Advances in Applied Geophysics

Advances in Applied Geophysics

Guest Editors

Filippos Vallianatos
Vassilis Sakkas

MDPI

Basel • Beijing • Wuhan • Barcelona • Belgrade • Novi Sad • Cluj • Manchester

Guest Editors
Filippos Vallianatos
Department of Geology
and Geoenvironment
National and Kapodistrian
University of Athens
Athens
Greece

Vassilis Sakkas
Department of Geology
and Geoenvironment
National and Kapodistrian
University of Athens
Athens
Greece

Editorial Office
MDPI AG
Grosspeteranlage 5
4052 Basel, Switzerland

This is a reprint of the Special Issue, published open access by the journal *Applied Sciences* (ISSN 2076-3417), freely accessible at: https://www.mdpi.com/journal/applsci/special_issues/Advances_in_Applied_Geophysics.

For citation purposes, cite each article independently as indicated on the article page online and as indicated below:

Lastname, A.A.; Lastname, B.B. Article Title. *Journal Name* **Year**, *Volume Number*, Page Range.

ISBN 978-3-7258-7340-1 (Hbk)
ISBN 978-3-7258-7341-8 (PDF)
https://doi.org/10.3390/books978-3-7258-7341-8

Contents

applied sciences

MDPI

Article

Empirical Correlation between Standard Penetration Resistance (SPT-N) and Shear Wave Velocity (V_S) for Soils in Metro Manila, Philippines

Arturo S. Daag *, Oliver Paul C. Halasan, Arielle Anne T. Magnaye, Rhommel N. Grutas and Renato U. Solidum, Jr.

Department of Science and Technology—Philippine Institute of Volcanology and Seismology, C.P. Garcia Ave., Diliman, Quezon City 1101, Philippines

* Correspondence: arturo.daag@phivolcs.dost.gov.ph

Abstract: The measurement of the shear wave velocities (V_S) of soils is an important aspect of geotechnical and earthquake engineering, due to its direct relation to the shear modulus (G), which in turn influences the stress–strain behavior of geomaterials. V_s can be directly measured or estimated using a variety of onsite tests or in a laboratory. Methods such as downhole PS logging require boreholes and may not be logistically and economically feasible in all situations. Many researchers have estimated V_s from other geotechnical parameters, such as standard penetration test resistance (SPT-N), by means of empirical correlations. This paper aimed to contribute to this subject by developing an empirical relationship between V_s and SPT-N. Data from twenty sites in Metro Manila were obtained from geotechnical investigation reports. V_s profiles of the same sites were also acquired using the refraction microtremor method. New empirical relationships were developed for all, sandy, and clayey soil types, using a non-linear regression method that is applicable for Metro Manila soils. Statistical evaluation and comparison of the proposed correlations with other previous works suggested the viability of the empirical model.

Keywords: shear wave velocity; standard penetration test; refraction microtremor; empirical correlation; Metro Manila

1. Introduction

The measurement of the shear wave velocities (V_S) of soils is an important aspect in the application of geotechnical and earthquake engineering. This geophysical parameter is related to the shear modulus (G) through the elastic theory, in the equation $G = \rho V_s^2$, where ρ is the density of material. The stress–strain behavior of materials is then dictated by the shear modulus, as the strain is directly affected by this parameter [1]. This relationship is also important in ground response analysis, as it is used to evaluate the dynamic response of soils during earthquakes. The seismic shear wave velocity of the upper 30-m soil layer (V_{s30}) largely influences the ground motion amplification and is therefore considered an important parameter in earthquake engineering [2]. Additionally, the shear wave velocity of geomaterials is also used in studies on soil stratigraphy, liquefaction analysis, and site-specific subsurface modeling [3–5].

Shear wave velocities can be measured or estimated using a variety of laboratory or field tests. Bender element and resonant columns are some of the laboratory techniques that measure V_s. The accuracy of these techniques is greatly dependent on the degree of disturbance of the soil samples [6]. In situ field measurements include invasive methods such as down-hole, cross-hole, and up-hole tests. These techniques require the drilling of boreholes to a particular depth and, thus, may not be economically feasible in all situations. On the other hand, non-invasive geophysical methods, such as spectral analysis of surface waves (SASW), multichannel analysis of surface waves (MASW), refraction microtremor

(ReMi), among others, can also estimate V_s on site. These techniques do not require drilling equipment and only use acoustic sensors to determine the subsurface shear wave velocities.

There have been attempts to correlate this geophysical parameter to other physically measured soil properties. These empirical correlations between the geotechnical parameters of soil and V_s were established in order to aid in soil behavior estimation and seismic demand analysis (e.g., Azam Ghazi et al., 2015 [7]; Gautam, 2017 [2]). However, these relationships are only accepted in localized data. In Japan, many works regarding this area were done in the 1970s, utilizing datasets from different regions (i.e., Imai and Yoshimura, 1970 [8]; Ohba and Toriumi, 1970 [9]; Shibata, 1970 [10]; Ohta et al., 1972 [11]; Fujiwara, 1972 [12]; Ohsaki and Iwasaki, 1973 [13]; Imai, 1977 [14]). These works mainly evaluated the dynamic response of soils (e.g., liquefaction) due to earthquakes. This continued into the 1980s, as Imai and Tonouchi (1982) [15] proposed a correlation with the most data points (n = 1654) based on the relative density of the subsurface soil (N-value), soil type, and geologic age. The work of Imai and Tonouchi (1982) also indicated that clayey layers have higher V_s values in comparison to sandy soils. In the United States, a similar correlation was proposed by Seed and Idriss (1981) [16], to determine liquefaction susceptibility of alluvium and marine sands. Their correlation utilizes corrected SPT parameters such as the N_{60} and $(N_1)_{60}$. This proposal is similar with the works by Seed et al., (1985) [17]; Sykora and Stokoe (1983) [18], Rollins et al. (1998) [19], and Brandenberg et al., (2010) [20].

Other countries have developed similar empirical correlations. Over the past decades, works by Kalteziotis et al. (1992) [21], Pitilakis et al. (1999) [22], Athanasopoulos (1995) [23], Raptakis et al. (1995) [24], Tsiambaos and Sabatakakis (2011) [25] have been completed in Greece. From Turkey, works by Iyisan (1996) [26], Kayabali (1996) [27], Hasancebi and Ulusay (2007) [28], Koçkar and Akgün (2008) [29], Dikmen (2009) [30], and Akin et al. (2011) [31] also explored the empirical correlations between V_s and SPT-N values and other geotechnical parameters. A considerable number of works on this topic have also been done in India: Anbazhagan and Sitharam (2010) [32], Anbazhagan et al. (2013) [33], Hanumantharao and Ramana (2008) [34], and Uma Maheswari et al. (2010) [35]. Kuo et al. (2011) [36] correlated V_s in terms of soil characteristics such as depth, soil type, and SPT-N value, using datasets from Taiwan.

Due to differences in soil type and properties from region to region, which arise from differences in geologic settings and active geomorphic processes, the proposed empirical correlations are local and were only extrapolated to areas with a similar lithology. The application was generally constrained to similar soil conditions and geomorphic settings. As such, despite the numerous proposed empirical correlations, there is still a great importance in correlating V_s and geotechnical parameters using datasets from local settings. Additionally, a variety of reasons, such as lack of drilling and geophysical equipment, personnel, and/or budget, can hinder the carrying out of downhole measurements of V_s, especially in smaller projects in the Philippines. Direct V_s measurements may be preferable, but in most cases where this is not feasible, empirical correlations between V_s and geotechnical parameters can be useful for site investigations.

The main objective of this paper was to develop an empirical relationship between V_s values from a refraction microtremor to N-values from standard penetration test drillings, using statistical regression analysis. Twenty sites (20) in Metro Manila, Philippines were surveyed using the refraction microtremor technique to establish one-dimensional V_s profiles correlated to N-values from SPT. The proposed empirical relationship between V_s and the N-value of soils in Metro Manila represents an important development in site characterization techniques and seismic microzonation efforts in the study area.

2. The Study Area

Metropolitan Manila (Metro Manila) is the capital region of the Philippines (Figure 1), which is also the center of the economy and the most populated areas. The region is composed of several cities and municipalities, sprawled over an area of around 620 square kilometers. It lies in the southern extension of the Luzon Central Valley. In the early stages of the development of the Central Valley, rapid sedimentation from highlands persisted until the Late Cretaceous, intercalated with tuff and other volcanic products [37]. During the Late Tertiary and Quaternary periods, complex tectonic events led to the basic geologic structure of Metro Manila at present. This geologic evolution has produced three distinct terranes in the metropolis: the Coastal Lowland, Central Plateau, and the Marikina Valley. East of the metropolis, the uplands of Rizal province with basement-ophiolite complexes can be found. To the west, this region is bounded by Manila Bay.

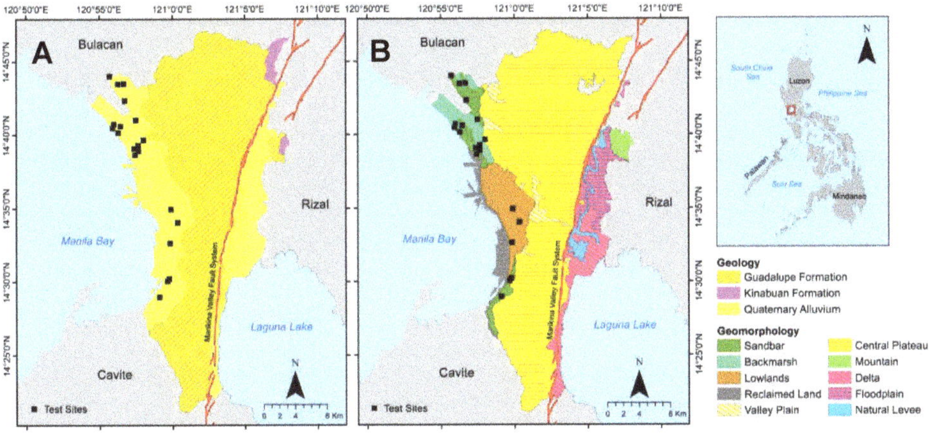

Figure 1. (**A**) Geological and (**B**) geomorphologic map of Metro Manila, Philippines, and the locations of the refraction microtremor surveys. Geologic map modified from BMG (1983) [38] and geomorphologic map obtained from MMIERS (2004) [39]. Locations of the testing sites are also depicted in black squares.

The Coastal Lowland is the flat and low plain section of Metro Manila, facing the Manila Bay. This section can be further subdivided into deposits from sand bars, backmarshes, tidal flats, the Pasig River delta, and some reclaimed land. This Quaternary Alluvium deposit has a thickness that can reach up to 40 m [40]. Sixty percent of the Metro Manila area is underlain by the Central Plateau, which is the geomorphic manifestation of the bedrock of the region, namely the Diliman Tuff member of the Pleistocene Guadalupe Formation. This lithologic member is made of flat-lying sequences of medium to thin-bedded, fine-grained vitric tuff and welded pyroclastic breccias. Minor fine to medium-grained tuffaceous sandstone was also reported to be part of this unit [41]. The Marikina Valley is the flat section on the eastern bounds of the metropolis, with lower elevations than the Central Plateau. This is produced by a pull-apart basin associated with the West Valley Fault (WVF) and East Valley Fault (EVF), collectively called the Marikina Valley Fault System [42]. Quaternary alluvium deposits in this section are associated with the floodplain of the Marikina River and the delta along the Laguna de Bay. Surface deposits on this sector are reported to reach up to 50 m in thickness [40].

Geologic field mapping and different subsurface investigations have revealed that most of the Metro Manila soils and sediments are composed of thick sequences of pyroclastic and epiclastic deposits. Pyroclastic deposits, primarily tuff materials, are derived from eruptions of nearby volcanoes in the region. The epiclastic deposits come from pyroclastic materials that were reworked by fluvial and lacustrine processes. Tan (1983) [43] reported that these epiclastic fluvial deposits are generally made of a sequence of loose to firm silty fine sand at the surface, underlain by very soft clayey silt or silty clay, further underlain by stiff to very stiff clay, and finally hard clay and silt, sometimes intercalated with very dense sand and/or gravel that grade into bedrock. The thickness and lateral continuity of these layers is highly variable, such that during borehole investigations on a particular site, it is normal to encounter different thicknesses and sequences of soil and sediment deposits.

3. Field Investigations

3.1. Geotechnical Investigations

The most common type of geotechnical site investigation in the country is through the standard penetration test (SPT). The National Structural Code of the Philippines [44] requires at least one borehole for two-story structures and greater, with the number of boreholes determined depending on the footprint area of the building. This in situ geotechnical test involves the driving of a hollow tube into the soil, while noting the number of blows needed to push the tube sampler down a specific vertical distance. A 63.5 kg hammer is repeatedly dropped from a height of 76 cm to advance the tube, in three successive increments of 150 mm. The first 150 mm is considered as a seating penetration and is disregarded, while the total number of blows needed to penetrate the second and third 150-mm iterations is the determined N-value of that layer. This procedure is described in the ASTM D1586 [45] guideline. The number of blow counts or N-value can be affected by a variety of factors. The most important factors include the amount of energy delivered to the tube in each impact of the hammer, the type of hammer, diameter of the borehole, and the length of the rod.

Standard penetration test reports of twenty sites in Metro Manila (Figure 1) were obtained from the Department of Education, and Department of Public Works and Highways of the Philippines. These SPT investigations were typically carried out as a prerequisite before the construction of school buildings and other structures. The sites selected to be included in this study were identified as having sufficient space, where geophysical surveys can be conducted and geotechnical data are available. Moreover, the spatial distribution was intended so that there are representatives of the various major geomorphic units identified in the coastal portion of the area. These sites are located along the western coast of Metro Manila, atop the Coastal Lowlands, and are expected to have relatively thick soil deposits. Specifically, the sites were built on top of sandbars, backmarshes, and lowland deposits.

Most of the boreholes have depths that reach up to 15 m, the typical depth requirement for low-story buildings in the country. In some cases, the depth can reach up to 30 m, particularly for sites with thicker soil profiles. On the other hand, there are also SPT investigations where the penetration only reaches up to five to ten meters, for sites with a shallower bedrock depth. In such cases, a coring procedure is employed to penetrate the target depth. The obtained soil samples from the SPT investigation were subjected to typical laboratory tests, such as grain size analysis, moisture content determination, Atterberg Limit test, soil classification, and unconfined compression test. A summary of the measured N-values and soil types identified for each depth in the twenty sites is presented in Table 1.

Table 1. Summary of the SPT from the twenty sites, indicating the location, water table depth, and Unified Soil Classification System (USCS) soil types at varying depths. The USCS scheme is adapted as the standard practice for soil classification for engineering purposes [46]. SM—silty sand, GM—silty gravel, ML—silt, CL—low plasticity clay, MH—high plasticity silt, SC—clayey sand, SP—poorly-graded sand, GP—poorly-graded gravel, CH—high plasticity clay, SW—well-graded sand, and GC—clayey gravel.

Site No.	Name	City	Long. (°)	Lat. (°)	Water Table Depth (m)	Depth (m)	Soil Type (USCS)
1	Wawang Pulo Elementary School	Valenzuela	120.928018	14.733015	0.2	0–1 1–2 2–5 5–6 6–15	SM GM SM ML CL
2	Pio Valenzuela Elementary School	Valenzuela	120.946090	14.707048	0.3	0.5 0.5–4.5 4.5–10 10–11.5 11.5–13 13–15	GM SM CL ML SM CL
3	Tagalag Elementary School	Valenzuela	120.938120	14.724660	0.7	0–15	SM
4	Coloong Elementary School	Valenzuela	120.944549	14.725007	4.0	0–10.0 10–15	ML SM
5	San Roque Elementary School	Navotas	120.938040	14.669389	0.8	0–6 6–12 12–25	SM MH ML
6	Tangos National High School	Navotas	120.932246	14.674792	0.6	0–1 1–2 2–3 3–6 6–7.5 7.5–10.5 10.5–12 12–13.5 13.5–22.5 22.5–25	SM SC SM SP-SM SC MH ML SC MH SM
7	Kapitbahayan Elementary School	Navotas	120.957491	14.643937	0.6	0–4.5 4.5–9 9–15	SM CL SC
8	Tanza Elementary School	Navotas	120.933461	14.679129	2.0	0–3 3–4.5 4.5–6 6–7.5 7.5–9 9–10.5 10.5–12 12–13.5 13.5–15 15–18 18–20	SC ML GP CL SC MH CL ML CL SM GC-GM
9	Dagat-Dagatan Elementary School	Navotas	120.956282	14.651251	0.3	0–1.5 1.5–6 6–12	CH SC-SM CH

Table 1. *Cont.*

Site No.	Name	City	Long. (°)	Lat. (°)	Water Table Depth (m)	Depth (m)	Soil Type (USCS)
10	Malabon Elementary School	Malabon	120.940551	14.677009	1.0	0–4.5 4.5–12 12–15	SP-SM ML CL
11	Panghulo Elementary School	Malabon	120.957800	14.684380	0.5	0–1.5 1.5–6 6–7.5 7.5–9 9–10.5 10.5–13.5 13.5–16.5 16.5–18 18–20	SP-SM SM CH CL CH MH CH MH CH
12	Imelda Elementary School	Malabon	120.960990	14.648665	1.2	0–3 3–9 9–15	SM CL CH
13	Ninoy Aquino Elementary School	Malabon	120.960700	14.654450	0.2	0–3 3–7.5 7.5–15	SC SM SC
14	Epifanio Delos Santos Elementary School	Malabon	120.992414	14.575557	3.0	0–4.5 4.5–6	SC SM
15	Manuel A. Roxas High School	Manila	120.999027	14.582204	0.9	0–1.5 1.5–4.5 4.5–6 6–10.5 10.5–12 12–13.5 13.5–15	SP-SM CH ML SM CH SM CH
16	Rafael Palma Elementary School	Manila	121.005917	14.567431	2.5	0–3 3–4.5 4.5–7.5 7.5–9 9–10.5 10.5–12 12–13.5 13.5–15	CL SC SM SC SW-SM SM SW-SM SP-SM
17	Jose Rizal Elementary School	Pasay	120.997418	14.543973	3.5	0–1.5 1.5–3 3–6 6–7.5 7.5–12 12–13.5 13.5–16.5 16.5–18 18–19.5 19.5–21 21–22.5 22.5–24 24–25.5 25.5–30	GC SP SW SP-SM SM SC CH SM ML MH CH MH CH MH

Table 1. *Cont.*

Site No.	Name	City	Long. (°)	Lat. (°)	Water Table Depth (m)	Depth (m)	Soil Type (USCS)
18	Paranaque Science High School	Parañaque	120.994677	14.500793	0.7	0–1.5	GM
						1.5–3	GP
						3–4.5	SP-SM
						4.5–6	SM
						6–7.5	MH
						7.5–12	SM
						12–13.5	CL
						13.5–15	MH
19	Sto. Nino Elementary School	Parañaque	120.996088	14.503082	No data	0–1.5	SC
						1.5–3	SP-SM
						3–4.5	SM
						4.5–6	ML
						6–7.5	MH
						7.5–9	GC
						9–10.5	ML
						10.5–12	MH
						12–13.5	CL
						13.5–15	CH
20	Manuyo Elementary School	Las Piñas	120.985444	14.482353	1.0	0–1.5	SP-SM
						1.5–4.5	SW-SM
						4.5–6	SP
						6–7.5	SM
						7.5–9	SP-SM
						9–10.5	CL
						10.5–12	MH
						12–13.5	CL
						13.5–15	SC

3.2. Geophysical Investigations

In the country, the most popular method among geotechnical engineers for assessing the strength of soil at a site is through standard penetration testing. Cross- or downhole profiling is also common to obtain in situ measurements of shear wave velocity with depth. These techniques may be precise and can provide measurement at, typically, one-meter resolutions. However, these methods can also rather be difficult and expensive to carry out in urban areas, as they require borehole drilling. Hence, non-invasive geophysical techniques are also utilized to overcome the problems posed by the practicality of borehole drilling. Seismic exploration techniques, e.g., multichannel analysis of surface waves (MASW), spectral analysis of surface waves (SASW), and refraction microtremors (hereafter "ReMi") are relatively more economical, faster, and can be performed in a variety of sites, including urban areas. This study utilized ReMi surveys to derive one-dimensional shear wave velocity profiles of the surveyed sites in Metro Manila.

The ReMi technique was first developed by Louie (2001) [47], to demonstrate the application of ambient noise in determining the shallow shear wave velocity structure of a site. The idea behind this technique is that the vertical component of ambient noise, primarily dominated by Rayleigh waves that are caused by human activities, can be picked up by a geophone array.

In this study, ReMi surveys were performed at the twenty sites in Metro Manila where the SPT investigations were performed. The surveys are done by coupling a straight linear array of twelve vertical geophones into the ground. These geophones have a frequency of 4.5 Hz and were arranged to have a spacing that ranged from four to eight meters. The array was then connected to a DAQLink II multichannel recorder to gather thirty iterations of thirty seconds unfiltered noise. The sampling rate of the recording was set at

2 ms (500 Hz). In the surveyed sites, typical sources of noise include pedestrians, vehicle traffic, construction, and wind. To provide high frequency noise for the recordings, a 10 lbs. sledgehammer was introduced, to hit a 6 inches × 6 inches × 1 inch steel plate placed at each end of the array, and which was utilized as an active noise source.

Data processing was done using ReMi Vspect and ReMi Disper modules. The Rayleigh waves were identified from other wave arrivals using slowness-frequency (p-f) transforms of the noise recording, using the ReMi Vspect module. The fundamental mode phase velocity was chosen along the minimum velocity envelope of energy within the p-f domain. Louie (2001) [47] demonstrated that selecting along this trend is the best procedure for choosing a dispersion curve to obtain the best estimate of the true phase velocities of shear waves (Figure 2A). The ReMi Disper module allows the modelling of a dispersion curve with multiple layers and shear wave velocities, to match the dispersion curve of the field data. The modeler varies the layer thickness and velocities until the resulting dispersion curve matches the previously selected dispersion points (Figure 2B,C).

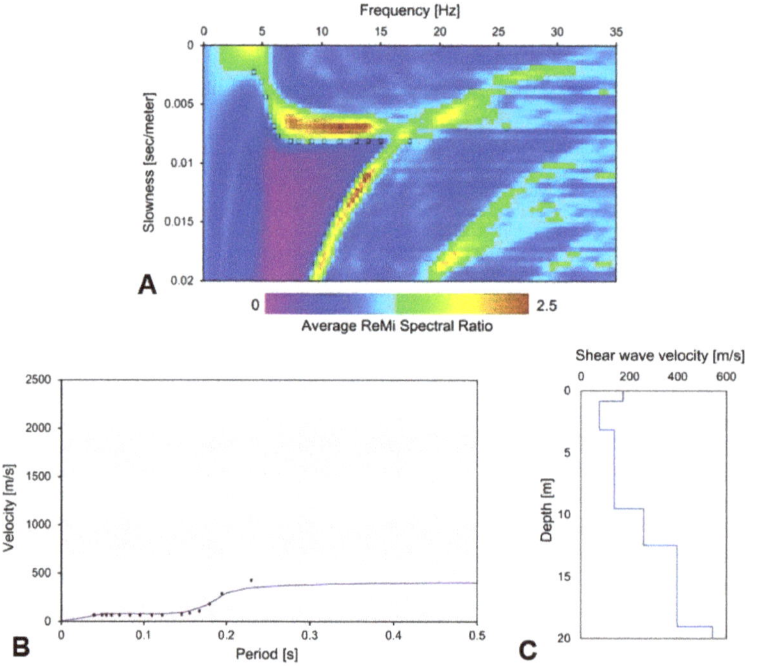

Figure 2. (**A**) Sample of a generated p-f image from the ReMi surveys and the identified dispersion curve and picks (black box) in the ReMi Vspect module. (**B**) Fitting of dispersion curve model into the dispersion picks in the ReMi Disper module. (**C**) Corresponding one-dimensional shear wave velocity profile from the dispersion curve model.

Vertical variations of shear wave velocities on the upper 20 m of the sites are illustrated in Figure 3. For most of the sites, the shear wave velocity of the upper 20 m generally ranged from ~100 m/s to ~220 m/s, corresponding to SPT-N values typically ranging from 2–10. For sites with much stiffer strata, the V_s at some depths can reach up to ~250 m/s to ~500 m/s. This was in areas with higher recorded N-values, that ranged from 15 to 50. In terms of the average V_s of the upper 30 m of the sites (V_{s30}), most of the western portion of Metro Manila was classified into Site Class D, based on the scheme of the National Structural Code of the Philippines [48]. This indicates that the area is generally underlain by stiff soil material, with a V_{s30} ranging from 180–360 m/s. This was also observed in the data obtained from the refraction microtremor surveys, as the V_{s30} of the sites mostly fell in the Site D range.

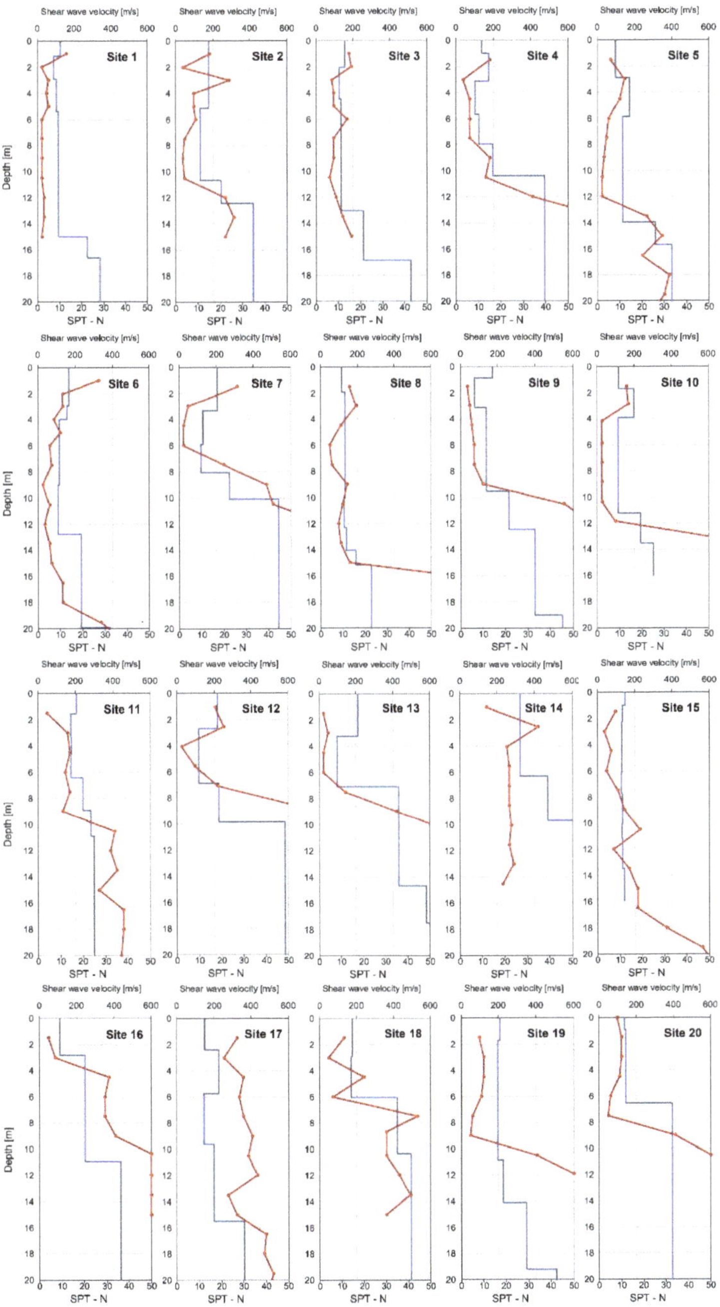

Figure 3. Variation of shear wave velocity (V_s) and penetration resistance (SPT-N) with depth for all twenty sites. Legend: blue lines for shear wave velocities; red lines for SPT-N values.

4. Methodology

A regression analysis is a statistical technique employed for analyzing empirical relationships between a dependent variable and one or more independent variables. This technique can be primarily applied to infer a relationship between the dependent and independent variable in a fixed dataset. It can also help in the visualization of variation of the values of the dependent variable, when one of the independent variables is adjusted while keeping the other independent variables constant. A particular type of regression analysis, the non-linear regression technique involves the modelling of the observed dependent variables using a mathematical function that combines one or more independent variables in a non-linear function. Power functions, exponential functions, logarithmic functions, and trigonometric functions are several typical examples of non-linear regression functions.

In this research, a power function was employed to correlate the estimated shear wave velocities, to measure the SPT-N values. This non-linear function is widely utilized in rapid evaluation of V_s and associated geotechnical parameters. The power function has a basic form of

$$E(x,y) = ax^b \tag{1}$$

where x and y are two variables, with x as the independent variable and where y is the response variable, while a and b are parameters of the power function [49]. a is a constant controlling the amplitude of the function, and b controls the relationship curve. This research examined the correlation based upon corrected and uncorrected SPT-N values and the V_s, irrespective of the depth, overburden pressure, geologic age of the deposit, and fines content.

Applying the natural logarithm function to Equation (1) transforms the non-linear function into a linear regression equation with the form

$$Y' = b_0 + b_1 X' \tag{2}$$

where $Y' = \ln y$, $b_0 = \ln a$, $b_1 = b$, and $X' = \ln x$. The parameters of regression b_0 and b_1 are determined using the least squares method

$$b_1 = \frac{\sum_{i=1}^{n} x_i y_i - \frac{1}{n}\sum_{i=1}^{n} x_i \sum_{i=1}^{n} y_i}{\sum_{i=1}^{n} x_i^2 - \frac{1}{n}\left(\sum_{i=1}^{n} x_i\right)^2} \tag{3}$$

$$b_0 = \frac{1}{n}\sum_{i=1}^{n} y_i - \frac{b_1}{n}\sum_{i=1}^{n} x_i \tag{4}$$

with x_i substituted by $\ln x$, y_i by $\ln y$, and n as the number of data points. When the values of b_0 and b_1 are determined, then the original parameters of the power function at Equation (1), a and b, can be also be calculated.

To measure the quality of the non-linear regression analysis, various statistical parameters need to be determined as well. The residual (ε_i) of the regression analysis is the difference from the observed value (y_i) of the dependent variable in comparison to its predicted value (y_i^*), using the regression function (Equation (5)).

$$\varepsilon_i = y_i - y_i^* \tag{5}$$

The total variability (S_{YY}) of the dependent variable (y_i) is determined from two components: the (1) variability in the observations of the y_i, which is accounted for by the regression function (SS_R); and the (2) residual variation, which is left unexplained by the regression function (SS_E) [49]. The method of determining S_{YY} is shown in Equations (6) and (7). In these equations, $\overline{y}$ is the average of the observed values of the dependent variable.

$$S_{YY} = SS_R + SS_E \tag{6}$$

$$\sum_{i=1}^{n} (y_i - \overline{y})^2 = \sum_{i=1}^{n} (y_i^* - \overline{y})^2 + \sum_{i=1}^{n} (y_i - y_i^*)^2 \tag{7}$$

The coefficient of regression (R^2) is the fraction of the total variability in the dependent variable that is accounted for by the regression function. This has a value that ranges from 0 to 1 and is expressed as

$$R^2 = \frac{SS_R}{S_{YY}} = \frac{\sum_{i=1}^{n} (y_i^* - \overline{y})^2}{\sum_{i=1}^{n} (y_i - \overline{y})^2} \tag{8}$$

To define the strength and nature of relationship between the dependent and independent variables, the coefficient of correlation (r) is also determined [49,50]. This value is expressed as

$$r(x,y) = \frac{\frac{1}{n}\left[\sum_{i=1}^{n} x_i y_i - \frac{\sum_{i=1}^{n} x_i \sum_{i=1}^{n} y_i}{n}\right]}{\sqrt{\frac{\sum_{i=1}^{n} x_i^2}{n} - \left(\frac{\sum_{i=1}^{n} x}{n}\right)^2}\sqrt{\frac{\sum_{i=1}^{n} x_i^2}{n} - \left(\frac{\sum_{i=1}^{n} x}{n}\right)^2}} \tag{9}$$

Another method to examine the quality of the obtained correlation is to determine Spearman's rank correlation coefficient (r_s) [50,51]. This non-parametric measure of correlation utilizes ranks to measure the correlation between an independent and response variable. The ordered datasets are replaced by rankings and the correlation coefficient (r_s) is calculated on the ranks, measuring the strength of association between the two ranked variables. This coefficient indicates how closely two sets of rankings agree with each other. r_s is obtained using the equation

$$r_s = 1 - \frac{6\Sigma d_i^2}{n(n^2 - 1)} \tag{10}$$

where d_i is the difference between the ranks and n is the number of members for each parameter. For the purposes of this work, the smallest value of each parameter (i.e., SPT N-values, shear wave velocities) was assigned with rank 1, and the rest of the dataset were ranked subsequently, in increasing order. Identical values in a dataset were assigned a rank equal to the average of their positions in the ascending order of values.

Aside from calculating different statistical parameters, graphical analysis of residuals can also be used to evaluate the proposed correlations. Histograms of standardized residuals can identify potential outliers and can determine whether the variance is normally distributed. A probability–probability (p-p) plot compares the empirical cumulative distribution function with a theoretical cumulative distribution function. p-p plots can determine if a model follows the assumed normality, as the points must follow the 1:1 diagonal line to be normally distributed data.

5. Results and Discussion

Several empirical relations of shear wave velocity (V_S) and penetration resistance (N) have been developed by a variety of researchers. In some references, corrected N-values were used in the regression analysis. These empirical correlations are summarized in Table 2. These correlations have a basic power law form $V_s = a \times N^b$, where a and b are coefficients that govern the shape of the empirical function. From Table 2, it is observable that for all previously proposed empirical correlations, a typically ranges from 19 to 121 and b ranges from 0.24 to 0.85.

Table 2. Existing correlation between shear wave velocity and standard penetration resistance.

Author	All Soils	Sand	Silt	Clay
Kanai (1966) [52]	$V_s = 19N^{0.6}$	-	-	-
Shibata (1970) [10]	-	$V_s = 31.7N^{0.54}$	-	-
Imai and Yoshimura (1970) [8]	$V_s = 76N^{0.33}$	-	-	-
Ohba and Toriumi (1970) [9]	$V_s = 84N^{0.31}$	-	-	-
Ohta et al. (1972) [11]	-	$V_s = 87.2N^{0.36}$	-	-
Fujiwara (1972) [12]	$V_s = 92.1N^{0.337}$	-	-	-
Ohsaki and Iwasaki (1973) [13]	$V_s = 81.4N^{0.39}$	-	-	-
Imai et al. (1975) [53]	$V_s = 89.9N^{0.341}$	-	-	-
Imai (1977) [14]	$V_s = 91N^{0.337}$	$V_s = 80.6N^{0.331}$	-	$V_s = 80.2N^{0.292}$
Ohta et al. (1978) [54]	$V_s = 85.35N^{0.348}$	-	-	-
Seed and Idriss (1981) [16]	$V_s = 61.4N^{0.5}$	-	-	-
Imai and Tonouchi (1982) [15]	$V_s = 96.9N^{0.314}$	-	-	-
Seed et al. (1983) [55]	-	$V_s = 56.4N^{0.5}$	-	-
Sykora and Stokoe (1983) [18]	-	$V_s = 100.5N^{0.29}$	-	-
Fumal and Tinsley (1985) [56]	-	$V_s = 152 + 51N^{0.27}$	-	-
Tonouchi et al. (1983) [57]	$V_s = 97N^{0.314}$	-	-	-
Jinan (1987) [58]	$V_s = 116\ (N + 0.3185)^{0.6}$	-	-	-
Okamoto et al. (1989) [59]	-	$V_s = 125N^{0.3}$	-	-
Lee (1990) [60]	-	$V_s = 57.4N^{0.49}$	$V_s = 105.64N^{0.319}$	$V_s = 114.43N^{0.31}$
Athanasopoulos (1995) [23]	$V_s = 107.6N^{0.36}$	-	-	$V_s = 76.55N^{0.445}$
Yokota et al. (1981) [61]	$V_s = 121N^{0.27}$	-	-	$V_s = 114N^{0.31}$
Kalteziotis et al. (1992) [21]	$V_s = 76.2N^{0.24}$	-	-	-
Pitilakis et al. (1992) [62]	-	$V_s = 162N^{0.17}$	-	-
Raptakis et al. (1995) [24]	-	$V_s = 100.7N^{0.24}$	-	-
Sisman (1995) [63]	$V_s = 32.8N^{0.51}$	-	-	-
Iyisan (1996) [26]	$V_s = 51.5N^{0.516}$	-	-	-
Kayabali (1996) [27]	-	$V_s = 175 + (3.75N)$	-	-
Jafari et al. (1997) [64]	$V_s = 22N^{0.85}$	-	-	-
Pitilakis et al. (1999) [22]	-	$V_s = 145\ (N_{60})^{0.178}$	-	$V_s = 132\ (N_{60})^{0.271}$
Kiku et al. (2001) [65]	$V_s = 68.3N^{0.292}$	-	-	-
Jafari et al. (2002) [66]	-	-	$V_s = 22N^{0.77}$	$V_s = 27N^{0.73}$
Hasancebi and Ulusay (2007) [28]	$V_s = 90N^{0.309}$	$V_s = 90.82N^{0.319}$	-	$V_s = 97.89N^{0.269}$
Hasancebi and Ulusay (2007) [28]	$V_s = 104.8\ (N_{60})^{0.26}$	$V_s = 131\ (N_{60})^{0.205}$	-	$V_s = 107.6\ (N_{60})^{0.237}$
Dikmen (2009) [30]	$V_s = 58N^{0.39}$	$V_s = 73N^{0.33}$	$V_s = 60N^{0.36}$	$V_s = 44N^{0.48}$
Uma Maheswari et al. (2010) [35]	$V_s = 95.64N^{0.301}$	$V_s = 100.53N^{0.265}$	-	$V_s = 89.3N^{0.358}$
Fauzi (2014) [67]	$V_s = 105.03N^{0.286}$	-	-	-

5.1. All Soils

This study utilized 265 data pairs from 20 sites in Metro Manila to propose new V_s–N empirical correlations for all soil types (Figure 4A). The empirical correlations were developed using non-linear regression analysis to fit the power law form. The proposed correlation from the data of this study is

$$V_s = 56.82N^{0.4861} \tag{11}$$

This relation has a regression coefficient $R^2 = 0.7515$, which indicates a strong correlation between the datasets [68]. Empirical correlations for the soil types proposed by other researchers were also compiled and compared to the results of this study (Figure 4C). In this compilation, for correlations that follow the form $V_s = a \times N^b$, the values of a range from 19 to 121 and b ranges from 0.24 to 0.85 can be observed. Among the compiled research that published empirical correlations, only Hasancebi and Ulusay (2007) [28] utilized corrected SPT-N values. On the other hand, Jinan (1987) [58] also proposed a correlation with a form $V_s = a\ (N + b)^c$.

It can be observed that the correlation by Kanai (1966) [52] can be used as the lower bound, while the work of Athanasopoulos (1995) [23] and Jafari et al. (1997) [64] provide a suitable upper bound for the proposed empirical relation of this study. The other twenty-one correlations were constrained by these three functions. Furthermore, the proposed curve of the present study lies closest to that proposed by Iyisan (1996) [26].

This study also explored correlating the shear wave velocity to corrected SPT-N values by applying several correction factors (Figure 4B). These correction factors are for the rod length (C_R), sampling method (C_S), borehole diameter (C_B), energy ratio (C_E), and overburden pressure (C_N). There are different equations for the correction of SPT-N values $(N_1)_{60}$, but the equation applied in this work is as follows:

$$N_{60} = N \times C_R \times C_S \times C_B \times C_E \tag{12}$$

$$(N_1)_{60} = N_{60} \times C_N \tag{13}$$

For overburden pressure correction, we adapted the equation proposed by Kayen et al. (1992) [69], which is as follows:

$$C_N = \frac{2.2}{1.2 + \frac{\sigma'_{vo}}{P_a}} \tag{14}$$

σ'_{vo} is the effective overburden pressure at the depth in kPa, and P_a is the reference pressure equal to 100 kPa. Rod length correction (C_R) was applied using the factors provided by Youd et al. (2001) [70]. A sampling method (C_S) correction factor of 1 was used, as most SPT drillings in the Philippines utilize standard samplers. Borehole diameter (C_B) and energy ratio (C_E) factors were applied following the scheme of Skempton (1986) [71]. After the application of these correction factors, non-linear regression analysis was employed again, to determine the empirical function to correlate V_s and corrected SPT-N values.

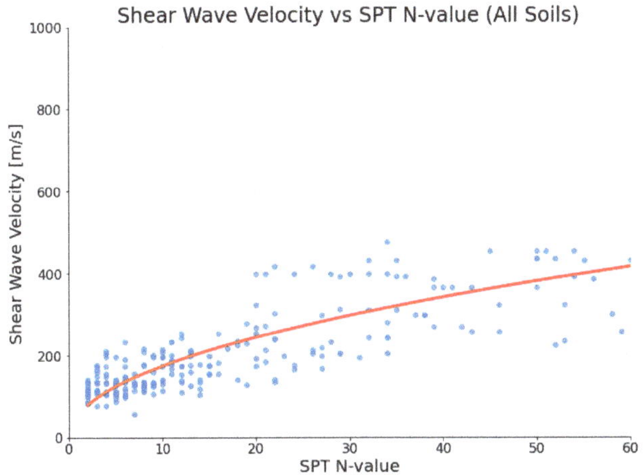

(A)

Figure 4. *Cont.*

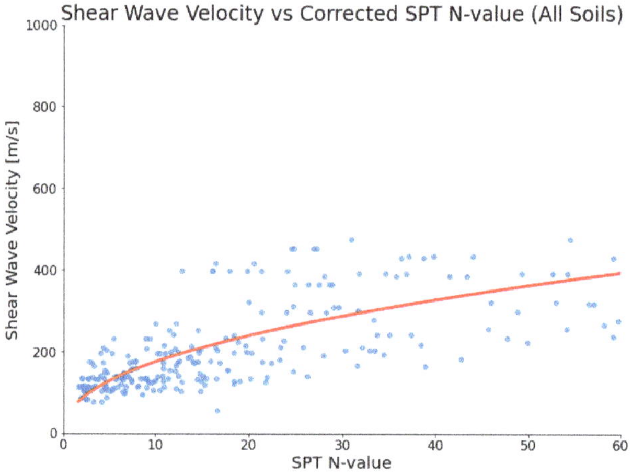

(**B**)

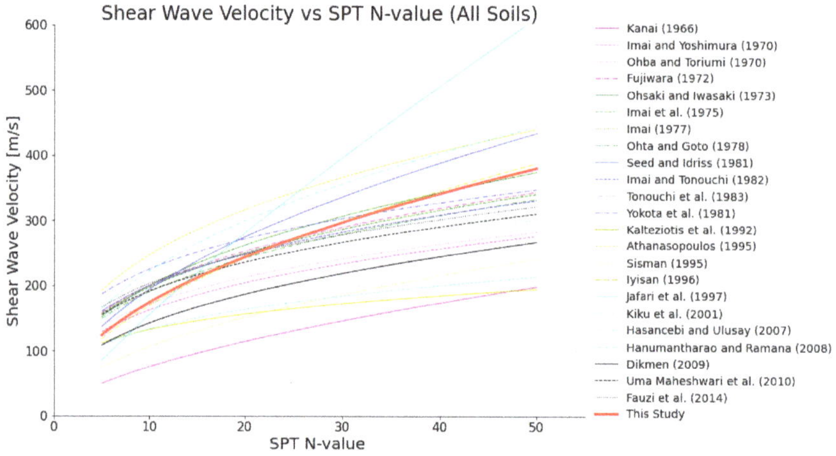

(**C**)

Figure 4. (**A**). Uncorrected SPT-N and V_s pairs for all soils and the determined empirical correlation. (**B**). Corrected SPT-N and V_s pairs for all soils and the determined empirical correlation. (**C**). Comparison of the proposed empirical correlation for all soil types and other published works [8,9,12–16,21,23,26,28,30,34,35,52,53,57,61,63–65,67,72].

The determined correlation from the data is

$$V_s = 61.77(N_1)_{60}{}^{0.4528} \tag{15}$$

This relation has a regression coefficient $R^2 = 0.5685$, indicating a moderate correlation, which is slightly lower than the regression coefficient when uncorrected SPT-N values are used in the analysis. This suggests that uncorrected SPT-N values correlate shear wave velocity better than corrected values.

5.2. Sandy Soils

A similar non-linear regression analysis was applied for sandy soil materials. Out of the 265 data points, 139 were identified to be of sandy soil. The proposed correlation for sandy soil types is

$$V_s = 45.07N^{0.5534} \tag{16}$$

This proposed correlation has a coefficient of regression value of $R^2 = 0.7922$. This value is relatively higher than the coefficient for all soil types and can be regarded as a strong correlation [68]. Figure 5 shows the plotted V_s and N-value for sandy soil and the corresponding proposed empirical relation. The empirical correlations for sandy soils proposed by other researchers were also compiled. Thirteen correlations follow the power law form $V_s = a \times N^b$ while utilizing the uncorrected SPT-N values in the analysis, while Pitilakis et al. (1999) [22] and Hasancebi and Ulusay (2007) [28] utilized corrected N-values for their statistical analysis. From the compiled curves, the a coefficient ranges from 31.7 to 162, while the b coefficient ranges from 0.17 to 0.54. On the other hand, Fumal and Tinsley (1985) [56] and Kayabali (1996) [27] proposed an empirical relation with the form $V_s = a(N + b)^c$.

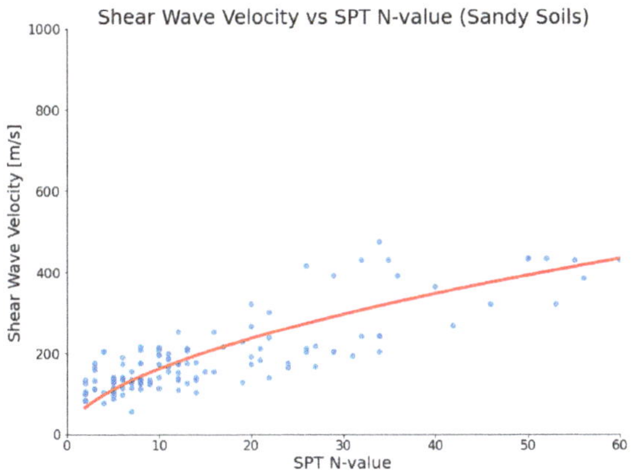

(A)

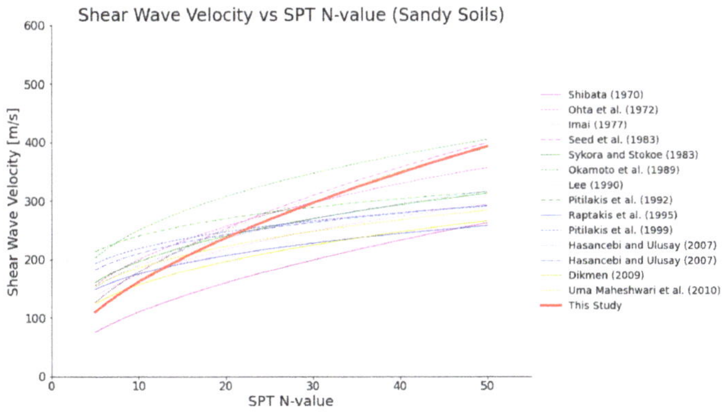

(B)

Figure 5. (**A**). Uncorrected SPT-N and V_s pairs for sandy soils and the determined empirical correlation. (**B**). Comparison of the proposed empirical correlation for sandy soil types and other published works [10,11,14,18,22,24,28,30,35,55,59,60,62].

When the proposed empirical relation is compared to the works of other researchers, it can be observed that the relations proposed by Okamoto et al. (1989) [59] and Shibata (1970) [10] can act as the upper and lower limits for empirical relationships for sandy soil types, respectively. The other eleven published empirical relations lie between these constraints, including the one proposed by this study. As the curve of this present study lies in the middle of all the other curves, this indicates a very good agreement with respect to the previous works.

5.3. Clayey Soils

A non-linear regression analysis was also applied to clayey soil types. A total 120 data pairs of V_s-SPT-N values were identified and utilized for the statistical analysis. The proposed correlation for clayey soil in this study is

$$V_s = 70.26N^{0.4220} \tag{17}$$

This proposed correlation has a coefficient of regression value of $R^2 = 0.73$. This value is relatively lower than the coefficient for all soil types but can still be regarded as having a strong correlation [68]. Figure 6 illustrates the plotted data pairs of V_s and N-values for clayey soil and the corresponding empirical relations. Empirical relations proposed by other authors were also compiled and compared. Eight correlation functions used the basic power law function while incorporating uncorrected SPT-N values in the analysis. In this compilation, coefficient a ranges from 27 to 114.43, and coefficient b ranges from 0.217 to 0.73.

It can be observed from the compiled empirical curves for clayey soils that the functions by Pitilakis et al. (1999) [22] and Athanasopoulos (1995) [23] can serve as the upper bound of the proposed relations, with the curve of Pitilakis et al. (1999) [22] for N values less than 22 and N values greater than 22 for Athanasopoulos (1995) [23]. For the lower bound, the curve by Dikmen (2009) [30] can be applied for N-values less than 24. Further constraints by Imai (1977) [14] can be applied as a lower bound for N-values greater than 24. The other empirical relations lie within these boundaries, including the one proposed in this study for clayey soils. As this curve lies in the middle of the other published curves, this is an indication of the good agreement between the proposed relation and the other published works. It can also be noticed that the curve proposed by this study lies close to the work by Uma Maheshwari et al. (2010) [35].

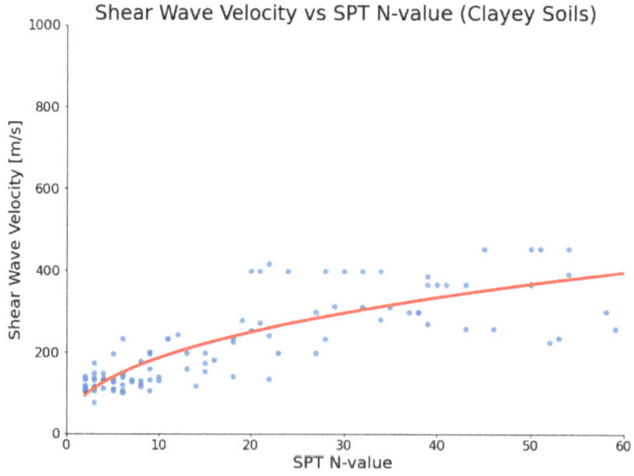

(**A**)

Figure 6. *Cont.*

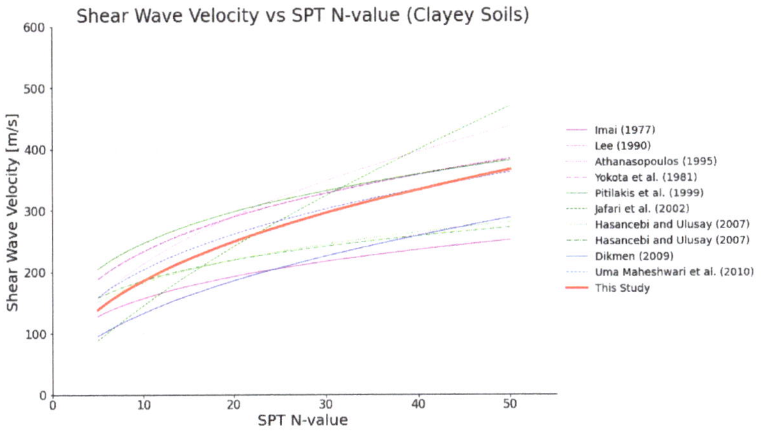

(B)

Figure 6. (**A**). Uncorrected SPT-N and V_s pairs for clayey soils and the determined empirical correlation. (**B**). Comparison of the proposed empirical correlation for clayey soil types and other published works [14,22,23,28,30,35,60,61,66].

5.4. Statistical Evaluation

Several methods were used to evaluate the proposed V_s-SPT-N values empirical relationships for all, sandy, and clayey soil types (Table 3). From the regression analysis, evaluated regression lines have an R^2 that ranges from 0.73 to 0.80, suggesting a moderate to strong fit of the proposed empirical function to the obtained dataset. Evaluation using the coefficient of correlation (r) and the Spearman rank coefficient (r_s) suggested that there is moderate to strong correlation between the V_s and SPT-N values from the dataset.

Table 3. Summary of the proposed correlations from this study.

Soil Type	Proposed Correlation	Coefficient of Regression (R^2)	Coefficient of Correlation(r)	Spearman Rank Correlation (r_s)
All soils	$V_s = 56.82N^{0.4861}$	0.7515	0.8767	0.8424
Sandy soil	$V_s = 45.07N^{0.5534}$	0.7922	0.9050	0.8192
Clayey soil	$V_s = 70.26N^{0.4220}$	0.7334	0.8453	0.8493

An alternative method applied to evaluate the validity of the proposed empirical relations was through an analysis of the residuals. Graphical methods for analyzing the residuals applied in this research included the development of residual histograms and probability–probability (p-p) plots (Figure 7). Histogram plots of the residuals determined whether the variance of the datasets was normally distributed. An ideal histogram of the residuals, which follows a bell-shaped distribution and which is distributed around zero, suggests that the normality assumption is likely true. In this analysis, histograms were produced from standardized residual values, which is solely the raw residual divided by its standard error. The method of standardizing residuals is applied to transform data, such that its mean centers at zero and the standard deviation equals one. This technique also identified potential outliers, such that residuals with standard deviations less than −3 or greater than 3, could be identified. In this analysis, histograms from the datasets suggested that the residuals generally followed a normal distribution, and outliers with standard deviations greater than $|3|$ were not detected.

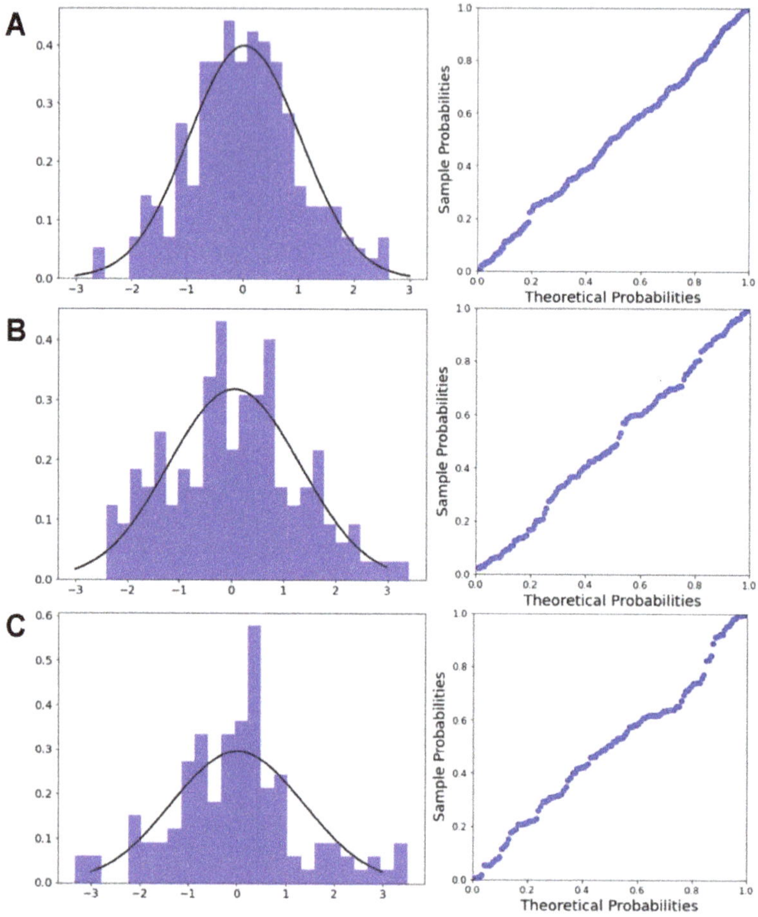

Figure 7. Histogram of the standardized residuals and the probability–probability (p-p) plots for (**A**) all soils, (**B**) sandy soils, and (**C**) clayey soils.

The second graphical technique applied was a p-p plot. This type of plot compares the empirical cumulative distribution function with the theoretical cumulative distribution function. Moreover, this plot can also show whether the residuals or errors follow the assumed normality. Points on the p-p plot must follow the 1:1 diagonal line for normally distributed data.

Finally, the measured V_s and estimated V_s from each empirical correlation were compared, in order to assess the performance of the regression models (Figure 8). Data pairs mostly lay in the region between the lines 1:1.50 and 1:0.65.

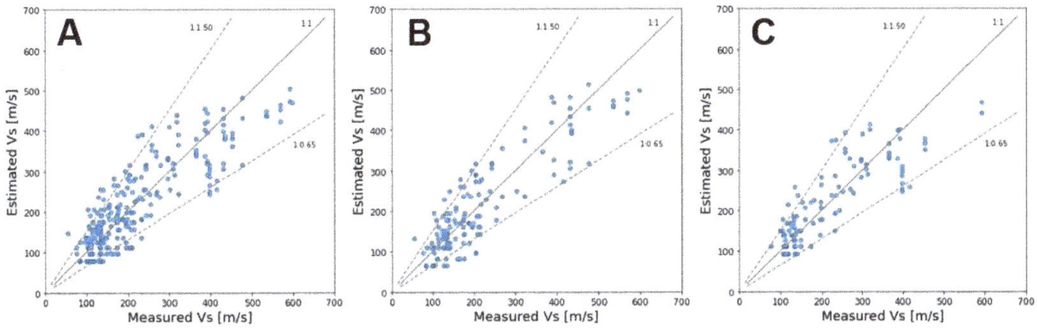

Figure 8. Measured versus estimated V_s for (**A**) all soils, (**B**) sandy soils, and (**C**) clayey soils with lines of 1:1.5, 1:1 and 1:0.65.

5.5. Application for Site-Specific Characterization

One of the main applications of these empirical correlations is the estimation of the average shear wave velocity of the upper 30 m of a site (V_{s30}). This metric is generally used for the classification of sites into the general characteristic of the subsurface, for earthquake engineering design. In the Philippines, V_{s30} is the most important parameter needed for the site classification scheme stipulated by the National Structural Code of the Philippines (2015) [44]. This scheme classifies a site into six categories used for earthquake engineering design (Table 4).

Table 4. Site classification scheme according to the National Structural Code of the Philippines, employing the V_{s30} metric to determine the general soil description and its typical corresponding N-values.

Soil Profile Type	Soil Profile Description	Shear Wave Velocity, V_s, (m/s)	SPT-N (Blows/30 cm)
S_A	Hard Rock	>1500	-
S_B	Rock	760–1500	-
S_C	Very Dense Soil and Soft Rock	360–760	>50
S_D	Stiff Soil	180–360	15–50
S_E [1]	Soft Soil	<180	<15
S_F	Soil requiring site-specific evaluation		

[1] Soil profile type S_E also includes any soil profile with more than 3.0 m of soft clay defined as soil with plasticity index > 20, moisture content ≥ 40%, and undrained shear strength < 24 kPa.

The magnitude of V_{s30} can be computed in accordance with the expression

$$V_{s30} = \frac{30}{\sum_{i=1}^{n} \frac{h_i}{v_i}} \quad (18)$$

where h_i and v_i are the layer thickness (m) and shear wave velocity (m/s) of the ith layer, respectively, in a total of n layers existing in the top 30-m of a site. The magnitude of the V_s is estimated from SPT-N values using the proposed empirical relations applicable for all soil types. Sample boreholes with penetration depths reaching 30 m were examined, and the proposed empirical correlation was applied to the datasets (Table 5). The calculated V_{s30}s highlight the site classifications obtained from SPT data. Site Class D classification is dominant in Metro Manila, particularly on the western sector of the region along the coast. This suggests that the sites are generally underlain by stiff soil deposits. This is further supported by the Quaternary Alluvium deposits mapped along the western coast, proximal to Manila Bay.

Table 5. Typical average shear wave velocity to a 30-m depth (V_{s30}) of all soils at various locations in Metro Manila.

Location	Coordinates	Generalized Soil Profile Description	V_{s30} Using Proposed Correlation for All Soil Types	Site Class Per NSCP (2015)
J. A. Santos High School, Manila City	120.972674° E, 14.597179° N	poorly graded to silty sand, silt	187.30	Class D
T. Paez Elementary School, Manila City	120.966613° E 14.630679° N	silt, well-graded and silty sand	243.10	Class D
Amang Rodriguez Elementary School, Malabon City	120.949242° E 14.671821° N	silty gravel, silty sand, clays	320.08	Class D
Tangos I Elementary School, Navotas City	120.932371° E 14.673729° N	poorly graded to silty sand, silt	274.71	Class D
Jose Rizal Elementary School, Pasay City	120.997418° E 14.543973° N	well graded to silty sand, silt	330.05	Class D
V.E. Fugoso Memorial Elementary School, Marikina City	121.118325° E 14.666042° N	high plasticity silt, low plasticity clay	225.04	Class D

An additional utility of these correlations is in the refinement of the corresponding N-values for a specific range of shear wave velocities. Based on the outcomes of this study, a simplified correlation can be established for basic soil strength classification (i.e., soft soils, stiff soils, and very dense soils) (Table 6).

Table 6. Simplified correlation of V_s and SPT-N for different soil strength classifications in this study.

Soil Description	V_s (m/s)	SPT N-value
Very soft to soft soils	<180 m/s	<10
Stiff soils	180–360 m/s	10–45
Very dense soils	360–760 m/s	>45

This demonstrates the utility of the proposed empirical correlations for geotechnical description at site-specific scales, as well as seismic microzonation efforts in the study area. Specific ranges of V_s have a corresponding range of N-values, where geophysical measurements of V_s can complement point borehole profiles and establish the lateral continuity of subsurface material between drilling points. In situations where SPT boreholes might not be feasible, due to economic or logistic reasons, refraction microtremor or other geophysical tests that can reliably measure shear wave velocities and can serve as a proxy to determine the relative strength of soils, especially for smaller construction projects. This was demonstrated by Pancha and Apperley (2021) [73], through utilizing refraction microtremors in two-dimensional shear wave velocity mapping, thereby extending subsurface characterization and mapping lateral variations between spot boreholes.

Alternatively, shear wave velocity surveys might not be feasible in some urban areas, due to a lack of availability of open space to lay the linear geophone array, which is a technical requirement of some geophysical tests. These sites instead might have robust borehole survey results acquired before the building's construction. Compilation of these SPT datasets can be utilized and converted to shear wave values, to produce or refine seismic microzonation maps. The derived empirical correlations of V_s and reliable static field data such as the SPT N-value can be an alternative way to determine these seismic parameters, such as the V_{s30}.

However, an important caveat, before the application of the proposed empirical relations, is to ensure the reliability of obtained SPT data to estimate the corresponding V_s values. On the other hand, the processing of geophysical data to acquire V_s must also be performed with utmost caution, to minimize uncertainty before applying it for basic soil strength classification. Furthermore, these empirical correlations must be applied with caution and with consideration to the geologic, geomorphologic, and geotechnical setting of an area. Areas with a very different environment must utilize other empirical functions that are more suitable to the locality.

6. Conclusions

By utilizing 265 pairs of SPT-N values and V_s, empirical relations between the two variables were developed for the area of Metro Manila, Philippines. Data from field surveys using standard penetration tests and refraction microtremor were obtained in 20 sites in Metro Manila, all situated along the western coast of the region and underlain by unconsolidated Quaternary Alluvium deposits. These datasets were mainly grouped into three: all soils, sandy soils, and clayey soil types. New empirical relations were determined using non-linear regression analysis. To evaluate the statistical validity of the empirical models, parameters such as coefficient of regression (R^2), coefficient of correlation (r), and Spearman rank correlation (r_s) were determined for each soil grouping. The determined values from these coefficients (R^2: 0.73–0.79, r: 0.84–0.90, r_s: 0.81–0.85) suggested a moderate to strong correlation between the two parameters. Moreover, residual histograms and probability–probability plots were also produced, to graphically analyze the datasets and evaluate the proposed empirical functions. This analysis showed the presumed normality distribution of the residuals and therefore supported the validity of the empirical models. The newly developed empirical correlations for the three categories of soil are comparable to other works on this subject by researchers around the world. This suggests that the new correlations can be applied for site-specific seismic characterizations. This was demonstrated by estimating magnitudes of V_{s30} from SPT-N values, which were consistent with pre-identified site classifications. This shows the utility of the proposed correlations for seismic microzonation efforts in the study area, as long as there are reliable SPT values.

Author Contributions: Conceptualization, A.S.D., O.P.C.H., A.A.T.M. and R.N.G.; methodology, A.S.D., O.P.C.H., A.A.T.M. and R.N.G.; formal analysis, O.P.C.H. and A.A.T.M.; investigation, O.P.C.H. and A.A.T.M.; writing—original draft preparation, O.P.C.H. and A.A.T.M.; writing—review and editing, A.S.D., R.N.G. and R.U.S.J. All authors have read and agreed to the published version of the manuscript.

Funding: This research was funded by the Republic of the Philippines—Department of Science and Technology (DOST) and Philippine Council for Industry, Energy and Emerging Technology (DOST-PCIEERD) Grants-in-Aid (GIA).

Institutional Review Board Statement: Not applicable.

Informed Consent Statement: Not applicable.

Data Availability Statement: Not applicable.

Acknowledgments: The authors would like to thank A.T. Serrano, M.J.V. Reyes, A.O. Amandy L.E.G. Aque, M.C.Dela Cruz, K.S. Sochayseng, E.J.M. Arnoco, and O.S. Locaba for assistance in data acquisition; M.I.T. Abigania, M.P. Dizon, D. Buhay, and E. Mitiam for rendering further technical support; the Specific Earthquake Project of DOST-PHIVOLCS for lending the Refraction Microtremor data processing software. We also thank the Department of Education for their participation in this research.

Conflicts of Interest: The authors declare no conflict of interest.

References

1. Duan, W.; Cai, G.; Liu, S.; Puppala, A.J. Correlations between Shear Wave Velocity and Geotechnical Parameters for Jiangsu Clays of China. *Pure Appl. Geophys.* **2019**, *176*, 669–684. [CrossRef]
2. Gautam, D. Empirical Correlation between Uncorrected Standard Penetration Resistance (*N*) and Shear Wave Velocity (V_S) for Kathmandu Valley, Nepal. *Geomat. Nat. Hazards Risk* **2017**, *8*, 496–508. [CrossRef]
3. Andrus, R.D.; Stokoe, K.H.; Hsein Juang, C. Guide for Shear-Wave-Based Liquefaction Potential Evaluation. *Earthq. Spectra* **2004**, *20*, 285–308. [CrossRef]

4. Kayen, R.; Moss, R.E.S.; Thompson, E.M.; Seed, R.B.; Cetin, K.O.; Kiureghian, A.D.; Tanaka, Y.; Tokimatsu, K. Shear-Wave Velocity–Based Probabilistic and Deterministic Assessment of Seismic Soil Liquefaction Potential. *J. Geotech. Geoenviron. Eng.* **2013**, *139*, 407–419. [CrossRef]
5. Mohamed, A.M.E.; Abu El Ata, A.S.A.; Abdel Azim, F.; Taha, M.A. Site-Specific Shear Wave Velocity Investigation for Geotechnical Engineering Applications Using Seismic Refraction and 2D Multi-Channel Analysis of Surface Waves. *NRIAG J. Astron. Geophys.* **2013**, *2*, 88–101. [CrossRef]
6. Sasitharan, S.; Robertson, P.K.; Sego, D.C. Sample Disturbance from Shear Wave Velocity Measurements. *Can. Geotech. J.* **1994**, *31*, 119–124. [CrossRef]
7. Ghazi, A.; Moghadas, N.H.; Sadeghi, H.; Ghafoori, M.; Lashkaripur, G.R. Empirical Relationships of Shear Wave Velocity, SPT-N Value and Vertical Effective Stress for Different Soils in Mashhad, Iran. *Ann. Geophys.* **2015**, *58*, 2. [CrossRef]
8. Imai, T.; Yoshimura, Y. Elastic Wave Velocity and Soil Properties in Soft Soil. *Tsuchi-ToKiso* **1970**, *18*, 17–22. (In Japanese)
9. Ohba, S.; Toriumi, I. Dynamic Response Characteristics of Osaka Plain. In Proceedings of the Annual Meeting, Architectural Institute of Japan, Tokyo, Japan, 5 October 1970.
10. Shibata, T. *The Relationship between the N-Value and S-Wave Velocity in the Soil Layer*; Disaster Prevention Research Laboratory, Kyoto University: Kyoto, Japan, 1970.
11. Ohta, T.; Hara, A.; Niwa, M.; Sakano, T. Elastic Shear Moduli as Estimated from N-Value. In Proceedings of the 7th Annual Convention of Japan Society of Soil Mechanics and Foundation Engineering, Tokyo, Japan, 10–15 July 1972; pp. 265–268.
12. Fujiwara, T. Estimation of Ground Movements in Actual Destructive Earthquakes. In Proceedings of the Fourth European Symposium on Earthquake Engineering, London, UK, 5–7 September 1972; pp. 125–132.
13. Ohsaki, Y.; Iwasaki, R. On Dynamic Shear Moduli and Poisson's Ratios of Soil Deposits. *Soils Found.* **1973**, *13*, 61–73. [CrossRef]
14. Imai, T. P and S Wave Velocities of the Ground in Japan. In Proceedings of the IX International Conference on Soil Mechanics and Foundation Engineering, Tokyo, Japan, 10–15 July 1977; pp. 127–132.
15. Imai, T.; Tonouchi, K. Correlation of N-Value with S-Wave Velocity and Shear Modulus. In Proceedings of the 2nd European Symposium of Penetration Testing, Amsterdam, The Netherlands, 24–27 May 1982; pp. 57–72.
16. Seed, H.B.; Idriss, I.M. Evaluation of Liquefaction Potential Sand Deposits Based on Observation of Performance in Previous Earthquakes. In Proceedings of the ASCE National Convention 1981, St. Louis, MO, USA, 26–30 October 1981; pp. 481–544.
17. Seed, H.; Tokimatsu, K.; Harder, L.F.; Chung, R.M. Influence of SPT Procedures in Soil Liquefaction Resistance Evaluations. *J. Geotech. Eng.* **1985**, *111*, 1425–1445. [CrossRef]
18. Sykora, D.E.; Stokoe, K.H. Correlations of In-Situ Measurements in Sands of Shear Wave Velocity. *Soil Dyn. Earthq. Eng.* **1983**, *20*, 125–136.
19. Rollins, K.M.; Evans, M.D.; Diehl, N.B.; Iii, W.D.D. Shear Modulus and Damping Relationships for Gravels. *J. Geotech. Geoenviron. Eng.* **1998**, *124*, 396–405. [CrossRef]
20. Brandenberg, S.J.; Bellana, N.; Shantz, T. Shear Wave Velocity as Function of Standard Penetration Test Resistance and Vertical Effective Stress at California Bridge Sites. *Soil Dyn. Earthq. Eng.* **2010**, *30*, 1026–1035. [CrossRef]
21. Kalteziotis, N.; Sabatakakis, N.; Vassiliou, J. Evaluation of Dynamic Characteristicsof Greek Soil Formations. In Proceedings of the 2nd Hellenic Conference on Geotechnical Engineering, Thessaloniki, Greece, 28 September–1 October 1992; pp. 239–246.
22. Pitilakis, K.; Raptakis, D.; Lontzetidis, K.; Tika-Vassilikou, T.; Jongmans, D. Geotechnical and Geophysical Description of Euro-Seistest Using Field, Laboratory Tests and Moderate Strong Motion Recordings. *J. Earthq. Eng.* **1999**, *3*, 381–409. [CrossRef]
23. Athanasopoulos, G. Empirical Correlations Vs-N SPT for Soils of Greece: A Comparative Study of Reliability. In Proceedings of the 7th International Conference on Soil Dynamics and Earthquake Engineering, Crete, Greece, 7 May 1995; pp. 19–36.
24. Raptakis, D.G.; Anastasiadis, S.A.J.; Pitilakis, K.D.; Lontzetidis, K.S. Shear Wave Velocities and Damping of Greek Natural Soils. In Proceedings of the 10th European Conference on Earthquake Engineering, Vienna, Austria, 28 August–2 September 1994; pp. 477–482.
25. Tsiambaos, G.; Sabatakakis, N. Empirical Estimation of Shear Wave Velocity from in Situ Tests on Soil Formations in Greece. *Bull. Eng. Geol. Environ.* **2011**, *70*, 291–297. [CrossRef]
26. Iyisan, R. Correlations between Shear Wave Velocity and In-Situ Penetration Test Results. *Chamb. Civ. Eng. Turk. Tek. Dergi* **1996**, *7*, 1187–1199.
27. Kayabali, K. Soil Liquefaction Evaluation Using Shear Wave Velocity. *Eng. Geol.* **1996**, *44*, 121–127. [CrossRef]
28. Hasancebi, N.; Ulusay, R. Empirical Correlations between Shear Wave Velocity and Penetration Resistance for Ground Shaking Assessments. *Bull. Eng. Geol. Environ.* **2007**, *66*, 203–213. [CrossRef]
29. Koçkar, M.K.; Akgün, H. Development of a Geotechnical and Geophysical Database for Seismic Zonation of the Ankara Basin, Turkey. *Env. Geol.* **2008**, *55*, 165–176. [CrossRef]
30. Dikmen, Ü. Statistical Correlations of Shear Wave Velocity and Penetration Resistance for Soils. *J. Geophys. Eng.* **2009**, *6*, 61–72. [CrossRef]
31. Akin, M.K.; Kramer, S.L.; Topal, T. Empirical Correlations of Shear Wave Velocity (Vs) and Penetration Resistance (SPT-N) for Different Soils in an Earthquake-Prone Area (Erbaa-Turkey). *Eng. Geol.* **2011**, *119*, 1–17. [CrossRef]
32. Anbazhagan, P.; Sitharam, T.G. Relationship between Low Strain Shear Modulus and Standard Penetration Test N Values. *Geotech. Test. J.* **2010**, *33*, 102278. [CrossRef]

33. Anbazhagan, P.; Kumar, A.; Sitharam, T.G. Seismic Site Classification and Correlation between Standard Penetration Test N Value and Shear Wave Velocity for Lucknow City in Indo-Gangetic Basin. *Pure Appl. Geophys.* **2013**, *170*, 299–318. [CrossRef]
34. Hanumantharao, C.; Ramana, G.V. Dynamic Soil Properties for Microzonation of Delhi, India. *J. Earth Syst. Sci.* **2008**, *117*, 719–730. [CrossRef]
35. Uma Maheswari, R.; Boominathan, A.; Dodagoudar, G.R. Use of Surface Waves in Statistical Correlations of Shear Wave Velocity and Penetration Resistance of Chennai Soils. *Geotech. Geol. Eng.* **2010**, *28*, 119–137. [CrossRef]
36. Kuo, C.-H.; Wen, K.-L.; Hsieh, H.-H.; Chang, T.-M.; Lin, C.-M.; Chen, C.-T. Evaluating Empirical Regression Equations for Vs and Estimating Vs30 in Northeastern Taiwan. *Soil Dyn. Earthq. Eng.* **2011**, *31*, 431–439. [CrossRef]
37. Gervacio, F. The Geology, Structures, and Landscape Development of Manila and Suburbs. *Philipp. Geol.* **1968**, *21*, 178–192.
38. Bureau of Mines and Geosciences. *Geological Map of Manila and Quezon City Quadrangle—Sheet 3263-IV [Map]*; Bureau of Mines and Geosciences: Quezon City, Philippines, 1983.
39. Japan International Cooperation Agency; Metropolitan Manila Development Agency; Philippine Institute of Volcanology and Seismology. *Earthquake Impact Reduction Study for Metro Manila, Republic of the Philippines*; Japan International Cooperation Agency: Makati City, Philippines, 2004.
40. Miura, H.; Midorikawa, S.; Fujimoto, K.; Pacheco, B.M.; Yamanaka, H. Earthquake Damage Estimation in Metro Manila, Philippines Based on Seismic Performance of Buildings Evaluated by Local Experts' Judgments. *Soil Dyn. Earthq. Eng.* **2008**, *28*, 764–777. [CrossRef]
41. Teves, J.S.; Gonzales, M.L. The Geology of the University Site, Balara Area, Quezon City. *Phil. Geol.* **1950**, *4*, 1–10.
42. Rimando, R.E.; Knuepfer, P.L.K. Neotectonics of the Marikina Valley Fault System (MVFS) and Tectonic Framework of Structures in Northern and Central Luzon, Philippines. *Tectonophysics* **2006**, *415*, 17–38. [CrossRef]
43. Tan, R.C. *Engineering Properties of Manila Subsoils*; University of the Philippines: Quezon City, Philippines, 1983.
44. Association of Structural Engineers of the Philippines, Inc. *National Structural Code of the Philippines*, 7th ed.; Association of Structural Engineers of the Philippines, Inc.: Quezon City, Philippines, 2015; Volume 1.
45. *ASTM D1586/D1586M-18e1*; Test Method for Standard Penetration Test (SPT) and Split-Barrel Sampling of Soils. ASTM International: West Conshohocken, PA, USA, 2022.
46. *ASTM D2487-17e1*; Practice for Classification of Soils for Engineering Purposes (Unified Soil Classification System). ASTM International: West Conshohocken, PA, USA, 2022.
47. Louie, J.N. Faster, Better: Shear-Wave Velocity to 100 Meters Depth from Refraction Microtremor Arrays. *Bull. Seismol. Soc. Am.* **2001**, *91*, 347–364. [CrossRef]
48. Philippine Institute of Volcanology and Seismology. *The Philippine Earthquake Model*; Philippine Institute of Volcanology and Seismology: Quezon City, Philippines, 2017.
49. Chatterjee, K.; Choudhury, D. Variations in Shear Wave Velocity and Soil Site Class in Kolkata City Using Regression and Sensitivity Analysis. *Nat. Hazards* **2013**, *69*, 2057–2082. [CrossRef]
50. Rezaei, S.; Shooshpasha, I.; Rezaei, H. Empirical Correlation between Geotechnical and Geophysical Parameters in a Landslide Zone (Case Study: Nargeschal Landslide). *Earth Sci. Res. J.* **2018**, *22*, 195–204. [CrossRef]
51. Lin, J.; Cai, G.; Liu, S.; Puppala, A.J.; Zou, H. Correlations Between Electrical Resistivity and Geotechnical Parameters for Jiangsu Marine Clay Using Spearman's Coefficient Test. *Int. J. Civ. Eng.* **2017**, *15*, 419–429. [CrossRef]
52. Kanai, K. *Conference on Cone Penetrometer*; The Ministry of Public Works and Settlement: Ankara, Turkey, 1966.
53. Imai, T.; Fumoto, H.; Yokota, K. The Relation of Mechanical Properties of Soil to Pand S-Wave Velocities in Japan. In Proceedings of the 4th Japan Earthquake Engineering Symposium, Tokyo, Japan, 6 September 1975.
54. Ohta, Y.; Goto, N.; Kagami, H.; Shiono, K. Shear Wave Velocity Measurement during a Standard Penetration Test. *Earthq. Eng. Struct. Dyn.* **1978**, *6*, 43–50. [CrossRef]
55. Seed, H.B.; Idriss, I.M.; Arango, I. Evaluation of Liquefaction Potential Using Field Performance Data. *J. Geotech. Eng.* **1983**, *109*, 458–482. [CrossRef]
56. Fumal, T.E.; Tinsley, J.C. Predicting Area Limits of Earthquake Induced Land Sliding Based on Seismological Parameters. In *Evaluation of Earthquake Hazards in the Los Angeles Region–An Earth Science Prospective*; U.S. Geological Survey: Los Angeles, CA, USA, 1985; pp. 127–150.
57. Tonouchi, K.; Sakayama, T.; Imai, T. S Wave Velocity in the Ground and the Damping Factor. *Bull. Int. Assoc. Eng. Geol.* **1983**, *26–27*, 327–333.
58. Jinan, Z. Correlation between Seismic Wave Velocity and the Number of Blow of SPT and Depth. *Sel. Pap. Chin. J. Geotech. Eng.* **1987**, *1*, 92–100.
59. Okamoto, T.; Kokusho, T.; Yoshida, Y.; Kusuonoki, K. Comparison of Surface versus Subsurface Wave Source for P–Slogging in Sand Layer. In Proceedings of the 44th Annal Conference JSCE, October 1989; Volume 3, pp. 996–997.
60. Lee, S.H. Regression Models of Shear Wave Velocities in Taipei Basin. *J. Chin. Inst. Eng.* **1990**, *13*, 519–532. [CrossRef]
61. Yokota, K.; Imai, T.; Konno, M. *Dynamic Deformation Characteristics of Soils Determined by Laboratory Tests.* OYO Technical Report. 1981, p. 26. Available online: https://www.oyo.co.jp/oyocms_hq/wp-content/uploads/2014/12/1981_03.pdf (accessed on 20 June 2022).
62. Pitilakis, K.D.; Anastasiadis, A.; Raptakis, D. Field and Laboratory Determination of Dynamic Properties of Natural Soil Deposits. In Proceedings of the 10th World Conference Earthquake Engineering, Rotherdam, The Netherlands, 19 July 1992; pp. 1275–1280.

63. Sisman, H. An Investigation on Relationships between Shear Wave Velocity, and SPT and Pressuremeter Test Results. Master's Thesis, Ankara University, Geophysical Engineering Department, Ankara, Turkey, 1995.
64. Jafari, M.K.; Asghari, A.; Rahmani, I. Empirical Correlation between Shear Wave Velocity (Vs) and SPT-N Value for South Tehran Soils. In Proceedings of the 4th International Conference on Civil Engineering, Tehran, Iran, 4–6 May 1997.
65. Kiku, H.; Yoshida, N.; Yasuda, S.; Irisawa, T.; Nakazawa, H.; Shimizu, Y.; Ansal, A.; Erkan, A. In-Situ Penetration Tests and Soil Profiling in Adapazari, Turkey. In Proceedings of the ICSMGE/TC4 Satellite Conference on Lessons Learned From Recent Strong Earthquakes, Istanbul, Turkey, 25 August 2001; pp. 259–265.
66. Jafari, M.K.; Shafiee, A.; Razmkhah, A. Dynamic Properties of Fine Grained Soils in South of Tehran. *J. Seismol. Earthq. Eng.* **2002**, *4*, 25–35.
67. Fauzi, A. Empirical Correlation of Shear Wave Velocity and N-SPT Value for Jakarta. *Geomate* **2014**, *7*, 980–984. [CrossRef]
68. Moore, D.S.; Notz, W.; Fligner, M.A. *The Basic Practice of Statistics*, 6th ed.; W.H. Freeman: New York, NY, USA, 2013; ISBN 978-1-4641-0254-7.
69. Kayen, R.E.; Mitchell, J.K.; Seed, R.B.; Lodge, A.; Nishio, S.; Coutinho, R. Evaluation of SPT-, CPT-, and Shear Wave-Based Methods for Liquefaction Potential Assessment Using Loma Prieta Data. In Proceedings of the 4th Japan-U.S. Workshop on Earthquake Resistant Design of Lifeline Facilities and Countermeasures for Soil Liquefaction, Honolulu, Hawaii, 27–29 May 1992; Volume 1, pp. 117–204.
70. Youd, T.L.; Idriss, I.M.; Andrus, R.D.; Arango, I.; Castro, G.; Christian, J.T.; Dobry, R.; Finn, W.D.L.; Harder, L.F.; Hynes, M.E.; et al. Liquefaction Resistance of Soils: Summary Report from the 1996 NCEER and 1998 NCEER/NSF Workshops on Evaluation of Liquefaction Resistance of Soils. *J. Geotech. Geoenviron. Eng.* **2001**, *127*, 817–833. [CrossRef]
71. Skempton, A.W. Standard Penetration Test Procedures and the Effects in Sands of Overburden Pressure, Relative Density, Particle Size, Ageing and Overconsolidation. *Géotechnique* **1986**, *36*, 425–447. [CrossRef]
72. Ohta, Y.; Goto, N. Empirical Shear Wave Velocity Equations in Terms of Characteristic Soil Indexes. *Earthquake Engng. Struct. Dyn.* **1978**, *6*, 167–187. [CrossRef]
73. Pancha, A.; Apperley, R.A. Multidisciplinary Site Investigations: Refraction Microtremor Surveys. *J. Seism.* **2021**, 1–14. [CrossRef]

applied sciences

MDPI

Article

Subduction Evolution Controlled Himalayan Orogenesis: Implications from 3-D Subduction Modeling

Weiling Zhu [1,2,3], Lin Ding [1,2,3], Yingfeng Ji [1,2,3,*], Rui Qu [1,3,*], Ye Zhu [1,3], Chaodi Xie [4] and Deng Zeng [1,3]

[1] State Key Laboratory of Tibetan Plateau Earth System, Resources and Environment (TPESRE), Institute of Tibetan Plateau Research, Chinese Academy of Sciences, Beijing 100101, China; zhuweiling@itpcas.ac.cn (W.Z.); dinglin@itpcas.ac.cn (L.D.); zhuye@itpcas.ac.cn (Y.Z.); zengdeng@itpcas.ac.cn (D.Z.)
[2] Center of Excellence in Tibetan Plateau Earth Science, Chinese Academy of Sciences, Beijing 100101, China
[3] University of Chinese Academy of Sciences, Beijing 100049, China
[4] Department of Geophysics, Yunnan University, Kunming 650091, China; xiecd@ynu.edu.cn
* Correspondence: yingfengji@itpcas.ac.cn (Y.J.); qurui@itpcas.ac.cn (R.Q.); Tel.: +86-158-1009-4403 (Y.J.); +86-173-2343-5109 (R.Q.)

Featured Application: Resource and environmental effects of the Tibetan Plateau.

Abstract: Himalayan orogenesis remains enigmatic in terms of Tibetan Plateau geodynamics originating from the Cenozoic India–Eurasian continental collision. India underthrusts below Tibet to the Yarlung–Tsangpo suture, which has been identified as the northernmost boundary for underplating. However, the way in which the historical evolution of continental subduction induces plateau uplift and the way it controls the variation in uplift between outboard and inboard areas is still unclear. To interpret the evolutionary mechanisms involved in the Himalayan growth history, we constructed different 3-D dynamic models at important stages to address these questions related to the formation of the Himalayas on the basis of paleoenthalpy evidence encoded in fossil leaves from recently documented assemblages in southern Tibet. The results show that (1) the effect of crustal thickening was the predominant factor in the early evolution from the Paleocene to the early Eocene, which resulted in a moderate growth rate. (2) The consecutive slab break-off eastward from the western syntaxis and the associated slab rebound significantly accelerated orogenesis from the late Eocene to the Oligocene. The upwelling asthenospheric flow was a key control of increasing crustal buoyancy, which resulted in the fastest growth of the Himalayas during the early Miocene. (3) Thereafter, the gradually enhanced monsoon and surface erosion during accompanying the increasing mountain height resulted in a slowdown of the orogenic rate, which counterbalanced the buoyant force produced by asthenospheric flow driving continuous Himalayan growth.

Keywords: Himalayas; uplift; subduction; 3-D modeling

1. Introduction

Himalayan uplift is one of the most important orogenic and climate forcing events of the Cenozoic (e.g., [1]). As the largest and most active collisional orogen at present (e.g., [2]), the Himalayas are the result of the Indian plate colliding with Eurasia, which first occurred at the center portion of the Yarlung–Tsangpo suture zone (YTSZ) between ca. 65 Ma and 63 Ma (e.g., [3]), and India continues to plunge into Eurasia along the main boundary thrust (MBT) (Figure 1). The India–Eurasia collision is characterized by a significant level of uplift, particularly along the high Himalayas, and an integral rise of the Tibetan Plateau ranging across thousands of kilometers, constituting the highest landscape on Earth in northeastern India (Figure 1). However, the way in which the historical evolution of continental subduction induces plateau uplift and the way it controls the variation in uplift between outboard and inboard areas is still unclear.

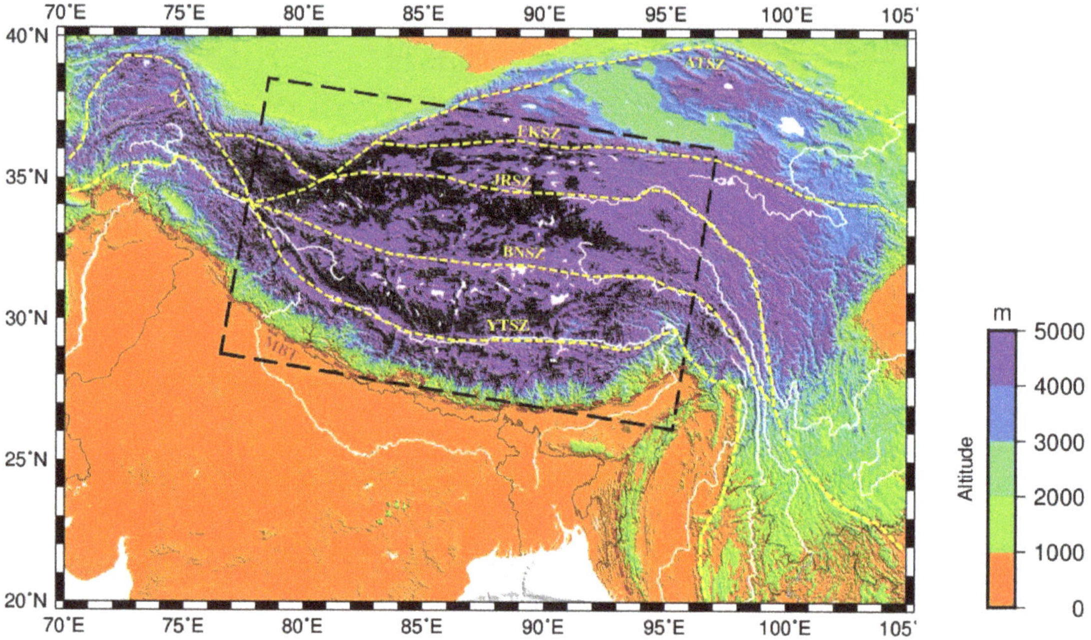

Figure 1. Tectonic map of the Himalayan orogen, including the major topographic features and sutures: YTSZ, Yarlung–Tsangpo suture zone; BNSZ, Bangong–Nujiang suture zone; JRSZ, Jinsha River suture zone; EKSZ, East Kunlun suture zone; ATSZ, Altyn Tagh suture zone; KF, Karakorum fault; MBT, main boundary thrust. Dashed lines indicate the main orogenic region focused by this study.

The growth of Himalayan orogen was reconstructed using a paleoaltimeter based on paleoenthalpy contained in fossil leaves from two newly reported assemblages in southern Tibet (Liuqu and Qiabulin) and four previously known floras from the Himalaya foreland basin, using climate leaf analysis, multivariate program analysis, and isotopic data [4]. Zircon U–Pb dating has constrained the Liuqu flora to the latest Paleocene (ca. 56 Ma) and the Qiabulin flora to the earliest Miocene (ca. 21 Ma), the latter of which marked the starting point for a sudden change to accelerated uplift in the Himalayas. The proto-Himalayas grew slowly from ≈1000 m in the late Paleocene to Eocene (>42 Ma) to ≈2.3 km at the beginning of the Miocene (23 Ma) and reached at least 5 km by approximately 15–11 Ma (e.g., [4]). The Himalaya–Tibet edifice and the Himalayan foreland basin have been gradually drying since 56 Ma, most likely due to the Himalayan orogen's uplift [4].

To assess the role of active tectonics and geometric variations in shaping the topography of the northwestern Himalayas, new constraints on deformation over geomorphic timescales are used, including morphometric analysis using high-resolution digital elevation models and field observations of rock type (e.g., [5]). The regional distribution of topographic growth indicates consistent active growth over million-year to millennial timescales in the Himalayas with a structural and/or tectonic control on topographic evolution. With a focus on deformation restoration, Li et al. [6] estimated that at least 1630 km of shortening (along 94° E) has occurred between India and Asia since 55 Ma, and Asia and the Himalayas have accommodated 1010 km and 620 km of the shortening, respectively. Wang et al. [7] suggested that slab break-off triggered lithosphere-asthenosphere interaction at the convergent margin. Jolivet et al. [8] showed that the closure of the former Tethys Ocean can best be explained by asthenospheric mantle flow transporting India northward, forming the Himalayas and the Tibetan Plateau, by comparing the tectonic and kinematic records from last 50 Ma with seismic tomography and anisotropy models.

The 3-D evolutionary models of Himalayan orogenesis were developed, and they revealed a substantial temporal relationship between the Indian slab's southward drift, lateral migration of slab detachment, and subsequent dynamic rebound with major alteration in Himalayas and coeval monsoon intensification (e.g., [9]). They further claimed that from 30 to 25 Ma, anchoring off the subducting Indian plate steepened the dip of the Himalayan thrust, resulting in crustal shortening deep within the Himalayan orogenic wedge. At around 13 Ma, slab detachment spread inward from both Himalayan syntaxes, causing a rapid rise in the mountain range. However, in their modeling, slab subduction evolved into slab rollback, the surface elevation is not constrained in detail in their focused dynamic processes, and the 3-D features of the heterogeneous uplift of the Himalayas remained unclear. Therefore, we tested several construction models at different evolutionary stages to quantify the Himalayan uplift history and mechanism in accordance with geological evidence (e.g., [4]) to provide insight into the Himalayan uplift on the basis of 3-D subduction modeling.

2. Materials and Methods

To estimate the multifactor-controlled Himalayan uplift evolution, we used a dynamic slab subduction model to simulate the effects on the uplift height of the controlling factors, including plate convergence thickening, slab rebounding, lithospheric flow, and surface erosion. Modeling reproduces the 1800 km width, 1600 km length, and 300 km depth viscoelastic domain along the Indian–Eurasian plate convergent margin beneath the Himalayas (Figures 1 and 2) with subduction velocity (Vs) and erosion rate (Ve) varying among different stages (Table S1). Horizontal resolution is approximately 20–22.5 km per grid. Continental collision occurs through the incoming plate slowly thrusting underneath the overriding plate along the plate boundary, extending to a depth of 100 km beneath the YTSZ and facilitating slab detachment and break-off, thus generating slab rebound forces upward and causing significant lithospheric flows beneath the tectonic complex of the lithosphere. Our simulation is inclusive of poloidal and toroidal mantle flows around the slab edge and lateral variation along the MBT during subduction-controlled orogenic processes, which are evolving self-consistently through time during the entire period of prescribed slab growth based on convergence rate [10,11].

The model is constructed on the basis of the finite difference method based on code Stag3D [11] and focuses on subduction regimes and geodynamics [10,12]. An anelastic liquid approximation and the equations of conservation of mass, momentum, and energy are used in this study (e.g., [12,13]):

$$\nabla \cdot \{\rho_s(z, T_s)\boldsymbol{v}\} = 0, \tag{1}$$

$$-\frac{\partial P}{\partial x_i} + \frac{\partial \tau_{ij}}{\partial x_j} - \delta_{i3}\rho_s g \alpha_0 (T - T_s) = 0, \tag{2}$$

$$\rho c_p \left(\frac{\partial T}{\partial t} + \boldsymbol{v} \cdot \nabla T \right) = k\nabla^2 T + \eta(\nabla v)^2 + \rho g \alpha T v_z + \rho H_r, \tag{3}$$

where P is the pressure deviation from hydrostatic pressure, α_0 is the reference thermal expansivity, τ_{ij} $(i, j = 1, 2, 3)$ is the stress tensor, and δ_{ij} is the Kronecker delta. The energy equation includes an advection term, thermal diffusion term, viscous dissipation term, adiabatic heating term, and radioactive heating term.

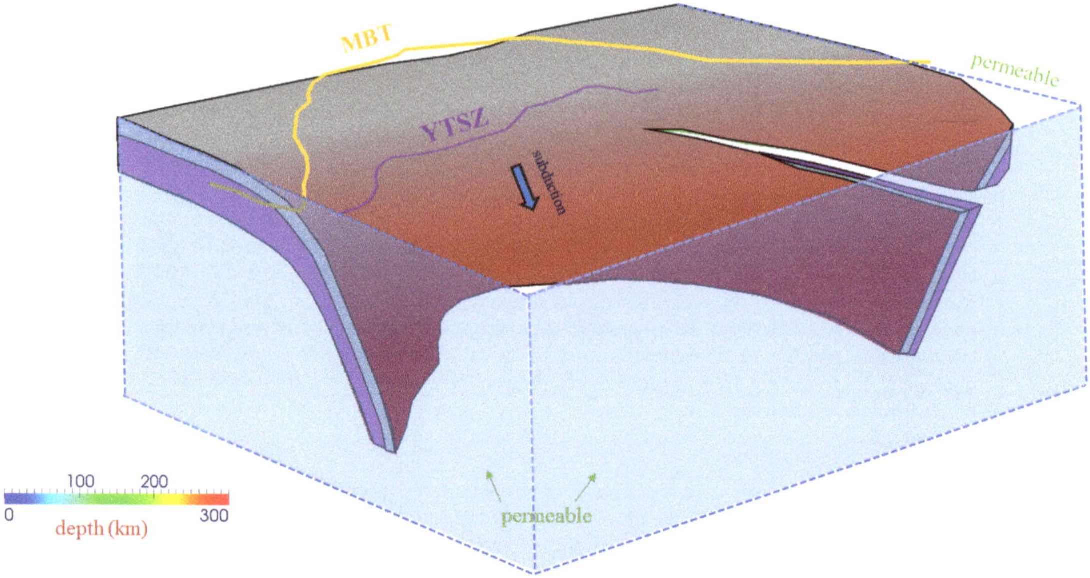

Figure 2. Conceptual figure of model domain and the slab geometry (splitting from west to east). The bottom and perpendicular planes are prescribed as adiabatic and permeable, and the top surface is set to be permeable. Detailed model settings are available in the Supplemental Information.

In this model, subducted plate geometry is kinematically prescribed according to Slab2.0 on the basis of seismic tomographic data [14] for depths of <60 km. The slab topography with a depth of 60–100 km adopts the extrapolation method and is assumed to feature steep subduction (45°) before 42 Ma (Figure 2) and shallow subduction (15°) after 42 Ma. The model dimensions are 72 × 72 × 72 grids. The Indian lithosphere is estimated to be >40 km thick according to Yoshii [15]. The temperature boundary condition agrees with the plate cooling model [16]. The slab's bottom and perpendicular planes are prescribed as adiabatic and permeable, while the top surface is set to be a constant temperature (0 °C) and permeable, allowing crustal extrusion and uplift. The composite upper mantle viscosity under a given condition is

$$\eta_{comp} = \frac{\eta_{df}\eta_{ds}}{\eta_{df} + \eta_{ds}}, \tag{4}$$

where η_{df} and η_{ds} are the diffusion creep and dislocation creep viscosities for olivine, respectively (model setting details are available in the Supplemental Information).

The thickness of the upper and lower crusts is prescribed to be 20 km. The upper crust is a pure elastic body, and the viscosity coefficient η is infinite; the lower crust is a maxwell body, and η is approximately 1.0×10^{19} Pa·s. The elastic modulus E of the crust is assumed to be 90–120 GPa (the reference value is 98 GPa), and the elastic modulus of the upper mantle is E = 170 GPa. Density ρ uses the global CRUST2.0 data from PREM and Stanford University (the upper crust is 2.6×10^3 kg/m^3, the lower crust is 2.9×10^3 kg/m^3, and the upper mantle is 3.3×10^3 kg/m^3) (Table S2). The kinematic plate subduction modeling method [12] is used to calculate subduction velocities inside a prescribed 3-D constrained volume of the oceanic lithosphere. The subduction velocity is set to vs. = 4.4–11.8 cm/year, as shown in Table S1. Current global MORVEL plate motion data have indicated a subduction velocity of 4–5 cm/year at the MBT [17,18], which is in accordance with our assumption.

The uplifted land surface has slowly been eroded, with erosion reducing the thickness of the continental crustal surface. The erosion rates reach the highest value in the High Himalayas at 2.7 ± 0.3 mm/a (1σ errors) and fall to <0.6 mm/a in the foothills to the south of the high mountains [19], which suggests an average erosion rate as low as <1 mm/a for the present-day Himalayas. Lenard et al. [20] further suggested that during the past six million years, the erosion rates were on average approximately 1 mm/a, which is close to the modern erosion rates in the Himalayas. During the Quaternary, there has been a marked increase in the erosion rate due to glacial erosion, especially in the High Himalaya (e.g., [21]). Considering these factors, we assumed the average erosion rate to be 0.2–0.4 mm/a (Table S1) in our modeling. The composite model setting, including model configuration, initial and boundary conditions, and physical parameters are illustrated in the Supplemental Information.

Slab thrusting, detachment, and subsequent possible delamination give a rise to the uplift of the overriding continental terrain surface. To simulate the steady subduction of the Indian plate, previously developed subduction models applied to various subduction zones have been utilized (SI). The evolution scenario is shown in Figure 3, and the surface uplift velocity is shown in Figure 4.

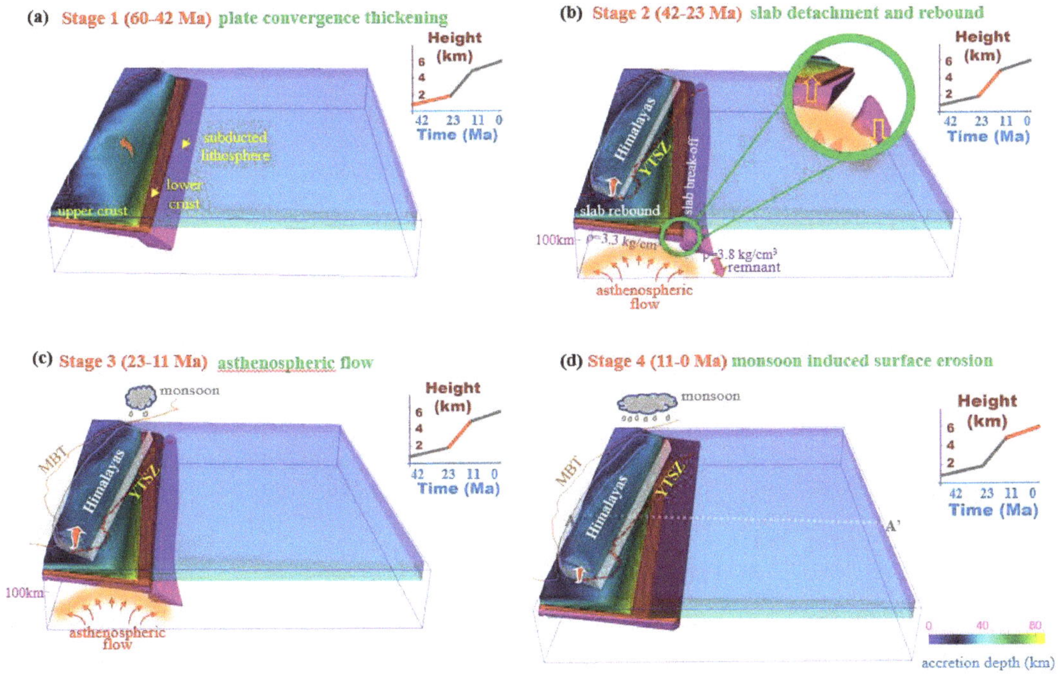

Figure 3. The four stages of the Himalayan uplift based on 3-D subduction modeling. Red arrows denote the elevation of the Himalayas. MBT, main boundary thrust; YTSZ, Yarlung–Tsangpo suture zone. The insets are modified from Ding et al. [4]. (**a**) Plate-convergence-induced thickening (60−42 Ma). (**b**) Progressive uplift of the Himalayas caused by slab detachment and rebound (42−23 Ma). (**c**) Upward asthenospheric flow causing rapid uplift of the Himalayas (23−11 Ma). (**d**) Slowed uplift of the Himalayas affected by surface erosion (11−0 Ma).

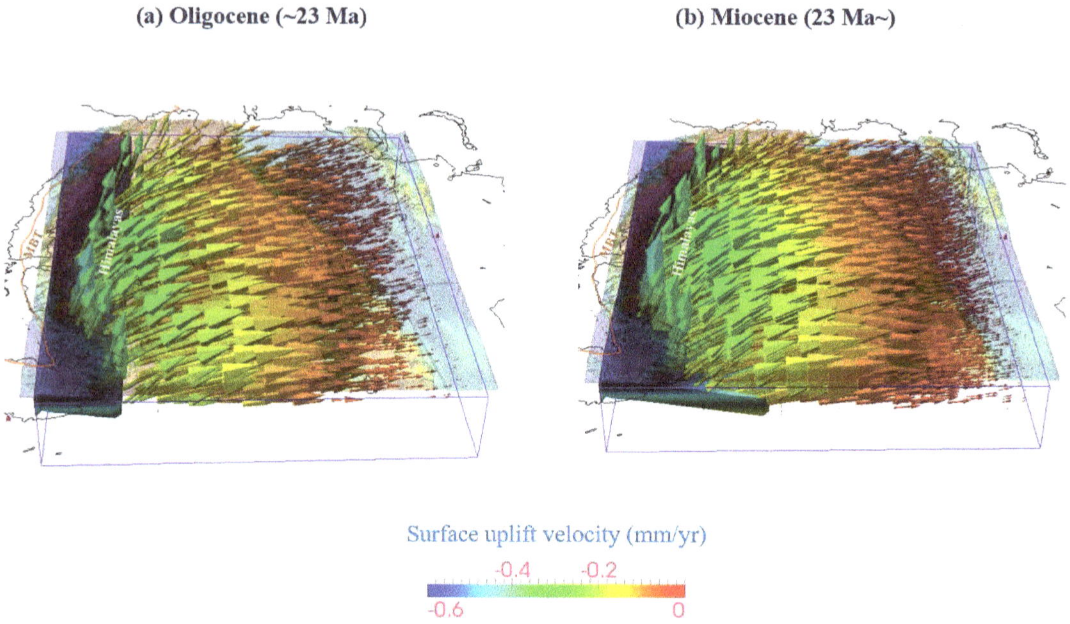

Figure 4. The vertical surface deformation field in and around the Himalayan orogen. (**a**,**b**) indicate the variation in vertical flow velocity (mm/year) in the Oligocene (42−23 Ma) and Miocene (23−11 Ma), respectively.

The first stage of the early plate collision period (60−42 Ma) is prescribed by a subduction velocity of 11.8 cm/year. The dominant uplift force is plate-convergence-induced thickening, and the Himalayan uplift rate is moderate (insets indicating the time-height covariation) until the region reaches a height of around 2 km. From 42 to 11 Ma, the Himalayas experienced a dramatic increase in elevation from 2 km to around 4-5 km. During this stage, slab break-off occurred, resulting in slab rebound and the occurrence of upward lithospheric flows, which are the prevailing driving forces of buoyancy; this mechanical mechanism explains such rapid uplift. Since 11 Ma, the Himalayan uplift slowed, probably due to the cessation of upward buoyant forces, and the monsoon-associated surface erosion became another crucial control affecting the surface height, which led to a gradually declining growth rate for the Himalayas. The subduction velocity during this period has passively decreased to an average of 4.4 cm/year since the late Miocene. The elevation of the surface topography increased continuously, reaching the current configuration characterized by significantly high topography with low relief.

3. Results

Considering the influence of crustal deformation, developed folds, and arc magmatism, the regions adjacent to the Yarlung–Zangpo suture zone are conjectured to have undergone rapid crustal thickening accompanied by structural shortening after the initial collision, which is evidenced by the formation of adakitic rocks (e.g., [22,23]). The Himalayan orogen is inferred to have featured a north-south compressional setting according to magmatic rock exhumation. The effect of crustal thickening was predominant in determining the uplift rate in the Paleocene and Eocene during the early evolution of the Himalayas (Figures 5a and 6a). The velocity field of the Himalaya orogen slows from south (>0.3 mm/year) to north (<0.1 mm/year), indicating strong shortening between India and Eurasia (Figure 4). In the second stage, the Himalayan uplift was linked to the detachment of the subducted Indian Plate. The mean compressive stress increased within the Indian plate and decreased the slab pull (e.g., [24]). Associated rebound occurred, facilitating uplift and increasing the elevation of the Himalayas by approximately 1.5 km prior to 23 Ma (Figures 5b and 6b). After slab break-off, the upwelling asthenospheric mantle flow also affected the acceleration of the uplift rate after 23 Ma in stage 3 (Figures 5c and 6c). Guo and Wilson [25], using Sr-Nd-Pb isotope data, indicated a change in the distribution of magmatism in Tibet, implying that asthenospheric mantle flow triggered further melting. Subducted Indian slab rollback could also have resulted in east-west-trending slab break-off ([26], shown in Figure 2). A large number of small tears within the underthrusting Indian plate are due to the upwelling of asthenospheric flow [27].

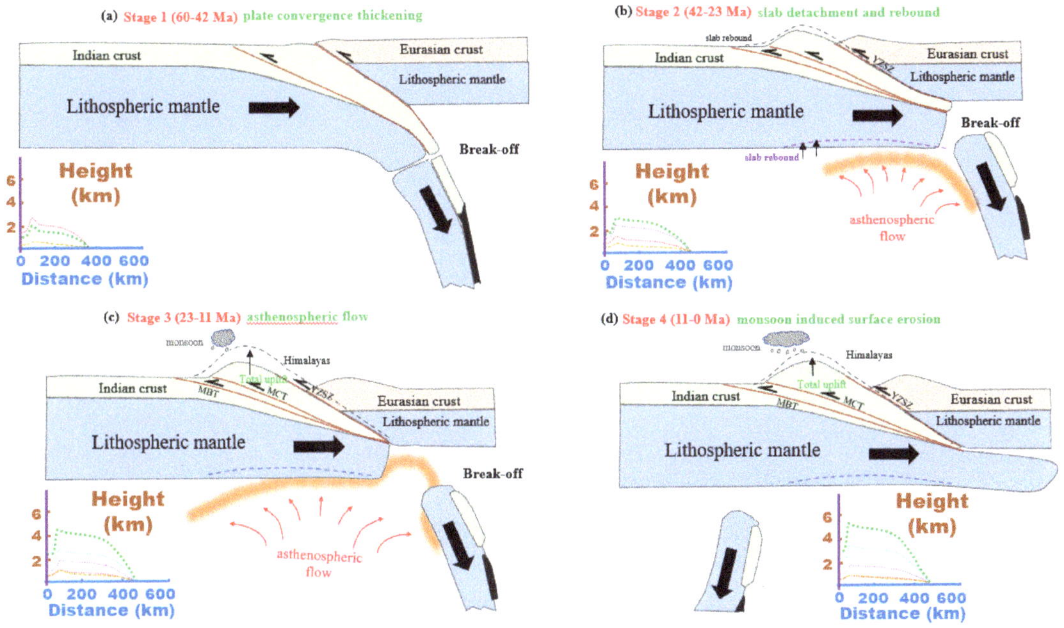

Figure 5. Schematic illustration of the 2-D tectonic evolution of the Himalayan orogeny corresponding to the four evolutionary stages of the Himalayas. The insets indicating variation of the height along the distance are specified in Figure 6. (**a**) Plate-convergence-induced thickening (60–42 Ma). (**b**) Progressive uplift of the Himalayas caused by slab detachment and rebound (42–23 Ma). (**c**) Upward asthenospheric flow causes rapid uplift of the Himalayas (23–11 Ma). (**d**) Slowed uplift of the Himalayas affected by surface erosion (11–0 Ma).

To explore the uplift rate of the Himalayas and its relationship to stress variation, we calculated the stress state in the underthrusting and overlying plates. Govin et al. [28] showed that the Himalayan uplift is caused by increased fault slip rates in response to stresses. Our results imply that the vertical flow velocity indicating the Himalayan uplift in the Miocene (23–11 Ma) was significantly higher than that in the Oligocene, which is a result of the bending of the Indian lithosphere due to lithospheric flow beneath the Himalayas (Figure 4). From the results, we found that the vertical flow velocity (uplift) was much greater than the average erosion rate (Figures 4 and 6). Although the strong monsoon climate results in a high erosion rate (e.g., [29]), erosion has not kept pace with the uplift of the Himalayas. The vertical velocity field reveals a phase of rapid uplift of the Himalayas in the Miocene (Figure 4b), and the range is still rising at a maximum rate of ~2 mm/year (e.g., [30]).

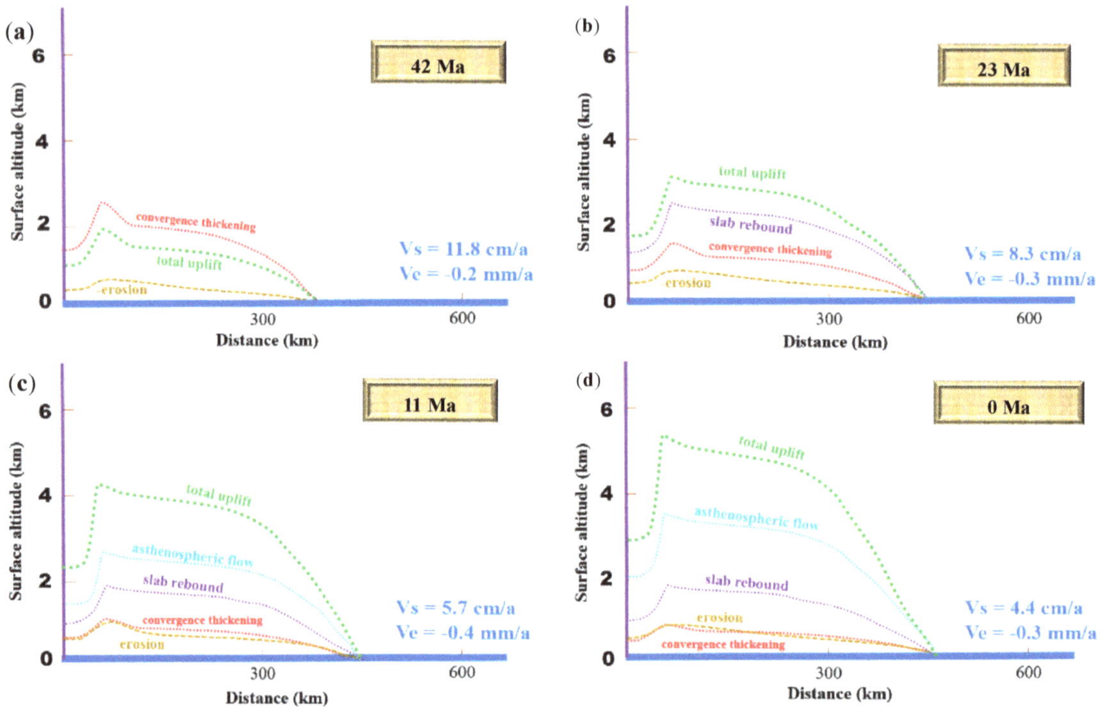

Figure 6. Model calculation of the surface height in the four periods along the profile AA′ in Fig. 3d. The curves indicate the net uplift effect for the Himalayas due to different tectonic processes, including convergence-induced thickening, slab rebound, asthenospheric flow, and surface erosion, at the end of the four evolutionary stages of the Himalayas. (**a**) 60−42 Ma. (**b**) 42−23 Ma. (**c**) 23−11 Ma. (**d**) 11−0 Ma.

The onset of slab detachment to reproduce surface morphometry results in significant surface uplift in southern Tibet (Figures 3d, 5d and 6d). The modeled uplift rates exhibit gradual along-subduction gradients (Figures 4 and 6), with faster uplift occurring predominantly in the regions adjacent to the Himalayas, where the plate interface is shallower and descends rapidly. The fact that the along-subduction variation in uplift rates has long wavelengths and low gradients suggests potential mechanisms responsible for plate-subduction-induced lithospheric flow. Hence, fast and steady uplift during the period from the late Eocene to the early Miocene (42-11 Ma) played a critical role in constructing the Himalayas into the highest mountain system on Earth.

4. Discussion

4.1. 3-D Numerical Modeling and Subduction Evolution

Dynamic models have been used to investigate Himalayan tectonic models, which are primarily focused on pro- and retro-plate interactions with contrasts in physical properties such as rigidity, density, rheology, widths, and lengths (e.g., [31]). The slab geometry determined in previous studies, however, is based on the model calculation and varies with time; thus, it is very difficult to reach the current tomography of the downgoing plate beneath the Himalayas. For this reason, a kinematically prescribed slab based on seismic tomography is helpful for high-resolution simulation of surface evolution. Hence, we aimed to construct such a 3-D, time-evolving kinematic model using seismic tomographic data to constrain the final geometry of the subducted Indian plate. With the modeling, we can explore lithospheric crustal-scale subduction processes and forearc surface development. The growth of the Himalayas is a complex process involving the collision of Gondwanan terranes and thickening of the crust close to the MBT first, instead of the rise of the Tibetan Plateau as a single entity (e.g., [32]). This indicates that Indian plate subduction is the major control for Tibetan uplift, especially for the southern part adjacent to the MBT, i.e., the Himalayas. According to estimates of the rise of the Tibetan Plateau (e.g., [4]), the uplift of the Himalayas (which at present reach elevations of at least 5.5 km) could be divided into four stages (insets in Figure 3). The proto-Himalayas grew slowly from ~1 km in the late Paleocene to ~2.5 km at the beginning of the Miocene (42−23 Ma) and produced significant uplift (e.g., from 2.5 to >5 km) at 23−11 Ma. Finally, the Himalayas reached the present-day elevation of ~6 km in the third stage (11 Ma to present). This proposal was adopted by this study for numerical modeling with further refinement of the subduction parameters. We consider that due to the comparison between the Himalaya–Tibet edifice and the Himalaya foreland basin from precipitation patterns, a characteristic feature is that the uplift of the Himalaya orogen may have generated progressive drying across southern Tibet over the past 56 Myr and intensified the monsoon over 11−0 Ma [33]. Hence, the surface erosion involving the monsoon effect was included in our modeling with the erosion rate prescribed in different stages (Table S1 and Figures 5 and 6).

4.2. Slab Detachment and Asthenospheric Flow

In this work, we calculated the Himalayan subduction and orogenic processes from the Eocene to the present day and related their covariation to an integrated subduction model that governs Himalayan elevation. In the modeling, we considered the steepened dip of the Himalayan thrust affected by the anchoring of the Indian subducted lithosphere (e.g., [9]). Slab detachment of the Indian crust subducted beneath the Himalayan orogen is the crucial factor for the rapid uplift in the Oligocene to early Miocene. In this context, slab detachment results in a buoyancy force on both the subducted and overlying crust and leads to the accelerated uplift of the Himalayas (Figures 5 and 6) (e.g., [34]). To explain this significant uplift, Husson et al. [35] suggested that the rise presumably occurred by means including convergence crustal thickening, and She and Fu [36] inferred that the upwelling of the mantle material is the dominant driver of the acceleration of uplift. This interpretation is supported by our modeling showing that the upwelling of asthenospheric flow facilitates the uplift of surface topography to ~3.5 km, and the effect of crustal thickening is comparatively weaker in our model (Figure 6). With continued convergence, we estimated that the sunken remnants of the detached lower lithosphere caused slab rebound and played a crucial role in elevating the Himalayas since 23 Ma (Figure 3b).

4.3. Orogenic Uplift and Surface Erosion

Moreover, the rise of the Himalayan orogen is thought to explain the development of the South Asian monsoon (e.g., [4,37]). In the fourth stage (11–0 Ma), the enhanced erosion and monsoon intensification contributed to shaping the peculiar topographic profile of the Himalayans we observed today (e.g., [38]). The variation in mantle convection and climate-induced erosion could modulate the crustal motion velocities (e.g., [39]) and affect the rise of the orogen (Figures 5d and 6d). The cold and dense lower lithosphere pushing into the hotter asthenosphere will cause convective instability in the mantle lithosphere (e.g., [40]). In this regard, the horizontal extension and thinning of the crust are a result of the thinning of the mantle lithosphere and thus slow acceleration in the Himalayas (Figure 5d). Furthermore, the enhanced climate change by uplifted mountains in turn influenced the precipitation and resultant surface erosion in the early and middle Miocene (e.g., [41]). Clift et al. [38] provide ample evidence by analyzing the chemistry and mineralogy of sediments record and suggest that Indian monsoon efficiently erodes the Himalayas. The potential counterbalancing relationship between orogenic uplift and surface erosion obtained from water, freeze-thaw, chemical weathering, and physical erosion observations from environmental measurements implies a nonlinear inverse relationship that is considered to predominantly control the current surface topography of the Himalayas and the Tibetan Plateau according to 3-D subduction modeling.

5. Conclusions

Through 3-D modeling of subduction beneath the Himalayas, we obtained the following results. During the early evolution of the Himalayas (60–42 Ma), the effect of crustal thickening was predominant in determining the surface uplift rate. The effects of slab break-off and associated slab rebound caused significant Himalayan uplift in the Oligocene (42–23 Ma). The upwelling asthenospheric mantle flow was the dominant driver of the rapid uplift of the Himalayas in the Miocene (23–11 Ma). The enhanced erosion and monsoon intensification counterbalance the further uplift of the Himalayas, and growth has slowed since 11 Ma, resulting in the present-day height of the Himalayas.

Supplementary Materials: The supplementary information is available for download: https://www.mdpi.com/article/10.3390/app12157413/s1. References [42–60] are mentioned in Supplementary Materials file. Table S1: Main model parameters; Table S2: Parameters for model domains.

Author Contributions: L.D. conceived the original idea and provided the funding. Y.J. designed the 3-D thermomechanical code, performed the numerical experiments, and interpreted the results. W.Z. and R.Q. elaborated the numerical study and wrote the manuscript. Y.Z., C.X. and D.Z. provided comments to improve the manuscript. All authors discussed the results and interpretations and participated in writing the paper. All authors have read and agreed to the published version of the manuscript.

Funding: This study benefited from financial support from the Second Tibetan Plateau Scientific Expedition and Research Program (2019QZKK0708) and the CAS Pioneer Hundred Talents Program.

Institutional Review Board Statement: Not applicable.

Informed Consent Statement: Not applicable.

Data Availability Statement: Data including the model code and results are available in figures or from the authors upon reasonable request.

Acknowledgments: We are thankful to P. Tackley for sharing the Stag3D code used in this study. We are grateful for the earthquake catalog from the IRIS [61]. Figures were created with Generic Mapping Tools, which included the coasts and borders [62], and with Paraview software developed by Kitware, Inc.

Conflicts of Interest: All the authors listed certify that they have no involvement in any organization or entity with any financial interest (including honoraria; educational grants; participation in speakers' membership, bureaus, consultancies, stock ownership, employment, or other equity interest; and patent-licensing arrangements or expert testimony) or nonfinancial interest (including professional or personal relationships, knowledge or beliefs, affiliations) in the subject materials or matter discussed.

References

1. DeCelles, P.G.; Kapp, P.; Gehrels, G.E.; Ding, L. Paleocene-Eocene foreland basin evolution in the Himalaya of southern Tibet and Nepal: Implications for the age of initial India-Asia collision. *Tectonics* **2014**, *33*, 824–849. [CrossRef]
2. Takada, Y.; Matsu'ura, M. A unified interpretation of vertical movement in Himalaya and horizontal deformation in Tibet on the basis of elastic and viscoelastic dislocation theory. *Tectonophysics* **2004**, *383*, 105–131. [CrossRef]
3. Ding, L.; Maksatbek, S.; Cai, F.; Wang, H.; Song, P.; Ji, W.; Xu, Q.; Zhang, L.; Muhammad, Q.; Upendra, B. Processes of initial collision and suturing between India and Asia. *Sci. China Earth Sci.* **2017**, *60*, 635–651. [CrossRef]
4. Ding, L.; Spicer, R.; Yang, J.; Xu, Q.; Cai, F.; Li, S.; Lai, Q.; Wang, H.; Spicer, T.; Yue, Y. Quantifying the rise of the Himalaya orogen and implications for the South Asian monsoon. *Geology* **2017**, *45*, 215–218. [CrossRef]
5. Dey, S.; Thiede, R.C.; Biswas, A.; Chauhan, N.; Chakravarti, P.; Jain, V. Implications of the ongoing rock uplift in NW Himalayan interiors. *Earth Surf. Dyn.* **2021**, *9*, 463–485. [CrossRef]
6. Li, Y.; Wang, C.; Dai, J.; Xu, G.; Hou, Y.; Li, X. Propagation of the deformation and growth of the Tibetan–Himalayan orogen: A review. *Earth-Sci. Rev.* **2015**, *143*, 36–61. [CrossRef]
7. Wang, W.; Pandit, M.K.; Zhao, J.-H.; Chen, W.-T.; Zheng, J.-P. Slab break-off triggered lithosphere-asthenosphere interaction at a convergent margin: The Neoproterozoic bimodal magmatism in NW India. *Lithos* **2018**, *296*, 281–296. [CrossRef]

8. Jolivet, L.; Faccenna, C.; Becker, T.; Tesauro, M.; Sternai, P.; Bouilhol, P. Mantle flow and deforming continents: From India-Asia convergence to Pacific subduction. *Tectonics* **2018**, *37*, 2887–2914. [CrossRef]
9. Webb, A.A.G.; Guo, H.; Clift, P.D.; Husson, L.; Müller, T.; Costantino, D.; Yin, A.; Xu, Z.; Cao, H.; Wang, Q. The Himalaya in 3D: Slab dynamics controlled mountain building and monsoon intensification. *Lithosphere* **2017**, *9*, 637–651. [CrossRef]
10. Ji, Y.; Yan, R.; Zeng, D.; Xie, C.; Zhu, W.; Qu, R.; Yoshioka, S. Slab dehydration in Sumatra: Implications for fast and slow earthquakes and arc magmatism. *Geophys. Res. Lett.* **2021**, *48*, e2020GL090576. [CrossRef]
11. Tackley, P.; Xie, S. STAG3D: A code for modeling thermo-chemical multiphase convection in Earth's mantle. In *Computational Fluid and Solid Mechanics 2003*; Elsevier: Amsterdam, The Netherlands, 2003; pp. 1524–1527.
12. Ji, Y.; Yoshioka, S.; Matsumoto, T. Three-dimensional numerical modeling of temperature and mantle flow fields associated with subduction of the Philippine Sea plate, southwest Japan. *J. Geophys. Res. Solid Earth* **2016**, *121*, 4458–4482. [CrossRef]
13. Yoshioka, S.; Murakami, K. Temperature distribution of the upper surface of the subducted Philippine Sea Plate along the Nankai Trough, southwest Japan, from a three-dimensional subduction model: Relation to large interplate and low-frequency earthquakes. *Geophys. J. Int.* **2007**, *171*, 302–315. [CrossRef]
14. Hayes, G.P.; Moore, G.L.; Portner, D.E.; Hearne, M.; Flamme, H.; Furtney, M.; Smoczyk, G.M. Slab2, a comprehensive subduction zone geometry model. *Science* **2018**, *362*, 58–61. [CrossRef] [PubMed]
15. Yoshii, T. Regionality of group velocities of Rayleigh waves in the Pacific and thickening of the plate. *Earth Planet. Sci. Lett.* **1975**, *25*, 305–312. [CrossRef]
16. Grose, C.J.; Afonso, J.C. Comprehensive plate models for the thermal evolution of oceanic lithosphere. *Geochem. Geophys. Geosystems* **2013**, *14*, 3751–3778. [CrossRef]
17. Argus, D.F.; Gordon, R.G.; DeMets, C. Geologically current motion of 56 plates relative to the no-net-rotation reference frame. *Geochem. Geophys. Geosystems* **2011**, 12. [CrossRef]
18. DeMets, C.; Gordon, R.G.; Argus, D.F. Geologically current plate motions. *Geophys. J. Int.* **2010**, *181*, 1–80. [CrossRef]
19. Vance, D.; Bickle, M.; Ivy-Ochs, S.; Kubik, P.W. Erosion and exhumation in the Himalaya from cosmogenic isotope inventories of river sediments. *Earth Planet. Sci. Lett.* **2003**, *206*, 273–288. [CrossRef]
20. Lenard, S.J.; Lave, J.; France-Lanord, C.; Aumaitre, G.; Bourles, D.L.; Keddadouche, K. Steady erosion rates in the Himalayas through late Cenozoic climatic changes. *Nat. Geosci.* **2020**, *13*, 448–452. [CrossRef]
21. Pedersen, V.K.; Egholm, D.L. Glaciations in response to climate variations preconditioned by evolving topography. *Nature* **2013**, *493*, 206–210. [CrossRef]
22. Guan, Q.; Zhu, D.-C.; Zhao, Z.-D.; Dong, G.-C.; Zhang, L.-L.; Li, X.-W.; Liu, M.; Mo, X.-X.; Liu, Y.-S.; Yuan, H.-L. Crustal thickening prior to 38 Ma in southern Tibet: Evidence from lower crust-derived adakitic magmatism in the Gangdese Batholith. *Gondwana Res.* **2012**, *21*, 88–99. [CrossRef]
23. Shellnutt, J.G.; Lee, T.-Y.; Brookfield, M.E.; Chung, S.-L. Correlation between magmatism of the Ladakh Batholith and plate convergence rates during the India–Eurasia collision. *Gondwana Res.* **2014**, *26*, 1051–1059. [CrossRef]
24. Mugnier, J.-L.; Huyghe, P. Ganges basin geometry records a pre-15 Ma isostatic rebound of Himalaya. *Geology* **2006**, *34*, 445–448. [CrossRef]
25. Guo, Z.; Wilson, M. Late Oligocene–early Miocene transformation of postcollisional magmatism in Tibet. *Geology* **2019**, *47*, 776–780. [CrossRef]
26. Replumaz, A.; Funiciello, F.; Reitano, R.; Faccenna, C.; Balon, M. Asian collisional subduction: A key process driving formation of the Tibetan Plateau. *Geology* **2016**, *44*, 943–946. [CrossRef]
27. Zou, Y.; Tian, X.; Yu, Y.; Pan, F.B.; Wang, L.; He, X. Seismic evidence for the existence of an entrained mantle flow coupling the northward advancing Indian plate under Tibet. *Earth Planet. Phys.* **2019**, *3*, 62–68. [CrossRef]
28. Govin, G.; Najman, Y.; Copley, A.; Millar, I.; Van der Beek, P.; Huyghe, P.; Grujic, D.; Davenport, J. Timing and mechanism of the rise of the Shillong Plateau in the Himalayan foreland. *Geology* **2018**, *46*, 279–282. [CrossRef]
29. Clark, M.K.; Bilham, R. Miocene rise of the Shillong Plateau and the beginning of the end for the Eastern Himalaya. *Earth Planet. Sci. Lett.* **2008**, *269*, 337–351. [CrossRef]
30. Liang, S.; Gan, W.; Shen, C.; Xiao, G.; Liu, J.; Chen, W.; Ding, X.; Zhou, D. Three-dimensional velocity field of present-day crustal motion of the Tibetan Plateau derived from GPS measurements. *J. Geophys. Res. Solid Earth* **2013**, *118*, 5722–5732. [CrossRef]
31. Huangfu, P.; Li, Z.-H.; Gerya, T.; Fan, W.; Zhang, K.-J.; Zhang, H.; Shi, Y. Multi-terrane structure controls the contrasting lithospheric evolution beneath the western and central–eastern Tibetan plateau. *Nat. Commun.* **2018**, *9*, 3780. [CrossRef]
32. Spicer, R.A.; Su, T.; Valdes, P.J.; Farnsworth, A.; Wu, F.-X.; Shi, G.; Spicer, T.E.; Zhou, Z. Why 'the uplift of the Tibetan Plateau' is a myth. *Natl. Sci. Rev.* **2021**, *8*, nwaa091. [CrossRef]
33. Kroon, D.; Steens, T.; Troelstra, S.R. Onset of monsoonal related upwelling in the western Arabian sea. *Proc. Ocean Drill. Program Sci. Results* **1991**, *117*, 257–263.
34. Chemenda, A.I.; Burg, J.-P.; Mattauer, M. Evolutionary model of the Himalaya–Tibet system: Geopoem: Based on new modelling, geological and geophysical data. *Earth Planet. Sci. Lett.* **2000**, *174*, 397–409. [CrossRef]
35. Husson, L.; Bernet, M.; Guillot, S.; Huyghe, P.; Mugnier, J.-L.; Replumaz, A.; Robert, X.; Van der Beek, P. Dynamic ups and downs of the Himalaya. *Geology* **2014**, *42*, 839–842. [CrossRef]
36. She, Y.; Fu, G. Uplift mechanism of the highest mountains at eastern himalayan syntaxis revealed by in situ dense gravimetry. *Geophys. Res. Lett.* **2020**, *47*, e2020GL091208. [CrossRef]

37. Boos, W.R.; Kuang, Z. Dominant control of the South Asian monsoon by orographic insulation versus plateau heating. *Nature* **2010**, *463*, 218–222. [CrossRef] [PubMed]
38. Clift, P.D.; Hodges, K.V.; Heslop, D.; Hannigan, R.; Van Long, H.; Calves, G. Correlation of Himalayan exhumation rates and Asian monsoon intensity. *Nat. Geosci.* **2008**, *1*, 875–880. [CrossRef]
39. Iaffaldano, G.; Husson, L.; Bunge, H.-P. Monsoon speeds up Indian plate motion. *Earth Planet. Sci. Lett.* **2011**, *304*, 503–510. [CrossRef]
40. Chen, Q.; Freymueller, J.T.; Wang, Q.; Yang, Z.; Xu, C.; Liu, J. A deforming block model for the present-day tectonics of Tibet. *J. Geophys. Res. Solid Earth* **2004**, 109. [CrossRef]
41. Beaumont, C.; Jamieson, R.A.; Nguyen, M.; Lee, B. Himalayan tectonics explained by extrusion of a low-viscosity crustal channel coupled to focused surface denudation. *Nature* **2001**, *414*, 738–742. [CrossRef]
42. Burkett, E.R.; Billen, M.I. Three-dimensionality of slab detachment due to ridge-trench collision: Laterally simultaneous boudinage versus tear propagation. *Geochem. Geophys. Geosyst.* **2010**, *11*, Q11012. [CrossRef]
43. Hirth, G.; Kohlstedf, D. Rheology of the upper mantle and the mantle wedge: A view from the experimentalists. *Geophys. Monogr.-Am. Geophys. Union* **2003**, *138*, 83–106.
44. Ranalli, G. *Rheology of the Earth*; Springer Science & Business Media: Berlin/Heidelberg, Germany, 1995.
45. Stein, C.A.; Stein, S. A model for the global variation in oceanic depth and heat flow with lithospheric age. *Nature* **1992**, *359*, 123–129. [CrossRef]
46. Müller, R.D.; Sdrolias, M.; Gaina, C.; Roest, W.R. Age, spreading rates, and spreading asymmetry of the world's ocean crust. *Geochem. Geophys. Geosyst.* **2008**, *9*, Q04006. [CrossRef]
47. Hacker, B.R.; Abers, G.A.; Peacock, S.M. Subduction factory 1. Theoretical mineralogy, densities, seismic wave speeds, and H_2O contents. *J. Geophys. Res. Solid Earth* **2003**, *108*, 2029. [CrossRef]
48. Ji, Y.; Yoshioka, S.; Manea, V.C.; Manea, M.; Matsumoto, T. Three-dimensional numerical modeling of thermal regime and slab dehydration beneath Kanto and Tohoku, Japan. *J. Geophys. Res. Solid Earth* **2017**, *122*, 332–353. [CrossRef]
49. Pollack, H.N.; Hurter, S.J.; Johnson, J.R. Heat flow from the Earth's interior: Analysis of the global data set. *Rev. Geophys.* **1993**, *31*, 267–280. [CrossRef]
50. Li, C.-F.; Lu, Y.; Wang, J. A global reference model of Curie-point depths based on EMAG2. *Sci. Rep.* **2017**, *7*, srep45129. [CrossRef]
51. Li, C.F.; Wang, J.; Lin, J.; Wang, T. Thermal evolution of the North Atlantic lithosphere: New constraints from magnetic anomaly inversion with a fractal magnetization model. *Geochem. Geophys. Geosyst.* **2013**, *14*, 5078–5105. [CrossRef]
52. Turcotte, D.L.; Schubert, G. *Geodynamics*; Cambridge University Press: Cambridge, UK, 2002.
53. Ji, Y.; Yoshioka, S.; Manea, V.C.; Manea, M. Seismogenesis of dual subduction beneath Kanto, central Japan controlled by fluid release. *Sci. Rep.* **2017**, *7*, 16864. [CrossRef]
54. Suenaga, N.; Ji, Y.; Yoshioka, S.; Feng, D. Subduction thermal regime, slab dehydration, and seismicity distribution beneath Hikurangi based on 3-D simulations. *J. Geophys. Res. Solid Earth* **2018**, *123*, 3080–3097. [CrossRef]
55. Ji, Y.; Yoshioka, S. Slab dehydration and earthquake distribution beneath southwestern and central Japan based on three-dimensional thermal modeling. *Geophys. Res. Lett.* **2017**, *44*, 2679–2686. [CrossRef]
56. Suenaga, N.; Yoshioka, S.; Ji, Y. 3-D thermal regime and dehydration processes around the regions of slow earthquakes along the Ryukyu Trench. *Sci. Rep.* **2021**, *11*, 11251. [CrossRef] [PubMed]
57. Ji, Y.; Yoshioka, S.; Banay, Y.A. Thermal state, slab metamorphism, and interface seismicity in the Cascadia subduction zone based on 3-D modeling. *Geophys. Res. Lett.* **2017**, *44*, 9242–9252. [CrossRef]
58. Wang, K.; Hyndman, R.D.; Yamano, M. Thermal regime of the Southwest Japan subduction zone: Effects of age history of the subducting plate. *Tectonophysics* **1995**, *248*, 53–69. [CrossRef]
59. Iwamori, H. Heat sources and melting in subduction zones. *J. Geophys. Res. Solid Earth* **1997**, *102*, 14803–14820. [CrossRef]
60. Christensen, U.R. The influence of trench migration on slab penetration into the lower mantle. *Earth Planet. Sci. Lett.* **1996**, *140*, 27–39. [CrossRef]
61. Trabant, C.; Hutko, A.R.; Bahavar, M.; Karstens, R.; Ahern, T.; Aster, R. Data products at the IRIS DMC: Stepping stones for research and other applications. *Seismol. Res. Lett.* **2012**, *83*, 846–854. [CrossRef]
62. Wessel, P.; Smith, W.H. New, improved version of Generic Mapping Tools released. *Eos Trans. Am. Geophys. Union* **1998**, *79*, 579. [CrossRef]

applied sciences

MDPI

Article

Non-Linear 3D Satellite Gravity Inversion for Depth to the Basement Estimation in a Mexican Semi-Arid Agricultural Region

Raúl Ulices Silva-Ávalos, Hugo Enrique Júnez-Ferreira *, Julián González-Trinidad and Carlos Bautista-Capetillo

Licenciatura en Ciencia y Tecnología del Agua, Universidad Autónoma de Zacatecas, Carretera Zacatecas-Guadalajara Km. 6, Ejido La Escondida, Zacatecas 98160, Mexico; raul.ulices.silva@gmail.com (R.U.S.-Á.); jgonza@uaz.edu.mx (J.G.-T.); baucap@uaz.edu.mx (C.B.-C.)
* Correspondence: hejunez@uaz.edu.mx; Tel.: +52-492-108-2170

Abstract: In Mexico, agriculture in semi-arid regions is highly dependent on groundwater resources, where most of the aquifers' characterization is a pending task. In particular, the depth to the basement is unknown for most of the Mexican territory. Hence, the development and performance of new techniques for the basement relief estimation is imperative for further hydrogeological studies. In this paper, we present a depth to the basement estimation using non-linear gravimetric inversion employing satellite data. Gravity forward modeling was implemented using both gravitational attraction due to juxtapositioned blocks and gravimetric non-linear inversion using conjugate gradient least squares to minimize the objective function in terms of a depth model. All of this took place under the sparse system framework. We present a synthetic result using the SEG-Bishop depth model taken for calibration purposes. Then, we recollected gravity data from The Satellite Geodesy group from SCRIPPS for the depth to the basement estimation of an unconfined aquifer in the northern-central semi-arid region of Zacatecas, Mexico. Both synthetic and satellite data were recovered, consistent depth models for both cases were presented, and a comparison with conventional gravimetric linear inversion for density estimation was performed.

Keywords: gravity; satellite; inversion; basement; conjugate gradient

1. Introduction

Mexico's population demands a water volume of 88,839.7 hm^3 annually for several needs [1]. Despite the wrongful belief that industry demands most of it, agriculture use 75.7% of the total water, where 63.4% of it is obtained on the surface and the other 36.6% from the subsoil. However, the major source of water for semi-arid regions corresponds to groundwater systems. Zacatecas state is one of such regions with large periods of drought, with a concession of 1688.4 hm^3 of water [2], where 85.3% is destined for agricultural activities to overcome the low precipitation problems in the region. Historically, the groundwater budget has been the principal tool to determine the availability of water within the aquifers in Mexico; however, several authors have demonstrated the need for the construction of groundwater flow models to support sustainable water management policies [3,4]. To achieve this goal, the understanding of groundwater flow systems needs to be increased, which requires a better knowledge of the hydrogeological properties of the subsoil, where data are usually scarce and sometimes outdated [5,6]. A fundamental task for further studies of groundwater functioning is the determination of basement geometry [7], which is necessary for the construction of conceptual and numerical groundwater models to achieve the sustainable management of water resources [8,9]. This task can be achieved using remote sensing, such as unmanned aerial vehicles [10–13] for near-shore studies. On the other hand, satellite data have become the most popular solution when equipment is not available, providing high resolution and faster results on a larger scale for multispectral imagery [14] or satellite gravity data.

Gravimetric studies have been widely used for depth to the basement estimation for geological mapping, exploring oil reservoirs, and modeling groundwater systems. The basement topography study is important for the understanding of the groundwater flow, which happens within sedimentary deposits above igneous or metamorphic rocks, having notorious density contrasts that produce gravity anomalies that provide structural information about the boundary between those layers. Such gravity interpretations can be grouped into two geophysical data inversion categories: density estimation and depth estimation. For density estimation, the geometry of the model is set to be known, and then the density of the cells or grid points is determined by linear data inversion [15,16], which can also be performed by incorporating a priori information and depth-dependence weights in the inversion to recover a density model with a more proper structure at different depths [17,18]. Alternatively, for depth estimation, a density contrast is assumed and non-linear data inversion is performed for the geometry of a parallelepiped. Such a gravity contrast can be considered constant [19,20] or varying with depth using exponential [21], quadratic [22], cubic [23], or hyperbolic [24] functions. These studies are well-tested for near-shore data in sedimentary basins where borehole data for oil exploration are available; however, such gravity constant functions for depth can be difficult to extrapolate for inland basin studies, where the geology is more complex.

With respect to the optimization, these works minimize an objective function performing Gauss–Newton optimization, which leads to a square matrix inversion that can be computationally expensive for high-resolution problems. Nowadays, additional optimization alternatives are being continuously developed for other geophysics areas, such as seismic methods, using conjugate gradient techniques, significantly reducing the computation cost of the inversion. In this context, the conjugate gradient least squares (CGLS) method was developed as an alternative for Gauss–Newton methods, reducing the computational time and storage.

We present a non-linear gravimetric inversion for depth to the basement estimation for an unconfined aquifer of the north-central region of Zacatecas, Mexico. The inverse problem was optimized by developing a CGLS algorithm for obtaining a depth model. A constant density contrast was assumed due to the lack of a priori information and the complexity of the region of interest.The algorithm was calibrated by computing synthetic gravity data using the Bishop model, recovering the depth model with the inversion scheme proposed. The procedure was repeated using gravity inversion with satellite data recollected from The Satellite Geodesy group from SCRIPPS. Finally, a comparison between density estimations and depth to the basement estimation was performed.

2. Geology and Previous Studies

The study area was located in the Mexican Altiplano, constituting part of the Central Terrane and most of the Guerrero Composite Terrane, formed between Late Jurassic and early Cretaceous [25]. This terrane is characterized by volcanic–sedimentary successions formed in a back-arc tectonic setting; we address Centeno-Garcia [26] for a complete compilation of works in this area. Locally, the study area was in Zacatecas State (Figure 1a), at the region of the Aguanaval (unconfined) aquifer (Figure 1b), situated in Fresnillo and Rio Grande municipalities. Historically, this area has been an important mine production source, mostly for silver exploitation, since the XVI century. The stratigraphy of this region is well-exposed and made up of shale, quartz-rich sandstone, scarce limestone, and basaltic pillow lavas [27]; however, basement rocks do not crop out in the Fresnillo area [28] and information on its depth remain unknown.

(a)

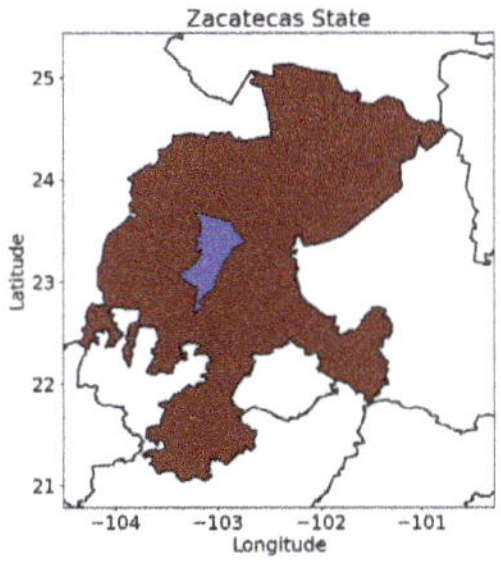

(b)

Figure 1. Location of Zacatecas State in Mexico (**a**) and the Aguanaval Aquifer polygon (**b**) for the study.

Agricultural and industrial activities predominate within the Aguanaval aquifer, which over-exploit the groundwater that, in itself, has low recoverability due to low precipitations. Other than geological, works in this region involve hydro-geological [29] and technical reports such as Comisión Nacional del Agua (CONAGUA) [2]. Despite the mining, industrial and agricultural importance of this region, there is a lack of geophysical academic works published—only AMT [30] and TEM [31] studies are found—where gravity and seismic interpretations need to be carried out for a better understanding of the basement relief; therefore, the importance of this study for depth to the basement estimation is emphasized.

3. Methodology

Newton's law of universal gravitation provides the vertical component of the gravity acceleration at a point $\vec{r_0} = (x_0, y_0, z_0)$ given by

$$g_z(x,y,z) = \gamma \int_{\Omega} \rho(x,y,z) \frac{z_0 - z}{\left[(x_0-x)^2 + (y_0-y)^2 + (z_0-z)^2\right]^{3/2}} \mathrm{d}x\mathrm{d}y\mathrm{d}z, \tag{1}$$

where $\gamma = 6.672 \times 10^{-11}\ \mathrm{m^3kg^{-1}s^{-2}}$ is the gravitational constant, ρ the density, and Ω the volume for the Cartesian coordinates of the body. We considered the gravitational attraction due to a rectangular prism of constant density [32,33], where the analytic solution of Equation (1) will be

$$g_z = \rho\gamma \left[z \tan^{-1}\left(\frac{xy}{zr}\right) - x\ln(y+r) - y\ln(x+r) \right] \Bigg|_{\Delta x_1}^{\Delta x_2} \Bigg|_{\Delta y_1}^{\Delta y_2} \Bigg|_{\Delta z_1}^{\Delta z_2}, \tag{2}$$

where $r = \sqrt{x^2+y^2+z^2}$, $\Delta x_k = x_0 - x_k$, $\Delta y_k = y_0 - y_k$, and $\Delta z = z_0 - z_k$ for $k = 1,2$ for the corners of the prism (Figure 2a). The gravity observation vector g_i ($i = 1, N$) was computed by adding the contribution for all of the j-th prisms ($j = 1, M$) of the density model, where N is the total of gravimetric stations measured along the surface and $M = n_x \times n_y \times n_z$ is the number prism of the model. Typically, the 3D gravimetric forward modeling is accomplished by computing Equation (2) for a considerable number of n_z grid points. For the depth of the basement estimation, only two prisms on the z-direction are considered (Figure 2b) for a large number of x, y grid points. The upper j-th prism will contain the density contrast ($\Delta\rho_1$) between the surface ($z_{\min}$) and depth z_j and the lower j-th prism will contain the density ($\Delta\rho_2$) for the bottom part of the model up to $z_{\max}$ for the maximum depth of study. In this case, z_j aimed for the depth of the basement, and the density contrast was considered to be known.

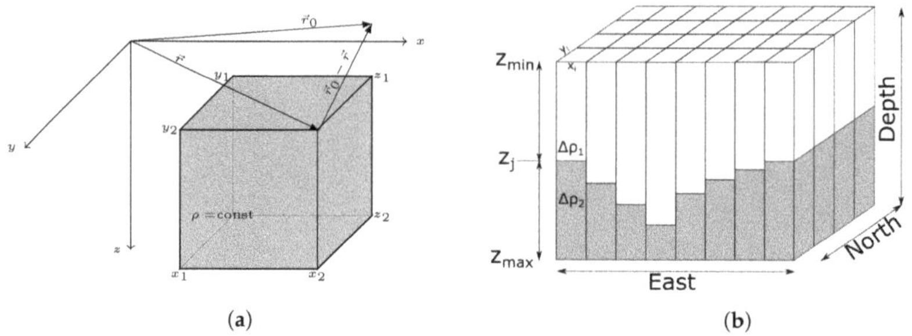

(a) (b)

Figure 2. (**a**) Single prism of constant density in Cartesian coordinates for gravity measurements and (**b**) an example of $n_x \times n_y \times 2 = 64$ juxtapositioned prisms for depth estimation.

For the inverse problem, the objective function Q proposed seeks to minimize the discrepancies between the observed gravity data $\mathbf{d}^{\text{obs}}$ and the synthetic $\mathbf{d}^{\text{cal}}$ using the L-2 norm as follows:

$$Q(\mathbf{m}) = ||\mathbf{C}_d(\mathbf{d}^{\text{obs}} - \mathbf{d}^{\text{cal}})||^2 + \alpha_{\text{reg}}^2||\nabla\mathbf{m}||^2 + ||\mathbf{C}_m(\mathbf{m} - \mathbf{m}_{\text{apr}})||^2, \tag{3}$$

where $\mathbf{m}$ is the model vector that minimizes Q and $\mathbf{C}_d$ is the diagonal matrix containing the covariance of the data. Here, a Tikhonov regularization constraint on the model was implemented, pondered by the parameter α_{reg}. The right term corresponds to the a priori information of the model, being $\mathbf{C}_m = \alpha_{\text{apr}}\mathbf{I}$ for a whole constraint on the model at the inversion. If borehole data are available, the matrix $\mathbf{C}_m$ will act as a weight on the regions of the model where the well is located, which will significantly reduce the non-uniqueness of the problem and will improve the accuracy of the inversion.

For density estimation of the model, the model parameter $\mathbf{m} \equiv \mathbf{m}^\rho$. It is well known that the inverse problem is linear for density estimation, as seen in Equation (2). The minimization of 3 with respect to the density model will result in the system

$$\left[\mathbf{A}^{\mathbf{T}}\mathbf{C}_{\mathbf{d}}^{\mathbf{T}}\mathbf{C}_{\mathbf{d}}\mathbf{A} + \alpha^2\mathbf{D}^{\mathbf{T}}\mathbf{D} + \mathbf{C}_{\mathbf{m}}^{\mathbf{T}}\mathbf{C}_{\mathbf{m}}\right]\mathbf{m}^\rho = \mathbf{A}^{\mathbf{T}}\mathbf{C}_{\mathbf{d}}^{\mathbf{T}}\mathbf{C}_{\mathbf{d}}\mathbf{d}^{\text{obs}} + \mathbf{C}_{\mathbf{m}}^{\mathbf{T}}\mathbf{C}_{\mathbf{m}}\mathbf{m}_{\text{apr}}^{\text{æ}}, \tag{4}$$

where $\mathbf{D}$ is the discrete operator of ∇ and $\mathbf{A}$ is the density kernel equal to the derivative of Equation (2) with respect to ρ (linear). For depth estimation of the model, the model parameter $\mathbf{m} \equiv \mathbf{m}^z$, where z is the depth of the prism (Figure 2b) and z does not denote exponent. By inspecting Equation (2), it is evident that the inversion by depth is a non-linear problem. This issue is solved by performing a linearization around a starting model $\mathbf{m}_0^{\mathbf{z}}$, resulting in the following system:

$$\left[\mathbf{J}^{\mathbf{T}}\mathbf{C}_{\mathbf{d}}^{\mathbf{T}}\mathbf{C}_{\mathbf{d}}\mathbf{J} + \alpha^2\mathbf{D}^{\mathbf{T}}\mathbf{D} + \mathbf{C}_{\mathbf{m}}^{\mathbf{T}}\mathbf{C}_{\mathbf{m}}\right]\mathbf{m}^z = \mathbf{J}^{\mathbf{T}}\mathbf{C}_{\mathbf{d}}^{\mathbf{T}}\mathbf{C}_{\mathbf{d}}(\mathbf{d}^{\text{obs}} + \mathbf{J}\mathbf{m}_0^z) + \mathbf{C}_{\mathbf{m}}^{\mathbf{T}}\mathbf{C}_{\mathbf{m}}\mathbf{m}_{\text{apr}}^{\mathbf{z}}. \tag{5}$$

In this case, the matrix $\mathbf{J}$ is the Jacobian corresponding to the derivative of Equation (2) with respect to z evaluated on the starting model $\mathbf{m}_0^z$; with this, Frechet derivatives are avoided, e.g, methods such as perturbations and sensitivity approaches, which can be computationally expensive. The solution of Equations (4) and (5) can be achieved by inverting the symmetric square matrix of the left size, commonly called $\mathbf{A}^{\mathbf{T}}\mathbf{A}$ inversion. However, this implementation requires the storage and computation of such a matrix, which results in a large computational time, and sometimes the memory is surpassed for large scale problems. One solution for this problem is to perform conjugate gradient least squares to solve a system in the form of $\mathbf{Gm} = \mathbf{d}$. This method internally computes the square matrix of Equation (4), allocating only the matrix **A** and the sparse matrix **D**, $\mathbf{C}_m$, and $\mathbf{C}_d$. For example, for depth inversion, the system $\mathbf{Gm} = \mathbf{d}$ is solved as

$$\begin{bmatrix} \mathbf{C_d J} \\ \alpha_{\text{reg}} \mathbf{D} \\ \mathbf{C_m} \end{bmatrix} \mathbf{m}^z = \begin{bmatrix} \mathbf{C_d}(\mathbf{d}^{\text{obs}} + \mathbf{J m}_0^z) \\ \mathbf{0} \\ \mathbf{C}_m \mathbf{m}_{\text{apr}}^z \end{bmatrix}, \tag{6}$$

In this case, the augmented matrix **G** were large and sparse due to the discrete operations for the Tikhonov regularization, and the model vector $\mathbf{m}^\rho$ was not modified. Notice that multiplication of Equation (6) by $\mathbf{G}^T$ yields to Equation (5). The algorithm to solve the system $\mathbf{Gm} = \mathbf{d}^{\text{CG}}$ is presented in Algorithm 1, modified from Nocedal & Wright [34]. Notice that an efficient implementation of this algorithm requires that all the matrices be stored in a sparse representation. We used coordinate list (COO) for storing such sparse matrices, given that the matrix and its transpose have to be accessed simultaneously [35].

Algorithm 1 CGLS algorithm to iteratively solve the problem $Gm = d^{CG}$

Require: m_0 (Starting model), G (Augmented kernel), d^{CG} (Data Vector), n (Iterations).
Ensure: $Gm = d^{CG}$
$d_0 \Leftarrow d^{CG}$
$r_0 \Leftarrow G^T d^{CG}$
$p_0 \Leftarrow r_0$
$t_0 \Leftarrow G p_0$
while $k < n$ or stop criteria is satisfied **do**
    $\alpha_k \Leftarrow ||r_{k-1}||^2 / ||t_{k-1}||^2$
    $m_k \Leftarrow m_{k-1} + \alpha_k p_{k-1}$
    $d_k \Leftarrow d_{k-1} - \alpha_k t_{k-1}$
    $r_k \Leftarrow G^T d_k$
    $\beta_k \Leftarrow ||r_k||^2 / ||r_{k-1}||^2$
    $p_k \Leftarrow r_k + \beta_k p_{k-1}$
    $t_k \Leftarrow G p_k$
    $k \Leftarrow k + 1$
end while
return m_k (Final model)

4. Results

We tested the algorithm developed on a single synthetic model for calibration purposes. First, a non-linear inversion was performed for the depth estimation of the basement. Finally, we present the results due to satellite gravity data for the depth to the basement estimation of the area of interest.

4.1. Synthetic Case: Bishop Model

In order to test the developed algorithm , we performed the inversion by depth for the Bishop model, a synthetic basement model created by upscaling and shifting the topographic data of an area of Bishop, CA, USA [36]. This model contains a variety of structures that resemble the complexity of real basement relief: two long faults, en echelon arrays of smaller faults, transfer zones between faults, and an unfaulted deep basin area. This features allows the model to be exploited by various authors to test new methods for depth to the basement estimation using magnetic data [36,37] and gravity data [38,39].

In the north-east region of the model, the shallower parts of the basement ($\approx$100 m) are presented and, in the south-west region of the model, the deepest parts of the basement are located up to $\approx$9300 m, as seen in the 3D view of Figure 3 and the 2D floor view of Figure 4a, which acts as an unknown true depth model in the inversion. A total of $n_x \times n_y \times 2 = 61 \times 61 \times 2 = 3162$ prisms are considered in a region of interest $0 \leq x \leq 380$ km for the east and $0 \leq y \leq 402$ km for the north, given a spacing of $\Delta x = 6.33$ km and $\Delta y = 6.7$ km (width and length) for each prism. The density for the upper prism is $\rho_1 = 2100\ \mathrm{g/cm^3}$, corresponding to sedimentary rocks above the basement of density $\Delta\rho_2 = 2750\ \mathrm{g/cm^3}$ for the lower prisms. 3D gravity forward modeling was performed, measuring along all of the grid points on the surface, giving a total of $n_x \times n_y \times 2 = 61 \times 61 = 1581$ gravimetric stations (length of the observed data vector d^{obs}). The gravity data show anomalies corresponding to the high and lower depths of the Bishop depth to the basement model (Figure 4b).

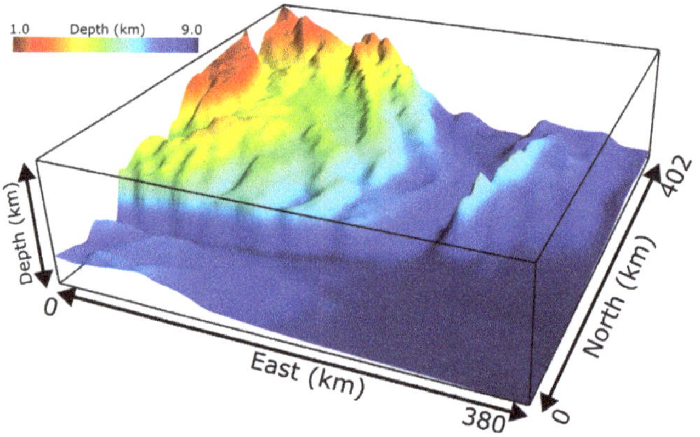

Figure 3. Bishop basement example for gravity data inversion by depth in kilometers.

For geophysical data inversion, *a priori* information was not considered; therefore, $\mathbf{C}_m = 0$ and the parameter of regularization were chosen using the L-curve method for Tikhonov regularization. We performed 10 iterations of the CGLS method (Equation (6)) starting from a constant depth model $\mathbf{m}_0^z$ of 1000 m. It is worth mentioning that the starting model does not affect the final solution, but it can help for a fast convergence of the problem. The depth model after 10 iterations of the CGLS is shown in Figure 4c, and the final synthetic response due to this model is shown in Figure 4d.

The result obtained clearly exhibits a successful recovery of the structure of the basement with the correct depths and, henceforth, the gravimetric anomalies were fitted properly, with few discrepancies. These differences can be observed in Figure 4e,f for the model error and the gravity residual, respectively. For the depth model, errors up to $\approx\pm0.7$ km appear, mainly located at regions where there is an abrupt change in lateral variation in density (see Figure 3); however, in regions where the basement behaves predominantly horizontal, the errors are almost negligible ($\approx\pm0.01$ km or $\approx\pm10$ m) as seen in the scale of Figure 4a,c. Although this error happens where lateral variation is presented, we infer that this is a product of the Tikhonov regularization constraint; the smoothness interposed does not recover parts where the total variation in the model is presented. We address Portniaguine & Zhdanov [40] for an optional regularization where a focused model is needed. Consequently, this error in the model also produces a gravity residual in such lateral-variation regions of up to 2.15 mGals, which is not harmful for the final results.

A single 1D depth to the basement profile is presented, located at $y = 201$ km for an x-variation as seen in Figure 5a. Again, the structure of the Bishop model is well-recovered in regions where there is not an abrupt change in density ($x = 280$ km). With respect to the convergence of the algorithm, the plot of the objective function (Misfit) for each iteration shows a fast convergence at the earliest iterations (Figure 5b). The CGLS algorithm encounters a stagnation in the misfit after 10 iterations, which is a sign that the minimum has been reached.

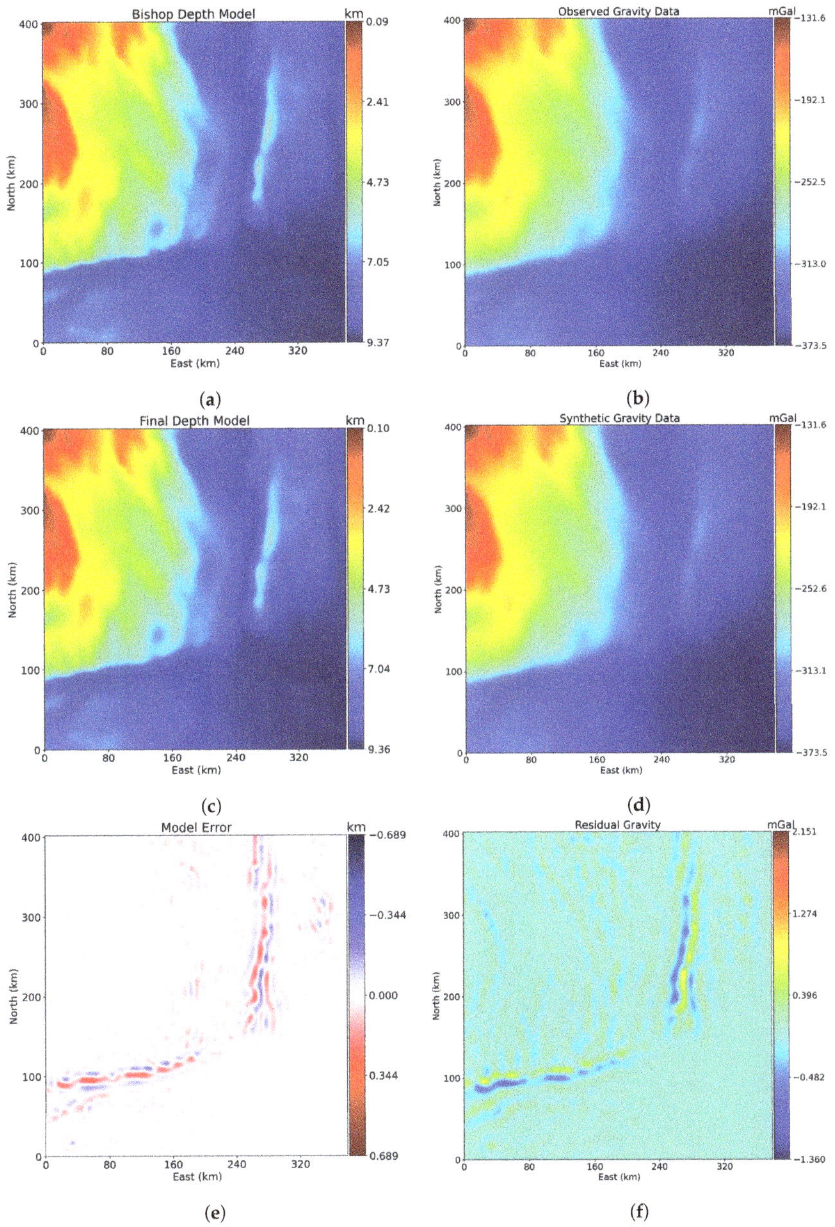

Figure 4. Inversion results for the Bishop true depth model (**a**) with gravimetric response (**b**) for the observed data vector. The final depth model after 10 iterations of CGLS (**c**) produces the gravimetric response of (**d**). The model error between the true and the final model (**e**) also produces error in the data (**f**) up to 2.15 mGal.

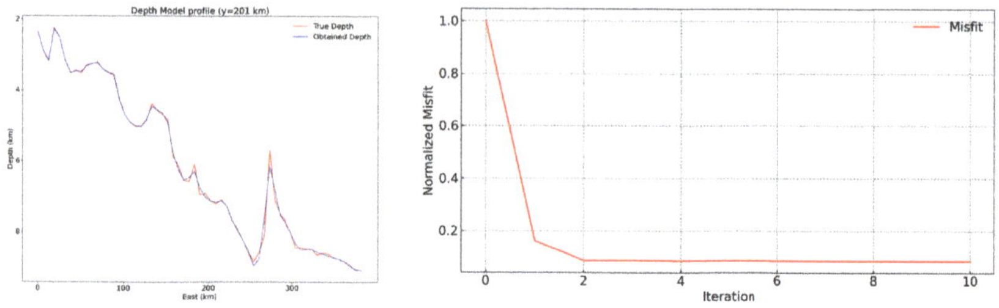

(**a**) (**b**)

Figure 5. 1D profile along x-direction (**a**) for the true (red) and final (blue) depth model at $y = 201$ km. (**b**) The objective function cost (Misfit) is plotted for each iteration (10).

4.2. Depth to the Basement Estimation for Real Gravity Data

The study area covers a minima and maxima latitude of 22.58 and 23.69 and minima and maxima longitude of −103.525 and −102.625, respectively; therefore, this area surrounds the aquifer polygon as seen in Figures 1 and 6a.

We recollected gravity data from The Satellite Geodesy research group at the Cecil H. and Ida M. Institute of Geophysics and Planetary Physics, Scripps Institution of Oceanography, University of California San Diego [41]. After geophysical processing, the observed gravity data (Figure 6a) of the area of study present consistencies with the geology and the topography information of the region of interest (Figure 6b). The interpretation of these gravity anomalies is straightforward: the sedimentary rock deposits, where the Aguanaval River streams, match with the lowest gravity anomalies, where the central part of the aquifer lies; on the other hand, the highest gravity anomalies are presented at the boundaries of the aquifer because these regions act as a source for the recharge of the aquifer due to high elevation.

The inversion result after 10 iterations of the algorithm is shown in Figure 6c. It is clear that the data have been fit for all zones, except for some regions with errors up to ≈2 mGal (see scale of Figure 6c), mostly at the regions where the gravity maximum and minimum are presented. Such errors can be reduced if the weight on the regularization is lowered; however, models with a higher roughness could be presented. As expected, the depth model presents a similar shape to the observed gravity data. A depth to the basement between 1.8 and 5.5 km is estimated, where the maximum depth is located at the central part of the aquifer limits as seen in Figure 6d for a 2D view and Figure 6e for a 3D view of the basement relief. The model does not present zero depths, which means that the basement does not crop out to the surface. Such depths are consistent with the topography and the geology of the study zone. The convergence of this result is well-behaved and fast in the same way as the synthetic example; therefore, the cost function will not be presented.

One-directional profiles are picked for interpretation, centered at the domain of the aquifer, horizontally and vertically. The north–south profile, along a constant latitude of 103°W, suggests the presence of an inland basin filled with sedimentary rocks according to the geology of the study zone. In the same way, the west–east profile covers the aquifer perpendicularly, where such a basin is narrower. This suggests that the aquifer is elongated in the north–south direction and slightly inclined in the same direction as the aquifer's polygon.

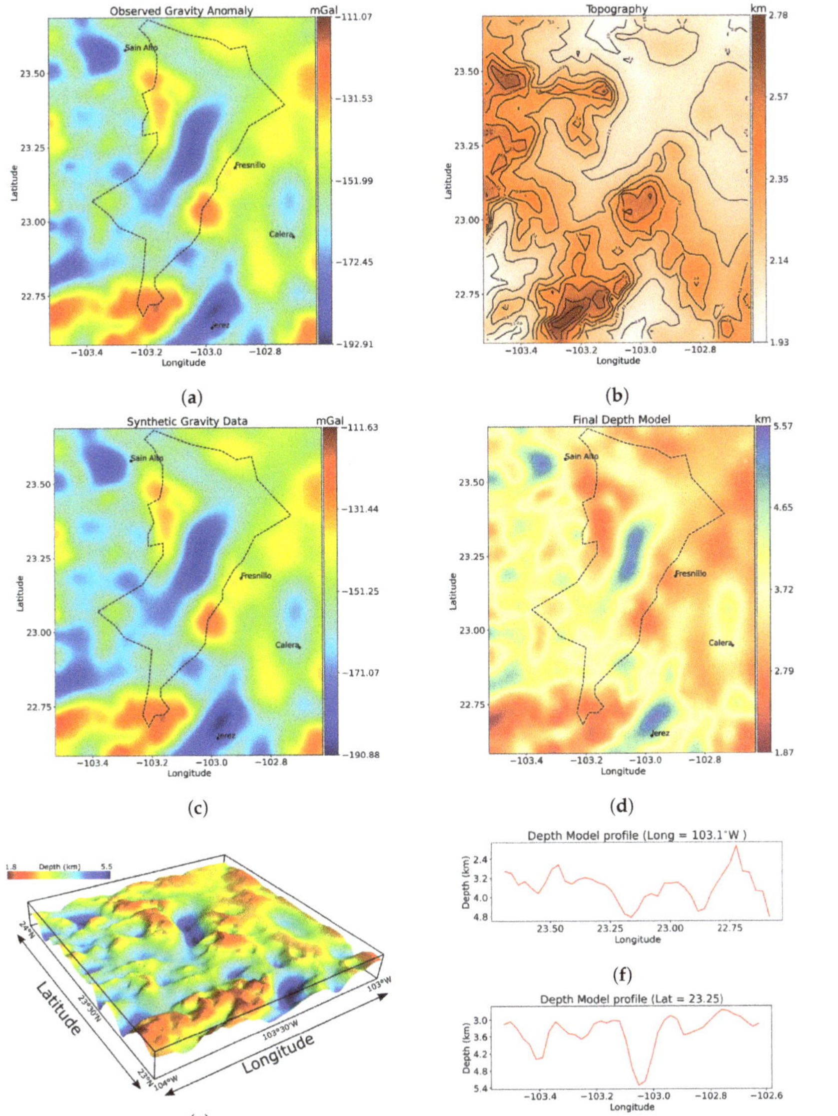

Figure 6. Observed gravimetric-satellite data for the area of interest (**a**) and its topography (**b**). The synthetic data after 10 iterations of CGLS method (**c**) recovers the observed data, obtaining the final depth model of (**d**). A 3D view of the basement (kilometers) is shown in (**e**). 1D profiles were picked for interpretation for N-S orientation at longitude 103° W (**f**) and for W-O orientation at latitude 23.25° N (**g**).

5. Conclusions

We developed an algorithm for depth to the basement estimation using the iterative CGLS method for the optimization, employing large sparse storage for the system of equations. The algorithm was calibrated using the Bishop synthetic model, successfully recovering the depth model and the observed gravity data, with neglectable errors due to Tikhonov regularization.

For satellite gravity data, the algorithm was applied for an unconfined aquifer located in the north-central region of Zacatecas, Mexico. The estimated relief has a depth between 1.8 and 5.5 km, where the deepest region is presented at the central part of the aquifer and shallower regions are approximately at the boundaries of the aquifer limits. This relief matches with the geology and the topography of the zone; however, the calibration remains as pending due to the lack of borehole data information (for an a priori model in the inversion) and the null acquaintance of other geophysical data available as an academic work or field report. Hence, we catalogued this depth to the basement model as an estimation, subject to modifications if a priori information could be obtained, but the relief will keep the same structure.

Finally, we conclude that this is an important step for a better understanding of the subsurface of the area of interest since the processes that occur within the aquifer need to be detailed, such as groundwater flow, for the sustainable management of this natural and vital resource.

Author Contributions: R.U.S.-Á. carried out the collection and processing of satellite gravity data. He also developed the algorithm for data inversion and performed the examples of the work. He mainly drafted the manuscript; H.E.J.-F. contributed with the data analysis and geological interpretation. He partially drafted the manuscript; J.G.-T. helped to draft the manuscript as a reviewer. He conceived the main goal of the project, supervising the objectives; C.B.-C. contributed to the review of the manuscript. He provided the sources needed for the research. All authors have read and agreed to the published version of the manuscript.

Funding: This research received no external funding.

Data Availability Statement: Not applicable.

Acknowledgments: To CONACyT for the postdoctoral fellowship granted to Raúl Ulices Silva Avalos under the CVU 546426. We also acknowledge *Consejo Zacatecano de Ciencia Tecnología e innovación* (COZCyT) for additional resources provided for this work.

Conflicts of Interest: The authors declare no conflict of interest.

References

1. CONAGUA. *Estadísticas del Agua en México;* Sistema Nacional de Información del Agua; Diario Oficial de la Federación, Government of Mexico: Mexico City, Mexico, 2019.
2. CONAGUA. *Actualización de la Disponibilidad Media Anual de Agua en el acuíFero Aguanaval Estado de Zacatecas;* Techreport 3214; Subdirección General Técnica, Gerencia de Aguas Subterráneas; Diario Oficial de la Federación, Government of Mexico: Mexico City, Mexico, 2020.
3. Bredehoeft, J.D.; Alley, W.M. Mining groundwater for sustained yield. *Bridge* **2014**, *44*, 33–41.
4. Konikow, L.F.; Reilly, T.E.; Barlow, P.M.; Voss, C.I. Groundwater modeling. In *The Handbook of Groundwater Engineering;* CRC Press: New York, NY, USA, 2006; pp. 815–866.
5. Carrillo-Rivera, J.J. Lack of a conceptual system view of groundwater resources in Mexico. *Hydrogeol. J.* **2003**, *11*, 519–520. [CrossRef]
6. Gleeson, T.; Cardiff, M. The return of groundwater quantity: A mega-scale and interdisciplinary "future of hydrogeology"? *Hydrogeol. J.* **2013**, *21*, 1169–1171. [CrossRef]
7. Hildenbrand, T.G.; Langenheim, V.E.; Mankinen, E.A.; McKee, E.H. *Inversion of Gravity Data to Define the Pre-Tertiary Surface and Regional Structures Possibly Influencing Groundwater Flow in the Pahute Mesa–Oasis Valley Region, Nye County, Nevada;* US Geological Survey Open-File Report; US Geological Survey: Reston, VA, USA, 1999; pp. 99–149.
8. Bredehoeft, J.D. The water budget myth revisited: Why hydrogeologists model. *Groundwater* **2002**, *40*, 340–345. [CrossRef] [PubMed]
9. Voss, C.I. Editor's message: Groundwater modeling fantasies—Part 1, adrift in the details. *Hydrogeol. J.* **2011**, *19*, 1281–1284. [CrossRef]
10. Bandini, F.; Lopez-Tamayo, A.; Merediz-Alonso, G.; Olesen, D.; Jakobsen, J.; Wang, S.; Garcia, M.; Bauer-Gottwein, P. Unmanned aerial vehicle observations of water surface elevation and bathymetry in the cenotes and lagoons of the Yucatan Peninsula, Mexico. *Hydrogeol. J.* **2018**, *26*, 2213–2228. [CrossRef]

11. Dabrowski, P.S.; Specht, C.; Specht, M.; Burdziakowski, P.; Makar, A.; Lewicka, O. Integration of multi-source geospatial data from GNSS receivers, terrestrial laser scanners, and unmanned aerial vehicles. *Can. J. Remote Sens.* **2021**, *47*, 621–634. [CrossRef]
12. Alevizos, E.; Oikonomou, D.; Argyriou, A.V.; Alexakis, D.D. Fusion of Drone-Based RGB and Multi-Spectral Imagery for Shallow Water Bathymetry Inversion. *Remote Sens.* **2022**, *14*, 1127. [CrossRef]
13. Wang, D.; Xing, S.; He, Y.; Yu, J.; Xu, Q.; Li, P. Evaluation of a New Lightweight UAV-Borne Topo-Bathymetric LiDAR for Shallow Water Bathymetry and Object Detection. *Sensors* **2022**, *22*, 1379. [CrossRef]
14. Specht, M.; Specht, C.; Lewicka, O.; Makar, A.; Burdziakowski, P.; Dabrowski, P. Study on the Coastline Evolution in Sopot (2008–2018) Based on Landsat Satellite Imagery. *J. Mar. Sci. Eng.* **2020**, *8*, 464. [CrossRef]
15. Last, B.J.; Kubik, K. Compact gravity inversion. *Geophysics* **1983**, *48*, 713–721. [CrossRef]
16. Barbosa, V.C.F.; Silva, J.B.C. Generalized compact gravity inversion. *Geophysics* **1994**, *59*, 57–68. [CrossRef]
17. Li, Y.; Oldenburg, D.W. 3-D inversion of gravity data. *Geophysics* **1998**, *63*, 109–119. [CrossRef]
18. Boulanger, O.; Chouteau, M. Constraints in 3D gravity inversion. *Geophys. Prospect.* **2001**, *49*, 265–280. [CrossRef]
19. Oldenburg, D.W. The inversion and interpretation of gravity anomalies. *Geophysics* **1974**, *39*, 526–536. [CrossRef]
20. Barbosa, V.C.F.; Silva, J.B.C.; Medeiros, W.E. Gravity inversion of basement relief using approximate equality constraints on depths. *Geophysics* **1997**, *62*, 1745–1757. [CrossRef]
21. Cordell, L. Gravity analysis using an exponential density-depth function—San Jacinto Graben, California. *Geophysics* **1973**, *38*, 684–690. [CrossRef]
22. Gallardo-Delgado, L.A.; Pérez-Flores, M.A.; Gómez-Treviño, E. A versatile algorithm for joint 3D inversion of gravity and magnetic data. *Geophysics* **2003**, *68*, 949–959. [CrossRef]
23. García-Abdeslem, J. The gravitational attraction of a right rectangular prism with density varying with depth following a cubic polynomial. *Geophysics* **2005**, *70*, J39–J42. [CrossRef]
24. Silva, J.B.; Costa, D.C.; Barbosa, V.C. Gravity inversion of basement relief and estimation of density contrast variation with depth. *Geophysics* **2006**, *71*, J51–J58. [CrossRef]
25. Campa, M.F.; Coney, P.J. Tectono-stratigraphic terranes and mineral resource distributions in Mexico. *Can. J. Earth Sci.* **1983**, *20*, 1040–1051. [CrossRef]
26. Centeno-Garcia, E. Tectonic Evolution of the Guerrero Terrane, Western Mexico. Ph.D. Thesis, The University of Arizona, Tucson, AZ, USA, 1994.
27. Centeno-Garcia, E.; Silva-Romo, G. Petrogenesis and tectonic evolution of central Mexico during Triassic-Jurassic time. *Rev. Mex. Cienc. Geol.* **1997**, *14*, 13.
28. De Cserna, Z. Geology of the Fresnillo area, Zacatecas, Mexico. *Geol. Soc. Am. Bull.* **1976**, *87*, 1191–1199. [CrossRef]
29. Júnez-Ferreira, H.E.; Dávila, F.M.; Bautista-Capetillo, C.; Steiner, J.; Ávila Carrasco, J.R. Quantitative and qualitative analysis of groundwater in aguanaval and chupaderos aquifers (Mexico). *J. Earth Sci. Eng.* **2013**, *3*, 425–436.
30. Wendt, C.J. *The Geology and Exploration Potential of the Juanicipio Property, Fresnillo District, Zacatecas*; Mexico Technical Report; A Division of Hart Crowser: Lakewood, CO, USA, 2002.
31. Krienen, L.; Cardona Benavides, A.; Lopez Loera, H.; Rüde, T.R. *Understanding Deep Groundwater Flow Systems to Contribute to a Sustainable Use of the Water Resource in the Mexican Altiplano*; Technical Report, Lehr-und Forschungsgebiet Hydrogeologie; RWTH Aachen University: Aachen, Germany, 2019.
32. Nagy, D. The gravitational attraction of a right rectangular prism. *Geophysics* **1966**, *31*, 362–371. [CrossRef]
33. Bhattacharyya, B.K.; Leu, L.-K. Spectral analysis of gravity and magnetic anomalies due to rectangular prismatic bodies. *Geophysics* **1977**, *42*, 41–50. [CrossRef]
34. Nocedal, J.; Wright, S.J. *Numerical Optimization*; Springer: Berlin/Heidelberg, Germany, 1999.
35. Press, W.H.; Teukolsky, S.A.; Vetterling, W.T.; Flannery, B.P. *Numerical Recipes in Fortran 90: Numerical Recipes in Fortran 77V. 2. Numerical Recipes in Fortran 90*; Cambridge University Press: Cambridge, MA, USA, 1996.
36. Williams, S.; Fairhead, J.D.; Flanagan, G. Realistic models of basement topography for depth to magnetic basement testing. In *SEG Technical Program Expanded Abstracts 2002*; Society of Exploration Geophysicists: Houston, TX, USA, 2002; pp. 814–817.
37. Fairhead, J.D.; Williams, S.E.; Flanagan, G. Testing magnetic local wavenumber depth estimation methods using a complex 3D test model. In *SEG Technical Program Expanded Abstracts 2004*; Society of Exploration Geophysicists: Houston, TX, USA, 2004; pp. 742–745.
38. Barnes, G.; Lumley, J. Processing gravity gradient data. *Geophysics* **2011**, *76*, I33–I47. [CrossRef]
39. Salem, A.; Green, C.; Cheyney, S.; Fairhead, J.D.; Aboud, E.; Campbell, S. Mapping the depth to magnetic basement using inversion of pseudogravity: Application to the Bishop model and the Stord Basin, northern North Sea. *Interpretation* **2014**, *2*, T69–T78. [CrossRef]
40. Portniaguine, O.; Zhdanov, M.S. Focusing geophysical inversion images. *Geophysics* **1999**, *64*, 874–887. [CrossRef]
41. Barlier, F.; Lefebvre, M. A new look at planet Earth: Satellite geodesy and geosciences. In *The Century of Space Science*; Springer: Berlin/Heidelberg, Germany, 2001; pp. 1623–1651.

applied sciences

MDPI

Article

Grouting Effect Detection within the Floor of a Coal Seam Using 3D Electric Resistivity Tomography (ERT) with Arbitrary Electrode Positions

Chuming Pang [1], Weifu Gao [2,*], Pengzheng Wu [2] and Lidong Wang [2]

1 College of Energy and Mining Engineering, Shandong University of Science and Technology, Qingdao 266590, China; 201981010010@sdust.edu.cn

2 College of Natural Resources, Shandong University of Science and Technology, Tai'an 271019, China; 202183300003@sdust.edu.cn (P.W.); 202183300018@sdkd.net.cn (L.W.)

* Correspondence: skd995703@sdust.edu.cn

Abstract: To prevent the occurrence of water inrush from the working face floor, explorations of water-rich floors are necessary. For a working face with a regular shape, a water-rich floor can be detected by laying electrodes and cables around the working face. However, the rectangular working face develops an irregular shape, and the exploration of water in irregular working faces is a difficulty in the study of 3D electric resistivity tomography (ERT). In this paper, an unconventional dipole–dipole array is used for data acquisition and the potential of a non-grid point electrode is replaced by the potential of the surrounding electrode, which identifies a water-rich floor using 3D ERT with arbitrary electrode positions. Taking the 8826 irregular working face of the Baizhuang coal mine as an example, the working face is explored by 3D ERT, the anomalous area is delineated, and targeted grouting is carried out in the anomalous area. After grouting, the grouting effect is detected by 3D ERT, and the safe mining of the working face is ensured. The results show that grouting effect detection within the floor of a coal seam using 3D electric resistivity tomography (ERT) with arbitrary electrode positions could be applied to solving the problem of water-rich exploration of floors in irregular working faces.

Keywords: water-rich; tomography; arbitrary; anomalous

1. Introduction

In recent years, with the increasing mining depth of coal seams in eastern China, more coal mines have the problems of increasing water pressure in confined aquifers and decreasing thickness of aquicludes, which leads to the occurrence of water inrush from the coal seam floor [1,2]. The aquifer in the working face is grouted to prevent water inrush and protect water resources [3,4]. The traditional uniform hole layout method [5] is adopted, but the grouting efficiency is low, and human and material resources are wasted [6]. The exploration of water-rich floors is essential before grouting, which can ensure the effective grouting of floor aquifers [7].

There are two main methods to identify a water-rich floor: the ERT method [8–11] and the (TEM) method [12–15]. At present, TEM is widely used in various projects [16], and this method is sensitive to low-resistance bodies [17]. However, it is easily affected by the surrounding environment, resulting in poor data acquisition quality, which affects the accuracy of detecting water. The ERT method is one of the most mature methods in theory and technology [18]. It is sensitive to low resistance and has strong anti-interference [19]. Therefore, it is widely used in the fields of tunnels [20,21], the environment [22], the control of water in coal mines, and [23] so on. The resistivity profiles of the roof and floor [24] can be displayed by 2D electric resistivity tomography (ERT), and a water-rich floor cannot be detected in the working face. Only when 3D ERT is adopted can the water-rich area of the floor be displayed in the working face. For a working face with a regular shape, electrodes and cables are buried around the working face, and the water-rich area of the working face is detected. A working face with an irregular shape is different. For an irregularly shaped working face, some of the electrodes are not on the model grid points, and the electrode data cannot be used. For the deep coal seam in North China Coalfield, the coal seam is affected by tectonic movement and fault development, which seriously affects the direction of the working face, resulting in the working face becoming a non-rectangular working face. The traditional 3D ERT can only detect the water-rich of the floor rock layer of the irregular working face, so there is a risk of water inrush from the floor rock layer for the non-rectangular working face.

In this paper, 3D ERT is realized with arbitrary electrode positions, then water-rich floors can be detected in irregular working faces. This 3D ERT approach is used to detect the water-rich area and design the location of the grouting [25] holes. After grouting, the method is used to detect the grouting effect again [26].

2. Materials and Methods

2.1. Overview of the Study Area

The Baizhuang coal mine is located in the western Feicheng minefield. The field is approximately 4.0 km in the east–west direction and 3.9–4.8 km in the north–south direction and covers an area of 15.6708 km^2. The coal type of the working face consists of gas and fat coal. The thickness of the coal seam is 1.9–2.4 m, with an average thickness of 2.1 m, which is a stable medium-thick coal seam.

The BF54-1 and BF54-2 faults are exposed in the transportation roadway (Figure 1C). To reduce the influence of faults on mining, coal pillars are reserved along the fault. Therefore, the transportation roadway of the 8826 working face is changed. The strike of the 8826 working face is 280 m, and the cutting span is 140 m. With the change in the transportation roadway, the span of the working face decreases gradually. Figure 1C shows that the shape of the 8826 working face is irregular. Therefore, the 3D ERT method cannot detect a water-rich floor.

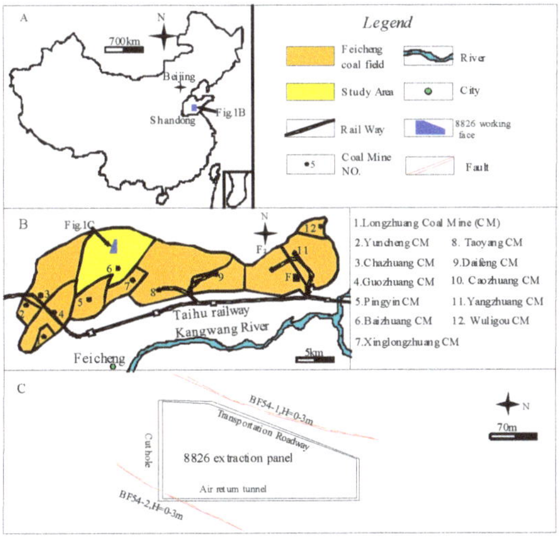

Figure 1. Location of the study area in China, including (**A**) the Feicheng coal field within China and (**B**) the location of the Baizhuang CM within the Feicheng coal field. (**C**) Schematic of the 8826 extraction panel.

As shown in Figure 2, the main aquifers of the 8826 working face are the NO. 4 LS, NO. 5 LS, and OLS aquifers. The NO.4 LS has developed fissures, but the water of NO. 4

LS has been drained in the process of mining other working faces. The NO.5 LS aquifer is contained within thin limestone, with an average thickness of 6.4 m, and the maximum head of NO. 5 LS is −153 m. The minimum thickness is 30.7 m from the floor to NO. 5 LS. The thickness of OLS is more than 800 m, and the water content in OLS is uneven. The water head height of OLS reaches 6.2 m. The minimum thickness is 35.7 m from the floor to OLS.

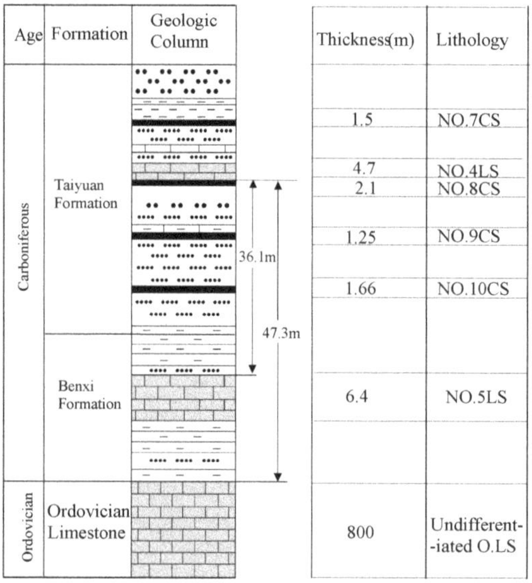

Figure 2. Stratigraphic section for the 8826 working face (CS = Coal Seam, LS = Limestone).

2.2. Water-Rich Evaluation and Data Acquisition

The water inrush coefficient method [27,28] is used to predict whether the floor of the working face is prone to water inrush events [29]. A water inrush coefficient less than 0.06 indicates that the working face is safe; relatively, water inrush may occur when the water inrush coefficient exceeds 0.06 [30,31]. The water inrush coefficients T1max and T2max of the NO. 5 LS and OLS aquifers, respectively, are 0.1 Mpa/m and 0.132 Mpa/m.

$$T = \frac{P}{M} \tag{1}$$

The water inrush coefficients of the NO.5 LS and OLS aquifers are greater than 0.06, so the 8826 working face cannot be mined. To prevent water inrush from the floor, the NO. 5 LS and OLS aquifers were grouted. To improve the accuracy and efficiency of grouting, water-rich floors were explored. The anomalous area was delineated and drilled to verify its water richness. After grouting, 3D ERT should be carried out again to detect the grouting effect.

In this paper, 3D ERT was used to explore water-rich floors. An unconventional dipole–dipole array was applied for data acquisition. As shown in Figure 3, current electrode A and potential electrodes MN were placed in the air return tunnel with an electrode spacing of 5.5 m according to the size of the air return tunnel, and current electrode B was arranged in another roadway.

The coordinates of current electrode B were (156, 0, 0). When electrode 1 and electrode B were used as current electrodes, other potential electrodes were simultaneously measured in the air return tunnel, the potential was measured at each point, and the potential difference between the potential electrodes was calculated. The measurement was completed with electrode 1 and electrode B as current electrodes and with electrode 2 and electrode

B as current electrodes and the other electrodes as potential electrodes. Then, electrode 3 and electrode B were used as current poles, and the other electrodes were used as potential poles for measurement. The above sequence was followed until the last electrode was measured as the current electrode.

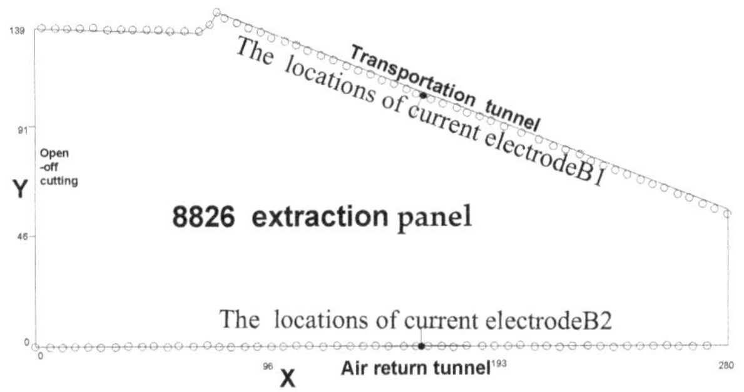

Figure 3. Layout plan of the electrodes; The black circle is the current electrode B; The circle of the return air roadway is the current electrodes A1; The circle in the transport roadway is the current electrodes A2.

Then, the electrodes and cables were laid out along the transportation roadway, and the current electrode B was laid in the air return tunnel (156, 107.8, 0). The abovementioned step could be repeated to realize the overall measurement. The data acquisition was accomplished with an FLSH-64 DC method instrument. The instrument had high acquisition efficiency with one point current electrode and the other 61 electrodes for measurement. After the data collection was completed, the collected data of the two roadways were combined and processed.

2.3. Data Processing of Arbitrary Electrode Positions

Due to the influence of faults, the orientation of the transportation roadway changed; as a result, the positions of numerous electrodes in the transportation roadway were inconsistent with the mesh of the model [32] (Figure 4). To resolve this issue, as shown in Figure 5A, this paper replaced the electrode potential method with the four nearest interpolation potentials around the electrode. Nevertheless, subject to the constant refinement of the mesh, this process not only is time-consuming but also requires much memory for the calculation. To develop an approach to resolving the problem, the potential of the non-grid point electrode was replaced by the potential of the nearest node using a distorted finite element mesh (Figure 5B); this approach detected the water-rich floor using 3D ERT with arbitrary electrode positions.

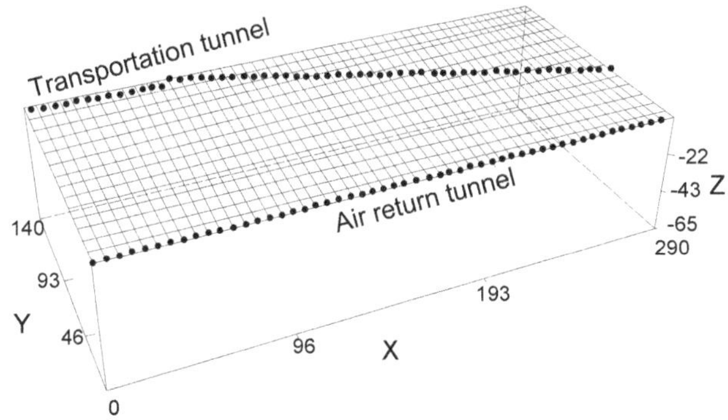

Figure 4. 3D mesh generation and electrode distribution.

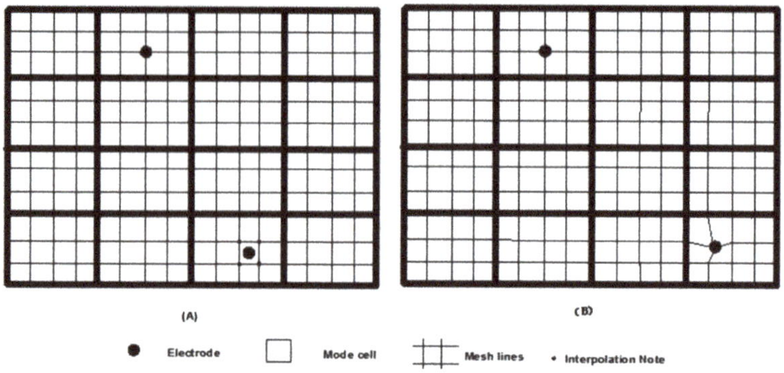

Figure 5. Non-grid electrode treatment methods: (**A**) Interpolation alternative method; (**B**) peripheral potential alternative method.

2.4. Synthetic Model

A 3D geoelectric model with dimensions of 29 m × 14 m × 6.5 m was established according to the proportionality of 10:1, and the forward and inverse calculation of the model was carried out to verify the feasibility of this method [33]. Electrodes were arranged in the two roadways of the working face, and their electrode spacing was 0.55 m. Furthermore, the resistivity of the surrounding rock was 100 Ω·m. An anomalous body with a resistivity of 10 Ω·m and a volume of 1.62 m × 2.16 m × 1.6 m was placed at coordinates of (11.6, 5.78, −1.6). Then, the horizontal results of forward and inverse calculations with the model are illustrated in Figure 6B [34].

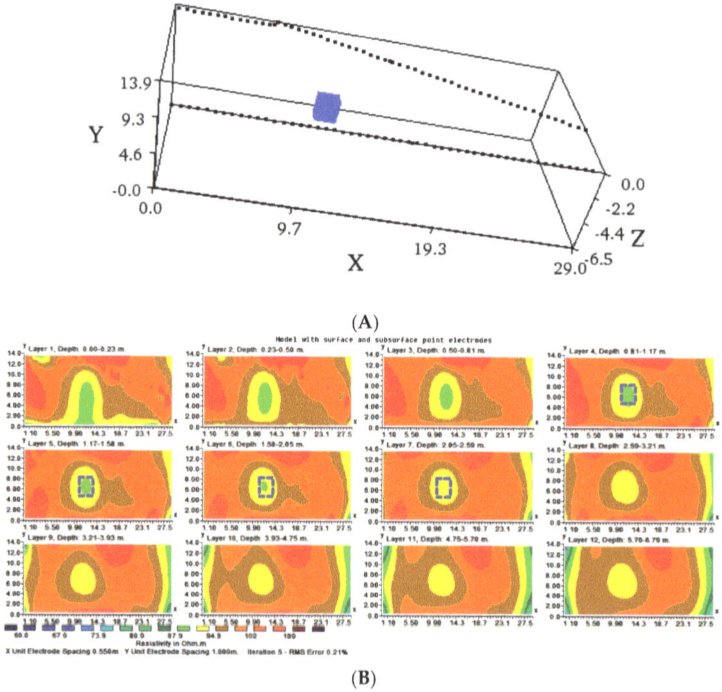

Figure 6. The model and the inversion results: (**A**) The synthetic model (**B**); The inversion results of the model, rectangle is the location of abnormal body.

3. Result and Discussion

This method can show the location of the low-resistivity body in the model, but the location of the low-resistivity area changes with the depth of the low-resistivity body. The low-resistivity anomaly area is the same as that in the model at a depth of 1.6 m. Moreover, the closer to the surface, the farther the position of the anomalous body extends in the negative Y direction. With the change in depth, the position of the low-resistance anomalous body hardly changes. Therefore, this method can accurately identify low-resistance bodies and can be used in field exploration.

Before grouting, data are acquired by laying electrodes and cables on both sides of the roadway. After data acquisition, first, the mutation points of resistivity are eliminated, and the resistivity curve is smoothed. Then, the preprocessed data are converted to the 3D inversion structure, and the data are inverted using RES3D commercial software. The Gaussian Newton method (The Newton–Gauss method consists of linearizing the model equation using a Taylor series expansion around a set of initial parameter values b0, also called preliminary estimates, whereby only the first-order partial derivatives are considered the linear theory is applicable to this linearized model) is used to invert the data. The data with large root mean square (RMS) errors are eliminated by observing inversion results after several inversions, and the data are then inverted once more to decrease the RMS error and promote the precision of continuous inversion.

After inversions, the 3D resistivity data volume of the floor of the 8826 working face is obtained. According to the hydrogeological conditions of the working face, −36.1 m in the floor is the position of the NO.5 LS aquifer. The horizontal resistivity profile of the NO. 5 LS aquifer is drawn by 3D Slicer Dicer visualization software. The anomalous resistivity areas of the NO. 5 LS aquifers are mainly situated in the ranges of 100 m–150 m and 200 m–240 m in the air return tunnel and 100 m–140 m in the transportation roadway (Figure 7A). Similarly, the horizontal resistivity map of the OLS aquifer is drawn by using Slicer Dicer software (Figure 7B). The areas with anomalous OLS resistivity are within the ranges of 100–130 m and 200–220 m in the air return tunnel and 100–130 m in the transportation roadway. Taking the exposure of drilling 1 as an example, the coal seam floor is 31.75m away from the NO. 5 LS, and the thickness of the NO. 5 LS is 4.89 m. The distance from the NO. 5 LS to the OLS is 3.61 m, and drilling 1 enters the OLS at 32.8 m.

A comparative analysis of charts A and B in Figure 7 reveals that the anomalous low-resistance areas between the NO. 5 LS aquifer and the OLS aquifer are similar. This indicates that the low-resistance anomalous areas may be caused by a single aquifer or multiple aquifers. To prevent water inrush from the floor, the anomalous area of the floor is verified by drilling.

As shown in Figure 7, Drillings 1–4 are designed in the return air roadway and drilling 5 is designed in the transportation roadway. According to the experience of the personnel involved with water prevention and control, when the water inflow of the NO. 5 LS aquifer is greater than 5 m^3/h, the aquifer should be grouted. Otherwise, grouting is not necessary.

Table 1 shows that the water inflow from the NO.5 LS aquifer is less than 5 m^3/h in drillings 1 and 5, and the NO. 5 LS aquifer is not grouted. The water inflow from the NO.5 LS aquifer is greater than 5 m^3/h in drillings 2–4. Therefore, the NO.5 LS aquifer needs to be grouted. The water inflow from the OLS aquifer is greater than 5 m^3/h; thus, all OLS drillings are grouted. The above situation implies that the NO. 5 LS aquifer is not rich in water, the OLS aquifer is rich in water, and the water abundance in the OLS is obviously different.

After grouting, 3D ERT is carried out on the floor of the working face again, and the results are shown in Figure 7D. Comparing the results before and after grouting, the resistivity of the OLS anomalous area increases obviously after grouting. It can be inferred that after grouting, the water-rich character of the OLS aquifer becomes weakened, indicating that the grouting effect is good. There is no obvious anomalous area in the floor of the working face, and the working face can be mined.

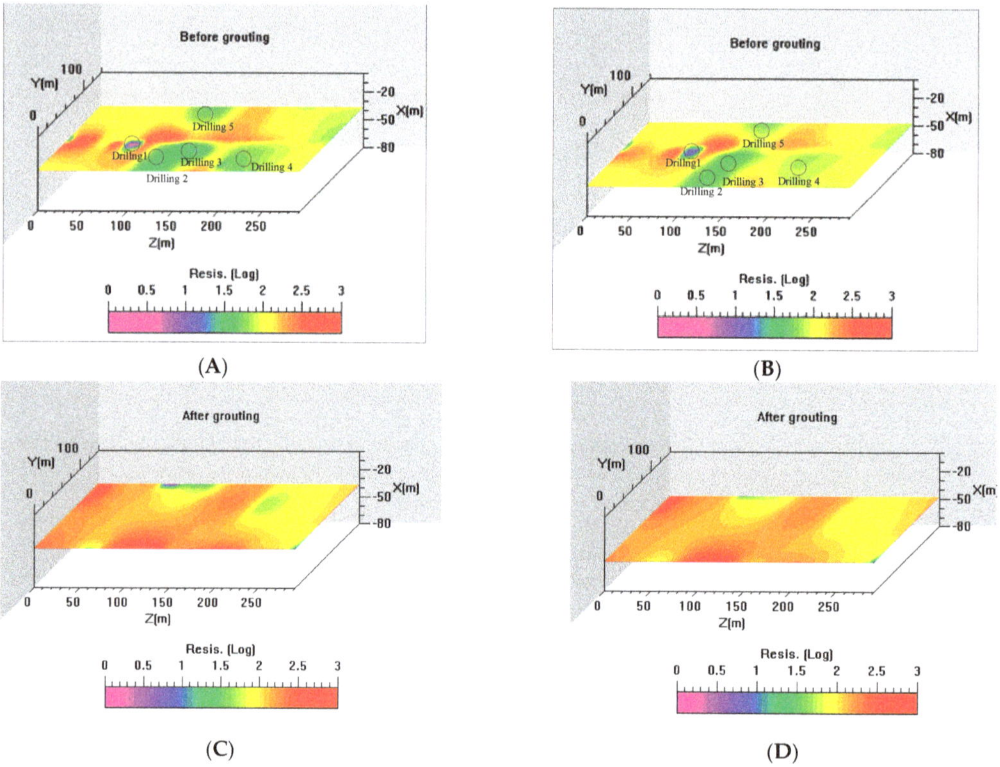

Figure 7. Resistivity profile of the working face floor aquifer before and after grouting: (**A**) Resistivity profile of the NO.5 LS aquifer before grouting; (**B**) Resistivity profile of the OLS aquifer before grouting; (**C**) Resistivity profile of the NO.5 LS aquifer after grouting; (**D**) Resistivity profile of the OLS aquifer.

Table 1. Statistical data of drilling water inflow.

Name of Drilling	The Water Inflow from NO. 5 LS (m^3/h)	The Water Inflow from OLS (m^3/h)
Drilling 1	2.0	60
Drilling 2	5.0	20
Drilling 3	7.0	25
Drilling 4	5.0	12
Drilling 5	0	10

4. Conclusions

In this paper, the boreholes reveal that the NO. 5 LS aquifer has the characteristics of thin thickness, poor water-rich but partial water-rich in the working face. The OLS aquifer with thick thickness is strong and water-rich but with great differences in water-rich. Therefore, the grouting of the OLS aquifer is an important aquifer in the grouting process.

In the process of detecting the water-rich of the floor for the working face, current electrode A and potential electrodes MN were placed in the air return tunnel and current electrode B was arranged in another roadway, an unconventional dipole–dipole array is used for data collection. The partial mesh subdivision method was adopted to realize arbitrary electrodes' measures to successfully delineate the water-rich area of the floor and provide target areas for subsequent grouting work. After grouting, this method is used to detect the water-rich of the floor again, so as to detect the grouting effect and ensure the safe mining of the working face.

Author Contributions: Conceptualization, W.G. and C.P.; methodology, C.P.; software, W.G.; validation, P.W. and L.W.; formal analysis, W.G.; investigation, C.P.; resources, W.G.; data curation, P.W. and L.W.; writing—original draft preparation, C.P.; writing—review and editing, W.G.; visualization, C.P.; supervision, W.G.; project administration, W.G.; funding acquisition, W.G. All authors have read and agreed to the published version of the manuscript.

Funding: The research was funded by the National Science Foundation (41807283), and Scientific Research Foundation of Shandong University of Science and Technology for Recruited Talent (2019RCJJ024), and special funds provided by the Taishan Scholars construction projects.

Institutional Review Board Statement: Not applicable.

Informed Consent Statement: Not applicable.

Data Availability Statement: Data available in a publicly accessible repository. This can be found at https://pan.baidu.com/s/1W1KFT6KDogi6TWe0xBiZIQ?pwd=292x (accessed on 5 May 2022).

Conflicts of Interest: No conflict of interest exists in the submission of this manuscript and the manuscript is approved by all authors for publication. I would like to declare on behalf of my co-authors that the work described was original research that has not been published previously, and is not under consideration for publication elsewhere, in whole or in part. All the authors listed have approved the manuscript that is enclosed.

References

1. Shi, L.; Han, J. *Floor Water Inrush Mechanism and Prediction*; China University of Mining and Technology Press: Xuzhou, China, 2004. (In Chinese)
2. Zhang, J. Investigations of water inrushes from aquifers under coal seams. *Int. J. Rock Mech. Min. Sci.* **2005**, *42*, 350–360. [CrossRef]
3. Zhang, W.; Zhu, X.; Xu, S.; Wang, Z.; Li, W. Experimental study on properties of a new type of grouting material for the reinforcement of fractured seam floor. *J. Mater. Res. Technol.* **2019**, *8*, 5271–5282. [CrossRef]
4. Miao, X.X.; Pu, H.; Bai, H.B. Principle of water-resisting key strata and its application in water-preserved mining. *J. China Univ. Min. Technol.* **2008**, *37*, 1–4.
5. Li, H.; Bai, H.; Wu, J.; Wang, C.; Ma, Z.; Du, Y.; Ma, K. Mechanism of water inrush driven by grouting and control measures—a case study of Chensilou mine, China. *Arab. J. Geosci.* **2017**, *10*, 468. [CrossRef]
6. Qian, D.; Zhang, N.; Zhang, M.; Shimada, H.; Cao, P.; Chen, Y.; Zhang, N. Application and evaluation of ground surface pre-grouting reinforcement for 800-m-deep underground opening through large fault zones. *Arab. J. Geosci.* **2017**, *10*, 285. [CrossRef]
7. Lu, P.; Hou, K. Current application status and development trend of curtain grouting in water-rich mine. *Mod. Min.* **2010**, *3*, 21–24.
8. Shi, L.; Wang, Y.; Qiu, M.; Gao, W.; Zhai, P. Application of three-dimensional high-density resistivity method in roof water advanced detection during working stope mining. *Arab. J. Geosci.* **2019**, *12*, 464. [CrossRef]
9. Schoor, M.V.; Binley, A. In-mine (tunnel-to-tunnel) electrical resistance tomography in South African platinum mines. *Near Surf. Geophys.* **2009**, *8*, 563–574. [CrossRef]
10. Bharti, A.K.; Pal, S.K.; Priyam, P.; Pathak, V.K.; Kumar, R.; Ranjan, S.K. Detection of illegal mine voids using electrical resistivity tomography: The case-study of Raniganj coalfield (India). *Eng. Geol.* **2016**, *213*, 120–132. [CrossRef]
11. Maillol, J.; Seguin, M.-K.; Gupta, O.; Akhauri, H.; Sen, N. Electrical resistivity tomography survey for delineating uncharted mine galleries in West Bengal, India. *Geophys. Prospect.* **2010**, *47*, 103–116. [CrossRef]
12. Tang, H.; Yang, H.; Lu, G.; Chen, S.; Yue, J.; Zhu, Z. Small multi-turn coils based on transient electromagnetic method for coal mine detection. *J. Appl. Geophys.* **2019**, *169*, 165–173. [CrossRef]
13. Yu, J.; Malekian, R.; Chang, J.; Su, B. Modeling of whole-space transient electromagnetic responses based on FDTD and its application in the mining industry. *IEEE Trans. Ind. Inform.* **2017**, *13*, 2974–2982. [CrossRef]
14. Fitterman, D.V.; Stewart, M.T. Transient electromagnetic sounding for groundwater. *Geophysics* **1986**, *51*, 995–1005. [CrossRef]
15. Buselli, G.; Lu, K. Groundwater contamination monitoring with multichannel electrical and electromagnetic methods. *J. Appl. Geophys.* **2001**, *48*, 11–23. [CrossRef]
16. Xue, G.Q.; Chen, W.Y.; Ma, Z.J.; Hou, D.Y. Identifying deep saturated coal bed zones in china through the use of large loop TEM. *J. Environ. Eng. Geophys.* **2018**, *23*, 135–142. [CrossRef]
17. Yue, J.; Yang, H.; Hu, B. 3D finite difference time domain numerical simulation for TEM in-mine. *Prog. Geophys.* **2007**, *6*, 1904–1909.
18. Gao, B.K. Research on the Forward of Slope Exploration in DC Based on ANSYS. Master's Thesis, Central South University, Changsha, China, 2008.
19. Gao, W.; Shi, L.; Zhai, P. Water Detection within the Working Face of an Underground Coal Mine Using 3D Electric Resistivity Tomography (ERT). *J. Environ. Eng. Geophys.* **2019**, *24*, 497–505. [CrossRef]

20. Liu, Z.G.; Qiao, D.H.; Li, Y.W.; Xu, J.Y. Tunnel Karst Disease Risk Estimation by Advance Geological Forecast. *Appl. Mech. Mater.* **2013**, *353–356*, 1689–1692. [CrossRef]
21. Gao, X.Q.; Zhu, Y.Q.; Ye, C.L. Survey and Analysis of the Regularity of the Surface Crack in the Loess Tunnel of the Special Passenger Line. *Appl. Mech. Mater.* **2011**, *90–93*, 2258–2264. [CrossRef]
22. Park, M.K.; Park, S.; Yi, M.-J.; Kim, C.; Son, J.-S.; Kim, J.-H.; Abraham, A.A. Application of electrical resistivity tomography (ERT) technique to detect underground cavities in a karst area of South Korea. *Environ. Earth Sci.* **2013**, *71*, 2797. [CrossRef]
23. Bazaluk, O.; Sadovenko, I.; Zahrytsenko, A.; Saik, P.; Lozynskyi, V.; Dychkovskyi, R. Forecasting Underground Water Dynamics within the Technogenic Environment of a Mine Field. Case Study. *Sustainability* **2021**, *13*, 7161. [CrossRef]
24. Gao, W.; Shi, L.; Han, J.; Zhai, P. Dynamic monitoring of water in a working face floor using 2D electrical resistivity tomography (ERT). *Mine Water Environ.* **2018**, *37*, 423–430. [CrossRef]
25. Zhai, M.; Bai, H.; Wu, L.; Wu, G.; Yan, X.; Ma, D. A reinforcement method of floor grouting in high-water pressure working face of coal mines: A case study in Luxi coal mine, North China. *Environ. Earth Sci.* **2022**, *81*, 28. [CrossRef]
26. Yu, X.; Pei, F.; Han, J.; Gao, W.; Wang, X. Ordovician limestone karst development law in Feicheng coal field. *Environ. Earth Sci.* **2018**, *77*, 781. [CrossRef]
27. Shi, L. Analysis of the origin of water inrush coefficient and its applicability. *J. Shandong Univ. Sci. Technol.* **2012**, *31*, 6–9.
28. Meng, Z.; Li, G.; Xie, X. A geological assessment method of floor water inrush risk and its application. *Eng. Geol.* **2012**, *143–144*, 51–60. [CrossRef]
29. Wu, Q.; Liu, Y.; Liu, D.; Zhou, W. Prediction of Floor Water Inrush: The Application of GIS-Based AHP Vulnerable Index Method to Donghuantuo Coal Mine, China. *Rock Mech. Rock Eng.* **2011**, *44*, 591–600. [CrossRef]
30. Zhu, Q.H.; Feng, M.M.; Mao, X. Numerical analysis of water inrush from working-face floor during mining. *Int. J. Min. Sci. Technol.* **2008**, *18*, 159–163. [CrossRef]
31. Rudakov, D.; Inkin, O. Evaluation of heat supply with maintaining a safe mine water level during operation of open geothermal systems in post-coalmining areas. *Min. Miner. Depos.* **2022**, *16*, 24–31. [CrossRef]
32. Rudakov, D.; Westermann, S. Analytical modeling of mine water rebound: Three case studies in closed hard-coal mines in Germany. *Min. Miner. Depos.* **2021**, *15*, 22–30. [CrossRef]
33. Wu, J.; Cai, J.; Zhao, D.; Chen, X. An analysis of mine water inrush based on fractal and non-darcy seepage theory. *Fractals* **2014**, *22*, 1440008. [CrossRef]
34. Loke, M.H.; Barker, R.D. Least-squares deconvolution of apparent resistivity pseudosections. *Geophysics* **1995**, *60*, 1682–1690. [CrossRef]

applied sciences

MDPI

Review

Paleomagnetic-Geodynamic Mapping of the Transition Zone from Ocean to the Continent: A Review

Lev V. Eppelbaum [1,2,*] and Youri I. Katz [3]

1 Department of Geophysics, Faculty of Exact Sciences, Tel Aviv University, Ramat Aviv, Tel Aviv 6997801, Israel
2 Azerbaijan State Oil and Industry University, 20 Azadlig Ave., Baku AZ1010, Azerbaijan
3 Steinhardt Museum of Natural History & National Research Center, Faculty of Life Sciences, Tel Aviv University, Ramat Aviv, Tel Aviv 6997801, Israel; kiyour@mail.ru
* Correspondence: levap@tauex.tau.ac.il

Abstract: The easternmost Mediterranean is a distinct transition zone from the ocean to the continent located at the junction between the largest Earth's lithospheric segments: Eurasian and African. The methodology of paleomagnetic mapping of such transition zones is based on integrating the mapping techniques for both continental and oceanic platforms: paleomagnetic reconstructions, results of radiometric dating of magnetized rocks, tectonic-structural reconstructions, biogeography, and utilization of the results of various geophysical surveys. The geodynamic-paleomagnetic mapping makes it possible to reveal the multilevel structural heterogeneity and display complex elements of the geodynamics of different ages inherent in this transition zone. The region of northern Israel is the most complex area in the easternmost Mediterranean. For the combined paleomagnetic mapping, well-studied paleomagnetically and radiometrically areas were selected: (1) the Carmel area, (2) the Atlit area (internal part of the Carmel area), (3) the Sea of Galilee with the adjoining zones (primarily, the Kinnarot Valley), and (4) the area of the Hula Basin with adjacent areas of the Golan Plateau, Hermon Mt., and Galilea uplift. The constructed paleomagnetic profiles for the Carmel area (on the top of the accumulative surface of the Lower Cretaceous traps) and the Kinnarot Valley—Sea of Galilee—Hula Basin, evidently indicate the complex history of the paleogeodynamic evolution of the region. These studies demonstrate the effectiveness of paleomagnetic mapping integrated with paleomagnetic profiles crossing these geologically complex areas.

Keywords: paleomagnetic mapping; paleomagnetic profiling; radiometric dating; tectonic-structural interpretation; integrated study

1. Introduction

The easternmost Mediterranean is a transition zone between the large tectonic-geodynamic structures of the Earth: Eurasian and African [1–5]. Unique structural features characterize the area that simultaneously expresses the elements of the tectonic collision associated with the evolution of the Tethys Ocean [6–8] and the initial spreading of the Red Sea rift system [9]. In the Cenozoic, four lithospheric plates intruded into the area: Nubian, Arabian, Aegean-Anatolian, and Sinai [10].

This tectonic-geodynamic setting is of extreme complexity and variegation of the developed tectonic elements and the rock complexes of the Earth's crust. The mentioned geological features form a very bizarre intersection, which combines heterogeneous continental and oceanic crust blocks, ophiolite belts, traps, folded island-arc, and terrane complexes, and the complex zones of strike-slip, thrusts, and underthrust faults [11–16].

The widespread development of the magmatic sequences of different ages and origins stimulated the employment of magnetic investigations in the region under study (e.g., [17–23]). Therefore, paleomagnetic methods are vital in considering the presence of a variety of arbitrary magnetized rocks. At the same time, the importance of paleomagnetic studies goes beyond just explaining the magnetic anomalies. Paleomagnetic data have successfully recommended itself for deciphering the structure of the intricate targets from the subsurface up to the large depths in the continents and oceans [24–28].

Paleomagnetic methods were applied partially in the transition zones from ocean to continent [29]. However, their initial application in the easternmost Mediterranean region was triggered by the work of Ben-Avraham et al. [30]. The previous experience of our studies in this region [16,18,31–33] indicates that paleomagnetic studies are effective tools for analyzing the geodynamic-tectonic features of the easternmost Mediterranean.

The use of paleomagnetic-radiometric mapping (based on the integrated use of magnetostratigraphic divisions of different orders and corresponding radiometric data) in this region began with the research of Eppelbaum et al. [18,19]. This combined mapping can be applied primarily in areas with comprehensive paleomagnetic study accompanied by detailed radiometric dating. Besides this, the mentioned mapping demands knowledge of tectonic-structural reconstructions, drilling results, and different geophysical data analyses. We have significantly expanded the techniques and methods of the paleomagnetic mapping in the Easternmost Mediterranean, combining it with the regional gravity, magnetic and seismic data, and the structural analysis of sedimentary formations and crustal blocks that are discussed below.

2. Tectono-Paleomagnetic Mapping and Its Role in Geological-Geophysical Integration

Paleomagnetic research is widely accepted as a powerful independent tool for geodynamic and tectonic analysis, studying in-depth features, searching for economic minerals, and archaeogeophysical analysis (e.g., [24,26,28]). In principle, the geological-geophysical conclusions reached from paleomagnetic data analysis cannot be obtained using other geophysical or geological methods [22,34]. Regional and local magnetostratigraphic charts are widely applied to unify the stratigraphic schemes and correlate the numerous local subdivisions. Paleomagnetic mapping (zonation) can be realized in well-studied areas with detailed paleomagnetic measurements, radiometric dating, and structural-tectonic examination. Paleomagnetic mapping enables to "jump" from the single measurements to some areal reconstructions that increase the paleomagnetic method's effectiveness. At the same time, integration of paleomagnetic data with the set of geophysical-geological approaches often strongly increases the reliability and the significance of the combined research.

The first paleomagnetic stratigraphic studies in sedimentary rocks are associated with investigations by Irving and Runcorn [35]. The authors detected normal and reversed magnetic polarities in the Proterozoic Torridonian Sandstones and rocks from Scotland's Devonian and Triassic ages. For the Torridonian Sandstones, an attempt was made to correlate the observed polarity zones in stratigraphic sequences (after [36]).

Khramov's [37] investigation focused on the magnetic polarity stratigraphy in Pliocene-Pleistocene sediments from western Turkmenistan (Central Asia). He carried out chronostratigraphic interpretations based on the equal time of polarity intervals. Picard [38] successfully applied magnetostratigraphic studies to Triassic red sandstones of the Chugwater Formation (Wyoming, USA). Irving [39] discovered the Kiama paleomagnetic zone of inverse polarity in the Upper Paleozoic in Australia and defined its interrelation with covering the Illawarra zone of normal polarity. This discovery has played a crucial role in subsequent paleomagnetic investigations [26].

The development of the first paleomagnetic maps was associated with mapping the ocean bottoms, where a series of direct and inverse magnetization zones were interpreted as spreading zones [40,41]. In the current stage, the paleomagnetic ocean reconstructions include stages from the Cretaceous and Cenozoic [42] to the Middle Jurassic [43,44]. McDougall et al. [45] demonstrated an effective integration of the paleomagnetic-radiometric data analysis to examine basaltic formations in western Iceland.

Several examples of paleomagnetic zonation of sedimentary deposits in the Volga-Ural and Caucasus regions (Russia) were displayed in [27,46]. Paleomagnetic mapping of the volcanogenic association of the Miocene traps of Transcarpathia was presented in [47].

Kissel and Laj [25] demonstrated the geodynamic paleomagnetic zonation with examples from the easternmost Mediterranean, New Zealand, the USA, and some other regions. Opdyke and Channel [26] presented various generalizations for paleomagnetic mapping. Kristjansson and Jonsson [48] significantly extended the perspectives of paleomagnetic mapping in Iceland. Goguitchaichvili et al. [49] have reliably mapped the Gilbert–Gauss geomagnetic reversal in the Pliocene volcanic sequences in the Lesser Caucasus.

Nur et al. [50] carried out the initial geodynamic paleomagnetic zonation of the transition zone from ocean to continent in northern Israel. The mentioned authors revealed several tectonic blocks of the predominant counterclockwise rotation. Granot et al. [51] employed paleomagnetic mapping in the Troodos gabbro massif (Cyprus) for estimation of the oceanic earth's crust block rotation. Rolland [52] has skillfully applied paleomagnetic mapping to unmask the Caucasian collisional history.

Eppelbaum et al. [18,19] have applied the integrated paleomagnetic-radiometric and magnetic analyses for the tectonic-structural reconstructions in the Sea of Galilee area. As a result, the first magnetic–paleomagnetic–radiometric map (scheme) for this area was developed.

The combined magnetic-paleomagnetic analysis enabled the discovery of the Kiama zone of inverse polarity in the easternmost Mediterranean [31]. This research triggered the development of the first paleomagnetic map of the easternmost Mediterranean based on the integrated interpretation of different geophysical fields and comprehensive analysis of the surrounding sedimentary and volcanogenic structures [22,32,33,53].

Paleomagnetic-geodynamic mapping [22,31] of the easternmost Mediterranean, located between the boundaries of Laurasia and Gondwana, where the Eurasian, Aegean-Anatolian, Nubian, Sinai, and Arabian plates converge, made it possible to clarify and substantiate the location and nature of their crust. For the first time, ophiolite exposures in the distal regions of the foreland adjacent to the southern Tethys zone were identified and mapped. A new tectonic map of the region has been compiled, where the Mesozoic Terrane Belt (MTB) with pre-collision, collision and post-collision traps have been recognized and contoured [22].

The specific complexity of the easternmost Mediterranean follows, first of all, from convergence here collision and spreading processes. A wide range of applied analytical, search and survey methods give a reason to believe that this area can become a reference for mapping techniques for transition zones from the ocean to the continent and geologically complex regions. It makes it possible to test a combination of paleomagnetic-radiometric studies and structural and tectonic analyses.

3. Paleomagnetic Mapping and Profiling in Northern Israel

The combined paleomagnetic mapping has been performed for the following geologically complex areas in northern Israel (see contoured areas in Figure 1): (1) Mt. Carmel and surrounding areas and (2) Atlit area as part of the Mt. Carmel paleomagnetic map (it was caused by the high complexity of the map mentioned above), (3) Sea of Galilee and its vicinity, and (4) Hula Basin and adjacent areas. The paleomagnetic profiles developed across the Carmel—Galilee region (***II–II′***) and across the Hula Valley—Sea of Galilee—Hula Basin (***I–I′–I″***) unmask nontrivial geological sections of these areas.

3.1. Paleomagnetic Maps of Carmel and Atlit Areas

The Mt. Carmel structure (see Figure 1) differs from other coastal plain structures of the easternmost Mediterranean by numerous geological-geophysical characteristics [17,54–58].

The Carmel structure is located at the boundary between the Galilee-Lebanon and Judea-Samaria terranes [6]. Neotectonically, this boundary coincides with the zone of seismic activity within the Yagur fault branch system; it continues onshore and divides the area into the southern and northern sectors [59]. The most recent analysis shows that this boundary is displaced a few km south of the Atlit fault zone [57].

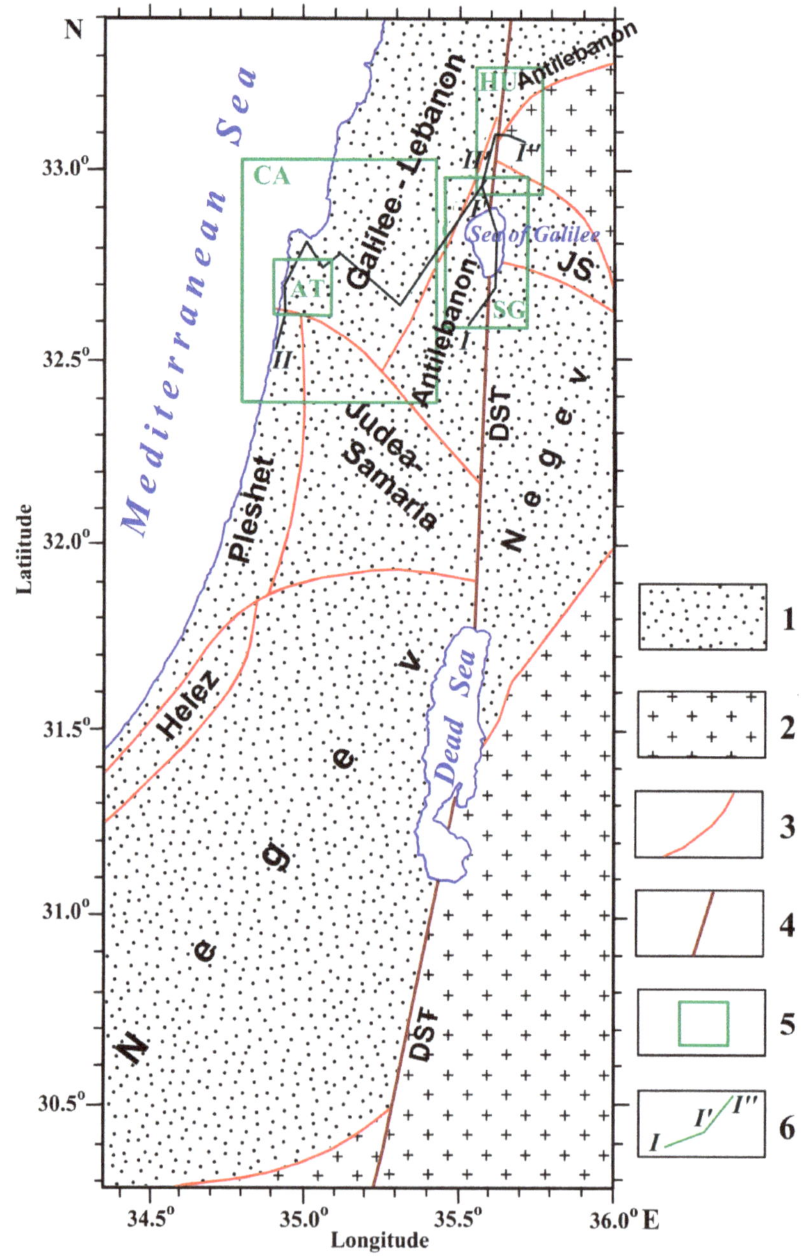

Figure 1. Very simplified structural map of with location of areas of paleomagnetic-radiometric generalization and examined magnetic anomalies. (1) Mesozoic Terrane Belt, (2) Neoproterozoic Belt, (3) intraplate faults, (4) interplate Dead Sea Transform (DST), (5) paleomagnetically investigated areas: CA, Carmel, AT, Atlit, SG, Sea of Galilee (Lake Kinneret), HU, Hula, (6) location of paleomagnetic profiles *I–I′–I″* and *II–II′*. JS, Judea-Samaria.

The Carmel tectonic system plays an essential role in estimating the spatial interaction of the continental and oceanic zones in this area. The oldest discovered deposits are Triassic strata, overlying series of the magmatic and sedimentary rocks of the Early Mesozoic (Asher Volcanics) [60]. The last associations covered by younger Mesozoic-Cenozoic deposits are affected by the prolonged complex deformation of the postaccretional tectonic stage (>130 Ma).

Gvirtzman and Steinitz [60] and Garfunkel [61] assumed that the Early Mesozoic formation of the Asher Volcanics was generated within the graben structure. Gvirtzman et al. [54] suggested that the formations composing this graben have a predominantly continental genesis, including soils and coal. According to other data [62,63], the Asher Volcanics relate to a deep basin with a possible ocean-like crust; the results of our investigation [22,31,32] partially agree with the last proposition.

Discovering numerous mantle minerals [58,64–69] also confirmed the mentioned suggestion. The anomalously high content of the Permian zircons was found in the xenoliths of the Cretaceous volcanoes [65]. Based on the analysis of the ratios of trace elements, the authors mentioned above believed that these zircons are associated with the magmatic complexes of the Earth's crust of the oceanic type.

In the current work, we have expanded the areas of paleomagnetic mapping of the northern part of the easternmost Mediterranean by the attraction of the well-studied geological-geophysical area of the Mt. Carmel and the Sea of Galilee and adjacent areas (Figure 2).

Figure 2 indicates a significant differentiation in the direction of tectonic-thermal processes from the Late Mesozoic to the Late Cenozoic. This testifies to the extreme geodynamic instability of this region and the variety of manifestations of tectonic movements, magmatic processes, and surface relief with the formation of faults, strike-slip valleys, and rotational structures. The dominant direction of the isopachs of the Lover Cretaceous traps of the Halal-1 paleomagnetic superzone corresponds to the NNE-SSW. It is 60° counterclockwise as opposed to the development field of the Upper Cretaceous volcanoes of the Halal-2 superzone—NNW-SSE. Meanwhile, it is 30° counterclockwise compared to the long axis of the Late Cretaceous traps [56] of Mt. Carmel. These data confirm the regional rotation of the deep mantle structure below the Eastern Mediterranean [16] over a long geological time. Diagonal and arc faults oppose the boundaries of larger tectonic blocks (terranes). These fractures, in recent times, have created fault and strike-slip valleys and coastal zones of marching.

The Atlit paleomagnetic map (Figure 3) occupies the central-northern part of the Carmel paleomagnetic map.

Tectonically, this area belongs mainly to the Galilee-Lebanon terrane and the marginal parts of the Pleshet, Judea-Samaria, and Antilebanon terranes (Figure 3). Geophysically, this area is significant in developing the Carmel regional magnetic-gravity anomaly [55]. The anomaly apical part almost coincides with the top part of the Carmel Plateau, with hypsometric elevations of up to 546 m. The Carmel anomaly marks the southern part of the Galilee-Lebanon terrane, where significant differences in the basement depth [55] and developed ophiolite sheets were found. The latter enclose the magmatic complexes and mélange [22].

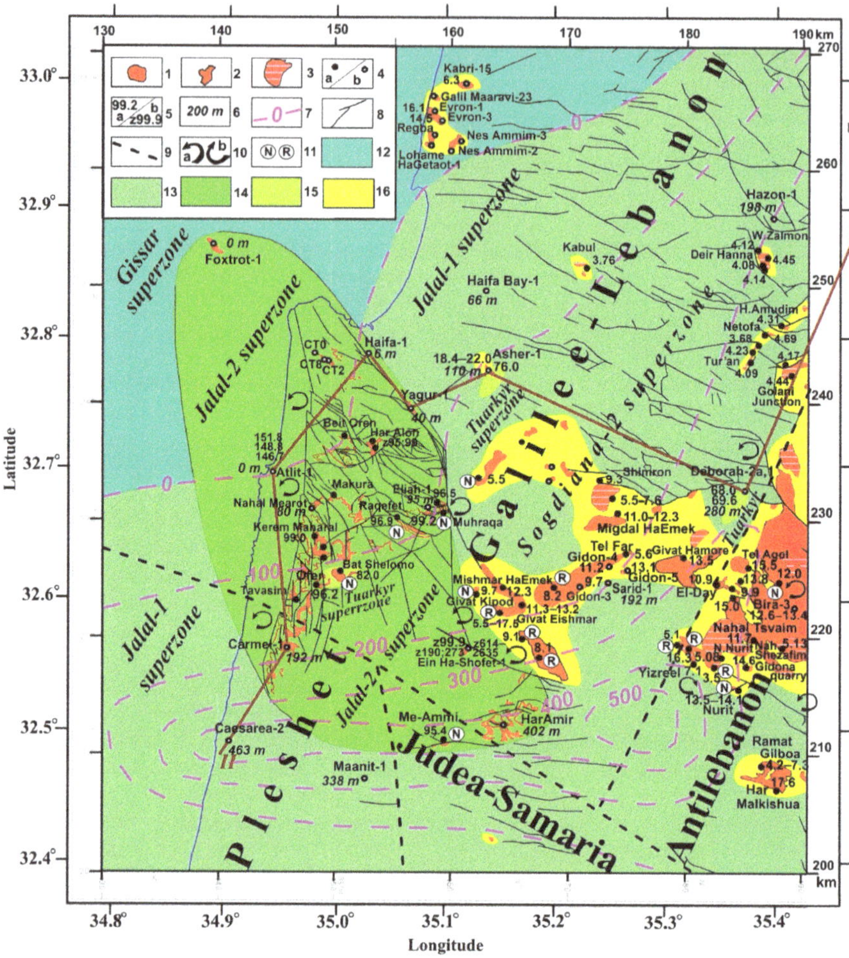

Figure 2. Geodynamic-paleomagnetic map of the Mt. Carmel—Galilee region, CA (see its location in Figure 1). (1) Cretaceous-Miocene basalts, (2) Miocene gabbroid intrusive, (3) Pliocene Cover basalts, (4) outcrops (a) and boreholes (b) with the Mesozoic-Cenozoic magmatic complexes, (5) radiometric age of magmatic rocks and minerals from K-Ar, Ar-Ar methods (a) and zircon geochronology (b), (6) thickness of the Lower Cretaceous traps (in m), (7) isolines of the Lower Cretaceous traps thicknesses (in m), (8) faults, (9) boundaries of terranes, (10) counterclockwise (a) and clockwise (b) rotation derived from tectonic and paleomagnetic data, (11) data of paleomagnetic measurements of magmatic rocks with normal *N* and reverse *R* polarities, (12–15) paleomagnetic zones: (12) Gissar, (13) Jalal-1, (14) Jalal-2, (15) Tuarkyr, (16) Sogdiana-2. The bold brown line shows the location of the paleomagnetic profile ***I–I′***. The following main works were used for this map construction: [22,23,50,57,64,65,70–88].

Applied paleomagnetic mapping (profiling) became possible due to new paleomagnetic, radiochronological, mineral, and petrological data and novel structural-tectonic reconstructions. The Gissar and Jalal-1 superzones were derived from the sections of multiple boreholes (e.g., [89]), where the thicknesses and the radiometric ages are accomplished. The geological boundaries of the Gissar and Jalal-1 superzones correspond to the pinching out of collisional-post-collisional effusive trap strata reaching a thickness of 463 m (Caesarea-3 borehole).

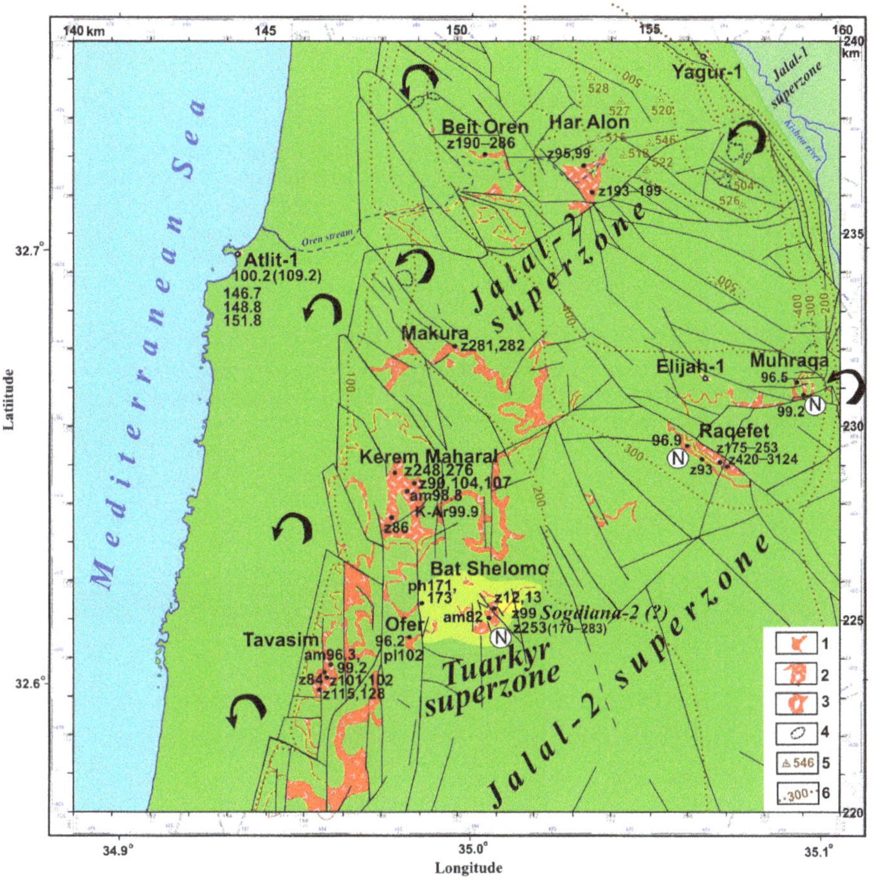

Figure 3. Geodynamic-paleomagnetic map of Atlit area (AT), the middle part of the Carmel area (see its location in Figure 1). This map was developed based on a geological map [57]. (1) basalt lava flows, (2) basalt tuffs, (3) tuffs and flows of basalt volcanic units, (4) landslide scars, (5) highest hypsometric marks, and (6) hypsometric lines.

Stratigraphic and radiometric data from the natural outcrops and Foxtrot-1, Eliah-1, Carmel-1, and Ein Ha-Shofet wells [57,79,90] indicate the development of the Upper Cretaceous (Cenomanian) diamondiferous volcanics [65] of the Jalal-2 superzone. It is pretty indicative that the outline and dimensions of the outcropping area of the paleomagnetic Jalal-2 superzone generally coincide with both magnetic [17,55,91] and gravity [17,55] anomalies of the Mt. Carmel and its outskirts.

The data mentioned above join the deep-geophysical mechanisms of different levels—the mantle plumes from the deep mantle and the movements of the basement of the Earth's crust, which are sources of the shallow anomalies in the gravity and magnetic fields [16,92]. The Tuarkyr paleomagnetic zone (confined in this area to the Senonian), was developed in the form of the insignificant domains at the surface as scattered outcrops of the Bat Shelomo effusive rocks in the south of the Carmel Plateau with the radiometric age of 82 Ma [90].

The other two finds in this superzone are associated with the radiometrically dated sills from the Asher-1 borehole [82] and tuffs from the Devorah-2 borehole [72]. These finds correspond to the Campanian and Maastrichtian, respectively. Paleomagnetic research [80] indicated that a normal polarity characterizes the Late Cretaceous volcanic rocks outcropping in a few areas. These rocks correspond to the Jalal-2 and Tuarkyr paleomagnetic superzones.

The Sogdiana-2 paleomagnetic superzone, concerning more ancient Mesozoic superzones, extends submeridionally and occurs discordantly. It ranges from the giant outcrop of the Cenozoic Ash Shaam traps diagonally stretching from the SE to the NW. In this area, the belt of the Cenozoic traps forms the scattered outcrops within the Cretaceous and Paleogene rocks and often combines with them in a narrow space (e.g., in the Asher-1 borehole [82]). The Cenozoic traps even used the ancient faults and volcanic apparatus to intrusion into the Mesozoic traps, as discovered [65] in the Santonian Bat Shelomo volcano field.

3.2. Paleomagnetic Profile across the Carmel-Galilee Region

The paleomagnetic profile crosses the area of the Mt. Carmel, Northern Galilee, and the Korazim Plateau of the DST region (north of the Lake Kinneret), from SW to NE from the Pleshet terrane through the Galilee-Lebanon terrane to the Antilebanon terrane (Figure 4; see location of this profile in Figure 1).

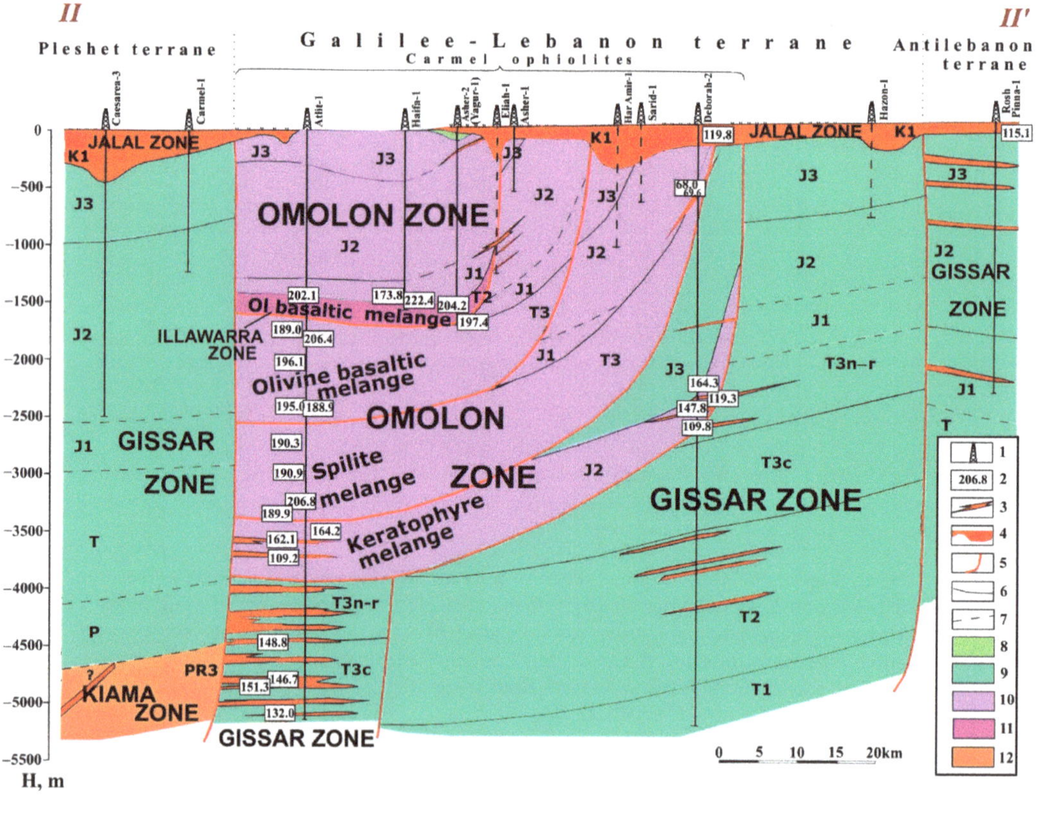

Figure 4. Paleomagnetic profile *II–II′* of the Carmel-Galilee region along the top of the accumulative surface of the Lower Cretaceous traps (location of this profile is shown in Figures 1 and 2). (1) borehole, (2) radiometric age, (3) dykes and sills, (4) traps, (5) faults, (6) established stratigraphic boundaries, (7) supposed position of the stratigraphic boundaries, (8–12) paleomagnetic zones: (8) Jalal, (9) Gissar, (10) Omolon, (11) Illawarra, (12) Kiama. The following main works were used for this map construction: [32,54,55,60,61,63,78,89].

The Pleshet and Antilebanon terranes are somewhat uplifted tectonically than the Galilee-Lebanon terrane. This profile is constructed along the top of the accumulative leveling surface of the Lower Cretaceous traps belonging to the Jalal-1 paleomagnetic zone (131–105 Ma).

The underlying Triassic-Jurassic sequences that make up the cover of the MTB carbonate platform contain dikes and sills of Tithonian-Neocomian basaltoids (132–153 Ma) and belong to the Gissar paleomagnetic zone. Concerning the terranes themselves, the carbonate platform represents an autochthonous complex; however, both carbonates and intrusive traps were formed in the pre-collision stage, when the terranes bordered the southern part of the Neotethys Ocean at the boundary with the Gondwana foreland. The profile testifies that an allochthonous complex of ophiolites, occurring above the carbonate platform, has developed in the Galilee-Lebanon terrane. It forms four sheets of different ages. The lowest of them, the youngest, is composed of keratophyre mélange aged 162–164 Ma (Omolon paleomagnetic zone) and is covered with relatively deep-water Upper Jurassic carbonates penetrated by intrusive traps of the Gissar zone. Two middle ophiolite plates are composed of spilite and olivine-basalt mélange (188–206 Ma) and deep-water Jurassic carbonates and generally belong to the Omolon paleomagnetic zone. Finally, the upper ophiolite plate, the oldest, 197.4–222.4 Ma, is composed of basalt mélange and covered with Jurassic carbonates (174 Ma). This age generally corresponds to the Illawarra-Omolon paleomagnetic zones (Figure 4).

The dynamics of sequential attachment of ophiolite sheets during collision and shear movements of oceanic plates and the Galilee-Lebanon terrane during the Levantine phase of tectogenesis at the boundary of the Early and Late Hauterivian was considered earlier [77]. These authors have shown that the amplitude of horizontal displacements of ophiolite plates could reach 120 km.

3.3. *Combined Paleomagnetic-Radiometric Scheme of the Sea of Galilee and Its Vicinity*

The region of the Sea of Galilee (Lake Kinneret) is a reference object for the paleomagnetic mapping transition regions from the ocean to the continent in terms of structural, historical-geological, and methodological approaches. The uniqueness of this target is because here developed the phenomena and structures of the collisional type caused by the closure of the Neotethys Ocean and the elements of the initial phases of the Red Sea—East African Rift system spreading.

It has long been used as the largest freshwater reservoir in the Middle East and as an etalon region for monitoring and accounting for seismicity in the area of active housing construction and areas of industrial and agricultural facilities development.

There are several fault systems in this area, the main ones being the N–S transform system and the E–W and NW–SE fault systems that break up the Galilee. The sea, and the plain to its south, are located in a depression bounded on the east and west by active fault scarps with steep gradients [93]. The superposition of vertical displacements perpendicular or oblique to the transform impedes structural interpretation of the investigated basin [94].

At the end of the twentieth century, various geological and geophysical surveys raised the question of developing a generalizing tectonic-geodynamic model (e.g., [94]). Further studies [18,19,22] showed that the optimal linkage consists of integrating the methods of magneto-geophysics (adopted in the study of the oceans) and paleomagnetic stratigraphy (assumed for the continents), combined with independent geological-geophysical investigations. The developed map (Figure 5) also contains the revised and generalized results of the quantitative interpretation of numerous magnetic anomalies in the Sea of Galilee [19].

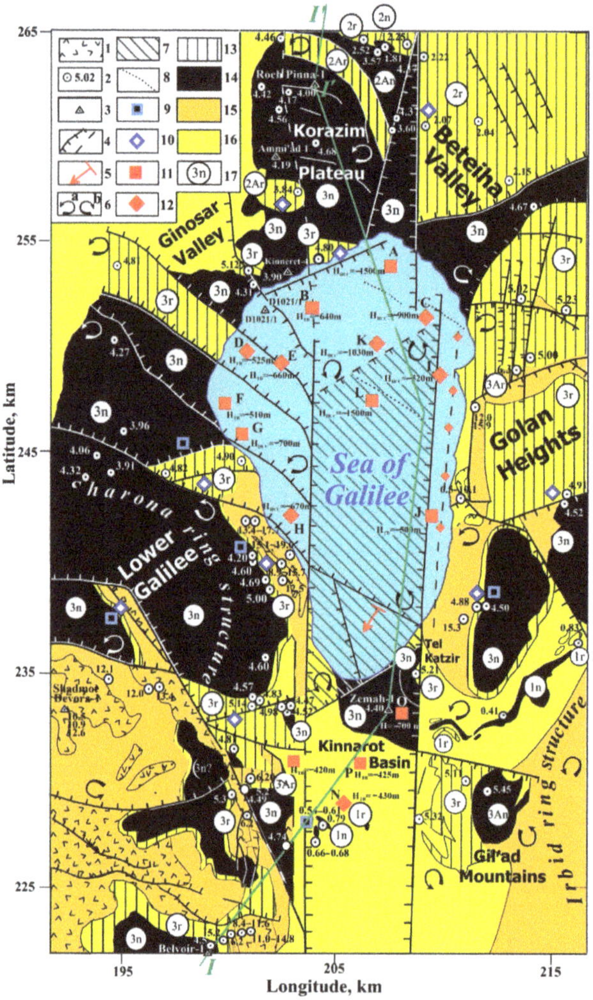

Figure 5. Integrated paleomagnetic scheme of the Sea of Galilee, SG (see its location in Figure 1). (1) outcropped Cenozoic basalts, (2) points with radiometric age of basalts (in Ma), (3) boreholes, (4) faults, (5) general direction of the proposed buried basaltic plate dipping in the southern part of the Sea of Galilee, (6) counterclockwise (a) and clockwise (b) rotation of faults and tectonic blocks, (7) pull-apart basin of the Sea of Galilee, (8) suggested boundaries of the paleomagnetic zones in the sea, data of land paleomagnetic measurements: (9 and 10) (9) reverse magnetization, (10) normal magnetization, (11 and 12) results of magnetic anomalies analysis: (11) normal magnetization, (12) reverse magnetization, (13) reversely magnetized basalt fields, (14) normal magnetized basalt fields, (15) Miocene basalts and sediments with complicated paleomagnetic characteristics, (16) Pliocene-Pleistocene basalts and sediments with complicated paleomagnetic characteristics, (17) index of paleomagnetic zonation. Tectonic setting after [76,94–98]. 1n, 2n, 3n, 1Ar, 2Ar, and 3Ar are the indexes of paleomagnetic zones. Radiometric data (K-Ar and Ar-Ar) after [73,75,97,99,100]. Paleomagnetic data after [70,73–75,85,95,97,99,101–103]. H_{TB} and H_{HCC} designate the calculated depths of basaltic bodies in the basin: H_{TB} is the upper edge for the model of the thin bed, H_{THP} is the upper edge for the model of the thin horizontal plate, and H_{HCC} is the center for the model of the horizontal circular cylinder. The bold green line designates the location of paleomagnetic profile ***I–I′***.

It was the first experience of such research in the transition areas from the ocean to the continent. In the context of the regional geological-geophysical studies of the African-Arabian region, paleomagnetic mapping of the area of the Sea of Galilee has been significantly expanded and detailed [16,23,32,53].

The supplemented edition of the paleomagnetic map of the Sea of Galilee (Figure 5) was extended to the south, where the Belvoir uplift has developed. Here, the data from a drilled deep well and numerous radiometric data of the Cenozoic traps were analyzed. The essential revision of this map is based on the attraction of new data from structural, radiometric, and paleomagnetic analyses, which expanded understanding of the studied area's structure and development.

In such a manner, we nailed the mapping of the transition region from the ocean to the continent to the methods widely used in the paleomagnetic mapping of the marine areas [22], and optimally compiled with the geodynamic principles [25]. A novel variant of the paleomagnetic map of the Sea of Galilee and adjacent areas (Figure 5) includes an anomalous pull-apart basin, areas of development of the circular structures [16,22], the arc faults, and the rotational markers of the crustal blocks identified from the structural and paleomagnetic data [94–96].

Here, two comparatively large ring structures are presented. The first is the Sharon trap depression located in the central-western part of the map. The second is the Irbid ring structure, bounded by conical dikes and partially presented in the southeastern area. Significantly, the largest amplitude of rotation is 58° counterclockwise, registered in the Sharona ring structure (Figure 5), which emphasizes the complexity of the geodynamics of the trap genesis zone. A different nature has an Irbid ring structure also prone to counterclockwise rotation (Figure 5).

3.4. Hula Paleomagnetic Map

In the extreme west of the district, in the Upper Galilee Heights, Upper Cretaceous and Paleogene sedimentary formations related to the Khorezm-Tuarkyr and Jalal paleomagnetic superzones were mainly developed. In the northeast, on the slope of the Hermon Mt., Jurassic and Cretaceous formations penetrated by pre-collisional, collisional, and post-collisional traps of the Gissar and Jalal superzones were developed (Figure 6).

Due to the paleomagnetic mapping methodology elaboration in the etalon area of the Sea of Galilee, it became possible to construct a paleomagnetic map of its northern extension—Hula Valley (Figure 6). The constructed map also includes areas of the Upper Galilee Heights, the northern slope of the Korazim Plateau, the northern part of the Golan Plateau, and the southern side of the Hermon Mts. The area comprises mainly the Pliocene-Quaternary traps and the sedimentary complexes belonging to the Brunhes-Matuyama zones. More ancient Pliocene trap complexes were developed to the south, in the Korazim Plateau and Beteiha Valley. These complexes belong to the Gauss and Gilbert paleomagnetic superzones.

In the extreme west of the district, in the Upper Galilee Heights, mainly Upper Cretaceous and Paleogene sedimentary formations developed. They relate to the Khorezm-Tuarkyr and Jalal paleomagnetic superzones. Jurassic and Cretaceous formations developed in the northeast on the slope of the Hermon Mt., penetrated by pre-collisional, collisional, and post-collisional traps. The latter relates to the Gissar and Jalal paleomagnetic superzones (Figure 6).

3.5. Paleomagnetic Profile Kinnarot Valley—Hula Basin

Without bringing it to Hermon, we slightly turned the profile line to show the structural contrast of the Antilebanon terrane with the Neoproterozoic base of Gondwana, which composes the lower structural levels of the northern part of the Golan Plateau. Here the Gissar, Jalal, and Sogdiana paleomagnetic superzones are sharply reduced in thickness compared to their analogs in the Antilebanon terrane. Within the blocks of the Gondwana oneself, there are almost no chalk traps developed. The complexity of the structure of the

Kinneret and Hula depressions becomes evident in the constructed paleomagnetic profile *I*–*I*′–*I*″ (Figure 7).

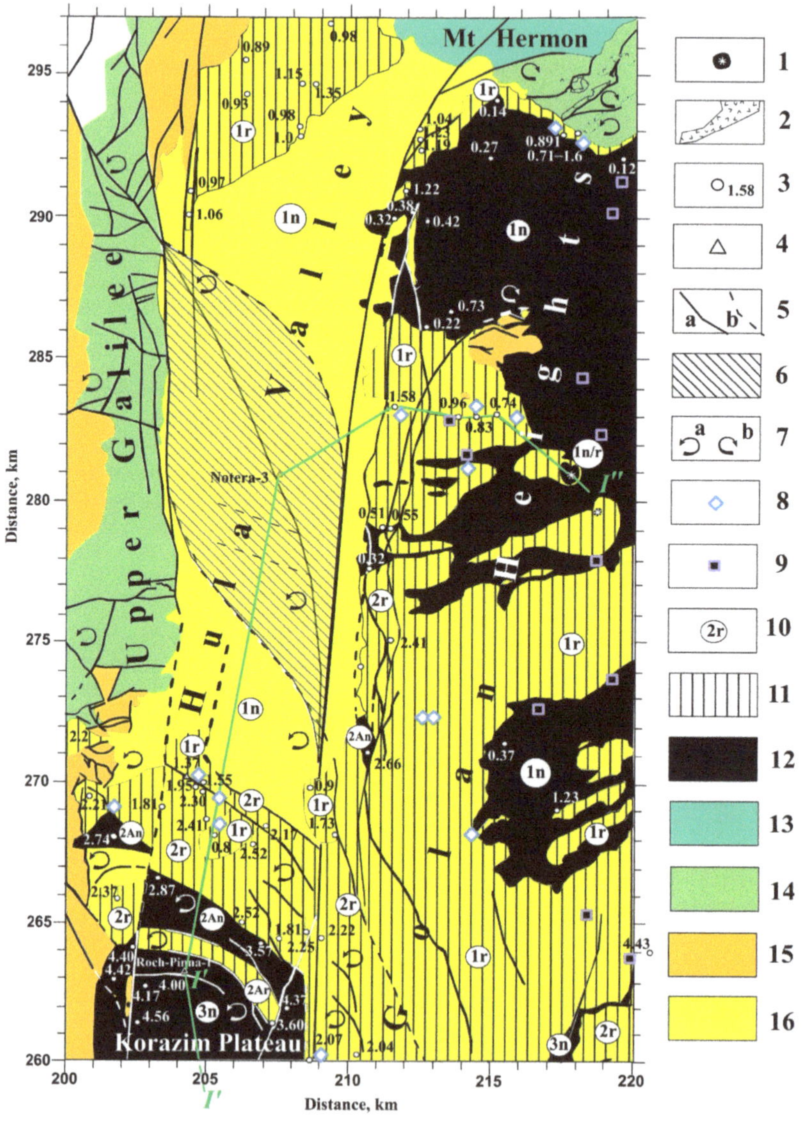

Figure 6. Paleomagnetic-geodynamic map of the Hula Basin and adjacent areas of the Golan Plateau, Hermon Mt., and Galilea uplift, HU (see its location in Figure 1). (1) volcanic cones, (2) Cretaceous traps, (3) outcrops with the radiometric ages of basalts (in Ma), (4) boreholes, (5) faults: (a) observed, (b) reconstructed, (6) pull-apart basin, (7) tectonic blocks rotations: (a) counterclockwise, (b) clockwise, (8) outcrops with the determined reversely magnetized basalts, (9) outcrops with the determined normally magnetized basalts, (10) chrons of Geomagnetic Polarity Timescale, (11) areas of the reversely magnetized basalts, (12) areas of the normally magnetized basalts, (13) Gissar superzone, (14) Jalal superzone, (15) Tuarkyr—Khorezm superzone, (16) Sogdiana superzone. Tectonic setting after: [76,97,104–106]. Radiometric data (K-Ar and Ar-Ar) after: [73,74,78,90,97,107]. Paleomagnetic data: [74,75,85,95,97,101]. The bold green line designates the location of paleomagnetic profile *I*′–*I*″.

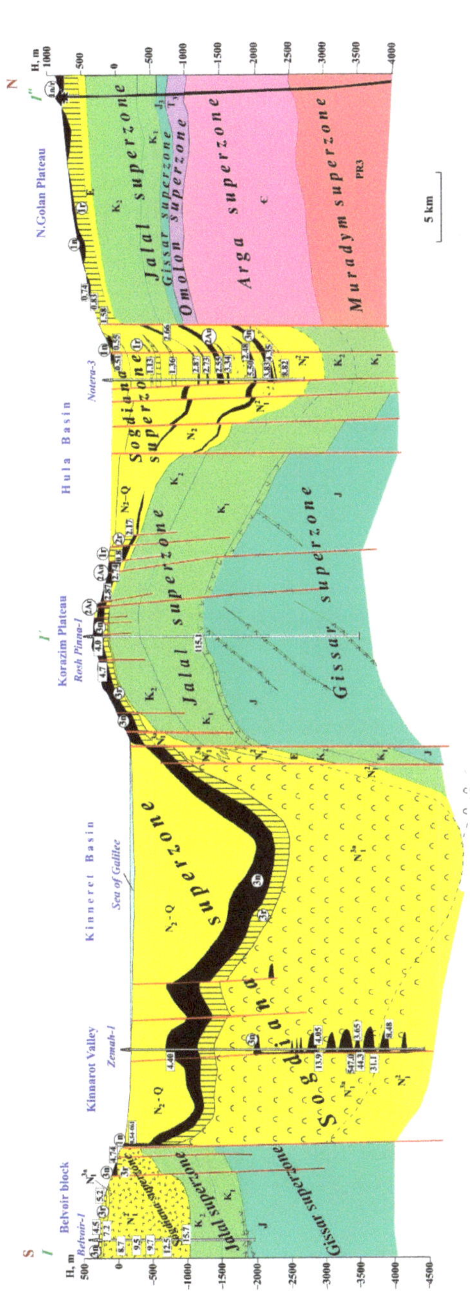

Figure 7. Paleomagnetic profile Kinnarot—Hula (its location is shown in Figures 1, 5 and 6). (1) basic traps, (2) basic dykes, (3) gabbro, (4) volcanic cones, (5) evaporite sediments, (6) faults, (7) radiometric dating, (8) reversal polarity, (9) normal polarity, (10) indexes of Plio-Pleistocene paleomagnetic scale, (11) Sogdiana superzone, (12) Jalal superzone, (13) Gissar superzone, (14) Omolon superzone, (15) Arga superzone, (16) Moradym superzone. The primary utilized sources (besides publications presented in captions for Figures 5 and 6) are: [33,100].

It stretches from the north to the south for more than 60 km. The southwestern and northeastern extremities of the profile are located about 20 km in latitude (Figures 5 and 6).

The deep structure is elicited from the regional tectonic constructions [15,32], results of seismic profiling [104,108,109], boreholes data [89,110], regional geological map [76], radiometric data [73,75,97,105,107]; and the results of the paleomagnetic zonation [16,18,19,22].

Structurally, the paleomagnetic profile corresponds to three uplifts: Belvoir, Korazim, and Golan, between which there are two depressions—Kinneret-Kinnarot and Hula. The first two uplifts and depressions are formed within the southern block of the Antilebanon terrane (Figure 1), shifted relative to the northern one uplifted block of the higher Hermon plateau along the Dead Sea Transform (DST) for about 100 km.

The Kinneret-Kinnarot and Hula are the pull-apart basins with the different amplitudes of subsidence of the Late Cenozoic sedimentary formations. Both the basins contain the traps related to the Sogdiana paleomagnetic superzone.

Without bringing it to Hermon, we slightly turned the profile line to show the structural contrast of the Antilebanon terrane with the Neoproterozoic base of Gondwana, which composes the lower structural levels of the northern part of the Golan Plateau. Within the blocks of the Gondwana oneself, there are almost no chalk traps developed. They are concentrated mainly in the MTB area [100,111].

4. Discussion and Conclusions

The presence of hard explained phenomena of magneto-geophysics prompted combined paleomagnetic-radiometric mapping to reveal the geodynamics history based on the paleomagnetic age reconstruction. The applied mapping includes the procedures for paleomagnetic determination of age and the typification of the magnetized objects, making it possible to widely use the plate geodynamics concepts. It enables the description of the stages of development of various forms of magmatism under transition from the ocean to the continent. The broad integration of paleomagnetic mapping with other geophysical and geological methods significantly increases the efficiency of its application.

Tectonic-paleomagnetic mapping as a new type of geological-geophysical survey contributed to the necessary amendment to the understanding of the nature and structure of the easternmost Mediterranean. The studied areas in northern Israel: Carmel, Atlit, Sea of Galilee, and Hula Basin, are unique objects from the geological point of view, and they are well investigated in various ways. However, their structure, geodynamics, and other critical questions of genesis are not clear and inexplicable. Therefore, we decided to apply the combined paleomagnetic mapping to investigate these intricate targets.

Interestingly, for the mapping examination, a wide diapason of the paleomagnetic data was used: from the Paleozoic (Permian—Carboniferous)—Kiama paleomagnetic superzone in the easternmost Mediterranean [31] to the Holocene—almost modern volcano with a lake in the Birkat Ram crater in the Golan Heights [101]. Without hesitation, this study became possible only thanks to the many years of painstaking work of Israeli geologists and geophysicists, who created a vast database of paleomagnetic (e.g., [50,70,73–75,85,95, 97,99,102,103]) and radiometric studies (e.g., [71–74,78,81,90,97,107]), data of geological mapping and tectonic reconstructions (e.g., [10,15,30,54,57,60–62,76,78,79,83,84,86–88,93, 94,97,100,104–106,108]) for the territory of Israel.

After changing the well-developed geosyncline theory to plate tectonics (convective geodynamics) and the decrease in standard geocartographic research worldwide, a methodological vacuum arose with the lack of mapping technology development. Our comprehensive studies in the African-Arabian region on the junction of Eurasia and Gondwana (e.g., [16]) with the application of paleomagnetic mapping techniques are the testing instrument for developing new kinds of combined investigations of geologically complex areas. Our comprehensive studies in the African-Arabian region on the junction of Eurasia and Gondwana (e.g., [16]) with the widespread application of paleomagnetic mapping techniques are the testing instrument for developing new kinds of combined investigations of geologically complex areas.

Paleomagnetic profiling is a very rarely used but effective research method. The paleomagnetic profile constructed along the top of the accumulative leveling surface of the Lower Cretaceous traps (Figure 4) clearly unmasks the geodynamic history of the Carmel-Galilee region. This procedure is practically impossible to implement using other geophysical methods. This profile makes it possible to reveal the significant tectonic and geodynamic differences between the Mesozoic and Cenozoic traps, differing in thickness and partly to the composition. The latter are usually associated not with the collision stage but with the spreading processes in East Africa and the Red Sea [9,15,16,112]. The beginning of the activation of this process 34 Ma ago is associated with a mantle plume near the Afar triangle. In the Narrat Ash Shaam volcanic field, the dikes of the most ancient traps have an age of 22–26 Ma [81], which corresponds to the boundary of the Oligocene and Miocene.

In the more southerly located Makhtesh Ramon area, integrated geological-geophysical data revealed the pre-collisional, collisional, and post-collisional traps [22], which are distinguished in the presented paleomagnetic profile (Figure 7). In the Korazim block (based on the Rosh Pinna-1 borehole data), these traps indicate the Levantine phase of the angular unconformity, which roughly corresponds to the boundary between the Gissar and Jalal paleomagnetic superzones.

Very indicative is the discovery of the xenoliths in the intrusions of Late Miocene—Early Pliocene gabbroids in the deep Zemah-1 well [100], with a radiometric age of 31.1 Ma (Lower Oligocene), 44.3 Ma (Middle-Upper Eocene), and 547 Ma (Precambrian—Lower Cambrian). These ages are also confirmed by the radiometric age of the basic rock pebbles—41.0 Ma from the Middle Miocene Um Sabune conglomerates in the Tabkha area [73] on the southwestern boundary of the Korazim plateau near the Lake Kinneret coast. Data on the older, Eocene trap magmatism in the Arabian Plate is associated with the beginning of the Red Sea rift opening and are novel.

The paleomagnetic profile (Figure 7) demonstrates a structural difference in the study region in the superzones' thicknesses and magmatism types in different structures. This profile also indicates the structural evolution manifested in the degree of activation of specific paleomagnetic units. The increase in these units' thicknesses relates to the boundary of the Jurassic and Cretaceous (Gissar superzone) and the Middle Miocene—Pliocene—Anthropogen (Sogdiana-2 superzone). Thus, the radiometric dating enables substantiating the Sogdiana-1 superzone trap assemblage and the upper part of the Khorezm superzone.

Overall, the paleomagnetic methods have established themselves as a powerful tool for geodynamic mapping. At the same time, optimal integration of paleomagnetic mapping with other geophysical and geological methods significantly increases its effectiveness and significance [22,23].

Based on the research conducted, we can draw the following main conclusions.

The side of the plate tectonics concept confirmed that the dominance of the geodynamics of the region in the Mesozoic and Cenozoic was carried out not from the western (Atlantic) but the eastern (Neothetical) direction. Paleomagnetic mapping (profiling) made it possible to evaluate the previously identified structures of different ranks. It is a novel unified approach due to the optimal combination of different applied geological-geophysical methods.

The area of development of the Sea of Galilee pull-apart basin has been identified, experiencing differentiation under conditions of the left DST shift and the formation of the faults and ring structures rotating counterclockwise in the adjacent zones. This phenomenon was explained at the level of the regional integrated geophysical studies using 3D modeling and structural-geodynamic analysis [92].

Integrated analysis of geological-geophysical data from the Sea of Galilee and Carmel (Atlit) areas indicates that both mentioned terranes and the MTB were moved in the Mesozoic from the east to the west in a counterclockwise direction. The paleomagnetic studies in the Carmel area showed that the Triassic-Jurassic ophiolite complex underlying the Aptian-Albian traps of the Mt. Carmel (within the Galilee-Lebanon terrane) contrasts sharply with the Mesozoic and sedimentary rocks penetrated by traps. These traps of the Late Jurassic-Neocomian age precede the Late Jurassic-Neocomian phase of joining MTB to the Gondwana.

The paleomagnetic profile constructed in the Carmel-Galilee region along the top of the accumulative leveling surface of the Lower Cretaceous traps is of essential importance: it unmasks key episodes of a complex geodynamic history of the easternmost Mediterranean.

The paleomagnetic profile Kinnarot Valley—Sea of Galilee—Hula Basin distinctly shows different amplitudes of subsidence of the Late Cenozoic sedimentary formations and traps related to the Sogdiana paleomagnetic superzone. First of all, this physical-geological model is essential for the geoecological monitoring of this significant source of fresh water in Israel.

The Cenozoic stage of development in the region looks to be discordant with the paleomagnetic structural zones of the Mesozoic and Cenozoic. This discordance was explained in light of the latest deep-geodynamic studies [16]. The axis of the discovered deep mantle structure [92] in the middle of the Cenozoic began to turn more intensively in the counterclockwise direction. This effect caused the development of the lithospheric plate disruptions, the formation of the topologically displaced dyke complex' zones, and younger effusive traps.

The proposed methodology of advanced paleomagnetic mapping can be effectively applied to solving different geological-geophysical problems in various geologically complex regions of the world.

Author Contributions: L.V.E. and Y.I.K.—equivalent contributions to all sections of this paper. All authors have read and agreed to the published version of the manuscript.

Funding: This research received no external funding.

Institutional Review Board Statement: Not applicable.

Informed Consent Statement: Not applicable.

Data Availability Statement: The data presented in this study are available on request.

Acknowledgments: The authors would like to thank three anonymous reviewers, who thoroughly reviewed the manuscript, and their critical comments and valuable suggestions were very helpful in preparing this paper.

Conflicts of Interest: The authors declare no conflict of interest.

References

1. McKenzie, D. Active tectonics of the Mediterranean region. *Geophys. J. Int.* **1972**, *30*, 109–185. [CrossRef]
2. Khain, V.E. *Tectonics of Continents and Oceans*; Scientific World: Moscow, Russia, 2001; 606p. (In Russian)
3. Muttoni, G.; Kent, D.V.; Garzanti, E.; Brack, P.; Abrahamsen, N.; Gaetani, M. Early Permian Pangea 'B' to Late Permian Pangea 'A'. *Earth Planet. Sci. Lett.* **2003**, *215*, 379–394. [CrossRef]
4. Stern, R.J.; Johnson, P. Continental lithosphere of the Arabian Plate: A geologic, petrologic, and geophysical synthesis. *Earth-Sci. Rev.* **2010**, *101*, 29–67. [CrossRef]
5. Faccenna, C.; Becker, T.W.; Auer, L.; Billi, A.; Boschi, L.; Brun, J.P.; Capitanio, F.A.; Funiciello, F.; Horvàth, F.; Jolivet, L.; et al. Mantle dynamics in the Mediterranean. *Rev. Geophys.* **2014**, *52*, 283–332. [CrossRef]
6. Ben-Avraham, Z.; Ginzburg, A. Displaced terranes and crustal evolution of the Levant and the eastern Mediterranean. *Tectonics* **1990**, *9*, 613–622. [CrossRef]
7. Le Pichon, X.; Kreemer, C. The Miocene-to-present kinematic evolution of the Eastern Mediterranean and Middle East and its implications for Dynamics. *Annu. Rev. Earth Planet. Sci.* **2010**, *38*, 323–351. [CrossRef]
8. Stampfli, G.M.; Hochard, C.; Vérard, C.; Wilhem, C.; von Raumer, J. The formation of Pangea. *Tectonophysics* **2013**, *593*, 1–19. [CrossRef]
9. Bosworth, W.; Huchon, P.; McClay, K. The Red Sea and Gulf of Aden basins. *J. Afr. Earth Sci.* **2005**, *43*, 334–378. [CrossRef]
10. Ben-Avraham, Z.; Schattner, U.; Lazar, M.; Hall, J.K.; Ben-Gai, Y.; Neev, D.; Reshef, M. Segmentation of the Levant continental margin, eastern Mediterranean. *Tectonics* **2006**, *25*, 1–17. [CrossRef]
11. Picard, L. Geology and oil exploration of Israel. *Bull. Res. Counc. Israel* **1959**, *G8*, 1–30.
12. Ben-Avraham, Z. The structure and tectonic setting of the Levant continental margin, Eastern Mediterranean. *Tectonophysics* **1978**, *46*, 313–331. [CrossRef]

13. Rotstein, Y.; Ben-Avraham, Z. Active Tectonics in the Eastern Mediterranean: The role of oceanic plateaus and accreted terranes. *Isr. J. Earth Sci.* **1986**, *35*, 23–39.
14. Robertson, A.H.F. Mesozoic-Tertiary Tectonic Evolution of the Easternmost Mediterranean Area: Integration of Marine and Land Evidence. In *Proceedings of the Ocean Drilling Program, Scientific Results*; Robertson, A.H.F., Eneis, K.C., Richter, C., Camerlenghi, A., Eds.; Texas A & M University: College Station, TX, USA, 1998; Volume 160, pp. 723–782.
15. Hall, J.K.; Krasheninnikov, V.A.; Hirsch, F.; Benjamini, C.; Flexer, A. Geological Framework of the Levant. Volume II: The Levantine Basin and Israel. In *Geological Framework of the Levant*; Historical Productions-Hall: Jerusalem, Israel, 2005.
16. Eppelbaum, L.V.; Ben-Avraham, Z.; Katz, Y.; Cloetingh, S.; Kaban, M. Giant quasi-ring mantle structure in the African-Arabian junction: Results derived from the geological-geophysical data integration. *Geotectonics* **2021**, *55*, 67–93. [CrossRef]
17. Ben-Avraham, Z.; Hall, J. Geophysical survey of Mount Carmel structure and its extension into the eastern Mediterranean. *J. Geophys. Res.* **1977**, *82*, 793–802. [CrossRef]
18. Eppelbaum, L.; Ben-Avraham, Z.; Katz, Y. Integrated analysis of magnetic, paleomagnetic and K-Ar data in a tectonic complex region: An example from the Sea of Galilee. *Geophys. Res. Lett.* **2004**, *31*, 1–4. [CrossRef]
19. Eppelbaum, L.V.; Ben-Avraham, Z.; Katz, Y.I. Structure of the Sea of Galilee and Kinarot Valley derived from combined geological-geophysical analysis. *First Break* **2007**, *25*, 21–28. [CrossRef]
20. Rybakov, M.; Voznesensky, V.; Ben-Avraham, Z.; Lazar, M. The Niklas anomaly southwest of Cyprus: New insights from combined gravity and magnetic data. *Isr. J. Earth Sci.* **2008**, *57*, 125–138. [CrossRef]
21. Eppelbaum, L.V. Comparison of 3D integrated geophysical modeling in the South Caucasian and Eastern Mediterranean segments of the Alpine-Himalayan tectonic belt. *ANAS Trans. Earth Sci.* **2015**, *3*, 25–45.
22. Eppelbaum, L.V.; Katz, Y.I. Paleomagnetic Mapping in Various Areas of the Easternmost Mediterranean Based on an Integrated Geological-Geophysical Analysis. In *New Developments in Paleomagnetism Research*; Series: Earth Sciences in the 21st Century; Eppelbaum, L., Ed.; Nova Science Publisher: Hauppauge, NY, USA, 2015; pp. 15–52.
23. Eppelbaum, L.V.; Katz, Y.I. Deep Tectono-Geodynamic Aspects of Development of the Nubian-Arabian Region. In *The Arabian Seas Biodiversity, Environment Challenges and Conservation Measures*; Jawad, L., Ed.; Springer: Berlin/Heidelberg, Germany, 2021; pp. 199–237. [CrossRef]
24. Khramov, A.N.; Goncharov, G.I.; Komissarova, R.A.; Pisarevsky, S.A.; Pogarskaya, I.A.; Rzhevsky, Y.S.; Rodionov, V.P.; Slauzitais, I.P. *Paleomagnetology*; Nedra: Leningrad, Russia, 1982; 312p. (In Russian)
25. Kissel, C.; Laj, C. *Paleomagnetic Rotations and Continental Deformation*; NATO ASI Series: Mathematical and Physical Sciences; Kluwer Academic Publishers: Dordrecht, The Netherlands; Boston, MA, USA; London, UK, 1989; 530p.
26. Opdyke, N.D.; Channell, J.E.T. *Magnetic Stratigraphy*; International Geophysical Series; Academic Press: Cambridge, MA, USA, 1996; Volume 64, 361p.
27. Molostovsky, E.A.; Khramov, A.N. *Magnetostratigraphy and Its Importance in Geology*; Saratov University: Saratov, Russia, 1997; 172p. (In Russian)
28. Tauxe, L. *Paleomagnetic Principles and Practice*; Kluwer Academic Publishers: Dordrecht, The Netherlands; Boston, MA, USA; London, UK, 2003; 300p.
29. Sholpo, L.E. (Ed.) *The Use of Rock Magnetism in Geological Survey*; Nedra: Leningrad, Russia, 1986; 182p. (In Russian)
30. Ben-Avraham, Z.; Ginzburg, A.; Makris, J.; Eppelbaum, L. Crustal structure of the Levant basin, eastern Mediterranean. *Tectonophysics* **2002**, *346*, 23–43. [CrossRef]
31. Eppelbaum, L.V.; Nikolaev, A.V.; Katz, Y.I. Space location of the Kiama paleomagnetic hyperzone of inverse polarity in the crust of the eastern Mediterranean. *Dokl. Earth Sci.* **2014**, *457*, 710–714. [CrossRef]
32. Eppelbaum, L.V.; Katz, Y.I. Newly Developed Paleomagnetic Map of the Easternmost Mediterranean Unmasks Geodynamic History of this Region. *Cent. Eur. J. Geosci. Open Geosci.* **2015**, *7*, 95–117. [CrossRef]
33. Eppelbaum, L.V.; Katz, Y.I. Eastern Mediterranean: Combined geological-geophysical zonation and paleogeodynamics of the Mesozoic and Cenozoic structural-sedimentation stages. *Mar. Pet. Geol.* **2015**, *65*, 198–216. [CrossRef]
34. Eppelbaum, L.V.; Khesin, B.E. *Geophysical Studies in the Caucasus*; Springer: Berlin/Heidelberg, Germany; New York, NY, USA, 2012; 411p.
35. Irving, E.; Runcorn, S.K. Analysis of the palaeomagnetism of the Torridonian sandstone series of north-west Scotland. *Philos. Trans. R. Soc. Lond.* **1957**, *250*, 83–99.
36. Laj, C.; Channell, J.E.T. Geomagnetic Excursions. In *Geomagnetism*; Kono, M., Ed.; Series: Treatise on Geophysics; Elsevier: Amsterdam, The Netherlands, 2007; Volume 5, pp. 373–416.
37. Khramov, A.N. *Paleomagnetic Correlation of Sedimentary Strata*; Australian National University: Canberra, Australia, 1960; 204p.
38. Picard, M.D. Paleomagnetic correlation of units within Chugwater (Triassic) formation, west-central Wyoming. *Am. Assoc. Pet. Geol. Bull.* **1964**, *48*, 269–291.
39. Irving, E. Paleomagnetism of some Carboniferous rocks from New South Wales and its relation to geological events. *J. Geophys. Res.* **1966**, *71*, 6025–6051. [CrossRef]
40. Vine, F.J.; Matthews, D.H. Magnetic anomalies over ocean ridges. *Nature* **1963**, *199*, 947–949. [CrossRef]
41. Pitman, W.G., III; Larson, L.R.; Herron, E.M. *Age of Ocean Basalts Determined from Magnetic Anomaly Lineation*; Geological Society of America: Boulder, CO, USA, 1974.

42. Cande, S.C.; Kent, D.V. A new geomagnetic polarity time scale for the Late Cretaceous and Cenozoic. *J. Geophys. Res.* **1992**, *97*, 13917–13951. [CrossRef]
43. Shreyder, A.A. Chronology of linear marine magnetic anomalies. *Izv. Phys. Solid Earth* **1993**, *9*, 305–319.
44. Tominaga, M.; Sager, W.W.; Tivey, M.A.; Lee, S.-M. Deep-tow magnetic anomaly study of the Pacific Jurassic Quiet Zone and implications for the geomagnetic polarity reversal timescale and geomagnetic field behavior. *J. Geophys. Res.* **2008**, *113*, B07110. [CrossRef]
45. McDougall, I.; Saemundssion, K.; Johannesson, H.; Watkins, N.D.; Kristjansson, L. Extension of the geomagnetic polarity time scale to 6.5 m.y.: K-Ar dating, geological and paleomagnetic study of a 3,500-m lava succession in western Iceland. *Geol. Soc. Am. Bull.* **1977**, *88*, 1–15. [CrossRef]
46. Molostovsky, E.A. Paleomagnetic Method for Sedimentary Rock Mapping. In *Application of Rock Magnetism for Geological Mapping*; Nedra: Leningrad, Russia, 1986; pp. 158–166. (In Russian)
47. Glevasskaya, A.M.; Mikhailova, N.P. Paleomagnetic Mapping of Volcanogenic Rocks. In *Application of Rock Magnetism for Geological Mapping*; Nedra: Leningrad, Russia, 1986; pp. 166–176. (In Russian)
48. Kristjansson, L.; Jonsson, G. Paleomagnetism and magnetic anomalies in Iceland. *J. Geodyn.* **2007**, *43*, 30–54. [CrossRef]
49. Goguitchaichvili, A.; Cervantes, M.A.; Calvo-Rathert, M.; Camps, P.; Sologashvili, J.; Maissuradze, G. Gilbert-Gauss Geomagnetic Reversal Recorded in Pliocene Volcanic Sequences from Lesser Caucasus: Revisited. *Earth Planets Space* **2009**, *61*, 71–82. [CrossRef]
50. Nur, A.; Ron, H.; Scott, O. Mechanics of Distributed Fault and Block Rotation. In *Paleomagnetic Rotations and Continental Deformation*; NATO ASI Series: Mathematical and Physical Sciences; Kissel, C., Laj, C., Eds.; Kluwer Acad. Publishers: Dordrecht, The Netherlands; Boston, MA, USA; London, UK, 1989; pp. 209–228.
51. Granot, R.; Abelson, M.; Ron, H.; Agnon, A. The oceanic crust in 3D: Paleomagnetic reconstruction in the Troodos ophiolite gabbro. *Earth Planet. Sci. Lett.* **2006**, *251*, 280–292. [CrossRef]
52. Rolland, Y. Caucasus collisional history: Review of data from East Anatolia to West Iran. *Gondwana Res.* **2017**, *49*, 130–146. [CrossRef]
53. Eppelbaum, L.; Katz, Y. Significant tectono-geophysical features of the African-Arabian tectonic region: An overview. *Geotectonics* **2020**, *54*, 266–283. [CrossRef]
54. Gvirtzman, G.; Klang, A.; Rotstein, Y. Early Jurassic shield volcano below Mount Carmel: New interpretation of the magnetic and gravity anomalies and implication for Early Jurassic rifting. *Isr. J. Earth Sci.* **1990**, *39*, 149–159.
55. Ginzburg, A.; Eppelbaum, L. *A Combined 3D Interpretation of the Carmel Gravity and Magnetic Anomalies*; Oilfields Ltd.: Tel Aviv, Israel, 1993; pp. 1–42.
56. Eppelbaum, L.; Katz, Y.; Ben-Avraham, Z. Mt. Carmel structure as a plate tectonics phenomenon. In Proceedings of the Annual Conference of the Israel Geological Society, Beit-Shean, Israel, 19–21 February 2006; p. 31.
57. Segev, A.; Sass, E. *The Geology of the Carmel Region, Albian-Turonian Volcano-Sedimentary Cycles on the Northwestern Edge of the Arabian Platform*; Report of the Israel Geological Soc.; Israel Geological Society: Jerusalem, Israel, 2009; pp. 1–77.
58. Lu, J.-G.; Griffin, W.L.; Huang, J.-X.; Dai, H.-K.; Castillo-Oliver, M.; O'Reilly, S.Y. Structure and composition of the lithosphere Mount Carmel, North Israel. *Contrib. Mineral. Petrol.* **2022**, *177*, 1–29. [CrossRef]
59. Garfunkel, Z.; Almagor, G. Geology and structure of the continental margin off northern Israel and the adjacent part of the Levantine basin. *Mar. Geol.* **1985**, *62*, 105–131. [CrossRef]
60. Gvirtzman, G.; Steinitz, G. *The Asher Volcanics—An Early Jurassic Event in the Northern Israel*; Current Research; Geological Survey of Israel: Jerusalem, Israel, 1982; pp. 28–33.
61. Garfunkel, Z. Tectonic setting of Phanerozoic magmatism in Israel. *Israel J. Earth Sci.* **1989**, *38*, 51–74.
62. Garfunkel, Z.; Derin, B. Reevaluation of the latest Jurassic-Early Cretaceous history of the Negev and the role of magmatic activity. *Isr. J. Earth Sci.* **1988**, *37*, 43–52.
63. Dvorkin, A.; Kohn, B.P. The Asher volcanics, northern Israel: Petrography, mineralogy, and alteration. *Isr. J. Earth Sci.* **1989**, *38*, 105–123.
64. Kaminchik, J.; Segev, A.; Katzir, Y. The Origin of Intraplate Alkaline Mafic Magmatism in Continental Shelves: Lavas and Xenoliths from the Upper Cretaceous Volcanos of Mt. Carmel. Master's Thesis, Beer-Sheva University, Beersheba, Israel, 2014, *unpublished*.
65. Griffin, W.L.; Gain, S.E.M.; Huang, J.-X.; Belousova, E.A.; Toledo, V.; O'Reilly, S.Y. Permian to quaternary magmatism beneath the Mt Carmel area, Israel: Zircons from volcanic rocks and associated alluvial deposits. *Lithos* **2018**, *314–315*, 307–322. [CrossRef]
66. Esperanca, S.; Garfunkel, Z. Ultramafic xenoliths from the Mt. Carmel area (Karem Maharal volcano), Israel. *Lithos* **1986**, *19*, 43–49. [CrossRef]
67. Mittlefehldt, D.W. Petrology of high pressure clinopyroxenite series xenoliths, Mount Carmel, Israel. *Contrib. Mineral. Petrol.* **1986**, *94*, 245–252. [CrossRef]
68. Apter, D.B. High pressure indicator minerals from the Rakefet magmatic complex (RMC). Mt. Carmel, Israel. In Proceedings of the Geological Society of South Africa Kimberley Diamond Symposium, Kimberley, South Africa, 11–13 September 2014.
69. Dobrzhinetskaya, L.; Mukhin, P.; Wang, Q.; Wirth, R.; O'Bannon, E.; Zhao, W.; Eppelbaum, L.; Sokhonchuk, T. Moissanite (SiC) with metal-silicide and silicon inclusions from tuff of Israel: Raman spectroscopy and electron microscope studies. *Lithos* **2018**, *310–311*, 355–368. [CrossRef]
70. Nur, A.; Helsey, C.F. Palaeomagnetism of Tertiary and Recent lavas of Israel. *Earth Planet. Sci. Lett.* **1971**, *10*, 375–379. [CrossRef]

71. Lang, B.; Mimran, Y. An Early Cretaceous volcanic sequence in central Israel and its significance to the absolute date of the base of the Cretaceous. *J. Geol.* **1985**, *93*, 179–184. [CrossRef]
72. Lang, B.; Steinitz, G. K-Ar dating of Mesozoic magmatic rocks in Israel: A review. *Isr. J. Earth Sci.* **1989**, *38*, 89–103.
73. Shaliv, G. Stages in the Tectonics and Volcanic History of the Neogene Basin in the Lower Galilee and the Valleys. Ph.D. Thesis, Hebrew University of Jerusalem, Jerusalem, Israel, 1991. (In Hebrew, Summary in English)
74. Mor, D. A time-table for the Levant Volcanic Province, according to K-Ar dating in the Golan Heights, Israel. *J. Afr. Earth Sci.* **1993**, *16*, 223–234. [CrossRef]
75. Heimann, A.; Steinitz, G.; Mor, D.; Shaliv, G. The Cover Basalt Formation, its age and its regional and tectonic setting: Implications from K-Ar and $^{40}Ar/^{39}Ar$ geochronology. *Isr. J. Earth Sci.* **1996**, *45*, 55–71.
76. Sneh, A.; Bartov, Y.; Rozensaft, M. *Geological Map of Israel, Scale 1:200,000*; Geol. Survey of Israel; Ministry of National Infrastructure: Jerusalem, Israel, 1998.
77. Katz, Y.I.; Eppelbaum, L.V. Levantine phase of tectonic-thermal activity in the Eastern Mediterranean. *Trans. Annu. Meet. Geol. Soc. Am. Sect. Planet. Geol.* **1999**, *31*, A119.
78. Segev, A. Synchronous magmatic cycles during the fragmentation of Gondwana: Radiometric ages from the Levant and other provinces. *Tectonophysics* **2000**, *325*, 257–277. [CrossRef]
79. Segev, A. $^{40}Ar/^{39}Ar$ and K-Ar geochronology of Berriasian-Hauteririan and Cenomanian tectomagmatic events in northern Israel: Implications for regional stratigraphy. *Cretac. Res.* **2009**, *30*, 818–828. [CrossRef]
80. Segev, A.; Sass, E.; Ron, H.; Lang, B.; Kolodny, Y.; McWilliams, M. Stratigraphic, geochronologic, and paleomagnetic constraints on Late Cretaceous volcanism in northern Israel. *Isr. J. Earth Sci.* **2002**, *51*, 297–309. [CrossRef]
81. Ilani, S.; Harlavan, Y.; Tarawneh, K.; Rabba, I.; Weinberger, R.; Ibrahim, K.; Peltz, S.; Steinitz, G. New K-Ar ages of basalts from the Harrat Ash Shaam volcanic field in Jordan: Implications for the span and duration of the upper mantle upwelling beneath the western Arabian plate. *Geology* **2001**, *29*, 171–174. [CrossRef]
82. Ilani, S.; Kafri, U.; Harlavan, Y. Campanian volcanism within the Asher-1 borehole. *Isr. J. Earth Sci.* **2005**, *54*, 179–181. [CrossRef]
83. Karcz, J.; Sneh, A. *Sheet 3-I, Haifa. Geological Map of Israel, Scale 1:50,000*; Geological Survey of Israel: Jerusalem, Israel, 2011; Volume 50.
84. Sass, E.; Dekel, A.; Sneh, A. *Sheet 5-II, Umm el Fahm. Geological Map of Israel, Scale 1:50,000*; Geological Survey of Israel: Jerusalem, Israel, 2013.
85. Dembo, N.; Hamiel, Y.; Granot, R. *Intraplate Rotational Deformation Induced by Faults: Carmel-Gilboa Fault System as a Case Study*; Report No. GSI/19/2015; Geological Survey of Israel: Jerusalem, Israel, 2015; pp. 1–32.
86. Sneh, A. *Sheet 3-II, Shefar'Am. Geological Map of Israel, Scale 1: 50,000*; Geological Survey of Israel: Jerusalem, Israel, 2013.
87. Sneh, A. *Sheet 3-IV, Nazerat. Geological Map of Israel, Scale 1:50,000*; Geological Survey of Israel: Jerusalem, Israel, 2018.
88. Sneh, A.; Sass, E.; Bein, A.; Arad, A.; Rosensaft, M. *Sheet 5-I, Hadera. Geological Map of Israel, Scale 1:50,000*; Geological Survey of Israel: Jerusalem, Israel, 2014.
89. Fleischer, L.; Varshavsky, A. *A Lithostratigraphic Data Base of Oil and Gas Wells Drilled in Israel*; Rep. OG/9/02; Ministry of National Infrastructures of Israel: Jerusalem, Israel, 2012.
90. Segev, A.; Lang, B. 40Ar/39Ar dating of Valanginian top Tayasir Volcanics in the Mount Hermon area, northern Israel. Israel Geological Survey. *Curr. Res.* **2002**, *13*, 100–104.
91. Folkman, Y. Magnetic and Gravity Investigations of the Crustal Structure in Israel. Ph.D. Thesis, Tel Aviv University, Tel Aviv, Israel, 1976. (In Hebrew, English Abstract)
92. Eppelbaum, L.V.; Ben-Avraham, Z.; Katz, Y.; Cloetingh, S.; Kaban, M. Combined Multifactor Evidence of a Giant Lower-Mantle Ring Structure below the Eastern Mediterranean. *Positioning* **2020**, *11*, 11–32. [CrossRef]
93. Garfunkel, Z.; Zak, I.; Freund, R. Active faulting in the Dead Sea Rift. *Tectonophysics* **1981**, *80*, 1–26. [CrossRef]
94. Ben-Avraham, Z.; ten-Brink, U.; Bell, R.; Reznikov, M. Gravity field over the Sea of Galilee: Evidence for a composite basin along a transform fault. *J. Geophys. Res.* **1996**, *101*, 533–544. [CrossRef]
95. Ron, H.; Freund, R.; Garfunkel, Z.; Nur, A. Block rotation by strike-slip faulting: Structural and paleomagnetic evidence. *J. Geophys. Res.* **1984**, *89P*, 6256–6270. [CrossRef]
96. Heimann, A.; Ron, H. Geometric changes of plate boundaries along part of the Northern Dead Sea transform—Geochronological and paleomagnetic evidence. *Tectonics* **1993**, *12*, 477–491. [CrossRef]
97. Heimann, A. The Development of the Dead Sea Rift and Its Margins in Northern Israel during the Pliocene and Pleistocene. Ph.D. Thesis, Hebrew University of Jerusalem, Jerusalem, Israel, 1990. (In Hebrew, Summary in English)
98. Hurwitz, S.; Garfunkel, Z.; Ben-Gai, Y.; Reznikov, M.; Rotstein, Y.; Gvirtzman, H. The tectonic framework of a complex pull-apart basin: Seismic reflection observations in the Sea of Galilee, Dead Sea transform. *Tectonophysics* **2002**, *359*, 289–306. [CrossRef]
99. Heimann, A.; Braun, D. Quaternary stratigraphy of the Kinarot Basin, Dead Sea Transform, northeastern Israel. *Isr. J. Earth Sci.* **2000**, *49*, 31–44. [CrossRef]
100. Segev, A. *Zemah-1, A Unique Deep Oil Well on the Dead Sea Fault Zone, Northern Israel: A New Stratigraphic Amendment*; Report GSI/21/2017; Geological Survey of Israel: Jerusalem, Israel, 2017; pp. 1–27.
101. Frank, U.; Schwab, M.J.; Negendank, J.F.W. A lacustrine record of paleomagnetic secular variations from Birkat Ram, Golan Heights (Israel) for the last 4400 years. *Phys. Earth Planet. Inter.* **2002**, *133*, 21–34. [CrossRef]

102. Freund, R.; Oppenheim, M.J.; Schulman, N. Direction of magnetization of some basalts in the Jordan Valley and Lower Galilee (Israel). *Isr. J. Earth Sci.* **1965**, *14*, 37–74.
103. Behar, G.; Shaar, R.; Tauxe, L.; Asefaw, H.; Ebert, Y.; Heimann, A.; Koppers, A.A.P.; Ron, H. Paleomagnetism and paleosecular variations from the Plio-Pleistocene Golan Heights volcanic plateau, Israel. *Geochem. Geophys. Geosyst.* **2019**, *20*, 4319–4334. [CrossRef]
104. Rotstein, Y.; Bartov, Y. Seismic reflection across a continental transform: An example from a convergent segment of the Dead Sea rift. *J. Geophys. Res.* **1989**, *94*, 2902–2912. [CrossRef]
105. Heimann, A.; Steinitz, G. Fault systems in the south-western Hula Valley and the eastern slopes of the Galilee—Dating and tectonic implications. In Proceedings of the Israel Geological Society Meeting, Ramot, Israel, 2–5 April 1989; pp. 73–75.
106. Schattner, U.; Weinberger, R. A mid-Pleistocene deformation transition in the Hula basin, northern Israel: Implications for the tectonic evolution of the Dead Sea Fault. *Geochem. Geophys. Geosyst.* **2008**, *9*, 1–18. [CrossRef]
107. Mor, D.; Steinitz, G. *K-Ar Age of the Cover Basalts Surrounding the Sea of Galilee*; Rep. Me/6/82; Geological Survey of Israel: Jerusalem, Israel, 1982; 14p.
108. Shulman, H.; Reshev, M.; Ben-Avraham, Z. The structure of the Golan Heights and its tectonic linkage to the Dead Sea Transform and the Palniyrides folding. *Israel J. Earth Sci.* **2004**, *53*, 225–237. [CrossRef]
109. Meiler, M.; Reshef, M.; Shulman, H. Seismic depth-domain stratigraphic classification of the Golan Heights, central Dead Sea Fault. *Tectonophysics* **2011**, *510*, 354–369. [CrossRef]
110. Marcus, E.; Slager, J. The sedimentary-magmatic sequence of the Zemah-1 well (Jordan–Dead Sea Rift, Israel) and its emplacement in time and space. *Isr. J. Earth Sci.* **1985**, *34*, 1–10.
111. Eppelbaum, L.; Katz, Y. Tectonic-Geophysical Mapping of Israel and eastern Mediterranean: Implication for Hydrocarbon Prospecting. *Positioning* **2011**, *2*, 36–54. [CrossRef]
112. Kazmin, V.G. *Rift Structures of the Eastern Africa—The Split of the Continent and the Emergence of the Ocean*; Nauka: Moscow, Russia, 1987; 210p. (In Russian)

applied sciences

MDPI

Article

Enhanced Discrimination of Seismic Geological Channels Based on Multi-Trace Variational Mode Decomposition

Jiaxuan Leng [1], Zhichao Yu [2,*] and Chaodong Wu [1]

[1] School of Earth and Space Sciences, Peking University, Beijing 100871, China; jxleng@pku.edu.cn (J.L.); cdwu@pku.edu.cn (C.W.)

[2] National Supercomputing Center in Shenzhen (Shenzhen Cloud Computing Center), Shenzhen 518005, China

* Correspondence: zcyu.426@163.com

Abstract: The spectral decomposition is a valuable tool for improving the resolution of seismic interpretation, and thus can improve the accuracy of the subtle geo-features (thin and narrow channels, thin reservoirs, etc.). Variational mode decomposition (VMD) is an adaptive signal decomposition algorithm that non-recursively decomposes multicomponent signals into several band-limited intrinsic mode functions, which is competitive in enhancing time-frequency resolution. However, discontinuity is normally caused by the trace-by-trace process, making the 3D seismic interpretation difficult. To address this issue, we present a novel seismic geological channel detection method for 3D seismic dataset based on multi-trace variational mode decomposition (MTVMD). The proposed method decomposes the broadband seismic data into several intrinsic mode functions and then computes seismic attributes from each component for geological feature analysis. Further tested by field seismic case, the proposed method demonstrates strength in depicting the detailed edges and sedimentary signatures of paleochannels. Overall, the proposed method provides an alternative approach to identifying seismic channels, especially for the detailed portrayals of subtle geo-features in low-quality seismic data.

Keywords: seismic interpretation; spectral decomposition; seismic geological channel; variational mode decomposition; multi-variate variational mode decomposition

1. Introduction

Channel sand bodies are of great importance in many petroliferous reservoirs and are targets in oil and gas explorations [1]. The appropriate distribution of sand bodies, particularly sand thicknesses, aids in the investigation of sedimentary facies [2]. Therefore, the precise interpretation of seismic geological channels is highly demanding. Ascribing to the continuous channel migration and frequent vertical superposition of sand bodies, sedimentary patterns of the channels in seismic records are always complicated. Doubly inhibited by varied physical properties and oil-bearing properties on seismic data, traditional attribute-based analytical methods have difficulties in recognition, especially for identifying the precise edges and geological patterns of paleo-channels [3,4].

With spectral decomposition techniques, initial broadband seismic data can be decomposed into a sequence of frequency-decomposed seismic volumes of various frequencies. Separating the frequency bands of seismic data is theoretically valid for extracting the information on varied geological scales [5]. The spectral characteristics of relatively narrow band components are usually used to delineate the anomalous geologic features that would otherwise be hidden within the full-bandwidth seismic response; thus, appropriate bands would effectively increase resolution and reliability in seismic interpretation [6,7]. Accordingly, spectral decomposition-based methods were developed and widely applied in seismic data analysis, including sequence stratigraphy analysis [8], reservoir characterization [9], thin-layer thickness estimation [10], and hydrocarbon detection [11]. Conventional spectral decomposition methods include short-time Fourier transform (STFT) [12], continuous wavelet transform (CWT) [13], and S transform (ST) [14]. The STFT, CWT, and ST are all bound by the Heisenberg uncertainty principle with a tradeoff between time and frequency localization [15]. In order to identify the intrinsic hidden properties contained in the nonlinear and non-stationary signals, the empirical mode decomposition (EMD) method is introduced by Huang et al. as a data-driven signal decomposition technique without using predefined wavelet libraries or base functions [16]. The EMD-based time-frequency analysis of seismic data shows the relative high resolutions in both the temporal and frequent realms, as it is effective on the instantaneous attributes of nonlinear and nonstationary signals. However, with a deficiency on the bandlimited assumption of the instinct mode function, the usage of the EMD method is limited. Further, the variational mode decomposition (VMD) method was developed, which can decompose the seismic signals into a sequence of quasi-orthogonal intrinsic mode functions, with the most compact bandwidth [17].

The VMD method was applied in seismic data analysis, including denoising [18,19], sequence stratigraphy analysis [7,20], seismic time-frequency analysis [21], and seismic facies analysis [22,23]. The VMD-based method, which calculates trace-by-trace, is unable to maintain the lateral continuity of 2D or 3D seismic data, causing difficulties in subtle geological feature discriminations [7], and consequently, including the seismic channels. Rehman and Aftab recently proposed the multivariate variational mode decomposition (MVMD) that is valid for multivariate or multichannel datasets [24]. In the MVMD method, the lateral continuity of the decomposed intrinsic mode functions (IMFs) is improved by applying lateral constraints, which are valid for extracting the spatial aspects of seismic reflection on different geological scales [25].

The tangled geo-features ascribed to the subtle discontinuities have brought difficulties in recognition by the original variational mode decomposition-based method. Taking into account the horizontal consistency of 3D seismic data, we propose a multi-trace variational mode decomposition-based seismic spectral decomposition method for geological feature discrimination in this study. First, we introduce the theory of variational mode decomposition and multivariate variational mode decomposition methods. Then, we illustrate the proposed MTVMD-based method for geological channel detection. Finally, we demonstrate the applicability of the proposed method to 3D field data.

2. Methods

In this section, we describe the variational mode decomposition method and multivariate variational mode decomposition method, then demonstrate the workflow of a multi-trace variational mode decomposition-based approach for 3D seismic data interpretation.

2.1. Variational Mode Decomposition

The variational mode decomposition method aims to decompose a real valued signal into a series of sub-signals with specific sparsity properties of its bandwidth in the spectral domain, while reproducing the input [17]. It is an adaptive and non-recursive signal decomposition approach in which each mode is assumed to be an amplitude modulation-frequency modulation (AM-FM) signal with a bandwidth and corresponding center frequency, i.e.:

$$u_k(t) = A_k(t)\cos(\phi_k(t)), \tag{1}$$

where, $A_k(t)$ and $\phi_k(t)$ are the instantaneous amplitude and phase of $u_k(t)$ respectively. The whole process of VMD can be considered as the construction and solution of a constrained variational problem, as described by Equation (2):

$$\begin{aligned} &\min\left\{\sum_{k=1}^{K} \|\partial_t\left[\left(\delta(t)+\frac{i}{\pi t}\right)\times u_k(t)\right]e^{-i\omega_k t}\|_2^2\right\}, \\ &s.t. \ \sum_{k=1}^{K} u_k(t) = x(t)\ , \end{aligned} \tag{2}$$

where u_k is the kth decomposed mode, ω_k is the center frequency of the corresponding kth decomposed mode, $\delta(t)$ is a Dirichlet function, $x(t)$ is the input signal. The term $(\delta(t) + i/\pi t) \times u_k(t)$ is the Hilbert transform of u_k.

A quadratic penalty term α and Lagrangian multipliers λ are adopted to render the variational problem unconstrained; the quadratic penalty term can guarantee better convergence properties, even in the presence of Gaussian noise, and the Lagrangian multiplier is used to enforce a strict constraint. The augmented Lagrangian is described as follows:

$$\begin{aligned} L(u_k, \omega_k, \lambda) &= \alpha \sum_{k=1}^{K} \left\| \partial_t \left[\left(\delta(t) + \frac{i}{\pi t} \right) \times u_k(t) \right] e^{-i\omega_k t} \right\|_2^2 \\ &+ \left\| x(t) - \sum_{k=1}^{K} u_k(t) \right\|_2^2 + \left\langle \lambda, x(t) - \sum_{k=1}^{K} u_k(t) \right\rangle . \end{aligned} \tag{3}$$

where α is a parameter balancing the variational energy term and the fidelity term. The alternating direction method of the multipliers (ADMM) can be employed to solve Equation (3). The modes and the center frequencies are updated iteratively by Equations (4) and (5), respectively:

$$\hat{u}_k^{n+1} = \frac{\hat{x} - \sum_{j \neq k} \hat{u}_j + (\hat{\lambda}/2)}{1 + 2\alpha(\omega - \omega_k)^2}. \tag{4}$$

$$\omega_k^{n+1} = \frac{\int_0^\infty \omega |\hat{u}_k(\omega)|^2 d\omega}{\int_0^\infty |\hat{u}_k(\omega)|^2 d\omega}. \tag{5}$$

where $(\hat{\cdot})$ indicates a Fourier-transformed signal. The Lagrangian multiplier will also be updated according to the following equation after each epoch:

$$\lambda^{n+1} = \lambda^n + \tau \left(\hat{x} - \sum_k \hat{u}_k^{n+1} \right). \tag{6}$$

Finally, the above iteration terminates when the convergence condition is satisfied, namely:

$$\sum_k \| \hat{u}_k^{n+1} - \hat{u}_k^n \|_2^2 / \| \hat{u}_k^n \|_2^2 < \varepsilon. \tag{7}$$

According to the preceding descriptions, there are four parameters that must be specified in advance: the mode number K, the balancing parameter α, the constraint intensity τ, and the tolerance of convergence criterion ε. The detailed complete algorithm of VMD can be found in [17]. It is important to note that VMD decomposes the input signal trace by trace, the lateral continuity is not considered.

2.2. Multivariate Variational Mode Decomposition

The goal of multivariate variational mode decomposition (MVMD) is to analyze a multivariate input signal containing M number of data channels into a predefined number K of multivariate components. As an extension of the VMD method, K multivariate components that will fully construct the input signal $\mathbf{x}(t)$ are searched:

$$\mathbf{x}(t) = \sum_{k=1}^{K} \mathbf{u}_k(t). \tag{8}$$

where the kth multivariate component. $\mathbf{u}_k(t) = [u_{k,1}(t), u_{k,2}(t), \ldots, u_{k,M}(t)]$ is a vector with M components.

MVMD takes the multivariate input signal as a whole and tries to seek K number of multivariate components from the input signal, with a minimum sum of bandwidths [24]. The optimization problem for MVMD can be naturally expanded from Equation (2) as follows:

$$\min_{u_k,\omega_k}\left\{\sum_{k=1}^{K}\sum_{m=1}^{M}\|\partial_t\left[\left(\delta(t)+\frac{i}{\pi t}\right)\times u_{k,m}(t)\right]e^{-i\omega_k t}\|_2^2\right\},\quad s.t.\ \sum_{k=1}^{K}u_{k,m}(t)=x_i(t), i=1,2,\ldots,M. \tag{9}$$

In Equation (9), the target function becomes the sum of the bandwidth of all of the modes over all of the channels, and the central frequencies ω_k in each mode are the same among different channels. The target function in Equation (8) can be reformulated as an augmented Lagrangian:

$$L(u_k,\omega_k,\lambda)=\alpha\sum_{k=1}^{K}\sum_{m=1}^{M}\|\partial_t\left[\left(\delta(t)+\frac{i}{\pi t}\right)\times u_{k,m}(t)\right]e^{-i\omega_k t}\|_2^2 + \sum_{m=1}^{M}\|x_m(t)-\sum_{k=1}^{K}u_{k,m}(t)\|_2^2+\sum_{m=1}^{M}\left\langle\lambda_m(t),x_m(t)-\sum_{k=1}^{K}u_{k,m}(t)\right\rangle. \tag{10}$$

where the Lagrangian multiplier and the penalty weighting factor play the same role as those in VMD. The ADMM procedure is also used for the solution of Equation (10), as formulated in Equations (10)–(12), which is slightly different from that in VMD:

$$\hat{u}_{k,m}^{n+1}=\frac{\hat{x}_m-\sum_{j\neq k}\hat{u}_{j,m}+(\hat{\lambda}_{l,m}/2)}{1+2\alpha(\omega-\omega_k)^2}, \tag{11}$$

$$\omega_k^{n+1}=\frac{\sum_{m=1}^{M}\int_0^\infty\omega|\hat{u}_{k,m}(\omega)|^2d\omega}{\sum_{m=1}^{M}\int_0^\infty|\hat{u}_{k,m}(\omega)|^2d\omega}, \tag{12}$$

$$\lambda^{n+1}=\lambda^n+\tau\left(x_m(\omega)-\sum_k u_{k,m}^{n+1}\right), m=1:M \tag{13}$$

Until convergence:

$$\sum_m\sum_k\|\hat{u}_{k,m}^{n+1}-\hat{u}_{k,m}^{n}\|_2^2/\|\hat{u}_{k,m}^{n}\|_2^2<\varepsilon. \tag{14}$$

The associated explanation can be located in the original work of Rehman and Aftab [24] and is not repeated here.

Figure 1 gives a schematic illustration of the algorithm of MVMD for a 2D seismic profile. In the algorithm, MVMD unifies the quantity of the decomposed component and their center frequency. It can not only increase the seismic signal similarity, but also the calculation efficiency by exchanging the time through memory space. The decomposition results reflect the seismic geological structures on various scales.

2.3. Multi-Trace VMD Based Seismic Channel Detection Method

Inspired by multi-variational mode decomposition [24] and quasi-bivariate variational mode decomposition [26] methods, we proposed a multi-trace variational mode decomposition-based (MTVMD) approach to deal with a 3D seismic dataset with only one primary dimension with discernible quasi-periodicity. However, instead of a simple concatenation of these intrinsic mode functions (IMFs) among the second non-decomposed dimension (inline and crossline direction are assumed here), the carrier frequency of IMFs with the same rank should be rebalanced among all of the traces. Such a consideration is based on the idea that there should be no big inter-trace discontinuity in each IMF; therefore, this step serves as an additional constraint to minimize the mode mixing problem.

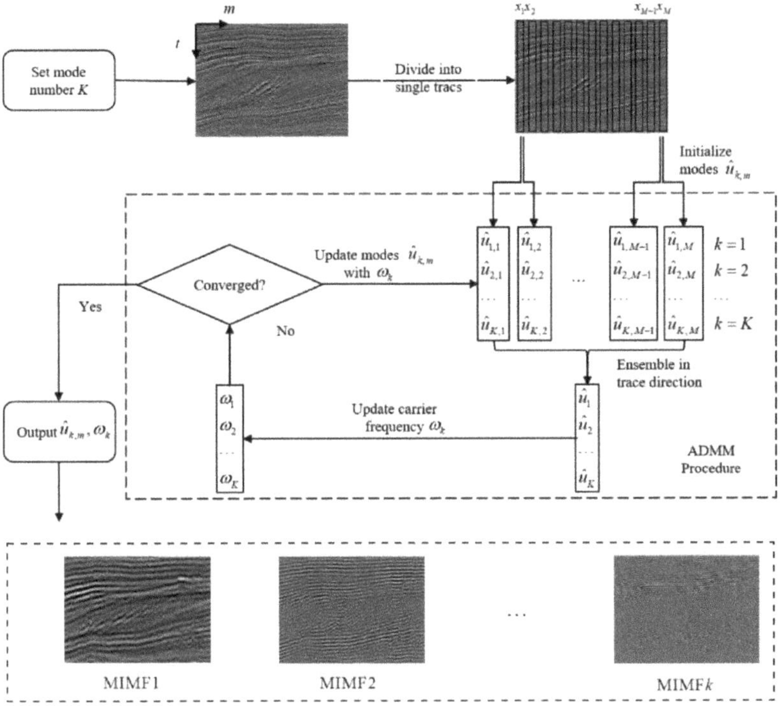

Figure 1. A flowchart of the MVMD algorithm for 2D seismic data.

The 3D seismic data $x_{l,m}(t)$ with L number of inline traces and M number of crossline traces is converted to be a quasi-2D seismic dataset after dimensionality reduction with the trace from both inline and crossline directions. We propose a novel seismic channel detection method, based on MTVMD in conjunction with seismic attributes. The detailed workflow (shown in Figure 2) can be summarized as follows.

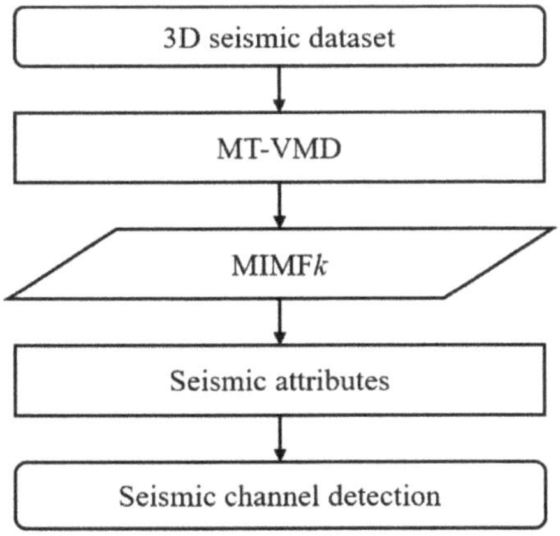

Figure 2. A flowchart of the proposed method for channel detection.

(1) Decompose the 3D seismic dataset into a series of band-limited components using the MTVMD method;

(2) The instantaneous frequency and amplitude of each component are utilized to obtain the seismic attributes, such as Root Mean Square amplitude, amplitude variance, and coherence;
(3) The attributes are involved in the analysis of seismic geological feature interpretation, such as seismic channels.

3. Field Data Application

We evaluate the performance of the proposed method by a 3D field seismic data. The dataset here is derived from the Chepaizi uplift in the western Junggar Basin, China. The Chepaizi Uplift was formed during the deposition of the Neogene Shawan Formation as a large-scale terrestrial slope with fluvial-delta sedimentation in the western basin, and it currently offers a potential exploratory perspective [27,28]. These data contain the sedimentary transition from the braided channel to the curved channel and the near-end delta, with frequent vertical superposition of sand bodies and continuous channel migration in the Shawan Formation. Discriminating subtle geological features of sedimentary channels from seismic data is difficult, due to the complicated patterns of the fluvial-delta deposition. Figure 3 shows the 3D field dataset, and time slices and vertical profiles are extracted from the seismic volume. The exacted time slice at t = 1665 ms (Figure 4a), the vertical profiles at inline No. 98520 (Figure 4b), and crossline No. 11272 (Figure 4c) are used as cases. As seen in the vertical profiles, pinch out on both sides of the seismic reflections are characteristic of the fluvial seismic facies (noted by the black arrows), however, hardly seen features of an entire channel in the horizontal time slice.

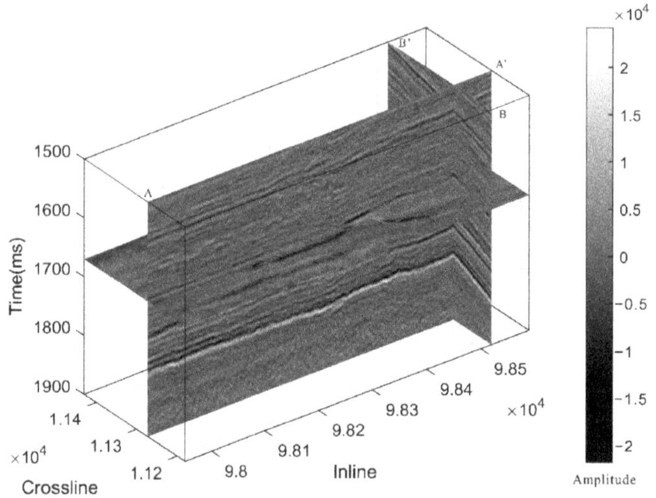

Figure 3. The 3D view of the field seismic dataset. AA′ and BB′ are through-well seismic profiles.

We can use the through-well seismic profiles to analyze more precisely the properties of the seismic geological channels. Figure 5 depicts the location of Well C1 on the southeastern channel of the selected real data. Through the lithological characteristics of the channel of Well C1, the grain size gradually becomes finer, the grains are coarser, and the lithology is gravel-bearing fine sandstone with a thickness of 2~9 m. The logging facies at the black ellipse dotted line is a zigzag box shape, which is in line with the logging facies type characteristics of the channel. On the well-passing seismic section, it can be seen that the seismic facies at the location of the black ellipse dotted line are continuous-moderately strong amplitude, which is consistent with the characteristics of the seismic facies type of the channel, and has a channel shape in the lateral direction.

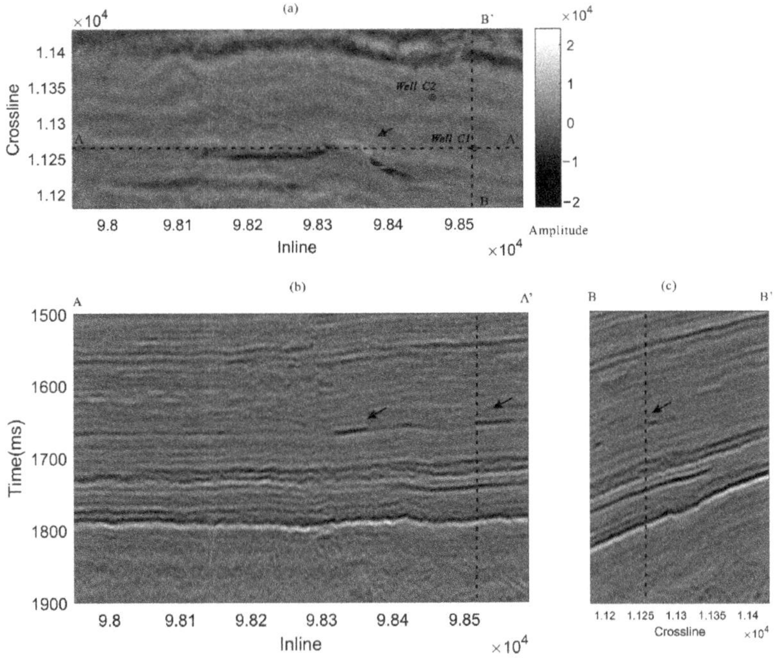

Figure 4. The location of wellbore and the through-well seismic profiles. (**a**) The time slice at t = 1650 ms; (**b**) the vertical seismic profile extracted at Crossline No. 98520; (**c**) the vertical seismic profile extracted at Crossline No. 11272. The dotted lines indicate the location of the through-well seismic profiles.

We use the full bandwidth dataset to calculate the Root Mean Square amplitude to analyze the seismic geological channels in this dataset, as shown in Figure 6. The figure shows that the extension direction and detailed description of the subtle channel facies are not clear. Apparently, spectral decomposition on the frequency bands is necessary in this case. We evaluate the performance of the proposed method by 3D field seismic data. To compare the performance on the resolution, we extracted band range frequency profiles and applied VMD- and MTVMD-based methods for comparison. Otherwise, RMS amplitude is else calculated. We demonstrate the superiority of MVMD over VMD using the through-well seismic profiles extracted from the seismic data. The input parameters K, α, τ and ε equal to 3, 2000, 0, and 10^{-4}, respectively. Figures 7 and 8 depict the decomposed components by using the MTVMD and VMD methods on the vertical seismic profile AA′ and BB′, respectively. The center frequencies of the decomposed components are from low frequency to high frequency. In Figures 7d–f and 8d–f, we can observe that the lateral consistency of the 2D data cannot be guaranteed, because the processing is calculated trace by trace. Even a slight oscillation can completely alter the outcome of the breakdown, indicated by the black arrows. Instabilities such as these are a common drawback of the mode decomposition methods, which make it harder to control noise and cause structural abnormalities. Figures 7a–c and 8a–c show the multivariate intrinsic mode functions (MIMFs) from the MVMD method, geological structural changes are more consistent and reasonable, which improves seismic interpretation.

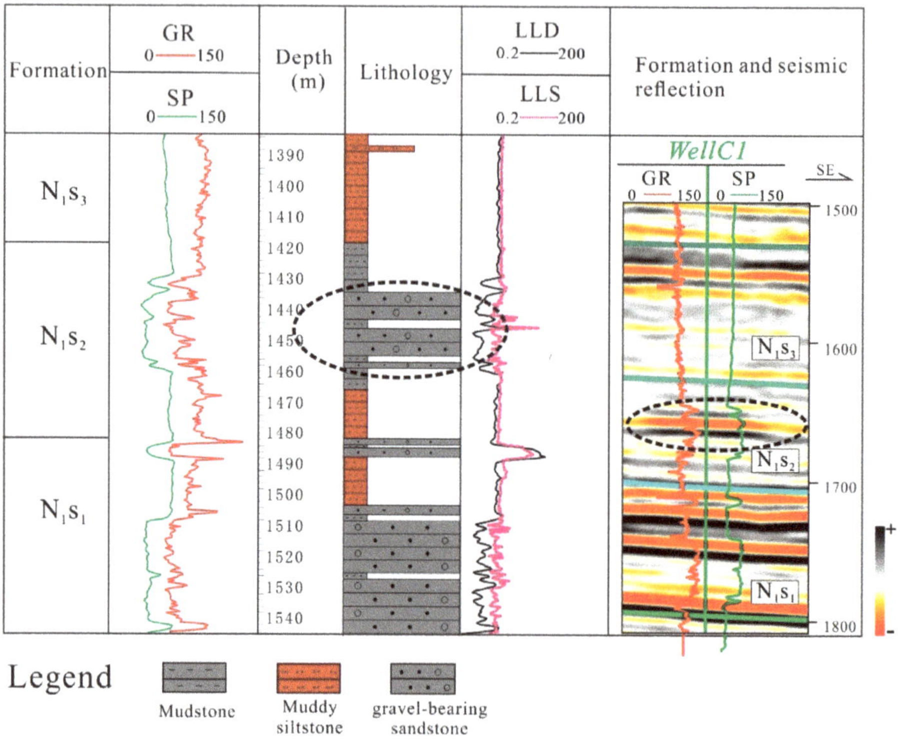

Figure 5. The well log recording of Well C1 in Figure 4. The dashed circle indicates the location of the seismic geological channel.

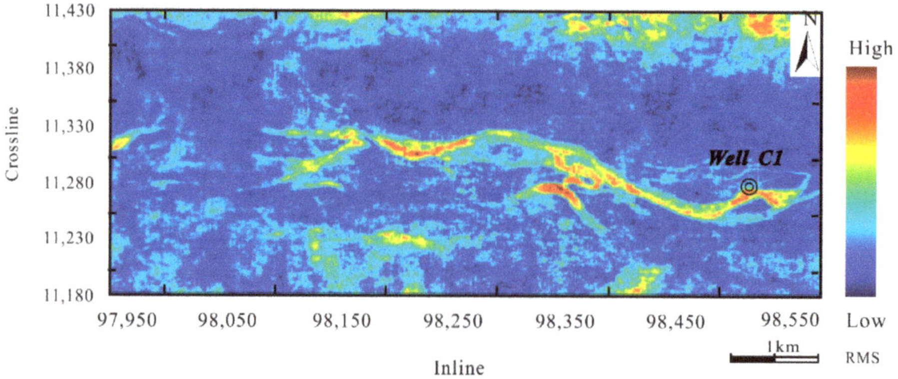

Figure 6. The RMS amplitude slices at t = 1650 ms calculated by full bandwidth seismic dataset.

For extracting the appropriate frequency bands, the dominant frequency of seismic data is considered, which is approximately 45 Hz on average with the effective bandwidth ranging from 10 to 81 Hz. Accordingly, three frequency bands of the RMS amplitude are chosen, including 10~33 Hz, 34~57 Hz, and 58~81 Hz, by the STFT method (Figure 9a–c). Consequently, three decomposed components are calculated by the VMD and MTVMD methods, as shown in Figures 10 and 11, respectively. Based on the VMD methods, channels of the fluvial-delta depositional system in the Shawan Formation are complete, verifying that both the VMD and MTVMD methods are valid for enhancing the geological features' discrimination. However, details of the geological features, especially for the subtle seismic geological channel features, are dissimilar.

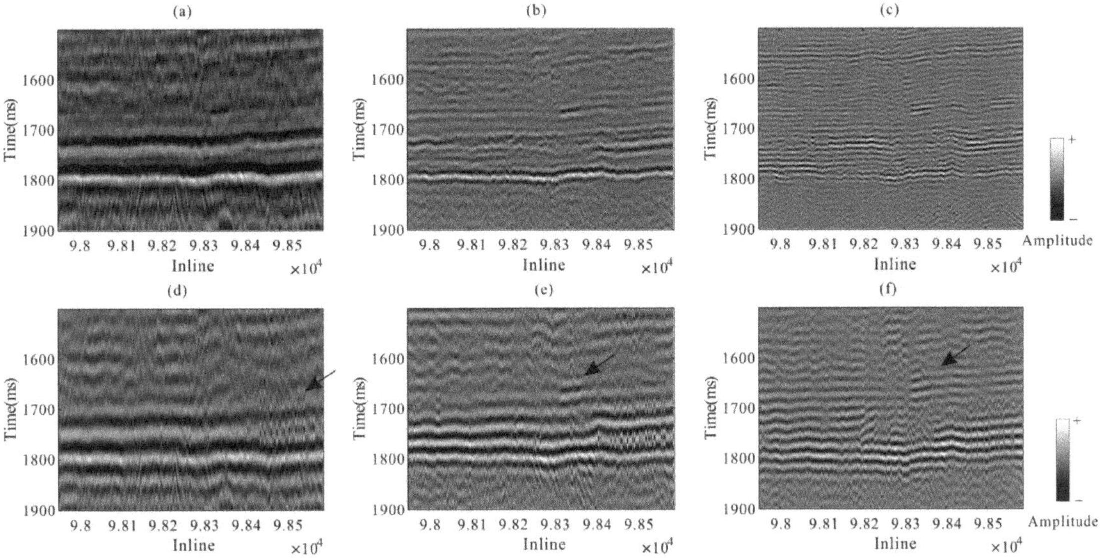

Figure 7. The decomposed components of the seismic vertical profile AA′ by using the MTVMD and the VMD methods. (**a–c**) are the components from the MTVMD method; (**d–f**) are the components from the VMD method. Black arrows indicate the lateral inconsistency in seismic profile.

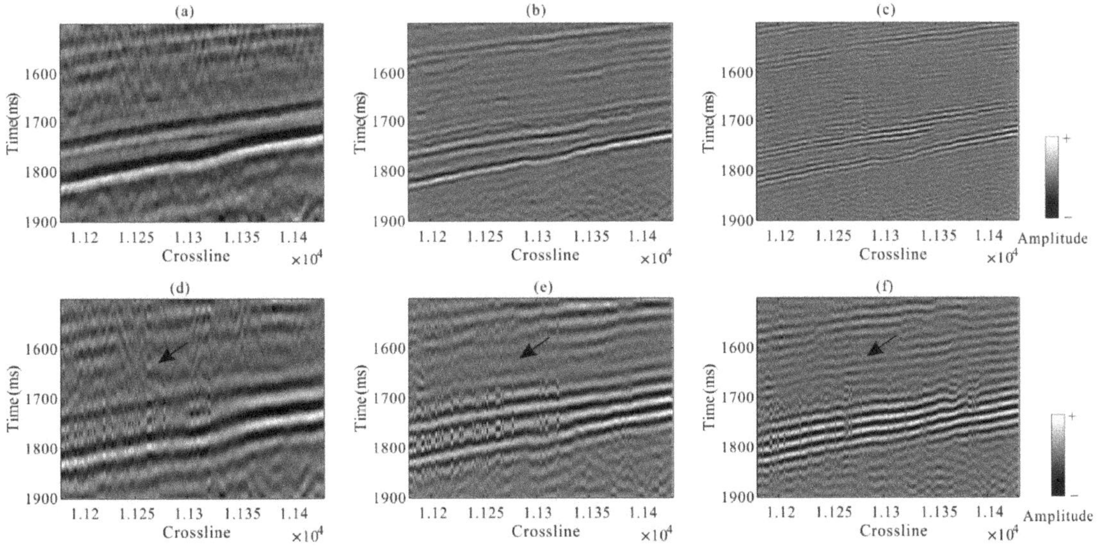

Figure 8. The decomposed components of the seismic vertical profile BB′ by using the MTVMD and the VMD methods. (**a–c**) are the components from the MTVMD method; (**d–f**) are the components from the VMD method. Black arrows indicate the lateral inconsistency in seismic profile.

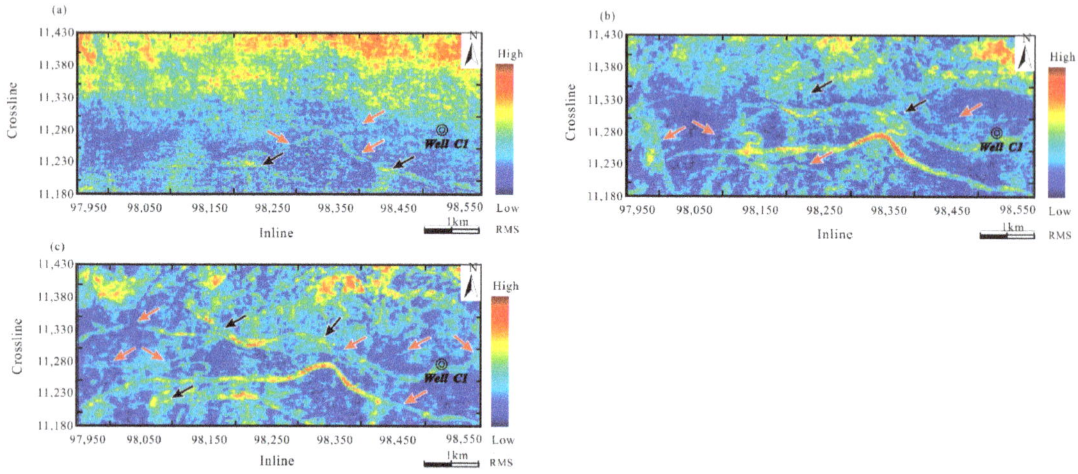

Figure 9. The RMS amplitude slices at t = 1650 ms. (**a**–**c**) show the RMS amplitude from the three frequency bands: (**a**) 10~33 Hz; (**b**) 34~57 Hz; (**c**) 58~81 Hz calculated by STFT method.

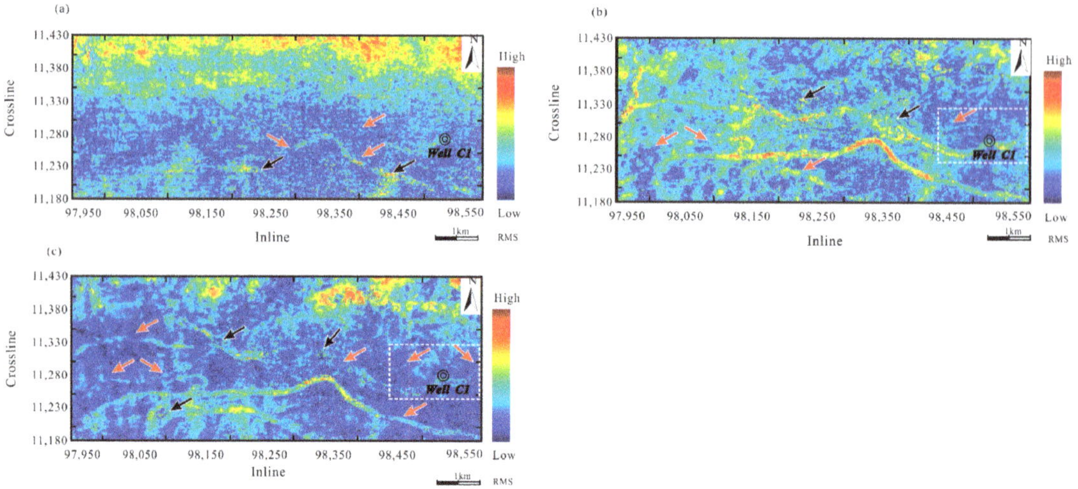

Figure 10. The RMS amplitude slices at t = 1650 ms. (**a**–**c**) show three decomposed components from the VMD method: (**a**) IMF1; (**b**) IMF2; (**c**) IMF3.

Recognized channels in fluvial-delta depositions are noted as red arrows in Figures 9–12. The presence of gaps and "jumps" in Figure 10a,c, marked as the black arrows, are evident due to the inconsistent center frequency. Overall, it is noticed that the high-frequency data (e.g., Figures 9c, 10c and 11c) are all presenting more sedimentary details compared to those of a low and medium frequency. However, it is always difficult to identify the spatial location, width, and flow direction of the narrow channels by using traditional time-frequency analysis methods. Although the high-frequency results are not as clear as the intermediate-frequency results in some locations, it is evident that the high-frequency spectrum is better described. For instance, some narrow channels (shown as the red arrow in Figure 11) in slices by the VMD method (Figures 10a and 11b) are expected but unrecognizable, however, these characteristics are significantly improved using the MTVMD based algorithm (Figure 11c,d).

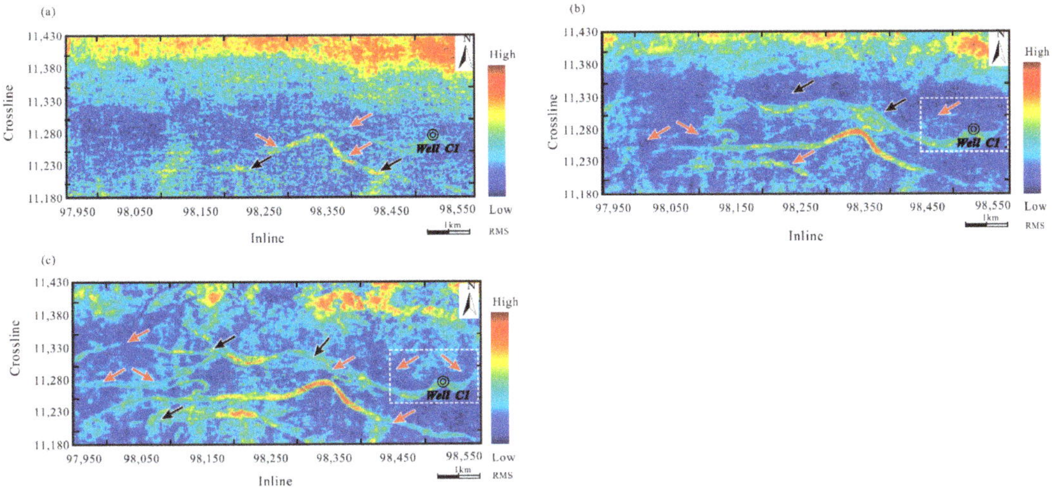

Figure 11. The RMS amplitude slices at t = 1650 ms. (**a**–**c**) shows three decomposed components from the MTVMD method: (**a**) IMF1; (**b**) IMF2; (**c**) IMF3.

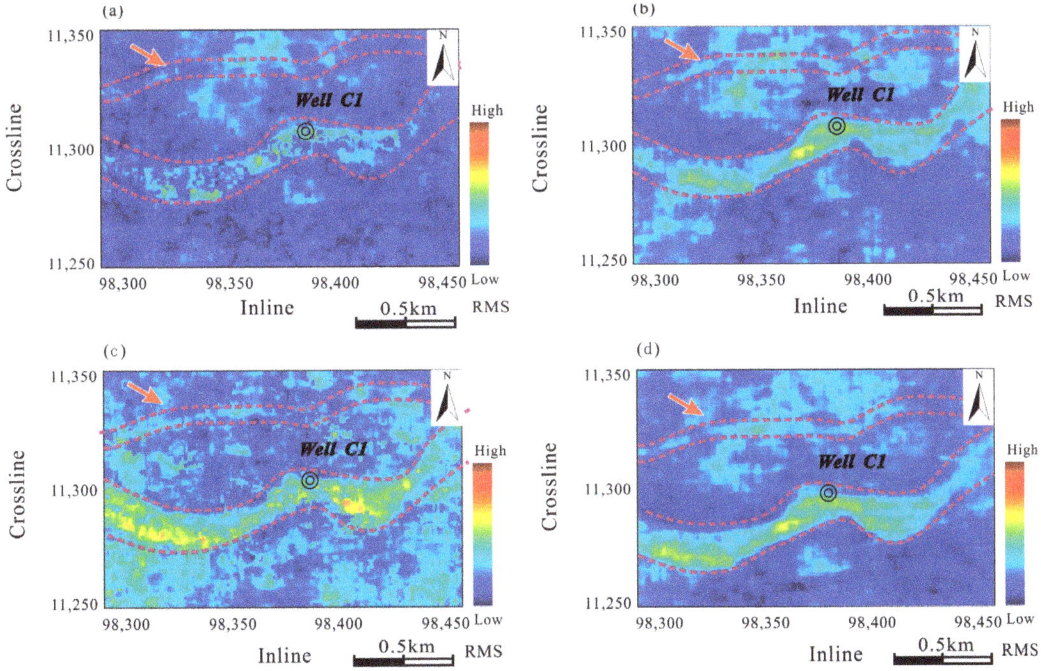

Figure 12. The detailed comparison in IMF2 and IMF3 components from the VMD and MT-VMD methods. Here, (**a**,**c**) are from Figure 10b,c, respectively; (**b**,**d**) are from Figure 11b,c, respectively. The original location of (**a**–**d**) are marked as white boxes in Figures 10 and 11.

Specifically, spatial spreading of the narrower branched channels is differentiated (Figure 10c, marked with a red arrow), except for the main stream of wide channels (Figure 9c). Subtle channels in lateral distributions are more distinct with the MTVMD method (Figure 11c, red dash lines). As estimated by scales, the resolution of channels by the MTVMD method can reach around 16 m (e.g., 16 m thickness and 20 m width, as shown in Figure 11). In addition, by the MTVMD method, overall clear depositional patterns and lateral consistency are evident (Figure 11), with the comparison of the results decomposed by STFT and VMD (Figures 9 and 10), which is expected to result in more concentrated energy and reduced shadow effect from the dip of seismic data by the MTVMD method.

4. Conclusions

A novel multi-trace variational mode decomposition-based seismic spectral decomposition method is presented for the purpose of better discrimination of the subtle seismic geological channels. The conclusions from this study are that:

(1) The variational mode decomposition method has the advantages of high resolution, complete decomposition, and avoidance of the mode stacking effect, which is suitable for enhancing the extraction of the target signal in seismic interpretation;
(2) Multi-trace variational mode decomposition provides an improved lateral consistency and resolution for recognizing the subtle geological features of channels, which enhances the quality and reliability of seismic interpretation;
(3) Tested by a 3D seismic dataset, the results have proved that, on data-driven decomposition approaches, a higher spectral-spatial resolution and clearer stratigraphy and lithology boundaries were obtained, compared to the traditional short-time Fourier transform method.

Although the computational efficiency was improved, a multi-core parallel computing strategy should be considered when processing 3D datasets by using the MTVMD method. With more big data input, the alternating direction method of multipliers for MVMD in the frequency domain may have high computation complexity. Data segmentation is also an effective strategy that divides the input data into a series of smaller ones, which can be considered for processing. In addition, subsequent research should also concentrate on the optimal selection of decomposition parameters.

Author Contributions: J.L. wrote the paper and performed the experiments; Z.Y. provided the original idea and wrote the program code; C.W. contributed to the conceptualization of the paper. All authors have read and agreed to the published version of the manuscript.

Funding: This research was financially supported by the National Science Foundation of China under Grant 42072125 and the National Science and Technology Major Project of China under Grant 2017ZX05008-001.

Institutional Review Board Statement: Not applicable.

Informed Consent Statement: Not applicable.

Data Availability Statement: Not applicable.

Acknowledgments: The calculations are supported by the High-performance Computing Platform of Peking University.

Conflicts of Interest: The authors declare no conflict of interest.

References

1. Li, W.; Yue, D.; Wang, W.; Wang, W.; Wu, S.; Li, J.; Chen, D. Fusing multiple frequency-decomposed seismic attributes with machine learning for thickness prediction and sedimentary facies interpretation in fluvial reservoirs. *J. Pet. Sci. Eng.* **2019**, *177*, 1087–1102. [CrossRef]
2. Zeng, H.L. Thickness imaging for high-resolution stratigraphic interpretation by linear combination and color blending of multiple-frequency panels. *Interpretation* **2017**, *5*, T411–T422. [CrossRef]
3. Zhang, J.H.; Liu, Z.; Zhu, B.H.; Feng, D.Y.; Zhang, M.Z.; Zhang, X.F. Fluvial reservoir characterization and identification: A case study from Laohekou Oilfield. *Appl. Geophys.* **2011**, *8*, 181–188. [CrossRef]
4. Anees, A.; Shi, W.; Ashraf, U.; Xu, Q. Channel identification using 3D seismic attributes and well logging in lower Shihezi Formation of Hangjinqi area, northern Ordos Basin, China. *J. Appl. Geophys.* **2019**, *163*, 139–150. [CrossRef]
5. Zeng, H.L. Frequency-dependent seismic-stratigraphic and facies interpretation. *AAPG Bull.* **2013**, *97*, 201–221. [CrossRef]
6. Li, F.; Qi, J.; Lyu, B.; Marfurt, K.J. Multispectral coherence. *Interpretation* **2018**, *6*, T61–T69. [CrossRef]

7. Lyu, B.; Qi, J.; Li, F.; Hu, Y.; Zhao, T.; Verma, S.; Marfurt, K.J. Multispectral coherence: Which decomposition should we use? *Interpretation* **2020**, *8*, T115–T129. [CrossRef]
8. Li, F.; Zhang, B.; Zhai, R.; Zhou, H.; Marfurt, K.J. Depositional sequence characterization based on seismic variational mode decomposition. *Interpretation* **2017**, *5*, SE97–SE106. [CrossRef]
9. Wang, Z.; Gao, J.; Wang, D.; Wei, Q. 3D seismic attributes for a tight gas sand reservoir characterization of eastern Sulige gas field, Ordos Vasin, China. *Geophysics* **2015**, *80*, B35–B43. [CrossRef]
10. Zhou, J.; Ba, J.; Castagna, P.; Guo, Q.; Yu, C.; Jiang, R. Application of an STFT-Based Seismic Even and Odd Decomposition Method for Thin-Layer Property Estimation. *IEEE Geosci. Remote Sens. Lett.* **2019**, *16*, 1348–1352. [CrossRef]
11. Liu, W.; Cao, S.; Wang, Z.; Kong, X.; Chen, Y. Spectral decomposition for hydrocarbon detection base VMD and Teager-Kasier energy. *IEEE Geosci. Remote Sens. Lett.* **2017**, *14*, 539–543. [CrossRef]
12. Lu, W.; Li, F. Seismic spectral decomposition using deconvolutive short-time Fourier transform spectrogram. *Geophysics* **2013**, *78*, V43–V51. [CrossRef]
13. Sinha, S.; Routh, P.S.; Anno, P.D.; Castagna, J.P. Spectral decomposition of seismic data with continuous-wavelet transform. *Geophysics* **2005**, *70*, P19–P25. [CrossRef]
14. Liu, N.; Gao, J.; Zhang, B.; Li, F.; Wang, Q. Time-frequency analysis of seismic data using a three parameters S transform. *IEEE Geosci. Remote Sens. Lett.* **2018**, *15*, 142–146. [CrossRef]
15. Tary, J.B.; Herrera, R.H.; Han, J.; van der Baan, M. Spectral Estimation-What is new? What is next? *Rev. Geophys.* **2014**, *52*, 723–749. [CrossRef]
16. Huang, N.E.; Shen, Z.; Long, S.R.; Wu, M.L.C.; Shih, H.H.; Zheng, Q.N.; Yen, N.C.; Tung, C.C.; Liu, H.H. The empirical mode decomposition and the Hilbert spectrum for nonlinear and non-stationary time series analysis. *Proc. R. Soc. A Math. Phys.* **1998**, *454*, 903–995. [CrossRef]
17. Dragomiretskiy, K.; Zosso, D. Variational mode decomposition. *IEEE Trans. Signal Process* **2013**, *62*, 531–534. [CrossRef]
18. Liu, W.; Cao, S.; Chen, Y. Application of variational mode decomposition in random noise attenuation and time-frequency analysis seismic data. In Proceedings of the EAGE 78th Annual International Conference & Print Exhibition, Vienna, Austria, 30 May–2 June 2016. [CrossRef]
19. Wu, G.N.; Liu, G.C.; Wang, J.X.; Fan, P.P. Seismic Random Noise Denoising Using Mini-Batch Multivariate Variational Mode Decomposition. *Comput. Intell. Neurosci.* **2022**, *2022*, 2132732. [CrossRef]
20. Xu, Z.; Zhang, B.; Li, F.; Cao, G.; Liu, Y. Well-log decomposition using variational mode decomposition in assisting the sequence stratigraphy analysis of a conglomerate reservoir. *Geophysics* **2018**, *83*, B221–B228. [CrossRef]
21. Xue, Y.J.; Cao, J.X.; Wang, D.X.; Du, H.K.; Yao, Y. Application of the variational-mode decomposition for seismic time-frequency analysis. *IEEE J. Sel. Top. Appl. Earth Obs. Remote Sens.* **2016**, *9*, 3821–3831. [CrossRef]
22. Zhao, T.; Li, F.; Marfurt, K.J. Constraining self-organizing map facies analysis with stratigraphy: An approach to increase the credibility in automatic seismic facies classification. *Interpretation* **2017**, *5*, T163–T171. [CrossRef]
23. Lyu, B.; Li, F.Y.; Qi, J.; Zhao, T.; Marfurt, K.J. Highlighting discontinuities with variational mode decomposition based coherence. In Proceedings of the 2018 SEG International Exposition and Annual Meeting, Anaheim, CA, USA, 13–18 October 2018. [CrossRef]
24. ur Rehman, N.; Aftab, H. Multivariate variational mode decomposition. *IEEE Trans. Signal Process.* **2019**, *67*, 6039–6052. [CrossRef]
25. Tian, Y.J.; Gao, J.H. Seismic depositional sequence characterization by using enhanced multichannel variational-mode decomposition. *Interpretation* **2021**, *10*, T103–T115. [CrossRef]
26. Wang, W.; Pan, C.; Wang, J. Quasi-bivariate variational mode decomposition as a tool of scale analysis in wall-bounded turbulence. *Exp. Fluids* **2018**, *59*, 1. [CrossRef]
27. Fan, X.Y.; Xue, G.Q.N.; Liu, B.; Huang, Y.T.; Qi, X.; Li, W.C. Recognition of Single Channel Sandbody of Meandering River With Logging-Seismic Combination. *Xinjiang Pet. Geol.* **2019**, *40*, 228. [CrossRef]
28. Lei, Q.I.N.; Li, X.G.; Qiu., Z.Y.; Li, Z.J.; Xu, S.H.; Li, Y.W.; Chen, C.; Wang, J.B.; Wen, S.Q. Characteristics and Formation Mechanism of Incised Valley on the Eastern Margin of Chepaizi Uplift, Junggar Basin. *Geoscience* **2021**, *35*, 832. [CrossRef]

applied sciences

MDPI

Article

Quantitative Seismic Geomorphology of Four Different Types of the Continental Slope Channel Complexes in the Canterbury Basin, New Zealand

Dicky Harishidayat [1,*] and Wasif Rehman Raja [2]

1 College of Petroleum Engineering and Geosciences, King Fahd University of Petroleum and Minerals, Dhahran 31261, Saudi Arabia
2 Department of Exploration, Pakistan Oilfields Limited, Rawalpindi 46602, Pakistan; wasif2283@gmail.com
* Correspondence: dicky.hidayat@kfupm.edu.sa or dickyharishidayat@gmail.com

Abstract: Modern marine seismic reflection data have allowed for the study of the morphology of submarine channels in a 3D perspective. This study presents a quantitative analysis of continental slope channel complexes morphology within the Canterbury Basin (New Zealand). It aims to characterize the morphology and seismic facies of submarine channels to better understand their formative sedimentary processes, paleoenvironment, and petroleum prospectivity. Submarine channel morphometric parameters are measured at 150 m intervals perpendicular to the axis of channels complexes. Based on the morphology analysis of erosional surfaces and seismic facies of channel complexes filling deposits, four types of continental slope channel complexes are found in the study area. These are vertical migrating channels, lateral migrating channels, V-shaped channels, and U-shaped channels. Furthermore, our work shows that channel morphometry varies over distances of 0.1 to 14 km in this continental slope system. These changes indicate a combination of submarine gravity flow process and channel wall collapse responsible for the development of continental slope channel complexes. Regionally, the evolution of the channels indicated less significant regional plate movement. This quantitative seismic geomorphology approach of characterizing submarine channels system has broader applications for better interpretation of paleoenvironment and petroleum prospectivity within frontier basins.

Keywords: seismic geomorphology; submarine channels; continental slope; channels complexes; morphometry; paleoenvironment; petroleum prospectivity; Canterbury; New Zealand

1. Introduction

Active submarine channels are the primary conduits for the transfer of sediment from the shelf to basin floor. They may become good hydrocarbon reservoirs when abandoned and infilled with sediment [1–4]. Submarine channels are located at the end of a shelf canyon system, down to the slope and basin floor regions, with dominant gravity-driven processes of the sediment transport mechanism [3,5–7]. Sediment delivery to submarine channels occurs when there is a connection between the canyon heads and the fluvial-deltaic system [8–10], longshore drift current in the shelf regions [9,11], and sediment mass movements on the canyon walls and shelf edge [4]. For this reason, factors that control the type, supply, and deposition of sediments such as basin tectonics, climate, and sea-level changes have substantial control over the genesis and evolution of submarine channels [4,8,12].

Recent advances in marine geophysical exploration have enabled the organization and comparison of architectural elements associated with submarine channels [4,6,13–16]. In addition, high-resolution 3D marine seismic reflection data have allowed the morphometric study of submarine channels to understand the sedimentary processes and paleoenvironments [17–20]. However, the interpretation of channel incision or channel isolated amplitude from the 3D marine seismic reflection data poses several challenges due to data artifacts, e.g., velocity push-downs issues [19,21,22]. Furthermore, statistical analysis of channel morphometry has become increasingly insightful in understanding sedimentary processes and the paleoenvironment [18,19,23–27]. A combination of the quantitative seismic geomorphology approach and a traditional seismic interpretation (seismic facies analysis, attribute analysis, and slicing technique) are used in this study to obtain a detailed 3D image of channel morphometric elements, which have implications for hydrocarbon prospectivity.

The study area is located in the southwestern area (continental slope) of the Canterbury Basin (Figure 1a), incorporating the offshore Waka 3D seismic reflection survey and several exploration and ocean drilling wellbore datasets. A previous study documented that the submarine channel initiation began during the Early Oligocene on the post-rift thermal sag phase (tectonically quiescent) with marine transgression to the Miocene—a recent time when the regression phase and prograding fore-sets occurred [28–30]. Furthermore, Zhao, Zhang, Li, and Marfurt [29] focused their research on the Miocene channels and Barrier, Nicol, Browne, and Bassett [30] on the Early Oligocene channels. In addition, a recent study by Kumar, et al. [31] revealed that pockmarks (circular depression in the seabed caused by fluid expulsion) occur in association with modern submarine canyons that were developed by turbidity currents, together with Southland current and regional tectonic. However, documentation and description of the Quaternary continental slope channel system in the study area are very limited. Thus, the study area's sedimentary process, paleoenvironment, and hydrocarbon implications are poorly understood. Therefore, the objectives of this study are as follows: (1) describe the morphology of several types of continental slope channel complexes and seismic facies of its filling deposits; (2) describe their morphometric elements and scaling relationship among them; (3) investigate their development and paleoenvironment; and (4) examine the impact of differential compaction, horizon interpretation pitfalls, and burial effect for petroleum prospectivity in the study area.

2. Geodynamic Evolution of the Canterbury Basin

The Canterbury Basin study area is located offshore of southeast New Zealand. It contains approximately 4 km of Cretaceous to Cenozoic sediments that follow a NE–SW trend, approximating the strike of the paleo margin (Figures 1 and 2). Before the breakup of Gondwana, the Australian margin was a convergent plate boundary with the westward subduction of the Phoenix plate, forming an extensive trench system [32–34]. Extensional tectonics created the early basin architecture through rapidly subsiding half grabens [35,36]. Sediments deposited in the syn-rift phase range from the mid-Cretaceous to the middle of the Late Cretaceous period include terrestrial conglomerate, sandstone, siltstone, carbonaceous mudstone, and coal [28,37,38]. However, local tectonism and thermally driven regional subsidence continued throughout the Cretaceous, producing a transgressive onlap of these terrestrially derived sediments [38], with subsequent post-rift tilting resulting in an angular unconformity during the Late Cretaceous (Figures 1–3). Furthermore, the Late Cretaceous to Early Oligocene post-rift deposits were associated with terrestrial and marine transgression in the western and eastern part of the Canterbury Basin, respectively [36,38].

The Canterbury Basin strata during the Late Oligocene to recent reflects the progressive uplift and erosion of the Southern Alps, with resulting marine regression [36,38–40]. An influx of terrigenous sediment from the west (Figure 1a), combined with strong oceanic currents, produced a stacked succession of a rapidly prograding clastic shelf system and drift prisms throughout the Pliocene—Pleistocene [36,38,41,42]. This resulted in a 2200 m thick succession of eastward fining interbedded conglomerates (alluvial fans), silt, shelf beds, and clay (Figures 1 and 2). During the last 5 million years, rapid uplift and erosion of the Southern Alps have greatly increased the supply of terrestrial coarse-clastic sediment to be deposited in the west and localized coarse-grained carbonates in the north, while fine-grained sediments continued to be deposited in the eastern parts (marine) of the Canterbury Basin [36,39,40]. Most of these sediments were transported to the deep water through a submarine canyon-channel system (Figure 1). In parallel with subsidence rates, sedimentation rates in the Canterbury Basin increased progressively during the Late Miocene to recent, with the peak of 320 mm/1000 years in the Pliocene-Pleistocene interval of Galleon-1 wellbore [36,42–44].

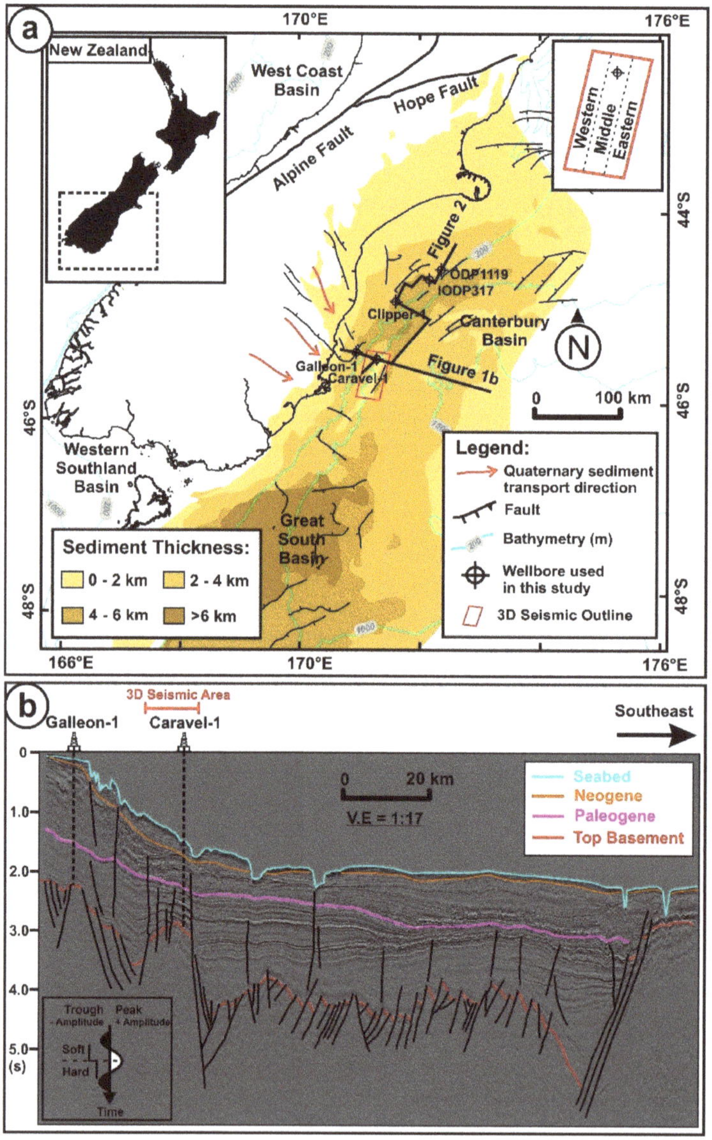

Figure 1. (**a**) Map of the study area with sediment thickness information from O'Leary and Mogg [38]. The red rectangle on the inset map is the western-middle-eastern zonation in the study area. (**b**) Regional NW-SE seismic and geological cross-section through the study area. See Figure 1a for section location.

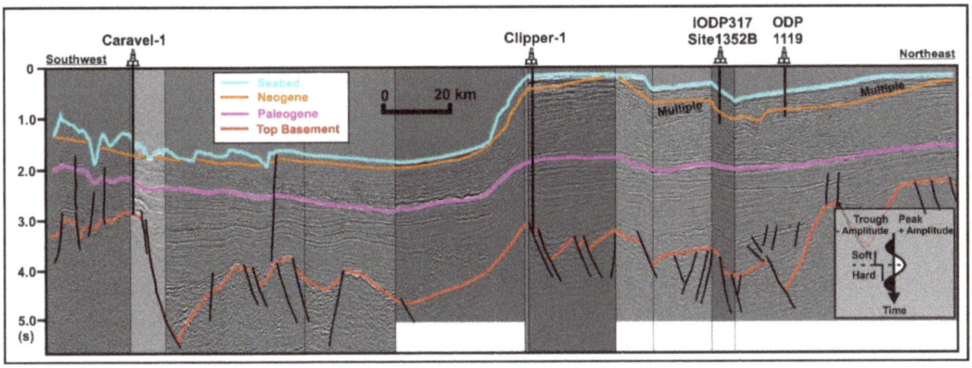

Figure 2. Regional SW-NE seismic-section through the study area. See Figure 1a for section location.

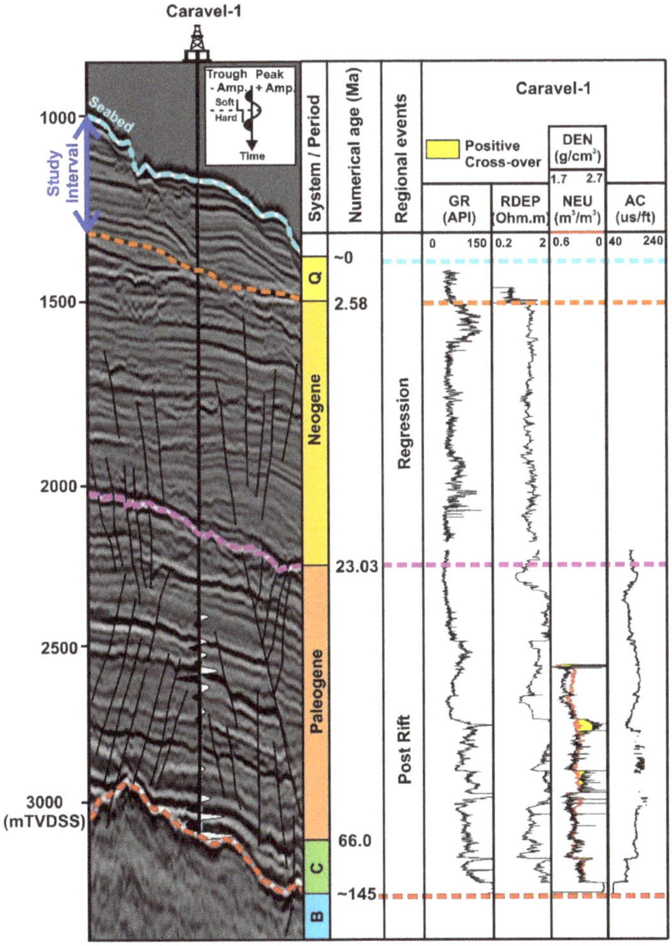

Figure 3. Stratigraphic column of the study area with seismic well-tie of Caravel-1 and regional events taken from O'Leary and Mogg [38] and Wilson, et al. [45]. Q denotes Quaternary; C denotes Cretaceous; B denotes basement.

3. Data and Method

3.1. Seismic Data

The subsurface investigation was based on the Waka 3D seismic reflection data located in the SE of the Canterbury Basin (Figure 1a). The Waka 3D seismic reflection survey was acquired by a 5100 m array of streamers, and it covers an area of 1150 km^2. The survey has a bin spacing of 12.5 by 25 m, and a 4 ms vertical sampling interval. The Waka 3D seismic reflection survey is the final stack time migration, which was compiled by the New Zealand Petroleum and Minerals department (NZP&M). The dominant frequency in the interested zone (1500–2200 ms) is around 35 Hz. The velocity at the seabed is 1500 m/s and at the Eocene level is 2500 m/s, therefore the average velocity in the zone of interest is about 2000 m/s. The vertical resolution is approximately 14.3 m in the Waka 3D seismic reflection survey after incorporating the values of velocity and frequency (using the quarter-wavelength criterion). The 2D seismic data (Figures 1b and 2a) has a similar dominance frequency (~35 Hz), but an average velocity of 2700 m/s, thus making the vertical resolution around 19.3 m.

The Waka 3D Seismic reflection survey uses the European or Australian polarity standards. The first hard kick in Figure 3 is the seabed, which depicts an increase in the acoustic impedance represented by white color or peak (positive amplitude value). In addition, this change also correlates with the different properties in the sediments due to density, porosity, and stiffness variations. The Waka 3D seismic reflection survey shows the black-white-black density contrast of the zero-phase data (Figure 3).

3.2. Wellbore Data

The wellbore used in the study (Caravel-1) was drilled inside the Waka 3D survey (Figure 1) by Anadarko New Zealand Company (ANZC) in 2014, and penetrated the basement to a total True Vertical Depth Sub-sea (TVDSS) of 3092 m. The log data were acquired with different logs suites of gamma-ray (GR), deep resistivity (RDEP), density (DEN), neutron (NEU), and sonic (AC). Unfortunately, gamma-ray and deep resistivity logs on the study interval (Quaternary) read unconsolidated formation (Figure 3). Meanwhile the other logs and drill cuttings start below the study interval, from Paleogene to the basement. In addition, biostratigraphy data and seismic horizon information from Galleon-1 [45], Clipper-1 [44], IODP317-Site 1352B [46], and ODP1119 [44,47] wellbores were used for regional correlation.

3.3. Methods

3.3.1. Interpretation of Lithology, Seismic Well-Tie, Horizons-Faults, and Seismic Attribute Analysis

The interpretation of lithology comes from gamma-ray and deep resistivity logs of Caravel-1 (Figure 3). This included check-shots data used for the seismic-well tie process to relate the seismic interpretation to their appropriate depth and stratigraphy. Since there are no DEN and AC logs in the interval of interest, synthetic seismograms are only available at the deeper part of Caravel-1 (Figure 3). The information on stratigraphy, ages, and regional events were obtained from Browne and Field [36], O'Leary and Mogg [38], Lu and Fulthorpe [44], Fulthorpe, Hoyanagi, Blum, Guèrin, Slagle, Blair, Browne, Carter, Ciobanu and Claypool [46], Carter, McCave, Richter and Carter [47], and Schiøler, et al. [48]. In addition, biostratigraphy data of the Galleon-1 wellbore, together with their seismic reflector, were used to constrain the geological age of the studied interval (Figure 1b). Other wellbores (including the wellbore from Ocean Drilling Program) are located far from the studied area, with no detail correlable seismic markers (Figure 2).

Seismic interpretation began with faults and regional correlation of the horizons. Seismic time to depth conversion was performed using a simple velocity model (interval velocity) that integrated the time–depth relationship from Caravel-1 wellbore data with regional velocity values from Mogg, et al. [49] within the interpreted horizons. This process is important for calculating the morphometric elements of submarine channels in the spatial rather than the temporal dimension. Manual interpretation of incision surface resulted in the characterization of 18 single submarine channels (Figure 4a) grouped into four different types (Figure 4b). Furthermore, the root mean square (RMS) of the seismic amplitude was analyzed for several depth slices of the four channel complex types. The RMS seismic attribute can assist in delineating geological features or channel bodies [50]. In addition, several other attributes that are well known for channel delineation (e.g., curvature, instantaneous frequency, sweetness, and variance) are also indicated as obvious channels features (Figure S1).

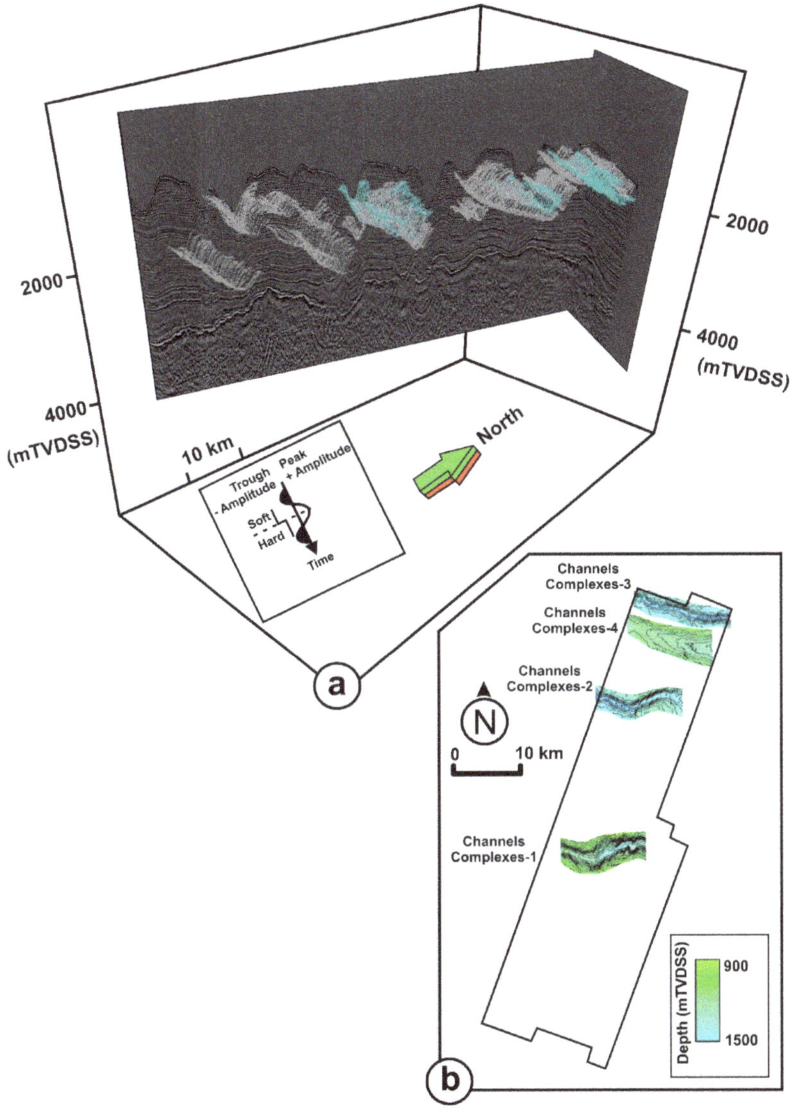

Figure 4. (**a**) Three-dimensional view of 18 continental slope channels complexes with inline and crossline seismic amplitude (the 4 represented channels complexes are marked by the blue color), and (**b**) Plan view of the 4 represented channels complexes. See Figure 1a for coordinate information.

3.3.2. Morphology of Channels Complexes Erosional Surface and Seismic Facies Analysis of Its Filling Deposits

In this study, the erosional surface of the channel complexes consists of flank, thalweg, terrace, and wall (Figure 5). The channel erosional surface is identified as a low area on the horizon, with the lowest point being the thalweg. The flank is an area outside the overspill point of the channel. Furthermore, the wall is a steep-sided area on an erosional surface, and the terrace is defined as a topographically flat area between a thalweg and wall (Figure 5). These morphologies of the erosional surface have been defined based on the works of Posamentier and Kolla [4], Qin, Alves, Constantine and Gamboa [17], and Harishidayat, Omosanya, Johansen, Eruteya and Niyazi [19].

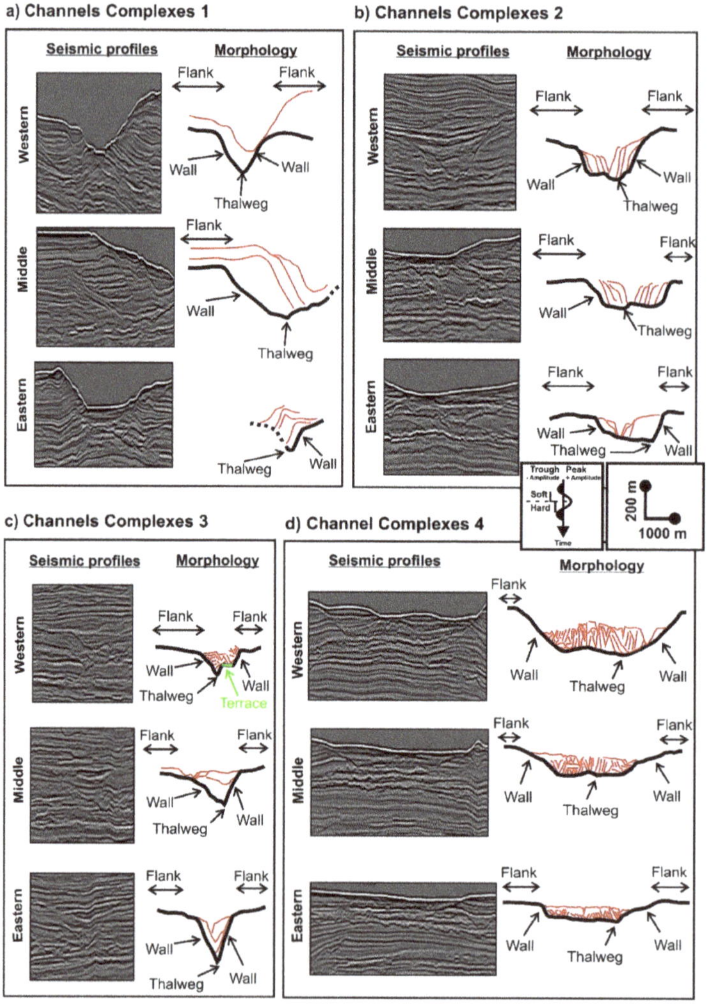

Figure 5. Erosional surface morphology of the channel's complexes and the seismic profile of each representative channel complexes in the study area represent sediment transport direction from west to the east. Note that the dashed line is the interpreted position of the erosional surface before it is cut out by the younger channel, while the red line is the younger channel.

After interpretation of the erosional surface, seismic facies analysis was performed to be able to interpret the depositional environment and lithologies from seismic reflection data using several seismic reflection parameters: configuration, continuity, amplitude, frequency, and interval velocity [51]. Figure 6 and Section 4.1 show detailed information about seismic reflection parameters leading to the identification and differentiation of submarine channel complexes in the study area.

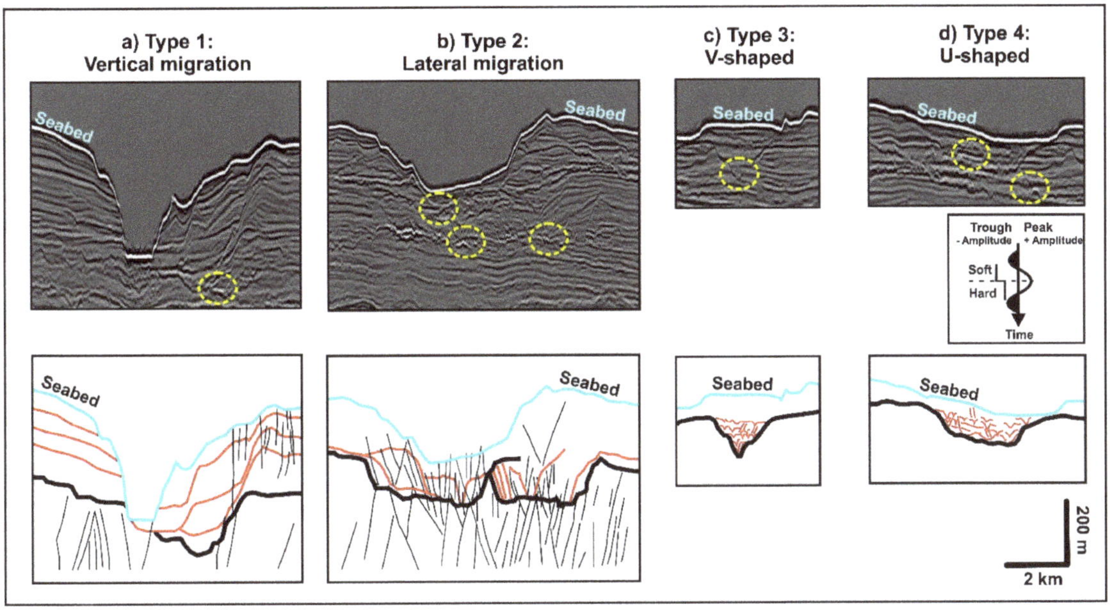

Figure 6. Seismic facies of filling deposits from four different channels complexes in this study. See Section 4.1 for detailed description. Note that the thick black line is the interpreted channel incision, the red line is the younger channel surface, the thin black line is the fault, and the yellow dashed line is isolated low to high seismic amplitude.

3.3.3. Morphometric Analysis of Submarine Channels

Morphometric dimension analysis (width, height, length, gradient, sinuosity, aspect ratio, and cross-sectional area) of genetically related channel complexes is an effective quantitative interpretation tool. It can reveal scaling relationships and infer the relation between sediment conduit morphology, sedimentary processes, and paleoenvironment interpretation [18,19,25].

The quantitative analyses of the channel system followed the methods of Qin, Alves, Constantine and Gamboa [17], and Harishidayat, Omosanya, Johansen, Eruteya and Niyazi [19] that are represented in Figure 7. These involved a combination of manual and auto-tracking procedures. Then, the incision interpretations were converted into the incision (channels) surfaces. The cross-sections are drawn perpendicular to these incision surfaces to calculate the morphometric parameters measured at an interval average of 150 m (Figure 7). Furthermore, the western-middle-eastern definition is based on the three zonation on the Waka 3D seismic reflection survey that consider the source of sediment in the west and the transport direction to the east (Figure 1a).

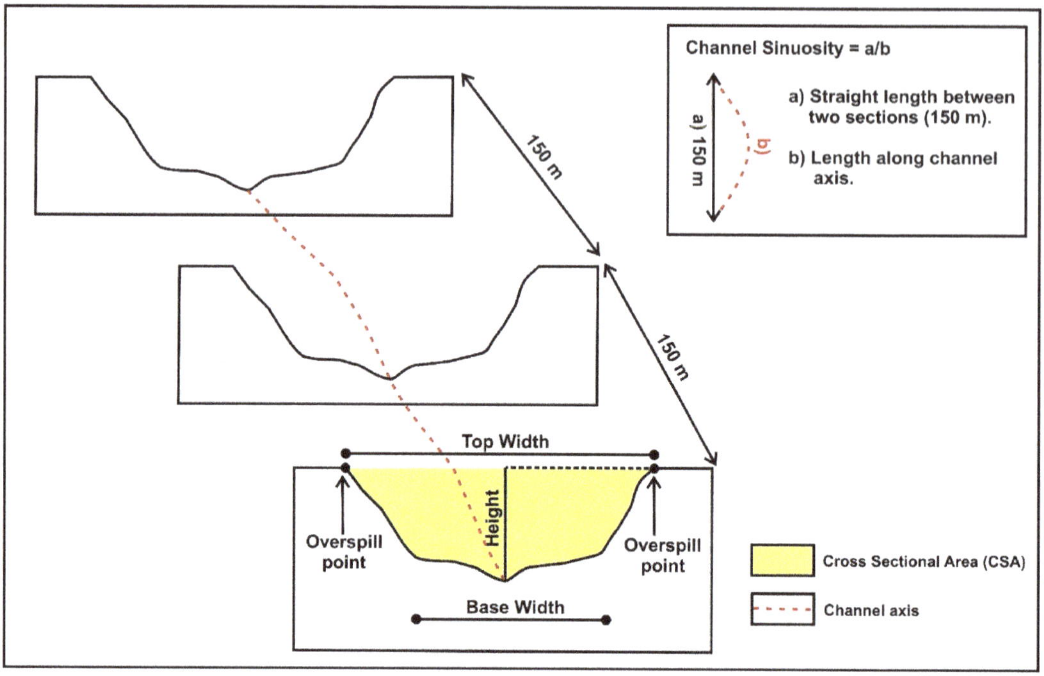

Figure 7. Morphometry elements of the channel's complexes that were analyzed in this study.

The morphometric measurements of the channel complexes investigated in the study area are as follow (Figure 7):

- Height is a vertical distance from thalweg to the overspill point.
- Top width is defined as the distance between overspill points on each channel complexes margin.
- Base width is the distance between two lower points in the channels floor of the channel complexes.
- Aspect ratio is defined as a ratio between top width and height of the channels complexes cross-sectional area (CSA).
- Sinuosity (i.e., the length between a reference point and along channels complexes axis) and channel gradient are calculated based on thalweg depth changes along the channel complexes' distance.

In addition, the overspill point is located between the channel flank and wall. Detailed sinuosity classification by Reimchen, Hubbard, Stright and Romans [27] was used in this study. The low sinuosity is defined between 1.05 and 1.20, whereas, for straight channels, complexes systems are between 1.00 to 1.05, and for high sinuosity systems, greater than 1.20. Moreover, all the morphometric elements on morphometry analysis are presented in the cross-plot. Among them is the scaling relationship, where R is the correlation coefficient that measures the strength and the direction of a linear relationship among two variables on the cross-plot. R^2 is a coefficient of determination that quantify the proportion of the variance (fluctuation) of one variable that is predictable from the other variable on the cross-plot [52].

4. Results

4.1. Morphology of Channels Complexes Erosional Surface and Seismic Facies of Its Filling Deposits

The main morphology of the channel's complexes erosional surface identified in the study area includes flank, wall, terrace, and thalweg (Figure 5). At the same time, seismic facies analysis is based on the reflection configuration of channel fills, and the shape of this erosional surfaces revealed four different types of channel complexes in the study area (Figure 6):

- Type 1 (Vertical migrating channels)
- Type 2 (lateral migrating channels)
- Type 3 (V-shaped channels)
- Type 4 (U-shaped channels).

In Type 1, the flank, wall, and thalweg features of channels complexes, the erosional surface is present in the western to eastern areas with proto flank existed in the eastern area of the channel complexes 1 (Figure 5a). The flank is characterized by low to high seismic amplitude with relatively continuous reflectors, while the erosional surface of channels complexes wall has discontinuous reflectors due to faulted and differential compaction (Figures 5a and 6a). In addition, the dip angle of the flank varied from horizontal to dipping flank with sediment wave in the flank. In contrast, the dip angle of the channel wall is relatively steep (>40°) in the western and eastern areas and less steep (<40°) in the middle area. Furthermore, the seismic facies of Type 1 are characterized by low to high amplitude, high continuity, and sub-parallel seismic reflections confined with a relatively V to U-shaped erosional surface. In addition, seismic facies inside the channel complexes show lateral and uniform fill reflection, which is mostly on-lapping into the erosional surface, and the exaggeration of the channel is in a relatively vertical direction (Figures 5a and 6a).

The morphology of Type 2 channels complexes consists of flank, wall, and thalweg that span from western to eastern areas (Figure 5b). The channels flank is characterized by medium to high seismic amplitude with relatively continuous reflectors, while the erosional surface of channels complexes wall has discontinuous low to high reflectors due to faulted and differential compaction (Figures 5b and 6b). The dip angle of the flank is mostly horizontal, with some dipping flank, especially in the western area (Figure 6b). In contrast, the dip angle of the channel wall is relatively steep (>40°) throughout the western to eastern areas. In Type 2 channel complexes, the seismic facies of the filling deposits consist of parallel to subparallel seismic reflections, variable low to high seismic amplitude, and continuity with a wider-shape erosional surface that is highly deformed (Figures 5b and 6b). Furthermore, the channels fills are present along the wall of the previous channels, which depicts that the channels are laterally migrating and on-lapping to the previous channel's erosional surface (Figure 6b). In addition, bright seismic amplitudes are present at the base of Type 2 on the western to eastern areas (Figures 5b and 6b).

Type 3 V-shaped channel complexes consist of flank, terrace (only found at this channel type), wall, and thalweg morphology presented in the western to eastern of study areas (Figure 5c). The channel flank is characterized by low to medium seismic amplitude with relatively continuous and dipping reflectors, while the terrace and wall have discontinuous low to medium reflectors due to faulting and differential compaction (Figures 5c and 6c). The dip angle of the terrace is relatively flat, whereas the wall of the channel is mostly steep, with more than 40° throughout the western to the eastern area (Figure 5c). The seismic facies of this Type 3 filling deposits consist of subparallel to parallel seismic reflectors, moderate amplitude, and variable continuity with V-shaped erosional surface. In addition, the reflections are on-lapping to the channel erosional surface. The upper part of channel-filled deposits is more continuous and parallel than the base deposits, indicating different lithologies filling inside Type 3 channels complexes (Figures 5c and 6c).

Lastly, Type 4 of the channels complexes in the study area has flank, wall, and thalweg morphology on their western, middle, and eastern areas, contributing to the U-shaped channel development (Figure 5d). The channel flank is characterized by a medium and continuous seismic reflector, with mostly horizontal layers and sediment waves. The wall itself has low to high amplitude, with relatively dipping and discontinuous reflectors (>40°) due to faulting and differential compaction (Figures 5d and 6d). In addition, the base of Type 4 has an irregular and discontinuous shape with relatively high seismic amplitude (Figures 5d and 6d). Furthermore, the seismic facies of Type 4 have variable amplitude values ranging from low to high amplitude, with a subparallel and hummocky reflection configuration. (Figure 6d). The channel-fill shows discontinuous reflections at the channel wall and main channel, indicating side wall collapses and cross-cutting channels, respectively (Figure 6d).

4.2. Morphological Analysis of Submarine Channels

Type 1 of the channel complexes narrows in a down-dip direction across the study area, where the top width decreases from 3821 to 2612 m (Table 1, Figures 4 and 8a). This linear trend is represented by relatively high correlation and determination coefficients (R and R^2), respectively (Table 1). These indicate that the decrease in the top width is proportional to the down dip distance. In parallel, the base width of Type 1 also decreases to the end of its western part, with an average base width of 436 m (Figure 8b). In contrast, the height of Type 1 ranges from 274 to 320 m and has a relatively flat trend through the channel distance (Figure 8c), indicated by low values of R and R^2 (0.42 and 0.18, respectively (Table 1). The CSA of Type 1 increases from 0.04 to 0.07 km^2 in a down-dip direction (Table 1 and Figure 8d). The Type 1 sinuosity values increase from straight (average 1) in the beginning of the western area to low sinuosity (average 1.05) at the end of the western area (Table 1 and Figure 8e). In parallel, the aspect ratio of Type 1 decreased from 12.90 to 8.90 down dip (Table 1 and Figure 8f).

Table 1. Average morphometry data of four channel complexes in the study area.

Channel Complexes	Height (m)			Top Width (m)			Width Average (m)			CSA (Km^2)		
	Max	Min	Avg.	Max	Min	Avg.	Top	Base	Ratio	Max	Min	Avg
1	320	274	298.50	3821	2612	3285	3285	436	7.50	0.07	0.04	0.06
2	242	208	228.80	2257	1690	1963	1963	327.60	5.99	0.07	0.04	0.05
3	308	151	218.70	2564	973	1796	1796	342.50	5.24	0.04	0.01	0.02
4	314	223	270.10	3551	1880	2879	2879	605.40	4.75	0.04	0.02	0.03
Channels Complexes	**Sinuosity**			**Gradient**			**Aspect Ratio (Width/Height)**			**Height**		**Top Width**	
	Max	Min	Avg	Max	Min	Avg	Max	Min	Avg	R^2	R	R^2	R
1	1.09	1.00	1.05	1.75	1.08	1.30	12.90	8.90	11.00	0.18	0.42	0.82	0.90
2	1.38	1.04	1.12	1.61	1.38	1.52	9.80	7.40	8.50	0.25	0.50	0.20	0.44
3	1.20	1.01	1.17	1.77	0.71	1.14	11.13	5.66	8.19	0.67	0.81	0.52	0.72
4	1.03	1.00	1.01	2.09	1.48	1.80	14.36	6.46	10.80	0.87	0.93	0.12	0.34

Type 2 has a top width ranging from 1690 to 2257 m with non-linear trends (Table 1, Figures 4 and 8a). This trend is represented by relatively low values of correlation and determination coefficients (R and R^2), respectively (Table 1). These indicate that decreases in top width are not proportional to down-dip distance. In parallel, the base width of Type 2 is also in a non-linear trend through the length of the channel (Figure 8b), with an average value of 327.6 m (Table 1 and Figure 8b). The height of Type 2 ranges from 208 to 242 m, with relatively low values of correlation coefficients and determination (R and R^2, respectively. See Table 1). The CSA of Type 2 ranges from 0.04 to 0.07 km^2, with no trend of the CSA through the channels' distance (Table 1 and Figure 8d). The sinuosity increases from low sinuosity (average 1.02) at the beginning of the western area to high sinuosity (average 1.30) at the end of the western area (Table 1 and Figure 8e). In addition, the aspect ratio of Type 2 has minimum and maximum values of 7.40 and 9.80, respectively, with no trend-down dip (Table 1 and Figure 8f).

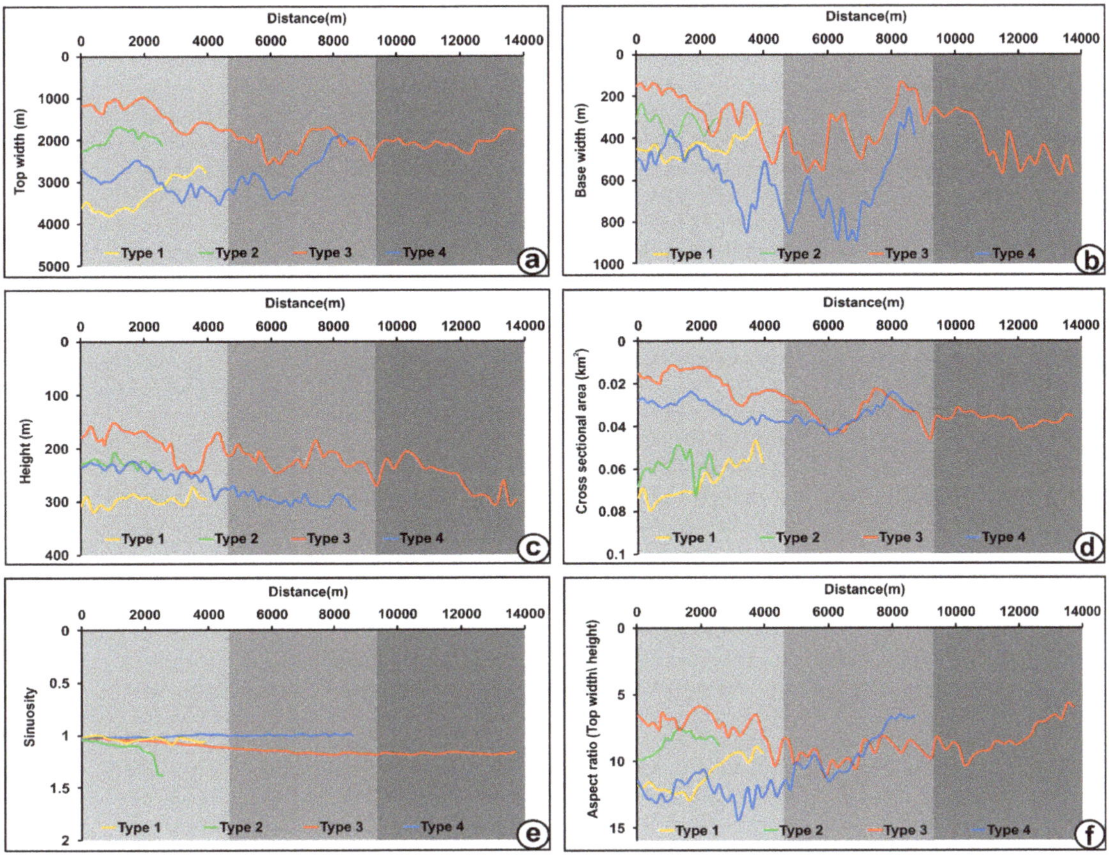

Figure 8. Morphological analysis of four different types of channels complexes in the western (light grey), middle (medium grey), and eastern (grey) areas. (**a**) Top width vs. distance, (**b**) Base width vs. distance, (**c**) Height vs. distance, (**d**) Cross-sectional area vs. distance, (**e**) Sinuosity vs. distance, (**f**) Aspect ratio vs distance.

Type 3 of the channel complexes narrows in a down-dip direction, where the top width is relatively decreased from 2564 to 973 m with better correlation trends (0.52 and 0.72 for R^2 and R, respectively) compared to the Type 2 channel complexes (Table 1 and Figure 8a). In parallel, the base width of Type 3 has an average value of 342.5 m with relatively linear trends down dip (Figure 8b). The height of Type 3 channels complexes increases down dip, starting from an average of 170 m in the western area to an average of 200 m and 250 m in the middle and eastern areas, respectively (Figure 8c). The height and channels distance relationships show a relatively good correlation indicated by high to mid values (0.81 and 0.67) of correlation and determination coefficients, respectively (Table 1). The CSA of Type 3 increases down dip, and it ranges from an average of 0.02 km^2 in the western area to an average of 0.03 km^2 in the middle and eastern areas, with relatively linear trends between the CSA and the distance of the channel (Table 1 and Figure 8d). The sinuosity itself increases down-dip from a straight channel (average sinuosity of 1.00) in the western area to a low sinuosity channel (average sinuosity of 1.10) in the middle and eastern areas (Table 1 and Figure 8e). It is characterized by a relatively linear trend in the channels distance and sinuosity relationship (Figure 8e). In addition, the aspect ratio has a set of minimum and maximum values of 5.60 and 11.10, respectively, with no trend-down dip (Table 1 and Figure 8f).

Type 4 has top width ranging from 1880 to 3551 m (Table 1 and Figure 8a) with a non-linear correlation that is represented by a very low correlation and determination coefficient (0.34 and 0.12, respectively) (Table 1). This indicated that the increase in channel distance is not proportional to the top width value of Type 4 channels complexes. In parallel, the base width value also has non-linear trends down dip (Figure 8b), with an average value of 605.4 m (Table 1 and Figure 8b). Meanwhile, the height of Type 4 channel complexes increased down dip, from 223 to 314 m (Figure 8c), with relatively high values of correlation and determination coefficients (Table 1). The minimum and maximum values of Type 4 CSA are 0.02 and 0.04, respectively, with relatively non-linear trends between the CSA and channels distance (Table 1 and Figure 8d). The sinuosity of Type 4 has an average of 1.01 (straight channel), while the aspect ratio has a minimum and maximum values of 6.46 of 14.36, respectively, with no trend (Table 1 and Figure 8f).

5. Discussion

5.1. Development of Submarine Channels

During the Quaternary period, channels (including canyons and channels complexes) acted as conduits for transferring sediment from the source area in the west (onshore Southern Alps) to the depositional area in the east (Figures 1 and 2). Furthermore, seismic facies, attributes, and morphological analysis in our zone of interest revealed some complexes of incisions with V and U shaped that are erosionally and constructionally confined (Figures 4 and 5). In addition, isolated bright amplitude with relatively linear and nonlinear correlations on the morphometric scaling relationships, also indicate submarine channels [4,6,19,53–56]. This observation is also supported by Sahoo, Kroeger, Thrasher, Munday, Mingard, Cozens and Hill [28], and Barrier, Nicol, Browne and Bassett [30], who concluded that submarine channels in Waka 3D seismic reflection survey are present from the Oligocene to Quaternary.

The first stage of the submarine channel's development is characterized by a V to U-shaped incision that cross-cuts the older deposits (truncated reflection termination pattern in Figures 5 and 6). The existence of isolated low to high seismic amplitude at the base of the channels (Figures 6 and 9) and the smaller base width distance (compared to the top width) of the channels (Figure 8b) are attributed to the coarse-grained basal confinement surface at the base of the turbidity flow. This contributed to the widening of the channel complexes [4,57–59] when a fall in the equilibrium profile reduced the slope accommodation [60]. In addition, the small values of the base width distance on the western area of the channels (<500 m, see Table 1 and Figure 8b), the V-shaped channel erosionally confined at the lower part of the channels (Figures 5 and 6), together with high correlation and determination coefficients (Table 1), shows a good example of intermittent turbidity flows. On the contrary, the high values of the base width distance of the channels (>500m, see Table 1 and Figure 8b) with the U-shaped constructionally confined channels indicate local factors of the channel's development (e.g., channel wall's collapse) or the existence of another turbidity flow widening the base of the channel complexes. These may be the product of sediment supply, transport regime, and gradient variations as the system was developing [5,6,10,12,14].

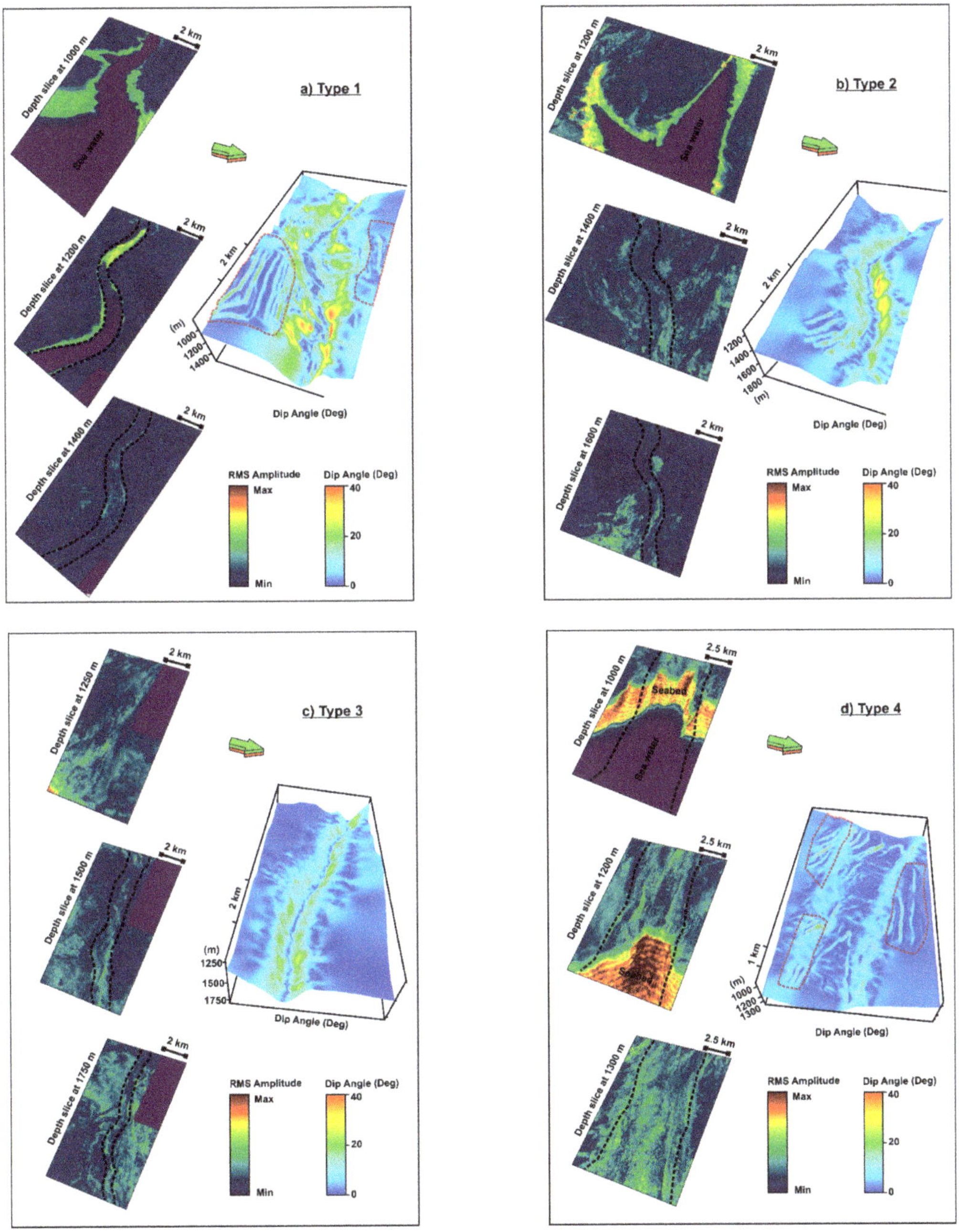

Figure 9. 3D view of dip angle (right images) and depth RMS Amplitude slice (left images) of (**a**) Type 1, (**b**) Type 2, (**c**) Type 3, and (**d**) Type 4 of channels complexes in this study. Dashed black and red lines represent channel outline and sediment waves, respectively. See Figure 4 for the channel's location.

The next stage is indicated lateral shifting of channel complexes (Types 2 and 4), and the vertically aggrading channel that associated with Types 1 and 3 (Figure 6). A relatively high value also indicates this lateral shifting on the top width of the channel (>1800 m, see Table 1, Figures 8a and 9) and cross-sectional area (>0.02 km^2, see Table 1 and Figure 8d). Some of the higher values of the top width and cross-sectional area from other types of channel complexes are attributed to local submarine channel wall failures. This lateral shifting is usually the product of channel avulsion or bend cut-off [5,12]. Furthermore, the vertically aggrading channels complexes in the study area are represented by the channel's height, where Type 1 has the highest average height of 298.5 m (Table 1 and Figure 8c). Once the channel equilibrium profile continues to rise, the submarine channel starts to be aggraded, with no significant lateral offset of depositional facies [6,60]. Overall, the morphometry of submarine channels has been influenced by local factors in this continental slope environments (e.g., submarine wall's collapse, different maturity of turbidite flows, and slope accommodation).

Lastly, the submarine channels in the study area have heights that range from 151 to 320 m. The top width ranges from 973 to 3821 m (Table 1), which indicates more deposition and late stage of filling after the channels were build-up. This is due to prograding fore-sets (Miocene to Recent) and the increase of sedimentation supply in the second stage caused by the strike-slip movement along the Alpine fault and tectonic uplift [30,36,41,43,44]. In addition, the accommodation was increased due to the rising equilibrium profile that allowed aggrading channel elements (more deposition) to develop a channel system fills or channel complexes at this last stage [6,60]. Furthermore, the depositional architecture and its morphometry in this continental slope system (mixed erosion-levee confined channels) is controlled by a cycle of accommodation variation in response to changes in the slope equilibrium profile [6,7,9,60].

5.2. *Implications of Submarine Channels' Morphologies for Paleoenvironment Interpretation*

The increase in morphometric parameters, together with the high correlation (R) and determination (R^2) coefficients (Table 1) from the western to the eastern areas, suggested that the movement of the channels is induced by gravity, and the net transport from these currents are primarily in a seaward direction [5,19,59]. Furthermore, based on the classification of Lemay, Grimaud, Cojan, Rivoirard and Ors [54], Jobe, Howes and Auchter [55], and Prerna and Mahender [61], the scaling relationship of the channel's morphometry elements in the study area is characterized as a submarine environment. In addition, biostratigraphy data from Galleon-1 wellbore at the corresponding intervals show the presence of *Euuvigerina rodleyi s.l.*, *Euuvigerina miozea s.l.*, *Notorotalia Hurupiensis*, and *Hopkinaina Mioinder*. This indicates a marine environment with an increase in water depths and turbid conditions followed by a rapid rate of sedimentation [45,48].

The U shaped submarine channels (Type 4 in particular), and its morphometry that lacked significant changes (Figures 5b,d, 8 and 9b,d), indicated that the extended and long process involving the movement of the fine-grained sediment to the submarine channels are without significant erosion [4,19,62]. Furthermore, the decrease in the top width of the channels in the eastern area (Figure 8a) is related to the contour current that shapes the sediment in a perpendicular direction to the direction of the channels. This is also supported by the presence of sediment waves in the flank area of the channels that represented contour current in the submarine environment (Types 1 and 4, see Figures 5a,d, 6a,d, and 9a,d). This current is known as a subtropical front with highly saline, nutrient-depleted, and warm surface water that flows in the S-N direction, perpendicular to the direction of the channels in the study area (northward-flowing southland current), along the New Zealand continental margin [35,40–42,63].

The lateral development of the channels' wall in the study area indicates the long processes of turbidity flows that occurred in the slope system of the submarine environment [2,5,6]. This is also supported by the relatively high channel top width value, compared with other morphometry elements (Figure 8). Furthermore, correlations between the increasing channel sinuosity and the widening channel floors inside the channel complexes have been observed in the study area (Figure 8e). This suggests a decrease in the dip angle of the submarine slope morphology with a relatively long history of channel development (Figure 9). This sinuosity behaviour is usually followed by channel cut-off in the later stages of channel development with channel complexes as an end product [5,6,12].

Furthermore, the provenance of the Cenozoic sediment in the Canterbury Basin was from the hinterland in the west (Figure 1a). It consists of volcanic, metamorphic (schist), and sedimentary rocks [32,35,36]. This relatively west to east sediment transport direction agrees with the late Cenozoic geodynamic evolution of the study area, where the Canterbury Basin that is located in the Eastern side of the New Zealand—South Island (sediment source area) was the depocenter in a submarine environment [44]. In addition, recent terrestrial rivers and their submarine counterparts display similar transport directions with Quaternary channels' complexes in the study area (Figure 10). This indicates less significant regional plate movement that could have deviated channel trajectory during the Quaternary in the study area (local processes might have occurred). Meanwhile, the northward-flowing southland current and the Antarctic Circumpolar Current contributed to the deposition of Pliocene–Pleistocene sediment drift in the study area [41,42].

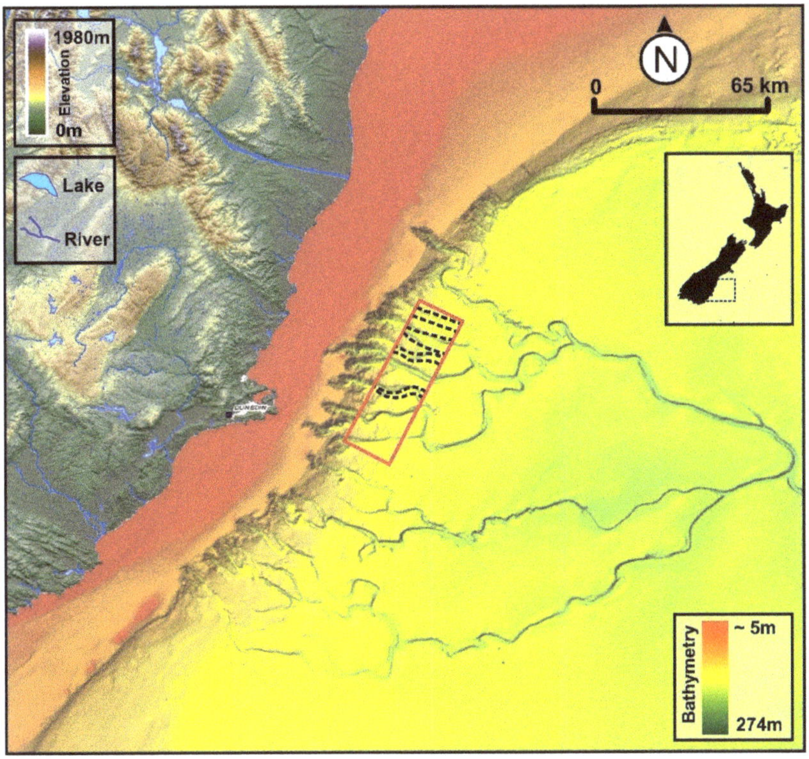

Figure 10. Recent New Zealand elevation and bathymetry (Source: NIWA Open Data License v1.0). Red rectangle is the study area that located in a marine environment (Continental shelf). See Figure 1a for coordinate location.

5.3. Implications of Differential Compaction, Horizon Interpretation and Burial Effects on the Petroleum Prospectivity

The presence of submarine channels in the study area (Figures 4, 9 and S1) with their fine- and coarse-grained sediments could have an impact on the differential compaction, horizon interpretation pitfalls, and burial effects that might influence petroleum prospectivity in the study area.

In the study area, low amplitude and continuous reflections elements of the channel-filled and surrounding deposits (Figures 5 and 6) indicate that the channels are mostly filled with a fine-grained suspended material [28,35,38,49]. Meanwhile, the high amplitude reflections depict the presence of coarse-grained sediments transported by high energetic flows [19,62]. Laterally, this lithological variation caused differential compaction between the stratigraphy below the channel features and its surroundings (Figure 11a). This could lead to hydrocarbon trap formation [64] and pore size reduction [65]. In addition, overpressure—fluid expulsion [66,67]—and sea-floor slope instability [68] could also be triggered. These overburden lateral variations, low-velocity anomalies (incised channels), and sea-floor incisions (causing variation on the water columns) on the final post-stack time-migrated 3D image (see Section 3.1) could also lead to a pitfall on horizon interpretation, namely velocity push-down (Figure 11a). This pitfall gives rise to an apparent feature that is mapped as channels (reservoir) or syncline (traps). However, post-stack depth-migration (PSDM) with careful velocity analysis will minimize these artificial features [69].

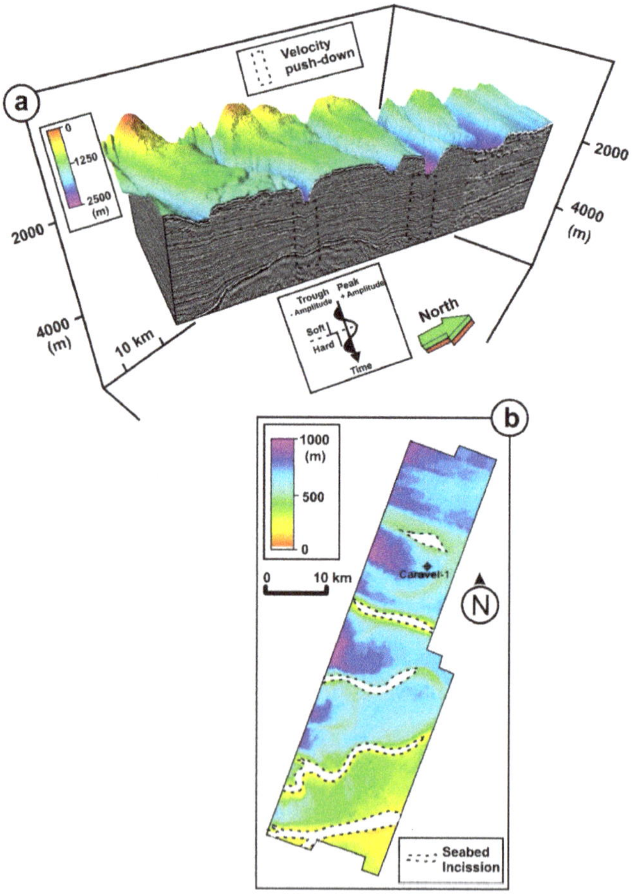

Figure 11. (**a**) Three-dimensional view of Seabed horizon and seismic amplitude, and (**b**) Thickness map between seabed and top Neogene (study interval). See Figure 1a for coordinate location.

Furthermore, the studied interval has a stratigraphic thickness ranging from a few meters to hundreds of meters with zero thickness when the seabed has incised the base horizon (Figure 11b). As sediments become buried, this thickness variation could also be contributing to the late-phase burial and generation of a petroleum system in the study area, underlying the Cretaceous and Paleocene coaly rocks [70], where the thinner area (south and SE) might provide less impact compared to other areas (Figure 11b). These implications could also be a petroleum prospectivity analogue for an area with a similar geological setting in a frontier basin worldwide.

6. Conclusions

This work has focused on the seismic geomorphology aspect of four different types of continental slope channels complexes in the Canterbury Basin, Offshore New Zealand. The conclusions from this research are that:

1. The channels complexes in the study area can be categorized into four main types, based on the shape of the erosional surfaces and the reflection configuration of channels fills. These are vertical migrating channels (Type 1), lateral migrating channels (Type 2), V-shaped channels (Type 3), and U-shaped channels (Type 4).
2. The morphometric analysis of the channels complexes in the study area shows an important aspect (geometry, facies, and scaling relationship) in the quantitative interpretation of its sedimentary processes and paleoenvironment. Thus, it is applicable to a similar setting worldwide.
3. The morphometry of the submarine channels is also influenced by local factors in these submarine environments (e.g., channel wall collapse, different maturity of turbidity flows, etc.)
4. The evolution of the channel's morphometry and trajectory in the study area indicates a less significant regional plate movement.
5. Seismic facies and morphometry analysis of channels complexes in the study area revealed that channels complexes are contributing to differential compaction, horizon interpretation pitfalls, and burial effects that might influence petroleum prospectivity in the study area.

Supplementary Materials: The following supporting information can be downloaded at: https://www.mdpi.com/article/10.3390/app12094386/s1, Figure S1: Seismic attribute of four types of the channels complexes in the study area; (a) curvature, (b) instantaneous frequency, (c) sweetness, and (d) variance with channel delineation (dashed line).

Author Contributions: Conceptualization, D.H.; methodology, D.H.; investigation, D.H. and W.R.R.; data curation, D.H. and W.R.R.; writing—original draft preparation, D.H. and W.R.R.; writing—review and editing, D.H. and W.R.R. All authors have read and agreed to the published version of the manuscript.

Funding: This research was funded by Department of Geosciences, College of Petroleum Engineering and Geosciences, King Fahd University of Petroleum and Minerals.

Institutional Review Board Statement: Not applicable.

Informed Consent Statement: Not applicable.

Data Availability Statement: Data are available in a publicly accessible repository that does not issue DOIs. Publicly available datasets were analyzed in this study. This data can be accessible here: www.nzpam.govt.nz/maps-geoscience/exploration-database. Last accessed on 24 April 2022.

Acknowledgments: The authors thank New Zealand Petroleum and Minerals (NZPM) for making this dataset into a public domain. We also thank King Fahd University of Petroleum and Minerals (KFUPM), as well as Abdullatif Al-Shuhail, Abdullah Alshuhail and Ståle Emil Johansen, for supporting this research. Schlumberger is thanked for PETREL software donated to KFUPM.

Conflicts of Interest: The authors declare no conflict of interest.

References

1. Mutti, E.; Normark, W.R. An Integrated Approach to the Study of Turbidite Systems. In *Seismic Facies and Sedimentary Processes of Submarine Fans and Turbidite Systems*; Weimer, P., Link, M.H., Eds.; Springer New York: New York, NY, USA, 1991; pp. 75–106.
2. Mayall, M.; Stewart, I. The Architecture of Turbidite Slope Channels. In *Deep-Water Reservoirs of the World*; Weimer, P., Ed.; SEPM Society for Sedimentary Geology: Tulsa, OK, USA, 2000; Volume 20.
3. Mayall, M.; Jones, E.; Casey, M. Turbidite channel reservoirs—Key elements in facies prediction and effective development. *Mar. Pet. Geol.* **2006**, *23*, 821–841. [CrossRef]
4. Posamentier, H.W.; Kolla, V. Seismic Geomorphology and Stratigraphy of Depositional Elements in Deep-Water Settings. *J. Sediment. Res.* **2003**, *73*, 367–388. [CrossRef]
5. Abreu, V.; Sullivan, M.; Pirmez, C.; Mohrig, D. Lateral accretion packages (LAPs): An important reservoir element in deep water sinuous channels. *Mar. Pet. Geol.* **2003**, *20*, 631–648. [CrossRef]
6. McHargue, T.; Pyrcz, M.; Sullivan, M.; Clark, J.; Fildani, A.; Romans, B.; Covault, J.; Levy, M.; Posamentier, H.; Drinkwater, N. Architecture of turbidite channel systems on the continental slope: Patterns and predictions. *Mar. Pet. Geol.* **2011**, *28*, 728–743. [CrossRef]
7. Pettinga, L.A.; Jobe, Z.R. How submarine channels (re)shape continental margins. *J. Sediment. Res.* **2020**, *90*, 1581–1600. [CrossRef]
8. Covault, J.A. Submarine fans and canyon-channel systems: A review of processes, products, and models. *Nat. Educ. Knowl.* **2011**, *3*, 4.
9. Covault, J.A.; Graham, S.A. Submarine fans at all sea-level stands: Tectono-morphologic and climatic controls on terrigenous sediment delivery to the deep sea. *Geology* **2010**, *38*, 939–942. [CrossRef]
10. Gong, C.; Steel, R.J.; Wang, Y.; Lin, C.; Olariu, C. Grain size and transport regime at shelf edge as fundamental controls on delivery of shelf-edge sands to deepwater. *Earth-Sci. Rev.* **2016**, *157*, 32–60. [CrossRef]
11. Posamentier, H.W.; Walker, R.G. Deep-Water Turbidites and Submarine Fans. In *Facies Models Revisited*; Posamentier, H.W., Walker, R.G., Eds.; SEPM Society for Sedimentary Geology: Tulsa, OK, USA, 2006.
12. Kolla, V. A review of sinuous channel avulsion patterns in some major deep-sea fans and factors controlling them. *Mar. Pet. Geol.* **2007**, *24*, 450–469. [CrossRef]
13. Niyazi, Y.; Eruteya, O.E.; Omosanya, K.; Harishidayat, D.; Johansen, S.E.; Waldmann, N. Seismic geomorphology of submarine channel-belt complexes in the Pliocene of the Levant Basin, offshore central Israel. *Mar. Geol.* **2018**, *403*, 123–138. [CrossRef]
14. Sylvester, Z.; Pirmez, C.; Cantelli, A. A model of submarine channel-levee evolution based on channel trajectories: Implications for stratigraphic architecture. *Mar. Pet. Geol.* **2011**, *28*, 716–727. [CrossRef]
15. Sylvester, Z.; Covault, J.A. Development of cutoff-related knickpoints during early evolution of submarine channels. *Geology* **2016**, *44*, 835–838. [CrossRef]
16. Harishidayat, D.; Johansen, S.E.; Batchelor, C.; Omosanya, K.O.; Ottaviani, L. Pliocene–Pleistocene glacimarine shelf to slope processes in the south-western Barents Sea. *Basin Res.* **2020**, *33*, 1315–1336. [CrossRef]
17. Qin, Y.; Alves, T.M.; Constantine, J.; Gamboa, D. Quantitative seismic geomorphology of a submarine channel system in SE Brazil (Espírito Santo Basin): Scale comparison with other submarine channel systems. *Mar. Pet. Geol.* **2016**, *78*, 455–473. [CrossRef]
18. Pettinga, L.; Jobe, Z.; Shumaker, L.; Howes, N. Morphometric scaling relationships in submarine channel–lobe systems. *Geology* **2018**, *46*, 819–822. [CrossRef]
19. Harishidayat, D.; Omosanya, K.O.; Johansen, S.E.; Eruteya, O.E.; Niyazi, Y. Morphometric analysis of sediment conduits on a bathymetric high: Implications for palaeoenvironment and hydrocarbon prospectivity. *Basin Res.* **2018**, *30*, 1015–1041. [CrossRef]
20. Omosanya, K.O.; Harishidayat, D. Seismic geomorphology of Cenozoic slope deposits and deltaic clinoforms in the Great South Basin (GSB) offshore New Zealand. *Geo-Mar. Lett.* **2019**, *39*, 77–99. [CrossRef]
21. Brown, A.R. *Interpretation of Three-Dimensional Seismic Data*, 7th ed.; Society of Exploration Geophysicists and the American Association of Petroleum Geologists: Tulsa, OK, USA; Houston, TX, USA, 2011.
22. Cox, D.R.; Newton, A.M.; Huuse, M. An introduction to seismic reflection data: Acquisition, processing and interpretation. In *Regional Geology and Tectonics*; Elsevier: Amsterdam, The Netherlands, 2020; pp. 571–603.
23. Hajek, E.A.; Heller, P.L.; Sheets, B.A. Significance of channel-belt clustering in alluvial basins. *Geology* **2010**, *38*, 535–538. [CrossRef]
24. Hofmann, M.H.; Wroblewski, A.; Boyd, R. Mechanisms Controlling the Clustering of Fluvial Channels and the Compensational Stacking of Cluster Belts. *J. Sediment. Res.* **2011**, *81*, 670–685. [CrossRef]
25. Shumaker, L.E.; Jobe, Z.R.; Johnstone, S.A.; Pettinga, L.A.; Cai, D.; Moody, J.D. Controls on submarine channel-modifying processes identified through morphometric scaling relationships. *Geosphere* **2018**, *14*, 2171–2187. [CrossRef]
26. Harishidayat, D.; Omosanya, K.O.; Johansen, S.E. 3D seismic interpretation of the depositional morphology of the Middle to Late Triassic fluvial system in Eastern Hammerfest Basin, Barents Sea. *Mar. Pet. Geol.* **2015**, *68*, 470–479. [CrossRef]
27. Reimchen, A.P.; Hubbard, S.; Stright, L.; Romans, B. Using sea-floor morphometrics to constrain stratigraphic models of sinuous submarine channel systems. *Mar. Pet. Geol.* **2016**, *77*, 92–115. [CrossRef]
28. Sahoo, T.R.; Kroeger, K.F.; Thrasher, G.; Munday, S.; Mingard, H.; Cozens, N.; Hill, M. Facies distribution and impact on petroleum migration in the Canterbury Basin, New Zealand. In Proceedings of the Eastern Australian Basin Symposium, Melbourne, Australia, 13–16 September 2015.
29. Zhao, T.; Zhang, J.; Li, F.; Marfurt, K.J. Characterizing a turbidite system in Canterbury Basin, New Zealand, using seismic attributes and distance-preserving self-organizing maps. *Interpretation* **2016**, *4*, SB79–SB89. [CrossRef]

30. Barrier, A.; Nicol, A.; Browne, G.H.; Bassett, K. Early Oligocene marine canyon-channel systems: Implications for regional paleogeography in the Canterbury Basin, New Zealand. *Mar. Geol.* **2019**, *418*, 106037. [CrossRef]
31. Kumar, P.C.; Alves, T.M.; Sain, K. Submarine canyon systems focusing sub-surface fluid in the Canterbury Basin, South Island, New Zealand. *Sci. Rep.* **2021**, *11*, 16990. [CrossRef] [PubMed]
32. Campbell, H.; Malahoff, A.; Browne, G.; Graham, I.; Sutherland, R. New Zealand Geology. *Episodes* **2012**, *35*, 57–71. [CrossRef]
33. Sutherland, R.; Spasojevic, S.; Gurnis, M. Mantle upwelling after Gondwana subduction death explains anomalous topography and subsidence histories of eastern New Zealand and West Antarctica. *Geology* **2010**, *38*, 155–158. [CrossRef]
34. Sutherland, R.; Dickens, G.; Blum, P.; Agnini, C.; Alegret, L.; Asatryan, G.; Bhattacharya, J.; Bordenave, A.; Chang, L.; Collot, J.; et al. Continental-scale geographic change across Zealandia during Paleogene subduction initiation. *Geology* **2020**, *48*, 419–424. [CrossRef]
35. Uruski, C.I. New Zealand's deepwater frontier. *Mar. Pet. Geol.* **2010**, *27*, 2005–2026. [CrossRef]
36. Browne, G.H.; Field, B.D. A review of Cretaceous-Cenozoic sedimentation and tectonics, East coast, South Island, New Zealand. *Can. Soc. Pet. Geol.* **1988**, *15*, 37–48.
37. Higgs, K.; Browne, G.; Sahoo, T. Reservoir characterisation of syn-rift and post-rift sandstones in frontier basins: An example from the Cretaceous of Canterbury and Great South basins, New Zealand. *Mar. Pet. Geol.* **2018**, *101*, 1–29. [CrossRef]
38. O'Leary, R.; Mogg, W.G. The Petroleum Potential of the Offshore Canterbury Basin: Insights from Petroleum Systems Modelling. In Proceedings of the Eastern Australian Basin Symposium III, Sydney, Australia, 14–17 September 2008.
39. King, D.J.; Newnham, R.M.; Gehrels, W.R.; Clark, K.J. Late Holocene sea-level changes and vertical land movements in New Zealand. *N. Z. J. Geol. Geophys.* **2020**, *64*, 21–36. [CrossRef]
40. Jiao, R.; Herman, F.; Seward, D. Late Cenozoic exhumation model of New Zealand: Impacts from tectonics and climate. *Earth-Sci. Rev.* **2017**, *166*, 286–298. [CrossRef]
41. Fulthorpe, C.; Carter, R.M. Continental-shelf progradation by sediment-drift accretion. *GSA Bull.* **1991**, *103*, 300–309. [CrossRef]
42. Lu, H.; Fulthorpe, C.; Mann, P. Three-dimensional architecture of shelf-building sediment drifts in the offshore Canterbury Basin, New Zealand. *Mar. Geol.* **2003**, *193*, 19–47. [CrossRef]
43. Lu, H.; Fulthorpe, C.S.; Mann, P.; Kominz, M.A. Miocene-Recent tectonic and climatic controls on sediment supply and sequence stratigraphy: Canterbury basin, New Zealand. *Basin Res.* **2005**, *17*, 311–328. [CrossRef]
44. Lu, H.; Fulthorpe, C.S. Controls on sequence stratigraphy of a middle Miocene-Holocene, current-swept, passive margin: Offshore Canterbury Basin, New Zealand. *GSA Bull.* **2004**, *116*, 1345–1366. [CrossRef]
45. Wilson, I.R.; Renton, P.H.; Mound, D.G.; Grant, J. *Well Completion Report of the Galleon-1 (PPL38203)*; BP SHELL TODD Canterbury Services Limited: Wellington, New Zealand, 1985; p. 1309.
46. Fulthorpe, C.S.; Hoyanagi, K.; Blum, P.; Guèrin, G.; Slagle, A.L.; Blair, S.A.; Browne, G.H.; Carter, R.M.; Ciobanu, M.C.; Claypool, G.E. Integrated ocean drilling program expedition 317 preliminary report canterbury basin sea level global and local controls on continental margin stratigraphy. *Integr. Ocean. Drill. Program. Prelim. Rep.* **2010**, 1–133. Available online: https://researchers.mq.edu.au/en/publications/integrated-ocean-drilling-program-expedition-317-preliminary-repo (accessed on 20 March 2022).
47. Carter, R.M.; McCave, I.N.; Richter, C.; Carter, L. Proceedings of the Ocean Drilling Program, Initial Reports Volume 181 (Site 1119); College Station, USA. 1999. Available online: http://www-odp.tamu.edu/publications/181_IR/VOLUME/CHAPTERS/IR181_01.PDF (accessed on 20 March 2022).
48. Schiøler, P.R.; Raine, J.I.; Crundwell, M.P.; Griffin, A.; Hollis, C.J.; Kulhanek, D.K.; Morgans, H.E.G.; Roncaglia, L.; Strong, C.P.; Uruski, C. *Revised Biostratigraphy and Well Correlation, Canterbury Basin, New Zealand (PR4365)*; Ministry of Economic Development: Wellington, New Zealand, 2011; p. 142.
49. Mogg, W.G.; Aurisch, K.; O'Leary, R.; Pass, G.P. The Carrack-Caravel prospect complex: A possible sleeping giant in the deep water Canterbury Basin, New Zealand. In Proceedings of the Eastern Australasian Basins Symposium III, Sydney, Australia, 14–17 September 2008.
50. Chopra, S.; Marfurt, K.J. Seismic attributes—A historical perspective. *Geophysics* **2005**, *70*, 3SO–28SO. [CrossRef]
51. Mitchum, R.M., Jr.; Vail, P.R.; Sangree, J.B. Seismic stratigraphy and global changes of sea level: Part 6. Stratigraphic interpretation of seismic reflection patterns in depositional sequences: Section 2. In *Application of Seismic Reflection Configuration to Stratigraphic Interpretation*; AAPG Memoir 26; American Association of Petroleum Geologists: Tulsa, OK, USA, 1977.
52. James, G.; Witten, D.; Hastie, T.; Tibshirani, R. *An Introduction to Statistical Learning*; Springer: Berlin/Heidelberg, Germany, 2013; Volume 112.
53. Kolla, V.; Posamentier, H.; Wood, L. Deep-water and fluvial sinuous channels—Characteristics, similarities and dissimilarities, and modes of formation. *Mar. Pet. Geol.* **2007**, *24*, 388–405. [CrossRef]
54. Lemay, M.; Grimaud, J.-L.; Cojan, I.; Rivoirard, J.; Ors, F. Geomorphic variability of submarine channelized systems along continental margins: Comparison with fluvial meandering channels. *Mar. Pet. Geol.* **2020**, *115*, 104295. [CrossRef]
55. Jobe, Z.R.; Howes, N.C.; Auchter, N.C. Comparing submarine and fluvial channel kinematics: Implications for stratigraphic architecture. *Geology* **2016**, *44*, 931–934. [CrossRef]
56. Howlett, D.M.; Gawthorpe, R.L.; Ge, Z.; Rotevatn, A.; Jackson, C.A. Turbidites, topography and tectonics: Evolution of submarine channel-lobe systems in the salt-influenced Kwanza Basin, offshore Angola. *Basin Res.* **2020**, *33*, 1076–1110. [CrossRef]

57. Paull, C.K.; Talling, P.J.; Maier, K.; Parsons, D.; Xu, J.; Caress, D.W.; Gwiazda, R.; Lundsten, E.M.; Anderson, K.; Barry, J.P.; et al. Powerful turbidity currents driven by dense basal layers. *Nat. Commun.* **2018**, *9*, 4114. [CrossRef]
58. Lowe, D.R. Sediment Gravity Flows: Their Classification and Some Problems of Application to Natural Flows and Deposits. In *Geology of Continental Slopes*; Doyle, L.J., Pilkey, O.H., Eds.; SEPM Society for Sedimentary Geology: Tulsa, OK, USA, 1979; Volume 27.
59. Middleton, G.V. Sediment Deposition from Turbidity Currents. *Annu. Rev. Earth Planet. Sci.* **1993**, *21*, 89–114. [CrossRef]
60. Kneller, B. The influence of flow parameters on turbidite slope channel architecture. *Mar. Pet. Geol.* **2003**, *20*, 901–910. [CrossRef]
61. Prerna, R.; Mahender, K. Geomorphometric comparison of submarine channel-levee complexes with fluvial river systems: Observations from the Indus. *Geo-Marine Lett.* **2020**, *40*, 573–592. [CrossRef]
62. Jobe, Z.R.; Lowe, D.R.; Uchytil, S.J. Two fundamentally different types of submarine canyons along the continental margin of Equatorial Guinea. *Mar. Pet. Geol.* **2011**, *28*, 843–860. [CrossRef]
63. Hillman, J.I.T.; Klaucke, I.; Pecher, I.A.; Gorman, A.R.; von Deimling, J.S.; Bialas, J. The influence of submarine currents associated with the Subtropical Front upon seafloor depression morphologies on the eastern passive margin of South Island, New Zealand. *N. Z. J. Geol. Geophys.* **2018**, *61*, 112–125. [CrossRef]
64. Ward, N.; Alves, T.M.; Blenkinsop, T.G. Differential compaction over Late Miocene submarine channels in SE Brazil: Implications for trap formation. *GSA Bull.* **2017**, *130*, 208–221. [CrossRef]
65. Chopra, S.; Marfurt, K.J. Seismic attribute expression of differential compaction. *Lead. Edge* **2012**, *31*, 1418–1422. [CrossRef]
66. Bjørlykke, K.; Høeg, K. Effects of burial diagenesis on stresses, compaction and fluid flow in sedimentary basins. *Mar. Pet. Geol.* **1997**, *14*, 267–276. [CrossRef]
67. Osborne, M.J.; Swarbrick, R.E. Mechanisms for Generating Overpressure in Sedimentary Basins: A Reevaluation. *AAPG Bull.* **1997**, *81*, 1023–1041. [CrossRef]
68. Georgiopoulou, A.; Krastel, S.; Masson, D.G.; Wynn, R.B. *Repeated Instability of The Nw African Margin Related to Buried Landslide Scarps*; Springer: Berlin/Heidelberg, Germany, 2007; pp. 29–36. [CrossRef]
69. Marfurt, K.J.; Alves, T.M. Pitfalls and limitations in seismic attribute interpretation of tectonic features. *Interpretation* **2015**, *3*, SB5–SB15. [CrossRef]
70. Sahoo, T.R.; Bland, K.J.; Arnot, M.J.; Boyes, A.F.; Funnell, R.; Griffin, A.; Kroeger, K.F.; Lawrence, M.J.F.; O' Brien, G.; Scadden, P.G.; et al. *Atlas of Petroleum Prospectivity, Southeast Province: ArcGIS Geodatabase and Technical Reports*; GNS Science: Lower Hutt, New Zealand, 2017; p. 50.

applied sciences

MDPI

Article

Surface Displacement and Source Parameters of the 2021 Bandar-e Genaveh, Iran, Earthquake Determined from InSAR Observations

Zeinab Golshadi [1,*], Nicola Angelo Famiglietti [2], Simone Atzori [2] and Annamaria Vicari [2]

1 Institute of Geophysics, University of Tehran, Tehran 141556466, Iran
2 Istituto Nazionale di Geofisica e Vulcanologia, 605, 00143 Rome, Italy; nicola.famiglietti@ingv.it (N.A.F.); simone.atzori@ingv.it (S.A.); annamaria.vicari@ingv.it (A.V.)
* Correspondence: z.golshadi@alumni.ut.ac.ir; Tel.: +98-919-2906-527

Abstract: On 18 April 2021, a M_W 5.8 earthquake occurred near the city of Bandar-e Genaveh, southwestern Iran. Four synthetic aperture radar (SAR) images, acquired from Sentinel-1 (ESA Copernicus project) satellites in ascending and descending orbits, were used to get two displacement maps, catching the surface co-seismic effects through the two-pass InSAR technique. Modeling the deformation patterns using equations for a shear dislocation in elastic half-space allowed the source parameters and the slip distribution of the seismogenic source to be determined. We calculated that the rupture occurred on a reverse fault extending NW-SE, gently dipping NE and with a maximum slip reaching about 1 m. The northeast and low-dip angle of this fault are also consistent with the tectonics of the region, which is subject to deformation and shortening along the northern margin of the Arabian plate. Our estimations of the fault parameters agree with the Zagros Foredeep reverse fault. We additionally processed four other SAR images to investigate the possibility that the Mw 5.0 aftershock, which occurred about one month later, induced surface effects visible with InSAR. This analysis, however, did not provide any clear conclusions.

Keywords: Genaveh earthquake; co-seismic deformation; Sentinel-1 A; interferometric synthetic aperture radar; geodetic modeling

1. Introduction

On 18 April 2021, at 6:41 UTC (11:11 Iran Standard Time), a M_W 5.8 earthquake occurred near the city of Bandar-e Genaveh, in the seismically active area of the Bushehr province, southwestern Iran (Figure 1). This earthquake occurred in the strongly active area, with structures oriented mostly northwest-southeast and dominated by right-slip and reverse fault mechanisms [1,2]. It did not cause casualties, extensive damage or surface ruptures in the affected area [3]. There were 200 aftershocks following this earthquake over the next 4 months, with 20 of a magnitude higher than 4 [4]. Among them, a large aftershock of M_W 5.0 occurred on May 21, potentially causing additional surface displacements.

A number of studies have investigated this earthquake using finite fault modeling (Table 1) based on different techniques. However, there are inconsistencies between the different fault models. According to the Harvard group and the U.S. Geological Survey (USGS), the Global Centroid Moment Tensor (GCMT) solution is a fault-plane solution that is consistent with nearly thrust-slip motion, but the USGS solution prefers a smaller seismic moment release.

In order to reconstruct the complex active fault system of the investigated area, we exploit the InSAR (Interferometry from Synthetic Aperture Radar) two-pass technique, a well-consolidated instrument to retrieve the surface deformations [5]. InSAR is, indeed, a geodetic technique capable of obtaining high resolution displacement maps with centimetric accuracy over a wide area [6]. These maps reveal the displacement in LoS (Line-of-Sight) direction, i.e., the line connecting the satellite to the pixel on the ground, and they can be used, through a geophysical model and optimization algorithms, to derive the parameters of the source responsible for the earthquake [7]. Given the peculiar geometry of the LoS direction, the combination of images acquired from ascending and descending orbits is generally preferred to better constrain the deformation field and, consequently, the source parameters.

Table 1. Source parameters of the 2021 Genaveh earthquake.

Reported by	Dataset	Magnitude	Plane No.	Strike (°)	Dip (°)	Rake (°)	Depth (km)	latitude (°)	longitude (°)
USGS	Teleseismic	5.8 (M_{WW})	1 2	308 139	28 62	80 95	8	29.753	50.678
GCMT	Teleseismic	5.9 (M_w)	1 2	300 132	26 65	79 95	12	29.61	50.60
IGUT	Regional seismic data	5.9 (M_w) 5 (M_n)	Mainshock aftershock	- -	- -	- -	9 9	29.73 29.79	50.67 50.57
IIEES	Regional seismic data	6 (M_L)	-	-	-	-	20	29.71	50.70
RHDRC	Local seismic data	6 (M_W)	-	-	-	-	10	50.79	29.79
EMSC	Regional seismic data	5.8 (M_W)	-	-	-	-	10	50.70	29.82
GFZ	Teleseismic	6 (M_W)	-	-	-	-	10	50.66	29.77

USGS: United States Geological Survey; GCMT: Global Centroid Moment Tensor; IGUT: Institute of Geophysics, University of Tehran; IIEES: International Institute for Earthquake Engineering and Seismology; RHDRC: Road, Housing and Development Research Center; EMSC: European-Mediterranean Seismological Centre; GFZ: German Research center for Geosciences.

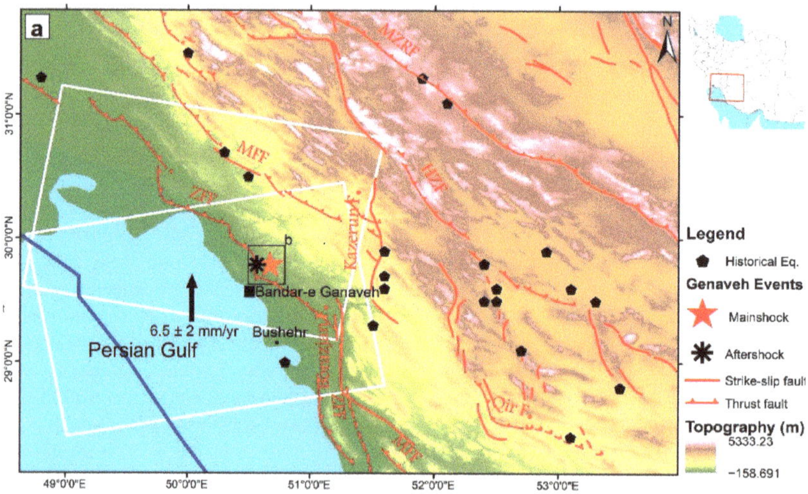

Figure 1. *Cont.*

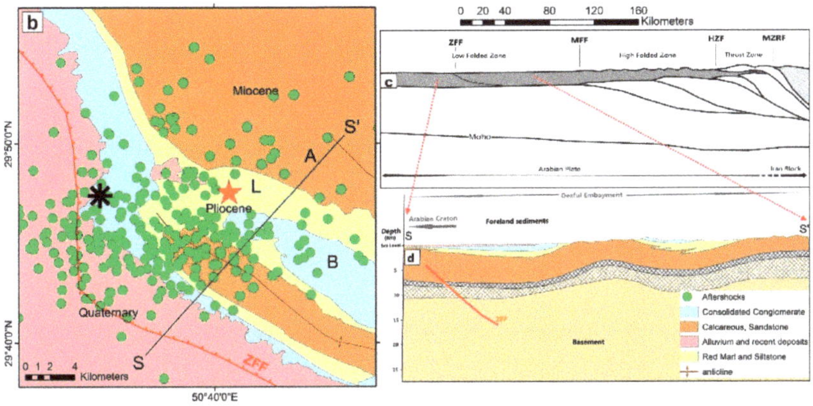

Figure 1. (**a**) Map of the investigated area plotted on shaded relief map of SRTM digital elevation model; red star represents the location of the 18 April 2021, M_W 5.8 mainshock and the black one indicates the 21 May 2021, M_W 5.0 aftershock [4]; black pentagons indicate historical earthquakes [8]; white rectangles mark the footprints of Sentinel-1 images used in this work; red lines are the active faults in the area [9,10]. Abbreviations are; ZFF-Zagros Foredeep Fault, HZF-High Zagros Fault, MFF-Mountain Front Fault and MZRF-Main Zagros Reverse Fault. Dark blue line show boundaries of cratons and tectonic belts as deformation front [11]. (**b**) Zoom on the investigated area, with the geological map [12]. The ZFF fault extends through sedimentary rocks and cuts simple folding of Bakhtiari, Lahbari, Aghajari formations (B, L and A labels in figure, respectively) [13]; Arrow shows the relative velocity of Arabia with respect to Eurasia in mm/year [14]; green circles indicate aftershocks up to 15 July 2021. (**c**) A simplified cross-section through northeast to southwest of (**a**) [15–17]. (**d**) Geological profile of SS' in (**b**) and enlarged (**c**) section [10,13].

In this study, we infer the source of the 2021, M_W 5.8, Genaveh earthquake through the modeling of ascending and descending InSAR displacement data, providing the fault geometries, kinematics and the slip distribution for the mainshock. We also investigate how the convergent tectonics of this region relate to the obtained causative fault. We additionally investigate the possibility that the M_W 5.0 aftershock induced some co-seismic surface effects by calculating and modeling two more InSAR displacement maps. The outcome of this analysis, however, did not provide an evident conclusion.

Geological and Tectonic Setting

This area is located in southwestern Iran, within the complex Neogene collision zone between the Arabian and Eurasian plates, with a tectonic shortening rate of approximately 6.5 ± 2 mm/yr [14] as a result of post-collisional crustal shortening that is transferred to the foreland structures and ophiolite obduction [18,19] (Figure 1).

This area is one of the most seismically active intra-continental fold-and-thrust belts on Earth. Zagros Mountains and their coupled foreland basin were evolved during a multi-episodic orogeny from the Late Cretaceous to recent times [20]. The estimated σ_1 axis in Zagros Folded and thrust Belt (ZFTB) trend rotates in anti-clockwise direction from NE-oriented trajectories to the southeast to ~N-oriented trends to the north [10]. There are four major active faults in ZFTB, which are the Zagros Foredeep Fault (ZFF), the High Zagros Fault (HZF), the Mountain Front Fault (MFF) and the Main Zagros Reverse Fault (MZRF). Most of these faults trend NW–SE and NNW–SSE, with dips of 30–60° and rakes of 60–120° along the plate boundary [9,21]. There is an older sequence of deformation in the inner parts of Zagros chain and a younger sequence in the outer Simply Folded Belt since the Late Cretaceous to Pliocene period [17].

In this region, several tectonic units are divisible from northeast to southwest with non-uniform deformation [10]. Tectonic slices of the highly deformed domain of thrust zone, including the Arabian margin, remains of island arcs, fragments of Cretaceous ophiolites and accretionary prisms, developed above a basal detachment in the latest Neoproterozoic–Lower Cambrian Hormuz evaporates or lateral equivalent [17,20,22]. In the Zagros belt, the deformation and tectonic process that reactivated the Late Cretaceous–Paleogene obduction belt is characterized by thin- and thick-skinned tectonics [17]. Tectonic deformation of the Zagros foreland basin was closely related to the propagation of faults from the Arabia–Eurasia suture zone to the southwestern side [22]. The Simply Folded Belt, limited by the HZF and the MFF, is characterized by many elongated anticlines. The low folded zone, located to the southwest of the MFF (Figure 1b), is characterized by a series of elongate folds of Miocene strata [23]. Multi-detachment folds with subordinate thrusting characterize the deformation of the sedimentary cover of the Arabian margin [24]. According to the progressive limb rotation model of fold kinematics, these formations are growth strata which accumulated during folding. As a result of folding, N- and NW-trending basement faults accommodated shortening [10]. Multiplicity of decollement levels promotes a partly decoupled deformations and limits the size of most faults in the area [19,24].

The Dezful embayment changes the obliquity of Arabian plate convergence, controls the slip rate of faults, and contributes to the formation of the ZFTB [10]. The study area is bounded by the Zagros Foredeep Fault (ZFF) and the Borazjan fault. Inverse ZFF is generally northwest–southeast with a slope to the northeast [1]. It separates the foredeep Zagros belt (north and northeast) from the Quaternary alluvium of the Persian Gulf coastal plain (south and southwest) and in the form of overthrust fault branches, has cut the folds in the region [1]. The strike-slip segment of the Kazerun Fault displaced the High Zagros Fault southward [10]. The ZFF has been extended as a dextral fault along the active Kazerun–Borazjan fault for 105 km and is composed of various hidden fault segments that partition the compressional structures within this range. This fault, like the MFF with a predominant thrust regime and the HZF with a combination of thrust and strike-slip regimes [10], is a deep fault that, after cutting the sedimentary cover of the upper crust, extends to the lowest parts of the crust and reaches the shear zone to Moho. As a result of the thrust Borazjan fault activation, the fold axis has curved, elongated, and dislocated across the Zagros belt as dextral [25] (Figure 1). Under NE dip low-angle basal thrusts, Arabian basement rocks are deformed by large crustal thrusts that actually merge at depth and correspond to the uppermost crystalline basement [24].

2. Methodology

2.1. InSAR Observations and Processing

In this work, we used Sentinel-1 (ESA Copernicus project) data in Interferometric Wide (IW) swath mode to retrieve the ground displacement field induced by the Genaveh earthquake; it acquires data with a 250 km swath at a 5 m by 20 m spatial resolution (single look). IW mode captures three sub-swaths using Terrain Observation with Progressive Scans SAR (TOPSAR). A white rectangle in Figure 1 indicates the footprints of SAR images covering the area of the event. We generated four interferograms, each with a temporal baseline of 12 days, to measure the mainshock and the M_W 5.0 aftershock deformations, both from ascending and descending orbits (Table 2).

Table 2. Sentinel-1 pairs used to detect the permanent deformations of the Genaveh mainshock and the aftershock.

Event	Interferogram Number	Master Date	Slave Date	Pass Direction	Orbit Number	Normal Spatial Baseline (m)	Temporal Baseline (Day)
Mainshock	1	14 April 2021	26 April 2021	Ascending	101	41.21	12
	2	10 April 2021	22 April 2021	Descending	35	39.18	12
Aftershock	1	20 May 2021	01 June 2021	Ascending	101	6.73	12
	2	16 May 2021	28 May 2021	Descending	35	39.92	12

Co-seismic processing was carried out with ENVI® SARscape® (Sarmap, CH, Version 5.5, distributed by HARRIS Geospatial Solutions, Switzerland, Retrieved from http://www.sarmap.ch/) software, using the two-pass InSAR method to produce the interferograms [26]. The Shuttle Radar Topography Mission (SRTM) digital elevation model (DEM) with 30 m resolution (Available online: https://earthexplorer.usgs.gov/ (accessed on 5 May 2021)) was used to remove the topographic phase contributions [27], precise orbits (Available online: https://qc.sentinel1.eo.esa.int (accessed on 19 May 2021)) were used for orbital corrections and the Goldstein filtering method was also applied to wrapped fringes [28] to reduce the interferometric phase noise. A phase unwrapping and displacement maps were created using the minimum cost flow (MCF) algorithm [29] and the slant range maps were eventually geocoded to UTM zone 39, with a pixel size of 30 m. The interferograms and the displacement maps are shown in Figure 2; due to the favorable conditions (dry soil, short temporal and spatial baselines), the interferometric coherence is very high and the coverage is continuous.

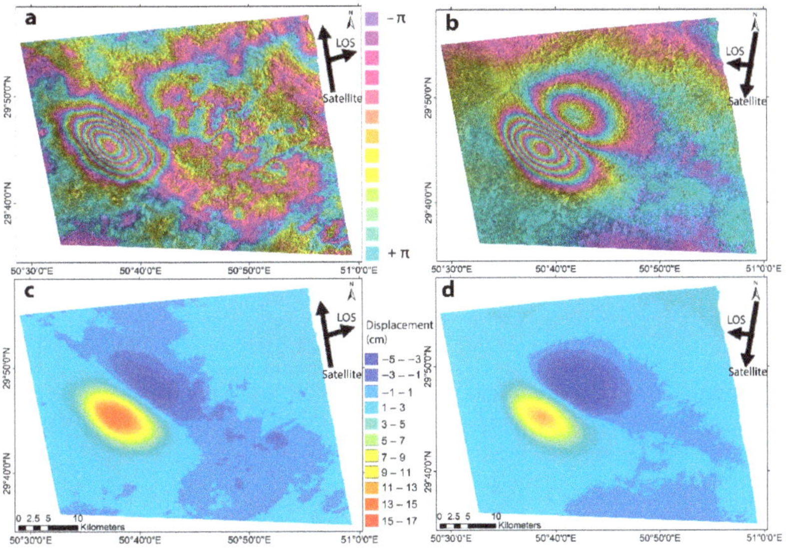

Figure 2. Interferometric fringes (**a**,**b**) and displacement maps (**c**,**d**) for ascending (left) and descending (right) orbits for the 2021 Genaveh mainshock. Positive displacement indicates a movement toward the satellite (e.g., uplift), whereas negative displacement indicates a movement away from the satellite (e.g., subsidence).

We also considered the possibility that the M_W 5.0 occurred on May 21st induced a detectable surface effect; the two displacement maps, obtained from the processing of images listed in Table 2, are shown in Figure 3. The two maps show a small, but clear, signal of a few centimeters that can be ascribed to several different causes, and we will investigate through their modeling in the next section.

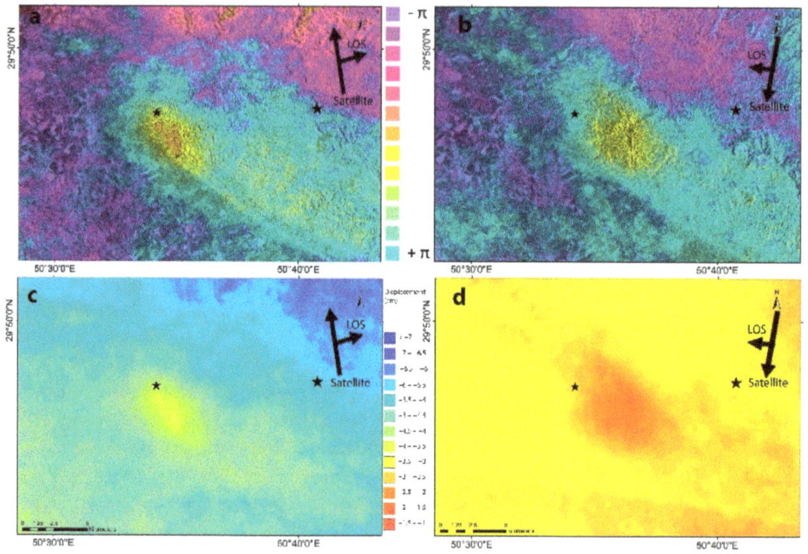

Figure 3. Ascending (**a**,**c**) and descending (**b**,**d**) pattern of Interferometric fringes and displacements including the May 21, M 5.0 aftershock. The biggest dark star indicates the location of the mainshock and the small dark star shows the location of May aftershock.

2.2. Elastic Dislocation Modelling Using InSAR Observations

To provide an accurate estimate of the earthquake source parameters based on a single rectangular fault plane and to determine where in-depth slip distributions are located, the following analysis of InSAR data and modelling was performed. After processing the four interferograms of Table 2, we started the modeling of displacement data to get the source parameters. In order to reduce the amount of data to handle in the inversion, particularly needed in this area where the radar coherence is maximized, we sampled the InSAR raster maps adopting a regular grid sampling of 2000 m in the far field and 500 m in the near field. With this sampling, we obtained a subset of 3500 points for the ascending and descending datasets. Then, co-seismic displacement modeling was conducted using a consolidated two-step inversion method: a non-linear inversion to solve for the geometry, location and uniform shear dislocation on a rectangular source, and a linear inversion to solve for the slip distribution over the constrained fault plane [30]. In both steps, the underlying model is the rectangular shear dislocation in a homogeneous and elastic half-space [31]. Further, we adopt a strategy to account for the effects of topography [32], ranging in this area from zero to around 600 m and playing a small but not negligible role. For every event, we jointly inverted the ascending and descending displacement maps. The non-linear inversion exploits the Levenberg–Marquardt minimization algorithm to find the global minimum of the cost-function based on the least-squares of the residual between the observed and predicted data [33]; it consists of a Gauss–Newton method, implemented with multiple restarts to guarantee catching of the global minimum. Details of the implementation can be found in [30].

After the definition of the source geometry with a uniform slip and rake, we calculated the slip distribution along the fault plane, by using the algorithm described in [34]. It consists of a least squares optimization based on linear inversion of the matrix:

$$\begin{bmatrix} d_a \\ d_d \\ 0 \end{bmatrix} = \begin{bmatrix} G \\ k \cdot \nabla^2 \end{bmatrix} \cdot m$$

where d_a and d_d are the observed ascending and descending data vectors, respectively, m is the vector of the unknown slip values, G is the Green's function matrix, ∇^2 is a Laplacian operator used to regularize the slip distribution, weighted by the empiric coefficient k [35,36]. To prevent the change of the slip direction (back-slip) along the fault, we used additional constraints of non-negativity of the parameters.

We remark that all the inversions (non-linear and linear, for the mainshock and the aftershock) are carried out with a joint contribution of ascending and descending InSAR datasets, automatically weighted according to the approach described in [37].Furthermore, for every inversion the parameters of a possible ramp that affects the InSAR data were simultaneously assessed together with the source parameters.

3. Results

The displacement maps show a peak of about 0.17 and 0.13 m, from ascending and descending orbits, respectively (Figure 2). Positive values indicate a displacement toward the satellite, in the LoS, which reflect the presence of the uplift typical of thrust earthquakes. In fact, a few centimeters of variation in LoS displacement measurements might also be related to the different viewing angles between the ascending and descending satellite observations, and the contribution of the horizontal (mainly E-W) motion component. In addition, the important difference between the two LoS, ascending and descending, witnesses the existence of an important horizontal component, as expected for this type of rupture that occurs along a low-dip plane. Table 3 shows the source parameters for the deformation constrained by the non-linear inversion of two maps containing the effects of the M_W 5.8 mainshock; it shows that the deformation was caused by a fault with 9.5 km in length, 5 km in width, 20° in dip, and 313° in strike with a prevalently inverse rupture mechanism with an average slip of about 0.7 m, consistent with the focal mechanism determined by USGS, as discussed in the next section.

Table 3. Fault parameters of the Genaveh mainshock after the non-linear inversion. Latitude and longitude refer to the source center, vertically projected to the surface; depth refers to the fault top edge. For every parameter, the 1-sigma uncertainty is reported within brackets.

Width (km)	Length (km)	Depth (km)	Strike (°)	Dip (°)	Longitude	Latitude	Rake (°)	Slip (m)	Focal Mechanism
5.0 (0.3)	9.5 (0.1)	4.0 (0.1)	313 (1)	20 (1)	50.650 (0.001)	29.773 (0.001)	100 (2)	0.7 (0.1)	

To calculate the slip distribution of the source, using the linear inversion described in the previous section, we firstly extended the fault planes so that the source could include the entire slip distribution, from the peak value to zero, at the edges; the fault was then subdivided into patches of 1 × 1 km. According to the slip distribution (Figure 4), the maximum slip of approximately 1 m was located at a depth of 4.5 km; the slip distribution is constrained between a depth of 2.5 km and 6.0 km, thus confirming that the rupture did not reach the surface. The comparison of the observed and modeled displacements after the linear inversion is shown in Figure 5, along with a map of residuals; observed and predicted displacements differ by less than a centimeter on most of the map, but with very localized fluctuations up to 3 cm, due to the unavoidable approximations introduced by the elastic model.

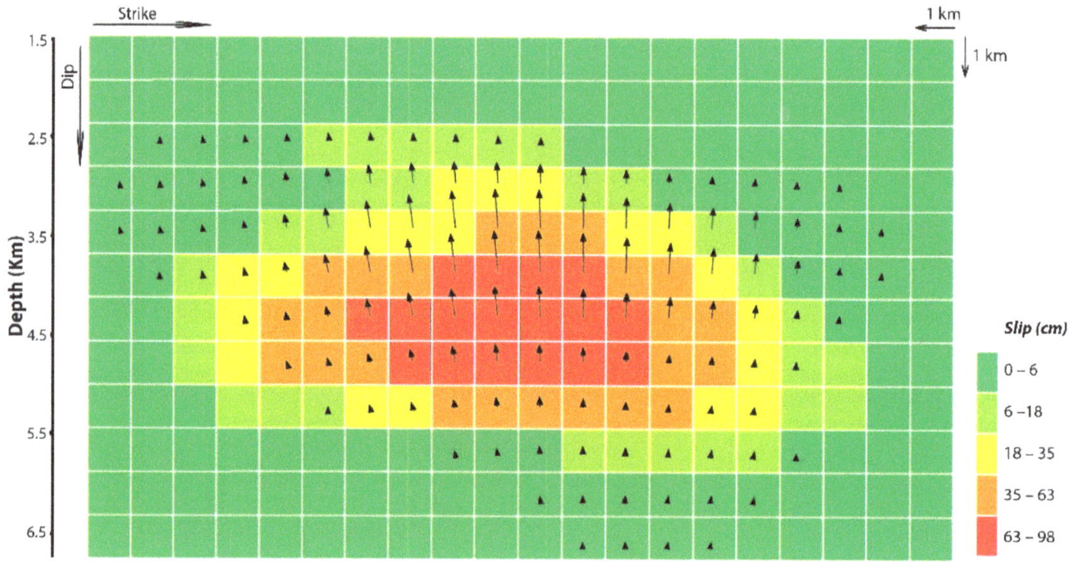

Figure 4. Slip distribution of the Mw 5.8 mainshock.

The geodetic moment corresponding to the slip distribution is 1.0×10^{18} Nm, close to the reported value of 6.7×10^{17} Nm by the USGS; the strike, dip and rake values retrieved through inversion (313°, 20°, 100°) are also in a good agreement with those calculated by the USGS (308°, 28°, 80°) and the Global CMT catalog (310°, 26°, 94°).

Additionally, we considered the hypothesis of constraining the source of the biggest aftershock, the M_W 5.0 event occurred May 21st, for which some deformation patterns were observed in both ascending and descending displacement maps (Figure 3). Though the displacement, from both orbits, has very low intensity (about 3 cm of maximum displacement), the location is close to the epicenter (Figure 3), suggesting a possible correlation of the signal with the seismic source. We therefore adopted the same modeling scheme used for the mainshock, in order to verify the compatibility of the detected signal with that predicted from a dislocation source. Unfortunately, no clear indications arose from the inversion: no fault configuration was found to unambiguously attribute the observed signal to a shear source. However, the presence of a similar pattern in both maps, obtained from completely independent image pairs, is unlikely to be caused by atmospheric artifacts, characterized by a very low correlation in time. It can be interpreted qualitatively as a post-seismic variation of the underlying volume, possibly caused by fluid migration.

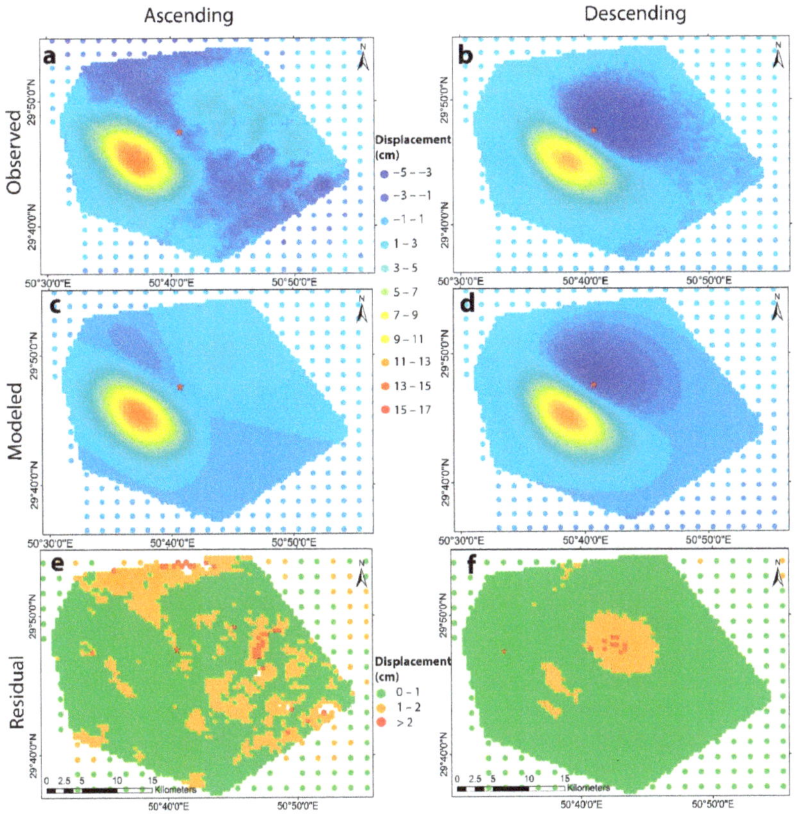

Figure 5. Observed, modeled and residual maps for displacements obtained from ascending and descending orbits for Mw 5.8 mainshock. The red star shows the location of mainshock.

4. Discussion and Conclusions

Geodetic observational data, such as InSAR measurements, are by now regularly used to infer information concerning sources of surface displacements and to understand the underlying processes. The significant co-seismic surface deformation signal of 18 April 2021, Bandar-e Genaveh earthquake, is clearly depicted by the InSAR images used in this study. The analysis of the mainshock co-seismic deformation demonstrated that the observed signal can be ascribed to a unique reverse fault with a 20° low dip to the northeast, across a broad ($\sim$9.5 $\times$ 5 km) rupture plane, corresponding to the pre-existing seismogenic thrust Zagros Foredeep Fault (Figure 6). In the southwest of Iran, it accommodates the convergence of low-depth values of density contrast discontinuity of the Arabian and Eurasian plates in the subduction zone [10,11].

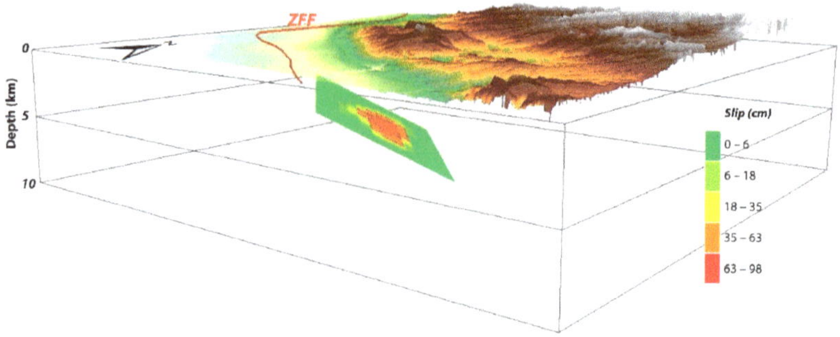

Figure 6. Fault slip distribution of Genaveh earthquake. ZFF are shown too. By extending the causative fault of the earthquake, we reach the ZFF fault line, which indicates that the activity in this seismically thrust active fault at a depth of about 5 km has triggered this earthquake through sedimentary rocks and the existence of alluvial deposits has caused rupture did not reach the surface.

The estimations of the causative fault agree with the previously known local structures, which cut simple sediment folding of Bakhtiari, Lahbari and Aghajari formations (Figure 1) in low folded zone in Dezful Embayment [17,22]. This fault can control the local fault-related folding and the frontal fold growth of large surface anticlines [21].

Despite the highly complex tectonic setting in the Zagros region [38] and the presence of large, active and hidden faults [39] with scattered seismicity [40], which sometimes do not follow the trend of the Zagros folds, we successfully demonstrated that the M_W 5.8 Genaveh mainshock occurred at a shallow dip angle subduction zone, located near the shortening Arabian–Eurasian Plate boundary on the Zagros Foredeep Fault. The used approach can help improve our understanding of the seismic potential of the Zagros region and refine the associated seismic hazards. It does not seem to be distinct from segments of the ZFF that generate steep range front topography elsewhere in the Zagros. According to [22], the deformation had propagated from MFF to Dezful Embayment in the late Miocene and based on our results it still continues.

The orientation of the retrieved fault plane is in good agreement with those published by other international Institutes (Table 1), and is in good agreement with the compressional tectonic regime in the area. The crustal seismicity occurring in the investigated area since 1900 with pure thrust faulting [21,41] is in agreement with our results. Its moment release rate does not greatly exceed those observed either instrumentally or historically and the formers show a seismic behavior similar to the one inferred for this zone [21]. The earthquakes along the ZFF mostly have ~20–60° dipping fault planes that is significantly similar to ours.

The results show that this moderate-size earthquake was located within the sedimentary cover in Zagros with a crustal thickness of ~10 km, a basement of ~25 km and a Moho depth of ~45 km [17,18,42], and does not reach to a Brittle/ Ductile transition zone with predominant dip of ~7° [17].

To get more detailed rupture slip distribution, the fault plane is subdivided into subfaults instead of just assuming single fault. As shown in Figure 4, the primary slip mechanism is dip–slip motion concentrated in a single main patch at a depth of 3.5–4.5 km, extending in the strike direction. There is no substantial slip shallower than 2.5 km, which confirms that this earthquake is associated with buried faults.

Even though InSAR is a very powerful instrument, it may not always be appropriate for estimating the effects of low to moderate events, such as earthquakes characterized by magnitudes ranging from 4.5 to <6. If the signature of the M_W 5.8 mainshock can be attributed to a blind rupture on a pre-existing and known fault, the analysis of the interferograms including the M_W 5.0 aftershock led to no evident conclusion such to attribute the earthquake to known structure. While the extent and amount of signal are compatible with an M 5.0 event at the expected hypocentral depth, as can be easily verified with a forward model, the deformation patterns cannot be satisfactorily reproduced with the inverse rupture, as expected in this area [1]. It is worth noting that these displacement patterns do not exceed 3 cm and can be significantly altered by the presence of common artifact InSAR sources that would obscure the pure co-seismic offset, or entirely due to surface effects not induced by a shear dislocation.

This earthquake highlights the need for more GPS coverage in this part of Zagros that could help differentiate strain accumulation on a regional low-angle structure and obtain the recurrence time of the main earthquake.

Author Contributions: Conceptualization, Z.G. and N.A.F.; methodology, S.A., Z.G. and N.A.F.; software, Z.G. and N.A.F.; validation, S.A., Z.G., N.A.F. and A.V.; formal analysis, N.A.F. and Z.G.; investigation, Z.G.; resources, N.A.F.; data curation, N.A.F. and Z.G.; writing—original draft preparation, Z.G.; writing—review and editing, Z.G., N.A.F., S.A. and A.V.; visualization, Z.G.; supervision, S.A.; project administration, Z.G.; funding acquisition, N.A.F. All authors have read and agreed to the published version of the manuscript.

Funding: This research received no external funding.

Institutional Review Board Statement: Not applicable.

Informed Consent Statement: Not applicable.

Data Availability Statement: Not applicable.

Acknowledgments: Sentinel data, from the ESA Copernicus Project, were obtained through the Open Access Hub. InSAR processing and data modeling were carried out with SARscape (sarmap, CH).

Conflicts of Interest: The authors declare no conflict of interest.

References

1. Berberian, M. Master 'blind' thrust faults hidden under the Zagros folds: Active basement tectonics and surface morphotectonics. *Tectonophysics* **1995**, *241*, 193–224. [CrossRef]
2. Agha Nabati, A. *Geology of Iran*; Geological Survey & Mineral Exploration of Iran: Tehran, Iran, 2004.
3. IIEES. International Institute of Earthquake and Engineering Seismology. Available online: http://www.iiees.ac.ir/fa/recentevents (accessed on 5 May 2021).
4. IGUT. Iranian Seismological Center (Institute of Geophysics, University of Tehran). Available online: http://irsc.ut.ac.ir (accessed on 5 May 2021).
5. Fialko, Y.; Sandwell, D.; Simons, M.; Rosen, P. Three-dimensional deformation caused by the Bam, Iran, earthquake and the origin of shallow slip deficit. *Nature* **2005**, *435*, 295–299. [CrossRef] [PubMed]
6. Zebker, H.A.; Rosen, P. On the derivation of coseismic displacement fields using differential radar interferometry: The Landers earthquake. *Int. Geosci. Remote Sens. Symp.* **1994**, *1*, 286–288. [CrossRef]
7. Massonnet, D.; Rossi, M.; Carmona, C.; Adragna, F.; Peltzer, G.; Feigl, K.; Rabaute, T. The displacement field of the Landers earthquake mapped by radar interferometry. *Nature* **1993**, *364*, 138–142. [CrossRef]
8. Ambraseys, N.N.; Melville, C.P. *A History of Persian Earthquakes*; Cambridge University Press: Cambridge, UK, 1982.
9. Hessami, K.; Jamali, F.; Tabasi, H. *Map of Major Active Faults of Iran at 1:2500000*; Seismology Research Center, International Institute of Earthquake Engineering and Seismology (IIEES): Tehran, Iran, 2003.
10. Malekzade, Z.; Bellier, O.; Abbassi, M.R.; Shabanian, E.; Authemayou, C. The effects of plate margin inhomogeneity on the deformation pattern within west-Central Zagros Fold-and-Thrust Belt. *Tectonophysics* **2016**, *693*, 304–326.
11. Eppelbaum, L.V.; Katz, Y.I. A New Regard on the Tectonic Map of the Arabian-African Region Inferred from the Satellite Gravity Analysis. *Acta Geophys.* **2017**, *65*, 607–626. [CrossRef]
12. Pollastro, R.M.; Persits, F.M.; Steinshouer, D.W. *Maps Showing Geology, Oil and Gas Fields, and Geologic Provinces of Iran*; U.S. Geological Survey: Reston, VA, USA, 1997. [CrossRef]
13. Setudehnia, A.; Perry, J.T. *Genaveh 1:100000 Geological Map*; National Iranian Oil Company: Tehran, Iran, 1967.
14. Vernant, P.; Nilforoushan, F.; Hatzfeld, D.; Abbassi, M.R.; Vigny, C.; Masson, F.; Nankali, H.; Martinod, J.; Ashtiani, A.; Bayer, R.; et al. Present-day crustal deformation and plate kinematics in the Middle East constrained by GPS measurements in Iran and northern Oman. *Geophys. J. Int.* **2004**, *157*, 381–398. [CrossRef]
15. Basilici, M.; Ascione, A.; Megna, A.; Santini, S.; Tavani, S.; Valente, E.; Mazzoli, S. Active deformation and relief evolution in the western Lurestan region of the Zagros mountain belt: New insights from tectonic geomorphology analysis and finite element modeling. *Tectonics* **2020**, *39*, e2020TC006402. [CrossRef]

16. Basilici, M.; Mazzoli, S.; Megna, A.; Santini, S.; Tavani, S. 3-D Geothermal Model of the Lurestan Sector of the Zagros Thrust Belt, Iran. *Energies* **2020**, *13*, 2140. [CrossRef]
17. Le Garzic, E.; Vergés, J.; Sapin, F.; Saura, E.; Meresse, F.; Ringenbach, J. Evolution of the NW Zagros Fold-and-Thrust Belt in Kurdistan Region of Iraq from balanced and restored crustal-scale sections and forward modeling. *J. Struct. Geol.* **2019**, *124*, 51–69.
18. Pirouz, M.; Avouac, J.-P.; Gualandi, A.; Hassanzadeh, J.; Sternai, P. Flexural bending of the Zagros foreland basin. *Geophys. J. Int.* **2017**, *210*, 1659–1680.
19. Tavani, S.; Parente, M.; Puzone, F.; Corradetti, A.; Gharabeigli, G.; Valinejad, M.; Morsalnejad, D.; Mazzoli, S. The seismogenic fault system of the 2017 Mw 7.3 Iran–Iraq earthquake: Constraints from surface and subsurface data, cross-section balancing, and restoration. *Solid Earth* **2018**, *9*, 821–831. [CrossRef]
20. Etemad-Saeed, N.; Najafi, M.; Vergés, J. Provenance evolution of Oligocene-Pliocene foreland deposits in the Dezful embayment to constrain Central Zagros exhumation history. *J. Geol. Soc.* **2020**, *177*, 799–817. [CrossRef]
21. Nissen, E.; Ghods, A.; Karasözen, E.; Elliott, J.R.; Barnhart, W.D.; Bergman, E.A.; Hayes, G.P.; Jamal-Reyhani, M.; Nemati, M.; Tan, F.; et al. The 12 November 2017 Mw 7.3 Ezgeleh-Sarpolzahab (Iran) earthquake and active tectonics of the Lurestan arc. *J. Geophys. Res. Solid Earth* **2019**, *124*, 2124–2152. [CrossRef]
22. Sun, J.; Sheykh, M.; Windley, B.F.; Talebian, M.; Cao, M.; Ahmadi, N.; Sha, J. Magnetostratigraphic age control of the timing of tectonic deformation and the shifting depositional environments in the Dezful Embayment, Iran. *Tectonics* **2022**, *41*, e2021TC006881. [CrossRef]
23. Najafi, M.; Vergés, J.; Etemad-Saeed, N.; Karimnejad, H.R. Folding, thrusting and diapirism: Competing mechanisms for shaping the structure of the north Dezful embayment, Zagros, Iran. *Basin Res.* **2018**, *30*, 1200–1229. [CrossRef]
24. Vergés, J.; Saura, E.; Casciello, E.; Fernàndez, M.; Villaseñor, A.; Jiménez-Munt, I.; García-Castellanos, D. Crustal-scale cross-sections across the NW Zagros belt: Implications for the Arabian margin reconstruction. *Geol. Mag.* **2011**, *148*, 739–761. [CrossRef]
25. Sheikholeslami, M.R.; Javadi, H.R.; Sarshar, M.A.; Hoseini, A.A.; Peyma, M.K.; Daneshmand, B.V. *Encyclopedia of Iranian Faults*; Geological Survey and Mineral Exploration of Iran: Tehran, Iran, 2013.
26. Massonnet, D.; Feigl, K.L. Radar interferometry and its application to changes in the earth's surface. *Rev. Geophys.* **1998**, *36*, 441–500. [CrossRef]
27. Farr, T.G.; Rosen, P.A.; Caro, E.; Crippen, R.; Duren, R.; Hensley, S.; Kobrick, M.; Paller, M.; Rodriguez, E.; Roth, L.; et al. The Shuttle Radar Topography Mission. *Rev. Geophys.* **2007**, *45*, RG2004. [CrossRef]
28. Goldstein, R.M.; Werner, C.L. Radar interferogram filtering for geophysical applications. *Geophys. Res. Lett.* **1998**, *25*, 4035–4038. [CrossRef]
29. Costantini, T.M. A novel phase unwrapping method based on network programming. *IEEE Trans. Geosci. Remote Sens.* **1998**, *36*, 813–821. [CrossRef]
30. Atzori, S.; Hunstad, I.; Chini, M.; Salvi, S.; Tolomei, C.; Bignami, C.; Stramondo, S.; Trasatti, E.; Antonioli, A.; Boschi, E. Finite fault inversion of DInSAR coseismic displacement of the 2009 L'Aquila earthquake (central Italy). *Geophys. Res. Lett.* **2009**, *36*, L15305. [CrossRef]
31. Okada, Y. Surface deformation due to shear and tensile faults in a half-space. *Bull. Seismol. Soc. Am.* **1985**, *75*, 1135–1154.
32. Williams, C.A.; Wadge, G. The effects of topography on magma chamber deformation models: Application to Mt. Etna and radar interferometry. *Geophys. Res. Lett.* **1998**, *25*, 1549–1552. [CrossRef]
33. Marquardt, D.W. An Algorithm for Least-Squares Estimation of Nonlinear Parameters. *J. Soc. Ind. Appl. Math.* **1963**, *11*, 431–441. [CrossRef]
34. Atzori, S.; Manunta, M.; Fornaro, G.; Ganas, A.; Salvi, S. Postseismic displacement of the 1999 Athens earthquake retrieved by the Differential Interferometry by Synthetic Aperture Radar time series. *J. Geophys. Res. Solid Earth* **2008**, *113*, B09309. [CrossRef]
35. Funning, G.J.; Parsons, B.; Wright, T.; Jackson, J.A.; Fielding, E. Surface displacements and source parameters of the 2003 Bam (Iran) earthquake from Envisat advanced synthetic aperture radar imagery. *J. Geophys. Res. Solid Earth* **2005**, *110*, B09406. [CrossRef]
36. Wright, T.J.; Lu, Z.; Wicks, C. Source model for the Mw 6.7, 23 October 2002, Nenana Mountain Earthquake (Alaska) from InSAR. *Geophys. Res. Lett.* **2003**, *30*, 18. [CrossRef]
37. Atzori, S.; Antonioli, A.; Tolomei, C.; De Novellis, V.; De Luca, C.; Monterroso, F. InSAR full-resolution analysis of the 2017–2018 M>6 earthquakes in Mexico. *Remote Sens. Environ.* **2019**, *234*, 111461. [CrossRef]
38. Alavi, M. Structures of the Zagros fold-thrust belt in Iran. *Am. J. Sci.* **2007**, *307*, 1064–1095. [CrossRef]
39. Karasözen, E.; Nissen, E.; Bergman, E.A.; Ghods, A. Seismotectonics of the Zagros (Iran) From Orogen-Wide, Calibrated Earthquake Relocations. *J. Geophys. Res. Solid Earth* **2019**, *124*, 9109–9129. [CrossRef]
40. Madahizadeh, R.; Mostafazadeh, M.; Ansari, A. Long-Term Seismicity Behavior of the Zagros Region in Iran. *Pure Appl. Geophys.* **2016**, *173*, 2637–2652. [CrossRef]
41. Palano, M.; Imprescia, P.; Agnon, A.; Gresta, S. An improved evaluation of the seismic/geodetic deformation-rate ratio for the Zagros Fold-and-Thrust collisional belt. *Geophys. J. Int.* **2018**, *213*, 194–209. [CrossRef]
42. Talebian, M.; Jackson, J. A reappraisal of earthquake focal mechanisms and active shortening in the Zagros mountains of Iran. *Geophys. J. Int.* **2004**, *156*, 506–526.

applied sciences

MDPI

Article

The Extended Data-Adaptive Probability-Based Electrical Resistivity Tomography Inversion Method (E-PERTI) for the Characterization of the Buried Ditch of the Ancient Egnazia (Puglia, Italy)

Marilena Cozzolino [1,*], Paolo Mauriello [1] and Domenico Patella [2]

[1] Department of Human, Social and Educational Science, University of Molise, Via De Sanctis, 86100 Campobasso, Italy; mauriello@unimol.it
[2] Faculty of Economics, Universitas Mercatorum, Piazza Mattei 10, 00186 Rome, Italy; domenico.patella@unimercatorum.it
* Correspondence: marilena.cozzolino@unimol.it

Abstract: A geoelectrical survey was carried out outside the walls of the ancient Egnazia (Puglia, Italy) with the aim of enriching the knowledge about its defense system. Nine Electrical Resistivity Tomographies (ERTs) were realized using the dipole-dipole (DD) electrode array, approximately transversal to the walls and equally spaced. The new Extended data-adaptive Probability-based Electrical Resistivity Tomography Inversion Method (E-PERTI) was applied, for the first time, to model the resistivity distribution of a large dataset. Considering some peculiar aspects of the general theory of probability, an optimization of results was reached giving major emphasis to one dataset portion rather than another and inspecting selectively vertical or lateral resistivity variations. In this way, sets of aligned resistivity lows attributable to the trace of an ancient ditch were found.

Keywords: archaeological prospection; resistivity tomography; E-PERTI; Egnazia

1. Introduction

The archaeological site of Egnazia is located on the Adriatic coast of Puglia region (Southern Italy) [1] (Figure 1). The coastal landscape is characterized by a sequence of sub-horizontal surfaces sloping from the Murge Plateau (locally about 400 m high) towards sea from an elevation of about 120 m. These surfaces are the result of marine abrasion during the Middle Pleistocene because of the superimposition of regional uplift and the glacio-eustatic sea level changes [2]. In addition, a relict drainage network characterized by deep valleys, locally named gravine, parallel to each other and perpendicular to the coastline, were shaped for sapping processes [3,4]. The stratigraphy of the Egnazia area is characterized at the base by the presence of the Calcare di Bari unit (Cretaceous), which represents the local basement, marked by karstic processes. Biocalcareous sandstone related to the Calcarenite di Gravina unit (Lower Pleistocene) overlies in transgression the limestone and outcrops extensively along the cost [5].

Mentioned by authors such as Pliny, Strabone, Horace, the city of Egnazia had great importance in the ancient world due to its geographical position. Thanks to the presence of the port and the Via Traiana, it was, in fact, an active center of trade and commerce.

The history of ancient Gnathia has unfolded over many centuries [6]. The first settlement, consisting of a village of huts on the hill of the acropolis, was built in the 15th century BC (Bronze Age). The village itself was rebuilt in the 14th century BC following a large fire [7]. Clayey beaten, post holes and a refractory cooking surface carved directly into the rock, attest to the presence of nuclei in different areas of the area dating back to the late Bronze Age [8].

Figure 1. Survey plan on a 2008 TerraItaly™ Orthophoto with indication of the 1–9 parallel 2D Electrical Resistivity Tomography (ERT) lines (in yellow).

In the 11th century BC (Iron Age), the invasion of populations from the Balkan area, the Iàpigi, is recorded, while with the 8th century BC begins the Messapian phase. From this period, besides the necropolis along the coast on the sides of the acropolis and in the southern area [9], the mighty defense walls remain, 7 m high, 2 km long, delimiting the three inland sides of a nearly trapezoidal urban area of about 40 hectares, being the fourth side coinciding with the shoreline (Figure 1). It enclosed a discontinuous residential fabric with non-straight road layouts, alternating with agricultural spaces, pastures, areas of worship and funerary ones, a typical arrangement of the Messapian cities of Salento [7]. The fortification dates back to the beginning of the 4th century BC and consists of a double retaining wall, an embankment of stones, a wall in square blocks and a ditch. At the end of the 4th century BC or at the beginning of the 3rd century BC the ditch was filled, a double row wall was built and a new and wider ditch was dug, as proved by an archaeological excavation in 1978 in the northernmost edge of the walls [10]. The new defensive structure had therefore covered and incorporated the old one, thus increasing the safety of the city.

Starting from the 3rd century BC, Egnazia will cease with the Roman occupation and the most important urban development occurs. The period of growth is accentuated with the assumption of the status of municipia after the end of the social war in 89 BC, a condition that suggests the need for an arrangement of administrative centers adapted to the new needs [6]. The main signs of this change concern the civil basilica, the forum, the enhancement of the port (starting with Agrippa), the creation of two piers and, probably, also the construction of the so-called cryptoporticus [7]. In the late Republican age and the early imperial age (2nd century BC–1st century AD), a more systematic organization took place. In fact, the arrangement of the road network, in particular from via Minucia, allows for a more defined subdivision of residential districts and public areas (Del Monte & Randino, 2014): the new insulae thus include houses, artisanal structures and commercials; the burial areas undergo transformations, with the expansion of the existing necropolis and the creation of the western necropolis that extends outside the walls; in the Augustan age there was the organization of the core of the forum with the construction of the public baths on the southern side [6,7].

A new functional renovation is promoted by Trajan between 108 and 110 AD; of all the initiatives, the most significant is the reorganization of the main artery which will take its name from the emperor and resumes the oldest road route of the Via Minucia.

Only towards the end of the 4th century AD, the city undergoes a period of pause in settlement following the destruction of numerous sectors caused by traumatic events of natural origin (earthquake with a tsunami of 365 AD). Already from the first years of the 5th century, new developments took place, including those of artisanal and commercial systems, the monumental strengthening of Christian buildings, the promotion of economic activities and commercial exchanges, architectural interventions [6,7].

In a similar context, thanks to the absence of modern overlays, geophysical prospection can be a useful tool for the subsurface mapping of buried features enhancing the knowledge of the archaeological sites. Thus, in order to ascertain the presence of the ditch along a portion of the southern side of the fortification, a geoelectrical survey has been performed, consisting of nine parallel 2D Electrical Resistivity Tomography (ERT) lines, located as in Figure 1.

In this paper, the Extended data-adaptive Probability-based Electrical Resistivity Tomography Inversion (E-PERTI) [11] has been applied for imaging the buried features. E-PERTI is the latest development within the probability tomography imaging approach as adapted to geoelectrical methods [12–15].

The probability tomography imaging was at the beginning used to characterize the occurrence probability of resistivity departures from a reference background resistivity. Although not capable of providing estimates of the true resistivity of the buried bodies, it proved useful to highlight the shape and position of anomalous resistivity highs and lows in different geological contexts [16–18].

The next step was the launch of the data-adaptive Probability-based ERT Inversion (PERTI) method, which was formulated with the aim of giving an estimation of the most probable resistivity values within a surveyed volume as weighted averages of the whole apparent resistivity dataset [19]. Using this approach, different near-surface prospections were successfully processed and interpreted [20–24].

In general, the PERTI method cannot be considered a standard tomographic inversion scheme. It does not make any reference to the geometry rank of the anomaly sources as an a priori constraint to the imaging algorithm, unlike other methods that were developed to work on this a priori characteristic of the expected geometry of the model such as the minimum gradient support inversion [25,26], the blocky inversion (L1 norm) [27], or the covariance-based inversion [28]. Conceptually, the PERTI method relates only to the pure physical aspects of the electric field interaction with the subsoil's structures as it is sensed on the ground. More precisely, the most probable resistivity values are estimated point by point in the surveyed volume, based exclusively on the field apparent resistivity dataset. The approach is thought independent of the true geometry of the bodies, the full resolution of which stands in principle well beyond the intrinsic illumination power of the geoelectrical method.

The E-PERTI method has at last been developed in order to enhance robustness to noise with respect to the original PERTI, as well as to optimize resistivity estimates. It consists of processing, with the PERTI algorithm, many different subsets of data extracted from the apparent resistivity dataset both randomly and by sequential vertical scanning and horizontal windowing across the datum space [11]. The result is a more or less dense cluster of resistivity values in each point of the surveyed volume, used at last to predict the best estimation of the most probable resistivity in the same point by an intrinsic linear regression model with ordinary least squares techniques.

In this paper, the first application of the E-PERTI scheme to a large apparent resistivity field dataset is illustrated. In the following, at first, a description of the ERT survey technique and a summary of the E-PERTI tool starting from the original PERTI theory are provided, then the explanation of the ERT plan designed in the Egnazia survey site and the presentation and discussion of the results reached from the ERT data inversion are shown.

2. The ERT Survey Technique

Geoelectrical surveys are among the most suitable geophysical tools for the study of the subsoil. The aim is to get information regarding the geometry and localization of buried structures in the light of their intrinsic electrical resistivities. A geoelectrical field survey consists of measuring on the ground the potential difference across a pair of electrodes, generated by a direct current injected in the subsoil through a second pair of electrodes. By changing the positions and spacing of the electrodes according to specific rules, horizontal and vertical resistivity changes in the subsoil can be investigated. Presently, the ERT approach is used routinely, thanks to the latest technological developments that make it a fast, versatile and cost-effective method for depicting resistivity structures. The dipole-dipole (DD) electrode configuration moved on the ground along a profile is the most suited array for gathering ERT data. A dipole is a paired electrode set with the two electrodes located relatively close to one another. The convention for a DD array is to maintain an equal amplitude for both the current (active) dipole and the potential (passive) dipole along the profile, with the distance between the inner electrodes being an integer multiple of the dipole amplitude (see Figure 2).

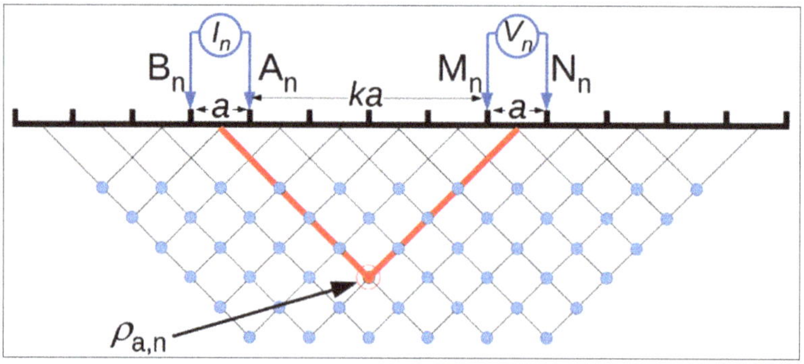

Figure 2. Schematic depiction of the dipole-dipole electrode array.

A primary advantage of the DD array is the ease of deployment in the field due to the short wire lengths needed to connect the active and passive dipoles to the measuring ammeter and voltmeter, respectively. Furthermore, it allows fast data acquisition as it supports multiple receiver channel measurements simultaneously. The DD layout is also largely more sensitive to lateral resistivity contrasts than other conventional arrays. For this reason, it can be particularly effective in archaeological contexts whose goal is often to highlight anthropogenic structures (walls, cisterns, tombs, trenches, etc.) that cause strong lateral resistivity contrasts at the boundary with the hosting subsoil. Nonetheless, it should be pointed out that the dipole-dipole array is very poor in mapping horizontal structures. Furthermore, the main disadvantage of this array is the decrease in signal strength as the distance between the dipole pair increases. However, in ERT prospecting for archaeology, both for the typical small size and closed shape of the target bodies and for the modest depths involved, these drawbacks appear largely minimized.

The first-instance result from a DD ERT survey is the so-called pseudosection, which is a straightforward representation of the well-known apparent resistivity parameter. On a not uniform subsoil, the apparent resistivity results to be a sort of complex weighted average of the true resistivities of the buried bodies, varying as the dipoles disposition along the profile. Referring to Figure 2, for a generic nth measurement with the current electrodes in the position A_n and B_n, and the potential electrodes in the position M_n and N_n, the apparent resistivity $r_{a,n}$ is calculated as $r_{a,n} = \pi k(k+1)(k+2)a(V_n/I_n)$, where I_n is the intensity of the direct current injected into the ground from the active dipole and V_n is the potential drop across the passive dipole. Each $r_{a,n}$ value is attributed at the midpoint between the two dipoles at a depth half their mutual distance, as in Figure 2. The plotted $r_{a,n}$ values are then contoured and colored by any conventional interpolation software. The resulting pseudosection represents a rough and sometimes severely distorted image of the true resistivity distribution of the subsurface. It thus needs to be converted by an inversion software into a true resistivity section in order to create a realistic image of the subsoil.

3. Outline of the E-PERTI Method

The E-PERTI method [11] is an extension of the PERTI method [19] that in turn was directly derived from the resistivity probability tomography approach [13]. The latter consisted in the determination of the resistivity anomaly occurrence probability function η_m calculated in the space below the surface composed of M contiguous cubic cells with equal volume ΔV, each identified by the location of its center (x_m, y_m, z_m) ($m = 1, 2, \ldots, M$) and its intrinsic resistivity ρ_m. In addition, the dataset was supposed to be composed of N apparent resistivity measurements, $\rho_{a,n}$, each identified by a running index n ($n = 1, 2, \ldots, N$) determined by the position of the electrode device on the ground. Thus, the η_m was defined as:

$$\eta_m = C_m \sum\nolimits_{n=1}^{N} \Delta\rho_{a,n} \left(\partial\rho_{a,n}^{ref} / \partial\rho_m\right), \quad (1)$$

where

$$C_m = \left[\sum_{n=1}^{N} (\Delta\rho_{a,n})^2 \sum_{n=1}^{N} \left(\partial\rho_{a,n}^{ref} / \partial\rho_m\right)^2\right]^{-1/2}. \quad (2)$$

In Equation (1), $\Delta\rho_{a,n}$ is the difference between $\rho_{a,n}$ and the reference apparent resistivity $\rho_{a,n}^{ref}$, computed at the same node as for $\rho_{a,n}$ using a reference model, and $\Delta\rho_m$ is the difference between the intrinsic resistivity ρ_m and the resistivity ρ_m^{ref} of the reference model in the mth cell.

At the base of the PERTI method, there is the assumption that the reference uniform resistivity is supposed to be the unknown value ρ_m that corresponds to a generic mth cell centered at (x_m, y_m, z_m), allowing to rewrite η_m as reported in [14]:

$$\eta_m = C_m \sum_{n=1}^{N} (\rho_{a,n} - \rho_m) \left(\partial\rho_{a,n}^{ref} / \partial\rho_m\right). \quad (3)$$

The rationale for the PERTI is that if at a point (x_m, y_m, z_m) it results $\eta_m = 0$, then in the cell centered at (x_m, y_m, z_m) the probability to find an increase or a decrease of the resistivity with respect to ρ_m is zero. In other words, in that cell, the intrinsic resistivity does not differ from ρ_m. Thus, referring to Equation (2), since C_m is always different from zero, the $\eta_m = 0$ condition leads to

$$\rho_m = \sum_{n=1}^{N} \rho_{a,n} \left(\partial\rho_{a,n}^{ref} / \partial\rho_m\right) / \sum_{n=1}^{N} \left(\partial\rho_{a,n}^{ref} / \partial\rho_m\right) \quad (4)$$

which represents the required solution for the application of the PERTI method. We only need to change repeatedly the coordinates (x_m, y_m, z_m) to retrieve, point by point, the resistivity pattern within V [19].

Unlike deterministic data inversion tools, starting from the pioneering work by Loke and Barker [29] onward, the PERTI method follows a probabilistic paradigm, totally independent of a priori information. Therefore, in principle, the two approaches are not comparable, despite the fact that in common practice, decisions are taken on the basis of the ERT reconstructions anyway produced. It is just to comply with this practical reason, that a comparison between PERTI and the well-known ERTLAB and RES2D/3DINV commercial software was made on synthetic models, showing that PERTI is generally as efficacious as the two previous tools in detecting, distinguishing and shaping the sources of the observed apparent resistivity anomalies [19].

Shortly, the main features of the PERTI method are [19]: (i) unnecessity of a priori reference models; (ii) full, unconstrained adaptability to any kind of dataset, acquisition technique and spatial regularity; (iii) drastic reduction of computing time, thus allowing for in-field real-time inversion.

It is now worth expressing a few words about the first of the PERTI peculiarities listed above. Specifically, we refer to the concept of depth of investigation (DOI), which is associated with an empirical index, based on at least two classic inversions of the dataset, attributing different values to the resistivities of the reference model [30–32]. The DOI is considered essential to provide information on how well the subsurface volumes are covered by the data and to discern between similar laterally correlated resistivity values. With PERTI, however, since it is a purely data-adaptive method not needing any reference model to start with, such a DOI index cannot be calculated.

Therefore, in order to improve the accuracy of the method, the extension of PERTI (E-PERTI) [11] was developed considering the Bernoulli conceptual scheme in probability theory [33]. According to this scheme, each single geoelectrical measurement is considered as a "test", that is, the realization of a specific set of conditions that can give rise to an elementary event of a space U of elementary events (test space). Grouping a set of N tests such as a resistivity dataset, each measurement is part of a new space of tests, called U_N, composed by arbitrary points of the space U. In addition, in ideal conditions, as every nth geoelectric measurement is an individual process, we can consider the N apparent resistivity tests as automatically satisfying the condition of mutual independence. Hence, if N given tests are independent, then any Q tests arbitrarily extracted from the given ones are still independent [33].

Assuming $N_q \leq N$ for $q = 1, 2, \ldots, Q$ and Q are arbitrarily large, a set of N_q data (tests) within the N available data is extracted. Applying the PERTI approach, for each N_q set Equation (3) becomes

$$\eta_{m,q} = C_{m,q} \sum_{k=1}^{N_q} (\rho_{a,k} - \rho_{m,q}) \left(\partial \rho_{a,k}^{ref} / \partial \rho_{m,q} \right). \tag{5}$$

Thus, letting

$$\eta_{m,q} = 0 \text{ for each } q = 1,\ 2,\ \ldots,\ Q, \tag{6}$$

we obtain Q estimates of ρ_m, say

$$\rho_{m,q} = \sum\nolimits_{k=1}^{N_q} \rho_{a,k} \left(\partial \rho_{a,k}^{ref} / \partial \rho_{m,q} \right) / \sum\nolimits_{k=1}^{N_q} \left(\partial \rho_{a,k}^{ref} / \partial \rho_{m,q} \right), \tag{7}$$

from which to derive the best estimator of ρ_m, using, for example, the standard least squares procedure for the determination of the slope of a linear equation of the type

$$y = \rho_{m,q} x, \tag{8}$$

where

$$x = \sum\nolimits_{k=1}^{N_q} \left(\partial \rho_{a,k}^{ref} / \partial \rho_{m,q} \right), \tag{9}$$

$$y = \sum\nolimits_{k=1}^{N_q} \rho_{a,k} \left(\partial \rho_{a,k}^{ref} / \partial \rho_{m,q} \right). \tag{10}$$

Some results are now worth being recalled from [11] about the main advantages of extending the PERTI method, that is, about the role of the E-PERTI linear least squares best fit expressed by Equation (8), which is directly based on the PERTI formula (4).

Using a DD ERT simulation, it was shown that by extracting many different sets of N_q data from the N-data original pseudosection, with $N_q < N$, more or less large spreads of the true resistivity values were observed in all of the common points of the N_q-data PERTI sections. In principle, such a spread was assumed as clear evidence of how much an inversion process may depend in an unpredictable way on the profile extent, array setup and data density. At the same time, it was proof of how important is to dispose of a ductile, reliable statistical tool, such as the E-PERTI, straight resulting from the adopted inversion process.

It was also shown that by corrupting the same N-data pseudosection with a random error in one sector, the E-PERTI section resulting from combining a set of N_q-data PERTI sections, was visibly a more realistic image than the original N-data PERTI section, as to position and true resistivity of the buried bodies. Such a result has been assumed as proof of the robustness of the proposed scheme against noise.

In conclusion, E-PERTI has been revealed to be a dynamic and objective tool, whose results can eventually be used as reference models in any other inversion algorithm.

4. The ERT Plan

Nine ERT profiles have been carried out using an MAE A3000™ resistivity meter (www.mae-srl.it, accessed on 25 October 2021). The profiles were arranged on the soil nearly orthogonal to the ancient walls at a mutual distance between 10 m and 16 m and with length from 22 m to 38.75 m, depending on the availability of areas free from obstacles. The apparent resistivity data were taken by a dipole-dipole configuration using an array of 23 up to 32 electrodes regularly spaced 1 m, 1.2 m or 1.25 m, according to the length of the profiles. For all profiles, a maximum of 15 steps were used as spacing between the inner electrodes of the two dipoles. Overall, about 2600 apparent resistivity data ranging between 5.66 Ωm and 324.38 Ωm were thus collected.

Table 1 lists all of the acquisition parameters of the ERT profiles, numbered as in Figure 1, and the minimum and maximum apparent resistivity along each profile.

Table 1. List of the acquisition parameters of all of the ERT profiles drawn in Figure 1.

Profile No.	Length (m)	Step (m)	Electrodes (No)	Datum Points (No)	$\rho_{a,min}$ (Ωm)	$\rho_{a,max}$ (Ωm)
1	33.6	1.2	29	285	32.3	293.44
2	26	1	27	255	5.66	310.43
3	22	1	23	195	6.24	178.8
4	26	1	27	255	10.63	198.31
5	31	1	32	330	7.71	209.71
6	34.8	1.2	30	330	8.6	175.42
7	37.2	1.2	32	330	10.6	213.44
8	38.75	1.25	32	330	5.74	167.3
9	38.75	1.25	32	330	13.71	324.38

5. Results and Discussion

In Appendix A, for each profile, the following pictures are shown:

- ERT field pseudosections (Figures A1, A4, A9, A10, A13, A18, A19, A22 and A25);
- PERTI sections obtained from data subsets made of N_q tests belonging to restricted data windows distributed along each pseudosection (left-side images in Figures A2, A5, A8, A11, A14, A17, A20, A23 and A26);
- PERTI sections obtained from data subsets made of N_q tests belonging to horizontal stripes of each pseudosection with progressively increasing thickness starting from the top (vertical scanning) (middle images in Figures A2, A5, A8, A11, A14, A17, A20, A23 and A26);

- PERTI sections obtained by averaging point by point 20 tomographies elaborated using randomly extracted N_q tests from the N starting dataset (right-side images in Figures A2, A5, A8, A11, A14, A17, A20, A23 and A26). The method of random selection considered also duplicate data when extracting subset;
- Final E-PERTI sections resulting from the estimation of ρ_m with a linear best-fit procedure applied point by point to the resistivity values belonging to the PERTI sections obtained by vertical scanning. In fact, particular emphasis was given to the vertical scanning approach with the ultimate goal of obtaining horizontal two-dimensional sections at gradually increasing depths compatibly with the acquisition parameters of the tomographies. For this type of objective, the procedure of the vertical scanning of the sections seemed to us the most suitable in order to show the horizontal trend of the ditch (Figures A3, A6, A9, A12, A15, A18, A21, A24 and A27).
- A common logarithmic scale has been used for visualization of both apparent resistivity and modeled real resistivity maps and, in order to enhance the resistivity contrasts, the same color sequence within the same $[\rho_{a,min}, \rho_{a,max}]$ interval has been assigned to each set.

Regarding the E-PERTI application, the rationale at the base of the new approach is to inspect in greater detail the electrical nature of the subsoil analyzing different subsets of an ERT dataset. To this respect, it is also worth recalling that with the original PERTI tool the resistivity values that result from using the whole ERT dataset were initially assumed as the most probable values, according to Equation (4). With the E-PERTI, instead, the best estimate of the most probable resistivity values, following Equation (8), is the result from a linear least-squares best-fit of a more or less spread distribution of resistivity values obtained in each single point by the PERTI tool applied to all selected subsets, the whole dataset included.

As regards the modality of selection of the subsets from the original ERT dataset, the horizontal stepwise windowing has the meaning of repeatedly applying the PERTI algorithm to different limited portions of a pseudosection [11]. The longer the pseudosection, the higher can be the number of windows opened along it, which can be contiguous or even overlapping. Of course, in each window, the relative subset vertically includes all the data from the surface down to the latest pseudodepth. The purpose is to get, from PERTI, resistivity estimates inside any window not influenced by points outside. This modality applied to the Egnazia case study has allowed us to draw the left-side images in Figures A2, A5, A8, A11, A14, A17, A20, A23 and A26.

The second conceived approach is the vertical scanning, which means that the PERTI algorithm is repeatedly applied to horizontal stripes of a pseudosection with progressively increasing thickness starting from top-down to the latest pseudodepth, that is, the whole pseudosection [11]. Therefore, the aim of the vertical scanning is just to observe how deeper points as stripes become thicker influence resistivity at shallow levels. For the case under study, the results of this approach are shown in the middle sections in Figures A2, A5, A8, A11, A14, A17, A20, A23 and A26. In the images, the last section obtained by a vertical scanning of the pseudosection has not always been processed using the entire data set so the result differs slightly from the final E-PERTI sections.

The last considered is the random extraction approach, consisting in applying the PERTI algorithm repeatedly to different subsets made of a varying number of N_q data (tests) extracted from the whole pseudosection dataset ($N_q < N$) [11]. The main purpose is to check for the stability of the resistivity nuclei in the various sections. Wandering or sporadic nuclei are likely to represent artifacts because of their variability in shape and intensity. At Egnazia, this last approach has provided the right-hand sequence of images in Figures A2, A5, A8, A11, A14, A17, A20, A23 and A26.

By means of these entire PERTI results, the last disposes of a cluster of resistivity values in each point of the section, which can even manifest some large dispersion, as amply shown in [11]. The spreading of the values is a consequence of the different roles that each partial PERTI can have in providing the most probable value of the true resistivity in a point. It follows that a statistical estimation of it is needed, using the best-fit linear regression tool, according to Equation (8). A more refined approach would be the use of a weighted best-fit regression, by assigning a damping or enhancing multiplier to all those resistivity values coming from visibly noise-corrupted PERTI, and/or from strongly depth-dependent PERTI sections if the research interest is, for example, addressed to the shallowest subsoil portion, as in this case study. Although a weighted best-fit may be seen as introducing some subjectivity into the process, in principle it does not conflict with the axioms of probability [11]. In field practice, a careful choice of different weights can help achieve a sounder interpretation of the results. Under this aspect, E-PERTI reveals to be a dynamic tool, which can also be of support in other inversion algorithms.

The final E-PERTI sections of the nine ERT profiles performed at Egnazia, shown in the Figures A3, A6, A9, A12, A15, A18, A21, A24 and A27, allow to single out the presence of a well-confined conductive body with resistivity ranging between 5 Ωm and 30 Ωm.

As the thickest stripes that have been processed by the vertical scanning approach coincide with the full pseudosections, the relative PERTI sections (bottom section of the central sequence in Figures A2, A5, A8, A11, A14, A17, A20, A23 and A26) can be used for comparison to evaluate the quality of the E-PERTI sections. At glance, the E-PERTI sections appear rather similar to the PERTI ones. However, a careful inspection allows one to recognize the ability of the new method to isolate and better define the limits of the buried conductive mass, as mainly in ERTs 1, 5, 6, 7, 9 (Figures A3, A15, A18, A21 and A27). Given that the 3D reconstruction of this shallow confined conductive body represents the final goal of the ERT survey at Egnazia, it can reasonably be assumed to correspond with the loose sandy-clayey sediments filling the ancient ditch.

In Figure 3, a 3D view of the E-PERTI sections is reported. As depicted by a dotted line, the high-conductivity volumes are distributed along a route that is parallel to the walls and that can be associated with the filling of the ancient ditch of the city, which had been supposed by archaeologists to lie close to that line.

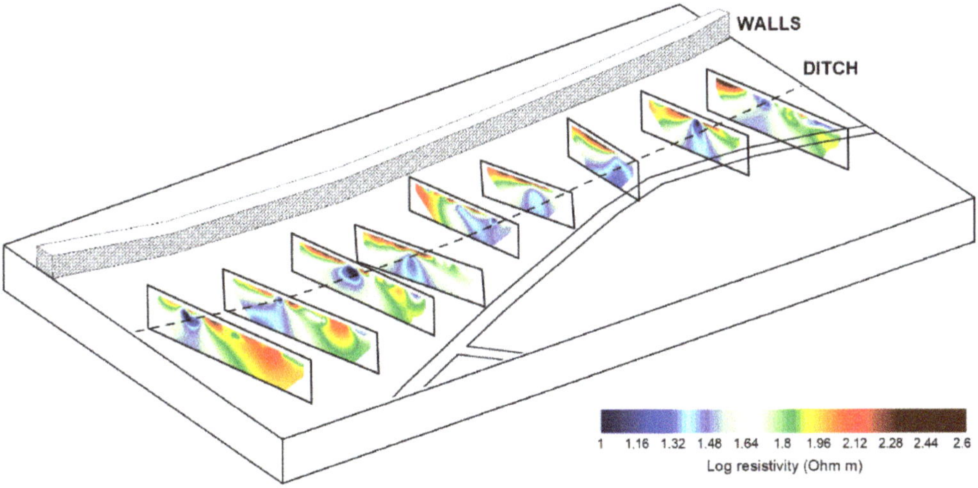

Figure 3. 3D assemblage of the E-PERTI sections.

Finally, by assembling all DD profiles, the E-PERTI solution of the entire apparent resistivity dataset has provided a 3D ERT model that in Figure 4 is shown as a sequence of horizontal slices at increasing depth from 2 m down to 5 m beneath the ground level, while in Figure 5 is depicted in full compact form. In both these last two figures, in order to better highlight the resistivity lows attributable to the presence of the ditch, the color scale has been modified, by compressing the left-hand half of the log scale below 1.8 and leaving the right-hand half colorless.

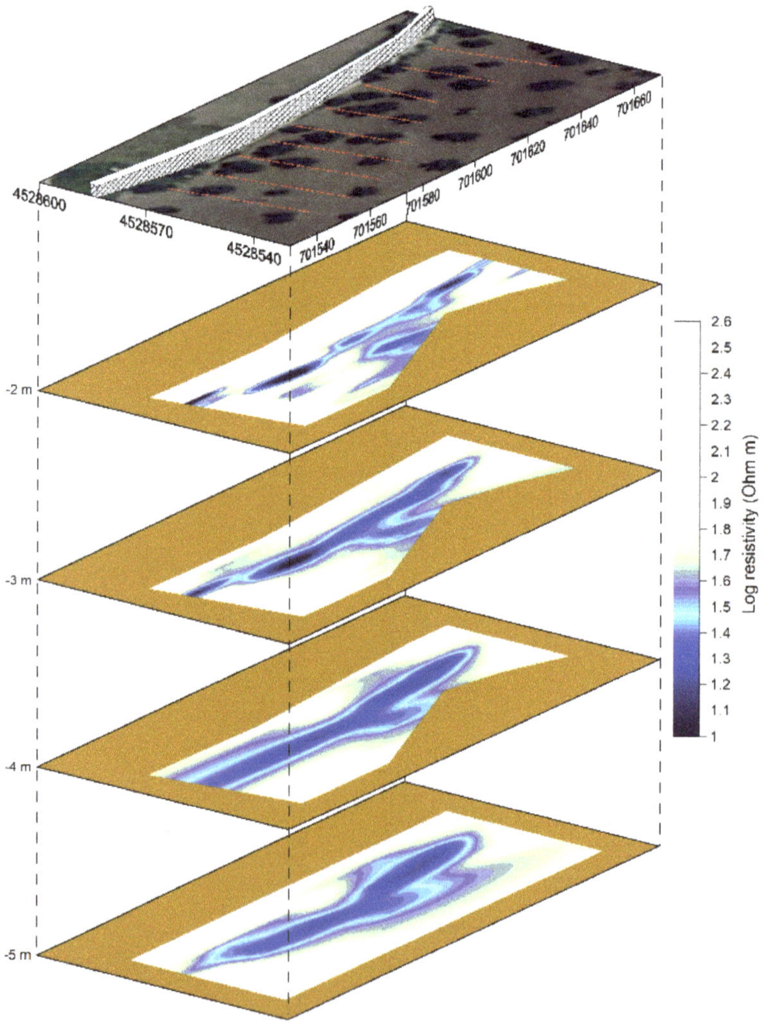

Figure 4. 3D view of the low resistivity body, obtained by the assemblage of the E-PERTI sections, ascribed to the ancient ditch as a sequence of horizontal slices at increasing depth (distances in meters).

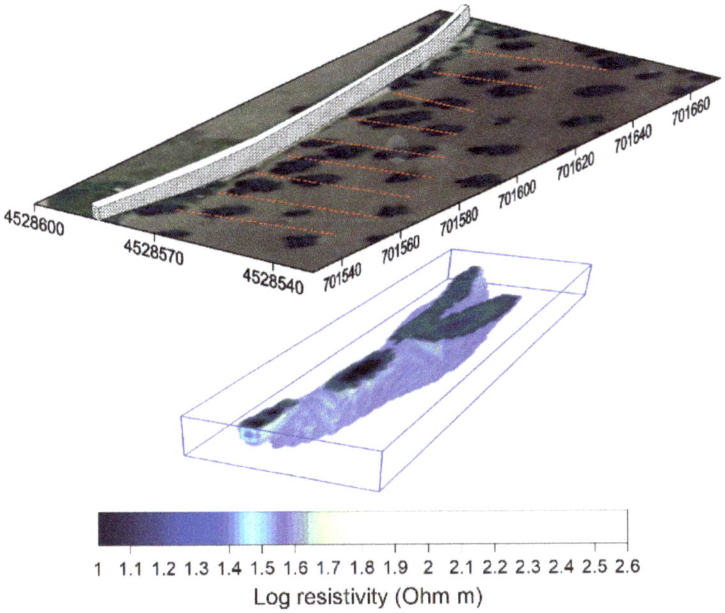

Figure 5. 3D view of the low resistivity body, obtained by the assemblage of the E-PERTI sections, ascribed to the ancient ditch in full compact form (distances in meters).

6. Conclusions

In this paper, for the first time, an ERT case study in archaeology in which the E-PERTI method was fully adopted to process and interpret the field survey dataset was shown. The purpose was to detect and model the course of the buried ancient ditch hypothesized to exist just outside a southern portion of the walls of the 4th century BC, built in defense of the ancient urban area of Egnazia (Puglia, Italy).

With the E-PERTI method, an interpretative model derived from the statistical determination of the most probable resistivity values within the surveyed space was proposed. This approach is theoretically justified by the probabilistic paradigm upon which the original PERTI method is based, which allows the extraction of a wealth of independent data subsets from the field dataset, following Bernoulli's conceptual scheme in probability theory.

In practice, the most probable model resistivity at each point of the surveyed space has been determined by a least-squares approach aimed at finding the slope of a linear equation that relates the multiplicity of the provisional resistivity estimates to the partial data subsets. Needless to say that the term "most probable" used to qualify the resistivity values of the model is a relative concept, as it intrinsically depends on amount and accuracy of the measures forming the entire initial database.

At last, the present field case history allows the confirmation of what was stated in the previous methodological study [11]. That is to say that E-PERTI is a robust approach to noise dejection and that the technique of extraction of subsets of the data is an easy viable technique for optimizing the modeling process in ERT applications. Ultimately, if necessary, the result deriving from the application of the E-PERTI method can be used as a valuable reference model in any standard iterative inversion process.

Author Contributions: Conceptualization, M.C., P.M. and D.P.; Data curation, M.C., P.M. and D.P.; Formal analysis, M.C., P.M. and D.P.; Methodology, M.C., P.M. and D.P.; Software, M.C., P.M. and D.P.; Validation, M.C., P.M. and D.P.; Writing—original draft, M.C., P.M. and D.P.; Writing—review & editing, M.C., P.M. and D.P. All authors have read and agreed to the published version of the manuscript.

Funding: This research received no external funding.

Institutional Review Board Statement: Not applicable.

Informed Consent Statement: Not applicable.

Data Availability Statement: Not applicable.

Conflicts of Interest: The authors declare no conflict of interest.

Appendix A

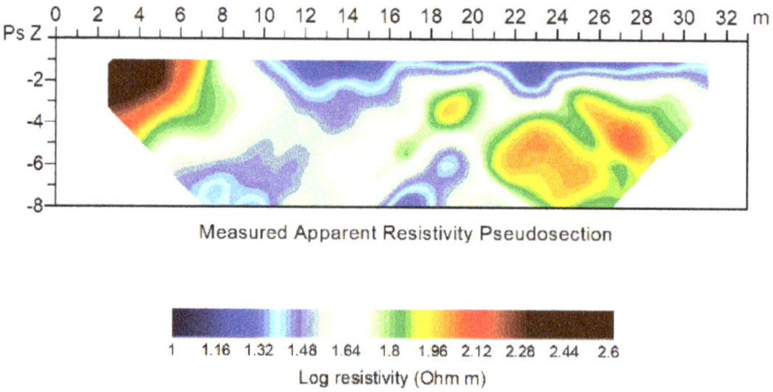

Figure A1. Profile 1: Measured apparent resistivity pseudosection.

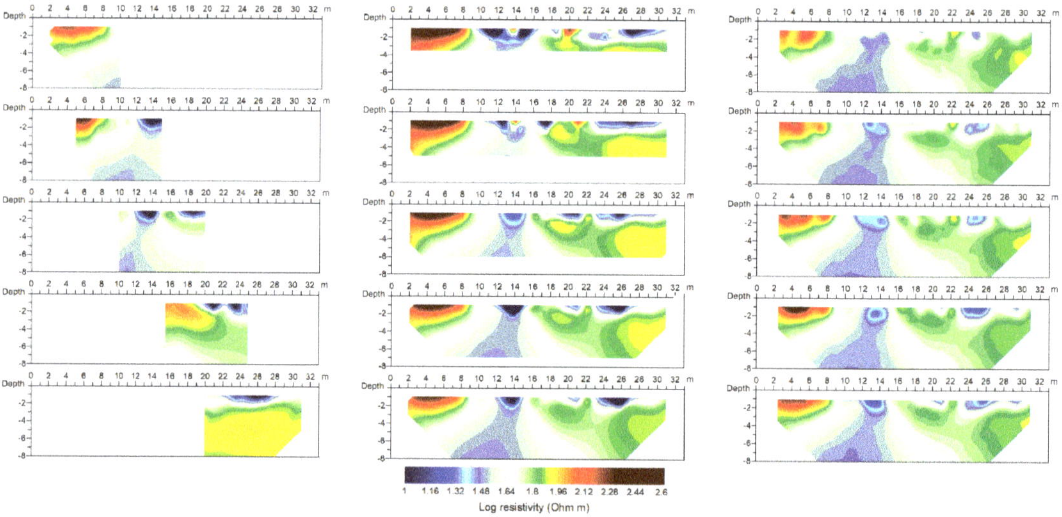

Figure A2. Profile 1: PERTI sections obtained by an irregular stepwise displacement of a data window with variable width along the pseudosection (**left** column); PERTI sections obtained by a vertical scanning of the pseudosection (**middle** column); average PERTI sections, obtained by averaging point by point 20 PERTI sections elaborated using randomly extracted N_q data (tests) from the $N = 285$ starting dataset. From top to bottom $N_q = 50$, $N_q = 100$, $N_q = 150$, $N_q = 200$ and $N_q = 250$ (**right** column).

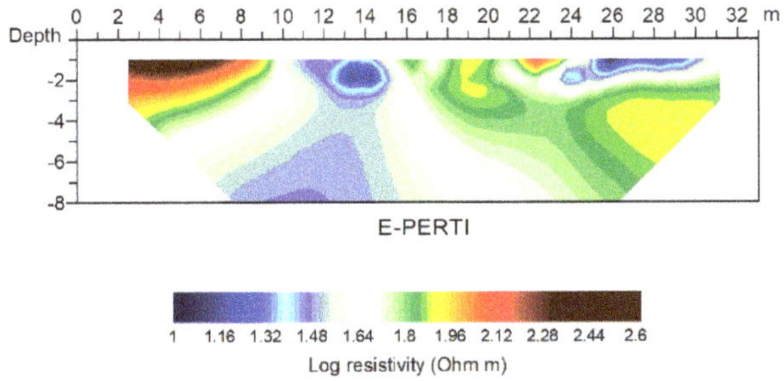

Figure A3. Profile 1: E-PERTI section resulting from the estimation of ρ_m with a linear best-fit procedure applied point by point to the resistivity values belonging to the PERTI sections in Figure A2.

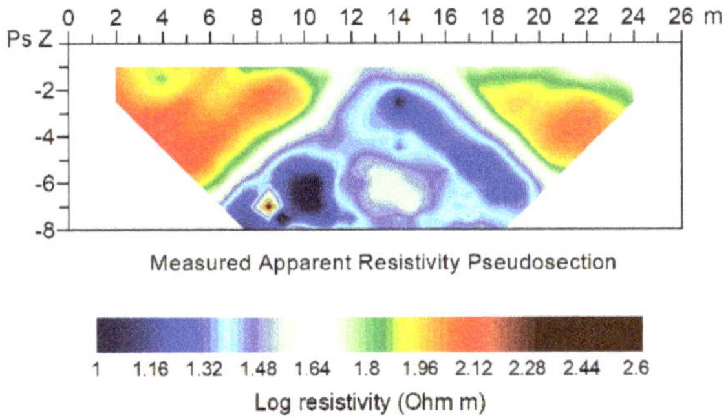

Figure A4. Profile 2: Measured apparent resistivity pseudosection.

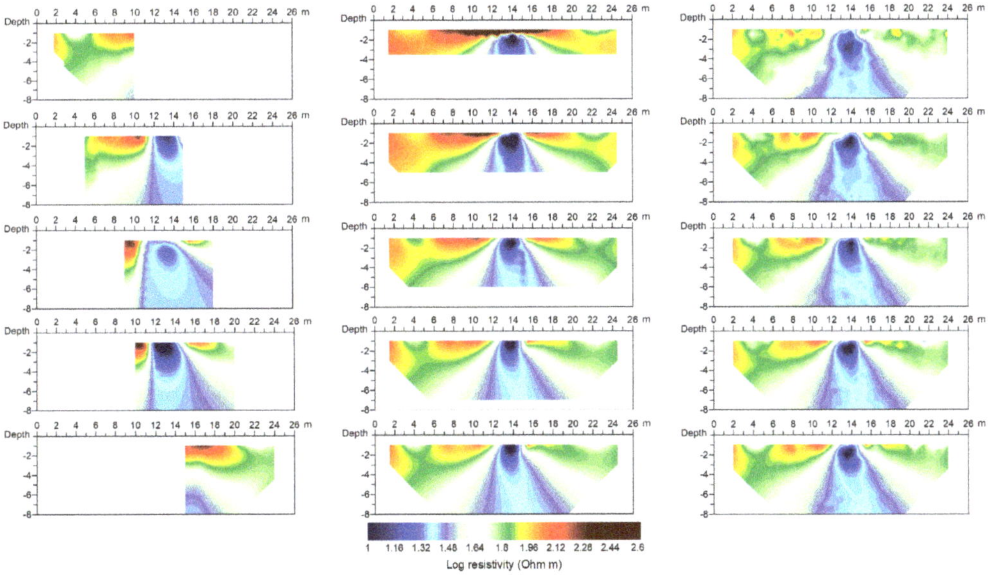

Figure A5. PERTI sections obtained by an irregular stepwise displacement of a data window with variable width along the pseudosection (**left** column); PERTI sections obtained by a vertical scanning of the pseudosection (**middle** column); average PERTI sections, obtained by averaging point by point 20 PERTI sections elaborated using randomly extracted N_q data (tests) from the N = 255 starting dataset. From top to bottom N_q = 50, N_q = 100, N_q = 150, N_q = 200 and N_q = 250 (**right** column).

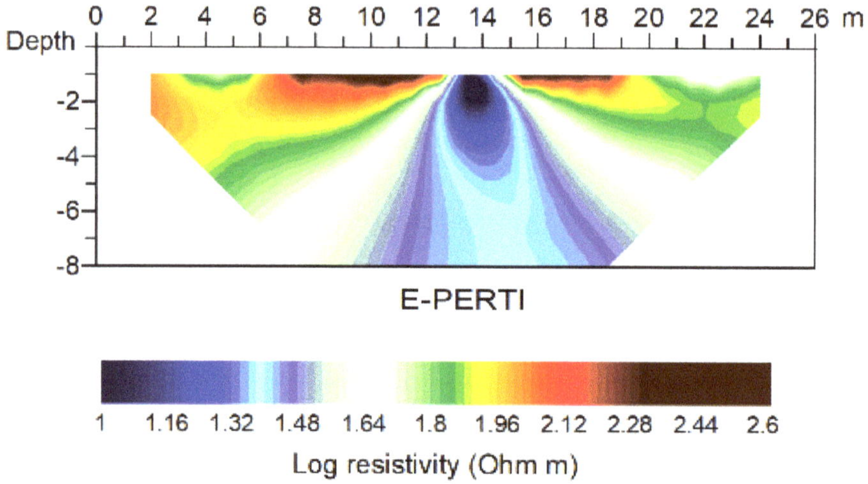

Figure A6. Profile 2: E-PERTI section resulting from the estimation of ρ_m with a linear best-fit procedure applied point by point to the resistivity values belonging to the PERTI sections in Figure A5.

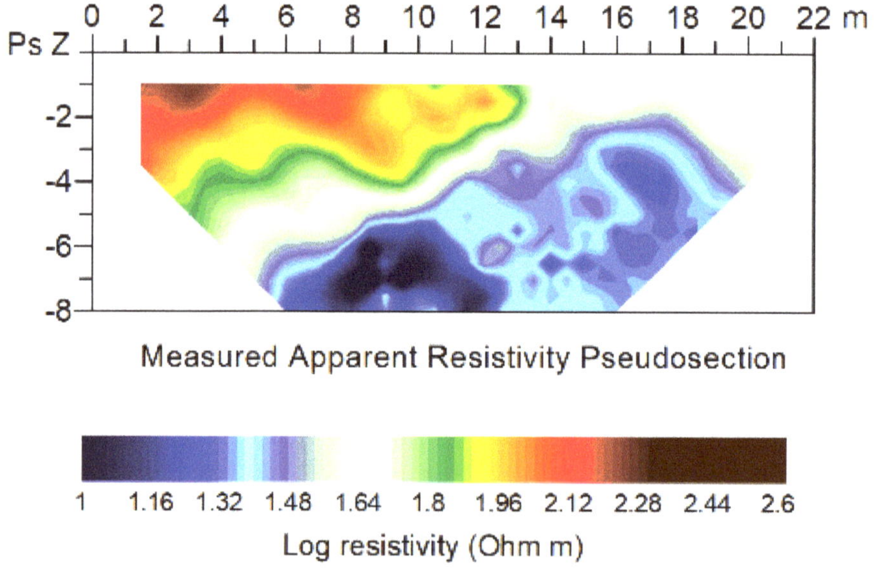

Figure A7. Profile 3: Measured apparent resistivity pseudosection.

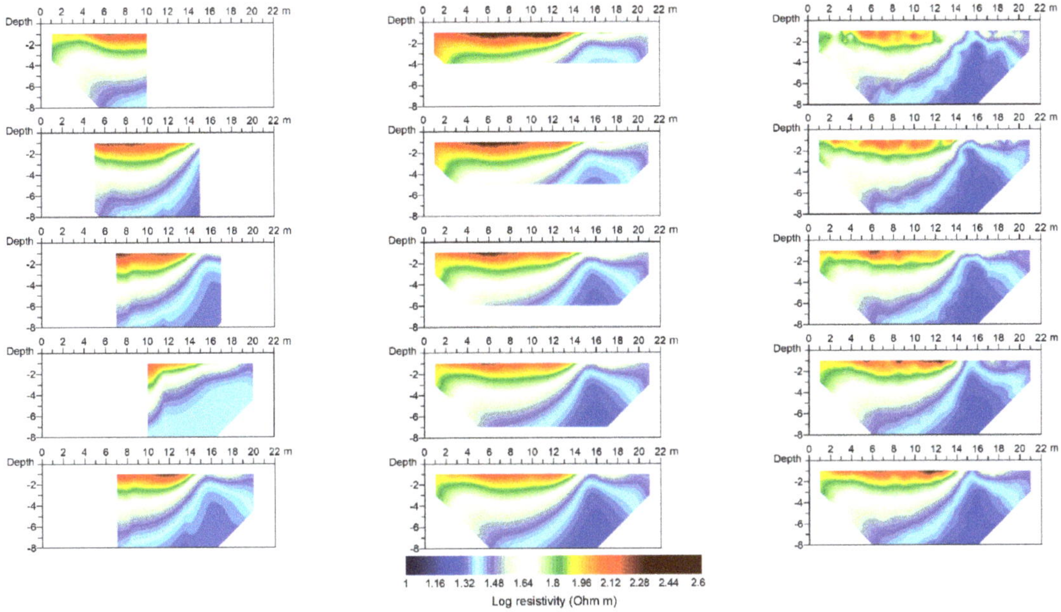

Figure A8. Profile 3: PERTI sections obtained by an irregular stepwise displacement of a data window with variable width along the pseudosection (**left** column); PERTI sections obtained by a vertical scanning of the pseudosection (**middle** column); average PERTI sections, obtained by averaging point by point 10 PERTI sections elaborated using randomly extracted N_q data (tests) from the $N = 195$ starting dataset. From top to bottom $N_q = 40$, $N_q = 80$, $N_q = 120$, $N_q = 160$ and $N_q = 180$ (**right** column).

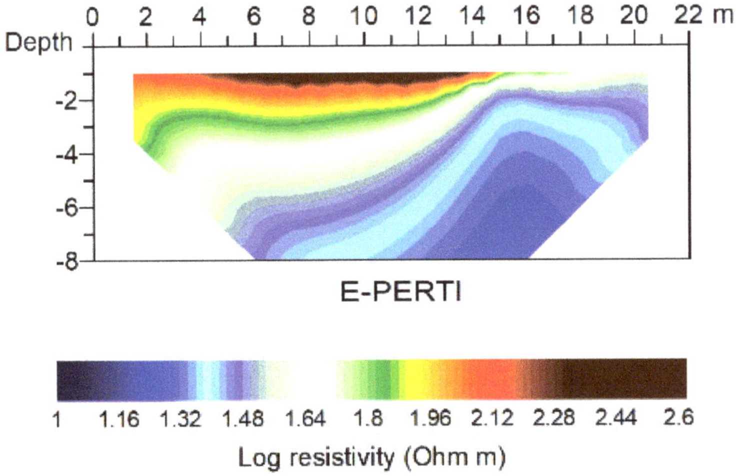

Figure A9. Profile 3: E-PERTI section resulting from the estimation of ρ_m with a linear best-fit procedure applied point by point to the resistivity values belonging to the PERTI sections in Figure A8.

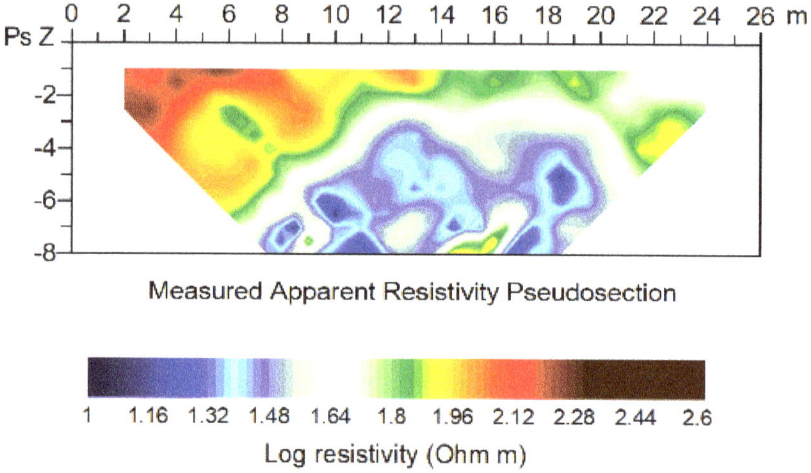

Figure A10. Profile 4: Measured apparent resistivity pseudosection.

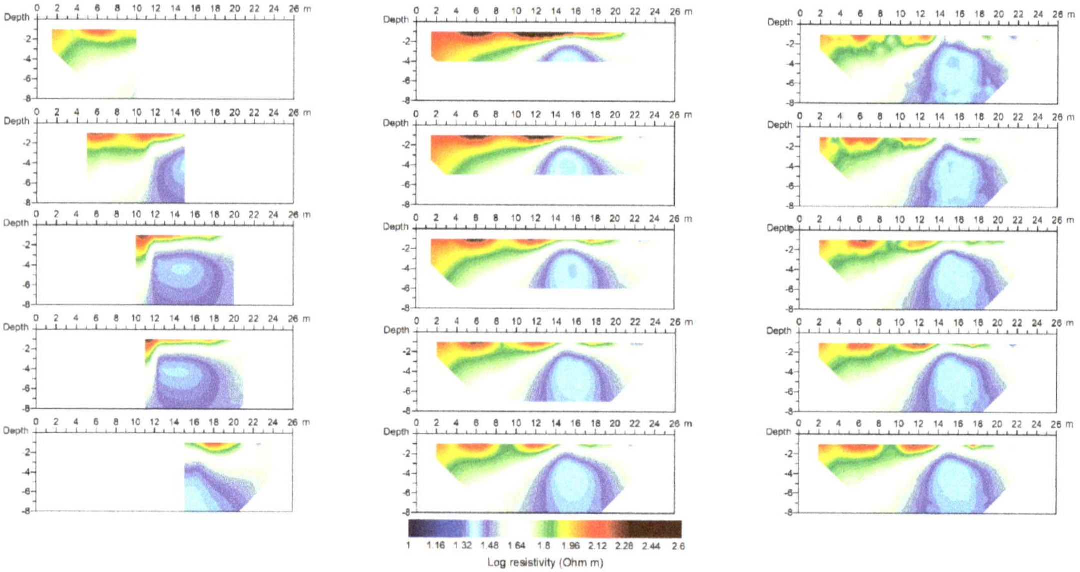

Figure A11. Profile 4: PERTI sections obtained by an irregular displacement of a data window with variable width along the pseudosection (**left** column); PERTI sections obtained by a vertical scanning of the pseudosection (**middle** column); average PERTI sections, obtained by averaging point by point 20 PERTI sections elaborated using randomly extracted N_q data (tests) from the $N = 255$ starting dataset. From top to bottom $N_q = 50$, $N_q = 100$, $N_q = 150$, $N_q = 200$ and $N_q = 240$ (**right** column).

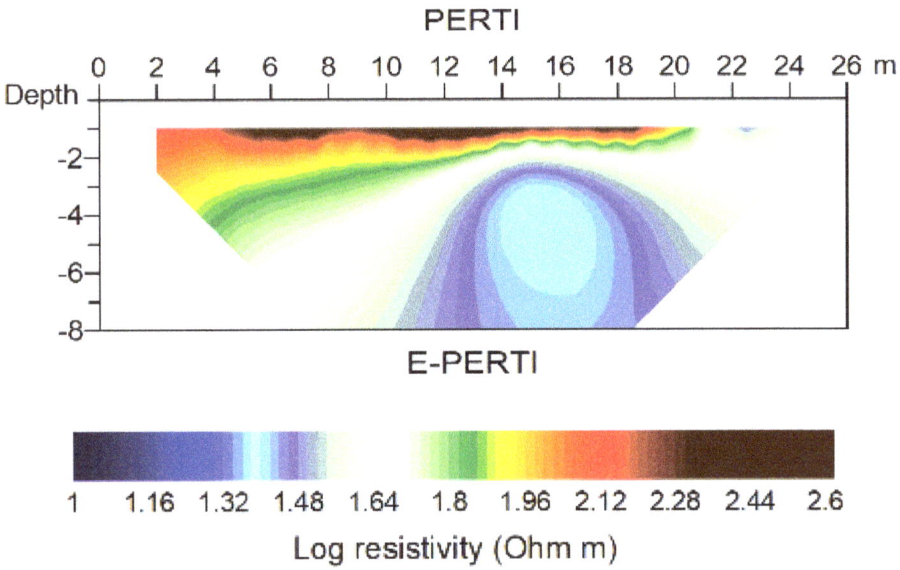

Figure A12. Profile 4: E-PERTI section resulting from the estimation of ρ_m with a linear best-fit procedure applied point by point to the resistivity values belonging to the PERTI sections in Figure A11.

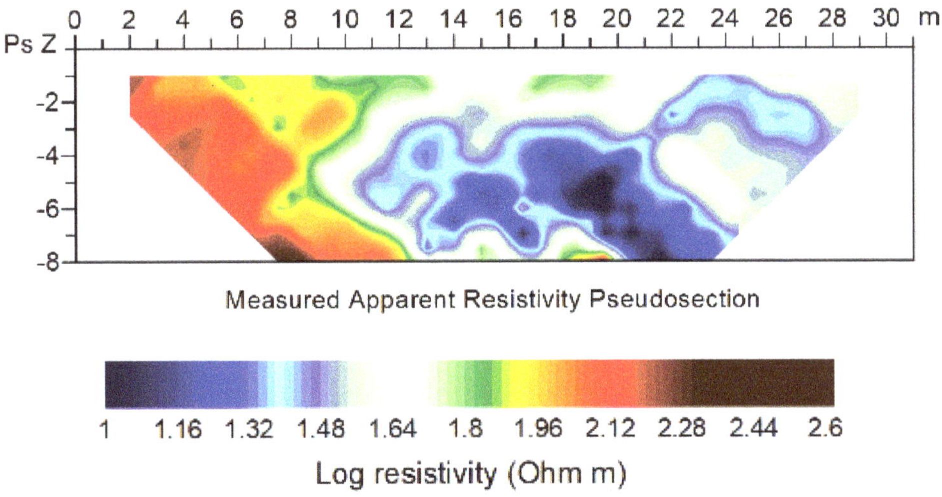

Figure A13. Profile 5: Measured apparent resistivity pseudosection.

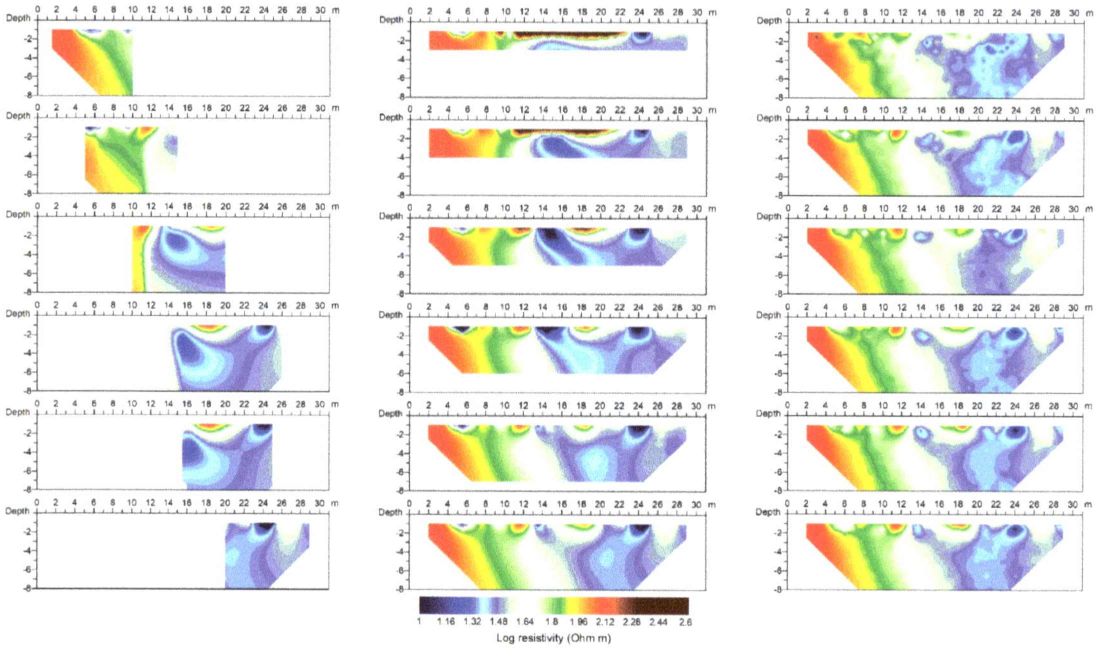

Figure A14. Profile 5: PERTI sections obtained by an irregular displacement of a data window with variable width along the pseudosection (**left** column); PERTI sections obtained by a vertical scanning of the pseudosection (**middle** column); average PERTI sections, obtained by averaging point by point 20 PERTI sections elaborated using randomly extracted N_q data (tests) from the $N = 330$ starting dataset. From top to bottom $N_q = 50$, $N_q = 100$, $N_q = 150$, $N_q = 200$, $N_q = 250$ and $N_q = 300$ (**right** column).

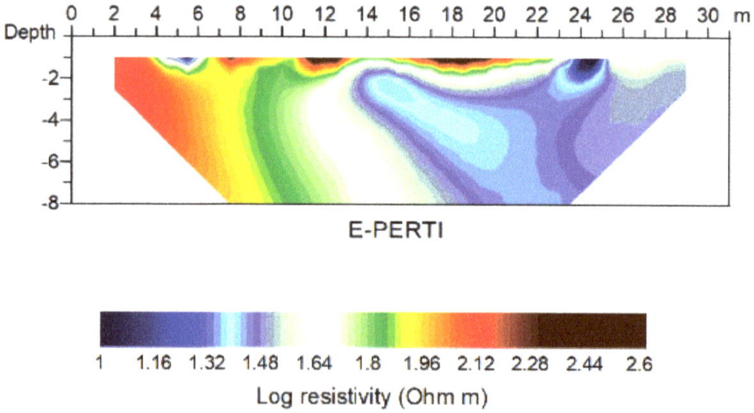

Figure A15. Profile 5: E-PERTI section resulting from the estimation of ρ_m with a linear best-fit procedure applied point by point to the resistivity values belonging to the PERTI sections in Figure A14.

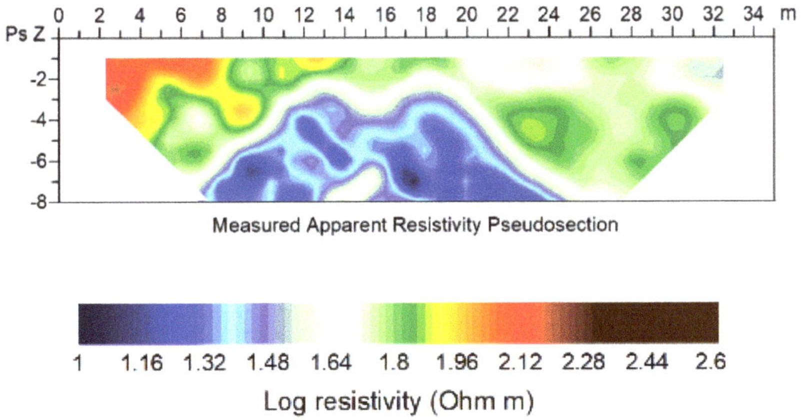

Figure A16. Profile 6: Measured apparent resistivity pseudosection.

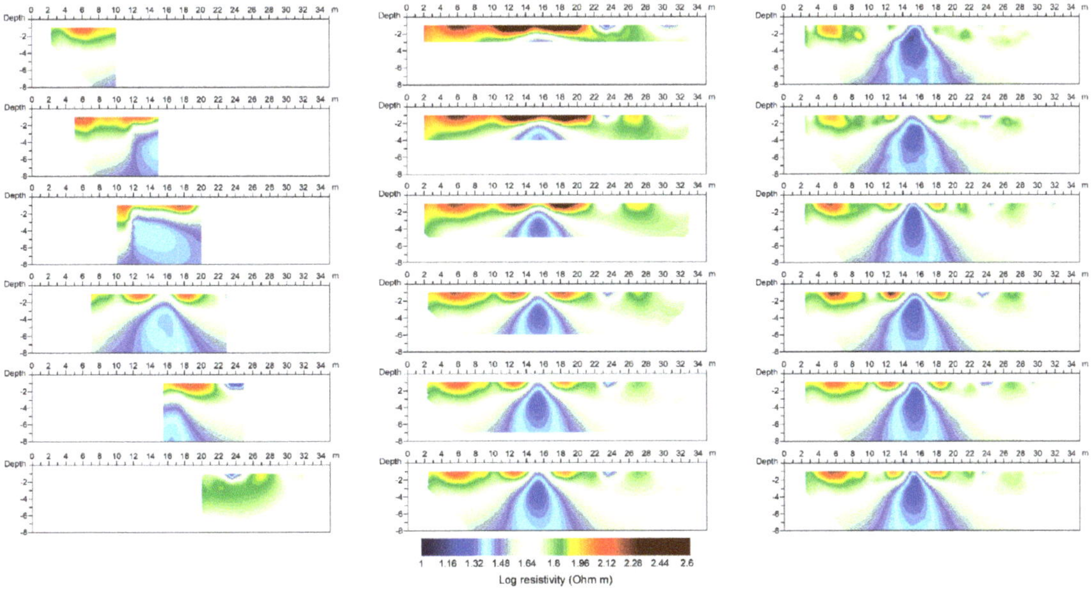

Figure A17. Profile 6: PERTI sections obtained by an irregular displacement of a data window with variable width along the pseudosection (**left** column); PERTI sections obtained by a vertical scanning of the pseudosection (**middle** column); average PERTI sections, obtained by averaging point by point 20 PERTI sections elaborated using randomly extracted N_q data (tests) from the $N = 330$ starting dataset. From top to bottom $N_q = 50$, $N_q = 100$, $N_q = 150$, $N_q = 200$, $N_q = 250$ and $N_q = 300$ (**right** column).

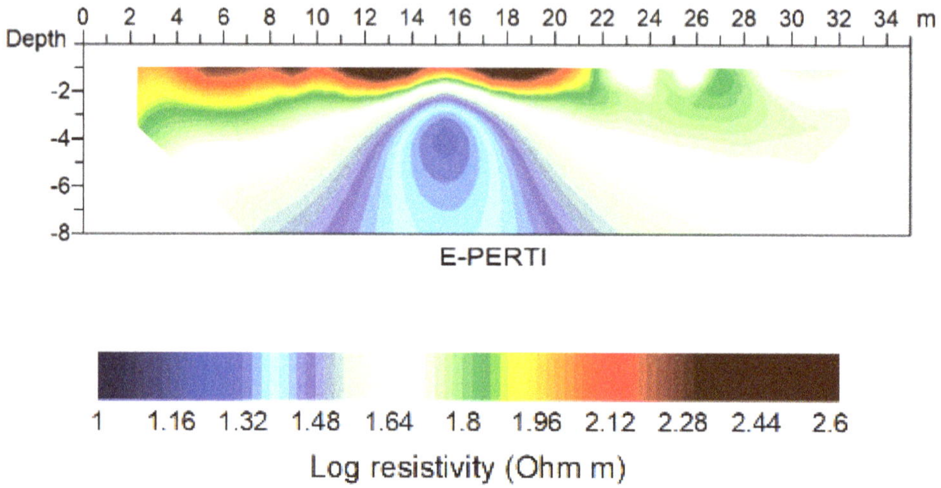

Figure A18. Profile 6: E-PERTI section resulting from the estimation of ρ_m with a linear best-fit procedure applied point by point to the resistivity values belonging to the PERTI sections in Figure A17.

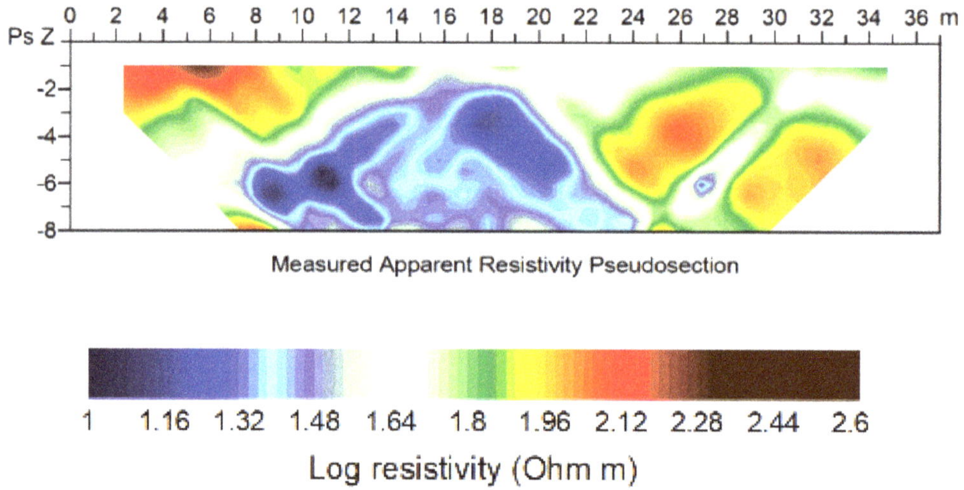

Figure A19. Profile 7: Measured apparent resistivity pseudosection.

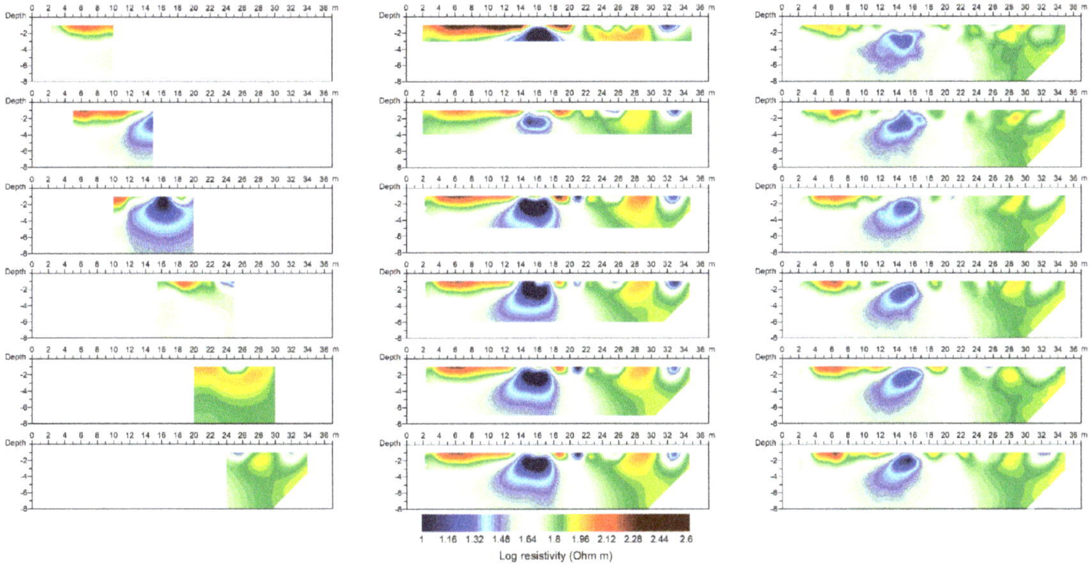

Figure A20. Profile 7: PERTI sections obtained by a displacement of 2 m of a 10 m wide data window along the pseudosection (**left** column); PERTI sections obtained by a vertical scanning of the pseudosection (**middle** column); average PERTI sections, obtained by averaging point by point 20 PERTI sections elaborated using randomly extracted N_q data (tests) from the N = 330 starting dataset. From top to bottom N_q = 50, N_q = 100, N_q = 150, N_q = 200, N_q = 250 and N_q = 300 (**right** column).

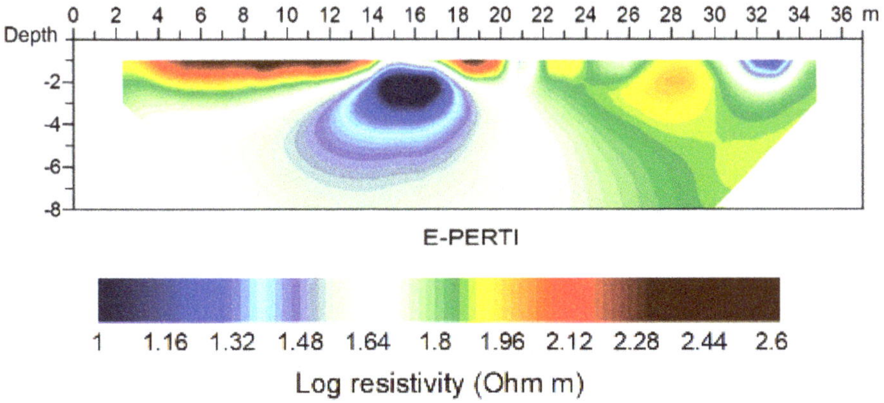

Figure A21. Profile 7 E-PERTI section resulting from the estimation of ρ_m with a linear best-fit procedure applied point by point to the resistivity values belonging to the PERTI sections in Figure A20.

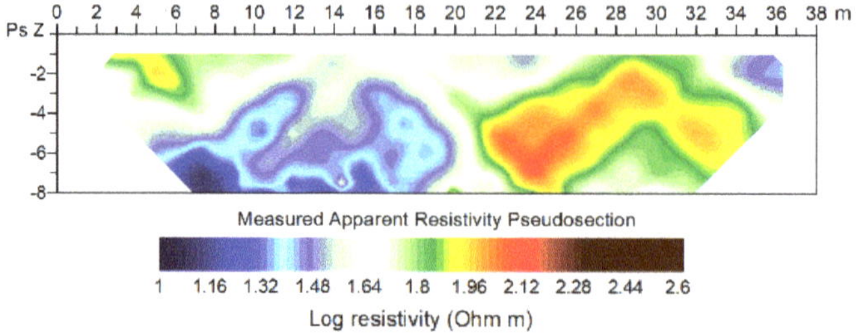

Figure A22. Profile 8: Measured apparent resistivity pseudosection.

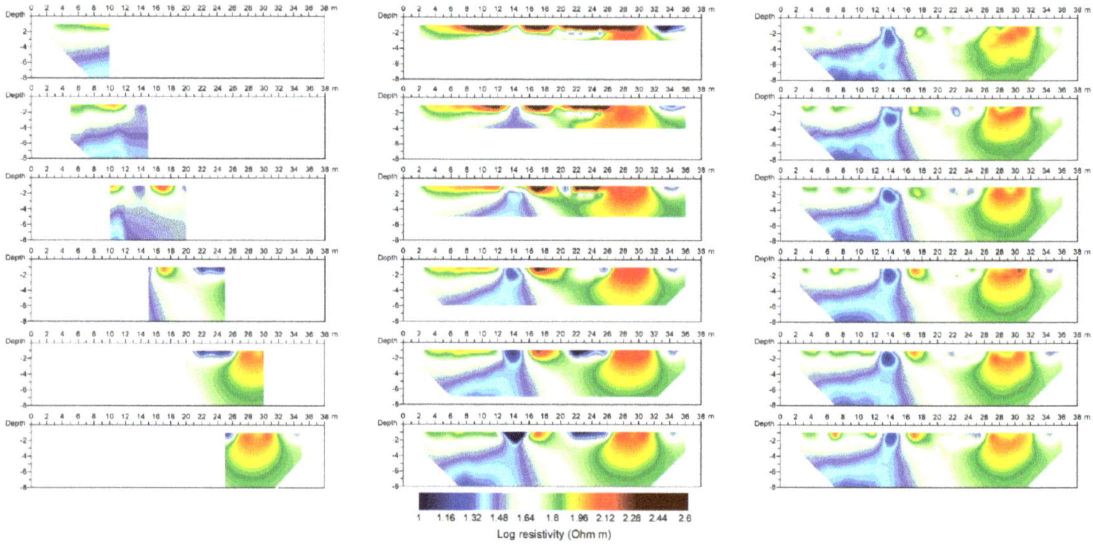

Figure A23. Profile 8: PERTI sections obtained by a displacement of 2 m of a 10 m wide data window along the pseudosection (**left** column); PERTI sections obtained by a vertical scanning of the pseudosection (**middle** column); average PERTI sections, obtained by averaging point by point 20 PERTI sections elaborated using randomly extracted N_q data (tests) from the $N = 330$ starting dataset. From top to bottom $N_q = 50$, $N_q = 100$, $N_q = 150$, $N_q = 200$, $N_q = 250$ and $N_q = 300$ (**right** column).

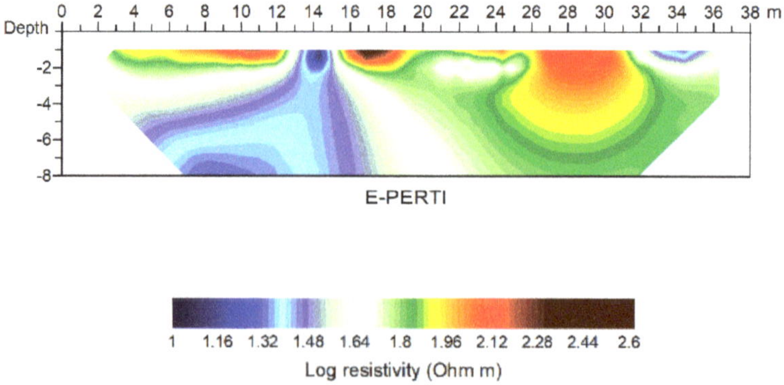

Figure A24. Profile 8 E-PERTI section resulting from the estimation of ρ_m with a linear best-fit procedure applied point by point to the resistivity values belonging to the PERTI sections in Figure A23.

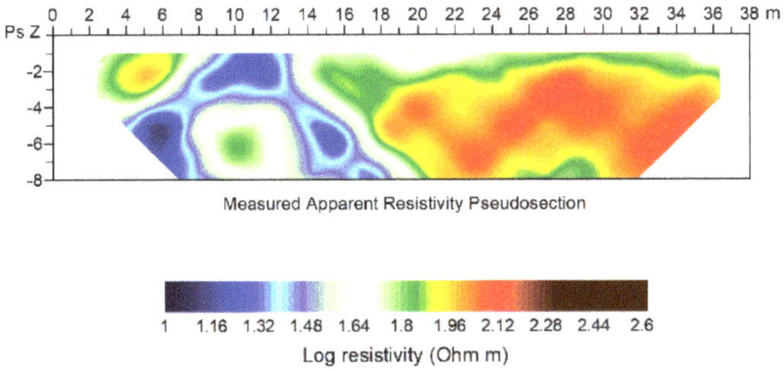

Figure A25. Profile 9: Measured apparent resistivity pseudosection.

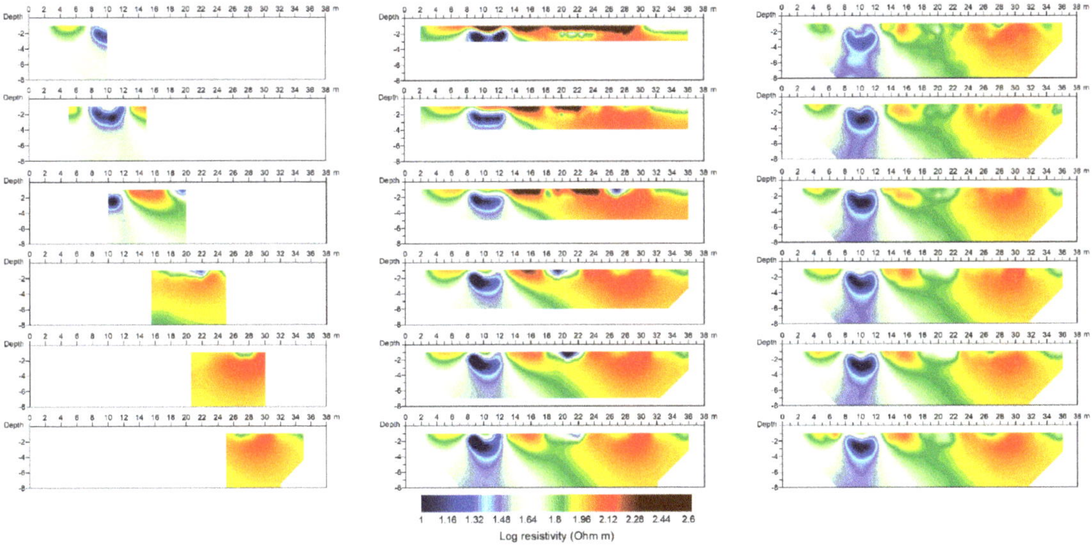

Figure A26. Profile 9: PERTI sections obtained by a displacement of 2 m of a 10 m wide data window along the pseudosection (**left** column); PERTI sections obtained by a vertical scanning of the pseudosection (**middle** column); average PERTI sections, obtained by averaging point by point 20 PERTI sections elaborated using randomly extracted N_q data (tests) from the N = 330 starting dataset. From top to bottom N_q = 50, N_q = 100, N_q = 150, N_q = 200, N_q = 250 and N_q = 300 (**right** column).

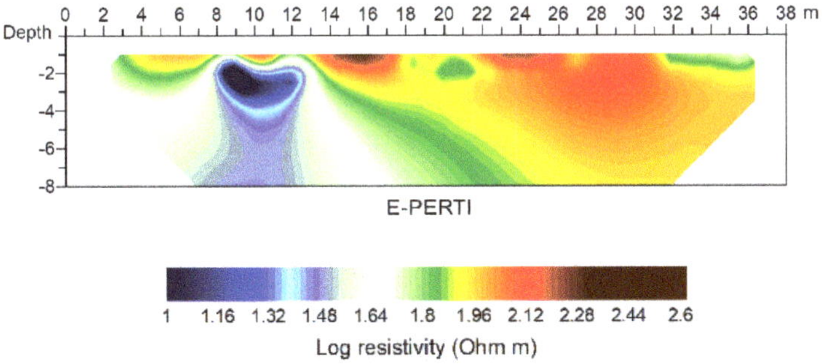

Figure A27. Profile 9 E-PERTI section resulting from the estimation of ρ_m with a linear best-fit procedure applied point by point to the resistivity values belonging to the PERTI sections in Figure A26.

References

1. Andreassi, G.; Cocchiaro, A.; Maruca, A. (Eds.) *Egnazia Sommersa: Dalla Terra al Mare*; Edipuglia: Bari, Italy, 2002.
2. ISPRA: Carta Geologica d'Italia 1:50.000. Available online: https://www.isprambiente.gov.it/Media/carg/puglia.html (accessed on 29 November 2021).
3. Mastronuzzi, G.; Sansò, P. Pleistocene sealevel changes, sapping processes and development of valley networks in the Apulia region (southern Italy). *Geomorphology* **2002**, *46*, 19–34. [CrossRef]
4. Mastronuzzi, G.; Sansò, P. Holocene coastal dune development and environmental changes in Apulia (southern Italy). *Sediment. Geol.* **2002**, *150*, 139–152. [CrossRef]
5. D'alessandro, A.; Iannone, A. Prime considerazioni sedimentologiche e paleoecologiche su alcune sezioni della Calcarenite di Gravina (Pleistocene) nei pressi di Monopoli. *Studi Geol. Geofis.* **1984**, *27*, 1–16.
6. Cinquepalmi, A.; Cocchiaro, A. *Egnazia Nel Tempo. Dal Villaggio Protostorico al Borgo Medievale*; C.D.S.: Valenzano, Italy, 2000.
7. Del Monte, R.; Randino, V. Egnazia. In *Archeologia delle Regioni d'Italia, Puglia*; Ceraudo, G., Ed.; Bradypus: Bologna, Italy, 2014; pp. 303–309.
8. Cinquepalmi, A. *Viaggio Nell'età del Bronzo. Egnazia tra Coppa Nevigata e Punta le Terrare*; C.D.S.: Valenzano, Italy, 2000.
9. Andreassi, G.; Cocchiaro, A. *Necropoli d'Egnazia*; C.D.S.: Fasano, Italy, 1988.
10. Cocchiaro, A.; Dell'Aglio, A. L'impianto difensivo. In *Mare di Egnazia*; Schena Editore: Fasano, Italy, 1983; pp. 45–56, 198.
11. Cozzolino, M.; Mauriello, P.; Patella, D. An Extension of the Data-Adaptive Probability-Based Electrical Resistivity Tomography Inversion Method (E-PERTI). *Geosciences* **2020**, *10*, 380. [CrossRef]
12. Patella, D. Introduction to ground surface self-potential tomography. *Geophys. Prospect.* **1997**, *45*, 653–681. [CrossRef]
13. Mauriello, P.; Patella, D. Resistivity anomaly imaging by probability tomography. *Geophys. Prospect.* **1999**, *47*, 411–429. [CrossRef]
14. Mauriello, P.; Patella, D. Resistivity tensor probability tomography. *Prog. Electromagn. Res. B* **2008**, *8*, 129–146. [CrossRef]
15. Mauriello, P.; Patella, D. Geoelectrical anomalies imaged by polar and dipolar probability tomography. *Prog. Electromagn. Res.* **2008**, *87*, 63–88. [CrossRef]
16. Alaia, R.; Patella, D.; Mauriello, P. Imaging multipole self- potential sources by 3D probability tomography. *Prog. Electromagn. Res. B* **2009**, *14*, 311–339. [CrossRef]
17. Compare, V.; Cozzolino, M.; Mauriello, P.; Patella, D. Resistivity probability tomography at the Castle of Zena (Italy). *EURASIP J. Image Video Process.* **2009**, *2009*, 1–9. [CrossRef]
18. Compare, V.; Cozzolino, M.; Di Giovanni, E.; Mauriello, P. Examples of resistivity tomography for cultural heritage management. In *Near Surface 2010-16th European Meeting of Environmental and Engineering Geophysics*; European Association of Geoscientists and Engineers, EAGE: Amsterdam, The Netherlands, 2010. [CrossRef]
19. Mauriello, P.; Patella, D. A data-adaptive probability-based fast ERT inversion method. *Prog. Electromagn. Res.* **2009**, *97*, 275–290. [CrossRef]
20. Amato, V.; Cozzolino, M.; De Benedittis, G.; Di Paola, G.; Gentile, V.; Giordano, C.; Marino, P.; Rosskopf, C.M.; Valente, E. An integrated quantitative approach to assess the archaeological heritage in highly anthropized areas: The case study of Aesernia (southern Italy). *Acta Imeco* **2016**, *5*, 33–43. [CrossRef]
21. Cozzolino, M.; Mauriello, P.; Patella, D. Resistivity Tomography Imaging of the substratum of the Bedestan Monumental Complex at Nicosia, Cyprus. *Archaeometry* **2013**, *56*, 331–350. [CrossRef]
22. Cozzolino, M.; Caliò, L.M.; Gentile, V.; Mauriello, P.; Di Meo, A. The Discovery of the Theater of Akragas (Valley of Temples, Agrigento, Italy): An Archaeological Confirmation of the Supposed Buried Structures from a Geophysical Survey. *Geosciences* **2020**, *10*, 161. [CrossRef]
23. Cozzolino, M.; Baković, M.; Borovinić, N.; Galli, G.; Gentile, V.; Jabučanin, M.; Mauriello, P.; Merola, P.; Živanović, M. The contribution of geophysics to the knowledge of the hidden archaeological heritage of Montenegro. *Geoscience* **2020**, *10*, 187. [CrossRef]
24. Di Giuseppe, M.G.; Troiano, A.; Fedele, A.; Caputo, T.; Patella, D.; Troise, C.; De Natale, G. Electrical resistivity tomography imaging of the near-surface structure of the Solfatara crater, Campi Flegrei (Naples, Italy). *Bull. Volcanol.* **2015**, *77*, 27. [CrossRef]
25. Blaschek, R.; Hördt, A.; Kemna, A. A new sensitivity-controlled focusing regularization scheme for the inversion of induced polarization data based on the minimum gradient support. *Geophysics* **2008**, *73*, F45–F54. [CrossRef]
26. Zhdanov, M.S.; Tolstaya, E. Minimum support nonlinear parametrization in the solution of a 3D magnetotelluric inverse problem. *Inverse Probl.* **2004**, *20*, 937–952. [CrossRef]
27. Loke, M.H.; Acworth, I.; Dahlin, T. A comparison of smooth and blocky inversion methods in 2D electrical imaging surveys. *Explor. Geophys.* **2003**, *34*, 182–187. [CrossRef]
28. Hermans, T.; Kemna, A.; Nguyen, F. Covariance-constrained difference inversion of time-lapse electrical resistivity tomography data. *Geophysics* **2016**, *81*, E311–E322. [CrossRef]
29. Loke, M.H.; Barker, R.D. Rapid least-squares inversion of apparent resistivity pseudosections by a quasi-Newton method. *Geophys. Prospect.* **1996**, *44*, 131–152. [CrossRef]
30. Oldenburg, D.W.; Li, Y. Estimating depth of investigation in dc resistivity and IP surveys. *Geophysics* **1999**, *64*, 403–416. [CrossRef]
31. Marescot, L.; Loke, M.H.; Chapellier, D.; Delaloye, R.; Lambiel, C.; Reynard, E. Assessing reliability of 2D resistivity imaging in mountain permafrost studies using the Depth of Investigation index method. *Near Surf. Geophys.* **2003**, *1*, 57–67. [CrossRef]

32. Deceuster, J.; Etienne, A.; Tanguy, R.; Nguyen, F.; Kaufmann, O. A modified DOI-based method to statistically estimate the depth of investigation of dc resistivity surveys. *J. Appl. Geophys.* **2014**, *103*, 172–185. [CrossRef]
33. Gnedenko, B.V. *Kurs Teorii Verojatnostej, Mir, Moscow*; Published in Italian as Teoria della Probabilità; Editori Riuniti: Rome, Italy, 1979.

applied sciences

MDPI

Article

Method of Eliminating Helicopter Vibration Interference Magnetic Field with a Pair of Magnetometers

Yongqiang Feng [1,2,3], Yaoxin Zheng [1,2,3], Luzhao Chen [4], Xiaodong Qu [5,*] and Guangyou Fang [1,2]

1 Aerospace Information Research Institute, Chinese Academy of Sciences, No. 9 Dengzhuang South Road, Haidian District, Beijing 100094, China; fengyq@aircas.ac.cn (Y.F.); zhengyaoxin17@mails.ucas.ac.cn (Y.Z.); gyfang@mail.ie.ac.cn (G.F.)
2 The Key Laboratory of Electromagnetic Radiation and Sensing Technology, Chinese Academy of Sciences, No. 19 North 4th Ring West Road, Haidian District, Beijing 100190, China
3 School of Electronic, Electrical and Communication Engineering, University of the Chinese Academy of Sciences, Beijing 100039, China
4 Beijing Automation Control Equipment Institute, Beijing 100074, China; chluzh@163.com
5 School of Information and Electronics, Beijing Institute of Technology, No. 5 South Street, Zhongguancun, Haidian District, Beijing 100081, China
* Correspondence: xdqu@bit.edu.cn

Abstract: The low-frequency electromagnetic fields and magnetic anomalies generated by ships and other underwater platforms are widely recognized as important features for ocean target detection. Low-frequency magnetic fields and anomalies are typically measured by optically pumped magnetometers installed on aircraft. However, the interference that is generated by the aircraft platform may significantly affect the detection performance. The traditional aeromagnetic compensation model has a good effect on eliminating the interference magnetic field that is caused by the carrier attitude variation. Usually, the magnetometer is fixed at the top of a long probe on the aircraft to avoid the influence from the main body in the aircraft. However, the probe is sensitive to external vibrations, and vibration-induced magnetic interference can occur in the measurements. The magnetometer is especially easily affected by the interference magnetic field, including the vibration frequency and harmonic frequency of the probe, in a moving platform, such as a helicopter. These interference fields usually have independent frequency characteristics that can be eliminated by compensation methods. In this paper, we propose a method based on the improved coherent noise suppression method with a pair of magnetometers to eliminate the effects from these magnetic field disturbances and improve the detection performance of the measurement system. The results of the flight experiment show that the method can effectively eliminate low-frequency vibration interference and improve the detection ability of weak signals from targets.

Keywords: vibration interference field; aeromagnetic compensation; coherent noise suppression method

1. Introduction

The electromagnetic fields generated by naval vessels and other underwater carriers have become important signals for tracking targets of interest in marine warfare [1,2]. When ships and underwater carriers are in seawater, in addition to the influence of ferromagnetic materials (such as iron, cobalt, and nickel), the metal material in the carrier composition (such as the steel shell and copper propeller) electrochemically reacts to generate a corrosive current around the vessel [3]. In order to protect the hull from corrosion, the impressed current cathodic protection system and the sacrificial anode cathodic protection system are widely used to generate corrosion protection currents in modern ships. These currents create electro-static and magneto-static anomalies around the ships. At the same time, the corrosion current and the protection current flow from the hull to the propeller through the seawater. Then, the currents return to the hull from the propeller through various bearings and mechanical wires [4]. The loop electrical impedance periodically changes as the drive shaft rotates, causing the current flowing through the seawater to be modulated, thereby forming an extremely low-frequency electromagnetic field in the seawater and propagating outward. The signal has frequency characteristics that are consistent with the propeller rotation frequency and its harmonic frequency. To a large extent, the signals can be transmitted several kilometers away. Therefore, they are considered important signal sources for detecting underwater military targets [5].

As a significant passive detection method, aeromagnetic exploration is widely used in underwater target detection [6–8]. The development and updating of aeromagnetic testing technologies is often concentrated in the military field. Tracing back to World War II, the British Army started to use aircraft to carry fluxgate sensors to detect underwater submarines. With the development of high-precision magnetometers, such as optically pumped magnetometers (OPMs), aeromagnetic surveying has become one of the most important means of underwater target detection. Given that the aircraft platform affects the accuracy of OPM measurements, W.E. Tolles proposed a classical aeromagnetic compensation model in the 1940s [9,10]. Paul Leliak proposed sinusoidal maneuvers to improve compensation performance in 1961 [11]. In 1980, B.W. Leach proposed a ridge regression algorithm to solve the aeromagnetic compensation parameters, and the experimental results showed that this method performed well in improving the compensation performance [12]. In 1995, Hardwick designed an innovative three-axis aeromagnetic gradiometry flown on a fixed-wing platform [13]. Around 2005, Nelson carried out a series of experiments with an experimental flight system [14,15]. In 2011 and 2013, Gerardo Noriega proved the viability and reliability of using the standard deviation of the signal as an assessment criterion through a large number of aeromagnetic trials [16,17]. The traditional aeromagnetic compensation model has a good effect on eliminating the interference field caused by aircraft attitude changes, which are generally less than 2 Hz. For time-varying interferences with an unknown signature, Sheinker proposed an adaptive magnetic interference cancelation method by using a pair of magnetometers, which obtained good results for highly correlated interference magnetic fields [18]. However, for magnetic measurement systems with the magnetometers on a probe of a helicopter, the probe vibrates when the helicopter is flying. In this case, there is a vibration frequency and its harmonic interference field that can be captured by the magnetometers. The harmonic interference magnetic field that is caused by aircraft vibration seriously affects the detection of target signals to a large extent.

In order to eliminate the electromagnetic interference field that is produced by the vibration of the helicopter platform, in this paper, we propose an interference cancelation algorithm based on the improved coherent noise suppression (ICNS) method for a detection system with a pair of magnetometers. The test for the calibration and the survey line proved that the proposed algorithm in this paper effectively eliminated the interference field. As a consequence, the system performance for detecting the target signal was improved.

2. Traditional Aeromagnetic Compensation Model

According to the classical Tolles–Lawson (T–L) aeromagnetic compensation model, the aircraft interference field $H_T(t)$ is a sum of the permanent magnetic field $H_P(t)$, the induced magnetic field $H_I(t)$, and the eddy current magnetic field $H_E(t)$, and is expressed as the following equation:

$$H_T(t) = H_P(t) + H_I(t) + H_E(t). \tag{1}$$

Usually, a three-axis magnetometer is used to measure the attitude of the platform during the aeromagnetic compensation flight. The quantities T, L, and V can be used as the three component outputs of the geomagnetic field $H_e(t)$. The cosine values of the angles between the platform and the geomagnetic field can be denoted as v_1, v_2, v_3, and they can be expressed as follows:

$$\begin{aligned} v_1 &= T(t)/H_e(t) = T(t)/\sqrt{T(t)^2 + L(t)^2 + V(t)^2} \\ v_2 &= L(t)/H_e(t) = L(t)/\sqrt{T(t)^2 + L(t)^2 + V(t)^2} \\ v_3 &= V(t)/H_e(t) = V(t)/\sqrt{T(t)^2 + L(t)^2 + V(t)^2} \end{aligned} . \quad (2)$$

Then, the interference field $H_T(t)$ can be displayed as follows:

$$H_T(t) = \begin{pmatrix} c_1v_1 + c_2v_2 + c_3v_3 \\ +H_e(t)\left(c_4v_1^2 + c_5v_1v_2 + c_6v_1v_3 + c_7v_2^2 + c_8v_2v_3 + c_9v_3^2\right) \\ +H_e(t)\begin{pmatrix} c_{10}v_1v_1' + c_{11}v_1v_2' + c_{12}v_1v_3' + c_{13}v_2v_1' + c_{14}v_2v_2' \\ +c_{15}v_2v_3' + c_{16}v_3v_1' + c_{17}v_3v_2' + c_{18}v_3v_3' \end{pmatrix} \end{pmatrix} = \sum_{i=1}^{18} A_i(t)c_i. \quad (3)$$

where the coefficients c_i, $I = 1,2, \ldots, 18$ represent the compensation parameter, and the time-varying quantities $A_i(t)$, $I = 1, 2, \ldots, 18$ can be calculated from v_1, v_2, and v_3 and their derivatives v_1', v_2', and v_3'.

Since the equations $v_1^2 + v_2^2 + v_3^2 = 1$ and $v_1v_1' + v_2v_2' + v_3v_3' = 0$, the parameters c_4, c_7, c_9, c_{10}, c_{14}, and c_{18} are not independent. To find the stable solutions for the model, Equation (3) can be simplified by removing the items $c_9v_3^2$ and $c_{18}v_3v_3'$. Thus, we obtain the replaced T–L model as follows:

$$H_T(t) = \sum_{i=1}^{16} A_i(t)c_i. \quad (4)$$

The corresponding matrix notation for Equation (4) can be expressed as follows:

$$\boldsymbol{H} = \boldsymbol{A}\boldsymbol{C}, \quad (5)$$

where $\boldsymbol{H}$ is the column vector constituted by the sampling points of $H_T(t)$. $\boldsymbol{C}$ is the column vector, and it can be expressed as $\boldsymbol{C} = [c_1c_2 \ldots c_{16}]^T$. $\boldsymbol{A}$ is the corresponding matrix constituted by the sampling points of attitude information, which can be expressed as follows:

$$\mathbf{A} = \begin{bmatrix} A_1(1) & A_2(1) & \cdots & A_{16}(1) \\ A_1(2) & A_2(2) & \cdots & A_{16}(2) \\ \vdots & \vdots & \ddots & \cdots \\ A_1(n) & A_2(n) & \cdots & A_{16}(n) \end{bmatrix}. \quad (6)$$

The compensation parameters can be obtained by a least square (LS) solution that can be expressed as follows:

$$\boldsymbol{C} = \left(\boldsymbol{A}^T\boldsymbol{A}\right)^{-1}\boldsymbol{A}^T\boldsymbol{H}. \quad (7)$$

Usually, after the calibration flight, the compensation coefficients can be obtained by Equation (7). Then, by taking the compensation coefficients back to Equation (5), the interference fields can be obtained and eliminated.

3. Vibration Interference Magnetic Field

The helicopter aeromagnetic system is shown in Figure 1. Two cesium OPMs (OPM-I and OPM-II) and a fluxgate magnetometer are fixedly connected to the probe. The fluxgate magnetometer is used to measure the attitude of the platform since it is vectorial. The noise performance of the OPMs is 0.6 pT/$\sqrt{\text{Hz}}$, and the fluxgate magnetometer is 6 pT/$\sqrt{\text{Hz}}$.

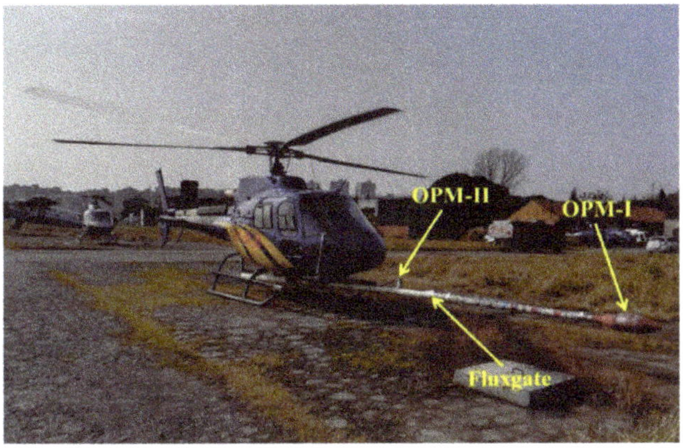

Figure 1. The helicopter platform and equipment.

The design of the flight path of the calibration compensation experiment is shown in Figure 2a. The sampling rate of the recording system was 40 Hz. The helicopter flew at a height of about 3000 m to avoid interference from the ground. The compensating flight executed standard maneuvers, including a series of ±10° rolls, ±5° pitches, and ±5° yaws maneuvers along each line in the box. Figure 2 details the flight path and information.

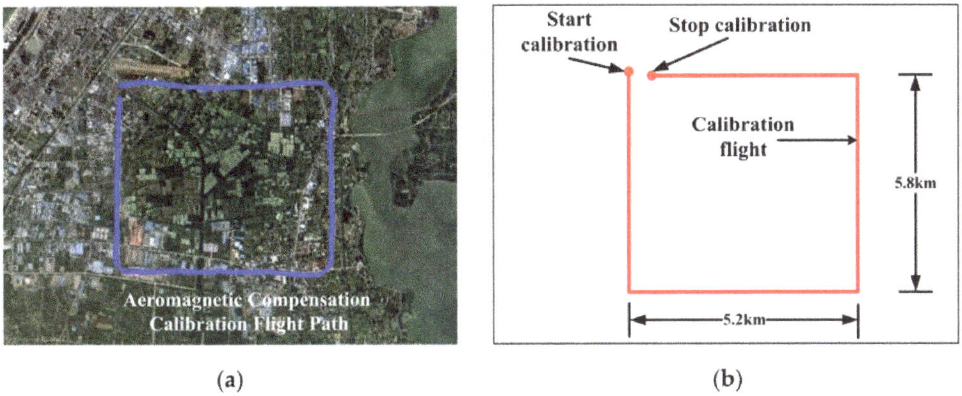

Figure 2. Calibration flight path in Google Earth (**a**) and its scale drawing (**b**).

After calculating by the LS solution, the compensation parameters of the OPM-I and OPM-II were calculated and separately listed in Tables 1 and 2. Consistent with the actual installation, we found that the compensation parameters of OPM-II were much larger.

Table 1. Compensation parameters of OPM-I.

Parameter	Value	Parameter	Value	Parameter	Value	Parameter	Value
c1	34.0nT	c2	5.57nT	c3	9.27nT	c4	5.61×10^5
c5	2.05×10^5	c6	4.04×10^6	c7	3.68×10^4	c8	1.78×10^4
c9	-4.96×10^5	c10	-8.73×10^5	c11	-5.31×10^6	c12	5.07×10^5
c13	-2.42×10^4	c14	-2.14×10^5	c15	-6.03×10^5	c16	-3.49×10^5

Table 2. Compensation parameters of OPM-II.

Parameter	Value	Parameter	Value	Parameter	Value	Parameter	Value
c1	302.7nT	c2	130.9nT	c3	−181.1nT	c4	1.59×10^4
c5	2.70×10^4	c6	-5.00×10^6	c7	3.20×10^3	c8	1.77×10^3
c9	-8.51×10^4	c10	-3.49×10^4	c11	-3.10×10^4	c12	4.51×10^4
c13	-1.53×10^3	c14	-1.37×10^3	c15	-3.31×10^4	c16	-2.91×10^4

Figures 3 and 4 show the compensation results of the calibration flight data using the compensation parameters of Tables 1 and 2. The blue dashed and solid lines represent the data before and after the compensation, respectively. Figures 3a and 4a present the results of the processing of the compensation data through a high-pass filter with a cut-off frequency of 0.5 Hz. The signals and the noise power spectral density of the uncompensated flight data and the compensated data are shown in Figures 3b and 4b, respectively. The aeromagnetic compensation had an obvious effect on the low-frequency electromagnetic interference, with a frequency below 2 Hz. Taking OPM-I as an example, the compensated interference field dropped from 45 $\text{pT}/\sqrt{\text{Hz}}$ at 1 Hz to 18 $\text{pT}/\sqrt{\text{Hz}}$ at 1 Hz. At higher frequencies, the traditional compensation model had essentially no effect. However, there were still many selective frequency interferences with the measurement magnetic field at 0.72 Hz, 1.445 Hz, 6.48 Hz, and so on. These multi-frequency interferences were mainly caused by the motion of the probe and the vibration interference field caused by the rotation of the propeller, which could not be eliminated by the basic aeromagnetic compensation operation.

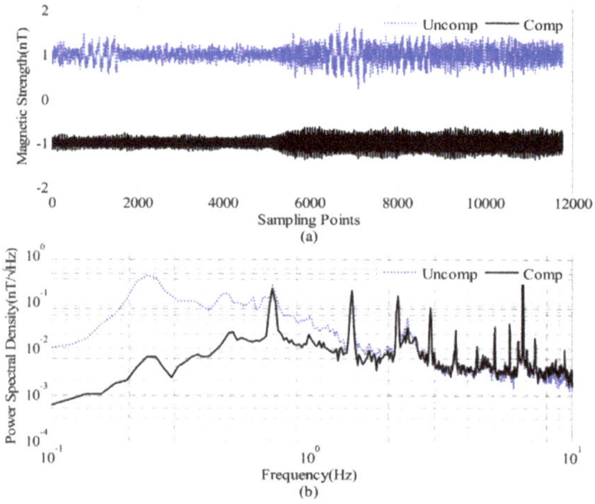

Figure 3. Compensation results of OPM-I: (**a**) calibration flight data and (**b**) the power spectral density of the calibration flight data.

The correlation coefficient of the residual interference magnetic field was calculated after the OPM-I and OPM-II magnetic compensation process, as shown in Figure 5. According to this figure, the correlation coefficient was high and reached or exceeded 0.8 in the fixed frequency points of the probe vibration and the rotation of the propeller, at 1.445 Hz, 2.15 Hz, 2.89 Hz, 6.48 Hz, and so on. The high correlation values indicate that the vibration interference field compensation based on the data of dual magnetometer had high feasibility to eliminate the interferences at the fixed frequencies.

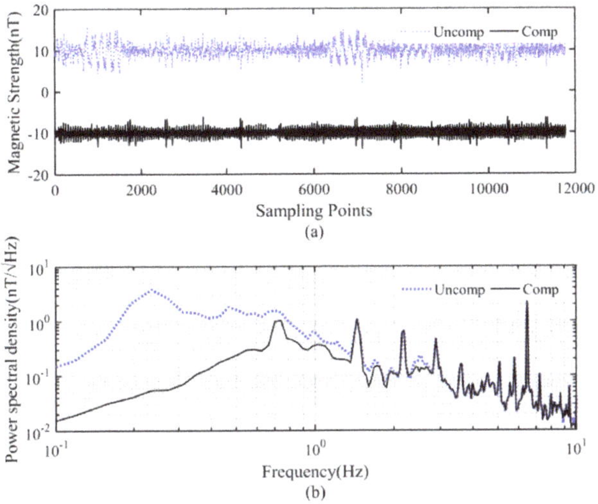

Figure 4. Compensation results of OPM-II: (**a**) calibration flight data and (**b**) the power spectral density of the calibration flight data.

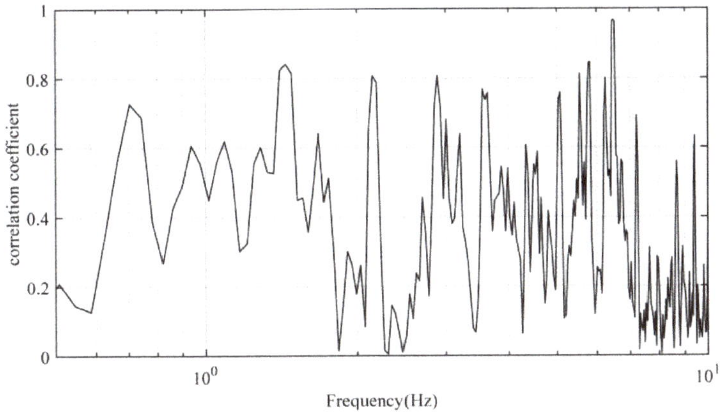

Figure 5. The correlation coefficient of the interference field in OPM-I and OPM-II.

4. The Proposed Method and Principle

After the aeromagnetic compensation, the interference field associated with the attitude of the aircraft was eliminated. However, the interference associated with the vibration in magnetometers was still present. These interference fields had many fixed frequency points that existed at fundamental frequencies and their harmonics. In order to eliminate the vibration interference magnetic field, we first introduced an adaptive cancelation method as a reference, and then proposed an improved coherent noise suppression (ICNS) method.

4.1. The Adaptive Interference Cancelation Method

Reference [18] introduced an adaptive method to eliminate the magnetic interference with a pair of magnetometers, and its theoretical model is shown in Figure 6, where $x(n)$ represents the ambient magnetic field, $e(n)$ is the magnetic interference, and $w_1(n)$ and $w_2(n)$ represent the intrinsic noises of OPM-I and OPM-II, respectively. The output readings of the two magnetometers are expressed as $y_1(n)$ and $y_2(n)$.

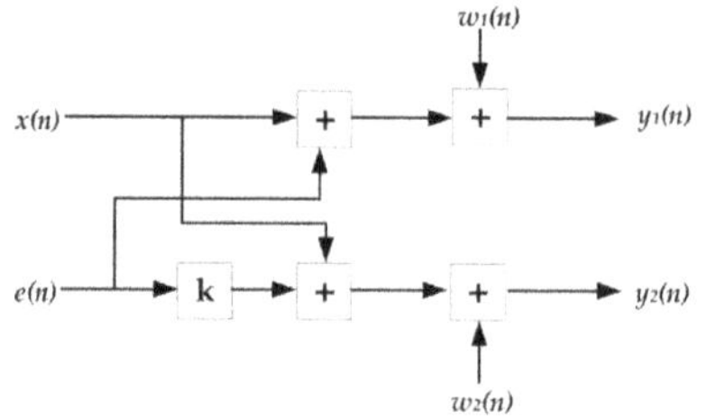

Figure 6. The model used for the adaptive interference cancelation method.

The model can be expressed as a set of equations, as shown in Equation (8). The interference magnetic field elimination is the solution of an adaptive coefficient k.

$$\begin{array}{l} y_1(n) = s(n) + e(n) + w_1(n) \\ y_2(n) = s(n) + ke(n) + w_2(n) \end{array}. \quad (8)$$

4.2. The Improved Coherent Noise Suppression (ICNS) Method

The CNS is a practical noise cancelation method. The distance between the two magnetometers is far less than the distance between the helicopter and the target, so it can be considered that the two magnetometers in the helicopter measure the common value of the magnetic field from the target and the different values from the interference field. In addition, it is known that the vibration interference field of the two magnetometers is highly correlated.

After the compensation, we obtained the magnetic field data sequences $y_1(n)$ and $y_2(n)$. The $y_1(n)$ involves the signal field $s(n)$ and interference field $d(n)$, while $y_2(n)$ involves the signal field $s(n)$ and stronger interference field $d_2(n)$. The n denotes the sample index, where $w_1(n)$ and $w_2(n)$ represent the intrinsic noises of OPM-I and OPM-II, respectively.

$$\begin{array}{l} y_1(n) = s(n) + d(n) + w_1(n) \\ y_2(n) = s(n) + d_2(n) + w_2(n) \end{array}. \quad (9)$$

Considering that the two magnetometers were subject to different interference levels from helicopters, the interference magnetic field that was measured by the closer magnetometer was larger and contained more interference components. In order to avoid the influence from the incoherent noise of the two magnetometers with the CNS method, it was noticed that the OPM-II had more interference field characteristics. Before the CNS processing, we needed to preprocess the measured data from the two magnetometers.

Here, we proposed a spectral subtraction method to reduce the quasi-stationary components from background interference based on a priori signal-to-noise ratio (PSNR) estimation, which was used to respectively process the measurement data from the two magnetometers. As an example, the magnetic field $y_1(n)$ measured by OPM-I can be expressed as follows:

$$y_1(n) = s(n) + d(n) + w_1(n). \quad (10)$$

In the process, this signal was evenly divided into short-time frames and analyzed frame by frame, assuming that its behavior was similar over the duration of all the analyses frames. For a short time frame, k defines the spectral component of the noisy signal as $Y_1(k, \omega)$, the signal as $S(k, \omega)$, and the interference and noise as $D(k, \omega)$. The interference reduction process was used for each short-time frequency spectrum, $Y_1(k, \omega)$, with a spectral gain, $G(k, \omega)$, which is a function of two parameters in the signal-to-noise ratio with a priori, $SNR_{prio}(k, \omega)$, and a posteriori, $SNR_{post}(k, \omega)$, as shown in Equations (11)–(13).

$$SNR_{prio}(k,\omega) = \frac{|Y_1(k,\omega)|^2}{E\{|D(k,\omega)|^2\}}. \quad (11)$$

$$SNR_{post}(k,\omega) = \frac{|S(k,\omega)|^2}{E\{|D(k,\omega)|^2\}}. \quad (12)$$

$$G(k,\omega) = \sqrt{\frac{SNR_{prio}(k,\omega)}{SNR_{prio}(k,\omega)+1}}. \quad (13)$$

The noise power spectral density $E\{|D(k,\omega)|^2\}$ can be estimated during the periods when the signal from the target is not present, where E is the expectation operator. The $SNR_{prio}(k,\omega)$ can be obtained through the following recursive relation, where α is the smoothing coefficient, which is generally a value between 0.8 and 1.

$$SNR_{prio}(k,\omega) = \alpha\frac{|S(k-1,\omega)|^2}{E\{|D(k,\omega)|^2\}} + (1-\alpha)\cdot max\left[SNR_{post}(k,\omega)-1,0\right]. \quad (14)$$

Then, $S(k,\omega)$ is the estimated signal spectrum at the previous frame and is obtained as follows:

$$S(k,\omega) = G(k,\omega)\cdot Y_1(k,\omega). \quad (15)$$

After the preprocessing, the outputs of OPM-I and OPM-II can be expressed as Equations (16) and (17), where $s(n)$, $e_1(n)$, and $e_2(n)$ represent the target signal, and the residual interference field in OPM-I and in OPM-II, respectively.

$$\begin{array}{l} y_1(n) = s(n) + e_1(n) \\ y_2(n) = s(n) + e_2(n) \end{array}. \quad (16)$$

By taking into account the coherence of the vibration interference field, the residual interference fields $e_1(n)$ and $e_2(n)$ can be expressed by Equation (17), where $h(n)$ is the transfer function. The coherent noise suppression method regards the detector as a 2×2 linear time-invariant system to suppress the residual interference field from the helicopter.

$$e_2(n) = h(n)\times e_1(n). \quad (17)$$

By applying a discrete Fourier transform (DFT) on both sides of Equation (18), they can be represented in the frequency domain as follows:

$$\begin{array}{l} Y_1(\omega) = S(\omega) + E_1(\omega) \\ Y_2(\omega) = S(\omega) + E_2(\omega) = S(\omega) + H(\omega)\cdot E_1(\omega) \end{array}. \quad (18)$$

Hence, the auto-spectra and cross-spectra of $Y_1(\omega)$ and $Y_2(\omega)$ are shown as Equation (19), where $P_{SS}(\omega)$ and $P_{EE}(\omega)$ are the auto-spectra of the signal and interference field, and $H(\omega)$ is the discrete Fourier transform of $h(n)$.

$$\begin{array}{l} P_{Y_1Y_1}(\omega) = P_{SS}(\omega) + P_{EE}(\omega) \\ P_{Y_1Y_2}(\omega) = P_{SS}(\omega) + H^*(\omega)P_{EE}(\omega) \\ P_{Y_2Y_1}(\omega) = P_{SS}(\omega) + H(\omega)P_{EE}(\omega) \\ P_{Y_2Y_2}(\omega) = P_{SS}(\omega) + H(\omega)H^*(\omega)P_{EE}(\omega) \end{array}. \quad (19)$$

The transfer function $H(\omega)$ can be expressed as Equation (20). Then, the signal $s(n)$ can be written again in a time domain by using the inverse Fourier transform, F^{-1}, as shown in Equation (21).

$$H(\omega) = \frac{P_{Y_2Y_2}(\omega) - P_{Y_2Y_1}(\omega)}{P_{Y_1Y_2}(\omega) - P_{Y_1Y_1}(\omega)}. \quad (20)$$

$$s(n) = F^{-1}\left(y_1(\omega) - \frac{y_2(\omega) - y_1(\omega)}{H(\omega) - 1}\right). \quad (21)$$

Finally, the structure of the aeromagnetic compensation and the improved coherent noise suppression algorithm is shown in Figure 7, including the compensation process and the interference suppression process.

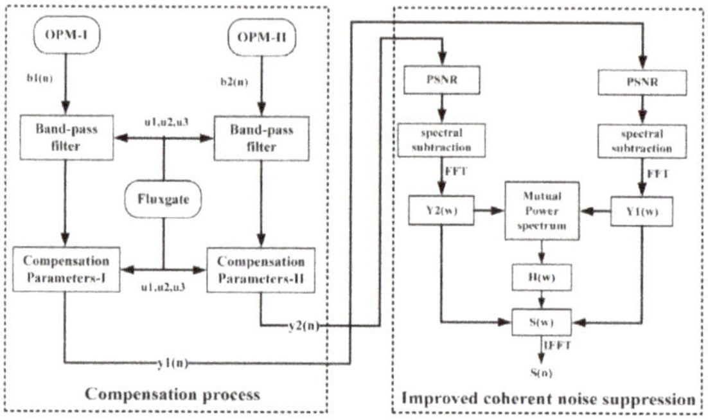

Figure 7. Block diagram of the improved coherent noise suppression (ICNS) method.

5. Results and Discussion

5.1. The Results of the Calibration Flight

5.1.1. Results of Adaptive Interference Cancelation (AIC)

As an important reference, the AIC method is often used in advance to eliminate the vibration interference magnetic field of a helicopter, and the results are calculated by using the method shown in Figure 8. The blue dot and black solid line represent the compensated field before and after the AIC method, respectively, and the black line is a clear offset of 1nT. In Figure 8, the vibration interference magnetic field is weakened to a certain extent, after using the AIC algorithm. In order to display the effect of elimination to a more visible extent, the algorithm effect is displayed using a time–frequency diagram, as shown in Figure 9.

In Figure 9, the interference fields at multi-frequency points still exist after the proposed method. Therefore, this AIC algorithm had a very limited effect on reducing the interference that was caused by the helicopter vibration.

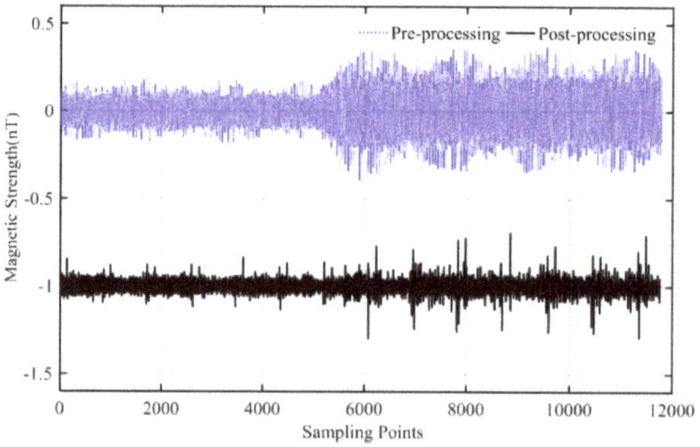

Figure 8. The pre-processing and post-processing results of the calibration flight data using AIC method.

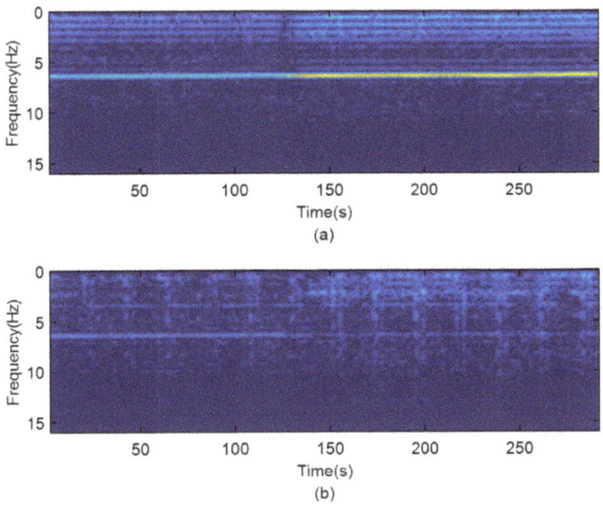

Figure 9. The time–frequency diagrams of the pre-processing (**a**) and post-processing (**b**) data using AIC method.

5.1.2. Results of the Improved Coherent Noise Suppression Method

For the same calibration flight data, the results using the ICNS method are shown in Figures 10 and 11. The blue dot and black solid line represent the compensated field before and after the ICNS method, respectively. In Figure 11, the multi-frequency point interference fields were mostly eliminated by the proposed method. However, some of the interference fields were still not eliminated, which may have originated from the random motion of the carrier.

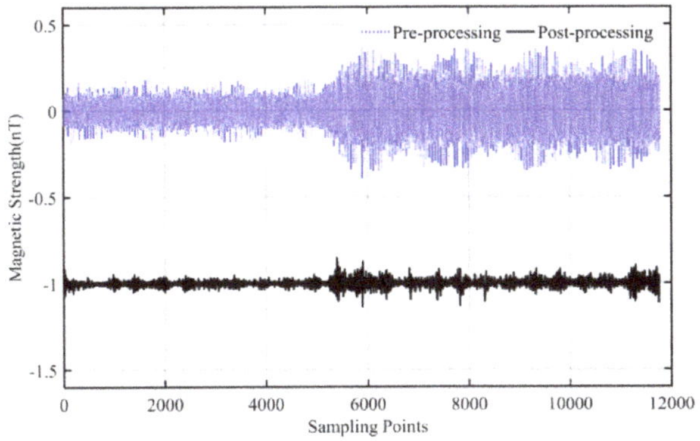

Figure 10. The pre-processing and post-processing results of the calibration flight data using ICNS method.

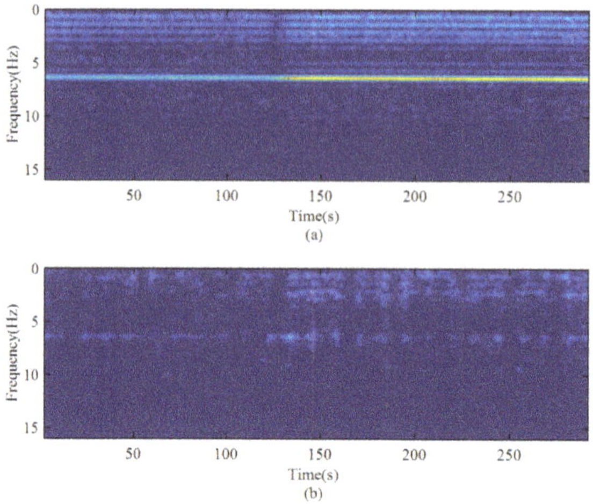

Figure 11. The time–frequency diagrams of the pre-processing (**a**) and post-processing (**b**) data using ICNS method.

By comparing the results in Figures 9 and 11, it can be seen that the ICNS method had a better suppression effect on the vibration interference magnetic field of the helicopter platform than the common AIC method if the two magnetometers were used. In order to further prove the effectiveness of the ICNS method proposed in this paper, Section 5.2 presents a simulation target design for the low-frequency magnetic field detection experiment to verify the effectiveness of this method.

5.2. The Results of the Detection Experiment of Target

The magnetic signal was produced by a square closed coil, which was placed at the position of (30°23′45″ N, 111°56′45″ E). The excitation current was in a square wave with a peak value of 4A at 2.08 Hz. The coil position and flight lines are shown in Figure 12. The flight test consisted of two lines above the analog signal source. The flight altitude of the survey line was about 500 m, as shown in Figure 12b.

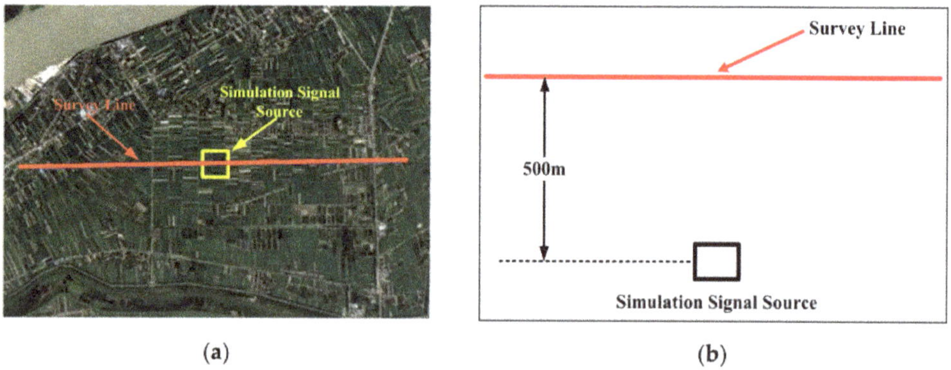

Figure 12. The flight path in Google Earth (**a**) and its scale drawing (**b**).

When the signal source was shut off, the measured magnetic field contained only the background field, and the measurement result of line 1 is shown in Figure 13. The interference field was eliminated by the proposed method. The blue dot line is the compensated data before the AIC method, the red dash line is the compensated data after the AIC method, and the black solid line is the compensated data after the ICNS method.

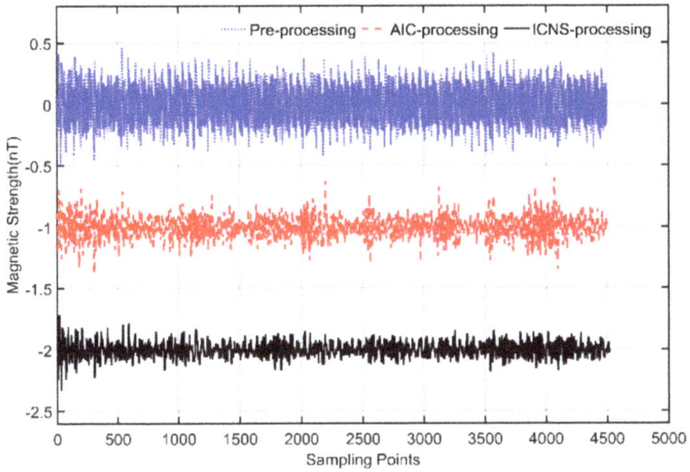

Figure 13. The pre-processing and post-processing results of the survey line without the target.

When the signal source was turned on, the measured magnetic field contained the background field and the low-frequency signal, and the measured result of line 1 is shown in Figure 14, where the blue dot line shows the compensated data before the AIC method, the red dash line shows the compensated data after the AIC method, and the black solid line is the compensated data after the ICNS method. It can be seen that the AIC method had little effect on the low-frequency signal. Meanwhile, the ICNS method eliminated most of the interference fields. In order to better show the processing results of the signal, the time–frequency diagram is given in Figure 15. It can be seen that the processed target signal was more obvious at 2.08 Hz and the harmonic frequency of 6.24 Hz than the original signal, as shown in Figure 15b.

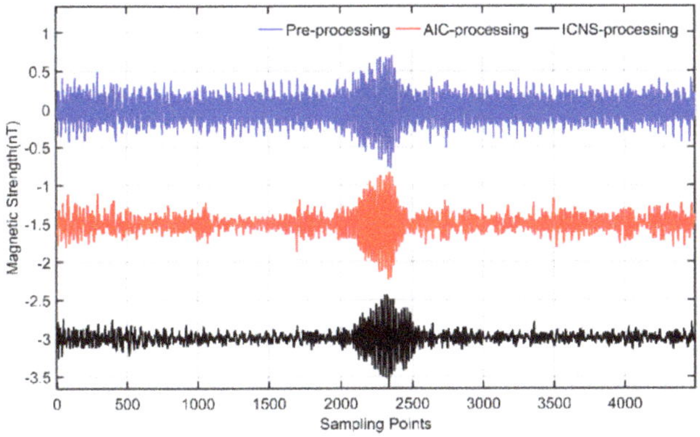

Figure 14. The pre-processing and post-processing results of the survey line with the target.

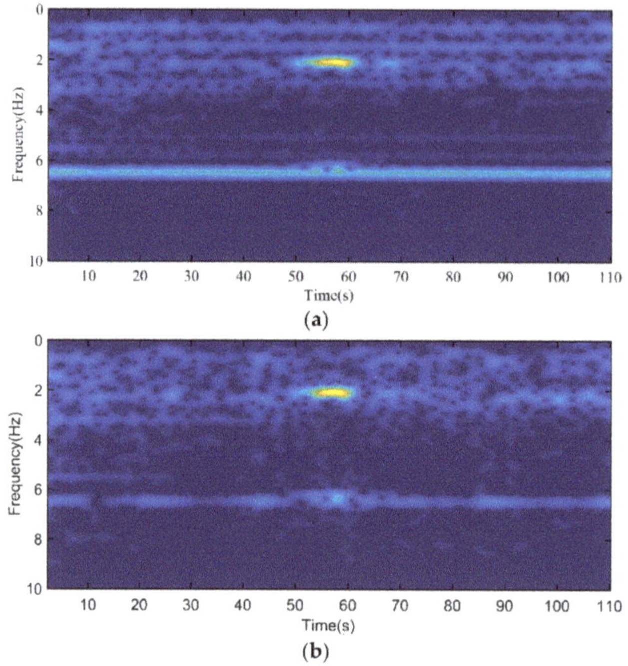

Figure 15. *Cont.*

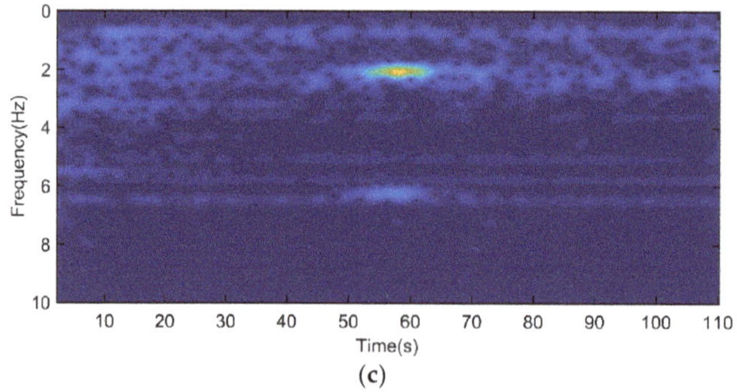

Figure 15. The time–frequency diagrams of the survey line with the target, the pre-processing data (**a**), post-processing data with AIC method (**b**) and post-processing data with ICNS method (**c**).

5.3. The Discussion of the Results

Through the processing of the compensation and detection survey flight data, the effectiveness of the algorithm proposed in this paper was well confirmed. Through the analysis of the results of the experiment, the conclusions can be summarized as follows:

(1) The traditional T–L aeromagnetic compensation model can only eliminate the interference magnetic field associated with the attitude of the manned helicopter, and its effective working frequency band is generally lower than 2 Hz.

(2) The ICNS-based method proposed in this paper can eliminate the interference field caused by the probe vibration and has little influence on the target signal, so it can

effectively improve the detection capability of the helicopter system for low-frequency target signals.

(3) The ICNS-based method can be used for the vibration magnetic field interference during the movement of the aircraft if the interference field has obvious correlation characteristics with a pair of magnetometers.

6. Conclusions

The vibration magnetic interference that is caused by the mechanical motion of the probe has a significant influence on the detection performance of low-frequency magnetic field of a helicopter aeromagnetic system. The traditional aeromagnetic compensation (T–L) model considers only the interference magnetic fields that are related to the attitude of the aircraft and is generally suitable for compensation with a frequency below 2 Hz. To eliminate these interference magnetic fields, a method based on the improved coherent noise suppression method (ICNS) with a pair of magnetometers is proposed in this paper. Since the distance between the two magnetometers and the helicopter is far less than the distance between the helicopter and the target, the ICNS method can adaptively eliminate the vibration interference fields and has little effect on the field to be measured. The results of the calibration and detection with a flight test show that the proposed method can effectively eliminate the interference field that is caused by the vibration in accordance with the approaching target. The proposed algorithm shows its effectiveness in detecting ocean targets that generate low-frequency magnetic field signals or anomalies.

Author Contributions: Conceptualization, Y.F.; methodology, Y.F. and L.C.; software, L.C. and Y.Z.; validation, Y.Z. and L.C.; formal analysis, X.Q.; investigation, Y.F. and G.F.; resources, G.F.; data curation, Y.F. and L.C.; writing—original draft preparation, Y.F.; writing—review and editing, Y.Z.; visualization, L.C.; supervision, G.F.; project administration, X.Q.; funding acquisition, G.F. All authors have read and agreed to the published version of the manuscript.

Funding: This research work was supported by the National Key R&D Program of China under grant 2019YFC1408102, and the National Natural Science Foundation of China under grant 61901441. The authors wish to express their gratitude to the editor and the anonymous reviewers.

Institutional Review Board Statement: Not applicable.

Informed Consent Statement: Not applicable.

Data Availability Statement: Not applicable.

Conflicts of Interest: The authors declare no conflict of interest.

References

1. Ioannidis, G. Identification of a Ship or Submarine from its Magnetic Signature. *IEEE Trans. Aerosp. Electron. Syst.* **1977**, *AES-13*, 327–329. [CrossRef]
2. Zolotarevskii, Y.M.; Bulygin, F.V.; Ponomarev, A.N.; Narchev, V.A.; Berezina, L.V. Methods of Measuring the Low-Frequency Electric and Magnetic Fields of Ships. *Meas. Tech.* **2005**, *48*, 1140–1144. [CrossRef]
3. Birsan, M. Measurement of the extremely low frequency (ELF) magnetic field emission from a ship. *Meas. Sci. Technol.* **2011**, *22*, 085709. [CrossRef]
4. Wu, Z.; Zhu, X.; Li, B. Modeling and measurements of alternating magnetic signatures of ships. *Sens. Transducers.* **2015**, *186*, 161–167.
5. Sun, Y.; Lin, C.; Jia, W.; Zhai, G. Analysis and measurement of ship shaft-rate magnetic field in air. *Prog. Electromagn. Res. M* **2016**, *52*, 119–127. [CrossRef]
6. Nabighian, M.N.; Grauch, V.J.S.; Hansen, R.O.; LaFehr, T.R.; Li, Y.; Peirce, J.W.; Phillips, J.; Ruder, M.E. The historical development of the magnetic method in exploration. *Geophysics* **2005**, *70*, 3361. [CrossRef]
7. Doll, W.E.; Gamey, T.J.; Bell, D.T.; Beard, L.P.; Sheehan, J.R.; Norton, J.; Holladay, J.S.; Lee, J.L.C. Historical Development and Performance of Airborne Magnetic and Electromagnetic Systems for Mapping and Detection of Unexploded Ordnance. *J. Environ. Eng. Geophys.* **2012**, *17*, 1–17. [CrossRef]
8. Hardwick, C.D. Non-oriented cesium sensors for airborne magnetometry and gradiometry. *Explor. Geophys.* **1984**, *2*, 266–267.
9. Tolles, W.E. Compensation of Aircraft Magnetic Fields. U.S. Patent 2692970A, 26 October 1954.
10. Tolles, W.E.; Mineola, N.Y. Magnetic Field Compensation System. U.S. Patent US2706801A, 19 April 1955.

11. Leliak, P. Identification and Evaluation of Magnetic-Field Sources of Magnetic Airborne Detector Equipped Aircraft. *IRE Trans. Aeronaut. Navig. Electron.* **1961**, *ANE-8*, 95–105. [CrossRef]
12. Leach, B.W. Aeromagnetic compensation as a linear regression problem. In *Information Linkage Between Applied Mathematics and Industry*; Academic Press: New York, NY, USA, 1980; pp. 139–161.
13. Hardwick, C.D. Aeromagnetic Gradiometry in 1995. *Explor. Geophys.* **1996**, *27*, 1–11. [CrossRef]
14. Nelson, J.B. *Aeromagnetic Noise during Low-Altitude Flights over the Scotian Shelf*; DRDC-ATLANTIC, Defence R&D Canada: Ottawa, ON, Canada, 2002.
15. Nelson, J.B. *Predicting In-Flight Mad Noise from Ground Measurements*; DRDC-ATLANTIC, Defence R&D Canada: Ottawa, ON, Canada, 2001.
16. Noriega, G. Performance measures in aeromagnetic compensation. *Lead. Edge* **2011**, *30*, 1122–1127. [CrossRef]
17. Noriega, G. Model stability and robustness in aeromagnetic compensation. *First Break* **2013**, *31*, 73–79. [CrossRef]
18. Sheinker, A.; Moldwin, M.B. Adaptive interference cancelation using a pair of magnetometers. *IEEE Trans. Aerosp. Electron. Syst.* **2016**, *52*, 307–318. [CrossRef]

applied sciences

MDPI

Article

Applications of Electromagnetic Induction and Electrical Resistivity Tomography for Digital Monitoring and Assessment of the Soil: A Case Study of Al-Ahsa Oasis, Saudi Arabia

Maged Mohammed [1,2,*], Ahmed El Mahmoudi [3] and Yousef Almolhem [4]

1. Date Palm Research Center of Excellence, King Faisal University, Al Hofuf 31982, Saudi Arabia
2. Agricultural and Biosystems Engineering Department, Faculty of Agriculture, Menoufia University, Shebin El Koum 32514, Egypt
3. Geology Department, Faculty of Science, Mansoura University, Mansoura 35516, Egypt; mahmoudi@mans.edu.eg
4. Department of Environment and Agricultural Natural Resources, College of Agricultural and Food Sciences, King Faisal University, Al Hofuf 31982, Saudi Arabia; yalmolhem@kfu.edu.sa

* Correspondence: memohammed@kfu.edu.sa

Abstract: Al-Ahsa Oasis is one of the main and oldest agricultural centers in Saudi Arabia and one of the largest natural oases in the world. It is the largest irrigated agricultural oasis worldwide. The cultivated area is about 12,000 hectares in the region of Al-Ahsa. Groundwater was considered the main water source, if not the only source, for the different purposes of Al-Ahsa Oasis. However, due to groundwater depletion from these aquifers, treated wastewater is currently used to meet the shortage of irrigation water at Al-Ahsa Oasis. Today, several problems affect the agricultural productivity at Al-Ahsa Oasis, and modern geophysical techniques could contribute effectively to solving these problems; such problems are increasing soil salinity and the existence of a hardpan layer in some parts of the Al-Ahsa oasis. The soil texture is considered a critical factor to be taken into consideration. Therefore, the objectives of this study were to evaluate the effectiveness of modern geophysical techniques to study the agricultural aspects of Al-Ahsa oasis. Ten profiles of 2-D electrical tomography using SuperSting R8/IP 8 channel multielectrode resistivity and IP imaging system with 112 electrodes at one-meter spacing were implemented. Moreover, twenty electrical conductivity profiles were measured of 2240 points using EM38-MK2 at vertical mode with transmitter–receiver coils of 0.5 and one meter spacing. The results indicated that the salinity variation along the measured profiles was changed vertically and laterally due to the changes in the soil type, texture, moisture content, sabkha, and the whole oasis. Analysis and interpretation of the interpreted resistively tomograms of the ten 2-D profiles indicate remarkably the different geoelectric units of different hydrogeological conditions and soil texture across the oasis. These results will be of good utility for farmers, development sectors, and research institutions, especially in agriculture development studies, at Al-Ahsa Oasis, to ensure the sustainability of agriculture in this important historical Oasis.

Keywords: agricultural geophysics; soil salinity; hardpan; soil texture; electrical resistivity tomography; electromagnetic induction; precision agriculture

1. Introduction

The methods of surface geophysics offer prompt and inexpensive means for subsurface characterization [1–3]. In addition, the surface geophysics technics provide information on subsurface properties, such as the layer's thickness and saturation zones, depth to bedrock, fracture zones, location and orientation of bedrock fractures, and faults. New developments in recent years are the use of 2-D electrical imaging/tomography surveys for mapping areas with moderately complex geology [4].

Agricultural geophysics is a modern discipline that applies noninvasive geophysical methods to obtain valuable soil surface and near-surface information. Geophysical techniques are becoming an important practical tool for many agricultural applications. Agricultural geophysical analyses are focused on delineating features over large areas within the soil profile (0 to 2 m depth) [5]. The most common geophysical methods predominantly used for agricultural purposes are direct current (DC) resistivity and electromagnetic induction. With the recent advent of precision agriculture, the ability to generate a large amount of information rapidly is becoming more critical than ever [5,6]. Site infrastructure assessment, hydrologic monitoring, and environmental investigations are just a few current and potential uses of near-surface geophysical methods in agriculture applications. In addition, modern geophysical techniques are being incorporated within site-specific agriculture to detect soil horizons, soil compaction, soil salinity, mapping of hardpan layer, and soil texture [6]. Today, with the advent of precision agriculture which employs data from multiple sources to enhance crop yields and assess soil quality to improve the cost-effectiveness of crop management strategies, it is necessary to advance the agricultural geophysics methods for clay-pan depth determination, soil hydrologic characterization, soil nutrient monitoring, crop/tree root biomass surveying, identification of subsurface flow pathways, subsurface irrigation, drainage systems infrastructure detection, and soil compaction evaluation [7–9].

Electromagnetic (EM) instruments have been used extensively to determine the spatial distribution of soil attributes, including clay content [10,11], moisture [12], and soil salinity [13–16]. In addition, different studies have used EM instruments to determine the depth of a clay layer [17], soil types, and organic carbon fraction [18]. Al-Ahsa Oasis is located 70 km to the west of the Arabian Gulf, and it is also one of the main agricultural areas in Saudi Arabia (Figure 1).

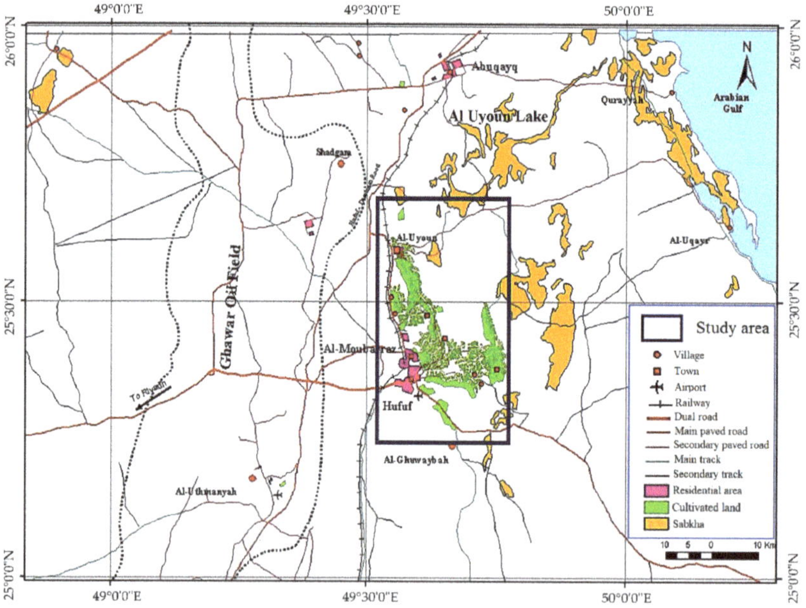

Figure 1. Location map of Al-Ahsa Oasis, Eastern Province, Saudi Arabia.

The cultivated area in the Al-Ahsa oasis is about 12,000 hectares. Groundwater was considered the main water source, if not the only source, for the different purposes of Al-Ahsa Oasis. As a consequence of the overexploitation of groundwater resources, a decline in groundwater levels is observed. The famous springs in the Al-Ahsa oasis are running dry. Furthermore, deterioration of groundwater quality is occurring, caused by upconing of deeper saline groundwater [19].

Due to groundwater depletion, treated wastewater is currently used to meet the shortage of irrigation water at Al-Ahsa Oasis. As a result, most groundwater is discharging from the Neogene aquifer. Today, due to excess overpumping, water depletion, and water deterioration, the signs of soil salinity at Al-Ahsa have become dominant [20,21].

Water and soil are the most critical factors that determine the success of agricultural activity in any location based on their characteristics. Some excavations at Al-Ahsa oasis indicated hardpan layers at different soil depths. Generally, hardpan is a term for a dense layer of soil below the uppermost topsoil layer. The negative impacts of the hardpan layer are that it is mainly impervious to water, restricts root growth, and makes it difficult for air, water, other gases, and organisms to move through the soil. Therefore, determining the existence of the hardpan layer nature and depth of its presence is very important in crop selection [22].

Al-Ahsa soils are underlain by an impermeable marl layer (mixtures of clay and calcium carbonate) at zero to 2 m depth below the ground surface. Excess irrigation water accumulates on top of this layer, forming a shallow water table beneath the soils of the Al-Ahsa Oasis. The water table depth is affected by varying depths to the marl layer, the soil texture, and depth and spacing between field drainage canals [23]. The irrigated arid zone of Al-Ahsa oasis faces twin problems of water scarcity and waterlogging accompanied by salinity. For more details about the studies related to the soil of Al Hassa, refer to [24–26]. Al-Dakheel [27] investigated the extent of soil salinity, irrigation water quality, and the relationship with the growth of vegetation, employing NDVI derived from Landsat satellite imagery. Al-Barrak and Al-Badawi [28] reported that salt-affected soils in Al-Ahsa region represent as much as 20 % of the total area.

El Najem [21] studied the degraded agricultural soils at Al-Ahsa Oasis and reported that the decline in the soil productivity distinguished the soil degradation at Al-Ahsa oasis. He attributed that to human activities and the marginal climatic conditions, which are considered the most important sources of soil degradation. The soil chemical properties exhibited slightly alkaline pH values, extending between 7.33 and 7.86, while the soil salinity data, as EC (electrical conductivity) values, varied across the Al-Ahsa Oasis. He found several values that exceeded 400 mS/m (i.e., the minimum value of saline soils), reaching 7.80 dS m^{-1}, implying the persistence of soil salinity conditions (i.e., saline soil) in the Oasis's agricultural soils. It was also noticed that the soil EC values increased with increasing soil depths, signifying the inefficient drainage systems and, hence, salinization of the soil profile in the Oasis [21].

Inland Sabkhas in Al-Ahsa area are salt-affected intermediate to dark brown soils found in relatively low relief areas. Sabkha in Al Hassa exists in very low relief areas, which are saturated with water (where groundwater rose and pushed salt to the surface).and may even have surface water [27].

Al-Ahsa drainage water is collected by earthen lateral open canals having depths of about 1 m. The collected drainage water in the eastern part of Al-Ahsa oasis flows from the lateral's canals to three sub-main drainage canals, which are connected to the main drainage canal D2 as shown in Figure 2. The collected drainage water in this drainage flows towards an evaporation lake in Al Asfar 13 km to the east of the Oasis, while the drainage water in the northern sector flows from the lateral to the sub-main drainage canals that were connected to the main drainage canal D1. The drainage water in D1 flows towards an evaporation lake in Al Uyoun (see Figure 2). The collected water in both lakes remains in the winter season to be a source for cattle to drink, as well for shallow aquifer recharge. From December until April, the agricultural drainage water overflows from Al Uyoun lake through small waterways into the Arabian Gulf [23].

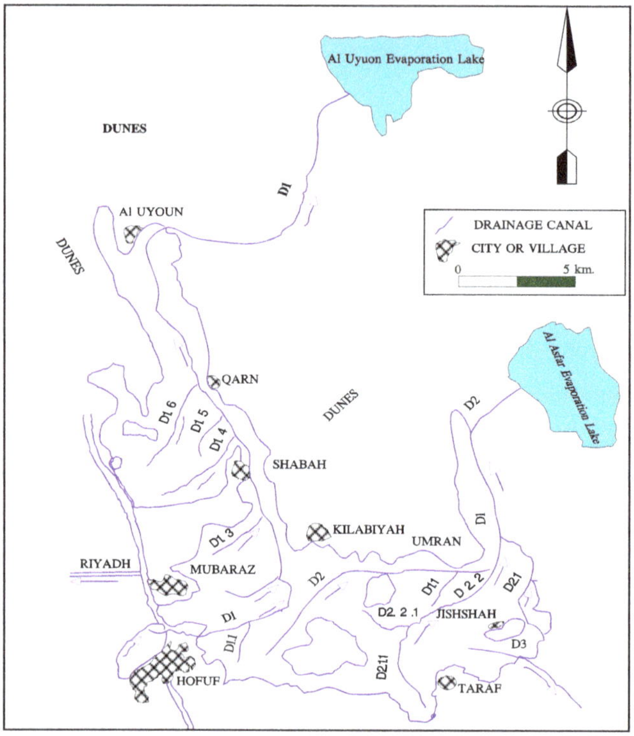

Figure 2. The main drainage system of Al-Ahsa [23].

For more details about the hydrogeology and the hydrochemical characterization of the groundwater of the Al-Ahsa area, refer to [19,27,29–32].

Soil surveys require quick and, when possible, nondisturbing estimations of numerous soil properties, such as salinity, texture, stone content, groundwater depth, and horizon sequence in soil profiles; however, conducting soil measurements with a high sampling density is costly and time-consuming. Conventional methods of soil analysis mostly require disturbing soil, removing soil samples, and analyzing them in a laboratory. So, the traditional methods used to acquire soil information involve intensive field surveys and laboratory analyses. As a consequence, only limited amounts of data are collected. In addition, the low resolution of the spatial data used to provide indicators of the distribution of water/soil properties and soil conditions may lead to interpretation errors and possibly errors in soil management. On the contrary, geophysical methods allow rapid measurement of soil electrical properties, such as electrical conductivity, resistivity, and potential, directly from the soil surface to any depth without soil disturbance. Modern geophysical techniques are a very efficient and effective tool for the nonintrusive gathering of continuous agricultural field data on many soil types, soil conditions, soil salinity, and underground topographies.

The objectives of this study were to evaluate the effectiveness of modern geophysical techniques to study the agricultural aspects of Al Hassa oasis.

2. Materials and Methods

2.1. The Geophysical Investigations

Electrical resistivity tomography (ERT) and Electromagnetic (EM) induction were implemented to achieve the objectives of this study. The DC resistively method of geophysical exploration is popular, has proved to be successful, and has many implications in agriculture, environmental studies, and hydrogeology. Moreover, the conductivity meter is an inexpensive, rapid, and effective means for mapping soil salinity. In this study, 2-D electrical resistivity measurements were carried out using a SuperSting R8/IP 8 channel multielectrode resistivity (Advanced Geosciences (AGI), Inc. Geoscience Dr., Austin Texas, TX, USA) and IP imaging system with 112 electrodes manufactured by Advanced Geosciences, Inc., (AGI), (Figure 3A), while the EM38-MK2 m (Geonics Limited, Meyerside Dr., Mississauga, Ontario, Canada) was used in vertical mode (Figure 3B) to measure the electrical conductivity (EC) along the study area.

(A) (B)

Figure 3. Geophysical data acquisition using the SuperSting R8/IP 8 resistivity (**A**) and the EM38-MK2system electrical conductivities (**B**) at the study area.

Ten profiles of 2-D (P-1 to P-10) electrical tomography using the SuperSting R8/IP 8 multielectrode channel resistivity and IP imaging system with 112 electrodes at one-meter electrode spacing were implemented. The locations of these profiles were chosen to cover different zones across the entire Al Hassa oasis. Moreover, 2240 measuring points using EM38-MK2 at vertical mode with the transmitter–receiver of 0.5 and one-meter spacing were measured with twenty electrical conductivity profiles. Figure 4 shows the base map of the ten conducted 2-D profiles along the study area. It is worth mentioning that the EM38-MK2 measured electrical conductivity was measured at the 10 exact locations of the measured 2-D profiles to correlate soil conductivity/resistivity at near-surface (These lines are Labeled Line One-1, Line Two-1, etc.).

In contrast, the other ten EM38-MK2 electrical conductivities, each of them, were measured five meters away from the location of the 2-D transects. These EC lines are labeled Line One-2, Line Two-2, etc., to distinguish them on the data logger files from those measured at the exact locations of the 2-D transects. These ten EC lines (labeled Line One-2, Line Two-2, etc.) were measured for correlation with those measured at locations of the 2-D transects.

For more details about the irrigation water quality refer to, for example, Hussein et al. [31]. For the soils of Al Hassa, refer to El Najem [21].

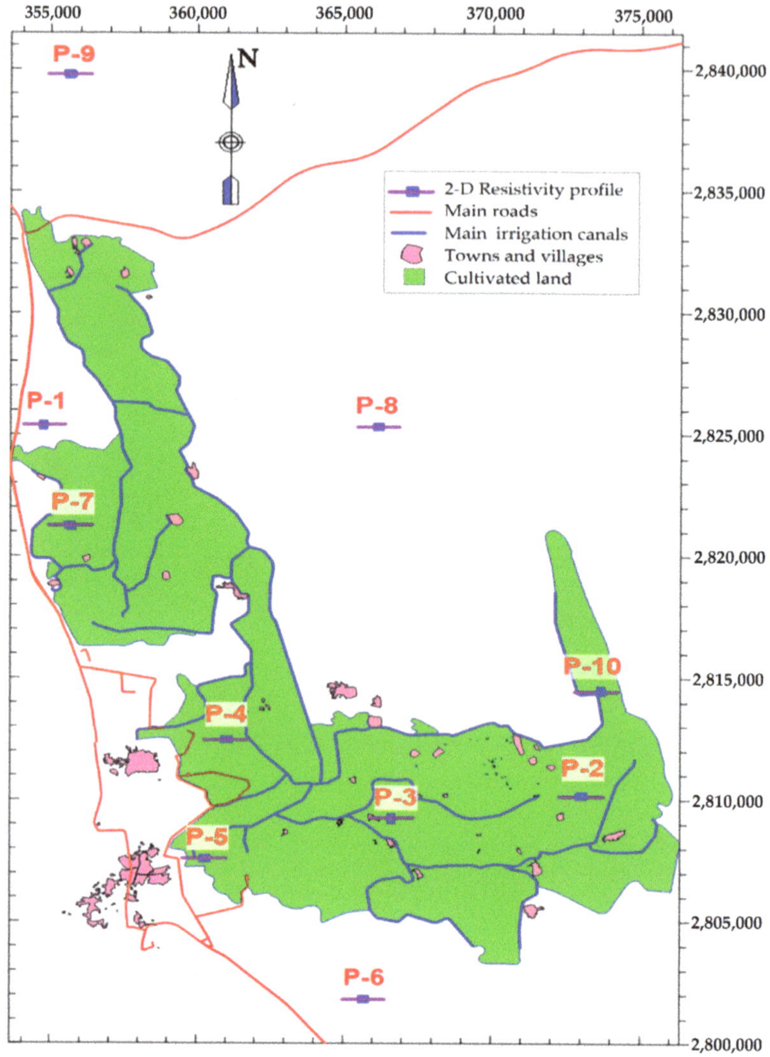

Figure 4. Base map showing the locations of the 2-D resistivity tomography profiles and EM38-MK2 at the study area.

2.2. Two-Dimensional Resistivity Description and Data Acquisition

The SuperSting R8/IP multielectrode resistivity provides apparent resistivity, induced polarization, and battery voltage. The measurement range is ±10 V, and the max measuring resolution is 30 nV, depending on the voltage level. The internal transmitter is 200 W and the external 15 kW. The output current ranges from 1 to 2000 mA continuous, and it can be measured with high accuracy. The output voltage is 800 V, and the actual electrode voltage depends on the ground resistivity and transmitted current. There are eight input channels.

The input gain ranges automatically and always uses the receiver's full dynamic range. The input impedance is more than 50 MΩ. The system calibration is conducted digitally by the microprocessor, based on the correction values stored in the memory. The supported configurations in manual mode are resistance, Wenner, Schlumberger, pole–dipole, dipole–dipole, and pole–pole, and any configuration can be programmed in automatic mode.

The operating system is stored in a reprogrammable memory. For each measurement, full-resolution readings of the mean value and error are stored along with the user-entered coordinates and time of the day.

The displayed and stored data in the memory are apparent resistivity (Ohm-m), current intensity (mA), and measured voltage (mV) for each measurement. In addition, the data can also be realtime displayed on an Android device with bright color, transmitter/receiver plot, IP curves, and contact resistance measurements. The internal SuperSting memory in the resistivity mode can store more than 79,000 measurements and can store more than 26,000 measurements in the combined resistivity/induced polarization mode. The RS-232C channel is available to dump the data from the instrument to a Windows-type computer based on the user command. The operating time of the SuperSting R8/IP depends on survey conditions and the used battery size. The internal circuitry in the auto mode can adjust the current to save energy. The operating temperature ranges from −5 to +50 °C.

The resistivity method depends on the injection of the electrical current (I) into the ground by using two current electrodes A (source) and B (sink), and two potential electrodes M and N, computing the resulting potential differences (ΔV) as illustrated in Figure 5. Increasing the electrode spacing results in a deep depth of investigation, where the electric current passes through the investigated subsurface targets. By considering the subsurface structure, the depth of investigation represents about 20% to 40% of the spacing of the current electrodes [33].

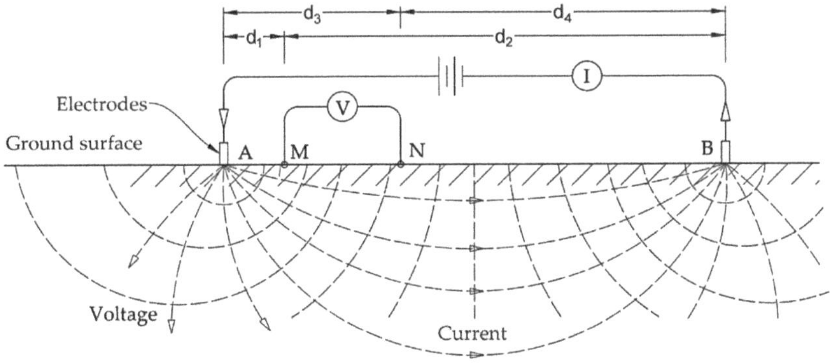

Figure 5. Schematic diagram illustrating the basic concept of electrical resistivity measurement.

The following Equations (1) and (2) give the apparent electrical resistivity of the subsurface

$$\rho_a = \frac{V_{MN}}{I} G \quad (1)$$

where G is the Geometrical Factor and is given by

$$G = \frac{2\pi}{1/d_1 - 1/d_2 - 1/d_3 - 1/d_4} \quad (2)$$

where d_1, d_2, d_3, and d_4 are the distances of the potential electrodes M and N from the current electrodes (A and B), see Figure 5.

For more details about resistivity and survey deployment basics, refer to Musset and Khan [34] and Reynolds [35]. In this study, the 2D electrical resistivity measurements were carried out using the SuperSting R8/IP 8 multielectrode channel resistivity and IP imaging system with 112 electrodes at one-meter spacing manufactured by AGI Advanced Geosciences, Inc., Austin, TX, USA (Figure 3A).

A combination of dipole–dipole, Wenner-Schlumberger, and pole–dipole configurations at each transect of the conducted ten 2-D resistivity profiles was implemented and merged during the processing of the interpretation of resistivity data for each transect.

The coordinates of the conducted 2-D profiles were measured using the GPS system. The quality of the data acquisition was entirely controlled by the computer software, which, before starting any measurement, checked the electrode connectivity to the soil and the magnitude of the contact resistances. The 1 m electrode spacing was chosen with an estimated investigation depth about 20 m below the ground. The apparent resistivity data were inverted to create a model of the resistivities of the subsurface using AGI Earthimager 2-D Software (after Advanced Geosciences, Inc., Austin, TX, USA). With this software, the data collected with the AGI SuperSting earth imaging resistivity instruments can be interpreted into easy-to-read 2-D earth sections (resistivity tomograms). A more accurate subsurface model is the two-dimensional (2-D) model, where the resistivity changes in the vertical and horizontal directions along the survey line are recorded. 2-D resistivity profiling was carried out by making many measurements at different profile locations and offsets. Finally, the 2-D resistivity profiling data were inverted using AGI Earthimager 2-D Software (after Advanced Geosciences, Inc., Austin, TX, USA) to create a tomogram-like resistivity model along a subsurface section that detected different salinity water zones [4]. AGI EarthImager 2-D software (after AGI Inc., Austin, TX, USA) was used for resistivity data inversion. AGI EarthImager 2-D software is a two-dimensional inversion modeling software for affordable resistivity and induced polarization (IP) imaging. Using this software, any array or mixed data from Schlumberger, pole–pole, pole–dipole, dipole–dipole, or Wenner electrode arrays can be inverted.

2.3. *EM38-MK2 Description and Data Acquisition*

EM38-MK2 provides the measurements of in-phase (magnetic susceptibility) and quad-phase (conductivity) components within two different depth ranges without any requirement to contact with or penetration of the instrument into the soil. The maximum practical depth of exploration is 1.5 m. Therefore, the EM38-MK2 applications in general soil and agriculture sciences are typical. The EM38-MK2 contains two receiver coils, divided to 1 m and 0.5 m from the transmitter, supplying the data from the effective depth ranges of 1.5 m and 0.75 m, respectively when placed in the vertical dipole orientation. On the other hand, it supplies the data from the effective depth ranges of 0.75 m and 0.375 m, respectively at horizontal dipole orientation. Once the EM38-MK2 is positioned within the collapsible calibration stand, the instrument can be calibrated automatically within seconds, without any necessity for iterative adjustments. In addition, the EM38-MK2 is supported by temperature equilibrium circuitry to improve temperature-related drift aspects. The data can be collected by any appropriate field computer or the DAS70-AR Data Acquisition System, which can be connected by either Bluetooth wireless technology or RS-232 serial cable. The DAS70-AR provides an improved level of support for data collection, and it includes a rugged field computer with 8 GB storage memory. For more details about this instrument, refer to Geonics Ltd., Mississauga, ON, Canada.

The diagram of the EM38 and its related electromagnetic fields is described in Figure 6. The primary magnetic field (Hp) is generated from the transmitting coil (Tx) at one end of the sensor, this field produces current loops in the ground below, and in turn, the current loops generate their own magnetic field (Hi). The generated secondary field is superimposed on the primary field, and both Hp and Hi are measured in a receiving coil (Rx) at the other end of the sensor [36]. The meter is intended to be placed or held in either the horizontal or vertical orientation; this orientation position controls the signal penetration depth and depth weighting response pattern. In a homogeneous soil profile, the vertical (EMV) signal penetrates to a depth of 1.5–2 m, while the horizontal (EMH) signal primarily reflects the shallower soil zone (0.75–1.0 m); Manual of EM38-MK2, (Geonics Limited).

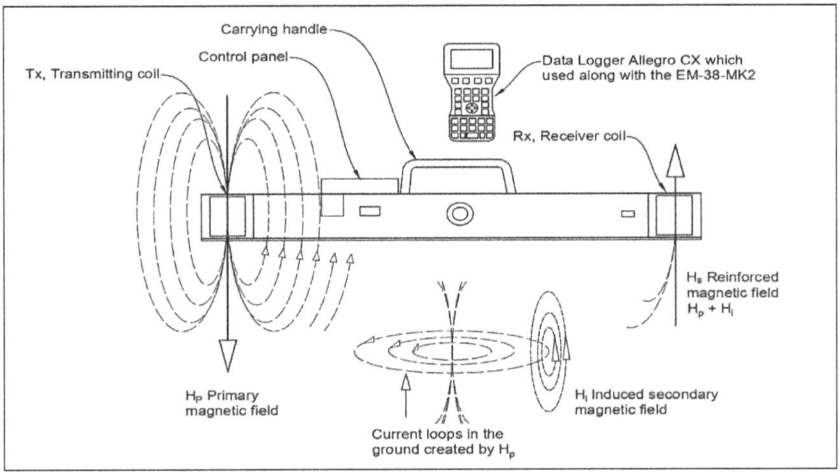

Figure 6. Diagram of an EM38 m showing the principles of operation.

Therefore, the instrument works by emitting an alternating current through a transmitter coil, which induces a primary magnetic field to pass perpendicular to the coil orientation and into the adjacent media. Eddy currents are generated as these currents pass through the coil. This strength is a function of the (a) amount of negative charge on the clay particle; (b) clay content; (c) concentrations of salts in the soil solution; and (d) soil moisture content. The more conductive the soil, the greater the secondary magnetic field produced. The ratio of the primary and secondary magnetic fields determines the apparent electrical conductivity (ECa).

Electromagnetic induction sensors, such as EM38-MK2, are becoming more widely used for mapping relative differences in soil characteristics in agricultural fields, including soil salinity; see [15]. In this study, EM38-MK2 was used in vertical mode to measure the electrical conductivity (EC) at 75 and 150 cm depths at about 2240 sites along the study area (see Figure 4). The acquired data using the EM38-MK2 conductivity meter were processed using the DAT38MK2 Ver. 1.05 package (Geonics Limited) and exported to XYZ format for the twenty EC measured profiles.

3. Results and Discussion

3.1. Two-Dimensional Resistivity Tomography Data

To tie the inverted 2-D resistivity tomograms with the available lithological and hydrogeological information, previous work conducted by different authors, for example [16,20,22,24,25], have been correlated with 2-D inverted resistivity tomograms. Using an iterative smoothness-constrained least-squares inversion method [37,38], apparent resistivity data collected by the 2D-DC resistivity system were inverted using AGI Earthimager 2D Software (after Advanced Geosciences, Inc.) to create a subsurface resistivity model that approximates the actual subsurface resistivity distribution [38]. Figure 7 shows the inversion of profile P-6 (see Figure 4 for location), where the resistivity field-data is presented on the upper part of the section, resistivity models are presented on the middle part of the section, and synthetic-data inversions are shown on the lower part of the section.

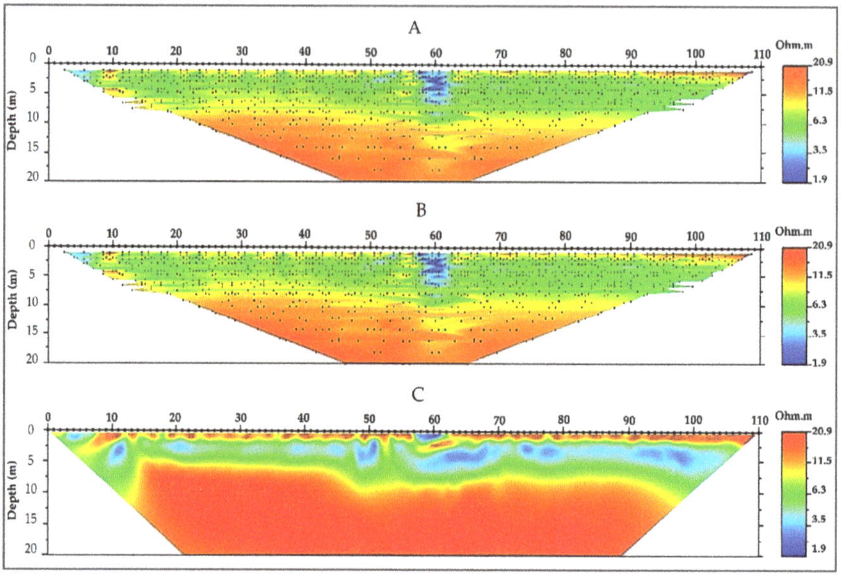

Figure 7. Two-dimensional resistivity section for the location P-6. The upper section is the measured resistivity (**A**), the middle section is the calculated resistivity (**B**), and the lower section is the inverted resistivity (**C**) (see Figure 4 for location).

Investigations of the constructed 2-D resistivity tomograms along with the data from the previous work conducted by different authors at Al-Ahsa Oasis, as well some soil pits dug by local farmers, helped in the interpretation of the resistivity tomograms and in the mapping of the near-surface condition of the soil along the study area which has its impacts on soil characterization.

Figure 8 shows the combined tomograms with their locations across the Al-Ahsa Oasis; vertical and lateral variations of different lithological units are recognized. Referring to Figure 8 and investigations of these 2-D Tomograms of the location of both of Al Asfar and Al Uyoun lakes, it is found that the main three identified geoelectric-lithologic units along these tomograms are:

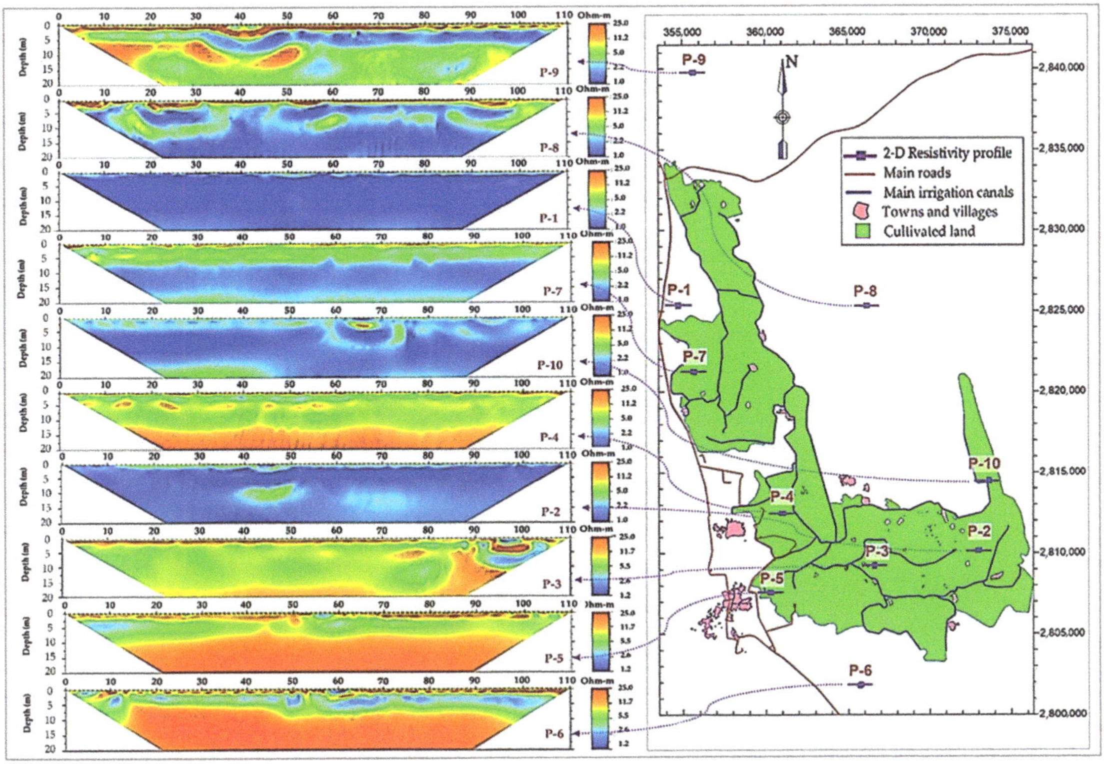

Figure 8. The Combined 2-D inverted tomograms show how the resistivity variations vertically and horizontally at the ten tested locations from 1 to 10 (P-1 to P-10), according to the positions, due to change in hydrogeological conditions and lithology and soil texture.

Layer-1 is a topsoil zone of loose to weakly consolidated sand. This zone corresponds to the upper part of the very resistive layer. The resistive nature of this layer is indicative of dry conditions in the upper part of the measured 2-D transects. However, and referring to Figure 8, it is noticed that resistivity values of this layer have relatively high values (above 10 Ohm-m) along 2-D transects of profiles 1, 10, and 7 (P-1 to P-10, see Figure 4 for locations); these are attributed to both dry conditions and the coarse grains sand at these locations. On the other hand, this layer at the 2-D transects of profiles 1, 10, and 2 (P-1, P-10, and P-2, Figure 8) have relatively low values due to the effect of Sabkha and clayey soil and the effects of the existence of both the Al Yuoun evaporation lake at the north of Al-Ahsa Oasis and of the Al Asfar Lake at the east of the Oasis. These match well with the results of Al-Dakheel et al. and El Mahmoudi et al. [27,39]. Referring to Figure 8, it is worth mentioning the alternating of high and low resistive parts at this layer (see, for example, P-9 at a distance of 30–50 m, Profile 10 at a distance of 60–75 m, and Profile 3 at a distance of 90–105 m. This indicates the inhomogeneity of this layer and variations of irrigation practices at the same farm (part of the farm is cultivated by vegetables and the other is date palm). Moreover, some farmers transport soils from other places to some locations at their farms.

Layer-2 has a moderate resistivity range of (2 to 10 Ohm-m). This layer exists at different depth intervals (for example, 2-D transects of profiles P-4, P-5, P-6, P-8, and P-9, Figure 8). This layer is interpreted as a partially saturated layer, which makes its resistivity values less than of the topmost soil layer.

Layer-3 is a thick zone below Layer-2. This layer has different resistivity ranges. The resistive nature of this third layer (above 10 Ohm-m) is along 2-D transects of profiles 6, 5, 3, and 4 (P-6, P-5, P-3, and P-4, Figure 8), which are located in the southwest of Al-Ahsa oasis. This layer at these locations is attributed to lithological changes and as well different hydrogeological conditions. The existence of Neogene limestone outcrops has been noticed at the surface and the near-surface in this part, while at the northern part, it does not exist. For more geological details of Al-Ahsa Oasis, refer to Hotzl et al. [40]. On the other hand, this layer has resistivities below 2 Ohm-m along with profiles 7, 1, 9, 8, 10, and 2 (P-7, P-1, P-9, P-8, P-10, and P-2, Figure 8) located in the north and east of Al-Ahsa oasis. However, and referring to Figure 8, it is noticed that this layer's resistivity values have relatively low values due to the effect of water saturation from both the Al Yuoun evaporation lake at the north of Al-Ahsa Oasis and of Al Asfar Lake at the east of the Oasis.

Figure 9 shows one of the dug pits beside 2-D profile 1 (P-1, Figure 8) and shows how the water table is very shallow and the saturation affects the resistivity values to be below 2 Ohm-m.

Figure 9. An image shows the dug pits and the shallow water table beside profile 1 (P-1, Figure 8).

A low resistivity layer was mapped along some 2-D transects, for example, profiles 6, 9, and 8 (P-6, P-9, as well as P-8, Figure 8). This layer is a hardpan layer composed of marl, clay, mudstone, or shale [22]. The layer directly affects the level of soil water and thus the accumulation of salt in some locations resulting in the presence of some of the sites affected by the salt composed sabkha, especially if it is very shallow and this, in turn, increased the soil conductivity measured by EM38-MK2, as will be discussed below. Figure 10 shows some excavation and the existing hardpan layer at different depths near profile 6 (P-6, Figure 8).

3.2. EM38-MK2 Results

Twenty profiles using the EM38-MK2 in the vertical mode were conducted to measure the apparent electrical conductivity (ECa) at 75 and 150 cm depths, based on implemented coil separation (0.5 and 1 m separation in vertical mode). The data from about 2240 measuring points sites acquired by EM38-MK2 were automatically logged by a special Data Logger Allegro CX. It is worth mentioning that ten of the EM38-MK2 profiles of measured electrical conductivity had the same locations of the measured 2-D profiles, while the other ten electrical conductivities lines measured by EM38-MK2 were measured five meters away from the location of the 2-D and the EM38-Lines-Ones of electromagnetic. The exported electrical conductivity (ECa) values for each EM38 measuring line were averaged for the values measured at 1 m coil separation and those measured at 0.5 coil separation. For every ten lines, the electrical conductivity (salinity indicator) was plotted for both two different coil separations (1 m and 0.5 m). Figure 11 shows the combined averaged measuring lines using EM38-MK2 across the Al-Ahsa Oasis. Investigations of these lines show how the variation of conductivity changed with depth, according to the positions, due to changes in soil salinity conditions, lithology, and soil texture at each line. The following remarks can be noticed from Figure 11.

Figure 10. Some excavation and the existing hardpan layer at different depths near profile 6 (P-6, Figure 8).

The measured conductivity at lines located at the southwest of the study area was characterized by relatively low conductivities (maximum was 140 mS/cm), see for example, Line 4 (Figure 12a) as compared with those lines measured at the north and east of the study area, see for example, line 7, Figure 12b (maximum was 750 mS/cm), which was about five times the maximum electrical conductivity at line 4 (P-4, Figure 12). These indicate the salinity of the soil at the north and east part of Al-Ahsa Oasis. The two evaporator lakes of Al Uyoun to the north and Al Afar to the east are the main reasons for the salinity increasing at these locations.

It is worth mentioning that there was a good match between the results of the 2-D profiles and the EM38-MK2 profiles for the soil zones. For example, the topsoil of resistivity profile 5 (P-5) was resistive, see Figure 8, which was confirmed by relatively low conductivities (maximum was about 150 mS/cm). On the other hand, compared with those lines measured at the north of the study area, see resistivity profile 1 and the EM38-MK2 for this profile for the soil zones; they are characterized by high conductivity (low resistivity).

Generally, the measured conductivity at most lines indicated high values of conductivities at a coil separation of 1 m (greater depth); see for examples Lines 7, 8, and 10 of Figure 12b, c, and d, which are located at the north and east of the study area, while only two lines (line 4, Figure 12a) and line 3 (P-3, Figure 12) showed high values of conductivities at a coil separation of 0.5 m (shallow depth) compared with those at a 1 m coil separation, located in the southwest of the study area.

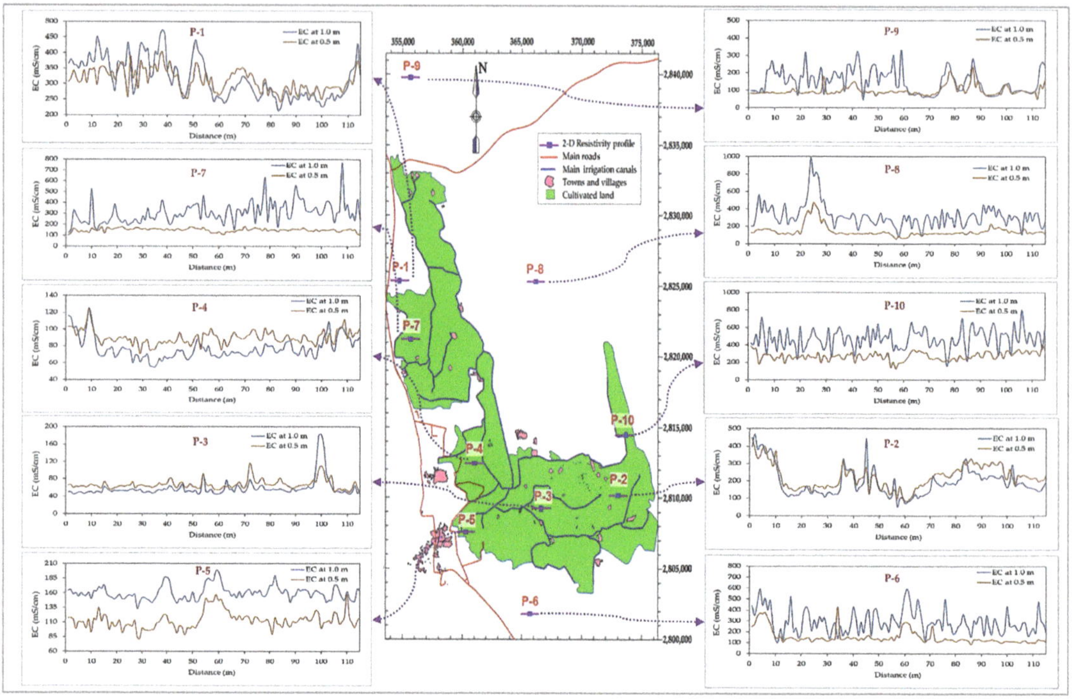

Figure 11. The combined averaged measuring lines at the ten tested locations from 1 to 10 (P-1 to P-10) using EM38-MK2 across the Al-Ahsa Oasis.

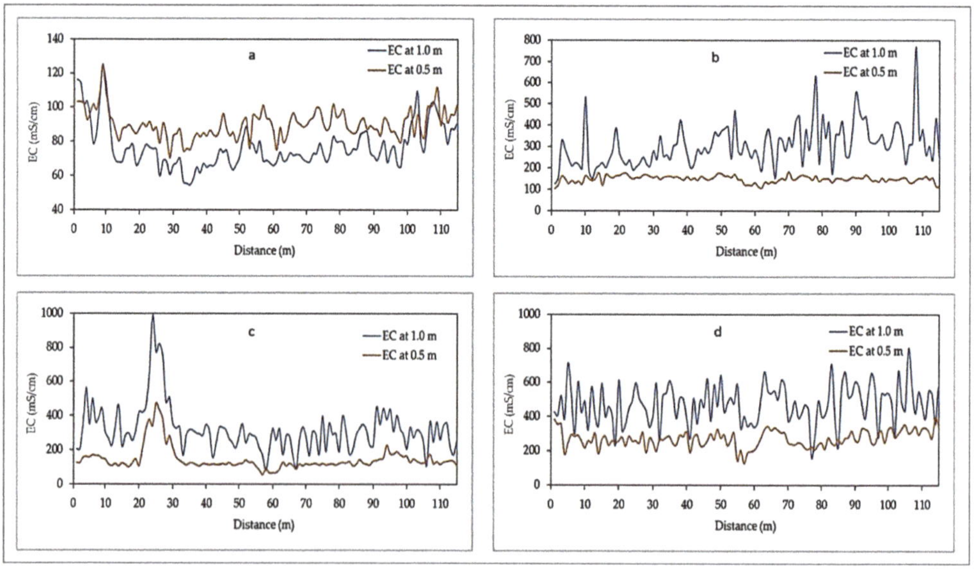

Figure 12. Electrical conductivities along Em38 at four tested locations (P-4, P-7, P-8, and P-10). Lines 4 at P-4 (**a**), Line 7 at P-7 (**b**), Line 8 at P-8 (**c**), and line 10 at P-10 (**d**) across the Al-Ahsa Oasis; for location see Figure 11.

These results are an excellent match with what was reported by El Najem [21]. El Najem [21] reported that soil chemical properties exhibited slightly alkaline pH values between 7.33 and 7.86, while, the soil salinity data, as EC values, varied across the Al-Ahsa Oasis. He found several values that exceeded 400 mS/ m (i.e., the minimum value of saline soils), reaching 7.80 dS/m, implying the persistence of soil salinity conditions (i.e., saline soil) in the Oasis's agricultural soils. Moreover, the author reported that the soil EC values increased with increasing soil depths, signifying the inefficient drainage systems and salinization of the soil profile in the Oasis.

4. Conclusions

Implementing electrical resistivity tomography (ERT) and modern electromagnetic induction techniques contributed to soil characterization salinity and mapping hardpan layer problems at Al-Ahsa Oasis, Eastern Province, Saudi Arabia. The tomograms of ten acquired 2-D resistivity profiles and data of the excavated soil pits judged by local farmers indicated the different geoelectrical layers. Information about the depth thickness of these geoelectric layers, including the topmost soil, was given. In some of the 2-D tomograms and at shallow depths, the hardpan layer was depicted, and this layer was composed of marl, clay, mudstone, or shale. This layer directly affects the level of soil water. Thus, the accumulation of salt in some locations results in some of the sites affected by the salt composed sabkha, especially if it is very shallow, which increases the soil conductivity. The 2-D tomographs measured at the southwest displayed resistivity values different to those those measured at the north and east of the study area, which were characterized by low resistivity values at greater depths due to effect of the sabkha and clayey soil and effects of both the Al Yuoun evaporation lake at the north of the Al-Ahsa Oasis and of the Al Asfar Lake at the east of the Oasis. The results of the twenty measured electrical conductivity profiles using the EM38-MK2 in the vertical mode were conducted at two different coil separations (1 m and 0.5 m) at the same locations of the measured 2-D profiles. The electrical conductivity results showed that the salinity variation along the measured profiles changed laterally and vertically due to changes in soil type/texture, soil moisture, and the existence of sabkha. The measured conductivity at most lines indicated high values of conductivities at coil separation of 1 m (greater depth); these profiles were located at the north and east of the study area, while only a few lines showed high values of conductivities measured at a coil separation of 0.5 m (shallow depth) compared to those at 1 m coil separation, these lines were located at the southwest of the study area. Moreover, it is found that the farms located in the southwest of the study area were characterized by low conductivities, while relatively high conductivities characterized the farms located to the north and east in the Al-Ahsa oasis. Thus, modern geophysical techniques proved to be efficient and effective tools for the nonintrusive gathering of continuous agricultural field data on many soil types, soil conditions, soil salinity, and near-surface conditions. Monitoring data about both soil and irrigation water was very important to prevent the degrading of agricultural soils in Al-Ahsa oasis. Degrading of agricultural soils will lead to poor agricultural productivity. The productivity limiting factors of these soils were mainly integrated into their chemical properties represented by the increased salinity. Such contribution provides technical support for planners, decision makers, and researchers in the field of agricultural sciences for ensuring agriculture sustainability in this Oasis.

Author Contributions: Conceptualization, A.E.M., M.M. and Y.A.; methodology, A.E.M.; software, A.E.M.; validation, M.M., A.E.M. and Y.A.; formal analysis, A.E.M.; investigation, A.E.M.; resources, M.M., A.E.M. and Y.A.; data curation, A.E.M.; writing—original draft preparation, M.M., A.E.M. and Y.A.; writing—review and editing, M.M., A.E.M. and Y.A.; visualization, M.M. and A.E.M. All authors have read and agreed to the published version of the manuscript.

Funding: The authors appreciate the Deanship of Scientific Research, King Faisal University, and extend their appreciation to the Deputyship for Research & Innovation, Ministry of Education in Saudi Arabia, for funding this study through project grant no. 150220.

Institutional Review Board Statement: Not applicable.

Informed Consent Statement: Not applicable.

Data Availability Statement: Not applicable.

Acknowledgments: The authors would like to express their sincere appreciation to the Deanship of Scientific Research, King Faisal University, for financial support of this study. Special thanks go to Mohamed Al Kulaib and Mohamed Al Firij for their assistance in the geophysical data acquisition.

Conflicts of Interest: The authors declare no conflict of interest.

References

1. Ellwood, B.B.; Harrold, F.B.; Marks, A.E. Site Identification and Correlation Using Geoarchaeological Methods at the Cabeço do Porto Marinho (CPM) Locality, Rio Maior, Portugal. *J. Archaeol. Sci.* **1994**, *21*, 779–784. [CrossRef]
2. Powers, C.J. *Surface-Geophysical Investigation of the University of Connecticut Landfill, Storrs, Connecticut*; US Department of the Interior, US Geological Survey: Reston, VA, USA, 1999; Volume 99.
3. Attwa, M.; El Mahmoudi, A.; Elshennawey, A.; Günther, T.; Altahrany, A.; Mohamed, L. Soil Characterization Using Joint Interpretation of Remote Sensing, Resistivity and Induced Polarization Data along the Coast of the Nile Delta. *Nat. Resour. Res.* **2021**, *30*, 3407–3428. [CrossRef]
4. Griffiths, D.H.; Barker, R.D. Two-dimensional resistivity imaging and modelling in areas of complex geology. *J. Appl. Geophys.* **1993**, *29*, 211–226. [CrossRef]
5. Allred, B.J.; Freeland, R.S.; Farahani, H.J.; Collins, M.E. Agricultural Geophysics: Past, Present, and Future. In Proceedings of the 23rd Symposium on the Application of Geophysics to Engineering and Environmental Problems, Keystone, CO, USA, 11–15 April 2010; pp. 190–202.
6. Allred, B.J.; Daniels, J.J.; Ehsani, M.R. *Handbook of Agricultural Geophysics*; CRC Press: Boca Raton, FL, USA, 2008; ISBN 9781420019353.
7. Friedl, M.A. 6.06-Remote Sensing of Croplands. In *Comprehensive Remote Sensing*; Liang, S., Ed.; Elsevier: Oxford, UK, 2018; pp. 78–95, ISBN 978-0-12-803221-3.
8. Mohammed, M.; Riad, K.; Alqahtani, N. Efficient iot-based control for a smart subsurface irrigation system to enhance irrigation management of date palm. *Sensors* **2021**, *21*, 3942. [CrossRef] [PubMed]
9. Mohammed, M.; Sallam, A.; Munir, M.; Ali-Dinar, H. Effects of deficit irrigation scheduling on water use, gas exchange, yield, and fruit quality of date palm. *Agronomy* **2021**, *11*, 2256. [CrossRef]
10. Williams, B.G.; Hoey, D. The use of electromagnetic induction to detect the spatial variability of the salt and clay contents of soils. *Aust. J. Soil Res.* **1987**, *25*, 21–27. [CrossRef]
11. Triantafilis, J.; Huckel, A.I.; Odeh, I.O.A. Comparison of statistical prediction methods for estimating field-scale clay content using different combinations of ancillary variables. *Soil Sci.* **2001**, *166*, 415–427. [CrossRef]
12. Tromp-van Meerveld, H.J.; McDonnell, J.J. Assessment of multi-frequency electromagnetic induction for determining soil moisture patterns at the hillslope scale. *J. Hydrol.* **2009**, *368*, 56–67. [CrossRef]
13. Hanson, B.R.; Kaita, K. Response of Electromagnetic Conductivity Meter to Soil Salinity and Soil-Water Content. *J. Irrig. Drain. Eng.* **1997**, *123*, 141–143. [CrossRef]
14. Lesch, S.M.; Corwin, D.L.; Robinson, D.A. Apparent soil electrical conductivity mapping as an agricultural management tool in arid zone soils. *Comput. Electron. Agric.* **2005**, *46*, 351–378. [CrossRef]
15. Yao, R.; Yang, J. Quantitative evaluation of soil salinity and its spatial distribution using electromagnetic induction method. *Agric. Water Manag.* **2010**, *97*, 1961–1970. [CrossRef]
16. El Mahmoudi, A.S.; Hussein, A.A.; Al-Molhem, Y.A. Mapping of soil salinity using electromagnetic Inductive Method at the Agricultural Research Station of KFU, AI Hassa, KSA. *Ecol. Environ. Conserv.* **2015**, *21*, 85–95.
17. Doolittle, J.A.; Sudduth, K.A.; Kitchen, N.R.; Indorante, S.J. Estimating depths to claypans using electromagnetic induction methods. *J. Soil Water Conserv.* **1994**, *49*, 572–575.
18. Jaynes, D.B.; Novak, J.M.; Moorman, T.B.; Cambardella, C.A. Estimating Herbicide Partition Coefficients from Electromagnetic Induction Measurements. *J. Environ. Qual.* **1995**, *24*, 36–41. [CrossRef]
19. Al Tokhais, A.S.; Rausch, R. The Hydrogeology of Al Hassa Springs. In Proceedings of the 3rd International Conference on Water Resources and Arid Environments, Riyadh, Saudi Arabia, 16–19 November 2008; Volume 1, pp. 16–19.
20. Almadini, A.M.; El-Garawany, M.M.; Al-Omran, A.A. *Evaluation of Soil Salinity and Soil Fertility at Al-Hassa Oasis, KSA*; Final Report of Research Project no. at-25–30; King Abdulaziz City for Science and Technology (KACST): Riyadh, Saudi Arabia, 2012.
21. El Najem, F.A.M. *Evaluating the Degraded Agricultural Soils at Al-Hassa Oasis, KSA*; King Fasisal University: Al Hofuf, Saudi Arabia, 2021.
22. El Mahmoudi, A.; Hussein, A.H.A.; Al-Molhem, Y. Modern Techniquesfor Mapping Soil Salinity and the Hard Pan Layer at the Agricultural and Research Station of KFU, Al Hassa, KSA. *Int. J. Water Resour. Arid Environ.* **2016**, *5*, 63–71.
23. Abderrhman, W.A.; Bader, T.A. Remote Sensing use in the management of Agricultural Drainage Water in Al Hassa Oasis of Saudi Arabia. *Remote Sens. Eval. Manag. Irrig.* **1992**, *42*, 237–246. [CrossRef]

24. Al-Barrak, S.A. Characteristics of some soils under date palm in Al-Hassa Eastern oasis, Saudi Arabia. *J. King Saud Univ. Agric. Sci* **1992**, *2*, 115–130.
25. Al-Barrak, S.A. *Soil Characteristic of Al-Hassa Oasis*; Fial Report; Water Studies Center, King Faisal University: Al-Hassa, Saudi Arabia, 1993.
26. El Prince, A.M. *National Fertilizer Program for Date Palm: Al-Hassa Phase*; Final Report; Water Studies Center, King Faisal University: Al-Hassa, Saudi Arabia, 2003.
27. Al-Dakheel, Y.Y.; Hussein, A.H.A.; El-Mahmoudi, A.S.; Massoud, M.A. Soil, water chemistry and sedimentological studies of Al Asfar evaporation lake and its inland sabkha, Al Hassa Area, Saudi Arabia. *Asian J. Earth Sci.* **2009**, *3*, 142–162. [CrossRef]
28. Al-Barrak, S.; Al-Badawi, M. Properties of some salt affected soils in Al-Ahsa, Saudi Arabia. *Arid Soil Res. Rehabil.* **1988**, *2*, 85–95. [CrossRef]
29. Almadini, A.M.; Al-Safarjalani, A.M.; Al-Naeem, A.A. Spatial variations in chemical properties of irrigation groundwater at Al-Hassa Oasis, Kingdom of Saudi Arabia. *Arab Gulf J. Sci. Res.* **2007**, *25*, 207–218.
30. Al Safarjalani, A.M.; Al-Dakheel, Y.Y.; El-Mahmoudi, A.S. Radionuclide levels in soil of Al Hassa. *J. Environ. Sci.* **2007**, *34*, 151–174.
31. Hussein, A.A.; El Mahmoudi, A.S.; Al Molhem, Y.A. Assessment of the validity of different types of irrigation water for agricultural purposes at Al Hassa Oasis, KSA (Kingdom of Saudi Arabia). *Ecol. Environ. Conserv.* **2016**, *22*, 1621–1627.
32. El-mahmoudi, A.S.; Hussein, A.A.; Hofouf, A. Hydrochemical Studies of Groundwater at Al Hassa Oasis, Eastern Region, Saudi Arabia. *Adsorption* **2017**, *6*, 20–32.
33. McDowell, P.W.; Barker, R.D.; Butcher, A.P.; Culshaw, M.G.; Jackson, P.D.; McCann, D.M.; Skipp, B.O.; Matthews, S.L.; Arthur, J.C.R. *Geophysics in Engineering Investigations*; CIRIA: London, UK, 2002; Volume 19.
34. Mussett, A.E.; Khan, M.A.; Button, S. *Looking into the Earth: An Introduction to Geological Geophysics*; Cambridge University Press: Cambridge, UK, 2000; Volume 83, ISBN 9780511810305.
35. Reynolds, J.M. *An Introduction to Applied and Environmental Geophysics*; John Wiley & Sons: Hoboken, NJ, USA, 2011; ISBN 0471968021.
36. McNeill, J.D. *Electromagnetic Terrain Conductivity Measurement at Low Induction Numbers*; Technical note TN-6; Geonics Limited: Mississauga, ON, Canada, 1980; Volume 6, p. 13.
37. Sasaki, Y. Resolution of resistivity tomography inferred from numerical SIMULATION1. *Geophys. Prospect.* **1992**, *40*, 453–463. [CrossRef]
38. Loke, M.H.; Barker, R.D. Rapid least-squares inversion of apparent resistivity pseudosections by a quasi-Newton method. *Geophys. Prospect.* **1996**, *44*, 131–152. [CrossRef]
39. El Mahmoudi, A.S.; Hussein, A.H.; Al-Dakheel, Y.Y.; Massoud, M.A. Evaluating the Degraded Agricultural Soils at Al-Hassa Oasis, KSA. *Int. J. Water Resour. Arid Environ.* **2014**, *3*, 84–95.
40. Hotzl, H.; Maurin, V. Geologic history of the Al Hasa area since Pliocene. In *Quaternary Period in Saudi Arabia*; Al-Sayari, S.S., Zotl, J.G., Eds.; Springer: Vienna, Austria, 1978.

applied sciences

MDPI

Article

Anisotropic Effective Elastic Properties for Multi-Dimensional Fractured Models

Shikai Jian [1,2], Liyun Fu [1,2,3,*] and Yifan Cheng [1,2]

1 Shandong Provincial Key Laboratory of Deep Oil and Gas, China University of Petroleum (East China), Qingdao 266580, China; jianshiakai@163.com (S.J.); 13261600217@163.com (Y.C.)
2 School of Geosciences, China University of Petroleum (East China), Qingdao 266580, China
3 Laboratory for Marine Mineral Resources, Qingdao National Laboratory for Marine Science and Technology, Qingdao 266071, China
* Correspondence: lfu@upc.edu.cn

Featured Application: The proposed method in the paper has significance in calculating the effective elastic properties of multi-dimensional fracture models and interaction mechanisms.

Abstract: The size, distribution, and orientation of fractures are generally multiscale and multi-dimensional in nature, leading to complex anisotropic characteristics. Theoretical or semi-analytical methods to determine the effective elastic properties depend on several assumptions, including the absence of the stress interaction and idealized fractures. On the basis of finite-element models, we conduct numerical oscillatory relaxation tests for determining the effective elastic properties of fractured rocks. The numerical approach for calculating equivalent stiffness tensors in two-dimensions is compared to the theoretical models for different fracture densities. Due to fracture interactions at high fracture densities, the suggested model makes a physical prediction. The effective elastic properties obtained from the application to a real fractured model, established from an outcrop, obviously disperse at different frequencies, which can be used to investigate fracture interactions and dynamic stress disturbances. The algorithm is extended to three-dimensional cases and also validated by using conventional effective medium theories. It is found that the fracture density obviously impacts the effective anisotropy properties, and the proposed method gives a reasonable prediction for high-fracture density. This work is significant because it enables the calculation of effective elastic properties of multi-dimensional fractured models and the fracture interaction mechanism.

Keywords: multi-dimensional fractured models; effective elastic properties; numerical oscillatory relaxation tests; anisotropy characteristics

1. Introduction

Cracks are the key factor in determining the deformability of rocks, properties of fluid movement, and tectonic history. We need to enhance ways for detecting and characterizing fractures. Due to the fact that natural fractures in geological formations frequently have a complicated shape and distribution, their interactions cause the complex anisotropic features of effective elastic properties. Thus, knowing the natural fractures' elastic behavior is essential for interpreting and predicting seismic responses in terms of rock properties and enables us to bridge the gaps in the spatial and temporal scale [1,2]. The overall effective elastic coefficient of a fractured rock is frequently composed of the stiffness tensor of the background host rock and fractures [3], which attract considerable attention because of their contribution to the elastic property. The theoretical medium models (TMM) have been widely applied to calculate the effective elastic properties of fractured rocks analytically [4–8]. However, due to the dilute assumption of fracture densities, these models ignore fracture interactions. The second-order Hudson's model takes into account the pair-wise interaction among cracks [9]. Chapman [10] proposed an equivalent-media theory that was sensitive to fracture length and studied the frequency-dependent anisotropy of fractured porous rocks using the squirt flow mechanism. Guo et al. [11] united linear-slip theory and anisotropic Gassmann equations to analyze equivalent stiffness coefficients. The self-consistent approximation (SCA) and differential effective medium (DEM) were applied to control the concentration of fractures to investigate the consequences of fracture interactions [12–14]. Jakobsen et al. [15] applied the T-matrix method to characterize the pores–fracture interaction. Meanwhile, the idealization and oversimplification of real rock severely limit the utility of theoretical or semi-analytical models.

In an attempt to circumvent certain assumptions inherent in effective medium theories [16–23], numerical simulations have been extensively applied for studying the elastic properties of complex rocks. Numerical methods can be used to characterize real fractures with different shapes, sizes, partial interconnections, and intersections, as well as provide independent validation of theoretical methods and laboratory measurements as an efficient means of determining their general applicability. Masson et al. [24] used numerical experiments to analyze the correlation between material structure and seismic attenuation anisotropy in porous media. Rubino et al. [25,26] employed 2D numerical oscillatory relaxation experiments to study the frequency-dependent effective stiffness matrix using the least-squares optimal approach. The method is used to characterize the frequency- and incidence-angle- dependent attenuation and velocity dispersion in cracked porous media for the P wave [27–29] and P-SV wave [30]. Jian et al. [31] expanded Rubino's relaxation tests to a three-dimensional (3D) case for computing the equivalent stiffness matrix.

We apply Rubino's numerical method to explore the elastic and seismic anisotropy characteristics of real fractures in this paper. Particular attention should be paid to the anisotropy variation of fractures generated by different frequencies and crack densities. The following is the outline. In the first part of the technique, we demonstrate the novel form of Rubino's 2D algorithm using three oscillatory relaxation tests with different fracture densities. Additionally, six oscillatory relaxation tests are used to spread Rubino's algorithm to three-dimensional fractured media. Following that, the effective stiffness of the 2D numerical method is compared with that of the classical Hudson's model and the Eshelby–Cheng model with different fracture densities. Then, 3D fractured media are compared with TMM in elastic responses to different fracture densities. Three 2D fracture models are presented to illustrate the primary mechanism of stress interreaction and their effects on the entire effective elasticity. Finally, the method is applied to a real fractured medium generated from an outcrop, demonstrating that frequency influences the stiffness tensor anisotropy. The anisotropic effective elastic properties have been discovered to have a direct effect on fracture density. This study aims to analyze the anisotropic effective elastic properties of multi-dimensional (2D and 3D) fracture models at different fracture densities.

2. Methodology

This section mainly introduces the methodology of studying anisotropic effective elastic properties of multi-dimensional fracture models. The entire modeling methodology is illustrated in Figure 1. First, we demonstrate the simulation equations and three oscillatory relaxation tests. Next, the 2D equivalent stiffness matrix is calculated by solving the simulation equations using the finite-element method (FEM). Following that, the effective stiffness of the 2D numerical method is compared with that of Hudson's model and the Eshelby–Cheng model with different fracture densities. Additionally, six oscillatory relaxation tests are used to spread to three-dimensional fractured media.

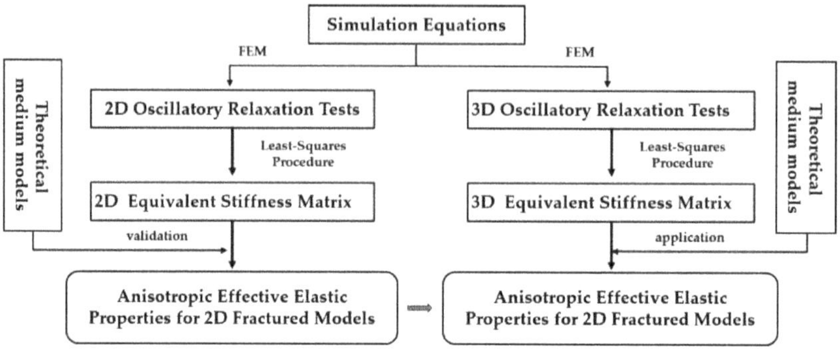

Figure 1. Flowchart of the modeling methodology.

2.1. *Anisotropic Effective Elastic Properties for 2D Fractured Models*

2.1.1. Simulation Equations

We have to use numerical methods because the TMM for predicting the elasticity of multi-scale real intersecting cracks is limited by its own assumptions. We here apply Rubino's 2D numeric upscaling procedure in an attempt to pattern natural cracks. Biot's consolidation system of equations in the space-frequency domain is expressed as [26].

$$\nabla \cdot \sigma = \mathbf{0}, \tag{1}$$

$$\nabla P_f = -i\omega \frac{\eta}{\kappa} \mathbf{W}. \tag{2}$$

where ∇ represents the spatial derivatives; $\sigma \equiv (\sigma_{ij})$ is total stress tensor; P_f denotes the fluid pressure, κ the permeability and ω the angular frequency; and $\mathbf{W}$ denotes average relative fluid–solid displacement tensor. η indicates the viscosity of pore fluid. Thus, the stress–strain relationships [32] state that

$$\sigma_{ij} = 2\mu e_{ij} + \delta_{ij}(\lambda_u \nabla \cdot \mathbf{U} - \alpha N \xi), \tag{3}$$

$$e_{ij} = \frac{1}{2}(\partial U_i/\partial x_j + \partial U_j/\partial x_x), \tag{4}$$

$$P_f = -\alpha N \nabla \cdot \mathbf{U} + N\xi, \tag{5}$$

where δ_{ij} is the Kronecker delta, and $\xi = -\nabla \cdot \mathbf{W}$ denotes the measurement of the local variations in the fluid content. $\mathbf{U}$ is the solid displacement vector; e_{ij} denotes strain tensor; and μ denotes the shear modulus of the rock with saturated water, which is equivalent to that of a dry frame μ_m. The Biot constant α, fluid storage coefficient N, and Lame parameter λ_u are expressed as

$$N = \left(\frac{\alpha - \phi}{K_s} + \frac{\phi}{K_f}\right)^{-1}, \tag{6}$$

$$\lambda_u = K_m + \alpha^2 N - \frac{2}{3}\mu_m, \tag{7}$$

$$\alpha = 1 - \frac{K_m}{K_s}, \tag{8}$$

where ϕ is the porosity; K_s denotes the bulk moduli of solid grains; K_m is the bulk moduli of dry frame; and K_f denotes the bulk moduli of fluid phase.

Substituting Equation (7) to Equation (3) and taking Equation (5), we have a new form of 2D/3D total stress tensor in porous fractured medium

$$\sigma_{ij} = 2\mu e_{ij} + \delta_{ij}\left(\lambda \nabla \cdot \mathbf{U} - \alpha P_f\right), \tag{9}$$

where $\lambda = K_m - 2\mu_m/3$. When $i, j = x, y$, it is the 2D case. When $i, j = x, y, z$, it is the 3D case. Substituting $\zeta = -\nabla \cdot \mathbf{W}$ into Equation (5), we have

$$P_f = -\alpha N \nabla \cdot \mathbf{U} - N \nabla \cdot \mathbf{W}, \quad (10)$$

Taking Equations (2) and (10), the P_f is given by

$$P_f = -\alpha N \nabla \cdot \mathbf{U} - N \nabla \cdot \left(\frac{ik}{\eta \omega} \nabla P_f \right). \quad (11)$$

In order to obtain the solid displacement fields, we conduct relaxation tests and solve Equations (1) and (11) with the corresponding boundary conditions applying the finite-element method (FEM).

2.1.2. 2D Oscillatory Relaxation Tests

As shown in Figure 2, we need to conduct three simulation oscillatory relaxation tests for calculating the elements of an equivalent stiffness tensor of a 2D sample of interest according to the 2D stress–strain correlation. Let $D = (0, L_x) \times (0, L_y)$ be an area that represents the interest model. Performing the first oscillatory relaxation test (see Figure 2a), a uniform vertical displacement is imposed on the bottom and top boundaries, with two lateral boundaries being confined and no shear stress being applied. As a result, the x- and y-directions' boundary conditions are as follows:

$$\begin{cases} \mathbf{U} \cdot \nu = -\Delta U, & (x, y) \in \mathbf{\Gamma}^b \cup \mathbf{\Gamma}^t, \\ \mathbf{U} \cdot \nu = 0, & (x, y) \in \mathbf{\Gamma}^l \cup \mathbf{\Gamma}^r, \\ (\sigma \nu) \cdot \chi = 0, & (x, y) \in \mathbf{\Gamma}, \end{cases} \quad (12)$$

where $\mathbf{U}$ is the solid displacement; ΔU is the displacement shift; $\boldsymbol{\sigma}$ represents the total stress tensor; $\mathbf{\Gamma}$ is the model boundary; $\mathbf{\Gamma}^b$, $\mathbf{\Gamma}^t$, Γ^l, and $\mathbf{\Gamma}^r$ are the bottom, top, left, and right boundary, respectively, on $\mathbf{\Gamma}$; χ is a unit tangent; and ν is the unit outer normal so that $\{\nu, \chi\}$ is an orthonormal system on $\mathbf{\Gamma}$.

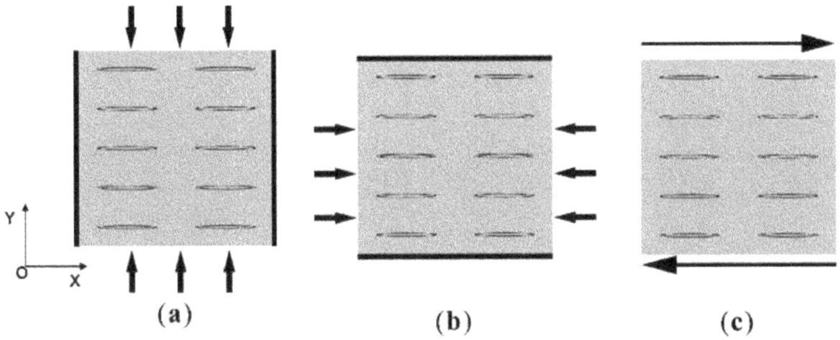

Figure 2. Sketch map of oscillatory relaxation simulation tests to determine 2D stiffness tensors. The back arrows denote the different orientations of solid displacements. Let the solid displacement vertical to the boundary on the black edge lines be zero. (**a**) the first oscillatory relaxation test, a uniform vertical displacement is imposed on the bottom and top boundaries, with two lateral boundaries being confined; (**b**) the second oscillatory relaxation test with a uniform horizontal displacement to the right and left boundaries and the two other boundaries confined; (**c**) the third test, where the top and bottom boundaries are loaded with displacement.

Figure 2b illustrates the second test that is analogous to the first, but with a uniform horizontal displacement to the right and left boundaries and the two other boundaries confined. Thus, the corresponding boundary conditions are expressed as

$$\begin{cases} \mathbf{U}\cdot\nu = -\Delta U, & (x,y) \in \Gamma^l \cup \Gamma^r, \\ \mathbf{U}\cdot\nu = 0, & (x,y) \in \Gamma^b \cup \Gamma^t \\ (\sigma\nu)\cdot\chi = 0, & (x,y) \in \Gamma, \end{cases}, \tag{13}$$

Figure 2c shows the third test, a shear test, applied to the interested model, where the top and bottom boundaries are loaded with displacement. The corresponding boundary conditions are

$$\begin{cases} \mathbf{U}\cdot\chi = -\Delta U, & (x,y) \in \Gamma^b \cup \Gamma^t, \\ \mathbf{U}\cdot\chi = 0, & (x,y) \in \Gamma^l \cup \Gamma^r \\ (\sigma\nu)\cdot\nu = 0, & (x,y) \in \Gamma. \end{cases}, \tag{14}$$

2.1.3. Calculation of 2D Equivalent Stiffness Matrix

After computing the displacement field, the stress and strain components of each element are computed using Equations (3) and (4). The corresponding stress and strain average over the model's volume can be calculated using these parameters

$$\overline{e_{ij}^m} = \frac{1}{S}\int_D e_{ij}^m dS, \tag{15}$$

$$\overline{\sigma_{ij}^m} = \frac{1}{S}\int_D \sigma_{ij}^m dS, \tag{16}$$

where m = 1–3 denotes the mth test, and S is the surface of domain D.

Taking the prerequisite that an equivalent homogeneous anisotropic elastic solid can stand for the average response of the interest model, the average stress and strain components can be related through the Voigt stiffness matrix $\mathbf{M} \equiv M_{ij}$. In the case of general anisotropy and under plane strain conditions, this relation is given by

$$\begin{pmatrix} \overline{\sigma_{xx}^m} \\ \overline{\sigma_{yy}^m} \\ \overline{\sigma_{xy}^m} \end{pmatrix} \equiv \begin{pmatrix} M_{11} & M_{12} & M_{16} \\ M_{12} & M_{22} & M_{26} \\ M_{16} & M_{26} & M_{66} \end{pmatrix} \begin{pmatrix} \overline{e_{xx}^m} \\ \overline{e_{yy}^m} \\ 2\overline{e_{xy}^m} \end{pmatrix} \tag{17}$$

where $\overline{\sigma_{ij}^m}$ and $\overline{e_{ij}^m}$ are cells of the averaging stress and strain vector, respectively, and m = 1–3 denotes the mth simulation test. It is worthy to note that the matrix $\mathbf{M}$ considers the greatest extent of anisotropy.

Thus, the six unknown components of the stiffness matrix possibly are calculated utilizing a least-squares procedure. That is, the components of equivalent stiffness matrix M_{ij} can be obtained by minimizing the cost function F_1 ($\mathbf{M}$) as follows:

$$\frac{\partial F_1(\mathbf{M})}{\partial M_{ij}} = 0. \tag{18}$$

We can solve the linear system of six equations yielded by Equation (18) to obtain the equivalent stiffness tensor M_{ij}; see Appendix A for more details.

2.2. Anisotropic Effective Elastic Properties for 3D Fractured Models

2.2.1. 3D Oscillatory Relaxation Tests

In this part, we apply Rubino's 2D method to a 3D cracked medium [30]. The governing Equations (1)–(11) would remain the same. As illustrated in Figure 3, we need to conduct six simulation oscillatory relaxation tests for calculating the elements of an equivalent stiffness matrix for a 3D probed model according to the 3D stress–strain correlation.

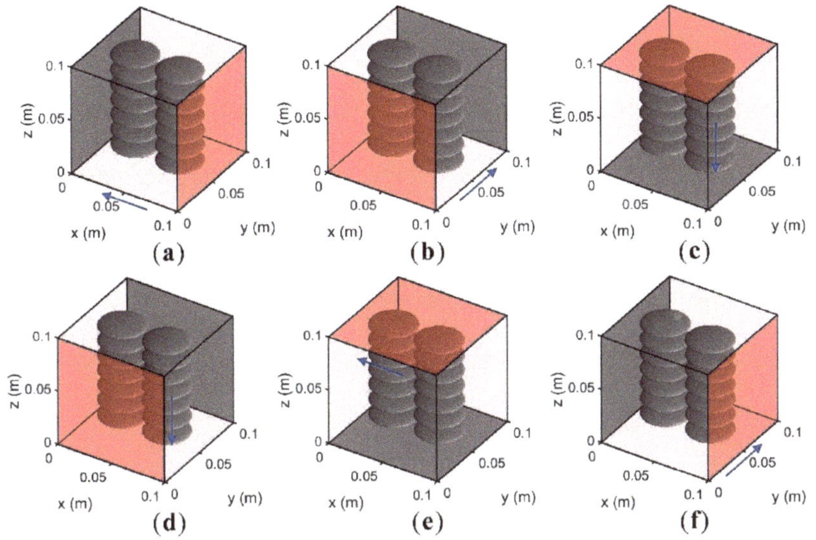

Figure 3. Schematic diagram of six oscillatory relaxation simulation tests to determine the equivalent stiffness matrix for a 3D model, including penny-shaped spheroid fractures. Giving every test, let a uniform solid displacement (blue arrow) be applied to the red side, the gray side be confined, and the periodical boundary conditions be imposed on the other sides. (**a**) the first oscillatory relaxation test; (**b**) the second oscillatory relaxation test; (**c**) the third oscillatory relaxation test; (**d**) the fourth oscillatory relaxation test; (**e**) the fifth oscillatory relaxation test; (**f**) the sixth oscillatory relaxation test.

Set $\Omega = (0, L_x) \times (0, L_y) \times (0, L_z)$ as a domain that represents a 3D interest model. Figure 3a displays the first oscillatory relaxation test in which a uniform vertical displacement along the x-direction (blue arrow) is applied to the model's red side, with the other sides employing periodic boundary conditions. The opposing side (gray side) is confined, and no shear stress is applied. Thus, the boundary conditions of the x-, y-, and z- directions are imposed as follows:

$$\begin{cases} \mathbf{U}\cdot\nu = -\Delta U, & (x,y,z) \in \boldsymbol{A}^r, \\ \mathbf{U}\cdot\nu = 0, & (x,y,z) \in \boldsymbol{A}^l, \\ (\boldsymbol{\sigma}\nu)\cdot\chi = 0, & (x,y,z) \in \boldsymbol{A}, \\ \mathbf{U}_{dst} = \mathbf{U}_{src}, & (x,y,z) \in \boldsymbol{A}^f \cup \boldsymbol{A}^k, \\ \mathbf{U}_{dst} = \mathbf{U}_{src}, & (x,y,z) \in \boldsymbol{A}^t \cup \boldsymbol{A}^b, \end{cases} \tag{19}$$

where σ represents the total stress tensor; $\mathbf{U}$ denotes the solid displacement; ΔU denotes the displacement shift; $\mathbf{U}_{src}$ denotes the source displacement field; $\mathbf{U}_{dst}$ denotes the displacement field of the target boundary; $\boldsymbol{A}$ is its side boundaries; $\boldsymbol{A}^f$, $\boldsymbol{A}^k$, $\boldsymbol{A}^l$, $\boldsymbol{A}^r$, $\boldsymbol{A}^t$, and $\boldsymbol{A}^b$ are the left, right, front, back, top, and bottom boundaries, respectively; χ is a unit tangent; and ν denotes the unit outer normal on $\boldsymbol{A}$, so that $\{\nu,\chi\}$ is an orthonormal system.

Figure 3b illustrates the second test that is analogous to the first, but with a uniform horizontal displacement applied to the red side from the y-direction and the opposing side fixed and periodical boundary conditions applied to the other surfaces. So, we have

$$\begin{aligned} &\mathbf{U}\cdot\chi = -\Delta U, && (x,y,z) \in \boldsymbol{A}^f, \\ &\mathbf{U}\cdot\chi = 0, && (x,y,z) \in \boldsymbol{A}^k, \\ &(\boldsymbol{\sigma}\nu)\cdot\nu = 0, && (x,y,z) \in \boldsymbol{A}, \\ &\mathbf{U}_{dst} = \mathbf{U}_{src}, && (x,y,z) \in \boldsymbol{A}^l \cup \boldsymbol{A}^r, \\ &\mathbf{U}_{dst} = \mathbf{U}_{src}, && (x,y,z) \in \boldsymbol{A}^t \cup \boldsymbol{A}^b, \end{aligned} \tag{20}$$

Figure 3c–f shows the third to sixth tests whose loaded direction of solid displacement and the relevant boundary conditions are given by

$$\begin{aligned}
&\mathbf{U}\cdot\nu = -\Delta U, && (x,y,z) \in \boldsymbol{A}^t,\\
&\mathbf{U}\cdot\nu = 0, && (x,y,z) \in \boldsymbol{A}^b,\\
&(\boldsymbol{\sigma}\nu)\cdot\chi = 0, && (x,y,z) \in \boldsymbol{A},\\
&\mathbf{U}_{dst} = \mathbf{U}_{src}, && (x,y,z) \in \boldsymbol{A}^f \cup \boldsymbol{A}^k,\\
&\mathbf{U}_{dst} = \mathbf{U}_{src}, && (x,y,z) \in \boldsymbol{A}^r \cup \boldsymbol{A}^l,
\end{aligned} \tag{21}$$

for the third test,

$$\begin{aligned}
&\mathbf{U}\cdot\chi = -\Delta U, && (x,y,z) \in \boldsymbol{A}^f,\\
&\mathbf{U}\cdot\chi = 0, && (x,y,z) \in \boldsymbol{A}^k,\\
&(\boldsymbol{\sigma}\nu)\cdot\nu = 0, && (x,y,z) \in \boldsymbol{A},\\
&\mathbf{U}_{dst} = \mathbf{U}_{src}, && (x,y,z) \in \boldsymbol{A}^l \cup \boldsymbol{A}^r,\\
&\mathbf{U}_{dst} = \mathbf{U}_{src}, && (x,y,z) \in \boldsymbol{A}^t \cup \boldsymbol{A}^b,
\end{aligned} \tag{22}$$

for the fourth test,

$$\begin{aligned}
&\mathbf{U}\cdot\chi = -\Delta U, && (x,y,z) \in \boldsymbol{A}^t,\\
&\mathbf{U}\cdot\chi = 0, && (x,y,z) \in \boldsymbol{A}^b,\\
&(\boldsymbol{\sigma}\nu)\cdot\nu = 0, && (x,y,z) \in \boldsymbol{A},\\
&\mathbf{U}_{dst} = \mathbf{U}_{src}, && (x,y,z) \in \boldsymbol{A}^f \cup \boldsymbol{A}^k,\\
&\mathbf{U}_{dst} = \mathbf{U}_{src}, && (x,y,z) \in \boldsymbol{A}^r \cup \boldsymbol{A}^l,
\end{aligned} \tag{23}$$

for the fifth test, and

$$\begin{aligned}
&\mathbf{U}\cdot\chi = \Delta U, && (x,y,z) \in \boldsymbol{A}^r,\\
&\mathbf{U}\cdot\chi = 0, && (x,y,z) \in \boldsymbol{A}^l,\\
&(\boldsymbol{\sigma}\nu)\cdot\nu = 0, && (x,y,z) \in \boldsymbol{A},\\
&\mathbf{U}_{dst} = \mathbf{U}_{src}, && (x,y,z) \in \boldsymbol{A}^f \cup \boldsymbol{A}^k,\\
&\mathbf{U}_{dst} = \mathbf{U}_{src}, && (x,y,z) \in \boldsymbol{A}^t \cup \boldsymbol{A}^b.
\end{aligned} \tag{24}$$

for the sixth test.

2.2.2. Calculation of 3D Equivalent Stiffness Matrix

We solve Equations (1) and (11) by applying the FEM to obtain the corresponding displacement field, which promotes the stress and strain components for every element to be obtained using Equations (3) and (4). Then, parameters allow us to estimate the corresponding average over the sample's volume:

$$\overline{e_{ij}^r} = \frac{1}{V}\int_\Omega e_{ij}^r dV, \tag{25}$$

And

$$\overline{\sigma_{ij}^r} = \frac{1}{V}\int_\Omega \sigma_{ij}^r dV, \tag{26}$$

where V is the volume of Ω, with r = 1–6 denoting the rth simulation oscillatory relaxation test described in Figure 3, respectively.

As in the case of 2D, we use a second-order equivalent Voigt stiffness matrix $\mathbf{M}\equiv M_{ij}$ to relate the averaging stress and strain components of 3D models. In the case of general anisotropy and under plane strain conditions, this relation is given by

$$\begin{pmatrix} \overline{\sigma_{xx}^r} \\ \overline{\sigma_{yy}^r} \\ \overline{\sigma_{zz}^r} \\ \overline{\sigma_{yz}^r} \\ \overline{\sigma_{zx}^r} \\ \overline{\sigma_{xy}^r} \end{pmatrix} \equiv \begin{pmatrix} M_{11} & M_{12} & M_{13} & M_{14} & M_{15} & M_{16} \\ M_{12} & M_{22} & M_{23} & M_{24} & M_{25} & M_{26} \\ M_{13} & M_{23} & M_{33} & M_{34} & M_{35} & M_{36} \\ M_{14} & M_{24} & M_{34} & M_{44} & M_{45} & M_{46} \\ M_{15} & M_{25} & M_{35} & M_{45} & M_{55} & M_{56} \\ M_{16} & M_{26} & M_{36} & M_{46} & M_{56} & M_{66} \end{pmatrix} \begin{pmatrix} \overline{e_{xx}^r} \\ \overline{e_{yy}^r} \\ \overline{e_{zz}^r} \\ 2\overline{e_{yz}^r} \\ 2\overline{e_{zx}^r} \\ 2\overline{e_{xy}^r} \end{pmatrix} \tag{27}$$

where M_{ij} is the equivalent stiffness components in the 3D media, and $\overline{\sigma_{ij}^{r}}$ and $\overline{e_{ij}^{r}}$ are cells of the average stress and strain vector, respectively. In general, the matrix **M** takes into account the highest degree of anisotropy for a 3D case. Like in the 2D case, we can obtain 21 unknown stiffness tensors by applying the least-squares procedure to minimize the following cost function F_2 (**M**),

$$\frac{\partial F_2(\mathbf{M})}{\partial M_{ij}} = 0. \tag{28}$$

We can solve the linear system of twenty-one equations yielded by Equation (28) to obtain the equivalent stiffness matrix M_{ij}, seeing Appendix B for more niceties.

3. Results

3.1. Stiffness Tensors for 2D Fractured Model

As a 2D case, a fractured model with sides of 0.1 m contains small, isolated, and penny-shaped elliptical cracks along the x-axis. Every elliptical fracture has an x-semiaxis of 1.5 cm and y-semiaxis of 0.08 cm. The following formula is used to define the fracture density [13].

$$\zeta = \frac{Na^2}{s} \tag{29}$$

where N and a denote the fracture numbers and major semiaxis, respectively; s is the area of the model. Five group parameters are designed with different fracture densities of 0.056, 0.11, 0.149, 0.23, and 0.49 in the horizontal direction. Table 1 shows the physical parameters of fluid parameters, fractures with saturated brine, and the embedding background matrix.

Table 1. Simulation parameters for multi-dimensional cracked models.

Parameters		Phase
	Matrix	
Grain Bulk moduli, k_s (GPa)		37.0
Grain Density, ρ_s (g/cm^3)		2.65
Porosity ϕ		0.1
Bulk modulus, k_m (GPa)		26
Shear modulus, μ or μ_m (GPa)		31.0
Permeability κ (m^2)		1×10^{-15}
	Fractures	
Grain Bulk moduli, k_s (GPa)		37.0
Grain Density, ρ_s (g/cm^3)		2.65
Porosity ϕ		0.8
Bulk modulus, k_m (GPa)		0.04
Shear modulus, μ or μ_m (GPa)		0.02
Permeability κ (m^2)		1×10^{-10}
	Brine	
Bulk modulus, k_f (GPa)		2.25
Density, ρ_s (g/cm^3)		1.09
Viscosity, η (Pa.s)		0.01

We solve Equations (1) and (11) using the relaxation experiments with FEM. Once the stress and strain fields are obtained, the stiffness tensor is calculated by the least-squares method. Figure 4 shows the resulting equivalent stiffness tensors M_{22} and M_{12}. With increasing fracture density, the equivalent stiffness tensors M_{22} and M_{12} have a similar decreasing trend.

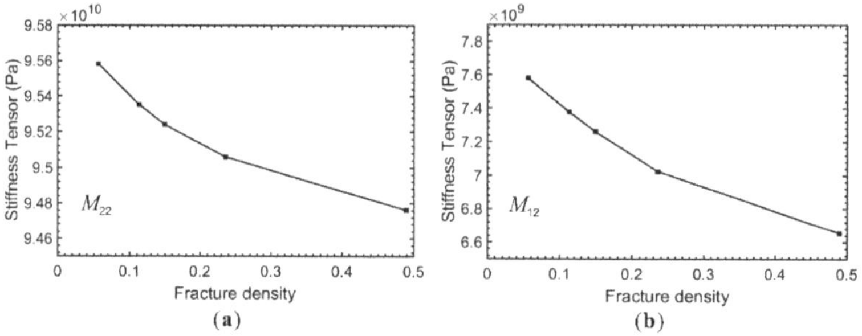

Figure 4. Effective stiffness tensors M_{22} (**a**) and M_{12} (**b**) for 2D fracture model.

3.2. Stiffness Tensors for 3D Fractured Model

In the 3D case, as shown in Figure 5, we build four models, each being a cube of 0.1 m on all sides for validating the 3D numerical oscillatory relaxation tests. These models include penny-shaped ellipsoidal cracks along the horizontal direction, and *x*-, *y*-, and *z*-semiaxes of 0.02 m, 0.02 m, and 0.003 m, respectively, with different fracture densities.

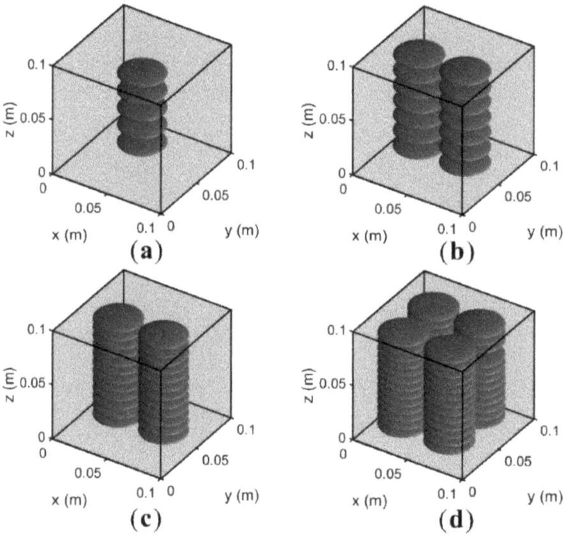

Figure 5. Four 3D models, each being a cube with 0.1 m on its sides, contain penny-shaped spheroid cracks with different fracture densities: 0.04 (**a**), 0.112 (**b**), 0.192 (**c**), and 0.416 (**d**).

As with 2D, the four fracture densities are set to 0.04, 0.112, 0.192, and 0.416, respectively, according to the basic definition,

$$\zeta = \frac{Na^3}{v} \tag{30}$$

where v is the volume of the cube. The same physical parameters in Table 1 are applied to the 3D models, and the relaxation tests are conducted at a frequency of 20 Hz.

Equations (1) and (11) are also solved by the relaxation experiments with FEM. Once the stress and strain fields are obtained, the stiffness tensor is calculated by the least-squares method. Figure 6 shows the resulting equivalent stiffness tensors M_{12} and M_{44}. With increasing fracture density, the equivalent stiffness tensors M_{12} and M_{44} tend to decrease.

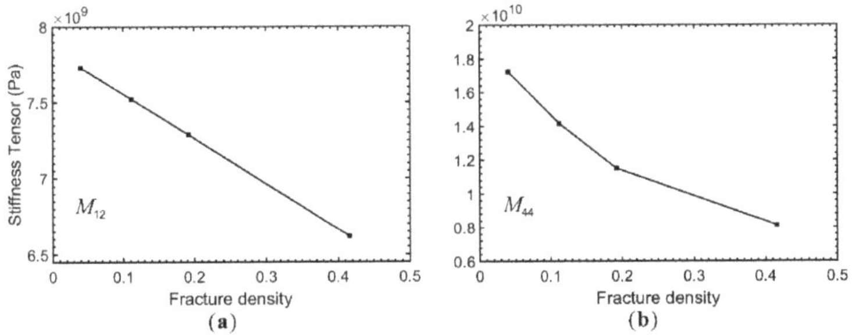

Figure 6. Effective stiffness tensors M_{12} (**a**) and M_{44} (**b**) for 3D fracture model.

3.3. Validation of Numerical Method

3.3.1. Validation of 2D Numerical Method

To validate the finite-element numerical simulation, the classical second-order Hudson's model and the Eshelby–Cheng model are used. Specific formulas are included in Appendix C.

Figure 7 compares the resulting equivalent stiffness tensors M_{22} and M_{12} calculated from the 2D numerical oscillatory relaxation tests for a frequency of 20 Hz with the Hudson model and the Eshelby–Cheng model. As expected, the solutions obtained using the different methods are almost comparable for low fracture densities but significantly different for high crack densities. We see that the second-order Hudson model for pair-wise interactions between cracks produces an unreasonable result at a high crack density that is over the dilution to make the assumption invalid. Both the equivalent stiffness matrix M_{11} and M_{12} by the Eshelby–Cheng model keep a good trend close to the numerical results, but their difference becomes large with increasing crack densities.

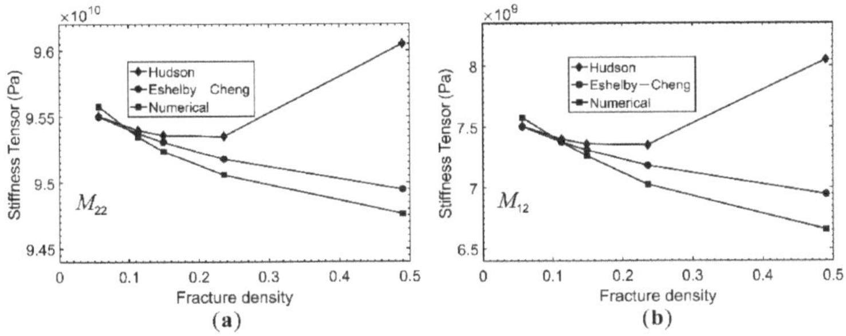

Figure 7. Comparison of numerical simulations and theoretical predictions for effective stiffness constants M_{22} (**a**) and M_{12} (**b**) for an isotropic background medium containing fractures saturated with brine.

3.3.2. Numerical Analysis of 3D Fractured Model

As illustrated in Figure 8, the resulting effective stiffness coefficients M_{12} and M_{44} are compared to the theoretical solutions obtained using the Hudson and Eshelby–Cheng models. As expected, these predictions exhibit a similar trend as facture densities increase. As illustrated in Figure 8a,b, there is a substantially better agreement between the numerical M_{12} and M_{44} and their theoretical counterpart solutions, especially at low crack densities. We also see that the numerical results closely match the Eshelby–Cheng model, with some minor departures between them at higher crack densities due to limited accuracies in the Eshelby–Cheng model. The simulation method provides a physical prediction at high fracture densities that control the overall elastic properties of rocks. We can analyze the fracture-induced anisotropy using these equivalent stiffness tensors.

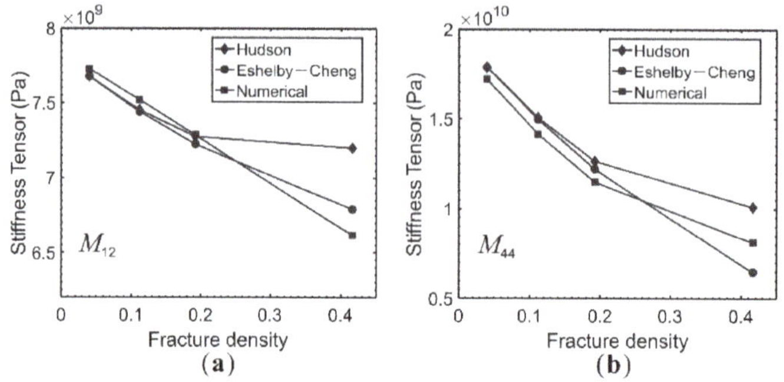

Figure 8. The effective stiffness coefficients M_{12} (**a**) and M_{44} (**b**) of the 3D models (Figure 5), including water-saturated penny-shaped cracks, set different fracture densities.

4. Discussion

4.1. Stress Distribution Induced on Crack Interaction

Stress interactions (stress amplification and dilution) produced by fractures have a substantial effect on the anisotropic characteristics of cracks. To better illustrate the stress interactions across cracks at different fracture densities, we did numerical tests with three 2D simple fracture models (Figure 9). As shown in Figure 9a, each model (0.01 m^2) contains two, six, and ten elliptic inclusions (fractures). Allow each fracture to have a main semiaxis and an aspect ratio of 0.015, and 0.053, respectively. The fractures are completely saturated with water. Table 1 contains the simulation parameters.

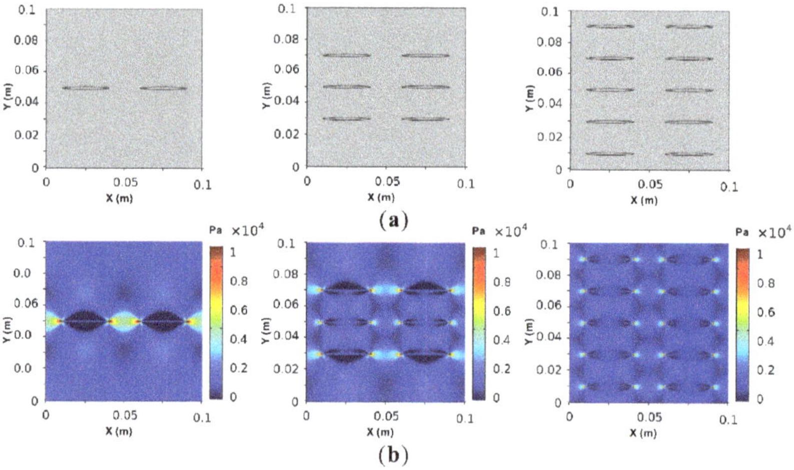

Figure 9. Stress distributions of 2D fractured models: (**a**) three 2D cracked models and (**b**) the stress (σ_{xx}) distributions of the models. The stress dilution and concentration occur near the faces and tips (central parts) of fractures, respectively.

We repeated the numerical test in Figure 2a with the corresponding boundaries, this time using the model in Figure 9a. As illustrated in Figure 9b, the stress distributions of the models vary. The fractures alter the stress distribution in their neighborhood. The stress dilution and concentration occurring near the face (central part) and tip of a fracture in Figure 9b. The stress distribution with a high fracture density contributes a broader range of effects to the background than the low fracture density. Additionally, when fracture density increases, the stress concentrations around the tips decrease. On the basis of the foregoing, it is possible to explain in Figures 7 and 8 that the elastic matrix with a high fracture density is lower than the theoretical method.

4.2. Anisotropy Properties for Real Fractures

Combined with Figures 7 and 8, we see that different fracture densities can induce elastic anisotropy, which can be quantitatively characterized using the equivalent stiffness tensors. We apply 2D anisotropic finite-element modeling to a real crack for anisotropy properties at this part. Figure 10a illustrates an outcrop from the Ordovician carbonate formation in the Tarim Basin, northwest China. Multistage tectonic movements result in a complicated fracture network in the formation [33]. From the Cambrian to the Middle Ordovician, strike-slip faults with numerous fractures were formed by the joint action of north–south and north–east stresses [34]. These secondary fractures are considered as both migration channels and fracture reservoir spaces for hydrocarbon. The outcrop is selected because its fracture characteristics are similar to those of subsurface strata over 7 km, and a complicated fracture network satisfies the simulation scale. Due to the complexity of natural fractures, modeling of irregular geometries and anisotropic features are challenging. The outcrop picture is simplified into a model consisting of background media and superimposed fractures with different shapes, sizes, and orientations to facilitate research on the anisotropy properties of fractures and their stress interaction, as illustrated in Figure 10b.

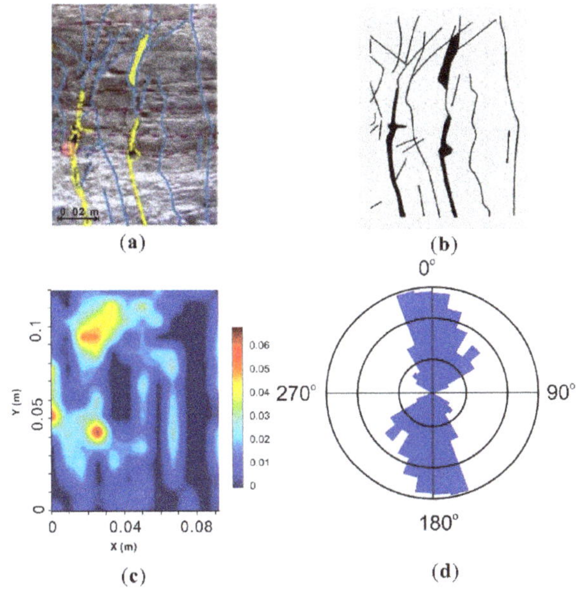

Figure 10. 2D real-fracture model and its geometric characteristics: (**a**) rock outcrop of Ordovician carbonates in Tarim Basin, northwest China; (**b**) the model can be composed of the background media and extracted fractures; (**c**) the model's fracture density distributions where the red and yellow areas in the maps indicate more fractures than others; (**d**) the equal-area rose diagrams of fracture length, which provide dip distribution.

We investigate the cracked model's density and dip-angle. For calculating fracture density, we use the method developed by reference [35,36], which employs a circular scanning window to calculate the number of total fractures by counting the number of endpoints of complete visible fractures within the window. An equal-area rose diagram instead of the histogram of fracture length is used since it provides an indirect orientation distribution (see reference [37] for more details). Figure 10c,d shows the geometric properties of the real model, which clearly exhibits a range of densities and orientations in different areas.

We use a numerical approach combined with finite-element modeling to study the anisotropic properties of the real fracture model at different frequencies. The same physical parameters in Table 1 are used, and oscillatory relaxation tests at frequencies of 0.01, 1, 10, 100, and 1000 Hz are applied.

The resultant real parts of the effective stiffness coefficients M_{11} are illustrated in Figure 11. As we see in Figure 11, the real parts of the stiffness matrix of M_{11} (red circle and red dotted line) show an increasing tendency with increasing frequency, which is not as regular as the simple models in reference [26]. The pictures indicated by the shallow green arrow represent the stress (σ_{xx}) distributions at different frequencies. Additionally, we observe that they change slowly at the beginning, then rapidly, and tend to be flat at high frequency, as the difference between protolith and faults affecting the stiffening of the fluid becomes smaller [38]. The stress (σ_{xx}) occurs an obvious difference between high frequency (100 and 1000 Hz) and low frequency, and the interaction becomes more intense as the frequency increases. The resultant anisotropies of effective elastic properties under the different frequencies can be used to characterize fracture interactions and dynamic stress disturbances. We can characterize the velocity anisotropy and seismic attenuation using these equivalent stiffness tensors.

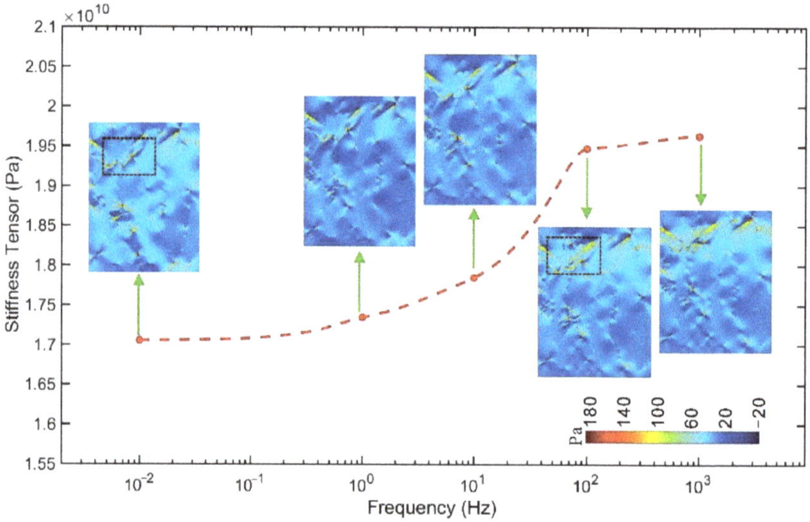

Figure 11. Real parts of the effective stiffness coefficients M_{11} (red circle) of the model with different frequencies indirectly describing the anisotropy characteristics of the model. The shallow green arrows are the stress (σ_{xx}) distributions under different frequencies. The stress (σ_{xx}) between high frequency (100 and 1000 Hz) and low frequency (0.01, 1 and 10 Hz) shows an obvious difference (black dotted rectangle). The magnitude of stress is denoted by the color bar.

The existence of fractures surely affects the overall elasticity, as shown in Figures 4 and 6. In the geophysical community, conventional analytical methods, such as the Hudson model and Eshelby–Cheng model, have been widely used to study the effective elastic properties of rocks, but they are limited to their assumption that fracture geometries are idealized [19,22] and hardly characterize the elastic properties of complex and irregular natural fracture networks. We see that the numerical results keep a good trend with the analytical methods at low crack density, while the difference between them becomes larger with the increase of fracture density (Figures 7 and 8) in that the numerical method explicitly simulates the influence of fractures distribution and their interaction (Figure 9).

It proves that our method can accurately simulate the equivalent stiffness tensors with different frequencies and low fracture density.

When the fracture density is very high, Figures 7 and 8 show that the decreasing trend may change because the solution of the simulation equation may be unstable. Some non-dilute crack theories should be used; for example, non-dilute crack theories established successfully based on the inclusion model DEM theory [39]. Additionally, we need to consider that fractures in nature can be too large to satisfy the simulation conditions. Therefore, further investigation on effective stiffness tensors of arbitrary complex multi-dimensional fracture should also consider these elements and laboratory measurements.

5. Conclusions

Numerical methods must be used to determine the effective elasticity of genuine fractures. We modify Rubino's numeric FEM for real fractures. For 2D simple fracture models with different fracture densities, the 2D technique is compared to theoretical models (Hudson's model and Eshelby–Cheng model). The results closely match the theoretical predictions at low fracture densities. The simulation method provides a physical prediction at high fracture densities that controls the overall elastic properties of rocks.

When applied to a sample of natural fractures extracted from an outcrop at different frequencies, the stress interactions between cracks become highly complex. The stiffness matrix of M_{11} shows a trend as the frequency increases. With increasing frequency, the stress concentration spreads quickly to cover the network's high fracture density, intensifying the interactions and becoming intense. The effective elastic properties of the resulting real fractured model can be altered to understand fracture interactions and dynamic stress disturbances.

The 2D finite-element numerical method is extended to 3D fractured media using the 3D stress–strain relationship defined by the six oscillatory relaxation simulation tests. A least-squares approach can be used to solvethe relevant equations for unknown stiffness tensors. The 3D fractured media are compared to effective medium theories in terms of their elastic responses to different fracture densities. The fracture density is found to have an obvious effect on the anisotropic effective properties, and the proposed method gives a physical prediction for high fracture densities. This work has significance calculating the effective elastic properties of multi-dimensional fracture models and fracture interaction mechanisms.

Author Contributions: Conceptualization, S.J. and L.F.; methodology, S.J.; software, S.J.; formal analysis, S.J.; investigation, S.J. and Y.C.; resources, S.J. and L.F.; data curation, S.J.; writing—original draft preparation, S.J.; writing—review and editing, S.J.; visualization, S.J.; supervision, L.F.; funding acquisition, L.F. All authors have read and agreed to the published version of the manuscript.

Funding: This research was funded by the Strategic Priority Research Program of the Chinese Academy of Sciences (Grant No. XDA14010303) and the National Natural Science Foundation of China (Grant Nos. 41821002).

Institutional Review Board Statement: Not applicable.

Informed Consent Statement: Not applicable.

Data Availability Statement: The data are available from the corresponding and the first author.

Acknowledgments: Thanks to the teachers and students of the FLY Team at the China University of Petroleum (East China). Shikai Jian wants to particularly thank the patience, care, and support from Xinxin Cao. Thanks to the anonymous reviewers for their constructive comments, and thanks to the editors for their hard work.

Conflicts of Interest: The authors declare no conflict of interest.

Appendix A. Least-Squares Procedure for 2D Equivalent Stiffness Tensors

The least-squares procedure to minimize the following cost function $F_1(\mathbf{M})$ for computing the equivalent stiffness matrix M_{ij} is described according to [26].

$$\begin{aligned} F_1(\mathbf{M}) = & \sum_{m=1}^{3}\Big(\|\overline{\sigma^m_{xx}} - M_{11}\overline{e^m_{xx}} - M_{12}\overline{e^m_{yy}} - 2M_{16}\overline{e^m_{xy}}\|^2 \\ & + \|\overline{\sigma^m_{yy}} - M_{12}\overline{e^m_{xx}} - M_{22}\overline{e^m_{yy}} - 2M_{26}\overline{e^m_{xy}}\|^2 \\ & + \|\overline{\sigma^m_{xy}} - M_{16}\overline{e^m_{xx}} - M_{26}\overline{e^m_{yy}} - 2M_{66}\overline{e^m_{xy}}\|^2\Big), \end{aligned} \tag{A1}$$

where m = 1–3 denotes the mth simulation test, and $\overline{\sigma^m_{ij}}$ and $\overline{e^m_{ij}}$ are the average stress and average strain, respectively. With the goal, the partial derivatives of the cost function related to the elements M_{ij} are set to be zero, resulting in the following linear equations:

$$\begin{aligned} & \frac{\partial F(\mathbf{M})}{\partial M_{11}} = 0 \\ & \sum_{m=1}^{3}\overline{\sigma^m_{xx}e^m_{xx}} = M_{11}\sum_{m=1}^{3}\overline{e^m_{xx}e^m_{xx}} + M_{12}\sum_{m=1}^{3}\overline{e^m_{yy}e^m_{xx}} + 2M_{16}\sum_{m=1}^{3}\overline{e^m_{xy}e^m_{xx}}, \end{aligned} \tag{A2}$$

$$\begin{aligned} & \frac{\partial F(\mathbf{M})}{\partial M_{12}} = 0 \\ & \sum_{m=1}^{3}\left(\overline{\sigma^m_{xx}e^m_{yy}} + \overline{\sigma^m_{yy}e^m_{xx}}\right) = M_{11}\sum_{m=1}^{3}\overline{e^m_{xx}e^m_{yy}} + M_{12}\sum_{m=1}^{3}\left(\overline{e^m_{xx}e^m_{xx}} + \overline{e^m_{yy}e^m_{yy}}\right) + 2M_{16}\sum_{m=1}^{3}\overline{e^m_{xy}e^m_{xx}} \\ & \qquad + 2M_{26}\sum_{m=1}^{3}\overline{e^m_{xy}e^m_{xx}} + M_{22}\sum_{m=1}^{3}\overline{e^m_{xx}e^m_{yy}}, \end{aligned} \tag{A3}$$

$$\begin{aligned} & \frac{\partial F(\mathbf{M})}{\partial M_{16}} = 0 \\ & \sum_{m=1}^{3}\left(2\overline{\sigma^m_{xx}e^m_{xy}} + \overline{\sigma^m_{xy}e^m_{xx}}\right) = 2M_{11}\sum_{m=1}^{3}\overline{e^m_{xx}e^m_{yy}} + 2M_{12}\sum_{m=1}^{3}\left(\overline{e^m_{yy}e^m_{xy}}\right) + M_{16}\left(\sum_{m=1}^{3}4\overline{e^m_{xy}e^m_{xy}} + \overline{e^m_{xx}e^m_{xx}}\right) \\ & \qquad + M_{26}\sum_{m=1}^{3}e^s_{yy}e^s_{xx} + 2M_{66}\sum_{m=1}^{3}\overline{e^m_{xx}e^m_{xy}}, \end{aligned} \tag{A4}$$

$$\begin{aligned} & \frac{\partial F(\mathbf{M})}{\partial M_{22}} = 0 \\ & \sum_{m=1}^{3}\overline{\sigma^m_{yy}e^m_{yy}} = M_{12}\sum_{m=1}^{3}\overline{e^m_{xx}e^m_{yy}} + M_{22}\sum_{m=1}^{3}\overline{e^m_{yy}e^m_{yy}} + 2M_{26}\sum_{m=1}^{3}\overline{e^m_{xy}e^m_{yy}}, \end{aligned} \tag{A5}$$

$$\begin{aligned} & \frac{\partial F(\mathbf{M})}{\partial M_{26}} = 0 \\ & \sum_{m=1}^{3}\left(2\overline{\sigma^m_{yy}e^m_{xy}} + \overline{\sigma^m_{xy}e^m_{yy}}\right) = 2M_{12}\sum_{m=1}^{3}\overline{e^m_{xx}e^m_{xy}} + 2M_{22}\sum_{m=1}^{3}\left(\overline{e^m_{yy}e^m_{xy}}\right) + 2M_{26}\sum_{m=1}^{3}\left(4\overline{e^m_{xy}e^m_{xy}} + \overline{e^m_{yy}e^m_{yy}}\right) \\ & \qquad + M_{16}\sum_{m=1}^{3}\overline{e^m_{yy}e^m_{xx}} + 2M_{66}\sum_{m=1}^{3}\overline{e^m_{xy}e^m_{yy}}, \end{aligned} \tag{A6}$$

and

$$\begin{aligned} & \frac{\partial F(\mathbf{M})}{\partial M_{66}} = 0 \\ & \sum_{m=1}^{3}\overline{\sigma^m_{xy}e^m_{xy}} = M_{16}\sum_{m=1}^{3}\overline{e^m_{xx}e^m_{xy}} + M_{26}\sum_{m=1}^{3}\overline{e^m_{yy}e^m_{xy}} + 2M_{66}\sum_{m=1}^{3}\overline{e^m_{xy}e^m_{xy}}. \end{aligned} \tag{A7}$$

Appendix B. Least-Squares Procedure for 3D Equivalent Stiffness Tensors

The least-squares procedure to minimize the following cost function $F_2(\mathbf{M})$ for computing the equivalent stiffness matrix M_{ij} is described as follows.

$$
\begin{aligned}
F_2(\mathbf{M}) = \sum_{r=1}^{6} \Big(& \|\overline{\sigma^r_{xx}} - M_{11}\overline{e^r_{xx}} - M_{12}\overline{e^r_{yy}} - M_{13}\overline{e^r_{zz}} - 2M_{14}\overline{e^r_{yz}} - 2M_{15}\overline{e^r_{zx}} - 2M_{16}\overline{e^r_{xy}}\|^2 \\
& + \|\overline{\sigma^r_{yy}} - M_{12}\overline{e^r_{xx}} - M_{22}\overline{e^r_{yy}} - M_{23}\overline{e^r_{zz}} - 2M_{24}\overline{e^r_{yz}} - 2M_{25}\overline{e^r_{zx}} - 2M_{26}\overline{e^r_{xy}}\|^2 \\
& + \|\overline{\sigma^r_{zz}} - M_{13}\overline{e^r_{xx}} - M_{23}\overline{e^r_{yy}} - M_{33}\overline{e^r_{zz}} - 2M_{34}\overline{e^r_{yz}} - 2M_{35}\overline{e^r_{zx}} - 2M_{36}\overline{e^r_{xy}}\|^2 \\
& + \|\overline{\sigma^r_{yz}} - M_{14}\overline{e^r_{xx}} - M_{24}\overline{e^r_{yy}} - M_{34}\overline{e^r_{zz}} - 2M_{44}\overline{e^r_{yz}} - 2M_{45}\overline{e^r_{zx}} - 2M_{46}\overline{e^r_{xy}}\|^2 \\
& + \|\overline{\sigma^r_{zx}} - M_{15}\overline{e^r_{xx}} - M_{25}\overline{e^r_{yy}} - M_{35}\overline{e^r_{zz}} - 2M_{45}\overline{e^r_{yz}} - 2M_{55}\overline{e^r_{zx}} - 2M_{56}\overline{e^r_{xy}}\|^2 \\
& + \|\overline{\sigma^r_{xy}} - M_{16}\overline{e^r_{xx}} - M_{26}\overline{e^r_{yy}} - M_{36}\overline{e^r_{zz}} - 2M_{46}\overline{e^r_{yz}} - 2M_{56}\overline{e^r_{zx}} - 2M_{66}\overline{e^r_{xy}}\|^2 \Big),
\end{aligned}
\tag{A8}
$$

where r = 1–6 denotes the rth simulation test, and $\overline{\sigma^r_{ij}}$ and $\overline{e^r_{ij}}$ represents the averaging stress and strain, respectively.

With the goal, the partial derivatives of the cost function related to the elements M_{ij} are set to be zero, resulting in the following linear equations:

$$
\begin{aligned}
& \frac{\partial F(\mathbf{M})}{\partial M_{11}} = 0 \\
& \sum_{r=1}^{6} \overline{\sigma^r_{xx} e^r_{xx}} = M_{11} \sum_{r=1}^{6} \overline{e^r_{xx} e^r_{xx}} + M_{12} \sum_{r=1}^{6} \overline{e^r_{yy} e^r_{xx}} + M_{13} \sum_{r=1}^{6} \overline{e^r_{zz} e^r_{xx}} + 2M_{14} \sum_{r=1}^{6} \overline{e^r_{yz} e^r_{xx}} + 2M_{15} \sum_{r=1}^{6} \overline{e^r_{zx} e^r_{xx}} + 2M_{16} \sum_{r=1}^{6} \overline{e^r_{xy} e^r_{xx}},
\end{aligned}
\tag{A9}
$$

$$
\begin{aligned}
& \frac{\partial F(\mathbf{M})}{\partial M_{12}} = 0 \\
& \sum_{r=1}^{6} \left(\overline{\sigma^r_{xx} e^r_{yy}} + \overline{\sigma^r_{yy} e^r_{xx}} \right) = M_{11} \sum_{r=1}^{6} \overline{e^r_{xx} e^r_{yy}} + M_{12} \sum_{r=1}^{6} \left(\overline{e^r_{xx} e^r_{xx}} + \overline{e^r_{yy} e^r_{yy}} \right) + M_{13} \sum_{r=1}^{6} \overline{e^r_{zz} e^r_{yy}} + 2M_{14} \sum_{r=1}^{6} \overline{e^r_{yz} e^r_{yy}} + 2M_{15} \sum_{r=1}^{6} \overline{e^r_{zx} e^r_{yy}} + 2M_{16} \sum_{r=1}^{6} \overline{e^r_{xy} e^r_{yy}} \\
& + M_{22} \sum_{r=1}^{6} \overline{e^r_{yy} e^r_{xx}} + M_{23} \sum_{r=1}^{6} \overline{e^r_{zz} e^r_{xx}} + 2M_{24} \sum_{r=1}^{6} \overline{e^r_{yz} e^r_{xx}} + 2M_{25} \sum_{r=1}^{6} \overline{e^r_{zx} e^r_{xx}} + 2M_{26} \sum_{r=1}^{6} \overline{e^r_{xy} e^r_{xx}},
\end{aligned}
\tag{A10}
$$

$$
\begin{aligned}
& \frac{\partial F(\mathbf{M})}{\partial M_{13}} = 0 \\
& \sum_{r=1}^{6} \left(\overline{\sigma^r_{xx} e^r_{zz}} + \overline{\sigma^r_{zz} e^r_{xx}} \right) = M_{11} \sum_{r=1}^{6} \overline{e^r_{zz} e^r_{xx}} + M_{12} \sum_{r=1}^{6} \overline{e^r_{yy} e^r_{zz}} + M_{13} \sum_{r=1}^{6} \left(\overline{e^r_{zz} e^r_{zz}} + \overline{e^r_{xx} e^r_{xx}} \right) + 2M_{14} \sum_{r=1}^{6} \overline{e^r_{yz} e^r_{zz}} + 2M_{15} \sum_{r=1}^{6} \overline{e^r_{zx} e^r_{zz}} + 2M_{16} \sum_{r=1}^{6} \overline{e^r_{xy} e^r_{zz}} \\
& + M_{23} \sum_{r=1}^{6} \overline{e^r_{yy} e^r_{xx}} + M_{33} \sum_{r=1}^{6} \overline{e^r_{zz} e^r_{xx}} + 2M_{34} \sum_{r=1}^{6} \overline{e^r_{yz} e^r_{xx}} + 2M_{35} \sum_{r=1}^{6} \overline{e^r_{zx} e^r_{xx}} + 2M_{36} \sum_{r=1}^{6} \overline{e^r_{xy} e^r_{xx}},
\end{aligned}
\tag{A11}
$$

$$
\begin{aligned}
& \frac{\partial F(\mathbf{M})}{\partial M_{14}} = 0 \\
& \sum_{r=1}^{6} \left(2\overline{\sigma^r_{xx} e^r_{yz}} + \overline{\sigma^r_{yz} e^r_{xx}} \right) = 2M_{11} \sum_{r=1}^{6} \overline{e^r_{xx} e^r_{yz}} + 2M_{12} \sum_{r=1}^{6} \overline{e^r_{yy} e^r_{yz}} + 2M_{13} \sum_{r=1}^{6} \overline{e^r_{zz} e^r_{yz}} + M_{14} \sum_{r=1}^{6} \left(4\overline{e^r_{yz} e^r_{yz}} + \overline{e^r_{xx} e^r_{xx}} \right) + 4M_{15} \sum_{r=1}^{6} \overline{e^r_{zx} e^r_{yz}} + 4M_{16} \sum_{r=1}^{6} \overline{e^r_{xy} e^r_{yz}} \\
& + M_{24} \sum_{r=1}^{6} \overline{e^r_{yy} e^r_{xx}} + M_{34} \sum_{r=1}^{6} \overline{e^r_{zz} e^r_{xx}} + 2M_{44} \sum_{r=1}^{6} \overline{e^r_{yz} e^r_{xx}} + 2M_{45} \sum_{r=1}^{6} \overline{e^r_{zx} e^r_{xx}} + 2M_{46} \sum_{r=1}^{6} \overline{e^r_{xy} e^r_{xx}},
\end{aligned}
\tag{A12}
$$

$$
\begin{aligned}
& \frac{\partial F(\mathbf{M})}{\partial M_{15}} = 0 \\
& \sum_{r=1}^{6} \left(2\overline{\sigma^r_{xx} e^r_{zx}} + \overline{\sigma^r_{zx} e^r_{xx}} \right) = 2M_{11} \sum_{r=1}^{6} \overline{e^r_{xx} e^r_{zx}} + 2M_{12} \sum_{r=1}^{6} \overline{e^r_{yy} e^r_{zx}} + 2M_{13} \sum_{r=1}^{6} \overline{e^r_{zz} e^r_{zx}} + 4M_{14} \sum_{r=1}^{6} \overline{e^r_{yz} e^r_{zx}} + M_{15} \sum_{r=1}^{6} \left(4\overline{e^r_{zx} e^r_{zx}} + \overline{e^r_{xx} e^r_{xx}} \right) + 4M_{16} \sum_{k=1}^{6} \overline{e^r_{xy} e^r_{zx}} \\
& + M_{25} \sum_{r=1}^{6} \overline{e^r_{yy} e^r_{xx}} + M_{35} \sum_{r=1}^{6} \overline{e^r_{zz} e^r_{xx}} + 2M_{45} \sum_{r=1}^{6} \overline{e^r_{yz} e^r_{xx}} + 2M_{55} \sum_{r=1}^{6} \overline{e^r_{zx} e^r_{xx}} + 2M_{56} \sum_{r=1}^{6} \overline{e^r_{xy} e^r_{xx}},
\end{aligned}
\tag{A13}
$$

$$
\begin{aligned}
& \frac{\partial F(\mathbf{M})}{\partial M_{16}} = 0 \\
& \sum_{r=1}^{6} \left(2\overline{\sigma^r_{xx} e^r_{xy}} + \overline{\sigma^r_{xy} e^r_{xx}} \right) = 2M_{11} \sum_{r=1}^{6} \overline{e^r_{xx} e^r_{xy}} + 2M_{12} \sum_{r=1}^{6} \overline{e^r_{yy} e^r_{xy}} + 2M_{13} \sum_{r=1}^{6} \overline{e^r_{zz} e^r_{xy}} + 4M_{14} \sum_{r=1}^{6} \overline{e^r_{yz} e^r_{xy}} + 4M_{15} \sum_{r=1}^{6} \overline{e^r_{zx} e^r_{xy}} + M_{16} \sum_{r=1}^{6} \left(4\overline{e^r_{xy} e^r_{xy}} + \overline{e^r_{xx} e^r_{xx}} \right) \\
& + M_{26} \sum_{r=1}^{6} \overline{e^r_{yy} e^r_{xx}} + M_{36} \sum_{r=1}^{6} \overline{e^r_{zz} e^r_{xx}} + 2M_{46} \sum_{r=1}^{6} \overline{e^r_{yz} e^r_{xx}} + 2M_{56} \sum_{r=1}^{6} \overline{e^r_{zx} e^r_{xx}} + 2M_{66} \sum_{r=1}^{6} \overline{e^r_{xy} e^r_{xx}},
\end{aligned}
\tag{A14}
$$

$$
\begin{aligned}
& \frac{\partial F(\mathbf{M})}{\partial M_{22}} = 0 \\
& \sum_{r=1}^{6} \overline{\sigma^r_{yy} e^r_{yy}} = M_{12} \sum_{r=1}^{6} \overline{e^r_{xx} e^r_{yy}} + M_{22} \sum_{r=1}^{6} \overline{e^r_{yy} e^r_{yy}} + M_{23} \sum_{r=1}^{6} \overline{e^r_{zz} e^r_{yy}} + 2M_{24} \sum_{r=1}^{6} \overline{e^r_{yz} e^r_{yy}} + 2M_{25} \sum_{r=1}^{6} \overline{e^r_{zx} e^r_{yy}} + 2M_{26} \sum_{r=1}^{6} \overline{e^r_{xy} e^r_{yy}},
\end{aligned}
\tag{A15}
$$

$$
\begin{aligned}
& \frac{\partial F(\mathbf{M})}{\partial M_{23}} = 0 \\
& \sum_{r=1}^{6} \left(\overline{\sigma^r_{yy} e^r_{zz}} + \overline{\sigma^r_{zz} e^r_{yy}} \right) = M_{12} \sum_{r=1}^{6} \overline{e^r_{xx} e^r_{zz}} + M_{22} \sum_{r=1}^{6} \overline{e^r_{yy} e^r_{zz}} + M_{23} \sum_{r=1}^{6} \left(\overline{e^r_{zz} e^r_{zz}} + \overline{e^r_{yy} e^r_{yy}} \right) + 2M_{24} \sum_{r=1}^{6} \overline{e^r_{yz} e^r_{zz}} + 2M_{25} \sum_{r=1}^{6} \overline{e^r_{zx} e^r_{zz}} + 2M_{26} \sum_{r=1}^{6} \overline{e^r_{xy} e^r_{zz}} \\
& + M_{13} \sum_{r=1}^{6} \overline{e^r_{xx} e^r_{yy}} + M_{33} \sum_{r=1}^{6} \overline{e^r_{zz} e^r_{yy}} + 2M_{34} \sum_{r=1}^{6} \overline{e^r_{yz} e^r_{yy}} + 2M_{35} \sum_{r=1}^{6} \overline{e^r_{zx} e^r_{yy}} + 2M_{36} \sum_{r=1}^{6} \overline{e^r_{xy} e^r_{yy}},
\end{aligned}
\tag{A16}
$$

$$
\begin{aligned}
& \frac{\partial F(\mathbf{M})}{\partial M_{24}} = 0 \\
& \sum_{r=1}^{6} \left(2\overline{\sigma^r_{yy} e^r_{yz}} + \overline{\sigma^r_{yz} e^r_{yy}} \right) = 2M_{12} \sum_{r=1}^{6} \overline{e^r_{xx} e^r_{yz}} + 2M_{22} \sum_{r=1}^{6} \overline{e^r_{yy} e^r_{yz}} + 2M_{23} \sum_{r=1}^{6} \overline{e^r_{zz} e^r_{yz}} + M_{24} \sum_{r=1}^{6} \left(4\overline{e^r_{yz} e^r_{yz}} + \overline{e^r_{yy} e^r_{yy}} \right) + 4M_{25} \sum_{r=1}^{6} \overline{e^r_{zx} e^r_{yz}} + 4M_{26} \sum_{r=1}^{6} \overline{e^r_{xy} e^r_{yz}} \\
& + M_{14} \sum_{r=1}^{6} \overline{e^r_{xx} e^r_{yy}} + M_{34} \sum_{r=1}^{6} \overline{e^r_{zz} e^r_{yy}} + 2M_{44} \sum_{r=1}^{6} \overline{e^r_{yz} e^r_{yy}} + 2M_{45} \sum_{r=1}^{6} \overline{e^r_{zx} e^r_{yy}} + 2M_{46} \sum_{r=1}^{6} \overline{e^r_{xy} e^r_{yy}},
\end{aligned}
\tag{A17}
$$

$$\begin{aligned}&\frac{\partial F(\mathbf{M})}{\partial M_{25}}=0\\&\sum_{r=1}^{6}\left(2\overline{\sigma_{yy}^{r}e_{zx}^{r}}+\overline{\sigma_{zx}^{r}e_{yy}^{r}}\right)=2M_{12}\sum_{r=1}^{6}\overline{e_{xx}^{r}e_{zx}^{r}}+2M_{22}\sum_{r=1}^{6}\overline{e_{yy}^{r}e_{zx}^{r}}+2M_{23}\sum_{r=1}^{6}\overline{e_{zz}^{r}e_{zx}^{r}}+4M_{24}\sum_{r=1}^{6}\overline{e_{yz}^{r}e_{zx}^{r}}+M_{25}\sum_{r=1}^{6}\left(4\overline{e_{zx}^{r}e_{zx}^{r}}+\overline{e_{yy}^{r}e_{yy}^{r}}\right)+4M_{26}\sum_{r=1}^{6}\overline{e_{xy}^{r}e_{zx}^{r}}\\&+M_{15}\sum_{r=1}^{6}\overline{e_{xx}^{r}e_{yy}^{r}}+M_{35}\sum_{r=1}^{6}\overline{e_{zz}^{r}e_{yy}^{r}}+2M_{45}\sum_{r=1}^{6}\overline{e_{yz}^{r}e_{yy}^{r}}+2M_{55}\sum_{r=1}^{6}\overline{e_{zx}^{r}e_{yy}^{r}}+2M_{56}\sum_{r=1}^{6}\overline{e_{xy}^{r}e_{yy}^{r}},\end{aligned}\tag{A18}$$

$$\begin{aligned}&\frac{\partial F(\mathbf{M})}{\partial M_{26}}=0\\&\sum_{r=1}^{6}\left(2\overline{\sigma_{yy}^{r}e_{xy}^{r}}+\overline{\sigma_{xy}^{r}e_{yy}^{r}}\right)=2M_{12}\sum_{r=1}^{6}\overline{e_{xx}^{r}e_{xy}^{r}}+2M_{22}\sum_{r=1}^{6}\overline{e_{yy}^{r}e_{xy}^{r}}+2M_{23}\sum_{r=1}^{6}\overline{e_{zz}^{r}e_{xy}^{r}}+4M_{24}\sum_{r=1}^{6}\overline{e_{yz}^{r}e_{xy}^{r}}+4M_{25}\sum_{r=1}^{6}\overline{e_{zx}^{r}e_{xy}^{r}}+M_{26}\sum_{r=1}^{6}\left(4\overline{e_{xy}^{r}e_{xy}^{r}}+\overline{e_{yy}^{r}e_{yy}^{r}}\right)\\&+M_{16}\sum_{r=1}^{6}\overline{e_{xx}^{r}e_{yy}^{r}}+M_{36}\sum_{r=1}^{6}\overline{e_{zz}^{r}e_{yy}^{r}}+2M_{46}\sum_{r=1}^{6}\overline{e_{yz}^{r}e_{yy}^{r}}+2M_{56}\sum_{r=1}^{6}\overline{e_{zx}^{r}e_{yy}^{r}}+2M_{66}\sum_{r=1}^{6}\overline{e_{xy}^{r}e_{yy}^{r}},\end{aligned}\tag{A19}$$

$$\begin{aligned}&\frac{\partial F(\mathbf{M})}{\partial M_{33}}=0\\&\sum_{r=1}^{6}\overline{\sigma_{zz}^{r}e_{zz}^{r}}=M_{13}\sum_{r=1}^{6}\overline{e_{xx}^{r}e_{zz}^{r}}+M_{23}\sum_{r=1}^{6}\overline{e_{yy}^{r}e_{zz}^{r}}+M_{33}\sum_{r=1}^{6}\overline{e_{zz}^{r}e_{zz}^{r}}+2M_{34}\sum_{r=1}^{6}\overline{e_{yz}^{r}e_{zz}^{r}}+2M_{35}\sum_{r=1}^{6}\overline{e_{zx}^{r}e_{zz}^{r}}+2M_{36}\sum_{r=1}^{6}\overline{e_{xy}^{r}e_{zz}^{r}},\end{aligned}\tag{A20}$$

$$\begin{aligned}&\frac{\partial F(\mathbf{M})}{\partial M_{34}}=0\\&\sum_{r=1}^{6}\left(2\overline{\sigma_{zz}^{r}e_{yz}^{r}}+\overline{\sigma_{yz}^{r}e_{zz}^{r}}\right)=2M_{13}\sum_{r=1}^{6}\overline{e_{xx}^{r}e_{yz}^{r}}+2M_{23}\sum_{r=1}^{6}\overline{e_{yy}^{r}e_{yz}^{r}}+2M_{33}\sum_{r=1}^{6}\overline{e_{zz}^{r}e_{yz}^{r}}+M_{34}\sum_{r=1}^{6}\left(4\overline{e_{yz}^{r}e_{yz}^{r}}+\overline{e_{zz}^{r}e_{zz}^{r}}\right)+4M_{35}\sum_{r=1}^{6}\overline{e_{zx}^{r}e_{yz}^{r}}+4M_{36}\sum_{r=1}^{6}\overline{e_{xy}^{r}e_{yz}^{r}}\\&+M_{14}\sum_{r=1}^{6}\overline{e_{xx}^{r}e_{zz}^{r}}+M_{24}\sum_{r=1}^{6}\overline{e_{yy}^{r}e_{zz}^{r}}+2M_{44}\sum_{r=1}^{6}\overline{e_{yz}^{r}e_{zz}^{r}}+2M_{45}\sum_{r=1}^{6}\overline{e_{zx}^{r}e_{zz}^{r}}+2M_{46}\sum_{r=1}^{6}\overline{e_{xy}^{r}e_{zz}^{r}},\end{aligned}\tag{A21}$$

$$\begin{aligned}&\frac{\partial F(\mathbf{M})}{\partial M_{35}}=0\\&\sum_{r=1}^{6}(2\overline{\sigma_{zz}^{r}e_{zx}^{r}}+\overline{\sigma_{zx}^{r}e_{zz}^{r}})=2M_{13}\sum_{r=1}^{6}\overline{e_{xx}^{r}e_{zx}^{r}}+2M_{23}\sum_{r=1}^{6}\overline{e_{yy}^{r}e_{zx}^{r}}+2M_{33}\sum_{r=1}^{6}\overline{e_{zz}^{r}e_{zx}^{r}}+4M_{34}\sum_{r=1}^{6}\overline{e_{yz}^{r}e_{zx}^{r}}+M_{35}\sum_{r=1}^{6}(4\overline{e_{zx}^{r}e_{zx}^{r}}+\overline{e_{zz}^{r}e_{zz}^{r}})+4M_{36}\sum_{r=1}^{6}\overline{e_{xy}^{r}e_{zx}^{r}}\\&+M_{15}\sum_{r=1}^{6}\overline{e_{xx}^{r}e_{zz}^{r}}+M_{25}\sum_{r=1}^{6}\overline{e_{yy}^{r}e_{zz}^{r}}+2M_{45}\sum_{r=1}^{6}\overline{e_{yz}^{r}e_{zz}^{r}}+2M_{55}\sum_{r=1}^{6}\overline{e_{zx}^{r}e_{zz}^{r}}+2M_{56}\sum_{r=1}^{6}\overline{e_{xy}^{r}e_{zz}^{r}},\end{aligned}\tag{A22}$$

$$\begin{aligned}&\frac{\partial F(\mathbf{M})}{\partial M_{36}}=0\\&\sum_{r=1}^{6}\left(2\overline{\sigma_{zz}^{r}e_{xy}^{r}}+\overline{\sigma_{xy}^{r}e_{zz}^{r}}\right)=2M_{13}\sum_{r=1}^{6}\overline{e_{xx}^{r}e_{xy}^{r}}+2M_{23}\sum_{r=1}^{6}\overline{e_{yy}^{r}e_{xy}^{r}}+2M_{33}\sum_{r=1}^{6}\overline{e_{zz}^{r}e_{xy}^{r}}+4M_{34}\sum_{r=1}^{6}\overline{e_{yz}^{r}e_{xy}^{r}}+4M_{35}\sum_{r=1}^{6}\overline{e_{zx}^{r}e_{xy}^{r}}+M_{36}\sum_{r=1}^{6}\left(4\overline{e_{xy}^{r}e_{xy}^{r}}+\overline{e_{zz}^{r}e_{zz}^{r}}\right)\\&+M_{16}\sum_{r=1}^{6}\overline{e_{xx}^{r}e_{zz}^{r}}+M_{26}\sum_{r=1}^{6}\overline{e_{yy}^{r}e_{zz}^{r}}+2M_{46}\sum_{r=1}^{6}\overline{e_{yz}^{r}e_{zz}^{r}}+2M_{56}\sum_{r=1}^{6}\overline{e_{zx}^{r}e_{zz}^{r}}+2M_{66}\sum_{r=1}^{6}\overline{e_{xy}^{r}e_{zz}^{r}},\end{aligned}\tag{A23}$$

$$\begin{aligned}&\frac{\partial F(\mathbf{M})}{\partial M_{44}}=0\\&\sum_{r=1}^{6}\overline{\sigma_{yz}^{r}e_{yz}^{r}}=M_{14}\sum_{r=1}^{6}\overline{e_{xx}^{r}e_{yz}^{r}}+M_{24}\sum_{r=1}^{6}\overline{e_{yy}^{r}e_{yz}^{r}}+M_{34}\sum_{r=1}^{6}\overline{e_{zz}^{r}e_{yz}^{r}}+2M_{44}\sum_{r=1}^{6}\overline{e_{yz}^{r}e_{yz}^{r}}+2M_{45}\sum_{r=1}^{6}\overline{e_{zx}^{r}e_{yz}^{r}}+2M_{46}\sum_{r=1}^{6}\overline{e_{xy}^{r}e_{yz}^{r}},\end{aligned}\tag{A24}$$

$$\begin{aligned}&\frac{\partial F(\mathbf{M})}{\partial M_{45}}=0\\&\sum_{r=1}^{6}\left(\overline{\sigma_{yz}^{r}e_{zx}^{r}}+\overline{\sigma_{zx}^{r}e_{yz}^{r}}\right)=M_{14}\sum_{r=1}^{6}\overline{e_{xx}^{r}e_{zx}^{r}}+M_{24}\sum_{r=1}^{6}\overline{e_{yy}^{r}e_{zx}^{r}}+M_{34}\sum_{r=1}^{6}\overline{e_{zz}^{r}e_{zx}^{r}}+2M_{44}\sum_{r=1}^{6}\overline{e_{yz}^{r}e_{zx}^{r}}+2M_{45}\sum_{r=1}^{6}\left(\overline{e_{zx}^{r}e_{zx}^{r}}+\overline{e_{yz}^{r}e_{yz}^{r}}\right)+2M_{46}\sum_{r=1}^{6}\overline{e_{xy}^{r}e_{zx}^{r}}\\&+M_{15}\sum_{r=1}^{6}\overline{e_{xx}^{r}e_{yz}^{r}}+M_{25}\sum_{r=1}^{6}\overline{e_{yy}^{r}e_{yz}^{r}}+M_{35}\sum_{r=1}^{6}\overline{e_{zz}^{r}e_{yz}^{r}}+2M_{55}\sum_{r=1}^{6}\overline{e_{zx}^{r}e_{yz}^{r}}+2M_{56}\sum_{r=1}^{6}\overline{e_{xy}^{r}e_{yz}^{r}},\end{aligned}\tag{A25}$$

$$\begin{aligned}&\frac{\partial F(\mathbf{M})}{\partial M_{46}}=0\\&\sum_{r=1}^{6}\left(\overline{\sigma_{yz}^{r}e_{xy}^{r}}+\overline{\sigma_{xy}^{r}e_{yz}^{r}}\right)=M_{14}\sum_{r=1}^{6}\overline{e_{xx}^{r}e_{xy}^{r}}+M_{24}\sum_{r=1}^{6}\overline{e_{yy}^{r}e_{xy}^{r}}+M_{34}\sum_{r=1}^{6}\overline{e_{zz}^{r}e_{xy}^{r}}+2M_{44}\sum_{r=1}^{6}\overline{e_{yz}^{r}e_{xy}^{r}}+2M_{45}\sum_{r=1}^{6}\overline{e_{zx}^{r}e_{xy}^{r}}+2M_{46}\sum_{r=1}^{6}\left(\overline{e_{xy}^{r}e_{xy}^{r}}+\overline{e_{yz}^{r}e_{yz}^{r}}\right)\\&+M_{16}\sum_{r=1}^{6}\overline{e_{xx}^{r}e_{yz}^{r}}+M_{26}\sum_{r=1}^{6}\overline{e_{yy}^{r}e_{yz}^{r}}+M_{36}\sum_{r=1}^{6}\overline{e_{zz}^{r}e_{yz}^{r}}+2M_{56}\sum_{r=1}^{6}\overline{e_{zx}^{r}e_{yz}^{r}}+2M_{66}\sum_{r=1}^{6}\overline{e_{xy}^{r}e_{yz}^{r}},\end{aligned}\tag{A26}$$

$$\begin{aligned}&\frac{\partial F(\mathbf{M})}{\partial M_{55}}=0\\&\sum_{r=1}^{6}\overline{\sigma_{zx}^{r}e_{zx}^{r}}=M_{15}\sum_{r=1}^{6}\overline{\sigma_{xx}^{r}e_{zx}^{r}}+M_{25}\sum_{r=1}^{6}\overline{\sigma_{yy}^{r}e_{zx}^{r}}+M_{35}\sum_{r=1}^{6}\overline{\sigma_{zz}^{r}e_{zx}^{r}}+2M_{45}\sum_{r=1}^{6}\overline{\sigma_{yz}^{r}e_{zx}^{r}}+2M_{55}\sum_{r=1}^{6}\overline{\sigma_{zx}^{r}e_{zx}^{r}}+2M_{56}\sum_{r=1}^{6}\overline{\sigma_{xy}^{r}e_{zx}^{r}},\end{aligned}\tag{A27}$$

$$\begin{aligned}&\frac{\partial F(\mathbf{M})}{\partial M_{56}}=0\\&\sum_{r=1}^{6}\left(\overline{\sigma_{zx}^{r}e_{xy}^{r}}+\overline{\sigma_{xy}^{r}e_{zx}^{r}}\right)=M_{15}\sum_{r=1}^{6}\overline{e_{xx}^{r}e_{xy}^{r}}+M_{25}\sum_{r=1}^{6}\overline{e_{yy}^{r}e_{xy}^{r}}+M_{35}\sum_{r=1}^{6}\overline{e_{zz}^{r}e_{xy}^{r}}+2M_{45}\sum_{r=1}^{6}\overline{e_{yz}^{r}e_{xy}^{r}}+2M_{55}\sum_{r=1}^{6}\overline{e_{zx}^{r}e_{xy}^{r}}+2M_{56}\sum_{r=1}^{6}\left(\overline{e_{xy}^{r}e_{xy}^{r}}+\overline{e_{zx}^{r}e_{zx}^{r}}\right)\\&+M_{16}\sum_{r=1}^{6}\overline{e_{xx}^{r}e_{zx}^{r}}+M_{26}\sum_{r=1}^{6}\overline{e_{yy}^{r}e_{zx}^{r}}+M_{36}\sum_{r=1}^{6}\overline{e_{zz}^{r}e_{zx}^{r}}+2M_{46}\sum_{r=1}^{6}\overline{e_{yz}^{r}e_{zx}^{r}}+2M_{66}\sum_{r=1}^{6}\overline{e_{xy}^{r}e_{zx}^{r}},\end{aligned}\tag{A28}$$

$$\begin{aligned}&\frac{\partial F(\mathbf{M})}{\partial M_{66}}=0\\&\sum_{r=1}^{6}\overline{\sigma_{xy}^{r}e_{xy}^{r}}=M_{16}\sum_{r=1}^{6}\overline{e_{xx}^{r}e_{xy}^{r}}+M_{26}\sum_{r=1}^{6}\overline{e_{yy}^{r}e_{xy}^{r}}+M_{36}\sum_{r=1}^{6}\overline{e_{zz}^{r}e_{xy}^{r}}+2M_{46}\sum_{r=1}^{6}\overline{e_{yz}^{r}e_{xy}^{r}}+2M_{56}\sum_{r=1}^{6}\overline{e_{zx}^{r}e_{xy}^{r}}+2M_{66}\sum_{r=1}^{6}\overline{e_{xy}^{r}e_{xy}^{r}}.\end{aligned}\tag{A29}$$

Appendix C. The Hudson's Model and Eshelby–Cheng Model

The second-order Hudson [6,7] equivalent stiffness matrix is given as

$$M_{ij}^{eff} = M_{ij}^{0} + M_{ij}^{1} + M_{ij}^{2} \tag{A30}$$

where M_{ij}^{0} is the elastic parameters of isotropy background, and M_{ij}^{eff} is the total stiffness tensor. Additionally, M_{ij}^{1} is the first-order correction of effects of the isolate crack, and M_{ij}^{2} are the second-order correction of fractures interaction. If the z-axis is perpendicular to the normal of cracks, such as the VTI (Vertical Transverse Isotropy) media, the Hudson model's first-order correction M_{ij}^{1} and second-order correction M_{ij}^{2} are given as follows:

$$\begin{cases} M_{11}^{1} = -\frac{\lambda}{\mu}\xi H_3, \\ M_{13}^{1} = -\frac{\lambda(\lambda+2\mu)}{\mu}\xi H_3, \\ M_{33}^{1} = -\frac{\lambda(\lambda+2\mu)^2}{\mu}\xi H_3, \\ M_{44}^{1} = -\mu\xi H_1 \\ M_{66}^{1} = 0, \end{cases} \tag{A31}$$

and

$$\begin{cases} M_{11}^{2} = \frac{q}{15}\frac{\lambda^2}{\lambda+2\mu}(\xi H_3)^2, \\ M_{13}^{2} = \frac{q}{15}\lambda(\xi H_3)^2, \\ M_{33}^{2} = \frac{q}{15}(\lambda+2\mu)(\xi H_3)^2, \\ M_{44}^{2} = \frac{2}{15}\frac{\mu(3\lambda+8\mu)\lambda^2}{\lambda+2\mu}(\xi H_3)^2 \\ M_{66}^{2} = 0, \end{cases} \tag{A32}$$

where $q = \frac{q}{15}\frac{\lambda^2}{\mu^2} + 28\frac{\lambda}{\mu} + 28$. λ and μ represents the lame coefficients, and ξ denotes the fracture density. In terms of dry rock, we have

$$H_3 = \frac{4(\lambda+2\mu)}{3(\lambda+\mu)}\frac{1}{1+B} \tag{A33}$$

and

$$H_1 = \frac{16(\lambda+2\mu)}{3(3\lambda+4\mu)}\frac{1}{1+k} \tag{A34}$$

where $B = \frac{\lambda+2\mu}{3(\lambda+\mu)}\frac{4\mu'}{\pi\alpha\mu}$, $k = \frac{\lambda+2\mu[K'+(4/3)\mu']}{\pi\alpha\mu(\lambda+\mu)}\frac{4\mu'}{\pi\alpha\mu}$, and K' and μ' denote the bulk and shear moduli of inclusions, respectively, and the aspect ratio of inclusions is α.

The first-order correction is calculated through exchanging M_{11}^{1} and M_{33}^{1}, M_{44}^{1} and M_{66}^{1}, as well as the second-order correction, if the x-axis is perpendicular to the normal of cracks, such as the HTI (Horizontal Transverse Isotropy) media.

For high fracture densities but small aspect ratios, the Eshelby–Cheng model can avoid some of the problems [4] given by

$$M_{ij}^{eff} = M_{ij}^{0}\frac{1-a_{ij}\xi}{1-b_{ij}\xi} \tag{A35}$$

where a_{ij} and b_{ij} represent the Padé coefficients, which are related to first- and second-order Hudson's models.

$$b_{ij} = -\frac{M_{ij}^{2}}{M_{ij}^{1}\xi},\ a_{ij} = -\frac{M_{ij}^{1}}{M_{ij}^{0}\xi} - b_{ij} \tag{A36}$$

References

1. Schultz, R. Introduction to Geologic Structural Discontinuities. In *Geologic Fracture Mechanics*; Cambridge University Press: Cambridge, UK, 2019; Chapter 1; pp. 1–26.
2. Brandes, C.; Tanner, D. Fault mechanics and earthquakes. In *Understanding Faults: Detecting, Dating, and Modeling*; Elsevier: Amsterdam, The Netherlands, 2020; Chapter 2; pp. 11–80. [CrossRef]
3. Grechka, V.; Kachanov, M. Effective elasticity of fractured rocks: A snapshot of the work in progress. *Geophysics* **2006**, *71*, W45–W58. [CrossRef]
4. Cheng, C.H. Crack models for a transversely isotropic medium. *J. Geophys. Res.* **1993**, *98*, 675–684. [CrossRef]
5. Hudson, J.A. Overall properties of a cracked solid. *Math. Proc. Camb. Phil. Soc.* **1980**, *88*, 371–384. [CrossRef]
6. Hudson, J.A. A higher order approximation to the wave propagation constants for a cracked solid. *Geophys. J. R. Astr. Soc.* **1986**, *87*, 265–274. [CrossRef]
7. Hudson, J.A. Overall elastic properties of isotropic materials with arbitrary distribution of circular cracks. *Geophys. J. Intern.* **1990**, *102*, 465–469. [CrossRef]
8. Eshelby, J.D. The determination of the elastic field of an ellipsoidal inclusion and related problems. *Proc. R. Soc. A Math. Phys. Sci.* **1957**, *241*, 376–396. [CrossRef]
9. Hudson, J.A. Overall properties of a material with inclusions or cavities. *Geophys. J. Intern.* **1994**, *117*, 555–561. [CrossRef]
10. Chapman, M. Frequency-dependent anisotropy due to mesoscale fractures in the presence of equant porosity. *Geophys. Prospect.* **2003**, *51*, 369–379. [CrossRef]
11. Guo, J.X.; Rubino, J.G.; Glubokovskikh, S.; Gurevich, B. Effects of fracture intersections on seismic dispersion: Theoretical predictions versus numerical simulations. *Geophys. Prospect.* **2016**, *65*, 1264–1276. [CrossRef]
12. Berryman, J.G. Single-scattering approximations for coefficients in Biot's equations of poroelasticity. *J. Acoust. Soc. Am.* **1992**, *91*, 551–571. [CrossRef]
13. Mavko, G.; Mukerji, T.; Dvorkin, J. *The Rock Physics Handbook*, 2nd ed.; Cambridge University Press: Cambridge, UK, 2009.
14. Ghosh, R.; Sen, M.K. Predicting subsurface CO2 movement: From laboratory to field scale. *Geophysics* **2012**, *77*, M27–M37. [CrossRef]
15. Jakobsen, M.; Hudson, J.A.; Johansen, T.A. T-matrix approach to shale acoustics. *Geophys. J. Intern.* **2003**, *154*, 533–558. [CrossRef]
16. Jing, L.; Hudson, J.A. Numerical methods in rock mechanics. *Intern. J. Rock Mech. Min. Sci.* **2002**, *39*, 409–427. [CrossRef]
17. Vlastos, S.; Liu, E.; Main, I.G.; Li, X.Y. Numerical simulation of wave propagation in media with discrete distributions of fractures: Effects of fracture sizes and spatial distributions. *Geophys. J. Intern.* **2003**, *152*, 649–668. [CrossRef]
18. Grechka, V.; Kachanov, M. Effective elasticity of rocks with closely spaced and intersecting cracks. *Geophysics* **2006**, *71*, D85–D91. [CrossRef]
19. Quintal, B.; Steeb, H.; Frehner, M.; Schmalholz, S.M. Quasi-static finite element modeling of seismic attenuation and dispersion due to wave-induced fluid flow in poroelastic media. *J. Geophys. Res.* **2011**, *116*, B01201. [CrossRef]
20. Zhao, L.X.; Yao, Q.L.; Han, D.H.; Yan, F.Y.; Nasser, M. Characterizing the effect of elastic interactions on the effective elastic properties of porous, cracked rocks. *Geophys. Prospect.* **2015**, *64*, 157–169. [CrossRef]
21. Zhang, W.H.; Fu, L.Y.; Zhang, Y.; Jin, W.J. Computation of elastic properties of 3D digital cores from the Longmaxi shale. *Appl. Geophys.* **2016**, *13*, 364–374. [CrossRef]
22. Lissa, S.; Barbosa, N.D.; Rubino, J.; Quintal, B. Seismic attenuation and dispersion in poroelastic media with fractures of variable aperture distributions. *Solid Earth* **2019**, *10*, 1321–1336. [CrossRef]
23. Song, Y.J.; Hu, H.S.; Han, B. Effective properties of a porous medium with aligned cracks containing compressible fluid. *Geophys. J. Intern.* **2020**, *221*, 60–76. [CrossRef]
24. Masson, Y.J.; Pride, S.R. On the correlation between material structure and seismic attenuation anisotropy in porous media. *J. Geophys. Res. Solid Earth* **2014**, *119*, 60–76. [CrossRef]
25. Rubino, J.G.; Ravazzoli, C.L.; Santos, J.E. Equivalent viscoelastic solids for heterogeneous fluid-saturated porous rocks. *Geophysics* **2009**, *74*, N1–N13. [CrossRef]
26. Rubino, J.G.; Caspari, E.; Müller, T.M.; Milani, M.; Barbosa, N.D.; Holliger, K. Numerical upscaling in 2-D heterogeneous poroelastic rocks: Anisotropic attenuation and dispersion of seismic waves. *J. Geophys. Res. Solid Earth* **2016**, *121*, 6698–6721. [CrossRef]
27. Guo, J.X.; Rubino, J.G.; Barbosa, N.D.; Glubokovskikh, S.; Gurevich, B. Seismic dispersion and attenuation in saturated porous rocks with aligned fractures of finite thickness: Theory and numerical simulations—Part I: P-wave perpendicular to the fracture plane. *Geophysics* **2018**, *83*, WA49–WA62. [CrossRef]
28. Favino, M.; Hunziker, J.; Caspari, E.; Quintal, B.; Krause, R. Fully-automated adaptive mesh refinement for media embedding complex heterogeneities: Application to poroelastic fluid pressure diffusion. *Computa. Geosci.* **2020**, *24*, 1101–1120. [CrossRef]
29. Cao, C.H.; Fu, L.Y.; Ba, J.; Zhang, Y. Frequency- and incident-angle-dependent *P*-wave properties influenced by dynamic stress interactions in fractured porous media. *Geophysics* **2019**, *84*, MR173–MR184. [CrossRef]
30. Cao, C.H.; Chen, F.Y.; Fu, L.Y.; Ba, J.; Han, T.C. Effect of stress interactions on anisotropic P-SV-wave dispersion and attenuation for closely spaced cracks in saturated porous media. *Geophys. Prospect.* **2020**, *68*, 2536–2556. [CrossRef]
31. Jian, S.K.; Fu, L.Y.; Cao, C.H.; Han, T.C.; Du, Q.Z. 3D finite-element modeling of effective elastic properties for fracture density and multiscale natural fractures. *J. Geophys. Eng.* **2021**, *18*, 567–582. [CrossRef]

32. Biot, M.A. Mechanics of deformation and acoustic propagation in porous media. *J. Appl. Phys.* **1962**, *33*, 1482–1498. [CrossRef]
33. Deng, S.; Li, H.; Zhang, Z.P.; Zhang, J.B.; Yang, X. Structural characterization of intracratonic strike-slip faults in the central Tarim Basin. *AAPG Bull.* **2019**, *103*, 109–137. [CrossRef]
34. Qi, L. Structural characteristics and storage control function of the shun I fault zone in the Shunbei region, Tarim Basin. *J. Petrol. Sci. Eng.* **2021**, *203*, 108653. [CrossRef]
35. Healy, D.; Rizzo, R.E.; Cornwell, D.G.; Farrell, N.; Smith, M. FracPaQ: A MATLAB toolbox for the quantification of fracture patterns. *J. Struct. Geol.* **2017**, *95*, 1–16. [CrossRef]
36. Mauldon, M.; Dunne, W.M.; Rohrbaugh, M.B. Circular scanlines and circular windows: New tools for characterizing the geometry of fracture traces. *J. Struct. Geol.* **2001**, *23*, 247–258. [CrossRef]
37. Nemec, W. The shape of the rose. *Sedimen. Geol.* **1988**, *59*, 149–152. [CrossRef]
38. Milani, M.; Rubino, J.G.; Müller, T.M.; Quintal, B.; Caspari, E.; Holliger, K. Representative elementary volumes for evaluating effective seismic properties of heterogeneous poroelastic media. *Geophysics* **2016**, *81*, D169–D181. [CrossRef]
39. Ghosh, R. Effective medium modeling of gas hydrate-filled fractures using the sonic log in the Krishna-Godavari basin, offshore eastern India. *J. Geophys. Res. Solid Earth* **2010**, *115*, B06101. [CrossRef]

applied sciences

MDPI

Article

An Improved Aeromagnetic Compensation Method Robust to Geomagnetic Gradient

Yongqiang Feng [1,2,3], Qimao Zhang [1,2], Yaoxin Zheng [1,2,3], Xiaodong Qu [4,*], Fang Wu [5] and Guangyou Fang [1,2]

[1] Aerospace Information Research Institute, Chinese Academy of Sciences, No. 9 Dengzhuang South Road, Haidian District, Beijing 100094, China; fengyq@aircas.ac.cn (Y.F.); theonezqm@126.com (Q.Z.); zhengyaoxin17@mails.ucas.ac.cn (Y.Z.); gyfang@mail.ie.ac.cn (G.F.)
[2] The Key Laboratory of Electromagnetic Radiation and Sensing Technology, Chinese Academy of Sciences, No. 19 North 4th Ring West Road, Haidian District, Beijing 100190, China
[3] School of Electronic, Electrical and Communication Engineering, University of the Chinese Academy of Sciences, Beijing 100039, China
[4] School of Information and Electronics, Beijing Institute of Technology, No. 5 South Street, Zhongguancun, Haidian District, Beijing 100081, China
[5] Naval Aviation University, No. 8 Honghou Road, Zhifu District, Yantai 264001, China; gaoqingwei1015@aliyun.com
* Correspondence: xdqu@bit.edu.cn

Abstract: Aeromagnetic surveys play an important role in many fields, for example, archaeology, anti-submarine warfare, and geophysical exploration. Being in the geomagnetic field, the aircraft generates a great deal of magnetic interference, resulting in bad performance during detection surveys. Thus, it is necessary and important to perform aeromagnetic compensation in advance. Conventional aeromagnetic compensation methods consider that the geomagnetic gradient is approximately zero after bandpass filtering, bringing about the inaccuracy of compensation coefficients. To address this issue, an improved aeromagnetic compensation method robust to geomagnetic gradient is proposed. In this study, the International Geomagnetic Reference Field (IGRF) model was employed to model the geomagnetic gradient. Then, the estimated geomagnetic gradient was subtracted from the measured data, which improved the accuracy of the compensation equations. Field experiments were conducted to verify the effectiveness of the proposed method. The experimental results show that compared to the traditional method, the compensation performance of the proposed method was improved by 152% to 329%. For the level flight, the standard deviation of residual noise after compensation can be as low as 3.3pT. The results indicate that the proposed method can significantly improve the compensation effect, showing great benefits for weak magnetic anomaly detection.

Keywords: aeromagnetic compensation; geomagnetic gradient; IGRF; magnetic anomaly detection

1. Introduction

Aeromagnetic surveying has been widely used in geophysical exploration, unexploded ordnance (UXO) detection, anti-submarine warfare, and many other fields [1–7]. The magnetic anomaly signal is measured by the magnetometer installed on the flight platform. However, the detection performance of the aeromagnetic survey system is significantly affected by the magnetic interference field that is generated by the flight platform. Therefore, it is important to estimate and eliminate the interference field [8–11], which method is called compensation.

The earliest aeromagnetic compensation model was established by Tolles [12,13]. The magnetic interference was divided into three categories, namely the permanent, the induced, and the eddy current field, respectively. In 1961, Leliak [14] proposed an effective calibration flight scheme and derived a method for solving the compensation coefficients. Since then, Leach [15] applied the least square method to solve the compensation coefficients. The multicollinearity in the compensation model was analyzed, and the ridge regression method was proposed to improve the accuracy of the estimated coefficients. Noriega [16,17] analyzed the stability of the aeromagnetic compensation algorithm and provided a method to evaluate the reliability of calibration flight results. Dou et al. [18,19] applied the wavelet analysis and recursive least squares methods to aeromagnetic compensation. The capability and the limitation of the two methods were also discussed. Wu and Chen et al. [20,21] have also conducted in-depth research on aeromagnetic data preprocessing, compensation methods, and the influence of fluxgate on the compensation performance. In addition, many scholars have proposed methods to improve the accuracy of aeromagnetic compensation, for example, an extended aeromagnetic compensation model was proposed in [22], which takes the non-maneuvering interference into account, and its effectiveness was verified by a series of experiments. Zhao et al. [23] proposed a multi-model aeromagnetic compensation method to mitigate the impact of multicollinearity and improve the quality of the measurements. A higher improvement ratio was obtained by the proposed method compared to the other two conventional methods, which means that this method has a better effect of interference suppression. An improved neural network method for aeromagnetic compensation was proposed by Yu et al. [24] that was based on the linear regression equations and the generalized regression neural network (GRNN). The generalization of the GRNN-based compensation model was improved, and the effect of overfitting was weakened by using proper smoothing factors and multiple training sets. Some researchers have studied magnetometer calibration methods based on the unmanned aerial vehicle (UAV) platforms, for example, Opromolla [25] proposed a method to improve the accuracy of heading angle estimation by using cooperative flight, and the determination of the magnetic biases' problem was formulated as a set of non-linear equations. An improved magnetometer calibration and compensation method based on the Levenberg–Marquardt algorithm for multi-rotor UAV was proposed in [26], and the experimental results indicated that the accuracy of the magnetometer in the yaw angle estimation had been greatly enhanced compared to the traditional ellipsoid fitting method.

The total magnetic field, measured by an optically pumped magnetometer (OPM), includes the geomagnetic field and the magnetic interference field. For the traditional compensation method, the geomagnetic gradient is usually considered a constant value, which is negligent after bandpass filtering. However, the geomagnetic field is a function of the spatial position and time, as described by the International Geomagnetic Reference Field (IGRF) model [27]. Based on the latest 13th generation IGRF model published by the International Association of Geomagnetism and Aeronomy (IAGA), the global geomagnetic intensity is shown in Figure 1. According to the requirements of standard calibration flight, the distance of profiles in each direction ranges from a few kilometers to more than ten kilometers, resulting in a geomagnetic gradient of several nanoTeslas or even hundreds of nanoTeslas. For the high-precision magnetic survey system, this is interference that cannot be ignored. Some previous studies have proved that considering the influence of the geomagnetic gradient can improve the accuracy of estimated compensation coefficients. Dou et al. [19] applied the World Magnetic Model 2010 to a generate geomagnetic field, which was then used to solve the compensation coefficients. The simulation results showed that the compensation performance can be improved. Ge et al. [28] proposed a relatively simple geomagnetic model to solve the compensation coefficients. The flight test results showed that the compensation performance of the two flights was improved by 27% and 22%, respectively. Zhou et al. [29] analyzed the influence of the geomagnetic gradient on the towed bird system and established the towed bird interference model; the feasibility of this compensation method was verified by actual flight tests.

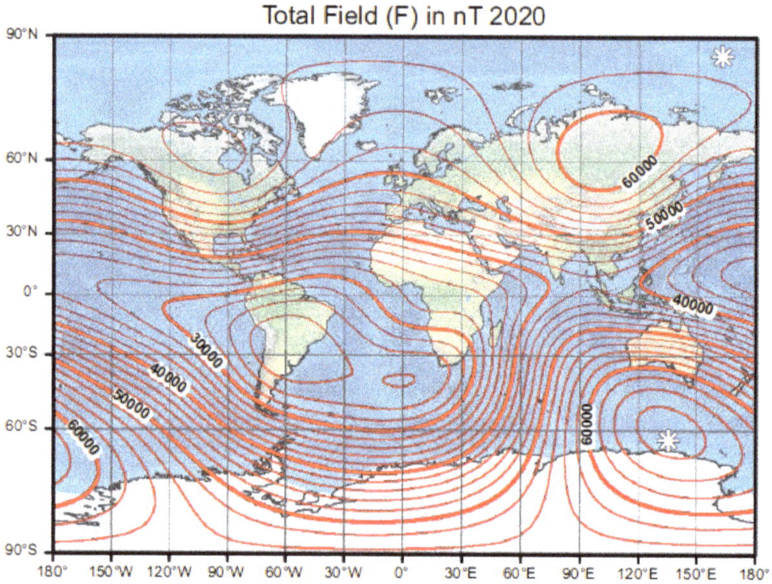

Figure 1. Map of the global geomagnetic total field intensity at the World Geodetic System-1984 Coordinate System (WGS84) ellipsoid surface for epoch 2020. White asterisks indicate locations of the magnetic dip poles.

To eliminate the residual interference field caused by the geomagnetic gradient as far as possible, an improved aeromagnetic compensation method robust to geomagnetic gradient is proposed in this paper. The application steps are as follows: first, the high-precision flight profiles are obtained by the Differential Global Positioning System (DGPS); secondly, the geomagnetic field information is obtained according to the IGRF model; finally, the background field, including the geomagnetic gradient, is substituted into the compensation model to obtain the compensation coefficients. Flight tests were carried out, and the results show that the residual noise of the proposed method was lower than that of the traditional method, which demonstrates that the geomagnetic gradient was effectively compensated; meanwhile, a higher detection ability was obtained for the aeromagnetic survey system.

The detailed study results are presented in the following four sections: Section 2 summarizes the principle of the traditional Tolles–Lawson model and the calibration flight. In Section 3, a new aeromagnetic compensation model robust to the geomagnetic gradient is extended by introducing the IGRF model, which was subsequently used to solve the compensation coefficients. An aeromagnetic survey system with low noise instruments and its data analysis are presented in Section 4. Section 5 presents the conclusion of this paper.

2. Principle of Aeromagnetic Compensation

2.1. Tolles–Lawson Model

To compensate for the platform-generated interference field, a three-axis fluxgate magnetometer is used to measure the attitude of the platform according to the classical Tolles–Lawson (T–L) model [12,13]. The reference coordinate system attached to the aircraft is established, as shown in Figure 2, where the origin O represents the location of the magnetometer, and the T, L, and V are the transverse, longitudinal, and vertical axis, respectively. H_e represents the geomagnetic field, and X, Y, and Z are the angles between the geomagnetic field and the three axes, respectively.

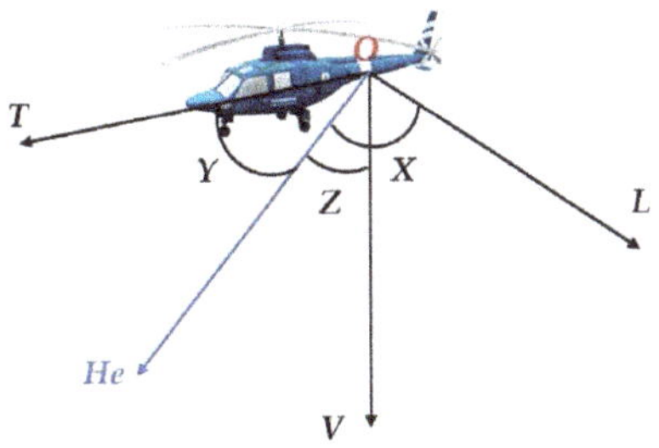

Figure 2. The definition of the reference coordinate system.

The direction cosines of the angles between the geomagnetic field and the platform are denoted as u_1, u_2, and u_3, and can be expressed as follows [21]:

$$\begin{aligned} u_1 &= \cos X(t) = \frac{T(t)}{\sqrt{T(t)^2+L(t)^2+V(t)^2}} \\ u_2 &= \cos Y(t) = \frac{L(t)}{\sqrt{T(t)^2+L(t)^2+V(t)^2}} \\ u_3 &= \cos Z(t) = \frac{V(t)}{\sqrt{T(t)^2+L(t)^2+V(t)^2}} \end{aligned}, \tag{1}$$

According to the T–L model, the platform-generated interference field can be grouped into three types, namely the permanent, the induced, and the eddy current magnetic fields. Define $H_d(t)$ as the total intensity of the interference field, then $H_d(t)$ can be expressed as follows:

$$H_d(t) = H_{PERM}(t) + H_{IND}(t) + H_{EDDY}(t). \tag{2}$$

The permanent magnetic field $H_{PERM}(t)$ caused by the permanent magnetism of the aircraft's ferromagnetic parts can be expressed as follows:

$$H_{PERM}(t) = c_1u_1 + c_2u_2 + c_3u_3 = \sum_{i=1}^{3} A_i(t)c_i, \tag{3}$$

where c_i (i = 1, 2, 3) is the compensation coefficient of the permanent magnetic field, and $A_i(t)$ (i = 1, 2, 3) is the variable constituted by the direction cosine.

The induced magnetic field $H_{IND}(t)$ is created in paramagnetic parts by the geomagnetic field, and it can be expressed as follows:

$$\begin{aligned} H_{IND}(t) &= H_e(t)(c_4u_1{}^2 + c_5u_1u_2 + c_6u_1u_3 + c_7u_2{}^2 + c_8u_2u_3 + c_9u_3{}^2) \\ &= \sum_{i=4}^{9} A_i(t)c_i \end{aligned}, \tag{4}$$

where c_i (i = 4, . . . , 9) is the compensation coefficient of the induced magnetic field, and $A_i(t)$ (i = 4, . . . , 9) is the variable constituted by the direction cosine and $H_e(t)$.

The eddy current, generated by the maneuvers of the metal components of the aircraft, is proportional to the time change rate of the flux through these components. The eddy current interference field $H_{EDDY}(t)$ can be expressed as follows:

$$\begin{aligned} &H_{EDDY}(t) = H_e(t)(c_{10}u_1u_1' + c_{11}u_1u_2' + c_{12}u_1u_3' + c_{13}u_2u_1' \\ &+c_{14}u_2u_2' + c_{15}u_2u_3' + c_{16}u_3u_1' + c_{17}u_3u_2' + c_{18}u_3u_3') = \sum_{i=4}^{9} A_i(t)c_i \end{aligned}, \tag{5}$$

where u_1', u_2', and u_3' are the time derivatives of u_1, u_2, and u_3, respectively, c_i (i = 10, . . . , 18) is the compensation coefficient of the eddy current magnetic field, and $A_i(t)$ (i = 10, . . . , 18) is the variable constituted by the direction cosine and $H_e(t)$.

The interference field generated by the platform can be expressed as follows:

$$H_d(t) = H_{PERM}(t) + H_{IND}(t) + H_{EDDY}(t) = \sum_{i=1}^{18} A_i(t)c_i. \tag{6}$$

The corresponding matrix notation of Equation (6) can be expressed as follows:

$$\boldsymbol{H}_d = \boldsymbol{AC}, \tag{7}$$

where $\boldsymbol{H}_d$ and $\boldsymbol{C}$ are both column vectors constituted by $H_d(t)$ and c_i ($i = 1, 2, \ldots, 18$), and $\boldsymbol{A}$ is the feature matrix that can be expressed as follows:

$$\boldsymbol{A} = \begin{bmatrix} A_1(1) & A_2(1) & \cdots & A_{18}(1) \\ A_1(2) & A_2(2) & \cdots & A_{18}(2) \\ \vdots & \vdots & \ddots & \ldots \\ A_1(n) & A_2(n) & \cdots & A_{18}(n) \end{bmatrix}. \tag{8}$$

The compensation coefficients can be obtained by the least square algorithm [15,30], as given by:

$$\boldsymbol{C} = (\boldsymbol{A}^T\boldsymbol{A})^{-1}\boldsymbol{A}^T\boldsymbol{H}_d. \tag{9}$$

Under the Tolles–Lawson model, the compensation coefficients are only related to the platform and the local background geomagnetic field. Thus, the compensation coefficients do not change unless the airplane platform system has changed. The time-varying $A_i(t)$ ($i = 1, 2, \ldots, 18$) can be obtained from the direction cosines u_i (i = 1, 2, 3), their time derivatives $u_i{}'$ (i = 1, 2, 3), and the magnitude of the geomagnetic field. It can be clearly demonstrated that the interference field generated by the platform would be well eliminated if the accurate compensation coefficients can be obtained.

2.2. Calibration Flight

To obtain the compensation parameters of the platform, the aircraft needs to conduct a set of maneuvers according to certain requirements, also known as calibration flight [14,31]. As shown in Figure 3, the calibration flight is carried out in four orthogonal headings, and the pitch, the yaw, and the roll maneuvers are completed successively in each direction, for which the standard amplitudes of the pitch, the roll, and the yaw are $\pm 5°$, $\pm 5°$, and $\pm 10°$, respectively. The altitude of the calibration flight is usually set at several kilometers to avoid being affected by the magnetic interference from the ground. Usually, the calibration flight is followed by a test flight that is used to evaluate the effectiveness of the estimated compensation parameters.

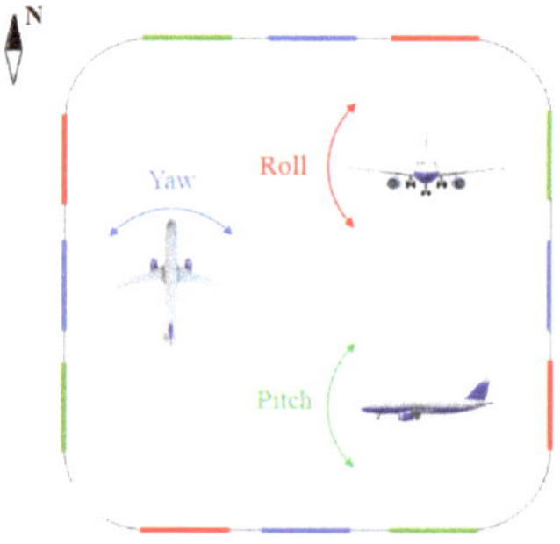

Figure 3. Attitude requirement during calibration flight.

3. Proposed Aeromagnetic Compensation Model

The total field intensity H_T measured by the optically pumped magnetometer is the superposition of the geomagnetic field H_e and the platform interference field H_d, hence the platform interference field H_d can be expressed as follows:

$$H_d = H_T - H_e. \tag{10}$$

It is difficult to obtain an accurate platform interference field in practice. A widely used method is to apply the bandpass filter to Equation (10), which can be expressed as follows:

$$bpf(H_d) = bpf(H_T) - bpf(H_e). \tag{11}$$

where *bpf*() is a bandpass finite impulse response (FIR) filter operator. The passband of the filter is determined by the frequency characteristics of the aircraft's maneuvers, which are measured by the three-axis fluxgate. The power spectrum of the direction cosines u_1, u_2, and u_3 are estimated, as shown in Figure 4. It can be seen that the maneuvering frequency of the aircraft during the calibration flight is concentrated between 0.15 Hz and 0.5 Hz, which provides a reference for the design of the bandpass filter in data preprocessing. On the other hand, the frequency band of magnetic anomaly detection (MAD) is about 0.04 Hz–0.4 Hz. Therefore, the cut-off frequency of the bandpass filter is set between 0.04 Hz and 0.5 Hz.

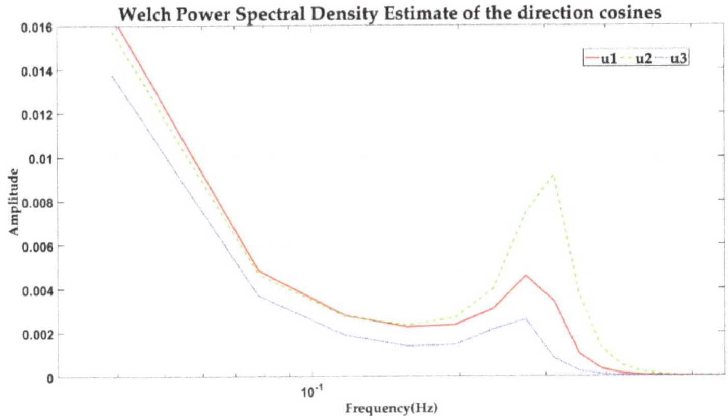

Figure 4. The power spectral density of the direction cosines.

For the traditional compensation method, the geomagnetic gradient is usually considered to be negligent, and the following equation can be obtained:

$$bpf(H_d) = bpf(H_T) - bpf(H_e) \approx bpf(H_T). \tag{12}$$

However, for a high-precision aeromagnetic survey, it is critical to consider the influence of the geomagnetic gradient when estimating the compensation coefficients. As suggested by the IGRF model, the geomagnetic field can be expressed by the gradient of a scalar potential, and the potential function can be described as a finite series expansion of spherical harmonic coefficients, as follows:

$$V(r,\theta,\phi,t) = a\sum_{n=1}^{N}\sum_{m=0}^{n}\left(\frac{a}{r}\right)^{n+1}[g_n^m(t)\cos m\phi + h_n^m(t)\sin m\phi]P_n^m(\cos\theta), \tag{13}$$

where r is the radial distance from the center of the Earth, θ and ϕ are the geocentric co-latitude and longitude, respectively, t represents time, α = 6371.2 km is the approximate mean radius of the Earth. ${g_n}^m(t)$ and ${h_n}^m(t)$ are the Gauss coefficients, ${P_n}^m(\cos\theta)$ is Schmidt semi-normalized associated Legendre functions of degree n and order m. N is the number of expansion terms of spherical harmonic function.

The spherical coordinate system is defined in Figure 5. By deriving the scalar potential of the internal magnetic field, the zonal component X, meridional component Y, and radial component Z of the geomagnetic field in the spherical coordinate system can be obtained as follows:

$$X = \frac{1}{r}\frac{\partial V}{\partial \theta}, Y = -\frac{1}{r\sin\theta}\frac{\partial V}{\partial \phi}, Z = \frac{\partial V}{\partial r}. \quad (14)$$

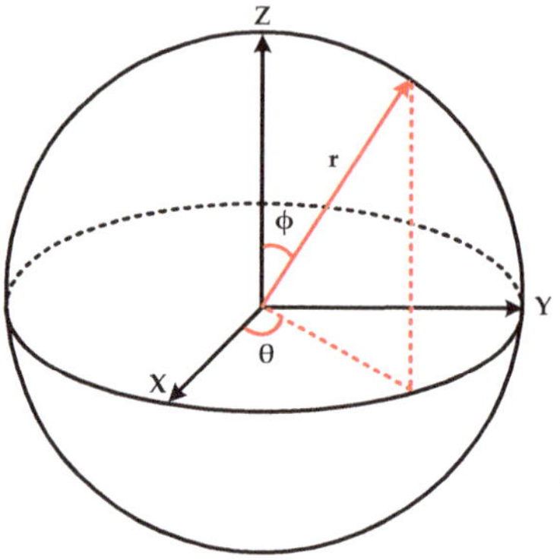

Figure 5. The definition of the spherical coordinate system.

The geomagnetic field H_e can be expressed as follows:

$$H_e = \sqrt{X^2 + Y^2 + Z^2}. \quad (15)$$

According to the aforementioned analysis, Equation (15) can be rewritten as a function related to the year, longitude, latitude, and height, as follows:

$$H_e = \psi(year, lat, lon, height), \quad (16)$$

where ψ is a model that can calculate the geomagnetic field by using the IGRF coefficients. Based on the magnetic field data from the magnetic survey satellites, observatories, and other magnetic survey programmers around the world, the IGRF coefficients are computed every five years by the IAGA. The 13th Generation IGRF, the latest version released in December 2019, represents the geomagnetic field from 2020 to 2025. The longitude, latitude, and height can be converted to the parameters r, ϕ, and θ in Equation (13) by coordinate transformation.

Based on Equations (9), (11), and (16), a compensation method robust to geomagnetic gradient is proposed, and the following equation can be established:

$$\begin{array}{l} \boldsymbol{C} = ((bpf(\boldsymbol{A}))^T bpf(\boldsymbol{A}))^{-1}(bpf(\boldsymbol{A}))^T(bpf(\boldsymbol{H}_T) - bpf(\boldsymbol{H}_e)) \\ \boldsymbol{H}_e = \varphi(year, lat, lon, height) \end{array}, \quad (17)$$

where $\boldsymbol{H}_T = (H_T(1), H_T(2), \ldots, H_T(n))^T$ is the column vector consisting of the intensities of the total field, $\boldsymbol{H}_e = (H_e(1), H_e(2), \ldots, H_e(n))^T$ is the column vector consisting of the geomagnetic field intensities, and $bps(\aleph)$ represents the bandpass filter that operates on each column of the matrix $\aleph$. The classic least squares method can be then used to estimate the compensation coefficients.

4. Field Experiments and Data Analysis

4.1. Aeromagnetic Survey System

As shown in Figure 6, the Cessna-208 fixed-wing aircraft was selected as the survey platform, and a 5-m-long probe rod was installed at the tail of the aircraft. The probe rod was equipped with an optically pumped cesium magnetometer and a three-axis fluxgate magnetometer. The data recording and compensation system was installed in the aircraft cabin. All these instruments were developed by the Aerospace Information Research Institute of the Chinese Academy of Sciences (AIRCAS). A DGPS antenna was attached to the top of the plane to better receive satellite signals.

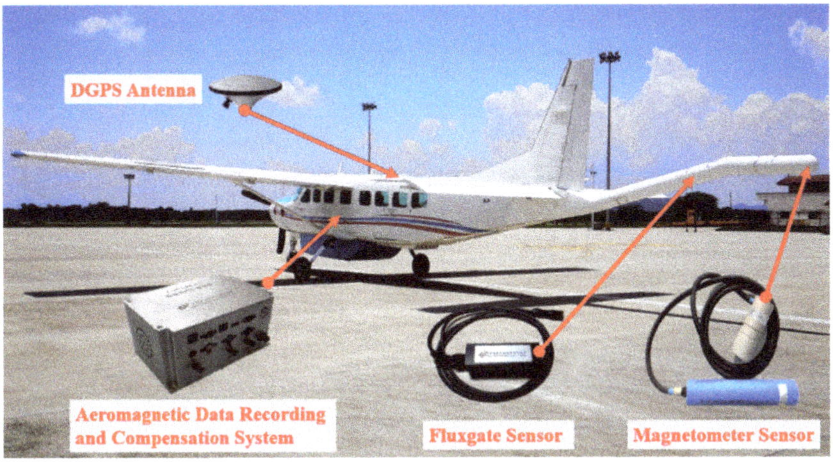

Figure 6. The aeromagnetic survey system.

The optically pumped cesium magnetometer was the most important sensor in the system, and the sensitivity was as low as 0.3 pT/$\sqrt{}$Hz at 1 Hz. The sensitivity of the fluxgate magnetometer was 6 pT/$\sqrt{}$Hz at 1 Hz. The technical specifications are shown in Tables 1 and 2.

Table 1. The technical specifications of the optically pumped cesium magnetometer.

Parameter	Value	Parameter	Value
Measurement range	10,000 nT to 105,000 nT	Operating temperature	−40 °C to +70 °C
Orientation range	Not less than 15° to 75° and 105° to 165°	Power consumption	12 W (24 V input)
Heading error	<±0.3 nT	Dimensions	Sensor: Φ45 mm × 130 mm; Electronics: 38 mm (H) × 50 mm (W) × 250 mm (D)
Noise	≤0.3 pT/$\sqrt{}$Hz	Weight	Sensor: 0.5 kg; Electronics: 0.5 kg
Gradient tolerance	40,000 nT/m		

Table 2. The technical specifications of the three-axis fluxgate magnetometer.

Parameter	Value	Parameter	Value
Measurement range	±100,000 nT	Offset error	±5 nT
Bandwidth	DC-1 kHz (−3 dB)	Operating temperature	−40 °C to +85 °C
Internal noise	≤6 pT/$\sqrt{\text{Hz}}$	Supply voltage	±12 to ±15 V DC
Scaling	100 mV/μT	Power consumption	0.4 W (±12 V input)
Scaling error	≤±0.5%	Dimensions	27 mm (H) × 41 mm (W) × 110 mm (D)

The aeromagnetic data recording and compensation system and the Differential Global Positioning System (DGPS) receiver were installed in the cabin in Figure 7. A software application was used to check the status of sensors and DGPS. All of the devices were powered by lithium batteries. The technical specifications of the recording system and the DGPS are shown in Tables 3 and 4.

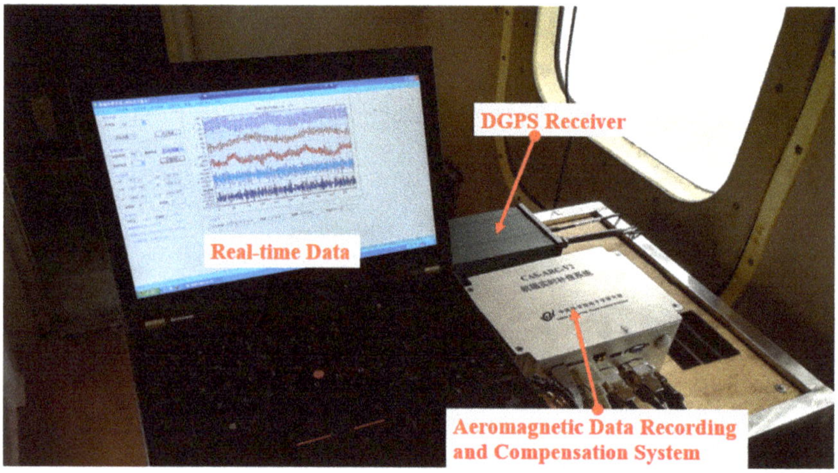

Figure 7. The equipment installed in the cabin.

Table 3. The technical specifications of the aeromagnetic data recording and compensation system.

Parameter	Value	Parameter	Value
Magnetometer input	Two optically pumped cesium magnetometers	Operating temperature	−40 °C to 70 °C
System noise	0.1 pT/$\sqrt{\text{Hz}}$	Input voltage	24 to 35 V DC
Sampling rate	160 Hz	Power consumption	10 W (24 V input)
Analog input	Three differential inputs for fluxgate, 24 bit converter	Receiver size	110 mm (H) × 220 mm (W) × 160 mm (D)
Data ports and interfaces	Ethernet, TCP/IP	Weight	1.6 kg

Table 4. The technical specifications of the DGPS receiver.

Parameter	Value	Parameter	Value
Constellations	GPS, GLONASS, BeiDou, Galileo and QZSS	Operating temperature	−15 °C to 55 °C
Navigation update rate	10 Hz	Input voltage	12 to 24 V DC
Horizontal position accuracy	5 cm at 2 σ (95%)	Power consumption	4 W (12 V input)
Elevation position accuracy	12 cm at 2 σ (95%)	Receiver size	90 mm (H) × 158 mm (W) × 227 mm (D)
Data ports and interfaces	3 Serial Connectors (EIA-232/422)	Weight	2.25 kg

The block diagram of the aeromagnetic data recording and compensation system is shown in Figure 8. Optically pumped magnetometer data and three-axis magnetometer data were recorded on the rising edge of the pulse per second (PPS), which was obtained from the DGPS. The sensitivity of the frequency counter channel and the analog to digital converter channel were 0.1 pT/$\sqrt{}$Hz at 1 Hz and 1 pT/$\sqrt{}$Hz at 1 Hz, respectively. In the meantime, the longitude, latitude, and altitude were collected for aeromagnetic data compensation in real time. The raw data and compensated data were all recorded in the flight.

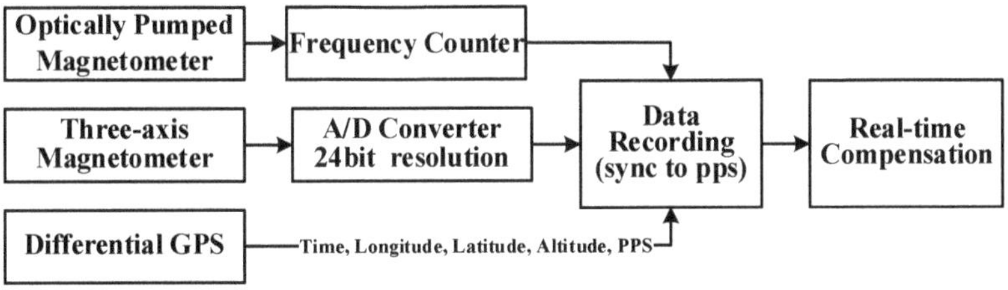

Figure 8. The block diagram of data recording and compensation system.

4.2. Data Evaluation Methods

The key data evaluation methods are the standard deviations of compensated and uncompensated signals and the improvement ratio (IR) derived from them. The IR can be expressed as follows [32]:

$$IR = \frac{\sigma_u}{\sigma_c}, \tag{18}$$

where σ_u and σ_c are, respectively, the standard deviation of the uncompensated and compensated signals. The larger the IR, the better the compensation result.

4.3. Compensation Flight and Level Flight

The compensation flight for this experiment was carried out in the South China Sea at an altitude of more than 3000 m and a flight speed of 180 km/h. Five cycles of maneuvers consisting of about ±5° rolls, ±5° pitches, and ±10° yaws are carried out in each heading. As it is shown in Figure 9, a calibration flight and a test flight were recorded, and these two flight paths were very similar. Usually, the test flight is conducted under the same magnetic background as the calibration flight to verify the performance of aeromagnetic compensation.

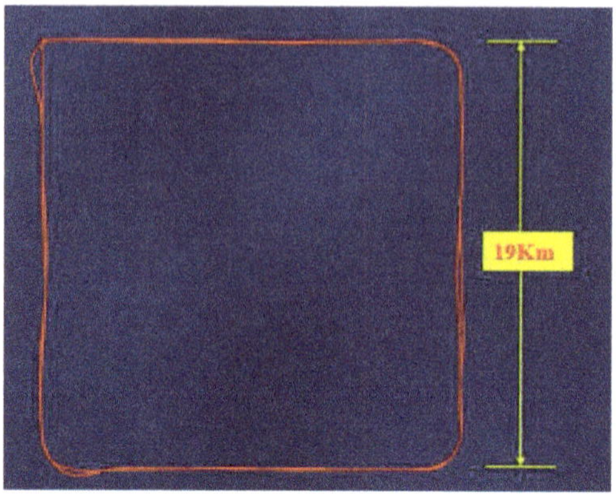

Figure 9. The compensation flight path on Google Earth.

The level flight path is shown in Figure 10. The level flight was about 30 km, the aim of which was to verify the best performance of aeromagnetic compensation.

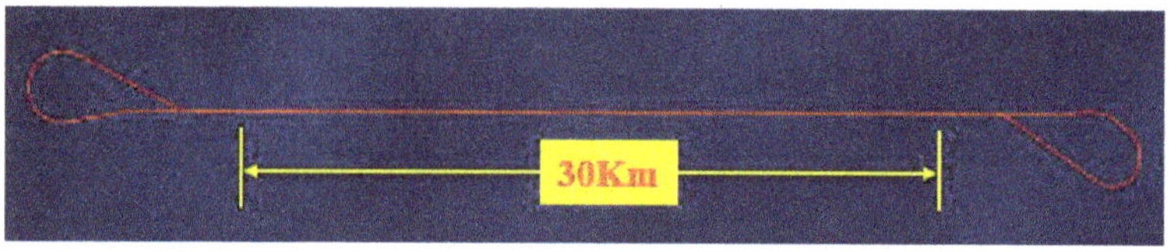

Figure 10. The level flight path on Google Earth.

4.4. Data Analysis

The compensation parameters were solved according to the process shown in Figure 11. The geomagnetic gradient was considered in the process of solving the compensation coefficients. The raw data were sampled at 160 Hz, then filtered by a lowpass filter with a cut-off frequency of 20 Hz. The geomagnetic field column $\boldsymbol{H}_e$ was computed with Equation (16), and the information matrix $\boldsymbol{A}$ was constituted by Equation (6). Finally, the compensation parameters were solved by Equation (9). After the calibration flight, the effectiveness of the proposed method was examined based on the test flight data.

The compensation coefficients could be obtained without taking into consideration the geomagnetic gradient (i.e., the traditional method), as shown in Table 5. Table 6 shows the compensation coefficients of the system considering the geomagnetic gradient (i.e., the method proposed in this paper). A comparison of the two tables shows that most of the parameters were very close.

Table 5. Compensation coefficients (traditional method).

Parameter	Value	Parameter	Value
c1	2.712	c2	1.552
c3	−0.178	c4	-1.322×10^6
c5	-3.084×10^6	c6	-8.296×10^6
c7	-8.197×10^6	c8	-5.655×10^6
c9	1.271×10^4	c10	7.522×10^4
c11	-1.408×10^6	c12	2.402×10^6
c13	-3.855×10^6	c14	7.457×10^4
c15	2.276×10^5	c16	1.605×10^6
c17	8.996×10^7	c18	6.976×10^4

Table 6. Compensation coefficients (proposed method).

Parameter	Value	Parameter	Value
c1	2.697	c2	1.580
c3	−0.428	c4	-4.884×10^7
c5	-3.632×10^6	c6	-8.316×10^6
c7	-6.011×10^6	c8	-1.781×10^5
c9	1.270×10^4	c10	8.491×10^4
c11	-1.534×10^6	c12	3.432×10^6
c13	-3.585×10^6	c14	8.434×10^4
c15	2.242×10^5	c16	1.355×10^6
c17	-1.104×10^6	c18	7.972×10^4

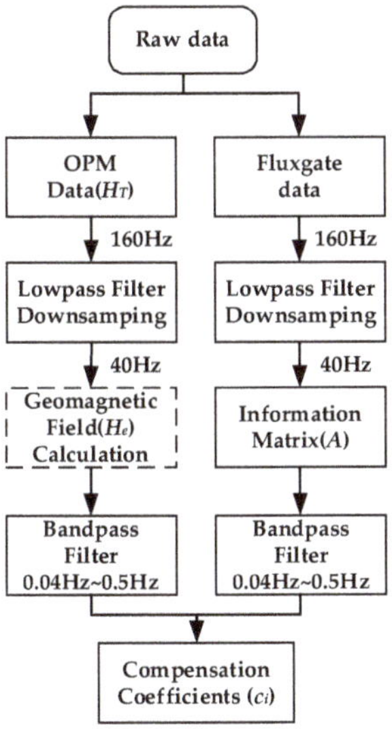

Figure 11. The process for solving compensation parameters.

The three-axis magnetometer data in the calibration flight are shown in Figure 12. It is obvious that there were four periods of time for the compensation maneuvers. At each stage, the pitch, yaw, and roll were executed in five cycles.

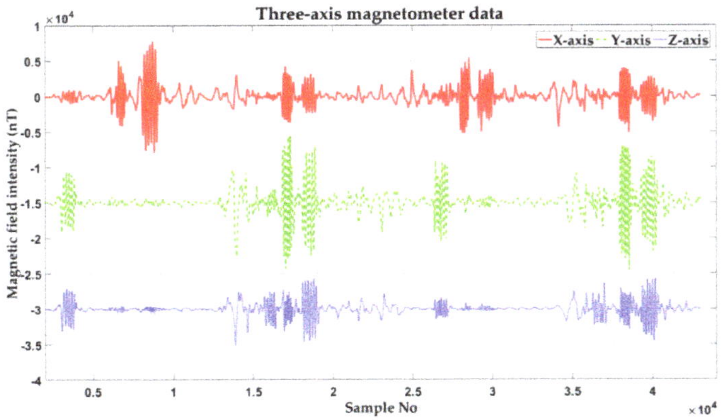

Figure 12. The three-axis magnetometer data in the calibration flight (after a bandpass filter, offset for clarity).

Figure 13 illustrates a typical performance of compensation in four orthogonal headings. The blue lines are the uncompensated data, and the red lines are the compensated data with the proposed method. The comparison of the compensation results between the traditional model, the Ge model [28], and the proposed model are shown in Figure 14, with the uncompensated signals in blue, the compensated signals of the traditional method in green, the compensated signals of the Ge method in black, and the compensated signals of the proposed method in red.

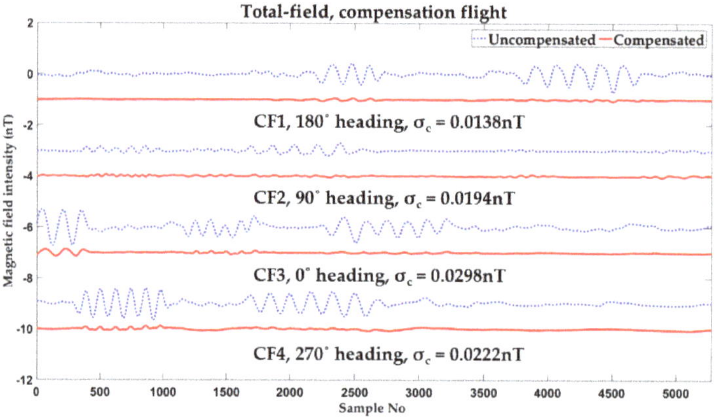

Figure 13. The compensation results of calibration flight using coefficients of Table 6 (after a bandpass filter, offset for clarity).

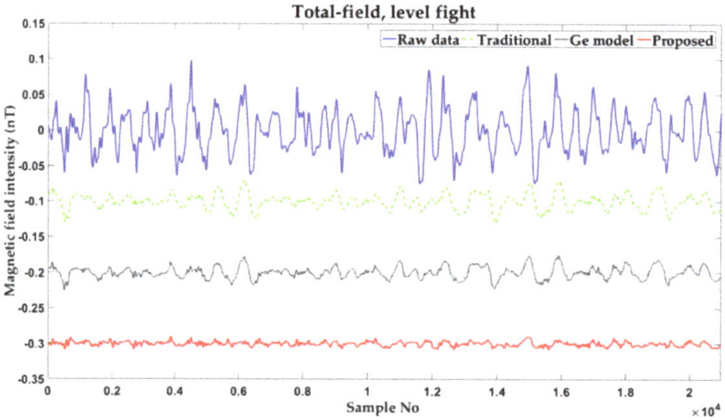

Figure 14. The compensation results of the traditional model, Ge model, and proposed model (after a bandpass filter, offset for clarity).

Five sets of data were compensated respectively. These include four sets of calibration flight data, CF1, CF2, CF3, and CF4, and one set of level flight data, marked as LF1. The compensation results obtained using the traditional method, the Ge method, and the proposed method are listed in Table 7.

Table 7. Comparison of compensation results.

Data	Method	σ_u	σ_c	IR
CF1	Traditional	0.1496	0.0309	4.8400
	Ge method	0.1496	0.0248	6.03
	Proposed	0.1496	0.0138	*10.8406*
CF2	Traditional	0.0699	0.0400	1.7469
	Ge method	0.0699	0.0311	2.24
	Proposed	0.0699	0.0194	*3.6031*
CF3	Traditional	0.1835	0.0464	3.9599
	Ge method	0.1835	0.0391	4.6931
	Proposed	0.1835	0.0298	*6.1577*
CF4	Traditional	0.1962	0.0339	5.7791
	Ge method	0.1962	0.0278	7.0576
	Proposed	0.1962	0.0222	*8.8378*
LF1	Traditional	0.0310	0.0109	2.8556
	Ge method	0.0310	0.0083	3.7349
	Proposed	0.0310	0.0033	*9.3939*

Based on the five sets of data shown in Table 7, it was concluded that the performance of the proposed method was better than that of the traditional method and the Ge method. For the four groups of calibrated flight data, compared to the traditional method, the improvement in IR after considering the geomagnetic gradient was between 152% and 224%, which is quite significant. The improvement in IR for the level flight data reached a significant 329%. As shown in Figure 15, the peak-to-peak value of the residual noise after compensation was 19.4 pT, which proves that the proposed method can well compensate the interference field of the aircraft and improve the ability of the system to detect weak magnetic anomalies.

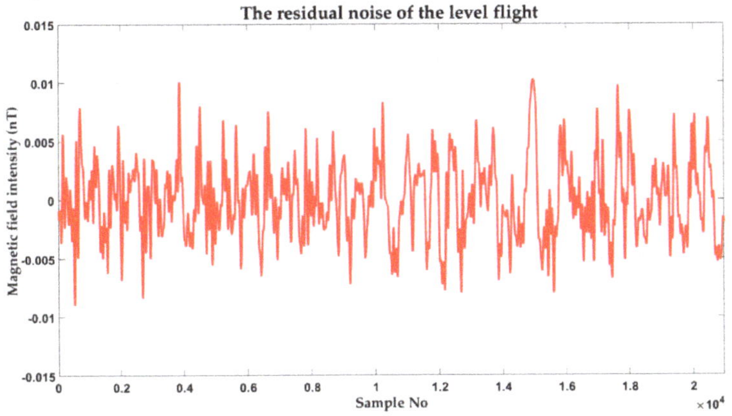

Figure 15. The residual noise of level flight after compensation (after a bandpass filter).

5. Conclusions

The traditional aeromagnetic compensation method does not take into consideration the influence of the geomagnetic gradient. The interference field obtained after bandpass filtering still contains the information of the geomagnetic field, which results in errors during the estimation of compensation parameters. In this paper, an improved compensation method that is robust to the geomagnetic gradient is presented. By introducing the IGRF model, the geomagnetic gradient is modeled as a function of longitude, latitude, and flight height. This significantly improved the accuracy and the stability of the estimated compensation parameters. The experiment results show that the proposed method greatly improved the compensation effect. All the results demonstrate that the proposed method could significantly improve the effect of aeromagnetic compensation and the weak magnetic anomaly detection. In the future, more aeromagnetic survey tests will be carried out to verify the effectiveness of the proposed method.

Author Contributions: Conceptualization, Y.F.; methodology, Y.F. and X.Q.; software, Q.Z. and Y.Z.; validation, Y.Z. and F.W.; formal analysis, X.Q.; investigation, Y.F. and G.F.; resources, G.F.; data curation, Y.F. and Q.Z.; writing—original draft preparation, Y.F.; writing—review and editing, Y.Z.; visualization, Q.Z.; supervision, G.F.; project administration, X.Q.; funding acquisition, G.F. All authors have read and agreed to the published version of the manuscript.

Funding: This research work was supported by the National Natural Science Foundation of China under grant 61901441 and the National Key R&D Program of China under grant 2019YFC1408102. The authors wish to express their gratitude to the editor and the anonymous reviewers.

Institutional Review Board Statement: Not applicable.

Informed Consent Statement: Not applicable.

Data Availability Statement: Not applicable.

Conflicts of Interest: The authors declare no conflict of interest.

References

1. Billings, S.; Wright, D. Interpretation of high-resolution low-altitude helicopter magnetometer surveys over sites contaminated with unexploded ordnance. *J. Appl. Geophys.* **2010**, *72*, 225–231. [CrossRef]
2. Doll, W.E.; Sheehan, J.R.; Gamey, T.J.; Beard, L.P.; Norton, J. Results of an airborne vertical magnetic gradient demonstration, New Mexico. *J. Environ. Eng. Geophys.* **2008**, *13*, 277–290. [CrossRef]
3. Pei, Y.; Liu, B.; Hua, Q.; Liu, C.; Ji, Y. An aeromagnetic survey system based on an unmanned autonomous helicopter: Development, experiment, and analysis. *Int. J. Remote Sens.* **2017**, *38*, 3068–3083. [CrossRef]
4. Walter, C.; Braun, A.; Fotopoulos, G. High-resolution unmanned aerial vehicle aeromagnetic surveys for mineral exploration targets. *Geophys. Prospect.* **2020**, *68*, 334–349. [CrossRef]
5. Zheng, Y.; Li, S.; Xing, K.; Zhang, X. Unmanned Aerial Vehicles for Magnetic Surveys: A Review on Platform Selection and Interference Suppression. *Drones* **2021**, *5*, 93. [CrossRef]
6. Schmidt, V.; Becken, M.; Schmalzl, J. A UAV-borne magnetic survey for archaeological prospection of a Celtic burial site. *First Break* **2020**, *38*, 61–66. [CrossRef]
7. Cunningham, M.; Samson, C.; Wood, A.; Cook, I. Aeromagnetic surveying with a rotary-wing unmanned aircraft system: A case study from a zinc deposit in Nash Creek, New Brunswick, Canada. *Pure Appl. Geophys.* **2018**, *175*, 3145–3158. [CrossRef]
8. Forrester, R.; Huq, M.S.; Ahmadi, M.; Straznicky, P. Magnetic signature attenuation of an unmanned aircraft system for aeromagnetic survey. *IEEE/ASME Trans. Mechatron.* **2013**, *19*, 1436–1446. [CrossRef]
9. Jirigalatu, J.; Krishna, V.; da Silva, E.L.S.; Døssing, A. Experiments on magnetic interference for a portable airborne magnetometry system using a hybrid unmanned aerial vehicle (UAV). *Geosci. Instrum. Methods Data Syst. Discuss.* **2021**, *10*, 25–34. [CrossRef]
10. Qiao, Z.; Ma, G.; Zhou, W.; Yu, P.; Zhou, S.; Wang, T.; Tang, S.; Dai, W.; Meng, Z.; Zhang, Z. Research on the comprehensive compensation of aeromagnetic system error of multi-rotor UAV. *Chin. J. Geophys.* **2020**, *63*, 4604–4612. (In Chinese)
11. Tuck, L.E.; Samson, C.; Laliberté, J.; Cunningham, M. Magnetic interference mapping of four types of unmanned aircraft systems intended for aeromagnetic surveying. *Geosci. Instrum. Methods Data Syst.* **2021**, *10*, 101–112. [CrossRef]
12. Tolles, W.; Lawson, J. *Magnetic Compensation of MAD Equipped Aircraft*; Airborne Instruments Lab. Inc.: Mineola, NY, USA, 1950; p. 201.
13. Tolles, W.E. Compensation of Aircraft Magnetic Fields. U.S. Patent 2692970A, 26 October 1954.
14. Leliak, P. Identification and evaluation of magnetic-field sources of magnetic airborne detector equipped aircraft. *IRE Trans. Aerosp. Navig. Electron.* **1961**, *ANE-8*, 95–105. [CrossRef]
15. Leach, B.W. Aeromagnetic compensation as a linear regression problem. In *Information Linkage between Applied Mathematics and industry II*; Academic: London, UK, 1980; Volume 2, pp. 139–161.
16. Noriega, G. Performance measures in aeromagnetic compensation. *Lead. Edge* **2011**, *30*, 1122–1127. [CrossRef]
17. Noriega, G.; Marszalkowski, A. Adaptive techniques and other recent developments in aeromagnetic compensation. *First Break* **2017**, *35*, 31–38. [CrossRef]
18. Dou, Z.; Han, Q.; Niu, X.; Peng, X.; Guo, H. An adaptive filter for aeromagnetic compensation based on wavelet multiresolution analysis. *IEEE Geosci. Remote Sens. Lett.* **2016**, *13*, 1069–1073. [CrossRef]
19. Dou, Z.; Han, Q.; Niu, X.; Peng, X.; Guo, H. An aeromagnetic compensation coefficient-estimating method robust to geomagnetic gradient. *IEEE Geosci. Remote Sens. Lett.* **2016**, *13*, 611–615. [CrossRef]
20. Wu, P.; Zhang, Q.; Chen, L.; Zhu, W.; Fang, G. Aeromagnetic Compensation Algorithm Based on Principal Component Analysis. *J. Sens.* **2018**, *2018*, 5798287. [CrossRef]

21. Chen, L.; Wu, P.; Zhu, W.; Feng, Y.; Fang, G. A novel strategy for improving the aeromagnetic compensation performance of helicopters. *Sensors* **2018**, *18*, 1846. [CrossRef]
22. Du, C.; Wang, H.; Wang, H.; Xia, M.; Peng, X.; Han, Q.; Zou, P.; Guo, H. Extended aeromagnetic compensation modelling including non-manoeuvring interferences. *IET Sci. Meas. Technol.* **2019**, *13*, 1033–1039. [CrossRef]
23. Zhao, G.; Han, Q.; Peng, X.; Zou, P.; Wang, H.; Du, C.; Wang, H.; Tong, X.; Li, Q.; Guo, H. An aeromagnetic compensation method based on a multimodel for mitigating multicollinearity. *Sensors* **2019**, *19*, 2931. [CrossRef]
24. Yu, P.; Zhao, X.; Jia, J.; Zhou, S. An improved neural network method for aeromagnetic compensation. *Meas. Sci. Technol.* **2021**, *32*, 045106. [CrossRef]
25. Opromolla, R. Magnetometer Calibration for Small Unmanned Aerial Vehicles Using Cooperative Flight Data. *Sensors* **2020**, *20*, 538. [CrossRef]
26. Wu, H.; Pei, X.; Li, J.; Gao, H.; Bai, Y. An improved magnetometer calibration and compensation method based on Levenberg–Marquardt algorithm for multi-rotor unmanned aerial vehicle. *Meas. Control* **2020**, *53*, 276–286. [CrossRef]
27. Alken, P.; Thébault, E.; Beggan, C.D.; Amit, H.; Aubert, J.; Baerenzung, J.; Bondar, T.N.; Brown, W.J.; Califf, S.; Chambodut, A.; et al. International geomagnetic reference field: The thirteenth generation. *Earth Planets Space* **2021**, *73*, 49. [CrossRef]
28. Ge, J.; Wang, W.; Dong, H.; Liu, H.; Zhang, X.; Luo, W.; Zhang, H.; Zhu, J. Modeling and reduction of the gradient effect of a magnetic sensor for an aeromagnetic compensator. *Rev. Sci. Instrum.* **2021**, *92*, 024502. [CrossRef]
29. Zhou, Z.; Liu, Z.; Li, W.; Wang, Y.; Wang, C. Analyze and improve the influence of geomagnetic gradient on aeromagnetic compensation in a towed bird. *Geosci. Instrum. Methods Data Syst.* **2021**, 1–12. [CrossRef]
30. Zhou, J.; Lin, C.; Chen, H. A method for aircraft magnetic interference compensation based on small signal model and LMS algorithm. *Chin. J. Aeronaut.* **2014**, *27*, 1578–1585. [CrossRef]
31. Tuck, L.; Samson, C.; Polowick, C.; Laliberte, J. Real-time compensation of magnetic data acquired by a single-rotor unmanned aircraft system. *Geophys. Prospect.* **2019**, *67*, 1637–1651. [CrossRef]
32. Jiang, D.; Zeng, Z.; Zhou, S.; Guan, Y.; Lin, T. Integration of an aeromagnetic measurement system based on an unmanned aerial vehicle platform and its application in the exploration of the Ma'anshan magnetite deposit. *IEEE Access* **2020**, *8*, 189576–189586. [CrossRef]

applied sciences

MDPI

Review

Time-Lapse Electrical Resistivity Tomography (TL-ERT) for Landslide Monitoring: Recent Advances and Future Directions

Vincenzo Lapenna * and Angela Perrone

Institute of Methodologies for Environmental Analysis, National Research Council of Italy (CNR-IMAA), C.da S.Loja, I-85050 Tito, PZ, Italy; angela.perrone@imaa.cnr.it
* Correspondence: vincenzo.lapenna@imaa.cnr.it

Abstract: To date, there is a growing interest for challenging applications of time-lapse electrical resistivity tomography (TL-ERT) in Earth sciences. Tomographic algorithms for resistivity data inversion and innovative technologies for sensor networks have rapidly transformed the TL-ERT method in a powerful tool for the geophysical time-lapse imaging. In this paper, we focus our attention on the application of this method in landslide monitoring. Firstly, an overview of recent methodological advances in TL-ERT data processing and inversion is presented. In a second step, a critical analysis of the main results obtained in different field experiments and lab-scale simulations are discussed. The TL-ERT appears to be a robust and cost-effective method for mapping the water-saturated zones, and for the identification of the groundwater preferential pathways in landslide bodies. Furthermore, it can make a valuable contribution to following time-dependent changes in top-soil moisture, and the spatio-temporal dynamics of wetting fronts during extreme rainfall events. The critical review emphasizes the limits and the advantages of this geophysical method and discloses a way to identify future research activities to improve the use of the TL-ERT method in landslide monitoring.

Keywords: time-lapse geophysical imaging; electrical resistivity tomography; landslides

1. Introduction

Electrical resistivity tomography (ERT) is a robust and well-consolidated method for near-surface geophysics, with a wide spectrum of applications in the geological, engineering and environmental sciences. This geoelectrical method provides 2D and 3D resistivity images of the subsurface; the basic principle is Ohm's law and the experimental data are the voltage signals produced by the direct electrical currents (f = 0 Hz) injected into the subsoil and filtered by the artificial noise. Technological advances (i.e., multi-channel arrays, innovative sensors) and novel tomographic algorithms for data inversion have rapidly transformed ERT in one of the most popular geophysical methods. Furthermore, the extreme variability of the resistivity parameter (different order of magnitude, from 1–10 Ohm·m of clayey materials to 10^5 Ohm·m of granite rocks) makes it possible to use the ERT method in many geological environments [1–4].

The ERT method can also be applied to study the time-dependent changes of the resistivity pattern of the subsoil: we used the analysis of 2D and 3D static resistivity images of the subsurface to the process the time-series of resistivity images (time-lapse approach or 4D tomography). The first laboratory and experimental works on time-lapse electrical resistivity tomography (TL-ERT or 4D-ERT) method date back to the period 1995–2005, when they were mainly devoted to monitor the injection of salt water in soil and aquifers [5–8]. The results of these controlled experiments stimulated the study of the complex relationship between resistivity values and hydraulic parameters.

Since these pioneering activities and seminal papers, the attention of the scientific community to the TL-ERT method has increased considerably, with exponential behavior in the yearly number of articles published in top-leading journals. In the last 20 years, more than 200 papers have been published on ISI journals (Figure 1).

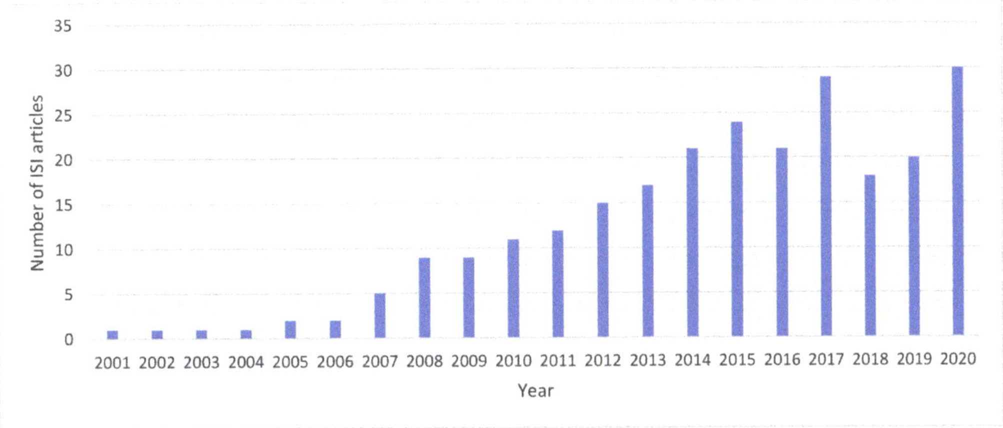

Figure 1. Yearly variations of the number of ISI articles concerning the topic "time-lapse ERT" spanning the period 2001–2020 (Source data: WoS database).

The TL-ERT method found an impressive number of applications crossing different disciplines, from hydrogeology [9–12] to agriculture [13–16], from engineering geology [17–20] to geohazards [21–24], from CO2 storage [25] to the study of the effects of the climate changes on soil and near subsurface [26–29].

Here, we focus our attention on the application of the TL-ERT method for landslide monitoring. In the last years, climate change and the occurrence of frequent extreme meteorological events have contributed to the triggering of landslide phenomena, with considerable effects on the economic and social life of the affected areas [30–32]. Indeed, the number of cases in which these phenomena involved strategic infrastructures (e.g., bridges, railways, highways) as well as civil housing and commercial activities [33–35], sometime causing fatalities [36,37], has considerably grown.

Taking this into account, the need to develop early warning systems allowing officials to monitor the variation of specific subsoil parameters responsible for triggering landslides has become more pressing [38–40]. Alongside the more classic monitoring systems, based on the use of purely geotechnical sensors, such as inclinometers, piezometers, etc., innovative systems based on indirect measures such as geophysical ones [41] have begun to be experimented with. Geophysical methods allow the measure of physical parameters, whose variation often depends on the variation of parameters directly related to the triggering of landslides (for example the water content, the porosity of the rock, etc.). The ERT method can be considered one of the most widely applied geophysical methods in the study of landslides. Many examples of 2D and 3D ERT applications (Figure 2) for the geometric characterization of the landslide body, the identification of the sliding surface, and the localization of areas with higher water content, are reported in the literature [42–44].

Recent technological improvements, which have allowed acquisition time to speed up in order to obtain many measurements and develop innovative algorithms for data inversion, have made it possible to test the ERT method and monitor landslides.

Indeed, considering the dependence of the electrical resistivity on the water content and the role that the latter plays in the triggering of landslides, repeated resistivity measurements over time along defined profiles (TL-ERT) can be carried out with the aim of monitoring, for example, the water content variations in the first subsoil levels and to try to define triggering thresholds.

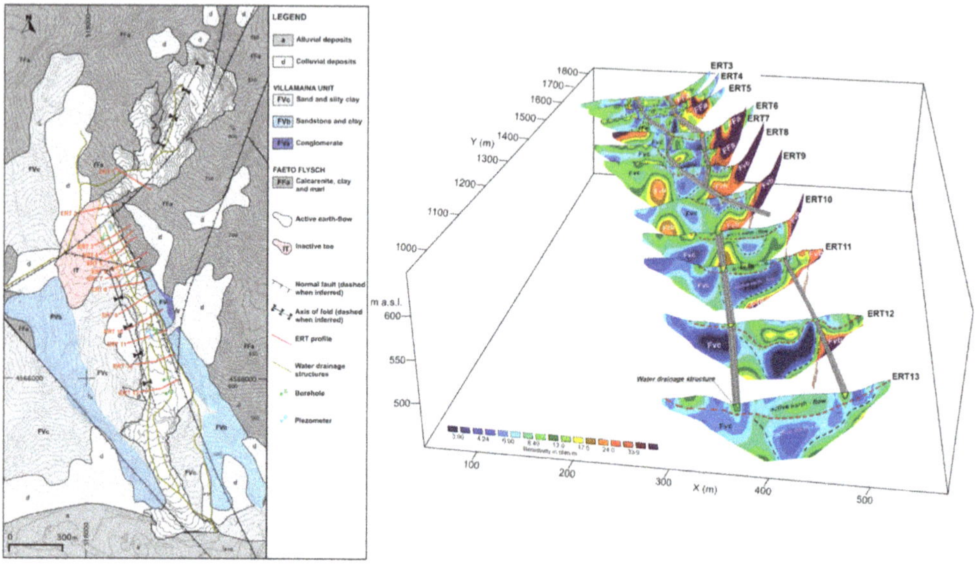

Figure 2. (**Left**) Geological map of the Montaguto earth-flow (Campania region, southern Italy) with the location of direct and indirect investigations. Red lines: ERT profiles. Green line: water drainage structures. Green dot: borehole. Blue triangle: piezometers. (**Right**) Fence diagram showing the locations of the 2D ERT (ERT 3–ERT 13) within a 3D space (after [44]).

Compared to other fields of application, the use of the TL-ERT in landslide monitoring can be considered recent. The geological complexity of the investigated phenomenon still requires additional experimentations and the development of petrophysical laws that allow quantitative correlation between the electrical resistivity values and the hydrologic parameters directly involved in the triggering of landslides.

In this paper, a selected list of papers extracted by the WOS database for the period 2010 up to now, concerning the TL-ERT method and its application for landslide monitoring, have been considered. The main goals of this work are: (i) to perform an overview of the recent methodological advances in TL-ERT data processing and inversion; (ii) to identify the limits and the advantages of this method when it is applied in monitoring hydrogeological phenomena; (iii) to perform a critical analysis with the identification of future directions for the research activity on this topic.

The work has been organized as follows. In the first step, the basic principles and the algorithms for data inversion of the TL-ERT method are briefly summarized. In a second step, a systematic overview of the main results obtained with this method in a selected list of field-scale experiments for landslide monitoring is carried out. Finally, some gaps and future methodological and technological trends are identified and discussed.

2. The TL-ERT Method: Data Processing and Inversion

2.1. Basic Principles

The TL-ERT method is based on the analysis of a dataset of 2D or 3D resistivity images obtained at different time intervals: it combines 2 or 3 spatial dimensions with a temporal dimension (4D ERT). The TL-ERT or 4D ERT method adopts multi-electrode arrays and the data acquisition procedures already used for the classical 2D and 3D ERT surveys, but the acquisition is generally controlled by a remote system, and it is repeated over time. The main experimental errors associated with the TL-ERT measurements are connected to the presence of man-made electrical noises, galvanic contact changes, electrode corrosion, and low signal-to-noise ratio in voltage signals (well-known sources of errors in 1D, 2D and

3D ERT measurements). The stacking techniques and algorithms for spectral analysis are generally used for extracting the weak voltage signals generated by the current injection into the subsurface, also in the condition of low signal-to-noise ratio [1,4,42]. However, it is necessary to consider other time-related errors connected to temperature and galvanic changes, such as rain precipitation and variations of electrode spacings due to surface ground movements [45].

As it concerns the interpretation and the inversion of the 4D ERT experimental data, the simpler and more common approach is based on the analysis of the differences between the resistivity images independently inverted during the acquisition phase (*independent approach*). However, the time-lapse of the resistivity images can easily produce an amplification of the inversion artifacts [46].

Another largely applied method is the *difference inversion* of the ERT data [47]. The Occam algorithm was modified to allow the inversion of the differences between the background or reference resistivity images and the other 3D resistivity images obtained at a different time. The resistivity obtained by the inversion of background data serves as a priori model in the difference inversion. The method allows a rapid convergence of the inversion process and reduces the errors due the field array configurations.

A *ratio inversion* was proposed by Daily 1992 [48], in which the inversion is performed using the ratio between two successive ERT measurements multiplied with a resistivity value representing a homogenous subsurface model. An alternative approach is the *cascaded inversion*: this method uses the inversion result of the initial dataset as the starting and reference model for the inversion of the subsequent datasets [49].

Another critical error affecting the interpretation of the TL-ERT data is the assumption of the time-invariant (static) subsurface model during the acquisition phase. This hypothesis is realistic for slow phenomena, but in field experiments we could observe rapid changes of the subsurface model (i.e., the migration of a fluid in a fractured material). To overcome this problem, Kim et al. (2009) [50] introduced a subsurface resistivity model that is continuously changed over the time, the model is completely defined in the space-time domains, and the inversion algorithm considers all data available (*Full 4D TL-ERT data inversion*). To reduce the computing time and to stabilize the inversion procedure, some regularization criteria are generally introduced.

2.2. Novel Algorithms for Data Processing and Inversion

After these seminal works, the TL-ERT method has found an impressive and large spectrum of applications, and many authors have proposed new methods for TL-ERT data analysis and inversion [4]. In past years, the number of papers on this topic has constantly grown. Here, an overview of selected papers published during the last ten years is presented.

Doetsch et al., 2010 [51], proposed a structural joint inversion using cross-gradient constraints of the time-lapse GPR and ERT data. They analyzed synthetic and experimental data collected during a water injection experiment. The structure is imposed when the updates of the model, with respect to the background, is in the same or opposite directions. Synthetic and experimental tests confirmed the capacity of the structural joint inversion to better describe the plumes geometry, with respect to the single individual inversion.

Herckenrath et al., 2011 [52] investigated the possibility to calibrate the hydrogeophysical inversion with the time domain electromagnetic (TDEM) and ERT data. A geophysical model was simultaneously inverted with a groundwater model by coupling the groundwater and geophysical parameters. Jardani et al., 2013 [53] used a stochastic joint inversion of resistivity and self-potential data integrated with in-situ measurements of hydrogeological parameters. The main results of the inversion revealed that incorporating the self-potential data improves the estimate of hydraulic conductivity.

Camporese et al., 2015 [54] analyzed the performance of coupled and uncoupled inversion approaches in the reconstruction of heterogeneous saturated hydraulic conductivity patterns. The comparison was carried out in the framework of a model inversion based on the ensemble Kalman filter. The results highlighted some advantages of the coupled approach, but it requires massive computational resources.

The joint inversion of geophysical data and the coupled inversion of hydrogeophysical data are extremely complex issues and challenging scientific problems, for this reason, there are not many papers about the joint inversion of the TL-ERT data with other geophysical measurements. Furthermore, there is a growing interest in the studies on hydrogeophysical and petrophysical relationships.

Many works are focused on improving and modifying the full 4D inversion algorithm based on the regularization in space and time [50]. Hayley et al., 2011 [55], introduced the concept of simultaneous inversion. They proposed two different approaches: in the first one, the inversion of two consecutive ERT datasets was performed under a single optimization procedure and constraints, to ensure that the changes from one time to another were smooth; in the second method, the inversions were performed simultaneously and constraints of smoothness and closeness to a reference model were applied to the difference image produced.

Karaoluis et al., 2011 [56] modified the method for the full 4D inversion based on the regularization in space and time. They introduced the concept of the 4D active time-constrained (4D-ATC) inversion. The main feature of the modified algorithm is that it allows the time regularization to vary between different time steps depending on the degree of spatial resistivity changes occurring among different monitoring stages. As a result, the 4D-ATC algorithm uses all the benefits of the 4D algorithm, but also introduces increased flexibility in terms of temporal regularization, which can help in focusing on the dynamic resistivity changes. In a successive work, Karaoluis et al., 2014 [57], proposed an inversion scheme based on distributed and non-uniform parameters, considering the time-related noise and the reduction of the sensitivity in ERT measurements as purely random.

Other improvements were based on the concept of the adaptive approach and the use of minimum gradient in TL-ERT data inversion. Wilkinson et al., 2015 [58] proposed, for the first time, an adaptive survey technique for geoelectrical monitoring. A static algorithm was modified for increasing the resolution within target zones; during the time-dependent period from the image analysis, the zones were interested by the resistivity changes, and the resolution within these zones in the successive ERT acquisition increased. Nguyen et al., 2016 [59] introduced the concept of the minimum gradient support (MGS) in the TL-ERT data inversion. This parameter was selected using the value of the TL-ERT data root mean square misfit and the first iteration of the inversion procedure.

Some recent works are mainly devoted to the development of innovative algorithms for considering and removing the influence of artifacts and errors in the TL-ERT inversion procedure.

Liu et al., 2017 [60] approached the problems related to the propagation of the false anomalies in the initial model and their propagation in the successive TL-ERT inversions, and the assumption of the uniform sensitivity to resistivity variations. They introduced a temporal gradient to enhance the vertical and/or horizontal variation of the model parameters, and normalized the observations before inversion.

Lesparre et al., 2017 [61] analyzed the problems related to the quantification of the errors in the TL-ERT data inversion. Their approach was based on the analysis of the differences between normal and reciprocal measurements (i.e., swapping of the energizing and measuring electrodes) acquired at different time. They extended the concept of the static error formed by a linear combination of the systematic and random components [7], the results emphasized the necessity to better evaluate the errors in the time-lapse resistivity monitoring independently by the algorithm inversion.

Tso et al., 2017 [62] organized an overview of the main methodologies for removing and minimizing the influence of the experimental errors in the ERT inversion algorithms. They extended the standard error model based on the linearity with the transfer resistance introducing a grouping variable based on the electrode number used for the experimental measurements. This new linear mixed model (LME) was tested using lab and field data, improving the quality of the ERT images.

Other works focus their attention on errors due to the presence of an irregular surface topography in the area interested by the field measurements [17], and the acquisition schemes with electrodes installed in the boreholes [63].

Furthermore, there are some studies based on the use of the classical Kalman Filter, Bayesian Markov chain, Monte Carlo simulation and clustering procedures for the TL-ERT data analysis and inversion. Notwithstanding these methods are largely applied in geophysics, there is a limited number of applications on the time-dependent geoelectrical monitoring. Saibaba et al., 2014 [64], introduced the Kalman filtering approach to analyze the TL-ERT data. This model is useful in practical applications in which data is acquired at a rapid rate, and when changes in subsequent states are small and can be approximated by a random process.

Oware et al., 2019 [65] developed a novel method for the TL-ERT data inversion using a Bayesian Markov chain Monte Carlo (McMC) technique. The approach was based on the parametrization of the Bayesain inversion model in terms of the optimal bases, which allows a sparse representation of the desired features leading to dimensionality reduction and Monte Carlo simulations tuned to the physics of the targeted process. The method was tested during a heat-tracer experiment. Delforge et al., 2021 [66] proposed the time clustering method as a tool for the post-inversion of the TL-ERT data. The authors tested three different algorithms (K-means, hierarchical agglomerative clustering, and Gaussian Mixture Model) on raw ERT data analyzing the role of the spatial constraints and evaluating the robustness of the method.

Recently, there has been growing attention on the development of open sources software and codes, and the use of parallel computing to reduce the time required for geophysical data inversion. As it concerns the TL-ERT data, Johnson et al., 2017 [67] used open-source codes to simulate spatio-temporal changes of subsurface resistivity. The module PFLOTRAN-E4D operating in parallel modality was applied to reconstruct the bulk electrical conductivity from a subsurface flow and transport simulation code. At each step of the iterative process, a tomographic image is simulated using the bulk electrical conductivity model. This multi-physics approach appears of great interest for the availability of open-source codes.

Rucker et al., 2017 [68] introduced an open-source framework for the inversion of a hydro-geophysical dataset. A pyGIMLi (Python Library for Inversion and Modelling in Geophysics) was implemented for solving the minimization problem with a Gauss–Newton algorithm for any forward operator. The method was used for incorporating hydrological and petrophysical constraints in the ERT data inversion. Blanchy et al., 2020 [69] introduced the open-source code RESIPy for the inversion of resistivity and induced polarization data. The code can also be used for the analysis of TL-ERT data. It is formed by three different layers: (i) the first one is composed of inversion codes R2, R3t and CR2 and a software to generate triangular meshes; (ii) the second layer is composed of Python API, containing processing routines for filtering and error modeling; (iii) the last is composed of routines for graphical application mainly devoted to data visualization.

Liu et al., 2020 [70] proposed a method for the fast inversion of the 4D ERT data that considers the problems related to the rapid changes in the resistivity subsurface model. Generally, the ERT inversion is time consuming, and during the period necessary for the data inversion a resistivity change in the subsurface could occur. The algorithm subdivided an entire data set into a multiple subset of data, that were dynamically substituted in the equations for data inversion. In this process, the regularization coefficients are adjusted; furthermore, a GPU parallel algorithm is used for the TL-ERT data inversion.

Finally, in Table 1 we summarize the different methods adopted for TL-ERT data inversion. At each class is associated a list of the selected works previously discussed. Of course, the list is not exhaustive, and does not include all of the published papers.

Table 1. Simplified classification of the different approaches and methods adopted in the TL-ERT data processing and inversion. The scheme is based on the analysis of a selected list of papers published in the last ten years.

Main Classes of the Methods for the TL-ERT Data Analysis and Inversion	List of the Papers
Joint-inversion with other geophysical data and hydro-geophysical data.	Doetsch et al., 2010 Herckenrath et al., 2013 Jardani et al., 2013 Camporese et al., 2015
Full 4D inversion, time and space active constraints, space and time regularization, algorithms based on the adaptive approaches and minimum gradients.	Hailey et al., 2011 Karaoulis et al., 2011 Karaoulis et al., 2014 Wilkinson et al., 2015 Nguyen et al., 2016
Methods for minimizing the errors and the artifacts.	Liu et al., 2017 Lesparre et al., 2017 Tso et al., 2017 Bievre et al., 2018 Perri et al., 2020
Clustering, Bayesian and other statistical approaches.	Saibaba et al., 2014 Oware et al., 2019 Delforge et al., 2021
Open-source codes and HPC techniques.	Johnson et al., 2017 Rucker et al., 2017 Blanchy et al., 2020 Liu et al., 2020

3. Landslide Monitoring

Landslide investigation is one of the more consolidated applications of the ERT method. The main contributions of the static 2D and 3D ERT (Figure 2) tomographic methods regard the delineation of the geometry of the landslide bodies and the identification of high water content areas [42,44,71]. As it concerns the *landslide monitoring*, the TL-ERT method represents an optimal tool to investigate the triggering factors of mass-movements related to water infiltration, and soil moisture's temporal and spatial changes. In fact, the time series of the resistivity images are strongly influenced by the time-dependent changes of water content in soil and subsoil.

Lebourg et al., 2010 [72] performed a field experiment at the Vence landslide (France), considered as a translational landslide, including 1.2×10^6 m^3 of material. They measured rainfall precipitation for three months, including water level and resistivity values, and investigated the possible correlations between these parameters. The study was carried out using raw data (apparent resistivity values) without any inversion, the results pointed out the strong negative correlations between the rainfall events and the resistivity correlation factor, and between the piezometric elevation and the resistivity average.

Travelletti et al., 2012 [73] carried out another field experiment at the Laval landslide (France). Starting from a non-saturated condition near the surface, the TL-ERT measurements were performed during water infiltration within clay-shale material. They observed a reduction of resistivity values (18%) with an asymptotic value in correspondence to steady state flow conditions. The inversion of the TL-ERT data was carried using different methods [2,49], and the results highlighted possible preferential flows near the landslide toe in the steady stationary flow (Figure 3). Other field experiments planned for landslide investigations in Taiwan and Italy are reported in Lee et al., 2012, Luongo et al., 2012, and Quagliarini et al., 2017 [74–76].

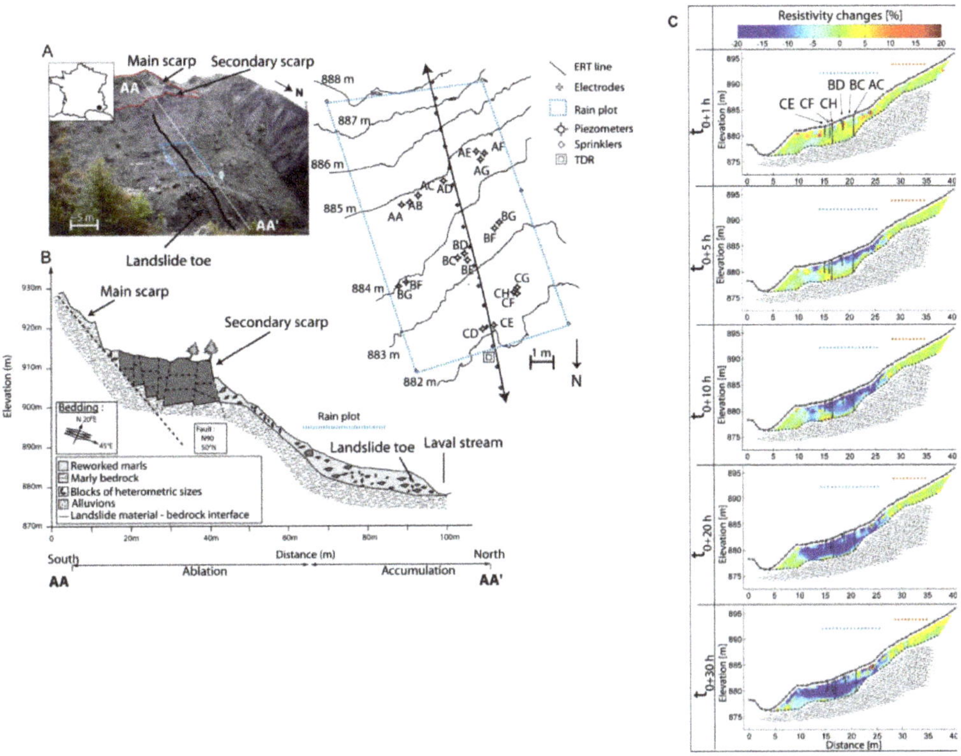

Figure 3. (**A**) Laval landslide (ORE Draix, South French Alps) with location of the ERT line and of the equipment used for the rain experiment. (**B**) Cross section of the landslide. (**C**) Resistivity changes during the rain experiment (after [73]).

Gunn et al., 2013 [77] investigated the complex geological environment of the Whitby Mudstone Formation, one of the highest landslide densities in the United Kingdom. They studied the landslide phenomena at Hollin Hill in North Yorkshire that includes shallow, retrogressive rotational, and translation movements. A permanent station able to perform time-lapse ERT observations was installed for monitoring the soil moisture pattern near the surface: the results allowed the authors to study the role of hydrogeological processes in triggering earth flows.

Gance et al., 2016 [78] installed a permanent 4D-ERT system at the Super-Sauze clayey landslide (France). They performed a statistical analysis to investigate possible relationships between the rainfall characteristics, the soil hydrological observations and the soil electrical resistivity response (Figure 4). The results highlighted the influence of the temperature in ERT measurements due to the heat exchange between the groundwater, the vadose zone water and the rainwater that hides the variations of resistivity due to variations in the soil water content. Another quite similar experiment was carried out in a landslide area in southwestern China, to map the pathways of rainwater infiltration in the slipping zones (Xu et al., 2016) [79].

Palis et al., 2017 [80] applying electrical resistivity tomography (daily acquisition), rainfall records (since 2006) and boreholes monitoring groundwater level (since 2009), carried out a 9.5-year multiparametric survey on the "Vence" (France) landslide. The aim of the survey was to understand fluid circulations within the unstable slope. ERT data were analyzed by using an automated clustering analysis enabled to isolate the ERT section into three groups of apparent resistivity values (Figure 5). The comparison between the variations of these cluster behaviors in time with the variations of groundwater levels on site, allowed the authors to locate distinct hydrogeological units. The authors showed that the qualitative information on the variability of the slope hydrogeological behavior could help improve the conversion of resistivity data into hydrologic quantities.

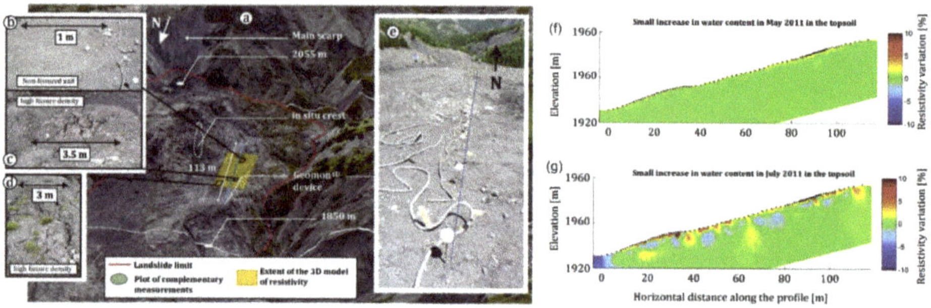

Figure 4. The map above (**a**) shows the location of the investigations carried out on the Super-Sauze landslide (Southeast France). The slope is characterized by non-fissured unit (**b**) and areas affected by a high fissure density (**c**,**d**). The dashed blue line (**e**) indicates the position of the GEOMON 4D system. The images (**f**,**g**) illustrate the soil cooling effect due to the rainwater. The percentage change in soil electrical resistivity after a small rainfall event could be due to the difference between rain and initial soil temperature (after [78]).

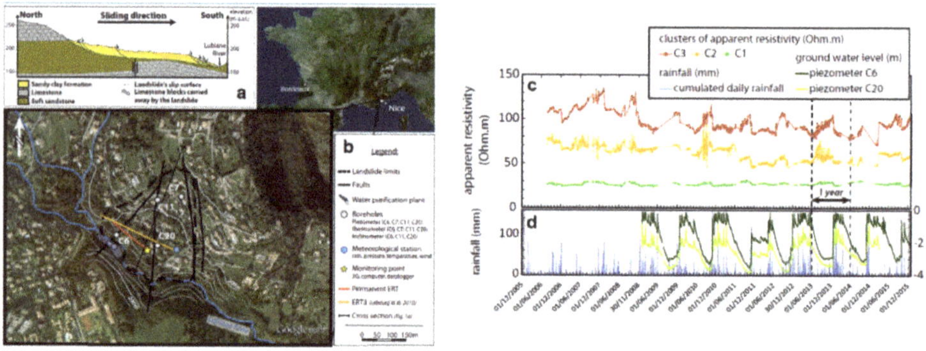

Figure 5. The map at the top reports the location of the survey carried out on the Vence landslide (south-eastern France), a simplified geological cross section (**a**) obtained along the sliding direction and the layout of the field survey carried out in the area. (**b**) Legend. The images below show (**c**) the variations of the daily median values of the three clusters of apparent resistivity measurements identified for the slope in the period 2006–2016, and (**d**) the accumulated rainfall index and groundwater level obtained by direct measurements in C6 and C20 for the same period (after [80]).

Zieher et al., 2017 [81] performed a spray irrigation experiment in a selected Alpine zone interested by shallow landslides. A sequence of 2D ERT images were obtained during the irrigation period. Furthermore, they installed a grid of time-domain reflectometry sensors at different depths for measuring water content, and used a linear diffusion model for simulating the time-dependent variations of the pore pressure. The results allowed them to detect the negative resistivity changes associated with the vertical progressing water. Unfortunately, the measurements and the model were not able to investigate the small-scale variations in pore pressure that may facilitate the triggering of shallow landslides.

Hojat et al., 2019 [82] carried out a laboratory-scale experiment to simulate the rainfall triggering of shallow landslides. They used the TL-ERT and time-domain reflectometry methods to monitor water infiltration and to evaluate the volumetric water content. After a calibration of the Archie's equation, the ERT images were transformed in water saturated maps. The results of the controlled experiment highlighted that the landslide body becomes unstable at zones where the water saturation exceeds 45%, and the instability phenomena could occur at the boundaries between areas with different water saturations.

Ivanov et al., 2020 [83] approached the study of shallow landslides in a mountain landscape mainly triggered by rainfall extreme events. They designed and implemented a landslide simulator, and a dataset of 20 land mass failures induced by water infiltration were described and analyzed. Both hydrogeological and TL-ERT measurements were performed. The time series of the ERT images were used for mapping the waterfront time-dependent changes during three different rainfall simulations: (i) temporal increase in the rainfall intensity; (ii) constant rainfall intensity and (iii) two sequences for rainfall events. The results of downscaled experiments confirmed that the rainfall intensity, the hydraulic conductivity and the soil moisture gradient controlled the stability, while rainfall intensity and duration were the main predictors of the shallow soil slips. Furthermore, their results could be useful for implementing procedures for the validation of the rainfall thresholds.

Boyd et al., 2021 [84] investigated one of the more critical problems that generally occur in the ERT monitoring of shallow landslides. It is common to observe low changes in the geometry of the electrode array due to ground movements. In a field experiment carried out at Hollin Hill site (UK), the variations of the distances between the electrodes were detected with topographic methods (UAV surveys and global position measurements) and considered in the TL-ERT inversion (Figure 6).

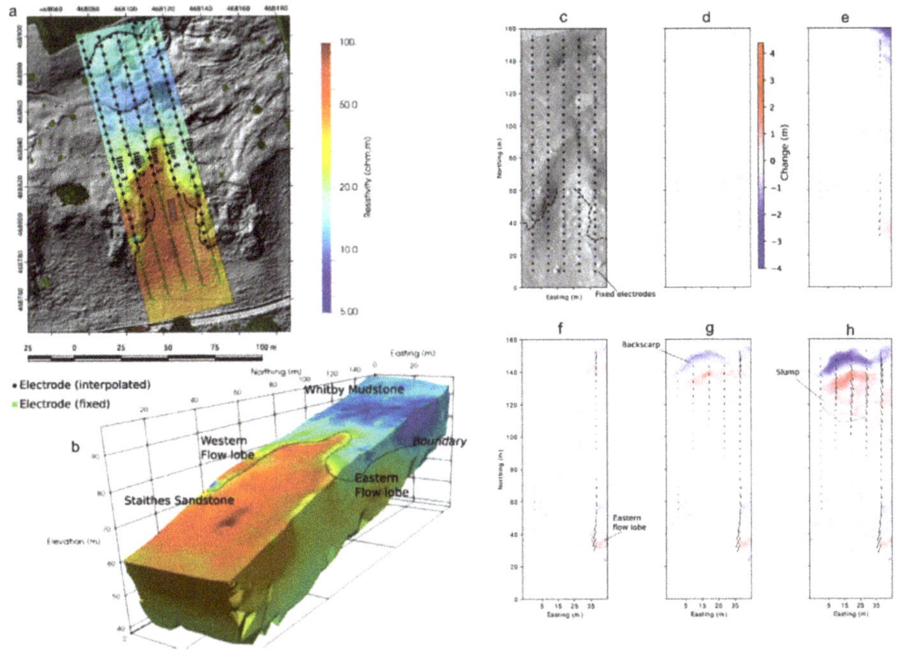

Figure 6. (**a**) The baseline inverted image of the Hollin Hill landslide (southern Howardian Hills, UK) measured in May 2016; (**b**) surface resistivity in relation to a hill-shaded relief map from a UAV survey in May 2016. The six images on the right (**c**–**h**) report an overview of elevation and electrode coordinate changes at Hollin Hill, as captured by terrestrial LiDAR (after [84]).

4. Discussion

This review helps to highlight the strengths and weaknesses of the TL-ERT method in the study of landslide areas, by considering both the technological aspects of instrumentation and data acquisition, and the phase of data analysis and processing.

As shown by the papers analyzed in the previous section, the main and robust contributions of the TL-ERT method in the study of landslides concern the monitoring of rainwater infiltration in the shallow subsurface, and the estimation of the spatial and temporal changes in soil moisture and water content in the top soils and shallow layers of the unstable slope, respectively. In many cases, authors tried to define significant correlations between rainfall, soil hydrological characteristics, and resistivity values [72,78,80]. The quantitative analysis of such correlations has always represented a challenge for the scientific community. Very often, the TL-ERT measurements have been integrated and compared with meteorological and hydrological data and, sometimes, with TDR measurements or other geophysical data [81,82]. The comparison and integration of data coming from different techniques allow the overcoming of some specific limitations of each single method. The results of the integration highlight the capability of TL-ERT to give quantitative information about the water content and to map a preferential pathway for water infiltration. This capability makes the TL-ERT method useful in the future for better evaluating the water saturation effects in the estimation of the rainfall thresholds in landslide early-warning systems.

The analyzed papers also confirmed that it is not possible to generalize the results obtained for every geological and geomorphological context. In order to make the landslide monitoring effective by applying the TL-ERT method, an appropriate and correct geological model must be adopted. As reported by other authors, the petrophysical models used to transform resistivity values in hydrological parameters are site-specific [81,85], and cannot be used to define resistivity thresholds applicable for the activation of landslides in any geological environment. The main evolution of the petrophysical relationships reported in the literature has regarded the change of the number of phases to be considered in the model [86]. Indeed, unlike what is claimed by the petrophysics, the rock formations are unconsolidated, and not composed of a continuous solid phase, but of different phases. So, starting from the one phase (pore water) model proposed by Archie's Law [87], to interpret the electrical properties acquired in different sedimentary contexts, multi-phase models have been considered [88].

Despite the attention paid by the scientific community, the numerous developed petrophysical relationships are still difficult to apply in the study of landslide areas. Often, they require calibration with other parameters (i.e., soil temperature, porosity, water content, etc.) that are not always available or measurable, requiring laboratory analysis for their definition. For this reason, authors working in landslide areas often prefer to consider the raw data and carry out a statistical analysis of them.

All the discussed applications were carried out almost exclusively in rural areas, i.e., in areas where the logistical conditions for data acquisition were favorable, and in which the quality of the data was generally high due to the absence of environmental noise sources. Working in rural areas offers some advantages, such as easier positioning of the instrumentation and planning of continuous acquisition over time (the electrodes can be embedded in the ground for long periods without creating problems for the normal activities of a community). The absence of buried objects and underground utilities, that could affect the flow of current in the subsoil, and the good electrode/terrain coupling mean that the acquired data are not very noisy, and do not require high filtering actions. Despite these, many authors, to avoid the appearance of the artifacts in the final image, prefer to analyse the apparent resistivity data, instead of applying inversion algorithms.

As described in paragraph two, in the last ten years an impressive number of papers concerning novel algorithms for resistivity data processing and inversion have been published. However, there are a limited number of cases in which these algorithms have been applied to analyze the TL-ERT data measured in field-experiments carried out for studying landslides. Probably, the use of these mathematical methods is quite complicated when applied to experimental data characterized by large variability in spatial and temporal domains, and affected by a high-level of anthropic noise. The TL-ERT data inversion must also consider the effect of the temperature in the first layers of the subsoil and, in active landslides, the effect of the electrode's movement. Corrective formulas have been introduced for both critical issues [78,84].

A gap between the quality of the algorithms that are currently available for the TL-ERT data inversion and the methods routinely applied in the interpretation of the experimental resistivity data is quite evident. The use of advanced methods for data processing and inversion will disclose the possibility to apply the TL-ERT method in the monitoring of landslides occurred in urban areas and complex geological slopes.

Except for a few studies [80,85], all the acquisitions were carried out along 2D longitudinal profiles. This still represents a limitation of the method, as the investigated object has a 3D structure. Unfortunately, making 3D acquisitions is not easy and requires the use of manageable instruments. Other reviews [4,42] have already highlighted this aspect, referring, as a possible solution, to the use of separate current injections and voltage measuring systems, but also to the experimentation of technologies that minimize the number of cables (smart sensors and web-based services for wireless sensor networks). Considering the difficulties that characterize 3D acquisition in landslide areas, it is well understood that these difficulties increase if we want to realize a 3D time-lapse ERT.

Other weaknesses identified by the critical analysis of the works presented in the previous paragraph concern the TL-ERT's difficulty in investigating deep subsoil layers in landslide bodies. The ERT method can investigate shallow (0–20 m) and deep layers (>50 m) in landslide bodies, but there are no experimental tests in which the time-lapse monitoring systems have been able to obtain information about the resistivity patterns in deep environments. Generally, the amplitude of the time-dependent changes of resistivity values are higher in shallow layers, and strongly decrease in deep environments; however, the possibility to detect weak perturbations in the resistivity patterns of deep environments (50–100 m) could be very interesting and challenging. In the future, the design of TL-ERT monitoring systems to obtain a time series of 3D resistivity images with an investigation range of 0–100 m could open new research opportunities.

5. Conclusions

Two very significant aspects emerge from the analysis of the papers published in the period 2010–2020 on the theoretical developments and related applications of the TL-ERT method. The first concerns the enormous development of innovative methodologies for the processing and interpretation of data acquired with the TL-ERT method, while the second resides in the wide and growing spectrum of applications of this method in landslide monitoring.

The TL-ERT method appears particularly suitable for studying the variations, both in the spatial and temporal domain, of water content in the most superficial layers of the soil in landslide bodies. It can also provide a useful contribution to the identification of preferential ways for the infiltration of fluids. In the future, this activity could make a significant contribution in early warning systems for landslides induced by extreme weather events by also applying the TL-ERT method to evaluate the water saturation effects in the estimation of the rainfall thresholds.

However, it will be necessary to investigate the complex relationships between resistivity and hydrogeological parameters. From the analysis of the results obtained, the use of innovative mathematical and statistical methods in interpreting the experimental results still appears to be limited. This aspect is fundamental in making the insertion of this method more robust in advanced and widespread monitoring systems. Finally, the TL-ERT method should also be applied for deeper geophysical investigations.

Author Contributions: Conceptualization, V.L. and A.P.; methodology, V.L. and A.P.; writing—original draft preparation, V.L. and A.P.; writing—review and editing, V.L. and A.P. All authors have read and agreed to the published version of the manuscript.

Funding: This research was carried out in the framework of the projects OT4CLIMA *"Advanced EO technologies for studying Climate Change impacts on the environment"* (D.D. 2261 of 6.9.2018, MUR PON R&I 2014–2020 program) and MITIGO *"Mitigazione dei rischi naturali per la sicurezza e la mobilità delle aree montuose del Sud Italia"* (ARS01_00964, MUR PON R&I 2014–2020 Program).

Institutional Review Board Statement: Not applicable.

Informed Consent Statement: Not applicable.

Acknowledgments: The authors thank Rocchino Caivano for the technical support in the bibliography classification and organization using the WoS database.

Conflicts of Interest: The authors declare no conflict of interest.

References

1. Koefoed, O. *Geosounding Principles 1: Resistivity Sounding Measurements*; Elsevier: Amsterdam, The Netherlands, 1979.
2. Loke, M.H.; Barker, R.D. Rapid least-squares inversion of apparent resistivity pseudosections using a quasi-Newton method. *Geophys. Prospect.* **1996**, *44*, 131–152. [CrossRef]
3. Loke, M.H.; Barker, R.D. Practical techniques for 3D resistivity surveys and data inversion. *Geophys. Prospect.* **1996**, *44*, 499–523. [CrossRef]
4. Loke, M.H.; Chambers, J.E.; Rucker, D.F.; Kuras, O.; Wilkinson, P.B. Recent developments in the direct-current geoelectrical imaging method. *J. Appl. Geophys.* **2013**, *95*, 135–156. [CrossRef]
5. Binley, A.; Henry-Poulter, S.; Shaw, B. Examination of solute transport in an undisturbed soil column using electrical resistance tomography. *Water Resour. Res.* **1996**, *32*, 763–769. [CrossRef]
6. Slater, L.; Binley, A.; Brown, D. Electrical imaging of fractures using groundwater salinity change. *Ground Water* **1997**, *35*, 436–442. [CrossRef]
7. Slater, L.; Binley, A.; Versteeg, R.; Cassiani, G.; Birken, R.; Sandberg, S. A 3D ERT study of solute transport in a large experimental tank. *J. Appl. Geophys.* **2002**, *49*, 211–229. [CrossRef]
8. Kemna, A.; Vanderborght, J.; Kulessa, B.; Vereecken, H. Imaging and characterisation of subsurface solute transport using electrical resistivity tomography (ERT) and equivalent transport models. *J. Hydrol.* **2002**, *267*, 125–146. [CrossRef]
9. Earon, R.; Riml, J.; Wu, L.W.; Olofsson, B. Insight into the influence of local streambed heterogeneity on hyporheic-zone flow characteristics. *Hydrogeol. J.* **2020**, *28*, 2697–2712. [CrossRef]
10. Paz, M.C.; Alcala, F.J.; Medeiros, A.; Martinez-Pagan, P.; Perez-Cuevas, J.; Ribeiro, L. Integrated MASW and ERT Imaging for Geological Definition of an Unconfined Alluvial Aquifer Sustaining a Coastal Groundwater-Dependent Ecosystem in Southwest Portugal. *Appl. Sci.* **2020**, *10*, 5905. [CrossRef]
11. Folch, A.; del Val, L.; Luquot, L.; Martinez-Perez, L.; Bellmunt, F.; Le Lay, H.; Rodellas, V.; Ferrer, N.; Palacios, A.; Fernandez, S.; et al. Combining fiber optic DTS, cross-hole ERT and time-lapse induction logging to characterize and monitor a coastal aquifer. *J. Hydrol.* **2020**, *588*, 125050. [CrossRef]
12. Palacios, A.; Ledo, J.J.; Linde, N.; Luquot, L.; Bellmunt, F.; Folch, A.; Marcuello, A.; Queralt, P.; Pezard, P.A.; Martinez, L.; et al. Time-lapse cross-hole electrical resistivity tomography (CHERT) for monitoring seawater intrusion dynamics in a Mediterranean aquifer. *Hydrol. Earth Syst. Sci.* **2020**, *24*, 2121–2139. [CrossRef]
13. Rao, S.; Lesparre, N.; Orozco, A.F.; Wagner, F.; Javaux, M. Imaging plant responses to water deficit using electrical resistivity tomography. *Plant Soil* **2020**, *454*, 261–281. [CrossRef]
14. Fishkis, O.; Noell, U.; Diehl, L.; Jaquemotte, J.; Lamparter, A.; Stange, C.F.; Burke, V.; Koeniger, P.; Stadler, S. Multitracer irrigation experiments for assessing the relevance of preferential flow for non-sorbing solute transport in agricultural soil. *Geoderma* **2020**, *371*, 114386. [CrossRef]
15. De Carlo, L.; Battilani, A.; Solimando, D.; Caputo, M.C. Application of time-lapse ERT to determine the impact of using brackish wastewater for maize irrigation. *J. Hydrol.* **2020**, *582*, 124465. [CrossRef]
16. Blanchy, G.; Watts, C.W.; Richards, J.; Bussell, J.; Huntenburg, K.; Sparkes, D.L.; Stalham, M.; Hawkesford, M.J.; Whalley, W.R.; Binley, A. Time-lapse geophysical assessment of agricultural practices on soil moisture dynamics. *Vadose Zone J.* **2020**, *19*, e20080. [CrossRef]
17. Bievre, G.; Oxarango, L.; Gunther, T.; Goutaland, D.; Massardi, M. Improvement of 2D ERT measurements conducted along a small earth-filled dyke using 3D topographic data and 3D computation of geometric factors. *J. Appl. Geophys.* **2018**, *153*, 100–112. [CrossRef]
18. Jodry, C.; Lopes, S.P.; Fargier, Y.; Sanchez, M.; Cote, P. 2D-ERT monitoring of soil moisture seasonal behaviour in a river levee: A case study. *J. Appl. Geophys.* **2019**, *167*, 140–151. [CrossRef]

19. Masi, M.; Ferdos, F.; Losito, G.; Solari, L. Monitoring of internal erosion processes by time-lapse electrical resistivity tomography. *J. Hydrol.* **2020**, *589*, 125340. [CrossRef]
20. Srivastava, S.; Pal, S.K.; Kumar, R. A time-lapse study using self-potential and electrical resistivity tomography methods for mapping of old mine working across railway-tracks in a part of Raniganj coalfield, India. *Environ. Earth Sci.* **2020**, *79*, 332. [CrossRef]
21. Di Giuseppe, M.G.; Troiano, A. Monitoring active fumaroles through time-lapse electrical resistivity tomograms: An application to the Pisciarelli fumarolic field (Campi Flegrei, Italy). *J. Volcanol. Geotherm. Res.* **2019**, *375*, 32–42. [CrossRef]
22. Tso, C.H.M.; Johnson, T.C.; Song, X.H.; Chen, X.Y.; Kuras, O.; Wilkinson, P.; Uhlemann, S.; Chambers, J.; Binley, A. Integrated hydrogeophysical modelling and data assimilation for geoelectrical leak detection. *J. Contam. Hydrol.* **2020**, *234*, 103679. [CrossRef] [PubMed]
23. Peskett, L.; MacDonald, A.; Heal, K.; McDonnell, J.; Chambers, J.; Uhlemann, S.; Upton, K.; Black, A. The impact of across-slope forest strips on hillslope subsurface hydrological dynamics. *J. Hydrol.* **2020**, *581*, 124427. [CrossRef]
24. Bouvier, C.; Adamovic, M.; Ayral, P.A.; Brunet, P.; Didon-Lescot, J.F.; Domergue, J.M.; Spinelli, R. Characterization of subsurface fluxes at the plot scale during flash floods in the Valescure catchment, France. *Hydrol. Processes* **2021**, *35*, e14144. [CrossRef]
25. Zhou, Q.L.; Yang, X.J.; Zhang, R.; Hosseini, S.A.; Ajo-Franklin, J.B.; Freifeld, B.M.; Daley, T.M.; Hovorka, S.D. Dynamic Processes of CO2 Storage in the Field: 1. Multiscale and Multipath Channeling of CO2 Flow in the Hierarchical Fluvial Reservoir at Cranfield, Mississippi. *Water Resour. Res.* **2020**, *56*, 2. [CrossRef]
26. Mollaret, C.; Hilbich, C.; Pellet, C.; Flores-Orozco, A.; Delaloye, R.; Hauck, C. Mountain permafrost degradation documented through a network of permanent electrical resistivity tomography sites. *Cryosphere* **2019**, *13*, 2557–2578. [CrossRef]
27. Conaway, C.H.; Johnson, C.D.; Lorenson, T.D.; Turetsky, M.; Euskirchen, E.; Waldrop, M.P.; Swarzenski, P.W. Permafrost Mapping with Electrical Resistivity Tomography: A Case Study in Two Wetland Systems in Interior Alaska. *J. Environ. Eng. Geophys.* **2020**, *25*, 199–209. [CrossRef]
28. Kellerer-Pirklbauer, A.; Avian, M.; Benn, D.I.; Bernsteiner, F.; Krisch, P.; Ziesler, C. Buoyant calving and ice-contact lake evolution at Pasterze Glacier (Austria) in the period 1998–2019. *Cryosphere* **2021**, *15*, 1237–1258. [CrossRef]
29. Scandroglio, R.; Draebing, D.; Offer, M.; Krautblatter, M. 4D quantification of alpine permafrost degradation in steep rock walls using a laboratory-calibrated electrical resistivity tomography approach. *Near Surf. Geophys.* **2021**, *19*, 241–260. [CrossRef]
30. Guha-Sapir, D.; CRED (Centre for Research on the Epidemiology of Disasters). EM-DAT: The Emergency Events Database, Brussels, Belgium. Available online: https://www.emdat.be/ (accessed on 12 January 2022).
31. Tiranti, D.; Cremonini, R. Editorial: Landslide Hazard in a Changing Environment. *Front. Earth Sci.* **2019**, *7*, 3. [CrossRef]
32. Mateos, R.M.; Lopez-Vinielles, J.; Poyiadji, E.; Tsagkas, D.; Sheehy, M.; Hadjicharalambous, K.; Liscak, P.; Podolski, L.; Laskowicz, I.; Iadanza, C.; et al. Integration of landslide hazard into urban planning across Europe. *Landsc. Urban Plan.* **2020**, *196*, 10374. [CrossRef]
33. Donnini, M.; Napolitano, E.; Salvati, P.; Ardizzone, F.; Bucci, F.; Fiorucci, F.; Santangelo, M.; Cadinali, M.; Guzzetti, F. Impact of event landslides on road networks: A statistical analysis of two Italian cases studies. *Landslides* **2017**, *14*, 1521–1535. [CrossRef]
34. Del Soldato, M.; Di Martire, M.; Bianchini, S.; Tomas, R.; De Vita, P.; Ramondini, M.; Casagli, N.; Calcaterra, D. Assessment of landslide-induced damage to structures: The Agnone landslide case study (southern Italy). *Bull. Eng. Geol.* **2019**, *78*, 2387–2408. [CrossRef]
35. Emberson, R.; Kirschbaum, D.; Stanley, T. New global characterisation of landslide exposure. *Nat. Hazards Earth Syst. Sci.* **2020**, *20*, 3413–3424. [CrossRef]
36. Petley, D. Global patterns of loss of life from landslides. *Geology* **2012**, *40*, 927–930. [CrossRef]
37. Froude, M.J.; Petley, D.N. Global fatal landslide occurrence from 2004 to 2016. *Nat. Hazards Earth Syst. Sci.* **2018**, *18*, 2161–2181. [CrossRef]
38. Stahli, M.; Sattele, M.; Huggel, C.; McArdell, B.W.; Lehmann, P.; Van Herwijnen, A.; Berne, A.; Schleiss, M.; Ferrari, A.; Kos, A.; et al. Monitoring and prediction in early warning systems for rapid mass movements. *Nat. Hazards Earth Syst. Sci.* **2015**, *15*, 905–917. [CrossRef]
39. Segoni, S.; Piciullo, L.; Gariano, S.L. Preface: Landslide early warning systems: Monitoring systems, rainfall thresholds, warning models, performance evaluation and risk perception. *Nat. Hazards Earth Syst. Sci.* **2018**, *18*, 3179–3186. [CrossRef]
40. Guzzetti, F.; Ariano, S.L.; Peruccacci, S.; Brunetti, M.T.; Marchesini, I.; Rossi, M.; Melillo, M. Geographical landslide early warning systems. *Earth-Sci. Rev.* **2020**, *200*, 102973. [CrossRef]
41. Jongmans, D.; Fiolleau, S.; Bievre, G. Geophysical Monitoring of Landslides: State-of-the Art and Recent Advances. In *Understanding and Reducing Landslide Disaster Risk*; Springer: Cham, Switzerland, 2021.
42. Perrone, A.; Lapenna, V.; Piscitelli, S. Electrical resistivity tomography technique for landslide investigation: A review. *Earth Sci. Rev.* **2014**, *135*, 65–82. [CrossRef]
43. Pazzi, V.; Morelli, S.; Fanti, R. A review of the advantages and limitations of geophysical investigations in landslide studies. *Int. J. Geophys.* **2019**, *2019*, 27. [CrossRef]
44. Bellanova, J.; Calamita, G.; Giocoli, A.; Luongo, R.; Macchiato, M.; Perrone, A.; Uhlemann, S.; Piscitelli, S. Electrical resistivity imaging for the characterization of the Montaguto landslide (southern Italy). *Eng. Geol.* **2018**, *243*, 272–281. [CrossRef]
45. Deceuster, J.; Kaufmann, O.; Van Camp, M. Automated identification of changes in electrode contact properties for long-term permanent ERT monitoring experiments. *Geophysics* **2013**, *78*, E79–E94. [CrossRef]

46. Kim, J.H. Four dimensional inversion of dc resistivity monitoring data. In *Proceedings of the Near Surface 2005-11th European Meeting of Environmental and Engineering Geophysics, Palermo, Italy, 4–7 September 2005*; European Association of Geoscientists and Engineers: Houten, The Netherlands, 2005; p. A006.
47. LaBrecque, D.J.; Yang, X. Difference inversion of ERT data: A fast inversion method for 3-D in situ monitoring. *J. Environ. Eng. Geophys.* **2001**, *6*, 83–89. [CrossRef]
48. Daily, W.; Ramirez, A.; Labrecque, D.; Nitao, J. Electrical resistivity tomography of vadose water movement. *Water Resour. Res.* **1992**, *28*, 1429–1442. [CrossRef]
49. Miller, C.R.; Routh, P.S.; Brosten, T.R.; McNamara, J.P. Application of Time-Lapse ERT Imaging to Watershed Characterization. *Geophysics* **2008**, *73*, G7–G17. [CrossRef]
50. Kim, J.H.; Yi, M.J.; Park, S.; Kim, J.G. 4D inversion of DC monitoring data acquired over a dynamically changing earth model. *J. Appl. Geophys.* **2009**, *68*, 522–532. [CrossRef]
51. Doetsch, J.; Linde, N.; Binley, A. Structural joint inversion of time-lapse crosshole ERT and GPR traveltime data. *Geophys. Res. Lett.* **2010**, *37*, L24404. [CrossRef]
52. Herckenrath, D.; Fiandaca, G.; Auken, E.; Bauer-Gottwein, P. Sequential and joint hydrogeophysical inversion using a field-scale groundwater model with ERT and TDEM data. *Hydrol. Earth Syst. Sci.* **2013**, *17*, 4043–4060. [CrossRef]
53. Jardani, A.; Revil, A.; Dupont, J.P. Stochastic joint inversion of hydrogeophysical data for salt tracer test monitoring and hydraulic conductivity imaging. *Adv. Water Resour.* **2012**, *52*, 62–77. [CrossRef]
54. Camporese, M.; Cassiani, G.; Deiana, R.; Salandin, P.; Binley, A. Coupled and uncoupled hydrogeophysical inversions using ensemble Kalman filter assimilation of ERT-monitored tracer test data. *Water Resour. Res.* **2015**, *51*, 3277–3291. [CrossRef]
55. Hayley, K.; Pidlisecky, A.; Bentley, L.R. Simultaneous time-lapse electrical resistivity inversion. *J. Appl. Geophys.* **2011**, *75*, 401–411. [CrossRef]
56. Karaoulis, M.C.; Kim, J.H.; Tsourlos, P.I. 4D active time constrained resistivity inversion. *J. Appl. Geophys.* **2011**, *73*, 25–34. [CrossRef]
57. Karaoulis, M.; Tsourlos, P.; Kim, J.H.; Revil, A. 4D time-lapse ERT inversion: Introducing combined time and space constraints. *Near Surf. Geophys.* **2014**, *12*, 25–34. [CrossRef]
58. Wilkinson, P.B.; Uhlemann, S.; Meldrum, P.I.; Chambers, J.E.; Carriere, S.; Oxby, L.S.; Loke, M.H. Adaptive time-lapse optimized survey design for electrical resistivity tomography monitoring. *Geophys. J. Int.* **2015**, *203*, 755–766. [CrossRef]
59. Nguyen, F.; Kemna, A.; Robert, T.; Hermans, T. Data-driven selection of the minimum-gradient support parameter in time-lapse focused electric imaging. *Geophysics* **2016**, *81*, A1–A5. [CrossRef]
60. Liu, B.; Liu, Z.Y.; Li, S.C.; Fan, K.R.; Nie, L.C.; Zhang, X.X. An improved Time-Lapse resistivity tomography to monitor and estimate the impact on the groundwater system induced by tunnel excavation. *Tunn. Undergr. Space Technol.* **2017**, *66*, 107–120. [CrossRef]
61. Lesparre, N.; Nguyen, F.; Kemna, A.; Robert, T.; Hermans, T.; Daoudi, M.; Flores-Orozco, A. A new approach for time-lapse data weighting in electrical resistivity tomography. *Geophysics* **2017**, *82*, E325–E333. [CrossRef]
62. Tso, C.H.M.; Kuras, O.; Wilkinson, P.B.; Uhlemann, S.; Chambers, J.E.; Meldrum, P.I.; Graham, J.; Sherlock, E.F.; Binley, A. Improved characterisation and modelling of measurement errors in electrical resistivity tomography (ERT) surveys. *J. Appl. Geophys.* **2017**, *146*, 103–119. [CrossRef]
63. Perri, M.T.; Barone, I.; Cassiani, G.; Deiana, R.; Binley, A. Borehole effect causing artefacts in cross-borehole electrical resistivity tomography: A hydraulic fracturing case study. *Near Surf. Geophys.* **2020**, *18*, 4. [CrossRef]
64. Saibaba, A.K.; Miller, E.L.; Kitandis, P.K. A fast kalman filter for time-lapse electrical resistivity tomography. In Proceedings of the 2014 IEEE Geoscience and Remote Sensing Symposium, Quebec City, QC, Canada, 13–18 July 2014. [CrossRef]
65. Oware, E.K.; Irving, J.; Hermans, T. Basis-constrained Bayesian Markov-chain Monte Carlo difference inversion for geoelectrical monitoring of hydrogeologic processes. *Geophysics* **2019**, *84*, A37–A42. [CrossRef]
66. Delforge, D.; Watlet, A.; Kaufmann, O.; Van Camp, M.; Vanclooster, M. Time-series clustering approaches for subsurface zonation and hydrofacies detection using a real time-lapse electrical resistivity dataset. *J. Appl. Geophys.* **2021**, *184*, 104203. [CrossRef]
67. Johnson, T.C.; Hammond, G.E.; Chen, X.Y. PFLOTRAN-E4D: A parallel open source PFLOTRAN module for simulating time-lapse electrical resistivity data. *Comput. Geosci.* **2017**, *99*, 72–80. [CrossRef]
68. Rucker, C.; Gunther, T.; Wagner, F.M. pyGIMLi. An open-source library for modelling and inversion in geophysics. *Comput. Geosci.* **2017**, *109*, 106–123. [CrossRef]
69. Blanchy, G.; Saneiyan, S.; Boyd, J.; McLachlan, P.; Binley, A. ResIPy, an intuitive open source software for complex geoelectrical inversion/modeling. *Comput. Geosci.* **2020**, *137*, 104423. [CrossRef]
70. Liu, B.; Pang, Y.H.; Mao, D.Q.; Wang, J.; Liu, Z.Y.; Wang, N.; Liu, S.H.; Zhang, X.X. A rapid four-dimensional resistivity data inversion method using temporal segmentation. *Geophys. J. Int.* **2020**, *221*, 586–602. [CrossRef]
71. Friedel, S.; Thielen, A.; Springman, S. Investigation of a slope endangered by rainfall-induced landslides using 3D resistivity tomography and geotechnical testing. *J. Appl. Geophys.* **2006**, *60*, 100–114. [CrossRef]
72. Lebourg, T.; Hernandez, M.; Zerathe, S.; El Bedoui, S.; Jomard, H.; Fresia, B. Landslides triggered factors analysed by time lapse electrical survey and multidimensional statistical approach. *Eng. Geol.* **2010**, *114*, 3–4. [CrossRef]
73. Travelletti, J.; Sailhac, P.; Malet, J.P.; Grandjean, G.; Ponton, J. Hydrological response of weathered clay-shale slopes: Water infiltration monitoring with time-lapse electrical resistivity tomography. *Hydrol. Processes* **2012**, *26*, 2106–2119. [CrossRef]

74. Lee, C.C.; Zeng, L.S.; Hsieh, C.H.; Yu, C.Y.; Hsieh, S.H. Determination of mechanisms and hydrogeological environments of Gangxianlane landslides using geoelectrical and geological data in central Taiwan. *Environ. Earth Sci.* **2012**, *66*, 1641–1651. [CrossRef]
75. Luongo, R.; Perrone, A.; Piscitelli, S.; Lapenna, V. A Prototype System for Time-Lapse Electrical Resistivity Tomographies. *Int. J. Geophysics.* **2012**, *2012*, 176895. [CrossRef]
76. Quagliarini, A.; Segalini, A.; Chelli, A.; Francese, R.; Giorgi, M.; Spaggiari, L. Joint Modelling and Monitoring on Case Pennetta and Case Costa Active Landslides System Using Electrical Resistivity Tomography and Geotechnical Data. In *Advancing Culture of Living with Landslides*; Mikoš, M., Casagli, N., Yin, Y., Sassa, K., Eds.; Springer: Cham, Switzerland, 2017. [CrossRef]
77. Gunn, D.A.; Chambers, J.E.; Hobbs, P.R.N.; Ford, J.R.; Wilkinson, P.B.; Jenkins, G.O.; Merritt, A. Rapid observations to guide the design of systems for long-term monitoring of a complex landslide in the Upper Lias clays of North Yorkshire, UK. *Q. J. Eng. Geol. Hydrogeol.* **2013**, *46*, 323–336. [CrossRef]
78. Gance, J.; Malet, J.P.; Supper, R.; Sailhac, P.; Ottowitz, D.; Jochum, B. Permanent electrical resistivity measurements for monitoring water circulation in clayey landslides. *J. Appl. Geophys.* **2016**, *126*, 98–115. [CrossRef]
79. Xu, D.; Hu, X.Y.; Shan, C.L.; Li, R.H. Landslide monitoring in southwestern China via time-lapse electrical resistivity tomography. *Appl. Geophys.* **2016**, *13*, 1–2. [CrossRef]
80. Palis, E.; Lebourg, T.; Vidal, M.; Levy, C.; Tric, E.; Hernandez, M. Multiyear time-lapse ERT to study short- and long-term landslide hydrological dynamics. *Landslides* **2017**, *14*, 4. [CrossRef]
81. Zieher, T.; Markart, G.; Ottowitz, D.; Romer, A.; Rutzinger, M.; Meissl, G.; Geitner, C. Water content dynamics at plot scale—Comparison of time-lapse electrical resistivity tomography monitoring and pore pressure modelling. *J. Hydrol.* **2017**, *544*, 195–209. [CrossRef]
82. Hojat, A.; Arosio, D.; Ivanov, V.I.; Longoni, L.; Papini, M.; Scaioni, M.; Tresoldi, G.; Zanzi, L. Geoelectrical characterization and monitoring of slopes on a rainfall-triggered landslide simulator. *J. Appl. Geophys.* **2019**, *170*, 103844. [CrossRef]
83. Ivanov, V.; Arosio, D.; Tresoldi, G.; Hojat, A.; Zanzi, L.; Papini, M.; Longoni, L. Investigation on the Role of Water for the Stability of Shallow Landslides-Insights from Experimental Tests. *Water* **2020**, *12*(4), 1203. [CrossRef]
84. Boyd, J.; Chambers, J.; Wilkinson, P.; Peppa, M.; Watlet, A.; Kirkham, M.; Jones, L.; Swift, R.; Meldrum, P.; Uhlemann, S.; et al. A linked geomorphological and geophysical modelling methodology applied to an active landslide. *Landslides* **2021**, *18*, 2689–2704. [CrossRef]
85. Mary, B.; Peruzzo, L.; Iván, V.; Facca, E.; Manoli, G.; Putti, M.; Camporese, M.; Wu, Y.; Cassiani, G. Combining Models of Root-Zone Hydrology and Geoelectrical Measurements: Recent Advances and Future Prospects. *Front. Water* **2021**, *3*, 767910. [CrossRef]
86. Mollehuara Canales, R.; Kozlovskaya, E.; Lunkka, J.P.; Guan, H.; Banks, E.; Moisio, K. Geoelectric interpretation of petrophysical and hydrogeological parameters in reclaimed mine tailings areas. *J. Appl. Geophys.* **2020**, *181*, 104139. [CrossRef]
87. Archie, G. The electrical resistivity log as an aid in determining some reservoir characteristics. *Trans. AIME* **1942**, *146*, 54–62. [CrossRef]
88. Glover, P.W.J. A generalized Archie's law for n phases. *Geophysics* **2010**, *75*, E247–E265. [CrossRef]

applied sciences

MDPI

Article

Improved Workflow for Fault Detection and Extraction Using Seismic Attributes and Orientation Clustering

Minki Kim [1], Jeongmin Yu [2,*], Nyeon-Keon Kang [2] and Byoung-Yeop Kim [2]

[1] Schlumberger Digital & Integration Division, Seoul 03056, Korea; mkim@slb.com
[2] Petroleum & Marine Research Division, Korea Institute of Geoscience and Mineral Resources, Daejeon 34132, Korea; nkkang@kigam.re.kr (N.-K.K.); bykim@kigam.re.kr (B.-Y.K.)
* Correspondence: yjmin@kigam.re.kr

Abstract: Faults represent important analytical targets for the identification of perceptual ground motions and associated seismic hazards. In particular, during oil production, important data such as the path and flow rate of fluid flows can be obtained from information on fault location and their connectivity. Seismic attributes are conventional methods used for fault detection, whereby information obtained from seismic data are analyzed using various property processing methods. The analyzed data eventually provide information on fault properties and imaging of fault surfaces. In this study, we propose an efficient workflow for fault detection and extraction of requisite information to construct a fault surface model using 3D seismic cubes. This workflow not only improves the ability to detect faults but also distinguishes the edges of a fault more clearly, even with the application of fewer attributes compared to conventional workflows. Thus, the computing time of attribute processing is reduced, and fault surface cubes are generated more rapidly. In addition, the reduction in input variables reduces the effect of the interpreter's subjective intervention on the results. Furthermore, the clustering method can be applied to the azimuth and dip of the fault to be extracted from the complexly intertwined fault faces and subsequently imaged. The application of the proposed workflow to field data obtained from the Vincentian oil field in Australia resulted in a significant reduction in noise compared to conventional methods. It also led to clearer and continuous edge detection and extraction.

Keywords: directional fault detection; fault extraction; seismic attributes; ant tracking; edge evidence; orientation clustering

1. Introduction

Fractures are formed when the magnitude of stress exceeds the strength of a rock, thereby causing the rock to diverge in a particular direction. A fracture refers to a fault in which the relative rock displacement is parallel to the fracture direction and/or a joint in which there is negligible relative rock displacement [1]. In particular, faults are important analytical targets for the identification of perceptual ground motion and associated seismic hazards [2–5]. It is also important to understand the sub-surface destructive connectivity in fluid flow path analysis [6–8]. The interpretation of the fault system is performed using structural and seismic attributes (including similarity, semblance, maximum curvature, and amplitude volumes) that can emphasize differences in seismic properties (amplitude, frequency, and phase) [9]. Seismic attributes are one of the most widely used means of detecting faults [10,11]. They provide analytical and quantitative measurements of data using 3D visualization [12] or machine learning techniques [13], thereby improving the analysis of information obtained from seismic data.

However, the process of detecting and characterizing faults from seismic amplitude data is time-consuming [14,15]. In general, several manual tasks are performed by interpreters to draw faults over raw seismic images. In addition, there are problems associated with significant variation in interpretation according to the interpreter's experience [16–20]. Therefore, several technologies and tools have been developed to reduce the time taken for data processing of raw seismic data and minimize the impact of the interpreters [21].

The coherence measurement method used for geometric and hierarchical discontinuous surface analysis was proposed by Bahorich and Farmer [14]. Subsequently, several such studies were performed based on different parameters such as semblance [22], eigenstructure [23,24], gradient structure tensor [25], similarity [10,26], variance [27], automated fault extraction [28], and directional structure–tensor coherence [29]. However, these methods are typically vulnerable to noisy data, which is problematic.

Image processing is the task of improving and extracting useful information from a given source. In seismic studies, amplitude images are primarily used, which are applied to fault detection through various attributes. Edge detection has been put forward to effectively locate discontinuities using differences between seismic traces [30]. Attributes such as ant tracking [31] and Hough transform [32] have been proposed for edge detection and enhancement.

A conventional fault-detection workflow using the aforementioned techniques is shown in Figure 1. A widely used method based on seismic attributes consists of the following parts: conditioning, edge detection, and edge enhancement. However, each part has several types of seismic characteristic parameters, whose selection depends on the researcher's subjective experience. As a result, differences in results are inevitable. Furthermore, obtaining a clear fault extraction output is difficult owing to noisy data. In addition, the extraction of individual faults from the cube requires intense manual work and might lead to human errors.

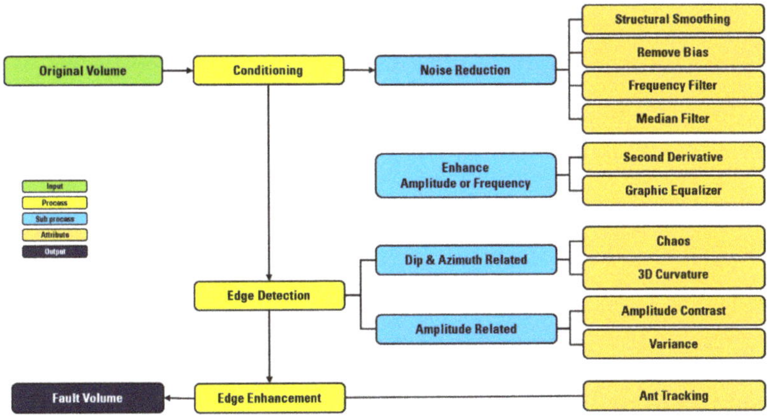

Figure 1. The conventional workflow for fault detection consists of conditioning, edge detection, and edge enhancement. It is set up to perform a significant number of sub-processes within each attribute.

In this study, we propose an effective workflow for fault detection and extraction to construct a fault surface model. This approach improves the workflow, enables faster and clearer fault detection, and leads to enhancement of the edges. In addition, individual fault surfaces can be extracted more easily by applying a clustering method to fault cubes. Furthermore, we demonstrate the applicability of the proposed workflow to field data.

2. Improved Workflow for Fault Detection

The workflow proposed in this paper consists of directional edge detection and edge enhancement. A flowchart of the improved workflow is shown in Figure 2, and the associated steps are summarized as follows:

(1) The amplitude contrast attribute was applied on the basis of the direction in which the edges are to be detected within the cube. This provides information about the major directions and edges of the fault to be investigated.
(2) Summations were performed by selecting cubes with high fault connectivity on the basis of the previously acquired amplitude contrast attribute. This reduces noise and improves connectivity between edges.
(3) The first edge enhancement was conducted using edge evidence. This attribute was applied sequentially to every direction to significantly reduce the noise and make the edges clearer.
(4) The second edge enhancement was performed by ant tracking. A combination of passive and aggressive modes was used to enhance the edges of major or detailed faults.

Upon comparing our method to the conventional workflow shown in Figure 1, it can be seen that the conditioning step is omitted in the studied method. This reduces the time required to process multiple processes, thereby rapidly providing the final results. Furthermore, a reduction in the processing step implies that the use of subjective input variable values is also reduced, which generates more objective results.

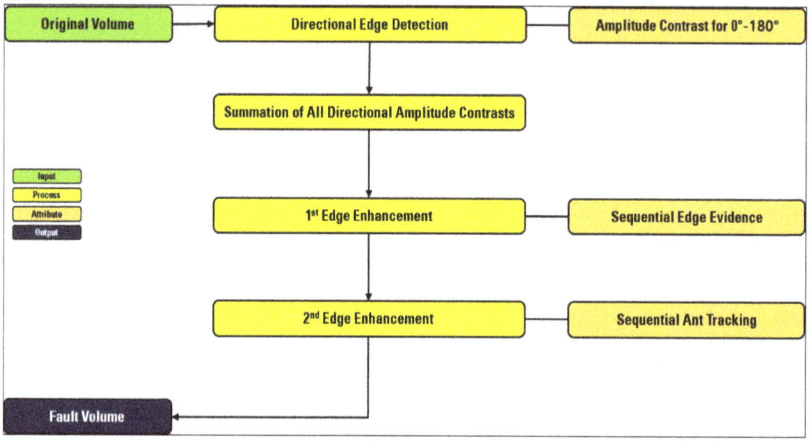

Figure 2. Workflow proposed in this study for fault detection consists of directional edge detection, summation, and edge enhancement. The workflow is significantly simplified, and the computational time is also reduced due to the application of reduced attributes.

3. Seismic Attributes and Application

3.1. Field Data in Vincent

We validated the proposed workflow by applying it to field data obtained from an offshore Vincent oil field adjacent to Western Australia, Australia. The Vincent field is a southern rift basin in the Canavon Basin, located in the Exmouth sub-basin with deposits including Jurassic to early Cretaceous sediments [33,34]. The formation structure of the basin was confirmed in the form of a three-way dip fault closure of Mooderong Shale.

3.2. Directional Edge Detection

In general, faults possess various directions (azimuth). However, a field or basin has few major directions depending on the geo-mechanical settings caused by their tectonic regime. Aqrawi et al. [35] proposed a technique for directional edge detection that takes into consideration of the major direction of the fault using amplitude contrast methods.

Amplitude contrast is a Sobel-based attribute computation of the amplitude derivatives between neighboring seismic traces, where nondiagonal neighbors are assigned double weights. The calculated differences were normalized, and the final value was calculated using three gradient calculations, where Sx, Sy, and Sz are the weighting operators in the corresponding dimensions. Figure 3 shows the amplitude contrast cubes when the azimuth direction increases from 0° to 170° in steps of 10°. It is evinced from the figure that the clarity of discontinuity varies with the azimuth.

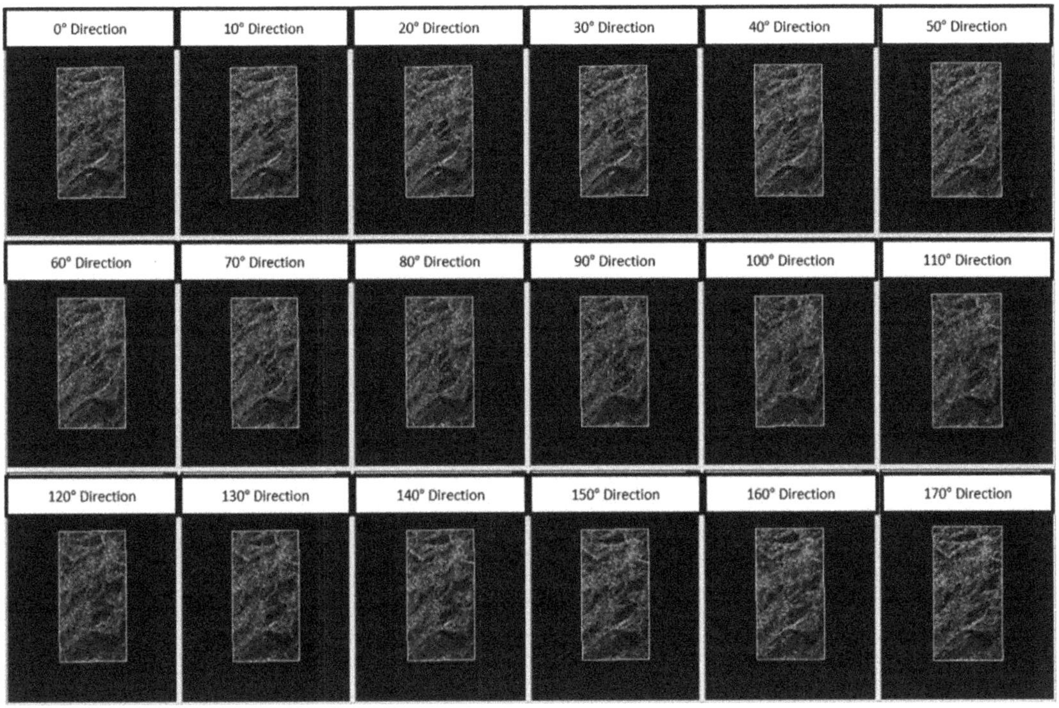

Figure 3. Amplitude contrast for specific azimuth direction. A change in amplitude contrast is shown when the azimuth direction increases from 0° to 170° at 10° intervals.

After calculating all directions, the cubes that display good continuity for faults were selected, and then the selected cubes were summed. Owing to this summation step, we can expect noise reduction of the background and enhancement of the edge (or fault) signal. This step is similar to the stacking process of seismic processing. The results of the directional edge detection and summation using the amplitude contrast are shown in Figure 4. It is evinced that, even with simple amplitude contrast attributes, the discontinuity of the formation is more prominent than that of the original cube section. Furthermore, it was observed that noise reduction after the application of both directional edge detection and summation improves the connectivity between discontinuities.

3.3. First Edge Enhancement

Even though the direction amplitude contrast cube showed good fault continuity, it was not sufficient to extract faults automatically. To enhance the continuity of faults, edge evidence of seismic attributes was applied to the directional amplitude contrast. Edge evidence uses integrals to detect faults in the image, which allows the delineation of the more enhanced fault edges from the results of amplitude contrast. This attribute retrieves lines with significantly different values in all directions, and it provides an estimate of the best line that passes the main point.

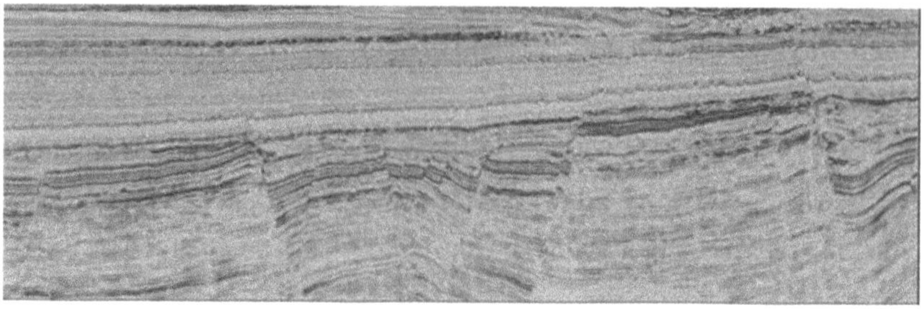

(**a**)

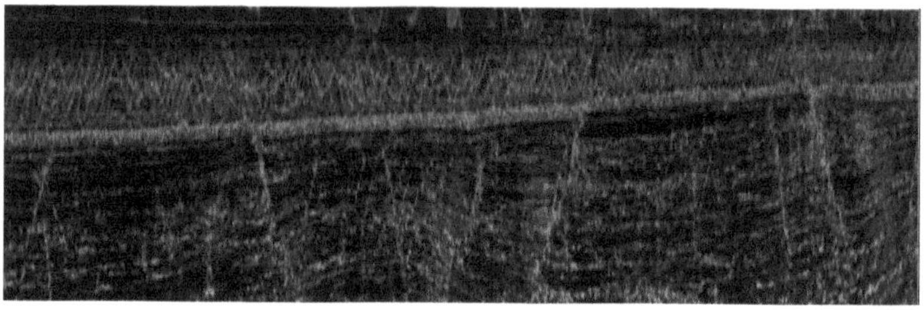

(**b**)

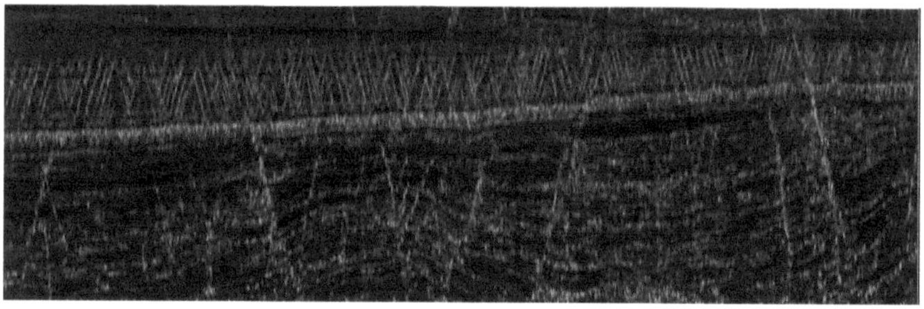

(**c**)

Figure 4. (**a**) Original cube section and the results of edge detection using amplitude contrast attribute with (**b**) nondirectional and (**c**) directional and summation steps.

In this study, the multiple edge evidence attribute was applied. As a result, noise was attenuated, owing to which the location of the discontinuity could be distinguished more clearly. It can be applied in both horizontal and vertical directions or along the direction of line orientation. A combination of these characteristics in sequence is expected to have a higher effect. Figure 5 shows the results of the sequential application of edge evidence considering the horizontal and vertical combinations. The results of horizontal application show that, although edge enhancements were generated, the overall noise remained strong. However, after the subsequent application of edge evidence in the vertical direction, most of the noise disappeared, and the discontinuity became more prominent.

3.4. Second Edge Enhancement

The second method, which follows edge evaluation, was applied to further enhance the discontinuity, and an ant tracking seismic attribute was used. Ant tracking is a patented technology by Schlumberger that performs edge enhancement to identify faults within the seismic data volume [36]. This method is based on the concept of swarm intelligence and is used to introduce a large number of agents (ants) in data volumes and evaluate swarm collective behavior. The analogy of the habit of ants to follow the shortest route for transporting food to their nests is applied to edge enhancement.

(a)

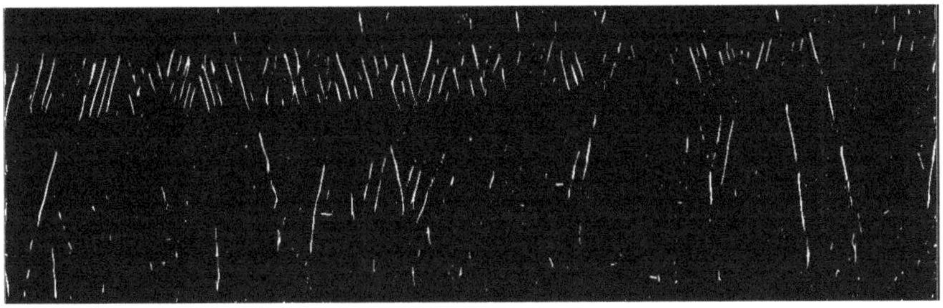

(b)

Figure 5. First edge enhancement with sequential edge evidence. When first applied along the (**a**) horizontal direction, the edges were sharper, and, when it was subsequently applied along the (**b**) vertical direction, the discontinuity was enhanced.

One of the issues in using ant tracking is that there are six parameters. Therefore, 720 variable combinations are possible, which indicates that different results can be obtained depending on the variable, even when applied to the same data. Another issue is that changing one of these values is difficult and requires a long computation time. Sequential ant tracking was used to avoid these issues. In ant tracking, there are two ant modes: passive and aggressive. Passive ant tracking is a common method for signals to move in a stronger direction and is effective in the identification of the overall trend of faults in large areas. The aggressive mode is more effective for detecting small faults in greater detail compared to the previous mode. In this study, these two modes were applied crosswise rather than individually to the cube to solve the abovementioned problems; the results are shown in Figure 6. The results were compared after cross-applying the passive–aggressive–passive and aggressive–passive–aggressive modes using the ant tracking method in the edge evidence cube. The difference between the two methods did not appear to be significant for this field data. However, the edges of the small faults were found to be slightly more enhanced in the aggressive–passive–aggressive sequence than in the passive–aggressive–passive sequence. In a multiple ant tracking experiment, most features of the first mode were found in the final cross-analysis result.

The results of fault detection using the conventional workflow and the proposed workflow are shown in Figure 7. The overall pattern of the fault was similar on both sides; however, the conventional workflow generated significantly poor continuity of the fault. On the other hand, the results using the proposed workflow significantly eliminated noise from the cube, and the fault edges appeared sharper and more continuous. Therefore, we demonstrate that the proposed workflow is applicable to field data, and it has the capacity to further improve fault detection.

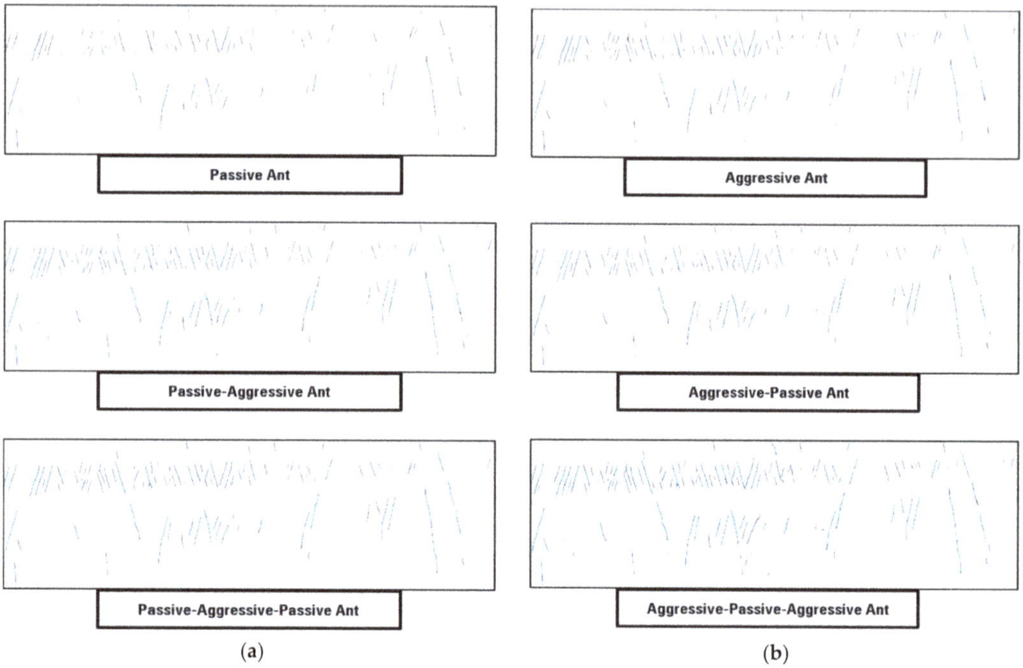

Figure 6. Second edge enhancement with cross-applying using the ant tracking attributes. In the (**a**) aggressive–passive–aggressive sequence, the edges of the small faults are more enhanced, and, in the (**b**) passive–aggressive–passive sequence, the edges of the major faults are enhanced.

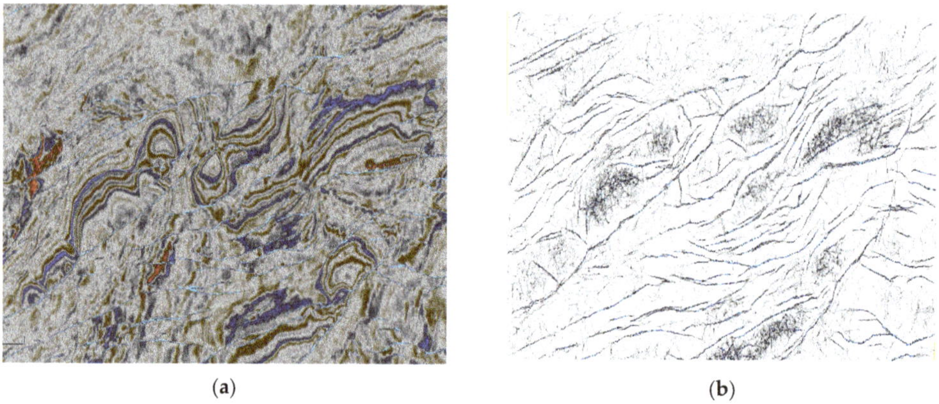

Figure 7. Fault detection using the (**a**) proposed and (**b**) conventional workflow. The proposed workflow generates clearer and simpler fault edges than conventional results.

4. Clustering and Fault Surface Extraction

The proposed workflow can be used to effectively detect faults, and it requires the processes of extracting and imaging fault surfaces from 3D cubes. The process of manual extraction of fault surfaces is influenced by the interpreter and requires an enormous amount of time. On the other hand, the automatic error extraction methods are quite complex and require several follow-up tasks with considerable computation time. In this workflow, we used orientation clustering to extract fault surfaces more easily.

To perform clustering, the fault surfaces were first converted to points, as shown in Figure 8a. At this point, it is important to transform the information of the direction of the surface using a sufficient number of points for effective description. In addition, information of the azimuth and dip at each point position was extracted at the time of conversion, as shown in Figure 8b,c, respectively.

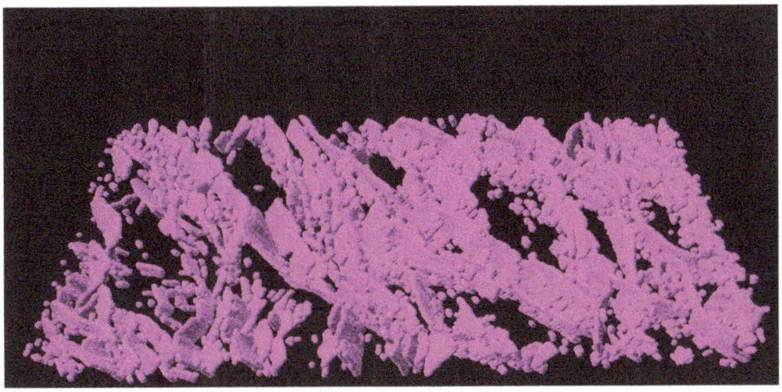

(**a**)

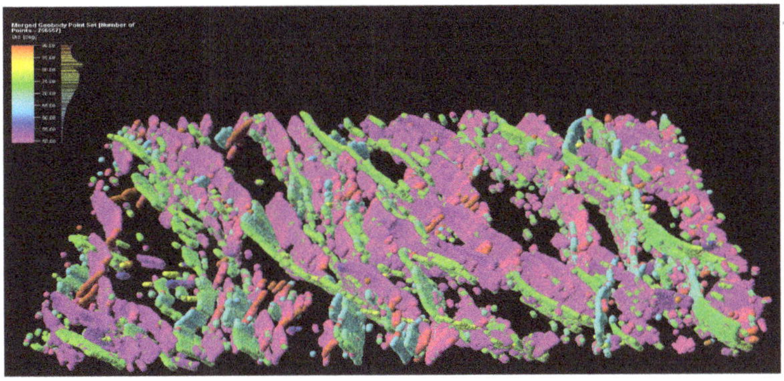

(**b**)

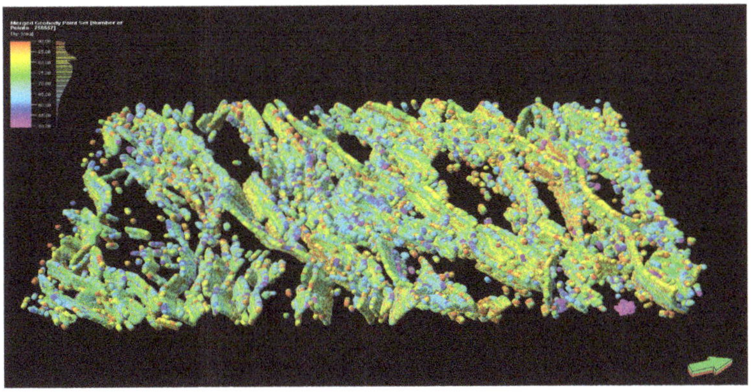

(**c**)

Figure 8. Extracted fault surfaces as (**a**) point type. To perform orientation clustering, (**b**) azimuth and (**c**) dip information for each point position was also extracted.

Clustering is an effective method to distinguish and extract similar groups by considering the characteristics of the data. The azimuth and dip information stored at each point was clustered using the density-based spatial clustering of applications with noise (DBSCAN) method [37,38]; subsequently, the fault was extracted separately. The DBSCAN algorithm expands clusters by finding data that are close to the initial data. The input parameters were the number of neighbors required to define the dense region (MinPts) and the distance representing the neighborhood (ε). Using these parameters, the data were classified into three types. If the number of points within a neighborhood with distance ε exceeded the minimum, then such a point was designated as a core point. After a core point was located, data within its neighborhood were defined as connected points or as other core points. If the data point was not connected to a neighbor, it was designated as a border point. Lastly, data in clusters that were neither core nor border points were designated as noise. Therefore, DBSCAN performs density-based clustering, and, unlike K-means, the number of final clusters is not required as an input parameter, which is an advantage. Figure 9 shows clustered fault point sets. Since DBSCAN was used, complexly intertwined faults could be identified individually according to the orientation, and the density-based attributes allowed the extraction of the connectivity of the fault surface. Furthermore, parts that lacked the number of points to be defined as a fault were considered outliers and removed.

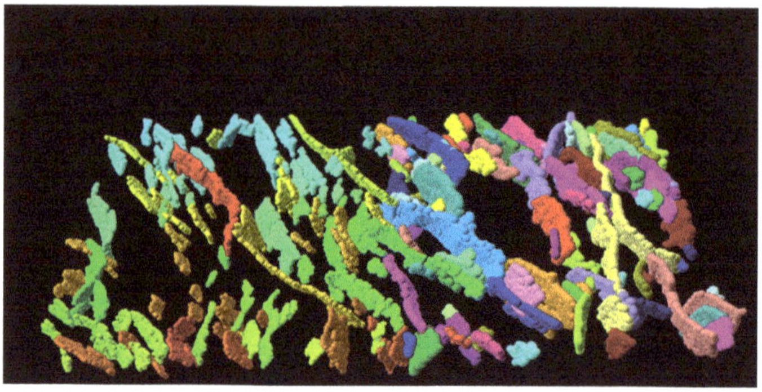

Figure 9. Fault point groups extracted individually by DBSCAN clustering from original point group data shown in Figure 8a.

Figure 10 shows the results of extracting the fault model from the original fault cube using the conventional automatic method and the proposed clustering method. The automated method showed significantly lower connectivity of the fault, and it seemed difficult to implement an accurate model because individual separation of the fault was not well generated at the intersection site of several intersections. On the other hand, using the proposed method, the generated model extracted individual fault surfaces with high connectivity. In addition, unnecessary outliers could be eliminated to obtain simpler and more visible images.

Comparing the time spent on the extracting method, it took at least two days to combine the calculation and editing processes when modeling faults using the automatic extraction method. If done manually, this would take several weeks or more. However, by applying the clustering method, the fault extraction could be completed within 4–5 h. Therefore, we demonstrate that the proposed method can obtain an effective fault model in a significantly shorter time than the conventional methods.

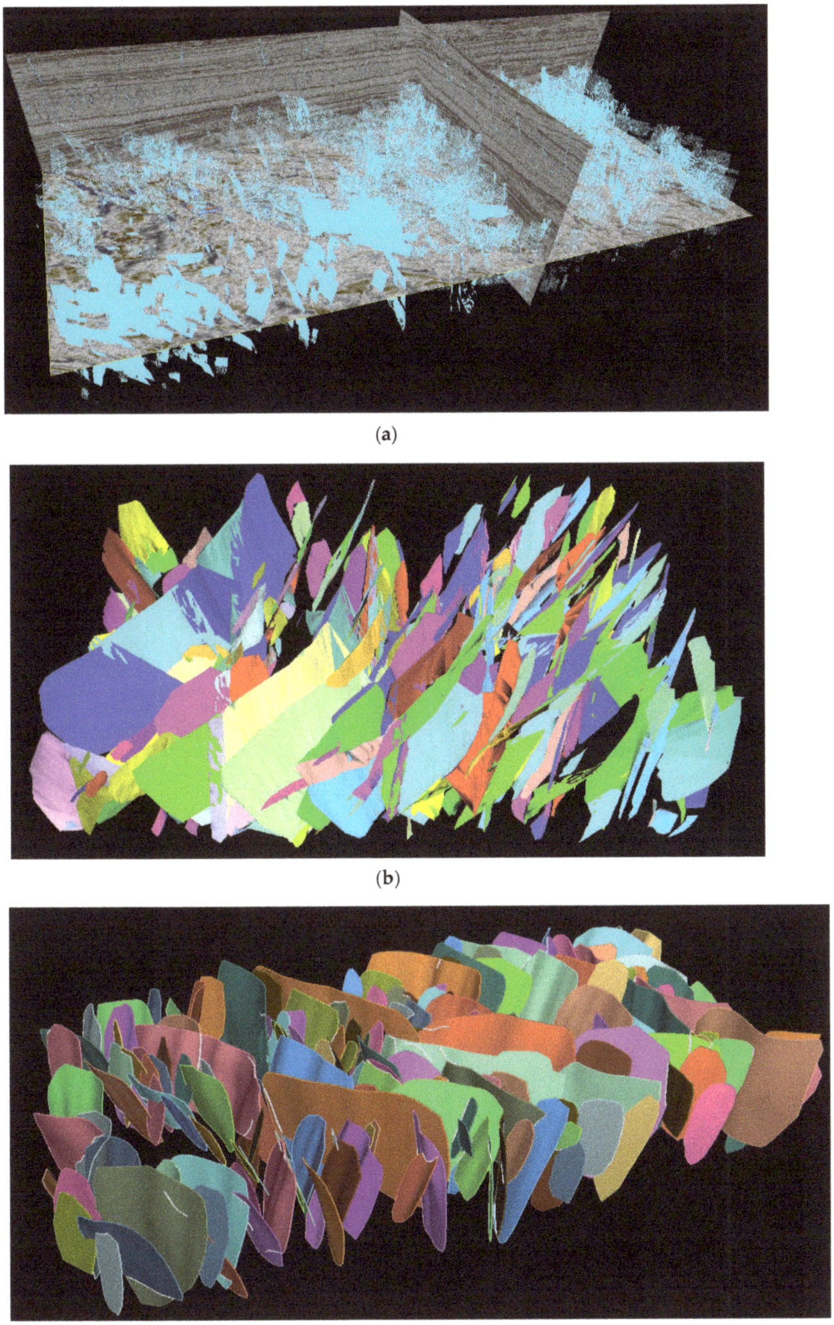

(a)

(b)

(c)

Figure 10. Results of fault modelling using (**a**) original fault cube, (**b**) conventional (automatic) approach, and (**c**) clustering (the proposed method). Automatic fault extraction methods show low connectivity between fault models; however, the proposed method shows high connectivity with the possibility of individual fault extraction through clustering.

5. Conclusions

We proposed an improved fault detection workflow consisting of directional edge detection, summation, and edge enhancement. Compared to the conventional fault detection workflow using seismic attributes, the proposed method is less cumbersome for controlling the parameters and does not require the selection of seismic attributes based on individual case. It mainly controls the results on the basis of a sequence of seismic attributes and displays consistent results regardless of the cases. In addition, the attributes used are smaller than those of the conventional method, and the proposed workflow requires lesser process time. Overall, we found that clustering effectively enables the extraction of individual fault surfaces.

In this method, the fault surface extracted from the edge detection cube was converted to a sufficient number of points to describe the orientation of the fault layer. Subsequently, clustering was performed on the azimuth and dip using DBSCAN, and individual fault surfaces with high continuity could be extracted from complex entangled fault structures.

The proposed workflow was applied to field data obtained from the Vincent oil field in Australia. The results showed that sharper and continuous edges could be detected with significant noise reduction compared to the conventional methods. Thus, we demonstrate that the proposed model is applicable to field data and is more effective for fault detection. The proposed workflow reduces the overall cycle of fault analysis and human-induced errors. This also reduces the effort and time spent during the manual interpretation of faults and allows the interpreter to further invest time in quality control of the extracted faults.

Author Contributions: Conceptualization, M.K. and N.-K.K.; methodology, J.Y. and N.-K.K.; investigation, B.-Y.K.; data curation, J.Y. and M.K.; writing—original draft preparation, J.Y. and B.-Y.K. All authors read and agreed to the published version of the manuscript.

Funding: This research was funded by the Korea Institute of Geoscience and Mineral Resources, grant number GP2020–023.

Institutional Review Board Statement: Not applicable.

Informed Consent Statement: Not applicable.

Data Availability Statement: Not applicable.

Acknowledgments: We wish to thank Joongmoo Byun and Rise Lab of Hanyang University for their support.

Conflicts of Interest: The authors declare no conflict of interest.

References

1. Mitcham, T.W. Fractures, joints, faults, and fissures. *Econ. Geol.* **1963**, *58*, 1157–1158. [CrossRef]
2. Bouchon, M.; Aki, K. Strain, tilt, and rotation associated with strong ground motion in the vicinity of earthquake faults. *Bull. Seismol. Soc. Am.* **1982**, *72*, 1717–1738. [CrossRef]
3. Passone, L.; Mai, P.M. Kinematic earthquake ground-motion simulations on listric normal FaultsKinematic earthquake ground-motion simulations on listric normal faults. *Bull. Seismol. Soc. Am.* **2017**, *107*, 2980–2993. [CrossRef]
4. Roberts, G.P.; Cowie, P.; Papanikolaou, I.; Michetti, A.M. Fault scaling relationships, deformation rates and seismic hazards: An example from the Lazio–Abruzzo Apennines, central Italy. *J. Struct. Geol.* **2004**, *26*, 377–398. [CrossRef]
5. Watkinson, I.M.; Hall, R. Fault systems of the eastern Indonesian triple junction: Evaluation of Quaternary activity and implications for seismic hazards. *Geol. Soc.* **2017**, *441*, 71–120. [CrossRef]
6. Raven, K.G.; Gale, J.E. Water flow in a natural rock fracture as a function of stress and sample size. *Int. J. Rock Mech. Min. Sci. Geomech. Abstr.* **1985**, *22*, 251–261. [CrossRef]
7. Kwicklis, E.M.; Healy, R.W. Numerical investigation of steady liquid water flow in a variably saturated fracture network. *Water Resour. Res.* **1993**, *29*, 4091–4102. [CrossRef]
8. Mayer, J.R.; Sharp, J.M. Fracture control of regional ground-water flow in a carbonate aquifer in a semi-arid region. *Geol. Soc. Am. Bull.* **1998**, *110*, 269–283. [CrossRef]
9. Carrasco, J.; Garcia, R. 3D seismic interpretation techniques for detecting oilfield exploration locations. In *SPE Latin America and Caribbean Petroleum Engineering Conference*; Society of Petroleum Engineers: Maracaibo, Venezuela, 2014; SPE-169452-MS.

10. Chopra, S.; Marfurt, K.J. *Seismic Attributes for Prospect Identification and Reservoir Characterization;* Society of Exploration Geophysicists: Tulsa, OK, USA, 2007.
11. Neves, F.A.; Zahrani, M.S.; Bremkamp, S.W. Detection of potential fractures and small faults using seismic attributes. *Lead. Edge* **2004**, *23*, 903–906. [CrossRef]
12. Chopra, S.; Marfurt, K.J. Emerging and future trends in seismic attributes. *Lead. Edge* **2008**, *27*, 298–318. [CrossRef]
13. Huang, L.; Dong, X.; Clee, T.E. A scalable deep learning platform for identifying geologic features from seismic attributes. *Lead. Edge* **2017**, *36*, 249–256. [CrossRef]
14. Bahorich, M.; Farmer, S. 3-D seismic discontinuity for faults and stratigraphic features: The coherence cube. *Lead. Edge* **1995**, *14*, 1053–1058. [CrossRef]
15. Qi, J.; Zhang, B.; Zhou, H.; Marfurt, K. Attribute expression of fault-controlled karst—Fort Worth Basin, Texas: A tutorial. *Interpretation* **2014**, *2*, SF91–SF110. [CrossRef]
16. Tversky, A.; Kahneman, D. Judgment under uncertainty: Heuristics and biases. *Science* **1974**, *185*, 1124–1131. [CrossRef] [PubMed]
17. Schaaf, A.; Bond, C.E. Quantification of uncertainty in 3-D seismic interpretation: Implications for deterministic and stochastic geomodeling and machine learning. *Solid Earth* **2019**, *10*, 1049–1061. [CrossRef]
18. Tannert, C.; Elvers, H.-D.; Jandrig, B. The ethics of uncertainty: In the light of possible dangers, research becomes a moral duty. *EMBO Rep.* **2007**, *8*, 892–896. [CrossRef] [PubMed]
19. Bond, C.E. Uncertainty in structural interpretation: Lessons to be learnt. *J. Struct. Geol.* **2015**, *74*, 185–200. [CrossRef]
20. Alcalde, J.; Bond, C.E.; Johnson, G.; Ellis, J.F.; Butler, R.W. Impact of seismic image quality on fault interpretation uncertainty. *GSA Today* **2017**, *27*, 4–10. [CrossRef]
21. Chahine, C.; Truelove, L.; Volcan, M.H. Enhanced delineation of reservoir compartmentalization from advanced preand post-stack seismic attribute analysis. *First Break* **2014**, *32*, 75–84. [CrossRef]
22. Marfurt, K.J.; Kirlin, R.L.; Farmer, S.L.; Bahorich, M.S. 3-D seismic attributes using a semblance-based coherency algorithm. *Geophysics* **1998**, *63*, 1150–1165. [CrossRef]
23. Kirlin, R.L. The relationship between semblance and eigenstructure velocity estimators. *Geophysics* **1992**, *57*, 1027–1033. [CrossRef]
24. Gersztenkorn, A.; Sharp, J.; Marfurt, K. Delineation of tectonic features offshore Trinidad using 3-D seismic coherence. *Lead. Edge* **1999**, *18*, 1000–1008. [CrossRef]
25. Bakker, P. Image structure Analysis for Seismic Interpretation. Ph.D. Thesis, Delft TU: Delft University of Technology, Delft, The Netherlands, 2012.
26. Tingdahl, K.M.; De Groot, P.; Heggland, R.; Ligtenberg, H. Semi-automated object detection in 3D seismic data. In Proceedings of the 2001 SEG Annual Meeting, San Antonio, TX, USA, 9–14 September 2001.
27. Van Bemmel, P.; Pepper, R. Seismic Signal Processing Method and Apparatus for Generating a Cube of Variance Values. U.S. Patent US6151555 A, 21 November 2000.
28. Dorn, G.A.; Kadlec, B.; Murtha, P. Imaging faults in 3D seismic volumes. In *SEG Technical Program Expanded Abstracts 2012;* Society of Exploration Geophysicists: Tulsa, OK, USA, 2012; pp. 1–5.
29. Wu, X. Directional structure-tensor-based coherence to detect seismic faults and channels. *Geophysics* **2017**, *82*, A-13. [CrossRef]
30. Luo, Y.; Higgs, W.G.; Kowalik, W.S. Edge detection and stratigraphic analysis using 3D seismic data. In *SEG Technical Program Expanded Abstracts 1996;* Society of Exploration Geophysicists: Tulsa, OK, USA, 1996; pp. 324–327.
31. Silva, C.C.; Marcolino, C.S.; Lima, F.D. Automatic fault extraction using ant tracking algorithm in the Marlim South Field, Campos Basin. In *SEG Technical Program Expanded Abstracts 2005;* Society of Exploration Geophysicists: Tulsa, OK, USA, 2005; pp. 857–860.
32. Wang, Z.; AlRegib, G. Automatic fault surface detection by using 3D Hough transform. In *SEG Technical Program Expanded Abstracts 2014;* Society of Exploration Geophysicists: Tulsa, OK, USA, 2014; pp. 1439–1444.
33. Loro, R.; Hill, R.; Jackson, M.; Slate, T. Technologies that have transformed the Exmouth into Australia's premier oil producing basin. *APPEA J.* **2015**, *55*, 233–246. [CrossRef]
34. Choi, J.; Kim, S.; Kim, B.; Byun, J. Probabilistic reservoir characterisation using 3D pdf of stochastic forward modelling results in Vincent oil field. *Explor. Geophys.* **2020**, *51*, 341–354. [CrossRef]
35. Aqrawi, A.; Weinzierl, W.A.; Daber, R.; Boe, T.H. Directional guided seismic attributes and their use in assisting structural, stratigraphic and lithological interpretation. In *SEG Technical Program Expanded Abstracts 2012;* Society of Exploration Geophysicists: Tulsa, OK, USA, 2012; pp. 1–5.
36. Pedersen, S.I.; Randen, T.; Sonneland, L.; Steen, Ø. Automatic fault extraction using artificial ants. In *SEG Technical Program Expanded Abstracts 2002;* Society of Exploration Geophysicists: Tulsa, OK, USA, 2002; pp. 512–515.
37. Ester, M.; Kriegel, H.P.; Sander, J.; Xu, X. A density-based algorithm for discovering clusters in large spatial databases with noise. *Knowl. Discov. Data Min.* **1996**, *96*, 226–231.
38. Yu, J.; Byun, J.; Seol, S.J. Imaging discrete fracture networks using the location and moment tensors of microseismic events. *Explor. Geophys.* **2021**, *52*, 42–53. [CrossRef]

applied sciences

MDPI

Article

Variations in Wedge Earthquake Distribution along the Strike Underlain by Thermally Controlled Hydrated Megathrusts

Rui Qu [1,2], Yingfeng Ji [1,2,*] and Weiling Zhu [1,3,*]

[1] State Key Laboratory of Tibetan Plateau Earth System, Resources and Environment (TPESRE), Institute of Tibetan Plateau Research, Chinese Academy of Sciences, Beijing 100101, China; qurui@itpcas.ac.cn
[2] College of Earth and Planetary Sciences, University of Chinese Academy of Sciences, Beijing 100049, China
[3] Department of Geophysics, Yunnan University, Kunming 650091, China
* Correspondence: yingfengji@itpcas.ac.cn (Y.J.); zhuweiling@itpcas.ac.cn (W.Z.)

Featured Application: Resource and environmental effects.

Abstract: Accretionary wedge earthquakes usually occur in the overriding crust close to the trench or above the cold nose of the mantle wedge. However, the mechanism and temperature properties related to the slab dip angle remain poorly understood. Based on 3D thermal models to estimate the subduction wedge plate temperature and structure, we investigate the distribution of wedge earthquakes in Alaska, which has a varying slab dip angle along the trench. The horizontal distance of wedge-earthquake hypocenters significantly increases from the Aleutian Islands to south–central Alaska due to a transition from steep subduction to flat subduction. Slab dehydration inside the subducted Pacific plate indicates a simultaneous change in the distances between the intraslab metamorphic fronts and the Alaskan Trench at various depths, which is associated with the flattening of the Pacific plate eastward along the strike. The across-arc width of the wedge-earthquake source zone is consistent with the across-arc width of the surface high topography above the fully dehydrated megathrust, and the fluid upwelling spontaneously influences wedge seismotectonics and orogenesis.

Keywords: wedge earthquakes; thermal regime; subduction process

1. Introduction

Subduction zone earthquakes are often detected in continental accretionary wedges close to trenches or in and around cold mantle wedges, while the distribution properties of continental/mantle wedge earthquakes (hereafter referred to as wedge earthquakes) related to slab dip variations remain poorly understood. The infiltration of fluids or melts derived from the subducted slab facilitates the generation of arc magma and earthquakes, including the Mélange diapirs that are physically mixed by sediments, fluids, and the mantle above the subducted slab [1]. Seismic scattering obstacles in the shallow mantle wedge of the subduction zone are reported to be delegated to the mélange diapirs or ascending melts or slab material directly associated with magma generation [1]. Plume generation is driven by the subduction of buoyant crustal rocks and the expulsion of aqueous slab fluids that cause the hydration and partial melting of the mantle wedge [2].

Instrumentally recorded arrival times to estimate an epicentral area centered at (−151.0°, 57.4°) offshore of Kodiak Island in Alaska suggest that active faults within the accretionary wedge may host various earthquakes [3]. The incoming deforming Pacific plate contacts an actively faulted overlying accretionary wedge, with megathrusts being capable of Mw > 9 earthquakes [3]. Plume development is enhanced by the hydration of the mantle wedge by volatiles released from the subducted slab, with the upward propagating hydration front that forms within the mantle wedge being a consequence of slab dehydration [4]. Tomographic results also show that low-velocity zones exist within the mantle wedge, with dimensional dynamic processes occurring in the mantle wedge, such as convection-controlled heterogeneous thermal structures and the nonuniform generation and migration of magmas [5].

According to observations, the distances of wedge-earthquake hypocenters from the trench are significantly different between the Aleutian Islands and south–central Alaska [6]. The Aleutian Islands are characterized by steep subduction, while south–central Alaska features quasi-flat subduction due to the presence of the Yakutat Block [7]. Slab dehydration occurring inside the incoming plate usually exhibits a variation in intraslab metamorphic front depths and loci separated from the convergent trench, as other subduction zones show and are commonly acknowledged [8]. Associated with the late flattening of the Pacific from west to east, the change in the slab dip angle and clustered zones of wedge earthquakes is noticeable, while the details are still elusive. It seems that the underlying thermally controlled hydrated Pacific plate plays a key role in determining the distribution of wedge earthquakes. However, the hydrothermal state of the megathrusts underlying the accretionary/mantle wedge itself is enigmatic. This becomes a primary issue to clarify.

Among various subduction zones, Alaska is typical because of its notably increasing dip angles along the trench, which present a good chance to study the variation in the subducted plate thermal state and distribution of wedge earthquakes along the trench. The contrasting topographies of the Pacific plate beneath south–central and southwest Alaska depend on the effects of variable factors, including potential fluid upwelling [6]. Therefore, in this paper, we use a 3D modeling method to investigate the distributed wedge-earthquake distribution underlain by particular hydrothermal state variables along the strike beneath trenchward Alaska.

2. Materials and Methods

In this study, our models were originally developed from stag3d [9]. An anelastic liquid approximation and the equations of conservation of mass, momentum, and energy [10] are used based on finite difference method. Based on the plate age and motion field mimicked for subducted curved plates [10,11], we construct three-dimensional (3-D) thermomechanical models to estimate the thermal state-based distribution of wedge earthquakes in Alaska. Subducted oceanic plate geometry was incorporated into the model using Slab2 [12]. The intraslab subduction velocity field used MORVEL [13,14] and interpolated the entire domain for the Pacific slab. The ocean seafloor ages of EarthByte [15] are used as an input of trenchward boundary conditions based on the plate cooling model [16,17] (Figure 1). The thickness of the marine lithosphere was specified according to oceanic lithosphere age [18]. We prescribe the subduction time to be 20 Myr to ensure that the oceanic lithosphere thrusts to reach a steady thermal state. The modeling results are constrained by the observations of surface heat flow from the global heat flow database [19] and the estimate from Curry depth [20] along the parallel profiles across the arc. The thermal model is prescribed with a size of 1600 $\times$ 800 $\times$ 400 km (width $\times$ length $\times$ depth); the model grids are 80 $\times$ 80 $\times$ 100. The model includes the updated 3D geometry [12] and subduction velocity of the Pacific plate underlying the North American plate. The mechanical boundary conditions are as follows: the top boundary is rigid (dirichlet) and the bottom boundary is permeable (neumann) in the vertical direction; lateral boundaries are also permeable (neumann) except for the lateral boundaries, which have prescribed a time-dependent subduction velocity. Viscous decoupling is not included on the megathrust since the frictional heating is not robustly identified from the observation of surface heat flow. The evolving thermal structure of the subducting lithosphere is obtained from the calculated result of thermal modeling, which depends on the properties of the slab geometry, the age of the subduction plate, and the history of the subduction zone.

Based on Omori et al. [21] and Hacker et al. [22], we use a P-T-wt %-facies database to estimate the fluid upwelling sources on Alaskan megathrusts. The pressure (GPa) at every grid point is obtained by converting its depths (km) through preliminary reference earth model (PREM) parameters. Using the temperature and pressure provided by the numerical simulation, we estimate each rock facies (the mid-ocean ridge basalt and ultramafic rock) domain and the corresponding water content (wt %) at every grid. Through the interpolation method, we obtain the intraslab water content distribution (wt %) at various depths. The water content value difference between two neighboring points divided by interval distance reflecting slab dehydration (wt %/km) is obtained at each point.

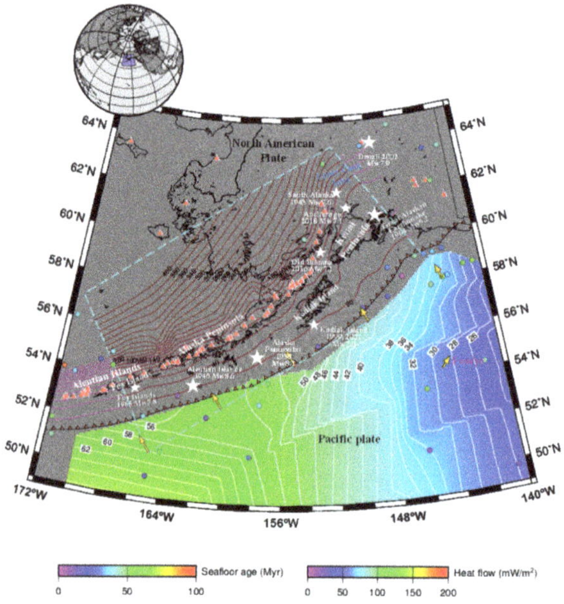

Figure 1. Study region in Alaska. Seafloor age and surface heat flow observations in Alaska are included. The colored solid circles represent the observed surface heat flow obtained from the global heat flow database [19]. The ocean seafloor age and its distribution are provided by EarthByte [15]. The light blue dotted line indicates the model boundary. The red (inside the model region) and purple (outside the model region) curves represent iso-depth contours of the upper surface of the subducted oceanic plate, with two adjacent lines separated by 20 km (Slab2) [12].

In this study, we focus on specific observations of wedge earthquakes and their underlying thermal and hydrous states in the Alaskan tectonic complex. To investigate the way the subduction regime controls or affects the distribution of continental wedge earthquakes, we focus on how the dip and slab geometry influence wedge quake distribution. The wedge earthquakes detected from 1 January 2000 to 31 December 2010 are employed in this study. The historical M > 7 earthquakes are also included.

3. Results

3.1. Constrained Thermal Regime Beneath the Alaskan Continental Wedge

Two-dimensional thermal models have been constructed with nonlinear mantle rheology to study 17 global subduction zones, including the Alaskan subduction zone [23]. The models incorporate slab mantle (or interplate) decoupling, which has been suggested to apply to NE Japan [24,25] based on the hypothesis that the observed surface heat flow above the cold forearc mantle is low in contrast to the hot arc and backarc mantle [23,

26]. However, this slab-mantle (or interplate) decoupling hypothesis relies on limited observations in NE Japan and Cascadia. The new development of surface heat flow measurements has included bottom-simulating reflectors [27,28], terrestrial boreholes and marine heat probes [29,30], and Hi-net boreholes [31], which are not used in Furukawa [24] but indicate the presence of forearc medium-high heat flow values (50–140 or 150 mW/m^2) located in NE Japan and Cascadia (Figure 2), which implies a much warmer forearc mantle than previously thought. Therefore, plate decoupling in thermal modeling, which significantly affects the subduction thermal regime, is still controversial. The thermal results are shown in Figure 3c,d. We found that the thermal models without plate decoupling result in a subduction regime via calculation, which better fits the observation of surface heat flow and seismicity distribution. Therefore, we suggest constraining the modeled thermal regime primarily by the integrated and updated observations for surface heat flow, which include the observations from bottom-simulating reflectors [27,28], terrestrial boreholes and marine heat probes [29,30], Hi-net boreholes [31], and Currie depths [20]. The Currie depth estimation [20] data cover global subduction zones and are a good reference for calculating the thermal regime on the shallow megathrusts.

(a) NE Japan

(b) Cascadia

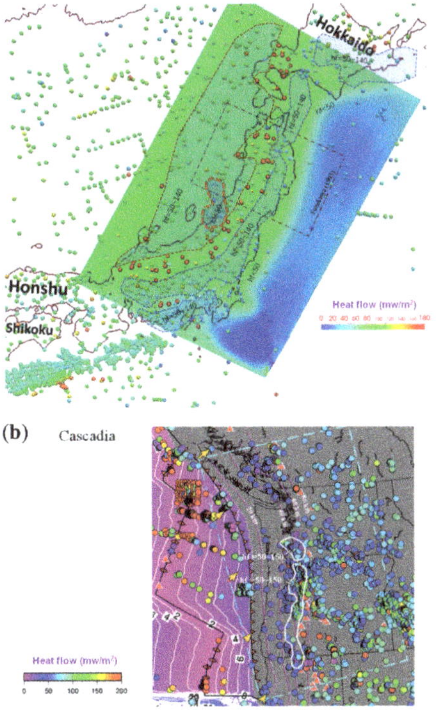

Figure 2. Integrated heat flow observations were based on the Global Heat Flow Database [19] and further combined with bottom-simulating reflectors [27,28], terrestrial boreholes, marine heat probes [29,30], and Hi-net boreholes [31]. The integrated observation of surface heat flow in NE Japan shows gradually changing values from the cold trench to the hot arc and then to the medium backarc. However, we found a forearc medium heat flow region (50–140/150 mW/m^2) in NE Japan and Cascadia, indicating that plate coupling occurs at a forearc depth of 40–80 km. The plate decoupling hypothesis is not robustly constrained by heat flow observations. (**a**) Northeast Japan. (**b**) Cascadia.

Compared with the result of temperature variation along the profile crossing the Augustine volcano (Figure 3) [23] in Alaska, we found that the slab geometry updated in Alaska by Slab2.0 [12] provides a high-resolution slab topography involving a slab depth of 100—115 km beneath the Augustine volcano, while the slab geometry [32] obtained by using seismic reflection and refraction data and earthquake distribution suggests a slab depth beneath Augustine volcano of approximately 85 km, as adopted by Wada and Wang [23] (Figure 3). This geometric difference greatly affects the calculated subduction thermal regime. In addition, the interplate temperature predicted by our models gradually rises from 300 to 800 °C with increasing depth, while the plate decoupling mode artificially creates an abrupt slab surface temperature increase from 400–600 °C to 800–1100 °C beneath the hot arc [23]. The abrupt increase in slab surface temperature makes it difficult to interpret the observation of forearc medium-high surface heat flow and the occurrence of subarc interplate earthquakes.

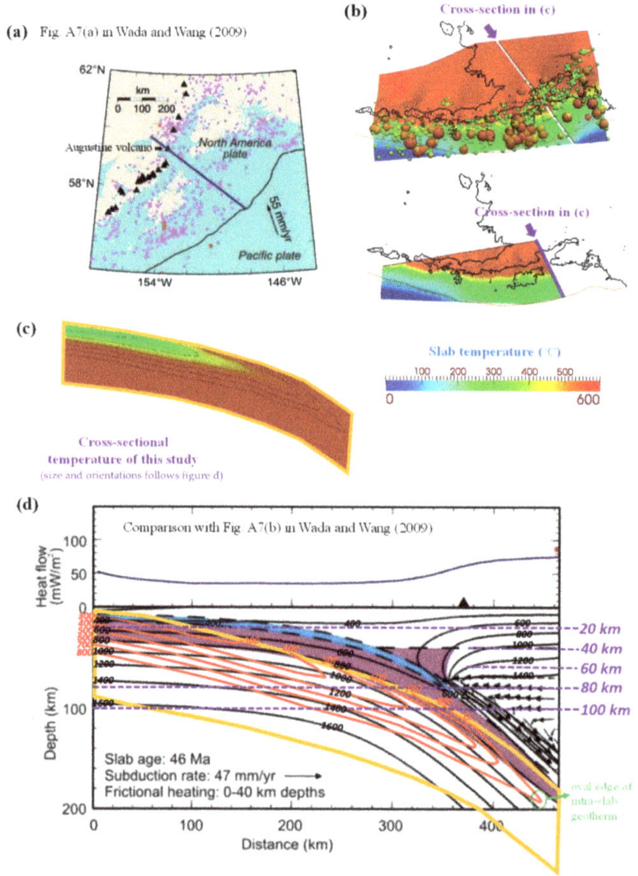

Figure 3. Comparison between our model result and Wada and Wang [23]. Black curves and symbols are from Wada and Wang [23], while yellow curves (slab surface) and red curves (intraslab thermal contours) are from this study. (**a**) The profile used to estimate the cross-section thermal structure in Alaska by 2-D model [23]. (**b**) Obtained the same profile from our 3-D model. (**c**) Cross-section thermal structure along the same profile in this study. (**d**) Slab geometry difference and modeled thermal result difference between Wada and Wang [23] and this study.

Furthermore, the intraslab temperature contours are quite different between the plate decoupling and plate-coupling models. The oval edge of the intraslab geotherm (Figure 3d) is usually observed in warm or plate coupling subduction zones, whereas the parallel layered intraslab thermal structure is often exhibited in cold or plate decoupling subduction zones (e.g., in [23]), which implies a very weak forearc mantle thermal effect on the slab due to the presence of a prescribed slab–mantle decoupling layer (Figure 3). The parallel layered intraslab thermal structure leads to slab dehydration over a wide range of depths (e.g., [23,33,34]), which in turn needs to focus the slab-produced fluid flow toward a subarc mantle at a depth of 100 km [35]. For example, compaction pressure gradients can be introduced to enhance the upslope flow within high-permeability layers in the slab [35]. However, the oval edge of the intraslab geotherm naturally leads to focused slab dehydration and fluid production at a narrower range of depths without having to modify the subarc upslab fluid migration pathways (e.g., Figure 3); i.e., the thermal front could focus the slab-derived fluids at a narrow range of depths without having to modify the subarc upslab fluid migration pathways.

In addition, the global range of subduction zone thermal structures from exhumed blueschists and eclogites is identified to be commonly hotter than the models in Syracuse et al. [36] and updated van Keken (pers. comm. by Penniston-Dorland, 2014 [37]) by more than 200 °C when Pmax < 2 GPa [37]. The reason is that the control of the slab-mantle decoupling setting in the aforementioned modeling is unrealistic, which decreases the forearc slab surface temperature by 100–300 °C. In contrast, the thermal models without plate decoupling predict the slab temperatures and better fit the observed temperature of exhumed blueschist and eclogite rocks recorded.

On the other hand, 3D models are further jointly constrained by a variety of morphologically controlling factors, such as along-strike plate motion and toroidal mantle flow variation, which are not included in 2D thermal models. The observed toroidal mantle flow in the case of oblique subduction changes the slab surface temperature at a scale of >0–20 °C, as we find in the comparison between the results of the 2D and 3D models.

Consequently, the 3D thermal regime constrained by the various observations of surface heat flow, including Currie depth estimations and the prescribed plate coupling setting, is better at predicting the thermal state of the underlain subducted megathrust beneath the wedge.

3.2. Variations in Wedge-Earthquake Distribution along the Strike

Based on the 3D thermal regime and subsequently calculated water content of the subducted Pacific plate beneath Alaska (Figure 4), we investigate how the slab dip change influences wedge quake distribution. The facies of MORB and ultramafic rocks represented by colors are shown in the figure. A comparison between the two wedge seismic events reveals a characteristic distribution: the wedge earthquakes are located primarily coinciding with the Pacific plate interface depth of <60–100 km (Figure 4). In the western part of the model, the Alaska Peninsula, the furthest wedge-earthquake hypocenter from the continental margin, is approximately atop the 60 km-deep Pacific plate surface, while that in the eastern part, southern Alaska, is above the 100 km-deep plate surface, although the clustered far inland crustal earthquakes there are partly (<60%) due to the aftershocks of some large earthquakes. Therefore, the difference between the western and eastern parts is attributable to the subduction angle decreasing eastward along the strike, and the distance of the slab dehydration front from the trench in the dipping direction increases gradually eastward. By modeling the mantle wedge in Alaska, the thermal structure calculated by other studies indicates that the mantle wedge near the subduction plate is widely hydrated (e.g., [38]), which is consistent with our estimation of the hydrated megathrust in Figure 4. Therefore, the dehydration of the subducted plate from the water-saturated updip portions on megathrust (blue zones in Figure 4 to lower water content downdip interface portions (yellow to red zones)) will certainly release water-rich fluids, which may in turn hydrate the mantle wedge and trigger the melting of the mantle wedge above the dehydrated megathrust. Then, the water-rich fluids carried away are transported to the surface by upwelling [39]. Accordingly, the wedge-earthquake distribution in Alaska may indicate a pathway for flux melt or fluid upwelling from the subduction channel, although the diapirs formed in accretionary wedges are complicated, and the infiltration of fluids or melts in Mélange diapirs are physically mixed by various components, such as sediments, fluids, altered oceanic crust, and mantle materials [1]. In particular, the upwelling angle is suggested to be primarily controlled by the oceanic plate subduction angle and then further affected by the inclination of the diapirs and faults inside the wedge, as we have demonstrated through detailed investigation of the distribution of wedge earthquakes in Alaska (Figure 4).

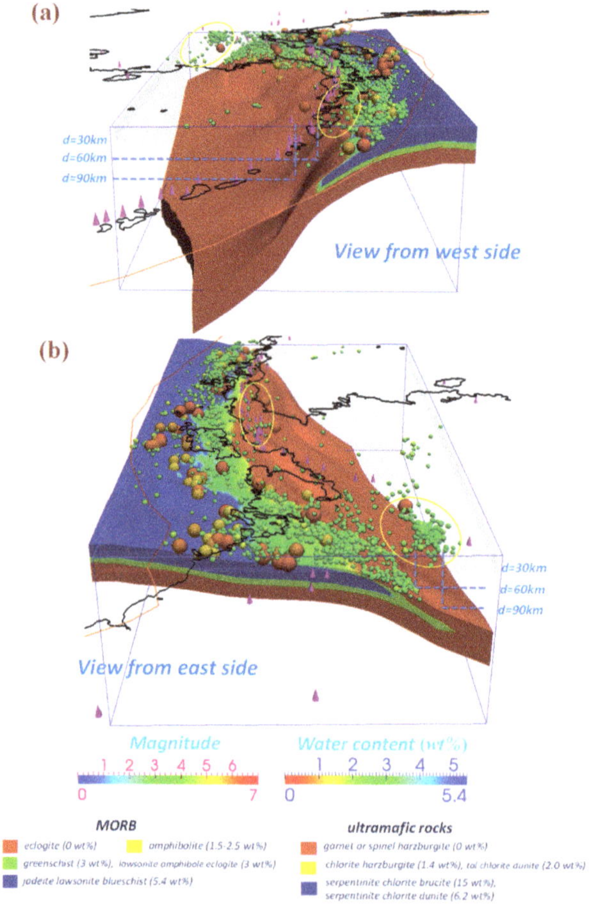

Figure 4. Wedge earthquakes in Alaska. Slab color indicates water content (wt %) distribution. The colored spheres indicate all M > 3 earthquakes (IRIS) in the North American plate from 1 January 2000 to December 2010 and all M > 5.5 earthquakes from 1 January 1900 to December 2000 (Centennial Database). The black curve on the surface represents the coast. Red cones indicate active arc volcanoes [40]. Yellow ellipses indicate clustered continental wedge earthquakes. (**a**) View of the 3D model from the west. (**b**) View of the 3D model from the east.

4. Discussion

Wedge earthquakes are also related to subduction-related surface topography formation (e.g., [41]) because plume/melt upwelling affects seismotectonics and orogenesis spontaneously. According to a numerical modeling study of a subduction zone, Horton [40] proposed that subduction at a shallow angle facilitates the uplift of a continental plateau, while subduction at a large angle contributes to the formation of a mountain or island. Previous studies also widely support that small angle subduction prefers the formation of a wide continental high topography, as a small portion of the continental crust closest to the margin possibly partitions due to the shallow subduction dip (e.g., [42]). In Alaska, shallow subduction in south–central Alaska results in fluid/molten upwelling landward, where the Alaska range uplifts and evolves landward. The steep subduction in the Alaska Peninsula causes mineral upwells limited in a narrow cross-strike width, which results in the formation of island arcs such as the Alaska Peninsula and the Aleutian Islands.

Small dip angle subduction of the cold lithosphere facilitates the fluid/melt rising in the continental wedge compared with the large dip angle oceanic plate. In addition, plate collision and subsequent upward extruding force both contribute to the formation of magma chambers or transcrustal magmatic that originates from melt processing in the deep crust [43], which facilitates the construction of an uplifted continental plateau. A broad forearc high, such as a large mountain system or continental plateau, an accretionary prism, and a shallow trench, is evidenced by south–central Alaska. On the other hand, when the slab dip angle increases to 20°–30°, mountains or island arcs are generated due to a limited cross-arc width for slab metamorphism and exhumation to the surface close to the trench, such as the Aleutian and Alaska Peninsulas. The frictional strength of the sediments and serpentinized peridotite exert a large control on the dip angle of the subduction interface at seismogenic depths [44]. In the high friction case, the subduction interface is steep, the trench is deeper, and the accretionary prism, forearc high, and basin are all absent [44]. In Alaska, the dip angle is greatly determined by the plate age (via the temperature and density comparison between two coupled plates), as shown in Figure 5. The reason that a broad forearc high is constructed by a shallow dip angle may be attributed to the extrusion effect brought by the incoming oceanic plate and the subsequent upwelling controlled by the dip angle. The Alaska Peninsula and the Fox Islands in the Aleutian Islands characterize an older oceanic plate age of approximately 50 Myr and a larger dip angle (12°–27°) at Moho depth compared with the Alaska range uplift in south-central Alaska, caused by a younger plate (approximately 40 Myr) and a smaller dip angle (7.7°).

Consequently, wedge earthquakes in Alaska are characterized by a distance to the trench transitioning from small (steep subduction) to large (flat subduction), which is likely determined by the slab surface depth of 60–100 km, where the slab is fully dehydrated and subsequently hydrates the above mantle wedge and crust, generating clustered wedge earthquakes in Alaska. Slab dehydration induces plume upwelling, which not only facilitates wedge seismic events but also alters the surface topography over a geological time scale. The farthermost point of the hypocenter of wedge earthquakes from the trench is determined by the slab dip angle and the slab surface depth of 60 km. Thus, the fluid upwelling pathway is decided by the inclination angle of faults and diapirs, contributing to fluid migration in the wedge, which facilitates fault slip and elevates in situ potential seismic risks. Therefore, the across-arc width of the wedge-earthquake source zone is consistent with the across-arc width of the surface high topography above the entirely dehydrated plate interface, because plume upwelling spontaneously influences forearc seismotectonics and orogenesis. The shallow slab dip facilitates the formation of a high forearc plateau, and wedge earthquakes widely span from the trench to the farthermost inland (with an interface depth of 100 km), which supports that wedge earthquakes are likely the by-product (energic and material circulating product) of the tectonic formation process of the forearc region of active plate subduction zones, which is evidenced by the observations in subduction zones.

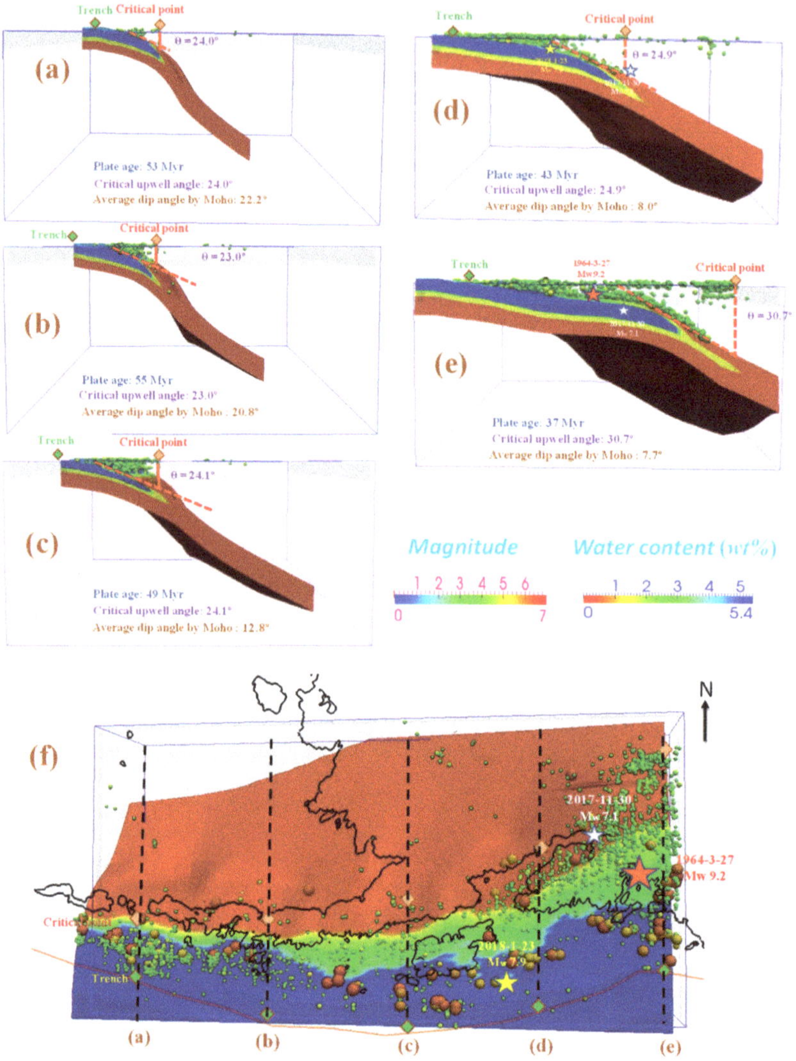

Figure 5. Wedge-earthquake distribution along five profiles (**a**–**e**) perpendicular to the margin from west to east. Slab color indicates the water content (wt %) distribution, which are the same with Figure 4. (**a**–**e**) Cross-section of wedge-earthquake distribution along a profile. (**f**) Distribution of profiles (**a**–**e**). The black dashed lines indicate the a–e profiles. Dimension scale and orientations of profiles are exhibited in (**f**).

5. Conclusions

Through modeling the thermal state of subducted Alaskan megathrusts and 3D comparison with the wedge earthquakes distribution, we concluded the following:

(1) The subduction thermal regime constrained by the observations of surface heat flow, including Currie depth estimations without plate decoupling, well interprets the thermal state of the underlain subducted plate beneath the wedge wherein earthquakes occurred.

(2) The distribution of wedge-earthquake variations in Alaska features a transition from steep subduction to flat subduction associated with different subduction hydrother-

mal states and fluids released. The farthermost point of the hypocenter of wedge earthquakes from the trench is determined by the slab dip angle and the slab surface depth of 60 km.

(3) The across-arc width of the wedge-earthquake source zone is consistent with the across-arc width of the surface high topography above the entirely dehydrated plate interface, and the fluid upwelling spontaneously influences forearc seismotectonics and orogenesis.

Author Contributions: Conceptualization, methodology, software, validation, resources, data curation, Y.J.; formal analysis, investigation, R.Q., W.Z.; writing-original draft preparation, R.Q.; writing-review and editing, visualization, supervision, project administration, and funding acquisition, Y.J. All authors have read and agreed to the published version of the manuscript.

Funding: This research was funded by financial support from the Second Tibetan Plateau Scientific Expedition and Research Program (2019QZKK0708) and the CAS Pioneer Hundred Talents Program.

Institutional Review Board Statement: Not applicable.

Informed Consent Statement: Not applicable.

Acknowledgments: We are grateful to P. Tackley for sharing the original code of stag3d developed in this study. We also thank IRIS for the earthquake catalog.

Conflicts of Interest: The authors declare no conflict of interest.

References

1. Lin, C.-H.; Shih, M.-H.; Lai, Y.-C. Mantle wedge diapirs detected by a dense seismic array in Northern Taiwan. *Sci. Rep.* **2021**, *11*, 1–12.
2. Gerya, T.V.; Connolly, J.A.; Yuen, D.A.; Gorczyk, W.; Capel, A.M. Seismic implications of mantle wedge plumes. *Phys. Earth Planet. Inter.* **2006**, *156*, 59–74. [CrossRef]
3. Tape, C.; Sykes, L.; Lomax, A. The 1900 M w 7.6–8.0 Earthquake Offshore of Kodiak Island, Alaska. *Bull. Seismol. Soc. Am.* **2021**, *111*, 1080–1109.
4. Gerya, T.V.; Stöckhert, B.; Perchuk, A.L. Exhumation of high-pressure metamorphic rocks in a subduction channel: A numerical simulation. *Tectonics* **2002**, *21*, 6-1–6-19. [CrossRef]
5. Tamura, Y.; Tatsumi, Y.; Zhao, D.; Kido, Y.; Shukuno, H. Hot fingers in the mantle wedge: New insights into magma genesis in subduction zones. *Earth Planet. Sci. Lett.* **2002**, *197*, 105–116. [CrossRef]
6. Martin-Short, R.; Allen, R.; Bastow, I.D.; Porritt, R.W.; Miller, M.S. Seismic imaging of the Alaska subduction zone: Implications for slab geometry and volcanism. *Geochem. Geophys. Geosyst.* **2018**, *19*, 4541–4560. [CrossRef]
7. Worthington, L.L.; Van Avendonk, H.J.; Gulick, S.P.; Christeson, G.L.; Pavlis, T.L. Crustal structure of the Yakutat terrane and the evolution of subduction and collision in southern Alaska. *J. Geophys. Res. Solid Earth* **2012**, *117*. [CrossRef]
8. Arcay, D.; Tric, E.; Doin, M.-P. Numerical simulations of subduction zones: Effect of slab dehydration on the mantle wedge dynamics. *Phys. Earth Planet. Inter.* **2005**, *149*, 133–153. [CrossRef]
9. Tackley, P.J.; Xie, S. STAG3D: A code for modeling thermo-chemical multiphase convection in Earth's mantle. In *Computational Fluid and Solid Mechanics*; Elsevier Science Ltd.: Amsterdam, The Netherlands, 2003; pp. 1524–1527.
10. Ji, Y.; Yoshioka, S.; Matsumoto, T. Three-dimensional numerical modeling of temperature and mantle flow fields associated with subduction of the Philippine Sea plate, southwest Japan. *J. Geophys. Res. Solid Earth* **2016**, *121*, 4458–4482. [CrossRef]
11. Ji, Y.; Yoshioka, S.; Manea, V.C.; Manea, M.; Matsumoto, T. Three-dimensional numerical modeling of thermal regime and slab dehydration beneath Kanto and Tohoku, Japan. *J. Geophys. Res. Solid Earth* **2017**, *122*, 332–353. [CrossRef]
12. Hayes, G.P.; Moore, G.L.; Portner, D.E.; Hearne, M.; Flamme, H.; Furtney, M.; Smoczyk, G.M. Slab2, a comprehensive subduction zone geometry model. *Science* **2018**, *362*, 58–61. [CrossRef]
13. DeMets, C.; Gordon, R.G.; Argus, D.; Stein, S. Current plate motions. *Geophys. J. Int.* **1990**, *101*, 425–478. [CrossRef]
14. DeMets, C.; Dixon, T.H. New kinematic models for Pacific-North America motion from 3 Ma to present, I: Evidence for steady motion and biases in the NUVEL-1A model. *Geophys. Res. Lett.* **1999**, *26*, 1921–1924. [CrossRef]
15. Müller, R.D.; Sdrolias, M.; Gaina, C.; Roest, W.R. Age, spreading rates, and spreading asymmetry of the world's ocean crust. *Geochem. Geophys. Geosyst.* **2008**, *9*. [CrossRef]
16. Stein, C.A.; Stein, S. A model for the global variation in oceanic depth and heat flow with lithospheric age. *Nature* **1992**, *359*, 123–129. [CrossRef]
17. Grose, C.J.; Afonso, J.C. Comprehensive plate models for the thermal evolution of oceanic lithosphere. *Geochem. Geophys. Geosyst.* **2013**, *14*, 3751–3778. [CrossRef]
18. Yoshii, T. Regionality of group velocities of Rayleigh waves in the Pacific and thickening of the plate. *Earth Planet. Sci. Lett.* **1975**, *25*, 305–312. [CrossRef]

19. Pollack, H.N.; Hurter, S.J.; Johnson, J.R. Heat flow from the Earth's interior: Analysis of the global data set. *Rev. Geophys.* **1993**, *31*, 267–280. [CrossRef]
20. Li, C.-F.; Lu, Y.; Wang, J. A global reference model of Curie-point depths based on EMAG2. *Sci. Rep.* **2017**, *7*, 1–9. [CrossRef]
21. Omori, S.; Kita, S.; Maruyama, S.; Santosh, M. Pressure–temperature conditions of ongoing regional metamorphism beneath the Japanese Islands. *Gondwana Res.* **2009**, *16*, 458–469. [CrossRef]
22. Hacker, B.R.; Abers, G.A.; Peacock, S.M. Subduction factory 1. Theoretical mineralogy, densities, seismic wave speeds, and H_2O contents. *J. Geophys. Res. Solid Earth* **2003**, *108*. [CrossRef]
23. Wada, I.; Wang, K. Common depth of slab-mantle decoupling: Reconciling diversity and uniformity of subduction zones. *Geochem. Geophys. Geosyst.* **2009**, *10*. [CrossRef]
24. Furukawa, Y. Depth of the decoupling plate interface and thermal structure under arcs. *J. Geophys. Res. Solid Earth* **1993**, *98*, 20005–20013. [CrossRef]
25. Peacock, S.M.; Wang, K. Seismic consequences of warm versus cool subduction metamorphism: Examples from southwest and northeast Japan. *Science* **1999**, *286*, 937–939. [CrossRef]
26. Wada, I.; Wang, K.; He, J.; Hyndman, R.D. Weakening of the subduction interface and its effects on surface heat flow, slab dehydration, and mantle wedge serpentinization. *J. Geophys. Res. Solid Earth* **2008**, *113*. [CrossRef]
27. Ashi, J.; Tokuyama, H.; Ujiie, Y.; Taira, A. Heat flow estimation from gas hydrate BSRs in the Nankai Trough: Implications for thermal structures of the Shikoku Basin. *EOS Trans. AGU* **1999**, *80*, F929–F930.
28. Ashi, J.; Tokuyama, H.; Taira, A. Distribution of methane hydrate BSRs and its implication for the prism growth in the Nankai Trough. *Mar. Geol.* **2002**, *187*, 177–191. [CrossRef]
29. Tanaka, A. Geothermal gradient and heat flow data in and around Japan (II) Crustal thermal structure and its relationship to seismogenic layer. *Earth Planets Space* **2004**, *56*, 1195–1199. [CrossRef]
30. Tanaka, A.; Yamano, M.; Yano, Y.; Sasada, M. Geothermal gradient and heat flow data in and around Japan (I): Appraisal of heat flow from geothermal gradient data. *Earth Planets Space* **2004**, *56*, 1191–1194. [CrossRef]
31. Matsumoto, T. Terrestrial heat flow distribution in Japan area based on the temperature logging in the borehole of NIED Hi-net. In Proceedings of the AGU Fall Meeting Abstracts, San Francisco, CA, USA, 10–14 December 2007. Abstract ID T23A-1217.
32. Oleskevich, D.; Hyndman, R.; Wang, K. The updip and downdip limits to great subduction earthquakes: Thermal and structural models of Cascadia, south Alaska, SW Japan, and Chile. *J. Geophys. Res. Solid Earth* **1999**, *104*, 14965–14991. [CrossRef]
33. Van Keken, P.E.; Kiefer, B.; Peacock, S.M. High-resolution models of subduction zones: Implications for mineral dehydration reactions and the transport of water into the deep mantle. *Geochem. Geophys. Geosyst.* **2002**, *3*, 1–20. [CrossRef]
34. Van Keken, P.E.; Wada, I.; Abers, G.A.; Hacker, B.R.; Wang, K. Mafic high-pressure rocks are preferentially exhumed from warm subduction settings. *Geochem. Geophys. Geosyst.* **2018**, *19*, 2934–2961. [CrossRef]
35. Wilson, C.R.; Spiegelman, M.; van Keken, P.E.; Hacker, B.R. Fluid flow in subduction zones: The role of solid rheology and compaction pressure. *Earth Planet. Sci. Lett.* **2014**, *401*, 261–274. [CrossRef]
36. Syracuse, E.M.; van Keken, P.E.; Abers, G.A. The global range of subduction zone thermal models. *Phys. Earth Planet. Inter.* **2010**, *183*, 73–90. [CrossRef]
37. Penniston-Dorland, S.C.; Kohn, M.J.; Manning, C.E. The global range of subduction zone thermal structures from exhumed blueschists and eclogites: Rocks are hotter than models. *Earth Planet. Sci. Lett.* **2015**, *428*, 243–254. [CrossRef]
38. Ponko, S.C.; Peacock, S.M. Thermal modeling of the southern Alaska subduction zone: Insight into the petrology of the subducting slab and overlying mantle wedge. *J. Geophys. Res. Solid Earth* **1995**, *100*, 22117–22128. [CrossRef]
39. Zhao, D.; Yu, S.; Liu, X. Seismic anisotropy tomography: New insight into subduction dynamics. *Gondwana Res.* **2016**, *33*, 24–43. [CrossRef]
40. Siebert, L.; Simkin, T.; Kimberly, P. *Volcanoes of the World*; University of California Press: Berkeley, CA, USA, 2010.
41. Horton, B.K. Tectonic regimes of the central and southern Andes: Responses to variations in plate coupling during subduction. *Tectonics* **2018**, *37*, 402–429. [CrossRef]
42. Schütt, J.M.; Whipp, D.M. Controls on continental strain partitioning above an oblique subduction zone, Northern Andes. *Tectonics* **2020**, *39*, e2019TC005886. [CrossRef]
43. Cashman, K.V.; Sparks, R.S.J.; Blundy, J.D. Vertically extensive and unstable magmatic systems: A unified view of igneous processes. *Science* **2017**, *355*, eaag3055. [CrossRef]
44. Tan, E.; Lavier, L.L.; Van Avendonk, H.J.; Heuret, A. The role of frictional strength on plate coupling at the subduction interface. *Geochem. Geophys. Geosyst.* **2012**, *13*. [CrossRef]

applied sciences

MDPI

Article

Two-Dimensional Full-Waveform Joint Inversion of Surface Waves Using Phases and Z/H Ratios

Chao Zhang [1,2,*], Ting Lei [2] and Yi Wang [3,*]

1 College of Oceanography, Hohai University, Nanjing 210098, China
2 Department of Earth Sciences, University of Toronto, Toronto, ON M5S 3B1, Canada; leiting@mail.utoronto.cn
3 State Key Laboratory of Geodesy and Earth's Dynamics, Innovation Academy for Precision Measurement Science and Technology, Chinese Academy of Sciences, Wuhan 430077, China
* Correspondence: czhang18@hhu.edu.cn (C.Z.); ghostzzz@mail.ustc.edu.cn (Y.W.); Tel.: +86-155-5546-2977 (C.Z.); +86-180-0842-4426 (Y.W.)

Abstract: Surface-wave dispersion and the Z/H ratio are important parameters used to resolve the Earth's structure, especially for S-wave velocity. Several previous studies have explored using joint inversion of these two datasets. However, all of these studies used a 1-D depth-sensitivity kernel, which lacks precision when the structure is laterally heterogeneous. Adjoint tomography (i.e., full-waveform inversion) is a state-of-the-art imaging method with a high resolution. It can obtain better-resolved lithospheric structures beyond the resolving ability of traditional ray-based travel-time tomography. In this study, we present a systematic investigation of the 2D sensitivities of the surface wave phase and Z/H ratio using the adjoint-state method. The forward-modeling experiments indicated that the 2D phase and Z/H ratio had different sensitivities to the S-wave velocity. Thus, a full-waveform joint-inversion scheme of surface waves with phases and a Z/H ratio was proposed to take advantage of their complementary sensitivities to the Earth's structure. Both applications to synthetic data sets in large- and small-scale inversions demonstrated the advantage of the joint inversion over the individual inversions, allowing for the creation of a more unified S-wave velocity model. The proposed joint-inversion scheme offers a computationally efficient and inexpensive alternative to imaging fine-scale shallow structures beneath a 2D seismic array.

Keywords: adjoint tomography; Rayleigh wave phase; Z/H ratio; joint inversion; 2D seismic array

1. Introduction

Surface waves are essential for resolving crust and upper mantle structures at various length scales. Surface waves have dispersive characteristics; that is, a surface wave travels with different velocities at different frequencies in the wave-propagation process. This feature has been well studied and widely used to image Earth structures, especially for the S-wave velocity model. Teleseismic surface-wave tomography was used in earlier studies, and it offered a good resolution for the upper mantle [1–3]. However, it is difficult to constrain structures in the shallow to middle crust due to the low frequency of such waves. In recent decades, ambient-noise tomography was developed, which can extract the Green's function of a surface wave from ambient-noise data, and it significantly improved the surface-wave-dispersion data coverage (mostly less than a 40 s period). Short- to intermediate-period surface waves have been widely used for imaging regional- or local-scale crustal structures [4–7].

However, the surface-wave-dispersion curve is only sensitive to the integral feature of the velocity structure of a depth range, resulting in ambiguities during inversion. Studies have shown that the amplitude information of a surface wave can also help improve the velocity structure. The Z/H ratio (or ellipticity) of a Rayleigh wave is always defined as the amplitude ratio of the vertical and horizontal components of the fundamental mode of the Rayleigh wave. The Z/H ratio of a Rayleigh wave is only controlled by the structure beneath the recording seismic station [8,9], and it is more sensitive to shallow structures than dispersion in the same period range. Therefore, the Rayleigh wave's Z/H ratio can provide extra constraints to reduce the ambiguity of surface wave tomography, and in particular, it can improve the imaging of shallow sedimentary structures. Because of the resolution gap caused by the different data sensitivities, independent inversions using the surface-wave-dispersion curve or the Z/H ratio cannot provide a unified S-wave velocity model. The joint inversion of the surface-wave-dispersion curve and the Z/H ratio has been proposed for imaging the complementary velocity structure. In particular, joint inversion based on ray theory has been widely used in imaging crust and upper mantle structures of various length scales [10]. However, this approach does not account for the complexities of wave propagation in complex media, and may lead to inaccurate tomographic results.

Recent advances in the numerical methods involving the wave equation combined with developments in high-performance computation (HPC) have enabled the routine simulation of seismic wave propagation in realistic 2D and 3D Earth models based on the spectral-element method (SEM) (Komatitsch and Tromp, 2002a, 2002b). Utilizing these seismic-wave simulations, adjoint-state methods can efficiently incorporate the full nonlinearity of the wave propagation into iterative seismic inversions [11–22]. Based on these advances, Chen et al. [23] proposed ambient-noise adjoint tomography for 3D crustal structures by matching the empirical Green's function (EGF) and the synthetic Green's function (SGF) calculated based on the SEM. In addition, full-wave 3D ambient-noise tomography has been developed, and has been used to image the structure of the Cascades, the Ontong Java Plateau, and the eastern North American margin [24–26]. Bao et al. [27] derived the equations for 3D full-wave sensitivity kernels of the Z/H ratio of a Rayleigh wave using the scattering-integral method [28–31].

However, the computational cost and storage requirement of 3D adjoint tomography are large. In contrast, 2D adjoint tomography is much more computationally efficient. Currently, an increasing number of dense linear arrays have been deployed around the world, which facilitates the high-resolution imaging of the 2D structures beneath the linear array. In global-scale tomography, Zhang et al. [32] presented a 2D ambient-noise adjoint tomography technique for a linear array with a significant reduction in computational cost, and applied it to an array in northern China. Tong et al. [33] developed 2D teleseismic body-wave adjoint tomography by an FK-SEM hybrid method. Zhang et al. [34] proposed a strategy of joint inversion for ambient-noise surface waves and teleseismic body waves recorded by linear arrays based on the adjoint-state method, which can be used to yield a fine S-wave velocity model and Moho topography. Small-scale tomography has been fully developed by the oil industry in the past three decades (well known as FWI), while 2D elastic FWI has recently been applied to shallow surface waves [35–38]. In particular, Forbriger et al. [39], Schäfer et al. [40], and Groos et al. [41] proposed a stable workflow for the 2D FWI of shallow-seismic Rayleigh waves not only for synthetic data, but also for real field-recorded data. However, including surface waves in FWI is a challenging issue, especially the cycle-skipping problems that occur due to these waves' higher amplitudes and different physical behaviors with respect to body waves. Benefiting from the fact that the sensitivity to the S-wave velocity is only concentrated near stations, the Z/H ratio information can constrain the shallow S-wave velocity structure in the near surface, which helps mitigate the cycle-skipping problem.

However, there is a lack of research about FWI using the Z/H ratio or the joint inversion of both data sets. In this paper, we present a systematic investigation of the 2D Rayleigh wave travel-time (phase) sensitivity kernel and the Z/H-ratio kernel via an adjoint-state method [19]. In addition, a full-waveform joint-inversion scheme for surface waves with phases and Z/H ratios is proposed to take advantage of their complementary sensitivities to the S-wave velocity structure. Finally, synthetic-imaging experiments are described to demonstrate the potential of the proposed method to yield a better complementary S-wave velocity model with a high-resolution.

2. Calculation of 2D Rayleigh Wave Phase and Z/H-Ratio Kernel

Tanimoto and Rivera [9] calculated the 1D depth-sensitivity kernel of the Rayleigh wave dispersion and the Z/H ratio either theoretically or numerically by perturbing the model parameters at various depths. The features of the 1D kernel include that both the dispersion and Z/H ratio are sensitive to changes in the S-wave velocity and the density, but they are less sensitive to the P-wave velocity. For the dispersion, the S-wave velocity sensitivity is positive at different depths, suggesting that the dispersion provides more constraints on the integral velocity feature. In contrast, the S-wave velocity sensitivity of the Z/H ratio changes from positive near a free surface to negative at deeper depths, implying that the Z/H ratio resolves the velocity contrasts better. To study the features of the 2D kernel, we derived equations for the 2D full-wave sensitivity kernels of the phases and the Z/H ratio based on the adjoint-state theory.

2.1. D Full-Wave Sensitivity Kernels of the Traveltime Equation

First, we briefly review the travel-time and amplitude-sensitivity kernels for single-component seismograms. Following the full-wave finite-frequency theory, the perturbations in the overall misfit $\delta\chi$ are linearly related to the relative perturbations in the density $(\delta ln\ \rho)$, P-wave velocity $(\delta ln\ \alpha)$, and S-wave velocity $(\delta ln\ \beta)$:

$$\delta\chi = \int \left[K_{\alpha}(\mathbf{r})\delta ln\alpha + K_{\beta}(\mathbf{r})\delta ln\beta + K_{\rho}(\mathbf{r})\delta ln\rho\right]d^2x, \tag{1}$$

In general, two primary measurements are used in adjoint tomography: the phase delay and the amplitude reduction. For the phase-delay measurements, the travel-time misfit perturbations (δt) can be represented as follows:

$$\delta t = -\frac{\int_{t_1}^{t_2} \partial\widetilde{s}(t)\delta s(t)dt}{\int_{t_1}^{t_2} |\widetilde{s}(t)|^2 dt} = \int \left[K_{\alpha}^{p}(\mathbf{r})\delta ln\alpha + K_{\beta}^{p}(\mathbf{r})\delta ln\beta\ + K_{\rho}^{p}(\mathbf{r})ln\rho\right]d^2x, \tag{2}$$

where δt is the travel-time anomaly; $\delta s(t)$ is the perturbation of the waveform displacement, $\widetilde{s}(t) - s(t)$; the tilde represents the synthetics based on the reference model; and t_1 and t_2 are the minimum and maximum values of the time window, respectively. ∂ is the time derivative, so δt is equal to the lag time at the maximum of the cross-correlogram between the observed and synthetic waveform. K_{α}^{p}, K_{β}^{p}, and K_{ρ}^{p} are the travel-time Fréchet kernels for the density, P-wave velocity, and S-wave velocity, respectively, which can be calculated using the adjoint-state method. The adjoint source of the cross-correlation travel-time is:

$$f_p^{\dagger}(t,\omega) = -\frac{w(t,\omega)\dot{s}(t,\omega)}{\int w(t,\omega)\dot{s}^2(t,\omega)dt}, \tag{3}$$

where $w(t,\omega)$ is the time window over which the measurement of the cross-correlation traveltime is made, s (t,ω) is the waveform recorded by a station, the dot represents the time derivative, and ω is the central frequency of the Gaussian filter.

2.2. D Full-Wave Sensitivity Kernels of the Z/H Ratio Equation

For the amplitude reduction measurements, the natural logarithm of the amplitude reduction (δln A) can be expressed as follows:

$$\delta \ln A = -\frac{\int_{t_1}^{t_2} \widetilde{s}(t)\delta u(t)dt}{\int_{t_1}^{t_2} |\widetilde{s}(t)|^2 dt} = \int \left[K_{\alpha}^{q}(\mathbf{r})\delta ln\alpha + K_{\beta}^{q}(\mathbf{r})\delta \ln\beta + K_{\rho}^{q}(\mathbf{r})\delta ln\rho\right]d^2x, \tag{4}$$

Here, A is defined as the ratio between (1) the amplitude of the cross-correlogram of the observed and synthetic waveforms at lag time δt and (2) the maximum amplitude of the auto-correlogram of the synthetic waveform. K_α^q, K_β^q, and K_ρ^q are the corresponding amplitude-reduction Fréchet kernels for the density, P-wave velocity, and S-wave velocity, respectively. The Z/H ratio (η) of a Rayleigh wave is always defined as the Z-component's amplitude (A_Z) over the H-component's amplitude (A_H) at a certain frequency ω:

$$\eta(\omega) = \frac{A_Z(\omega)}{A_H(\omega)}, \tag{5}$$

The perturbation of the Z/H ratio can be expressed as the difference between the Z-component and H-component of the amplitude perturbation:

$$\delta \ln \eta = \delta\left(\ln A^Z / A^H\right) = \delta \ln A^Z - \delta \ln A^H, \tag{6}$$

If we consider the fact that the variation in the measurements is only caused by the variation in the S-wave velocity (β), then

$$\delta \ln \eta = \int K_\beta^\eta \delta \ln \beta d^2x;\ \delta \ln A_Z = \int K_\beta^Z \delta \ln \beta d^2x;\ \delta \ln A_H = \int K_\beta^H \delta \ln \beta d^2x, \tag{7}$$

where K_β^η, K_β^Z, and K_β^H are the sensitivity kernels to the S-wave velocity of the Z/H ratio, the Z-component's amplitude, and the H-component's amplitude, respectively.

$$\int K_\beta^\eta \delta \ln \beta d^2x = \int K_\beta^Z \delta \ln \beta d^2x - \int K_\beta^H \delta \ln \beta d^2x, \tag{8}$$

This equation holds true for any volume V, so the Z/H-ratio kernel of the S-wave velocity can be expressed as:

$$K_\beta^\eta = K_\beta^Z - K_\beta^H, \tag{9}$$

Similarly, we can obtain the following equations for P-wave velocity (α) and density (ρ):

$$K_\alpha^\eta = K_\alpha^Z - K_\alpha^H, K_\rho^\eta = K_\rho^Z - K_\rho^H, \tag{10}$$

Thus, to generate the kernels above, we only need to calculate the sensitivity kernels of the Z-component's amplitude and the H-component's amplitude, which can also be solved by using the adjoint-state method. The adjoint source of the amplitude is:

$$f_q^\dagger(t,\omega) = \frac{w(t,\omega)s(t,\omega)}{\int w(t,\omega)s^2(t,\omega)dt}, \tag{11}$$

where $w(t,\omega)$ is the time window over which the measurement of the amplitude is made, $s(t,\omega)$ is the waveform recorded by a station, and ω is the central frequency of the Gaussian filter. The definition of Z/H ratio in this study is noteworthy. Generally, the previous studies [7,27] commonly used the maximum amplitude of the envelope to calculate Z/H ratio as follows:

$$\eta(\omega) = \frac{A_z(\omega)}{A_H(\omega)} = \frac{\max(E[s_z(\omega)])}{\max(E[s_H(\omega)])}, \tag{12}$$

where $s_z(\omega)$ and $s_H(\omega)$ are the Z-component and H-component of the seismogram processed using a Gaussian filter at a certain frequency ω, respectively, and E is their envelope. In this study, we used the following equation defined in [42]:

$$\eta(\omega) = \frac{A_z(\omega)}{A_H(\omega)} = \sqrt{\frac{\int_{t_1}^{t_2} s_Z^2 dt}{\int_{t_1}^{t_2} s_H^2 dt}}, \tag{13}$$

where t_1 and t_2 are the start and end of the measurement window, respectively. With respect to physical properties, both definitions in Equations (12) and (13) are robust and acceptable, and could result in very similar Z/H ratios for synthetic tests. In the real data application, they will be different, depending on the S/N level of the waveform data. Determining which is better would require a more detailed discussion regarding a different real data set. Otherwise, it should be noted that we need to guarantee that the same definition is used in the data processing and kernel calculation.

2.3. Sensitivity Calculation and Analysis

To illustrate the procedure for obtaining the travel-time and Z/H-ratio-sensitivity kernels, we used the 2D spectral-element method (SPECFEM2D) to calculate the 2D sensitivity kernels via the adjoint-state method. The numerical parameters of the modeling in the computation are shown in Table 1. The model measures 800 km in the horizontal direction, with 320 uniform mesh nodes (quadrangles in 2D), and 100 km in the vertical direction, with 40 uniform nodes. A total of (4 × 320 + 1) ∗ (4 × 40 + 1) = 206,241 unique grid points were used for the 2D spectral-element discretization with a four-degree polynomial (Komatitsch et al., 2005). The average grid spacing of the mesh was 2.5 km, which was sufficient to accurately simulate the surface waves at periods of greater than ~8 s (Komatitsch and Tromp, 2002). For simplicity and to minimize the possible interference of other arrivals on the Rayleigh waves, the P-wave velocity, the S-wave velocity, and the density were all held constant (Vp = 6000 m/s, vs. = 3500 m/s, and ρ = 2800 g/m^3). The time interval of the simulation was set to 0.02 s to satisfy the numerical stability, and the time step was set to 12,000. We placed a Gaussian-type source-time function with a dominant frequency of 0.2 Hz at location (650 km, 0 km) to generate the 2D synthetic waveform of the surface wave, which was computed using SPECFEM2D. A station was set at location (150 km, 0 km).

Table 1. Numerical parameters for kernel calculation in the computation.

Model Size (Nx × Nz)	**x-Direction: 800 km; z-Direction: 100 km (300 × 40 Elements)**
Material properties	Vp = 6000 m/s, vs. = 3500 m/s, ρ = 2800 g/m^3
Propagation time	12,000 time steps; Time discretization: 0.02 s
Type of source-time function	Gaussian
Dominant frequency of source	2 Hz
Source location	(650 km, 0 km)
Station location	(150 km, 0 km)

Figure 1 illustrates how to obtain the single-component phase sensitivity of the kernels ($K_\beta^{pZ}, K_\beta^{pH}$) to δVs on a vertical profile and horizontal profile in 2D media for surface waves with a period of 20 s. First, we obtained the seismograms of the Z-component (Figure 1a) and H-component (Figure 1b) recorded by the station, which were processed using a Gaussian filter with a central frequency of 0.5 Hz. Then, we obtained the adjoint source of the Z-component (Figure 1c) and H-component (Figure 1d) at the station using Equation (3). We performed one adjoint simulation by injecting the back adjoint source at the station into the model, which created adjoint wavefield $s^\dagger$. Finally, the phase-sensitivity kernels of the Z-component (Figure 1e) and H-component (Figure 1f) at periods of 20 s were obtained using the time integration of the interaction of $s^\dagger$ and the forward wavefield s. To study the features of the phase kernel in the depth and lateral directions at different periods, the Z-component kernels at three other periods of 10 s (Figure 1g), 30 s (Figure 1i), and 40 s (Figure 1j) also were calculated using the same procedure.

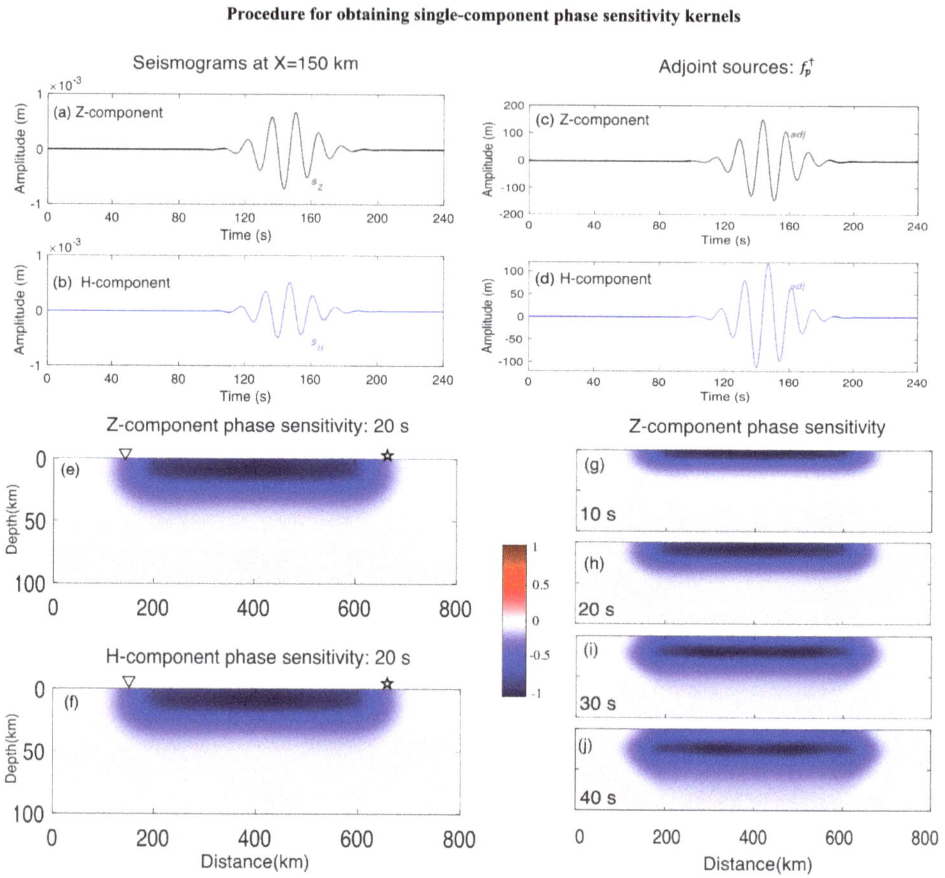

Figure 1. Procedure for obtaining the phase-sensitivity kernels for the surface waves with a period of 20 s, for a source (star) and station (triangle) pair. Seismograms were filtered using a Gaussian filter (central frequency: 0.5 Hz) recorded by the station for the (**a**) Z-component, (**b**) H-component, (**c**) adjoint source of the Z-component, and (**d**) adjoint source of the H-component. The shaded area represents the time window over which the measurement of the phase delay was made. (**e**) The phase-sensitivity kernels of the Z-component, and (**f**) the phase-sensitivity kernels of the H-component. The Z-component phase-sensitivity kernels at other periods of (**g**) 10 s, (**h**) 20 s, (**i**) 30 s, and (**j**) 40 s were also calculated to study the features of the phase kernel in the depth and lateral directions at different periods.

Figure 2 illustrates how to obtain the single-component amplitude kernels ($K_\beta^{pZ}, K_\beta^{pH}$) to δVs on a vertical profile and horizontal profile in 2D media, as well as their differential kernels ($K_\beta^{pZ} - K_\beta^{pH}$), which is called the Z/H-ratio kernel (K_β^{η}) in this paper. The procedure for obtaining the amplitude kernels was very similar to that for the phase kernels, except that the adjoint source was generated using different measurements followed by Equation (11). The amplitude-sensitivity kernels of the Z-component and H-component with periods of 20 s obtained using the adjoint-state method are shown in Figure 2e,f. Thus, the Z/H-ratio-sensitivity kernel at a period of 20 s (Figure 2g) could be obtained by subtracting the H-component amplitude-sensitivity kernel from the Z-component amplitude-sensitivity kernel. Similarity, the Z/H-ratio kernel at periods of 10 s (Figure 2h), 30 s (Figure 2j), and 40 s (Figure 2k) are also displayed and will be discussed later in this section.

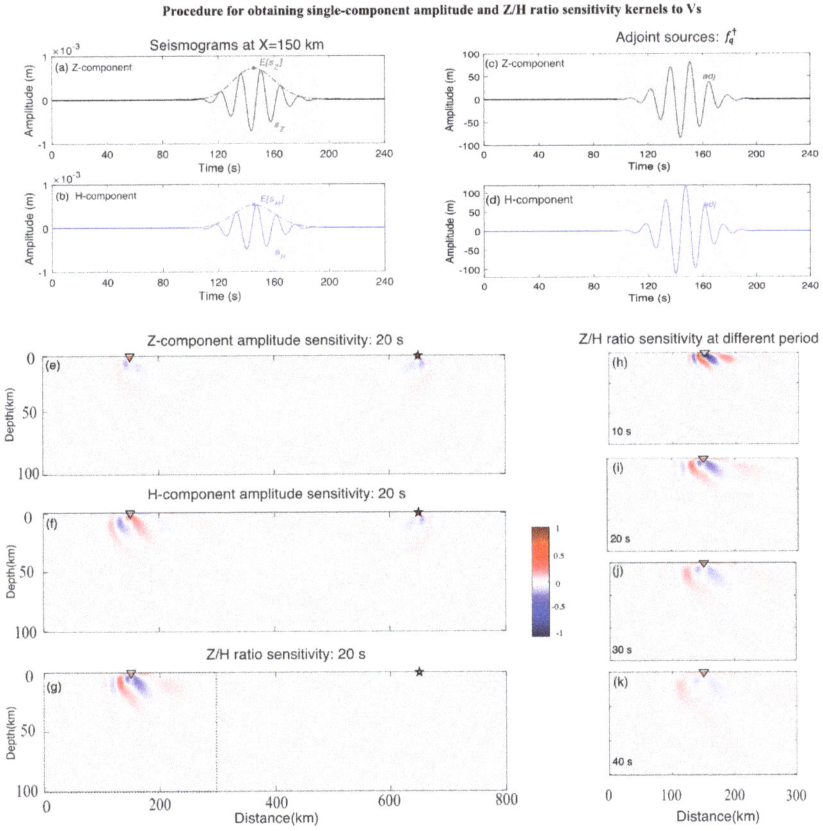

Figure 2. Procedure for obtaining the amplitude- and Z/H-ratio-sensitivity kernels for surface waves with a period of 20 s, for a source (star) and station (triangle) pair. Seismograms were filtered using a Gaussian filter (central frequency: 0.5 Hz) recorded by the station for the (**a**) Z-component, (**b**) H-component, (**c**) adjoint source of the Z-component, and (**d**) adjoint source of the H-component. The shaded area represents the time window over which the measurement of the amplitude reduction was made. (**e**) The amplitude-sensitivity kernels of the Z-component, and (**f**) the amplitude-sensitivity kernels of the H-component. (**g**) The Z/H-ratio-sensitivity kernels obtained by subtracting the H-component amplitude-sensitivity kernel from that of the Z-component. The Z-component Z/H-ratio kernels at other periods of (**h**) 10 s, (**i**) 20 s, (**j**) 30 s, and (**k**) 40 s were also calculated to study the features of the Z/H kernel in the depth and lateral directions at different periods. Note that only the kernels near the station (the region outlined by the dashed line in (**g**)) were plotted.

The experiment described above shows that both the 2D-phase and Z/H-ratio kernels provide more information than the 1D sensitivity kernels. The 1D kernels only had sensitivities in the vertical direction because they did not consider the source direction. In contrast, the 2D kernels varied with distance from the station and with depth, which indicated that they had strong resolving powers in both the depth and lateral directions.

Both kernels had different features in the vertical and lateral directions. The phase kernel was concentrated along the propagation path, with a much broader sensitivity range than the Z/H-ratio kernels, while the Z/H ratio sensitivity was only concentrated near the station. The reason for this is that the amplitude-sensitivity kernels of the Z-component and the H-component had the same pattern on the source side and a common propagation path, and the use of the component-differential calculation could remove the observational errors in the source and propagation path. This feature was consistent with previous studies of the Z/H ratio, especially for 1D cases; that is, the Z/H ratio was only sensitive to the structures beneath the station.

To compare the sensitivity to the S-wave velocity in depth for these two data sets, we reduced the spatial dimension of the phase and the Z/H-ratio kernels shown in Figures 1g–j and 2g–j from 2D to 1D by laterally summing the values on the horizontal plane in each vertical grid and plotting them as a function of depth. Figure 3 shows the reduced 1D kernels at periods of 10 s, 20 s, 30 s, and 40 s for the phase (Figure 3a) and Z/H ratio (Figure 3b). The results showed that their sensitivity kernels were different but complementary. The Rayleigh wave phase was more sensitive to the deeper structure than the Z/H ratio at the same period, while the Z/H ratio had a strong sensitivity to the shallow structure (e.g., the sedimentary layers). Moreover, the phase-sensitivity kernel was much broader than the Z/H-ratio kernel for the same period, which meant the phase information could be used to solve the smoothed velocity (background model), while the Z/H ratio could deal with the lateral heterogeneity (e.g., velocity anomaly or undulated interface) better.

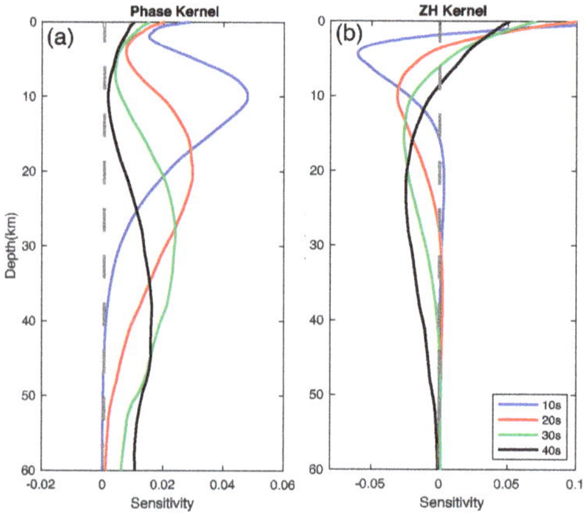

Figure 3. The laterally summed 1D sensitivity kernels to S-wave velocity in depth for (**a**) phase velocity and (**b**) Z/H ratio at four different periods: 10 s (blue line), 20 s (red line), 30 s (green line), and 40 s (black line).

The effect of the velocity anomaly on the Z/H ratio was obvious and easily reachable from Figure 2. Here, we designed another experiment to demonstrate the effect of undulated interface on the Z/H-ratio kernels as shown in Figure 4. We chose a typical structure containing interface variation, which followed a co-sinusoidal function with a maximum amplitude of 5.0 km. The results showed that the Z/H-ratio sensitivity was significantly affected by the undulated interface, and the maximum difference between the undulated interface and the flat interface was about 50%. The experiments described above suggested that the Z/H-ratio kernel was much more sensitive to the model with velocity anomalies and interface variations. This meant that the Z/H ratio could help us to constrain a shallow crust with a complex lateral heterogeneity in which the surface-wave phase cannot be resolved, as well as to improve the accuracy of the inversion.

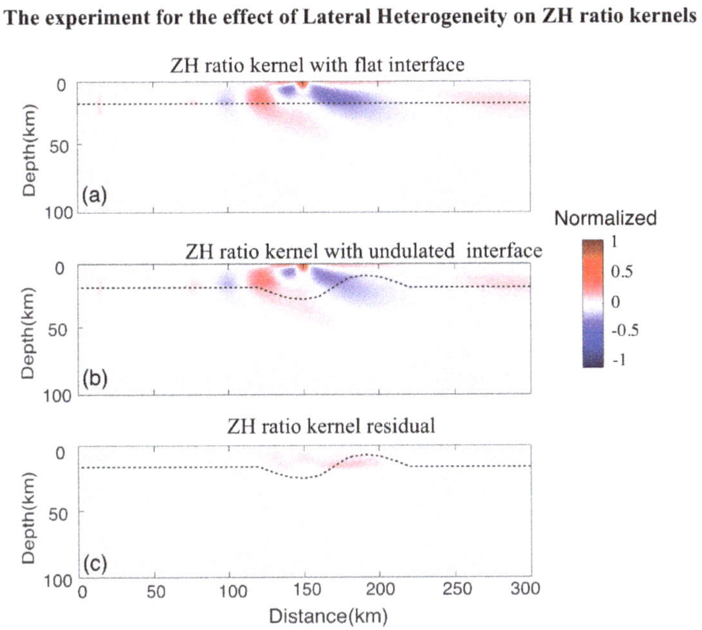

Figure 4. The effect of the local heterogeneities on the Z/H-ratio kernel: (**a**) Z/H-ratio kernel in the model containing a flat interface (dashed line); (**b**) Z/H-ratio kernel in the model containing an interface variation (dashed curve), which follows a cosine function with a maximum amplitude of 5.0 km; (**c**) differences between the Z/H-ratio kernels in (**a**,**b**).

3. Joint Inversion of the Rayleigh Wave Phase and the Z/H Ratio

As was demonstrated by the sensitivity analysis described in the previous section, the phase and the Z/H ratio of a Rayleigh wave had different sensitivities to the S-wave velocity structure, and they provided complementary constraints on the structure. Thus, joining the two data sets will help reduce the non-uniqueness of the inversion.

For adjoint tomography of the Rayleigh wave phase [32], we sought to minimize the travel-time misfits between the observed waveform and the synthetic waveform generated using SPECFEM2D. In this study, the frequency-dependent travel-time misfits can be written as:

$$\chi_{RP}(m) = \frac{1}{N_c}\sum_{i=1}^{N_c}\frac{1}{M_i}\sum_{j=1}^{M_i}\int\left(\frac{\Delta T_j\,(\omega,\,m)}{\sigma_{T,j}\,(\omega)}\right)^2 d\omega\,, \tag{14}$$

where m is the model vector, χ_{RP} is the misfit function for the Rayleigh wave phase, N_c is the number of periods, M_i is the total number of misfit measurements in the i-th period, $\Delta T_j\,(\omega)$ is the frequency-dependent phase delay between the jth pair of observed data and the synthetic data at frequency ω in the ith period, and $\sigma_{T,j}\,(\omega)$ is the corresponding measurement uncertainty. Similarity, the frequency-dependent Z/H ratio misfit is defined as follows:

$$\chi_{ZH}(m) = \frac{1}{N_c}\sum_{i=1}^{N_c}\frac{1}{M_i}\sum_{j=1}^{M_i}\int\left(\frac{\Delta zh_j\,(\omega)}{\sigma_{zh,j}\,(\omega)}\right)^2 d\omega\,, \tag{15}$$

where χ_{ZH} is the misfit function for the Rayleigh wave's Z/H ratio, and $\Delta zh_j\,(\omega)$ is the Z/H ratio variation measured between the jth pair of the observed data and synthetic data with an uncertainty of $\sigma_{zh,j}$.

Thus, the objective function of the joint inversion based on the adjoint-state method is defined as follows:

$$\chi(m) = \omega_{RP}\, \chi_{RP}(m) + \omega_{ZH}\, \chi_{ZH}(m)\,, \tag{16}$$

where ω_{RP} and ω_{ZH} are the weights used to balance the two data types to prevent the results from being dominated by either one. The joint-inversion scheme is described in Figure 5.

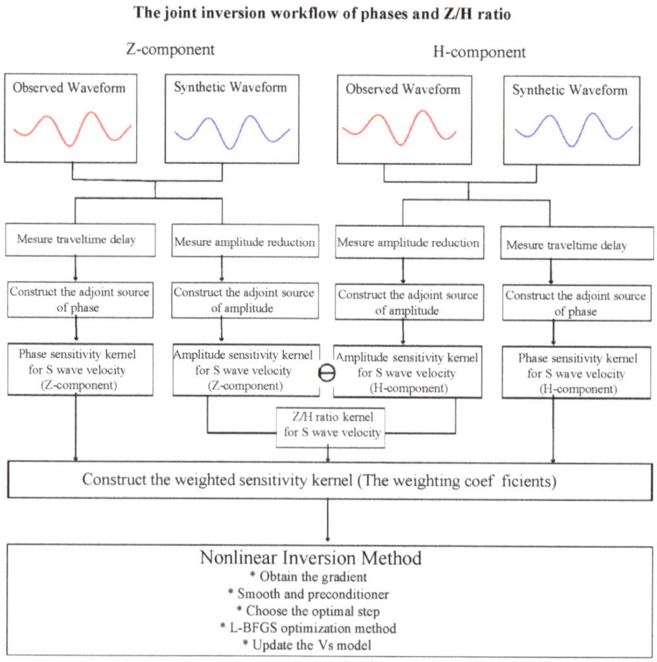

Figure 5. Flow chart of the 2D full-waveform joint inversion of the Rayleigh wave phase and the Z/H ratio.

4. Inversion Experiments

The complementary sensitivity of the phase and Z/H ratio to the vs. perturbation makes the joint inversion an important method of constraining S-wave velocity structures. Here, we designed two synthetic experiments to demonstrate the successful usage of this method on the large- and small-scale inversions.

4.1. Large-Scale Inversion Using Ambient Noise/Earthquake Surface Waves

Seismic linear arrays with dense station spacing have been deployed around the globe, which makes it possible to apply joint inversion to ambient-noise data or earthquake surface waves recorded by linear seismic arrays.

We set up a target model (Figure 6a) with alternating 6% box-shaped high- and low-Vs anomalies in the background model (also used as the initial model in this experiment; see material properties in Table 1), and no Vp or density perturbation was introduced into the target model. The size of the model and the number of uniform mesh nodes were the same as the parameter settings described in Section 2.3. All of the boxes with anomalies had lengths of 40 km, while the widths changed from 8 km in the upper row to 15 km in the middle row to 30 km in the lower row. These different sizes were designed to assess the spatial resolutions of the varying lengths for a Rayleigh wave with different periods. The source-time function was Gaussian, and the dominant frequency of the source was 0.2 Hz. Fifty stations were spaced 10 km apart and were located from 150 to 650 km, which could also be considered as a virtual source generating an empirical Green's function of the surface wave.

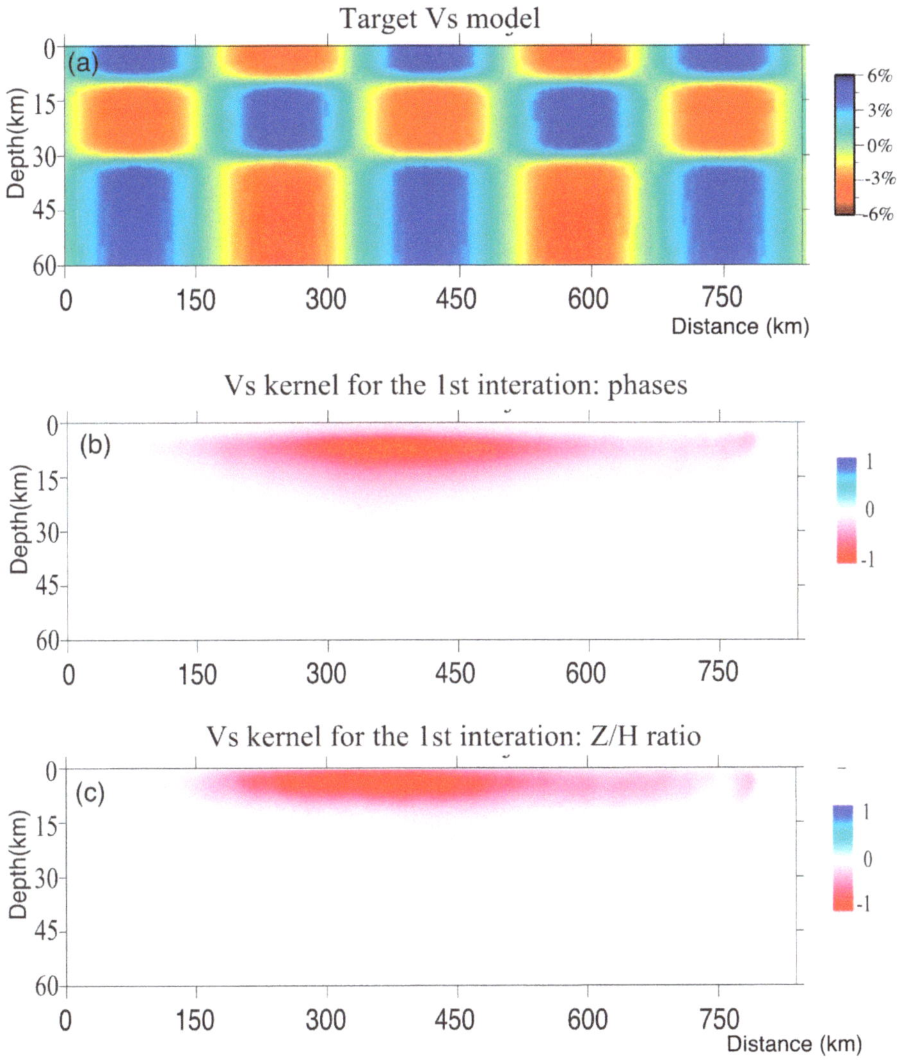

Figure 6. (**a**) Target model with alternating 6% box-shaped high- and low-Vs anomalies in the background model. The vs. sensitivity kernels for the first iteration of the (**b**) phase and (**c**) Z/H ratio.

We performed all of the forward and adjoint simulations using SPECFEM2D. Then, we filtered the synthetic and observed waveforms at four central period bands (10, 20, 30, and 40 s) using a Gaussian filter, which could cover the period band of 5 to 50 s quite well. Following the workflow of the joint inversion described in Figure 5, we measured the frequency-dependent travel-time and amplitude misfit between the synthetic and observed waveform and obtained the corresponding adjoint sources of the phase and amplitude, respectively. After that, we calculated the phase-sensitivity kernel and the amplitude kernel of the Z-component and H-component by injecting the adjoint sources at the stations based on the adjoint-state method. In addition, the Z/H-ratio kernels were calculated by subtracting the Z-component and H-component amplitude kernels. Finally, all of the event kernels were summed, smoothed, and preconditioned to obtain the final misfit gradient for updating the model. A 2D Gaussian function was used to smooth the summed kernel as a regularization procedure (Tape et al., 2010). The horizontal and vertical widths of the Gaussian function used to smooth the gradient were chosen as 30 km and 10 km, respectively, in the first 10 iterations. Then, they were reduced to smaller values of 10 km and 5 km, reflecting the incorporation of the shorter-period surface-wave measurements. Next, the smoothed kernel was preconditioned using a preconditioner, which was numerically calculated through the interaction between the forward and adjoint acceleration fields; namely, the pseudo-Hessian (Luo, 2012).

$$P(x) = H^2_{\rho\rho}(x,\ x) = \sum_{s=1}^{N_s} \int \partial_t^2 s(x,t) \cdot \partial_t^2 s^{\dagger}(x, T-t)dt\ , \tag{17}$$

where s and $s^{\dagger}$ are the forward and adjoint displacements, respectively, and N_s is the number of sources. Finally, we used the limited-memory Broyden–Fletcher–Goldfarb–Shanno (L-BFGS) method [43] (see detail shown in Appendix A) to update the model, and a line-search method was used to determine the optimal step length for updating the model. Once the optimal step length was determined from the minimum of the total function for all of the bands, the model parameter m was updated as follows:

$$ln\ (m_{i+1}/m_i) = \nu_i\ d_i\ , \tag{18}$$

where m_{i+1} and m_i are the models of the (i + 1)th iteration and the *i*th iteration, respectively; and ν_i and di are the optimal step length and the search direction (i.e., summed event kernels) of the *i*th iteration, respectively.

To demonstrate the advantage of our joint-inversion framework, we also conducted two additional separate inversions using either only the Rayleigh-wave phase data or only the Z/H-ratio data. The separate inversions also began with the background model and used the same inversion parameters as the joint inversion, including the smoothing width and step lengths. Both data sets were mostly sensitive to the shear-wave velocity. We only inverted for the vs. structures. In total, we conducted three inversions: (1) adjoint tomography of the Rayleigh-wave phase data (i.e., phase inversion); (2) adjoint tomography of the Rayleigh-wave Z/H-ratio data (i.e., Z/H-ratio inversion); and (3) the joint inversion using the two data sets (i.e., joint inversion).

Figure 6b,c show the gradients of the vs. for the first iteration of the phase and Z/H-ratio inversions. The gradients show that the two Rayleigh wave measurements had different and complimentary sensitivities to the crustal structure. The phase inversion was more sensitive to the deeper structure than the Z/H-ratio inversion, while the Z/H-ratio inversion had a strong sensitivity to the shallow structure. The total misfit evolution of the phase and Z/H-ratio inversions and the joint inversion are presented in Figure 7a,b. We can see that the joint inversion had a slower convergence rate and slightly larger misfits than those of the separate inversions, which was reasonable, since the joint-inversion scheme tried to fit both data sets simultaneously. The joint inversion converged after 32 iterations when the misfit changed over the last iteration because both the phase and Z/H-ratio data were less than 3%. Figure 7c,d show the waveform fitting of the observed Z-component and X-component data and the two synthetic computed data sets based on the initial and final models for the source–receiver pairs located at (750 km, 0 km) and (150 km, 0 km), and significant improvements in the travel-time and amplitude fits after joint inversion can be observed.

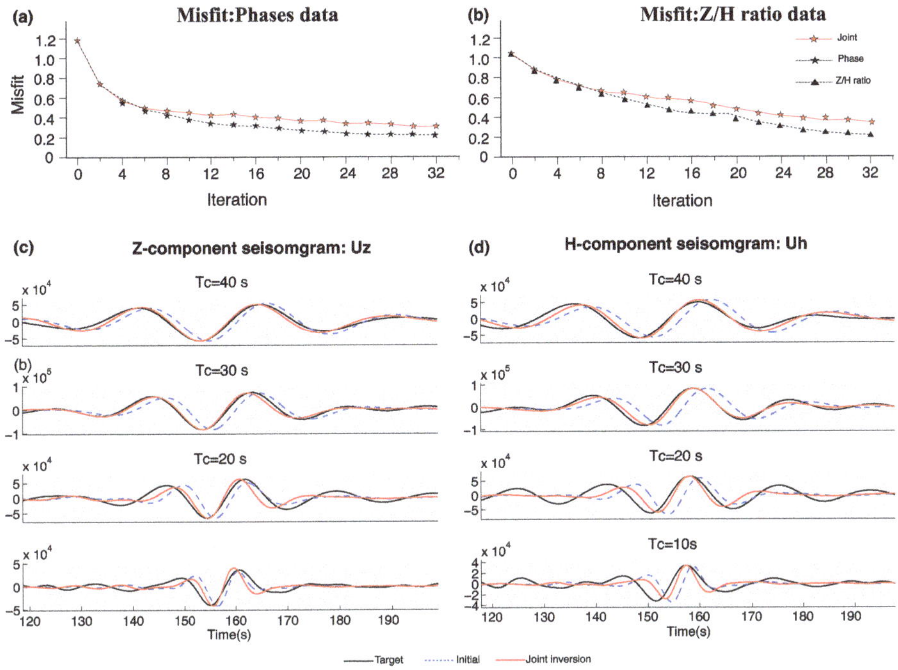

Figure 7. The total misfit evolution for the (**a**) phase and (**b**) Z/H ratio along with iterations in the phase inversion (black stars), the Z/H-ratio inversion (black triangles), and the joint inversion (red stars). The seismic traces obtained using joint inversion (red solid line) and the source–receiver pairs at (750 km, 0 km) and (150 km, 0 km) are shown together with the seismic traces of the initial (blue dashed line) and target (black solid line) models for the (c) Z-component and (**d**) H-component.

To compare the results of the three types of inversions, we show their final vs. models in Figure 8. The Z/H-ratio information for the surface was mostly sensitive to the vs. structures at shallow depths (<30 km), which was limited by the period range of the Rayleigh wave in this study. Thus, the recovered vs. model (Figure 8b) was better resolved in the shallow regions, and the upper and middle rows show the correct pattern with the anomaly close to target model. Compared with the Z/H-ratio inversion, the phase inversion was sensitive to much deeper vs. structures and could constrain the broader range of the vs. structures at depths of ~20 km to 60 km (Figure 8a). The pattern of the anomaly in the middle and lower rows is almost recovered; however, the recovered pattern is rather smoothed and lacks constraints in the shallow structure, which was caused by the sensitivity characteristic of the phases shown in Figure 3a. Benefiting from the sensitivity to the vs. structures concentrated near the station, the Z/H-ratio inversion helped to better resolve the structures in the shallow areas that were not well resolved by the phase inversion. Thus, the addition of the Z/H ratio in the joint inversion helped to constrain the crustal structure at shallow depths. At greater depths, the joint inversion had a better resolution than the phase inversion with the correct amplitude recovery. These tests demonstrated that the joint inversion combined the complementary sensitivities of the surface wave phases and Z/H ratio, and was capable of building a more-unified vs. model, thus outperforming the inversions based on the individual data sets.

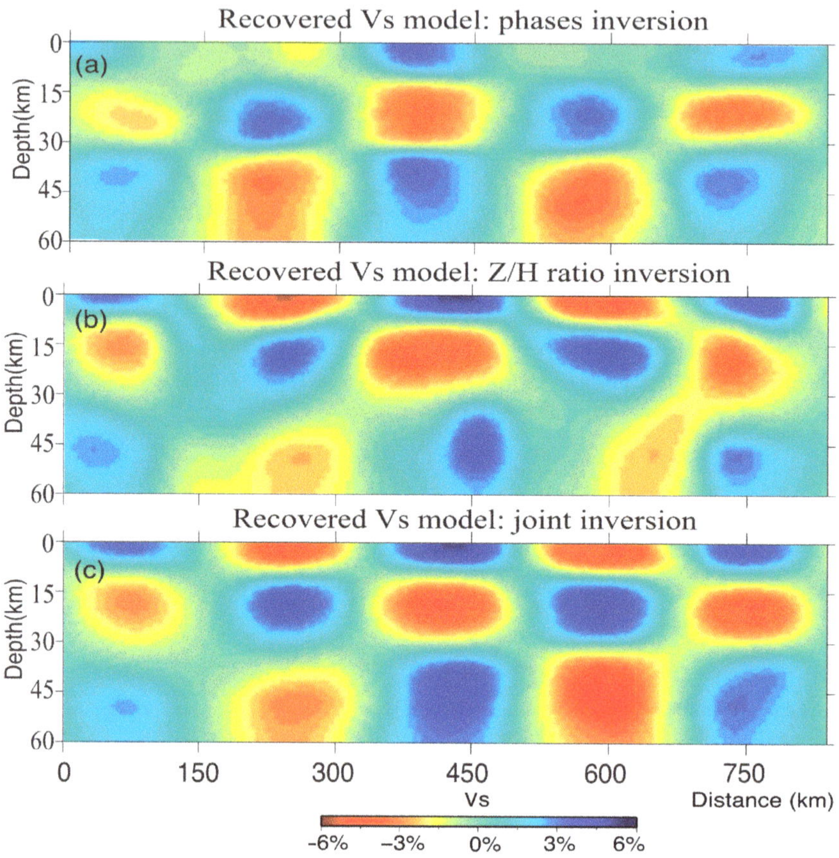

Figure 8. The recovered S-wave velocity models for the three types of inversions. (**a**) The final inverted model using only the Rayleigh wave phase. (**b**) The final inverted model using only the Rayleigh wave Z/H ratio. (**c**) The final inverted model using the joint inversion of the Rayleigh wave phase and Z/H ratio.

4.2. Small-Scale Inversion Using Active Sources

Another suitable application of this method is retrieving model parameters and characterizing the near-surface region, mainly the shear-wave velocity model using active sources, for engineering applications, geotechnical studies, and physical modeling tests. In this section, we used more realistic synthetic experiments to demonstrate the promising applications of this full-waveform joint-inversion method to small-scale inversions, with an improved resolution.

The 600 $\times$ 100 m target model for elastic S-wave and P-wave velocities used for the numerical experiments in this section is shown in Figure 9. With this model, we used 100 and 20 uniform mesh nodes in the horizontal and vertical directions, respectively, resulting in a total of $(4 * 100 + 1) * (4 * 20 + 1) = 32{,}481$ unique grid points. The target elastic model included heterogeneous Vp (Figure 9a) and vs. models (Figure 9b) with a homogeneous density of 1000 kg/m^3. Both the Vp and vs. increased with depth from a low-velocity layer, with the exception of one low-velocity zone (LVZ) and two high-velocity zones (HVZs) in the middle. We assumed an initial model with a homogeneous Vp of 1000 m/s, a homogeneous vs. of 500 m/s, and a homogeneous density of 1000 kg/m^3. We set up 50 shots located from 50 to 550 m at a depth 0.5 m below the model's surface, with an equal 10 m spacing. The stations were spaced 5 m apart and were located 5 m to 595 m at a depth of 0.5 m.

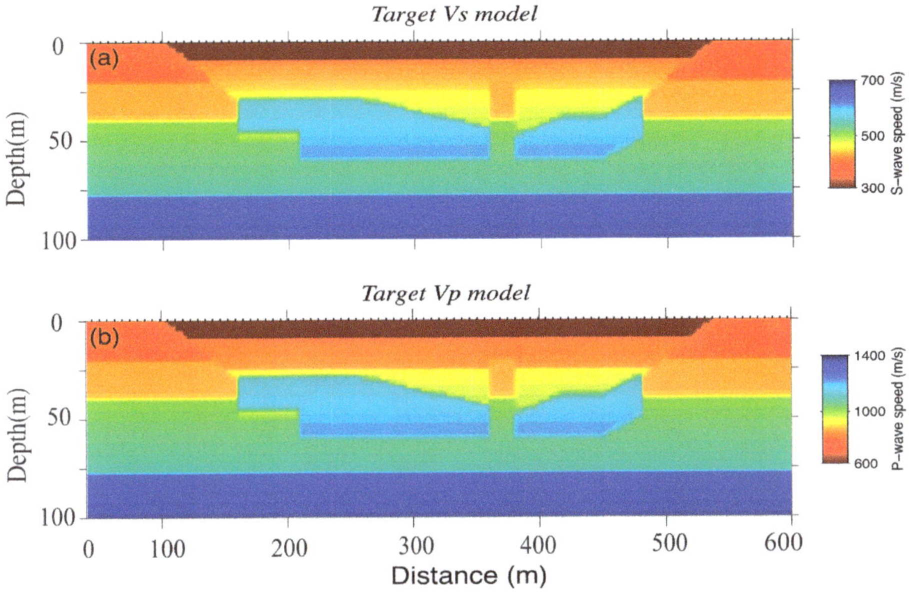

Figure 9. Target model with features in the near surface, specifying (**a**) S-wave and (**b**) P-wave velocities.

Synthetic data were computed for the target model and the initial model and were simulated using source-time functions with a dominant 10 Hz Ricker wavelet. Then, we conducted the joint inversion following the same inversion procedures described in Figure 5. To avoid the local minima and reduce the non-linearity of the waveform misfit function, we adopted a hierarchical strategy, in which we successively inverted from low to high frequency progressively through 20, 30, 40, 50, 60, and 80 Hz. We also conducted three types of inversions to further demonstrate the advantage of the joint inversion over the separate inversions. For each inversion, we simultaneously inverted both the S-wave and P-wave velocities.

First, we used the objective function in Equation (13) to update the model via the phases. The result of using the Rayleigh wave phase is shown in the inverted S-wave (Figure 10a) and P-wave (Figure 11a) velocity models. A better result for the vs. final model was expected, since the surface waves were more sensitive to this elastic parameter. Neither the S- nor the P-wave velocity model converged to a model near the target shown in Figure 9. The most probable reason for this was that the inversion suffered from strong cycle-skipping problems, which easily occur in surface waveform FWI, with incorrect updates leading to a secondary minimum. Next, we replaced the phases with the Z/H ratio in the objective function. That is, we minimized the discrepancy defined by the Z/H ratio in Equation (14). Figures 10b and 11b show the resulting vs. and Vp models. Because the envelope information using the Z/H ratio calculation in this paper mitigated the cycle-skipping effects, working only with the Z/H ratio led to a reasonably successful recovery of the vs. model of the shallow structure, but it did not constrain the deeper structure. Moreover, the capacity of the Z/H ratio to constrain the shallow crust with a complex lateral heterogeneity helped us obtain a significantly better low-velocity layer where the surface wave phase could not resolve the structure. The S-wave velocity structure was resolved, albeit with a low resolution, and only the very top portion of the P-wave velocity model showed signs of being updated in the desirable direction. The

Z/H ratio inversions appeared to converge at the price of losing resolution by discarding the phase information. Finally, we proposed a full-waveform joint inversion of the phase and Z/H ratio in the same multiscale framework. Figures 10c and 11c show the results obtained using the joint-inversion philosophy. The addition of the Z/H ratio to the joint inversion helped constrain the S-wave velocity structure in the near-surface region, so the low-velocity layer was correctly corrected and converged to the target model. Moreover, the mitigation of the cycle-skipping problems contributed to the Z/H ratio and could help the phase information better constrain the principal features of the deeper structure. The LVZ and HVZ in the middle were well covered with correct amplitudes. In addition, the interfaces between the velocity zones were better defined in the vs. model. In the Vp model, although only the very top portion of the P-wave velocity model was acceptably imaged, it can be seen that the inclusion of the phase information in the joint inversion provided more additional structure to the P-wave velocity model compared with the inversion results obtained using on obtained using only the Z/H ratio.

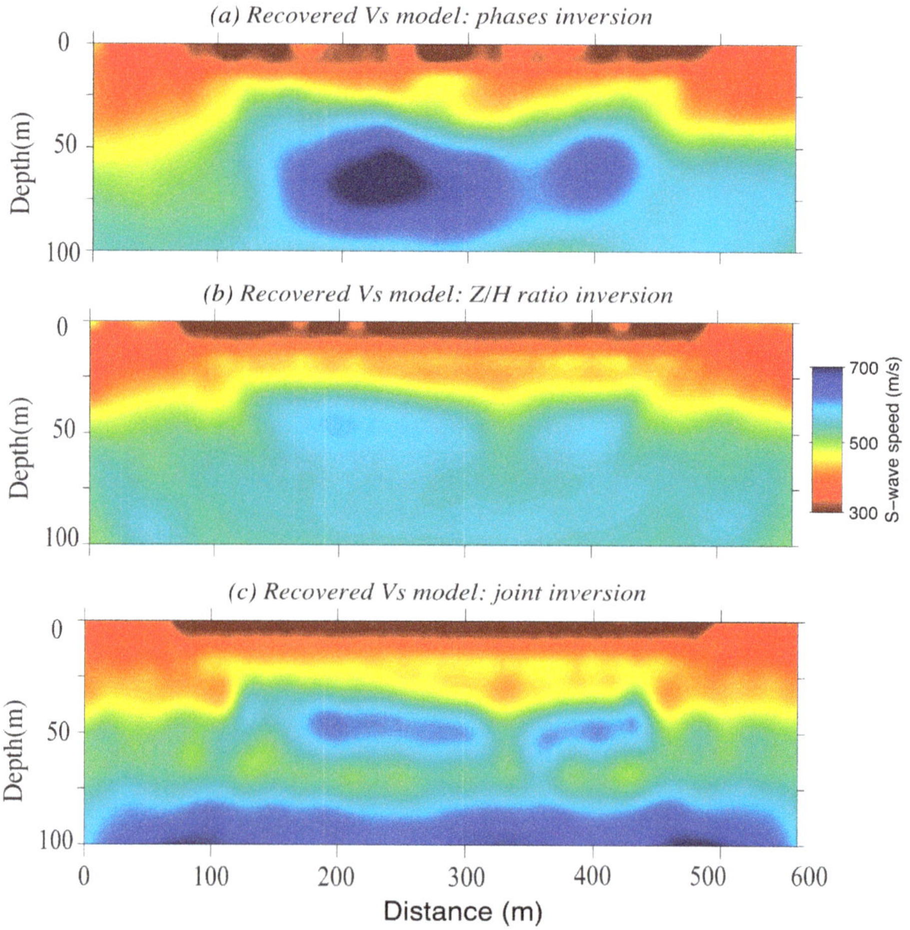

Figure 10. The recovered S-wave velocity models for the three types of inversion. (**a**) The final inverted model using only the Rayleigh wave phase. (**b**) The final inverted model using only the Rayleigh wave Z/H ratio. (**c**) The final inverted model using the joint inversion of the Rayleigh wave phase and Z/H ratio.

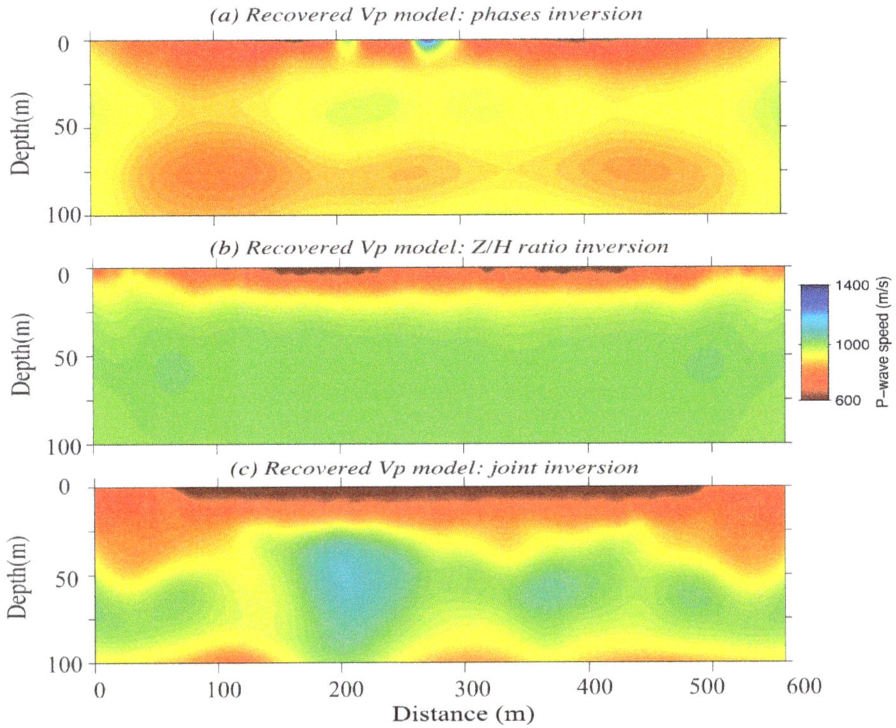

Figure 11. The recovered P-wave velocity models for the three types of inversions. (**a**) The final inverted model using only the Rayleigh wave phase. (**b**) The final inverted model using only the Rayleigh wave Z/H ratio. (**c**) The final inverted model using the joint inversion of the Rayleigh wave phase and Z/H ratio.

In Figure 12, we show how well the waveform data were fit under phase and joint inversion as described in this paper. The left column shows the vertical displacement components (Uz), and the right column shows the horizontal components (Uh) for three traces of stations at X = 150, 300, and 450 m from one shot gather located at X = 50 m. The black solid lines represent the target data, red dashed lines represent the initial predictions, and blue solid lines and green solid lines represent the final models obtained by phase inversion and joint inversion, respectively. We can see that the phase inversion suffered from a cycle-skipping problem, creating local minima in the FWI objective function, so the final traces were badly matched to the target traces (Figure 12a). Joint inversion incorporating Z/H ratio improved the vs. model by mitigating cycle-skipping issues, so the traces of joint inversion were matched to almost-perfect agreement with the target traces (Figure 12b).

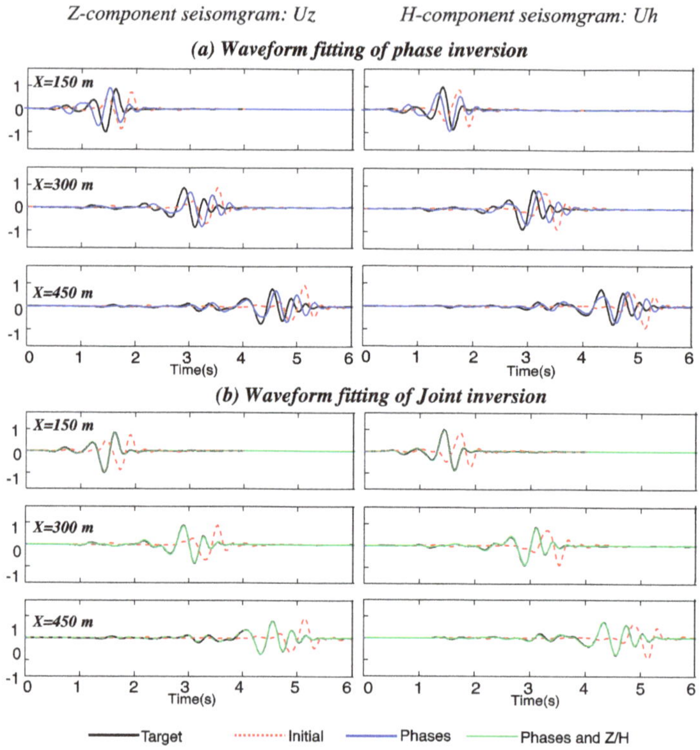

Figure 12. Waveform fitting for three traces of stations at X = 150, 300, and 450 m from one shot gather located at X = 50 m after inversion. Comparison of the target, initial, and final seismic traces obtained via (**a**) phase inversion; (**b**) joint inversion. Left: H-component. Right: Z-component.

5. Discussion

As the extension of 2D surface-wave adjoint tomography method [32], the computing of 2D joint inversion also kept the efficiency. Taking the small-scale inversion as an example, the model was 600 m in length and 100 m in depth with 100 × 20 elements, and the simulation contained 20,000 time steps with a 0.0003 s time interval. As shown in Table 2, the phase inversion included 50 forward-modelings and calculations of kernels; the number of processors for each source was 4, leading to a total cost of 200 CPU hours. The computation time was 6.25 h for 46 iterations. For joint inversion, the CPU hours cost was doubled (400 CPU hours total) due to the addition of the Z/H-ratio kernel calculation. However, the actual computation also showed relative efficiency, and 38 iterations could be completed in just 5.8 h, which was acceptable compared with the computational cost in the 3D case. On the basis of the time cost, the 2D joint inversion can be very efficient when computational resources are limited.

Table 2. Parameters of numerical modeling and computational time for inversion.

Type of Inversion	Phase Inversion	Z/H Ratio Inversion	Joint Inversion
Model size (Nx ∗ Nz)	x-direction: 600 m; z-direction: 100 m Nx = 100; Nz = 20		
Propagation time	20,000 time steps; Time discretization: 0.0003 s		
Number of processors for each source	4	4	8
Total number of CPUs/cores	200	200	400
Number of iterations	46	54	38
Time for inversion	6.25 h	6.72 h	5.80 h

The synthetic examples above provided us with two actual situations for the application of joint inversion. For large-scale inversion, we could use earthquake surface waves recorded by linear seismic arrays. In reality, it is difficult to find enough earthquakes along the profile. Fortunately, the approach has been developed over decades, in which the Z/H ratio can be extracted from ambient-noise cross-correlation data [7], which makes it the most suitable situation for applying the joint-inversion method. The tomographic experiment in Section 4.1 shows that both Rayleigh wave phase and Z/H ratio had sensitivity to S-wave velocity in the crustal structure, but their sensitivity patterns were different. Rayleigh wave phase velocity dispersion was sensitive to the average velocity of the depth range sampled. The Z/H ratio was more sensitive to the shallower structure compared to the Rayleigh wave phase velocity dispersion at the same frequency; thus, the Z/H ratio was more helpful to constrain the shallow structure, such as the sedimentary basin. The joint inversion combined the complementary sensitivities of the phases and Z/H ratio and was capable of building a more unified crustal structure.

For small-scale inversion, this method could be used effectively in retrieving model parameters and characterizing the near-surface region using active-source surface waves for engineering applications, geotechnical studies, and physical modeling tests. The tomographic experiment in Section 4.2 indicated that joint inversion has another important advantage; that is, it can deal with the cycle-skipping problem, which easily occurs in FWI of surface waves in near-surface cases. This contributes to envelope-information usage in the Z/H ratio calculation, which avoids incorrect vs. model updates, leading to a secondary minimum. Due to limited sensitivity to the P-wave velocity model, it was found that the P-wave model also seemed to be not well constrained by the joint inversion. This can be improved by inclusion of body-wave data sets, such as the receiver function for large-scale inversions and seismic refractions for small-scale inversions, which is sensitive to the P-wave velocity.

Before application to real data, there are some issues that require more study. One is the difficulty of extracting stable Z/H-ratio measurements from a single source–station pair. Similarly, using amplitude information from an ambient-noise cross-correlation is not always reliable. Another issue is the 3D structural effect on both phase and Z/H-ratio kernels, which were not properly modeled in the 2D case. For example, the ray path may not follow the profile. Thus, the requirements for the linear profile need to be considered before a field experiment is conducted. In addition, the real data FWI is indeed more difficult in near-surface application because of many practical problems, such as waveform data-signal processing problems in signal-to-noise ratio improvement, the effect of rugged topography beneath real irregular receiver array acquisition, a very complicated near-surface effect, the random or man-made noise effect, and heterogeneous distribution of ambient-noise sources. Ignoring any issue in the real data waveform inversion could cause the inversion result to be unexplained or unstable. This paper focused on the theoretical and technical aspects, and the real data application will be discussed and published in future works.

6. Conclusions

In this study, we developed a full-waveform joint-inversion scheme for the Rayleigh wave phase and the Z/H ratio simultaneously. First, we developed an approach for calculating the 2D sensitivity kernels of the Rayleigh wave phases and the Z/H ratio based on the adjoint-state method. A numerical experiment was carried out to investigate the sensitivities of the Rayleigh wave phase and the Z/H ratio to the S-wave velocity structure. The experiments showed that they were both mostly sensitive to the shear-wave velocity, but their sensitivity patterns were different. The phase kernel was concentrated along the propagation path, with a much broader sensitivity range than the Z/H-ratio kernels. The Z/H ratio was more sensitive to the shallower structure compared to the Rayleigh wave phase-velocity dispersion at the same frequency. In addition, the Z/H ratio could deal with the lateral heterogeneity better, which was more helpful in constraining shallow structures, such as sedimentary basins or near-surface structures.

The joint-inversion method was applied in two synthetic experiments involving large- and small-scale inversions. By comparing the tomographic results of the adjoint tomography of the Rayleigh wave phases, the adjoint tomography of the Z/H ratio, and the joint inversion, we demonstrated that the joint inversion outperformed the separate inversions because it combined the complementary sensitivities of both, resulting in a more S-wave velocity unified model. Also, joint inversion could deal with the cycle-skipping problem in FWI in near-surface cases. The proposed joint-inversion scheme offers a computationally efficient and inexpensive alternative to imaging fine-scale crustal structures or near-surface structures beneath a 2D seismic array.

Author Contributions: C.Z. made the main contributions to this research, providing the ideas of modeling and inversion; T.L. was responsible for the simulation; Y.W. provided suggestions on the feasibility and innovation of the solution. All authors have read and agreed to the published version of the manuscript.

Funding: This research was funded by financially supported by the Natural Science Foundation of Jiangsu Province, China, grant number BK20190499; the National Natural Science Foundation of China, grant number 42004037; and the Fundamental Research Funds for the Central Universities grant number 2019B0071428.

Institutional Review Board Statement: Not applicable.

Informed Consent Statement: Not applicable.

Data Availability Statement: The data of this study is available from the authors upon request.

Acknowledgments: We gratefully acknowledge Qinya Liu's constructive comments from university of Toronto.

Conflicts of Interest: The authors declare no conflict of interest.

Appendix A

Limited-Memory BFGS Algorithm

The main algorithm is: given an initial model m_0, objective function f, diagonal scaling D, memory value l, and stopping threshold $\delta > 0$, the L-BFGS algorithm is as follows:

(1) Evaluate $f_0 = f(m_0), g_0 = \nabla f(m_0)$;
(2) Set $p_0 = -g_0, k = 0$;
(3) If $k > 0$, compute p_k from recursion, below;
(4) Compute αk by line search and set $m_{k+1} = m_k + \alpha_k p_k$;
(5) Evaluate $f_{k+1} = f(m_{k+1}), g_{k+1} = \nabla f(m_{k+1})$;
(6) Set $s_k = m_{k+1} - m_k, y_k = g_{k+1} - g_k, k = k + 1$;

(7) Repeat (3–6), stopping immediately when $g_{k+1}^T g_{k+1} < \delta$;

Recursion

(1) Set $q = g_k, i = k-1, j = min(k, l)$;

(2) Perform j times: $\lambda_i = \frac{s_i^T q}{y_i^T s_i}, q = q - \lambda_i y_i, i = i-1$;

(3) Set $\gamma = \frac{s_{k-1}^T y_{k-1}}{y_{k-1}^T y_{k-1}}, r = \gamma D q, i = k-j$;

(4) Perform j times: $\mu = \frac{y_i^T r}{y_i^T s_i}, r = r + s_i(\lambda_i - \mu), i = i+1$;

(5) End with result $p_k = -r$;

L-BFGS's relatively modest memory usage rests on the fact that if k is the current iteration number and l is the memory value, then vector pairs prior to {sk−l,yk−l} are no longer needed and can be removed from storage. The scaling factor γ, which accounts for differences between the true Hessian and the approximation thereto, is essential to the good performance of the algorithm.

References

1. Trampert, J.; Woodhouse, J.H. Global phase velocity maps of Love and Rayleigh waves between 40 and 150 seconds. *Geophys. J. Int.* **1995**, *122*, 675–690. [CrossRef]
2. Simons, F.J.; Van Der Hilst, R.D.; Montagner, J.P.; Zielhuis, A. Multimode Rayleigh wave inversion for heterogeneity and azmuthal anisotropy of the Australian upper mantle. *Geophys. J. Int.* **2002**, *151*, 738–754. [CrossRef]
3. Shapiro, N.M.; Ritzwoller, M.H. Monte-Carlo inversion for a global shear-velocity model of the crust and upper mantle. *Geophys. J. Int.* **2002**, *151*, 88–105. [CrossRef]
4. Sabra, K.G.; Gerstoft, P.; Roux, P.; Kuperman, W.; Fehler, M.C. Surface wave tomography from microseisms in Southern California. *Geophys. Res. Lett.* **2005**, *32*, L14311. [CrossRef]
5. Shapiro, N.M.; Campillo, M.; Stehly, L.; Ritzwoller, M.H. Highresolution surface-wave tomography from ambient seismic noise. *Science* **2005**, *307*, 1615–1618. [CrossRef]
6. Yang, Y.; Ritzwoller, M.H.; Levshin, A.L.; Shapiro, N.M. Ambient noise Rayleigh wave tomography across Europe. *Geophys. J. Int.* **2007**, *168*, 259–274. [CrossRef]
7. Lin, F.C.; Moschetti, M.P.; Ritzwoller, M.H. Surface wave tomography of the western United States from ambient seismic noise: Rayleigh and Love wave phase velocity maps. *Geophys. J. Int.* **2008**, *173*, 281–298. [CrossRef]
8. Tanimoto, T.; Alvizuri, C. Inversion of the HZ ratio of microseisms for S-wave velocity in the crust. *Geophys. J. Int.* **2006**, *165*, 323–335. [CrossRef]
9. Tanimoto, T.; Rivera, L. The ZH ratio method for long- period seismic data: Sensitivity kernels and observational techniques. *Geophys. J. Int.* **2008**, *172*, 187–198. [CrossRef]
10. Chong, J.; Ni, S.; Zhao, L. Joint inversion of crustal structure with the Rayleigh wave phase velocity dispersion and the ZH ratio. *Pure Appl. Geophys.* **2015**, *172*, 2585–2600. [CrossRef]
11. Komatitsch, D.; Tromp, J. Spectral-element simulations of global seismic wave propagation—I. *Validation. Geophys. J. Int.* **2002**, *149*, 390–412. [CrossRef]
12. Komatitsch, D.; Tromp, J. Spectral-element simulations of global seismic wave propagation—II. Three-dimensional models, oceans, rotation and self-gravitation. *Geophys. J. Int.* **2002**, *150*, 303–318. [CrossRef]
13. Bozdag, E.; Peter, D.; Lefebvre, M.; Komatitsch, D.; Tromp, J.; Hill, J.; Podhorszki, N.; Pugmire, D. Global adjoint tomography: First-generation model. *Geophys. J. Int.* **2016**, *207*, 1739–1766. [CrossRef]
14. Chen, M.; Niu, F.; Liu, Q.; Tromp, J.; Zheng, X. Multi-parameter adjoint tomography of the crust and upper mantle beneath East Asia—Part I: Model construction and comparisons. *J. Geophys. Res.* **2015**, *120*, 1762–1786. [CrossRef]
15. Fichtner, A.; Kennett, B.L.N.; Igel, H.; Bunge, H.-P. Full seismic waveform tomography for upper-mantle structure in the Australasian region using adjoint methods. *Geophys. J. Int.* **2009**, *179*, 1703–1725. [CrossRef]
16. Liu, Q.; Gu, Y.J. Seismic imaging: From classical to adjoint tomography. *Tectonophysics* **2012**, *566–567*, 31–66. [CrossRef]
17. Tape, C.; Liu, Q.; Maggi, A.; Tromp, J. Adjoint tomography of the southern California crust. *Science* **2009**, *325*, 988–992. [CrossRef] [PubMed]
18. Tape, C.; Liu, Q.; Maggi, A.; Tromp, J. Seismic tomography of the southern California crust based on spectral-element and adjoint methods. *Geophys. J. Int.* **2010**, *180*, 433–462. [CrossRef]
19. Tromp, J.; Tape, C.; Liu, Q. Seismic tomography, adjoint methods, time reversal and banana-doughnut kernels. *Geophys. J. Int.* **2005**, *160*, 195–216. [CrossRef]
20. Zhu, H.; Tromp, J. Mapping tectonic deformation in the crust and upper mantle beneath Europe and the North Atlantic Ocean. *Science* **2013**, *341*, 871–875. [CrossRef]

21. Zhu, H.; Bozdağ, E.; Tromp, J. Seismic structure of the European upper mantle based on adjoint tomography. *Geophys. J. Int.* **2015**, *201*, 18–52. [CrossRef]
22. Zhu, H.; Bozdağ, E.; Peter, D.; Tromp, J. Structure of the European upper mantle revealed by adjoint tomography. *Nat. Geosci.* **2012**, *5*, 493–498. [CrossRef]
23. Chen, M.; Huang, H.; Yao, H.; van der Hilst, R.D.; Niu, F. Low wave speed zones in the crust beneath SE Tibet revealed by ambient noise adjoint tomograph. *Geophys. Res. Lett.* **2014**, *41*, 334–340. [CrossRef]
24. Gao, H.; Shen, Y. Upper mantle structure of the Cascades from full-wave ambient noise tomography: Evidence for 3D mantle upwelling in the back-arc. *Earth Planet. Sci. Lett.* **2014**, *390*, 222–233. [CrossRef]
25. Gao, H. Three-dimensional variations of the slab geometry correlate with earthquake distributions at the Cascadia subduction system. *Nat. Commun.* **2018**, *9*, 1204. [CrossRef] [PubMed]
26. Savage, B.; Covellone, B.M.; Shen, Y. Wave speed structure of the eastern North American margin. *Earth Planet. Sci. Lett.* **2017**, *459*, 394–405. [CrossRef]
27. Bao, X.Y.; Shen, Y. Full-waveform sensitivity kernels of component- differential traveltimes and ZH amplitude ratios for velocity and density tomography. *J. Geophys. Res. Solid Earth* **2018**, *123*, 4829–4840. [CrossRef]
28. Chen, P.; Jordan, T.H.; Zhao, L. Full three-dimensional tomography: A comparison between the scattering-integral and adjoint-wavefield methods. *Geophys. J. Int.* **2007**, *170*, 175–181. [CrossRef]
29. Zhang, W.; Shen, Y.; Zhao, L. Three-dimensional anisotropic seismic wave modelling in spherical coordinate by a collocated-grid finite-difference method. *Geophys. J. Int.* **2012**, *188*, 1359–1381. [CrossRef]
30. Zhang, Z.; Shen, Y. Cross-dependence of finite-frequency compressional waveforms to shear seismic wave speeds. *Geophys. J. Int.* **2008**, *174*, 941–948. [CrossRef]
31. Zhao, L.; Jordan, T.H.; Olsen, K.B.; Chen, P. Fréchet kernels for imaging regional Earth structure based on three-dimensional reference models. *Bull. Seismol. Soc. Am.* **2005**, *95*, 2066–2080. [CrossRef]
32. Zhang, C.; Yao, H.; Liu, Q.; Zhang, P.; Yuan, Y.O.; Feng, J.; Fang, L. Linear Array Ambient Noise Adjoint Tomography Reveals Intense Crust Mantle Interactions in North China Craton. *J. Geophys. Res. Solid Earth* **2018**, *123*, 368–383. [CrossRef]
33. Tong, P.; Chen, C.; Komatitsch, D.; Basini, P.; Liu, Q. High-resolution seismic arra-y imaging based on a SEM-FK hybrid method. *Geophys. J. Int.* **2014**, *197*, 369–395. [CrossRef]
34. Zhang, C.; Yao, H.; Tong, P.; Liu, Q.; Lei, T. Joint inversion of linear array ambient noise surface-wave and teleseismic body-wave data based on an adjoint-state method. *Chin. J. Geophys.* **2020**, *63*, 4065–4079.
35. Romdhane, G.; Grandjean, G.; Brossier, R.; Rejiba, F.; Operto, S.; Virieux, J. Shallow-structure characterization by 2D elastic full-waveform inversion. *Geophysics* **2011**, *76*, R81–R93. [CrossRef]
36. Nuber, A.; Manukyan, E.; Maurer, H. Enhancement of near-surface elastic full waveform inversion results in regions of low sensitivities. *J. Appl. Geophys.* **2015**, *122*, R192–R201. [CrossRef]
37. Tran, K.; McVay, M.; Faraone, M.; Horhota, D. Sinkhole detection using 2D full seismic waveform tomography. *Geophysics* **2013**, *78*, R175–R183. [CrossRef]
38. Groos, L. 2D Full Waveform Inversion of Shallow Seismic Rayleigh Waves. Ph.D. Thesis, Karlsruhe Institute of Technology, Karlsruhe, Germany, 2013.
39. Forbriger, T.; Groos, L.; Schäfer, M. Line-source simulation for shallow-seismic data—Part 1: Theoretical background. *Geophys. J. Int.* **2014**, *198*, 1387–1404. [CrossRef]
40. Schäfer, M.; Groos, L.; Forbriger, T.; Bohlen, T. Line-source simulation for shallow-seismic data. Part 2: Full-waveform inversion—A synthetic 2D case study. *Geophys. J. Int.* **2014**, *198*, 1405–1418. [CrossRef]
41. Groos, L.; Schäfer, M.; Forbriger, T.; Bohlen, T. Application of a complete workflow for 2D elastic full-waveform inversion to recorded shallow-seismic Rayleigh waves. *Geophysics* **2017**, *82*, R109–R117. [CrossRef]
42. Dahlen, F.; Baig, A. Fréchet kernels for body-wave amplitudes. *Geophys. J. Int.* **2002**, *150*, 440–466. [CrossRef]
43. Liu, D.; Nocedal, J. On the limited memory BFGS method for large scale optimization. *Math. Program.* **1989**, *45*, 504–528. [CrossRef]

applied sciences

MDPI

Article

Impact of Post-Earthquake Seismic Waves on the Terrestrial Environment

Xiangzhi Zeng [1,2] and Wencai Yang [2,3,*]

1 Institute of Geophysics & Geomatics, China University of Geosciences (Wuhan), Wuhan 430074, China; zeng.xz.rc@gmail.com
2 Institute of Geology, Chinese Academy of Geological Sciences, Beijing 100037, China
3 Institute of Geophysics, School of Earth Sciences, Zhejiang University, Hangzhou 310027, China
* Correspondence: yangwencai@cashq.ac.cn

Abstract: When an earthquake occurs, it is not only the crustal material in the seismic zone that moves violently; the seismic waves of the earthquake with certain level of energy can act on the crust over large areas, producing remote effects that affect the living environment. According to the records of the long-term observation station of Chinese Continental Scientific Drilling, the effects of near-surface crust caused by the post-earthquake seismic waves include the following four aspects: (1) the pore fissure loosening; (2) the pore pressure rising and groundwater upwelling; (3) gas releasing; and (4) exothermic reaction. The effects of groundwater upwelling, gas releasing and exothermic reaction may be superimposed on the process of global warming, which has a certain impact on the terrestrial environment and requires further studies.

Keywords: crustal movement; long-term observation; seismic waves; remote effects; terrestrial environment

1. Introduction

It has long been known that when an earthquake occurs, some crustal material in the seismic zones moves, and gas gushes out from the active faults [1–4]. However, at the same time, if large- and medium-sized earthquakes produce a certain amount of energy of seismic waves spreading widely in the land, will it affect the human living environment? If so, to what extent? Based on the analysis of long-term underground observation records, we briefly analyze the recorded data of the crustal movement observation station of Chinese Continental Scientific Drilling (CCSD), in order to understand the remote effects caused by vibration energy of the seismic waves.

2. CCSD Long-Term Crustal Movement Observations

The CCSD project was completed in 2005 by a borehole of 5158 m and achieved fruitful results [3–12]. In February 2007, the Institute of Geology of the Chinese Academy of Geological Sciences (CAGS) began to use this borehole to establish a long-term crustal movement observation station (referred to "CCSD-LOS" below). The CCSD-LOS is located in the eastern part of the Dabie-Sulu ultra-high metamorphic belt and is the first long-term observation experimental base with an observation depth of more than 4000 m in China. The corresponding author of this paper is the scientific consultant and planning master of the CCSD-LOS. Senior engineer Pi Jinyun is responsible for the installation of the instruments and equipment, and Zeng Xiangzhi, the first author of this paper, is responsible for data processing and management [11–15].

The CCSD-LOS is located at Donghai County in Jiangsu Province, with latitude of 34.4° N and longitude of 118.5° E (Figure 1a). In 2010, the CCSD-LOS installs two sets of high-precision digital seismographs in the borehole at four depths (544 m, 1559 m, 2545 m, and 4190 m) and on the ground as well. Compared with the single ground observation, deep-well seismic observations provide much better signals. A new 452 m hole was drilled at the CCSD-LOS in 2015. More than a dozen observational instruments, including high-precision measurements of stress–strain components, tilt, pore pressure, displacement, radon concentration, geomagnetic, temperature and others [14]. The station aims to achieve the long-term monitoring of crustal movement in eastern China. Because it is only 24 km apart from the Tanlu fault zone, its observation of overtime records may show the mechanical behavior of the active faults, and the changes of regional human living environment as well. The observation borehole of the CCSD-LOS is shown in Figure 1b, some underground probes are shown Figure 1c, and its supervisory control room is shown in Figure 1d.

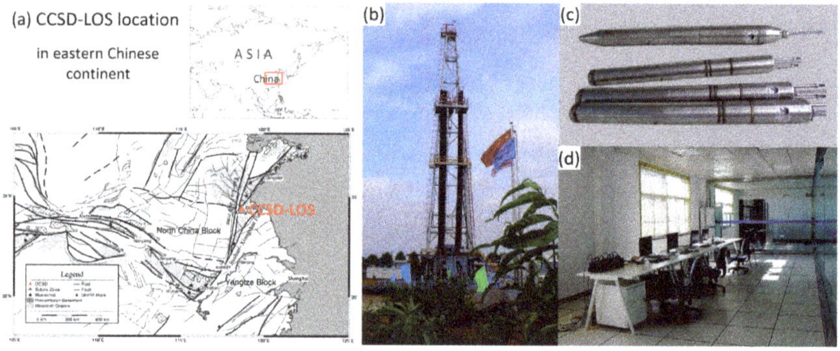

Figure 1. The CCSD long-term crustal movement observation station (CCSD-LOS). (**a**) The location of the station in China; (**b**) the observation borehole of the CCSD-LOS. The picture shows the derrick and drill pipes for the installation of instruments, with a 5158 m-deep hole below; (**c**) some underground probes installed in the boreholes. They have now been installed in 4 layers underground; and (**d**) the supervisory control room.

3. Gas Anomalies in the CCSD-LOS during the Magnitude-9 Earthquake in Sumatra

In situ fluid analysis was carried out during the CCSD drilling in 2002–2005. The drilling mud was degassed by using the freeze-drying method to remove steam from the gas, then it was extracted from vacuum pump to the OmniStar spectrometer for on-line analysis of gas components, in order to measure real-time concentration data of He, CH_4, CO_2, H_2, Ar, He, N_2 and other components [16,17]. Figure 2 shows the variations of methane, carbon dioxide and hydrogen with depths observed in the main borehole during the scientific drilling [11]. It can be seen from the records that a high concentration of methane gas rose abnormally in the underground fracture rocks from 2400 m to 3700 m. High concentrations of carbon dioxide and hydrogen are abnormal in a deeper formation below 4900 m. Since these high concentration anomalies of gases are located far from the surface, they seem not coming from the atmosphere. The observation records (Figure 2) prove that certain amount of some gases, such as methane, carbon dioxide and hydrogen, may originate from the deep Earth as their contents usually increases with the depth. The questions are, do they come to the near-surface strata and, if so, how?

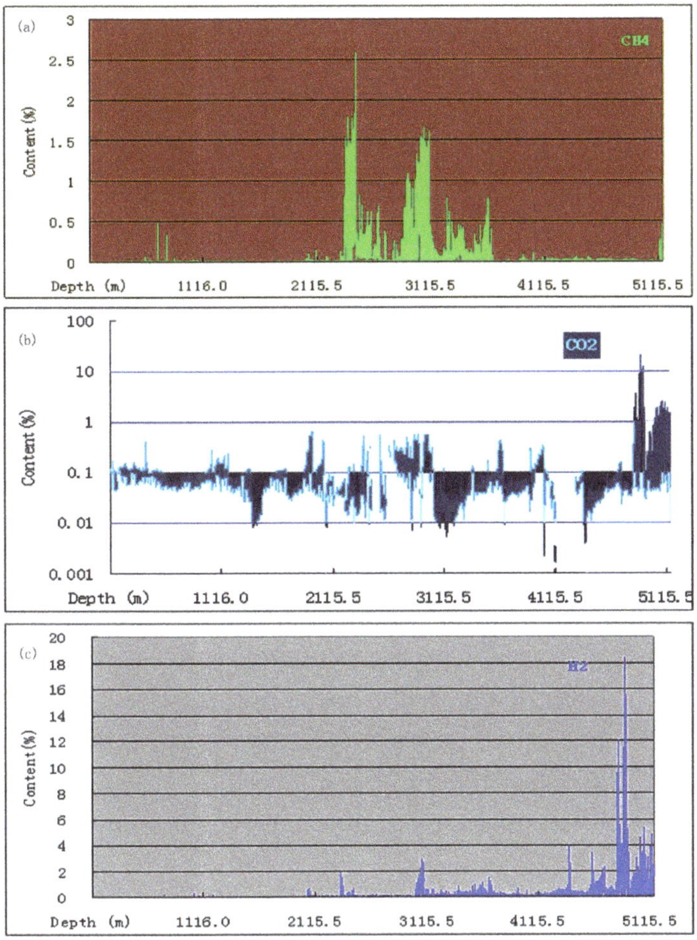

Figure 2. The gas anomaly records observed during the CCSD drilling: (**a**) methane, (**b**) carbon dioxide, and (**c**) hydrogen.

During the CCSD project performance, it is found that the gas contents in the borehole were changed significantly after big earthquakes. Records of gas contents during main earthquakes of magnitude 8 or larger occurred in Asia were obtained by the CCSD project; among them, the magnitude 8.7–9 Sumatra earthquake, occurring on December 26th of 2004, is outstanding. The earthquake occurred near the northern part of Sumatra, the focal coordinates are 2.30° N and 93.10° E with the epicenter depth of 20.0 km. Figure 3a shows a three-component record of the Sumatra earthquake observed on the CCSD-LOS during scientific drilling, the depth of the seismograph detector is at 1559 m. The strong underground vibrations caused by the seismic waves were clear at the drilling site, and the surface wave caused by the earthquake lasted for 4–5 days after the earthquake, then slow down.

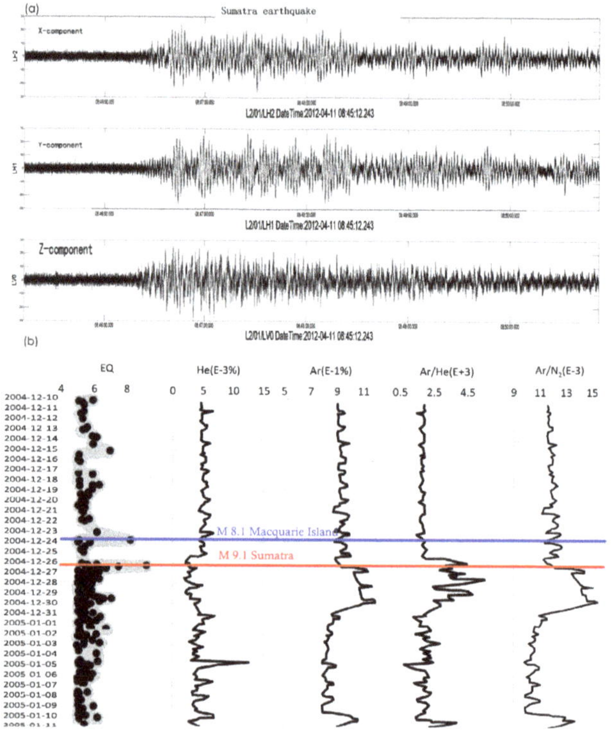

Figure 3. (**a**) A 3-component record of Sumatra earthquake observed. (**b**) Real-time gas anomalies observed during the CCSD drilling; the vertical ordinate is the date of observation; columns from the left to right are the seismic magnitude, records of helium, argon, argon/helium, and argon/nitrogen, respectively.

Figure 3b shows the gas content curves of the borehole fluids underground observed during the CCSD drilling from 7 December 2004 to 11 January 2005. The vertical ordinate is the date of observation; columns from the left to right are the seismic magnitude, records of helium, argon, argon/helium, argon/nitrogen, respectively. It shows that, from December 10 to 26, some earthquakes of magnitude 6 occurred in Sumatra. They caused fluctuations in the concentration of helium and argon slightly. After the magnitude 8.7–9 earthquake in Sumatra on December 26th, helium concentration slightly decreased, and the argon concentration sharply increased. The high argon anomaly lasted for about five days, as well as the ratio of Ar/He or Ar/N_2.

Positive anomalies of argon, argon/helium and argon/nitrogen ratios after the strong earthquakes indicate that there is a certain proportion of high Ar and low He fluid upwelling into the rocks below the CCSD-LOS after the earthquake. The distance between the CCSD-LOS and the epicenter of Sumatra earthquake is about 4000 km. This fluid upwelling at the CCSD-LOS after the earthquake could not come directly from the Sumatra seismic zone. However, seismic waves of the earthquake came about 17 min after the earthquake, consistent with the CCSD high Ar concentration anomalies. The fact indicates that these gas anomalies were triggered by the seismic waves triggered by this earthquake. Based on a theory we know, a sudden drop of the stress field in the seismic zones stimulates the crustal fluid upwelling and causes geoelectric, geomagnetic and other anomalies [18–20]. However, there have been no detailed studies on the long-distance effects of a strong earthquake, which cover thousands of kilometers from the epicenter. The surface waves from a big earthquake are relatively strong and slow to decay, and can spread in the global crust for several days. The energy of the seismic surface waves loosens the pore fissures exist in underground strata and promotes pore fracture connectivity, causing fluid release from the deep crust. The records of Figure 3 prove that there was a crustal gas-releasing effect after the strong earthquake occurred. If the post-earthquake seismic waves cause the corresponding degassing effect and fluid upwelling, it may have an impact on the terrestrial environment. However, the occurrence of earthquakes as strong as Sumatra is unusual; on average, this occurs once every a few years. Moderate earthquakes have much higher frequency of occurrence. If the post-earthquake seismic waves of a moderate earthquake also produce large-scale crustal effects and fluid movement, its impact on the living environment cannot be ignored. The question is, does a moderate earthquake also produce the large-scale crustal effects of fluid motion? Let us see once more a set of records of a typical moderate-intensity earthquake observed at the CCSD-LOS.

4. Multi-Parameter Anomalies during the Magnitude-5 Earthquake in Philippine

The most typical of moderate earthquakes was the magnitude-5 earthquake in the Philippines, which occurred on 17 June 2018. The coordinates of the quake epicenter are 20° N and 119.82° E. The distance between the CCSD-LOS and the epicenter is about 2000 km. As the installation of various geophysical monitoring instruments in 2015, we have rich comprehensive observation data about the Philippine earthquake, including the stress, magnetic field, pore pressure, temperature, water level, and gas radon.

Figure 4 shows the variation of some underground monitoring parameters before and after the earthquake in the Philippines, including the in situ stress components of multiple directions, the geomagnetic field components of six directions, the in situ tilt components of two directions, the groundwater level, the gas radon content, the underground temperature, pressure, and the pore pressure at depth of 500 m. One can see that, after the earthquake, the crustal stress in the CCSD-LOS had a sharp jump, lasting only a few hours. At the same time, the pore pressure also synchronously had a sudden jump. After the earthquake, the groundwater level, radon content, formation temperature and pressure all changed obviously. The underground water level in the shallow observation hole had been rising continuously for 4 months, and the rising amplitude has reached 0.4 m. Underground formation temperature rose for 2 months; the rise reached 25 °C, and then slowly decreased. Radon levels rose more than two days after the earthquake and then returned to a baseline; however, they rose again two months later. The formation pressure fluctuated obviously for 3 months before the earthquake, and then rose suddenly after the quake for another 3 months. The pore pressure of the formation began to rise slowly half a year before the earthquake, with a sudden jump of several hours after the earthquake, and then a slow rise of three months thereafter. The facts show that moderate-intensity earthquakes also produce remote effects, which must be caused by large-scale propagation of the post-earthquake seismic waves. The effects include a stress sharp and pore pressure rise for hours, gas radon content sudden jump for 2–3 days, and continuously delayed rise of groundwater level, as well as a rise in the formation temperature and pore pressure. According to these observation records, we find that the range of remote effects from moderate-intensity earthquakes can be over 2000 km.

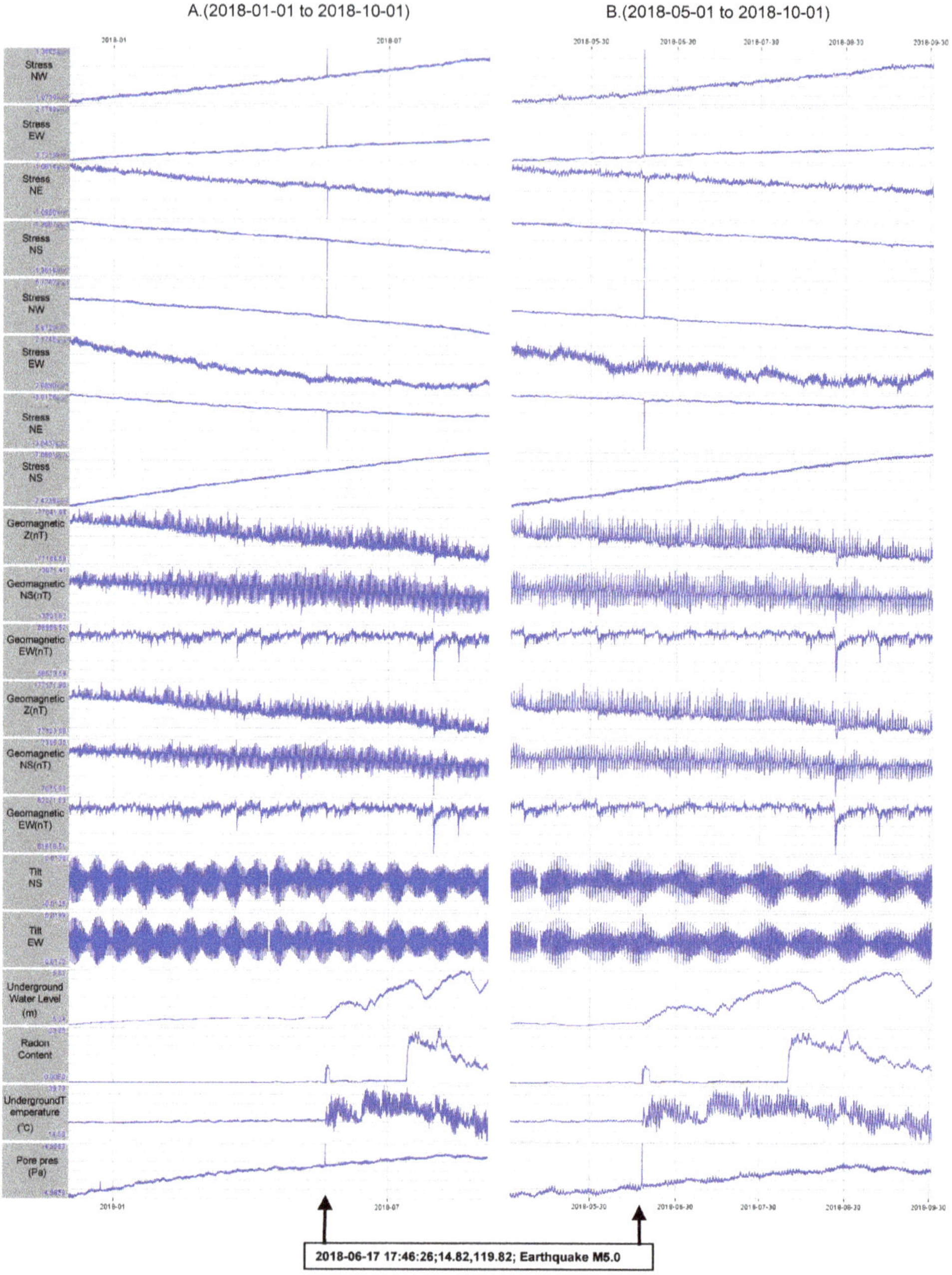

Figure 4. Records of underground monitoring parameters before and after the Philippine magnitude-5 earthquake. The monitoring parameters from the top to the bottom are as follows: the in situ stress components of eight azimuths, geomagnetic field components of six azimuths, the earth tilt components of two azimuths, underground water level, gas radon content, underground formation temperature and pressure, and pore pressure.

The monitoring parameters also include the earth tilt component and the geomagnetic field components in six directions, whose changes are not clear in Figure 4. In fact, they also have minor changes after the Philippines earthquake. Figure 5 shows these magnified parameter records. It shows that the geomagnetic component in the east–west direction increases obviously for 3 days after the earthquake, and then decreases obviously for about 100 days. Both ground tilt components had a small, serrated jump within 3 days after the earthquake. Thus, moderate-intensity earthquakes also have minor remote influences on the geomagnetic fields and ground tilt.

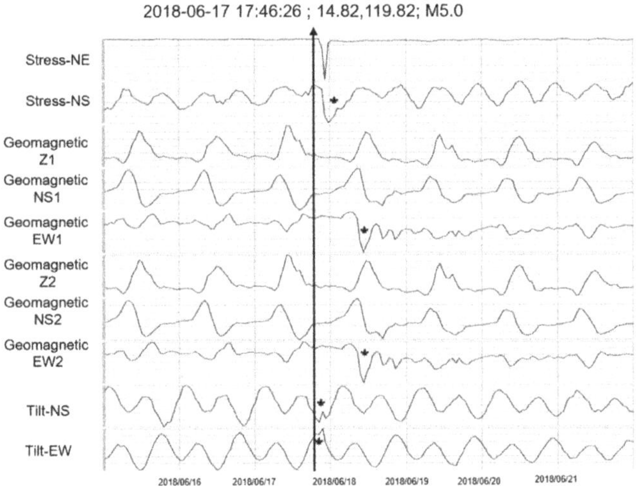

Figure 5. The magnified records of some parameters in Figure 4: the two stress components, six geomagnetic field components, and two earth tilt components.

5. Conclusions

In sum, earthquakes with moderate intensity or larger can generate remote effects on near-surface spaces in a large area of more than 2000 km. The effects are caused by the vibration energy of seismic waves, especially surface waves. The effects cause the following four types of changes in the near surface: (1) pore fissure loosening, i.e., vibration of the post-earthquake waves make pore fissure loosening and connecting; (2) the pore-pressure rising and groundwater upwelling. Because fractures connect, fluid motion makes pore pressure increase and fluid moves, resulting in groundwater upwelling; (3) gas releasing through the connected fractures from the deeper crust. The releasing gas include methane, carbon dioxide, hydrogen, helium, argon, and radon; and (4) exothermic reaction. Mass movement releases heat energy that may cause the exothermic reaction, which has a certain impact on the terrestrial environment. The degree of the effects requires further studies.

Author Contributions: X.Z., the first author, is responsible for data processing, management, as well as discovering and providing the data and figures used in this paper, joined determination of the conclusions. W.Y., finding scientific meanings in the data and writing the draft of this paper, joined determination of the estimation scheme and supplied the constrains. Both authors have read and agreed to the published version of the manuscript.

Funding: This research received no external funding.

Institutional Review Board Statement: Not applicable.

Informed Consent Statement: Not applicable.

Data Availability Statement: http://218.92.80.98:8005/GetData.html.

Acknowledgments: This study has been supported by Institute of Geology, Chinese Academy of Geolodical Sciense (CAGS) since 2007. We appreciate Pi Jinyun at the Institute of Geology for his great technical supports, and Xia Jianghai at Zhejiang University for constructive discussions and modifying the draft of the paper.

Conflicts of Interest: The authors declare that they have no conflict of interests.

References

1. Faber, E. Origin of hydrocarbon gases in the pump-test of the KTB pilot well. *Sci. Drill.* **1995**, *5*, 123–128.
2. Zhou, X.C.; Sun, F.X.; Chen, Z.; Lu, C.; Wu, K.; Du, J.; Li, J. Degassing of CO_2, CH_4, Rnand and Hg in the rupture zones produced by Wenchuan Ms8.0 earthquack. *Acta Petrol. Sin.* **2017**, *33*, 291–303. (In Chinese with English abstract) [CrossRef]
3. Zhiqin, X. The scientific goals and inversigation progresses of the Chinese Continental Scientific Drilling project. *Acta Petrol. Sin.* **2004**, *20*, 1–8. (In Chinese with English abstract)
4. Xu, Z.Q.; Li, H.B.; Wu, Z.L. Wenchuan earthquake and scientific drilling. *Acta Geol. Sin.* **2008**, *82*, 1613–1622. (In Chinese with English abstract)
5. Yang, W.C. Geophysical profiling across the Sulu ultra-high-pressure metamorphic belt, Eastern China. *Tectonophysics* **2002**, *354*, 277–288. [CrossRef]
6. Xu, Z.Q.; Yang, J.S.; Zhang, Z.M.; Liu, F.L.; Yang, W.C.; Jin, Z.M.; Wang, R.; Luo, L.; Huang, L.; Dong, H.L. Completion and achievement of the Chinese continental scientific drilling project. *Geol. China* **2005**, *32*, 177–183. (In Chinese with English abstract)
7. Yang, W.-C. Flat mantle reflectors in eastern China: Possible evidence of lithospheric thinning. *Tectonophysics* **2003**, *369*, 219–230. [CrossRef]
8. Yang, W.-C.; Yang, W.-Y.; Jin, Z.-M.; Cheng, Z.-Y. Lithospheric seismic fabrics of Sulu ultrahigh-pressure metamorphic belt. *Sci. China Ser. D* **2005**, *48*, 585–600.
9. Yang, W.-C.; Chen, Z.-D. Seismic multi-arch structures in East China. *Sci. China Ser. D* **2006**, *49*, 133–146. [CrossRef]
10. He, L.; Hu, S.; Huang, S.; Yang, W.; Wang, J.; Yuan, Y.; Yang, S. Heat flow study at the Chinese Continental Scientific Drilling site: Borehole temperature, thermal conductivity, and radiogenic heat production. *J. Geophys. Res.* **2008**, *113*, B02404. [CrossRef]
11. Yang, W.; Jin, Z.; Yu, C. Seismic response to natural gas anomalies in crystalline rocks. *Sci. China Ser. D* **2008**, *51*, 1726–1736. [CrossRef]
12. Xu, Z.; Yang, W.; Ji, S.; Zhang, Z.; Yang, J.; Wang, Q.; Tang, Z. Deep root of a continent-continent collision belt: Evidences from the CCSD deep borehole in the Sulu ultrahigh pressure metamorphic terrane in China. *Tectonophysics* **2009**, *475*, 204–219. [CrossRef]
13. Yang, W.-C. The crust and upper mantle of the Sulu UHPM belt. *Tectonophysics* **2009**, *475*, 226–234. [CrossRef]
14. Zhiqin, X.; Wencai, Y.; Jingsui, Y.; Zhisheng, A.; Chengshan, W.; Haibing, L.; Xiaoxi, Z.; Jiaqi, L.; Dechen, S.; Bizhu, H.; et al. 15 years of hardship and struggle history and the prospects for the future of the Chinese continental scientific drilling program (CCSD): In memory of the 15 year anniversary of CCSD and 20 year anniversary of ICDP. *Acta Geol. Sin.* **2016**, *90*, 2109–2122. (In Chinese with English abstract)
15. Yang, W.-C.; Xu, J.R.; Chen, Z.Y.; Hou, Z.Z. *Regional Geophysics and Crust-Mantle Interaction in Sulu-Dabie Orogenic Belt*; Geological Publishing House: Beijing, China, 2005; pp. 1–151. (In Chinese)
16. Luo, L.Q.; Sun, Q.; Zhan, X.C. 0–2000 m fluid profiles and sources in Chinese continental scientific drilling project. *Acta Petrol. Sin.* **2004**, *20*, 185–191. (In Chinese with English abstract)
17. Sun, Q.; Luo, L.Q.; Li, S.Q. N2-Ar-He compositions in the 0–2000 m mud of Chinese continental scientific drilling. *Acta Petrol. Sin.* **2004**, *20*, 179–184. (In Chinese with English abstract)
18. Mitrofanov, F.P.; Yakovlev, Y.N.; Ikorsky, S.V.; Yakovleva, A.K.; Vetrin, V.R.; Neradovsky, Y.; Tolstikhin, I.N.; Lanev, V.S.; Smirnov, Y.P.; Rusanov, M.S. A change on composition of rocks, mineral phases and trapped gases in the Kola super-deep borehole (SD-3) section of the Archean complex with depth. In *Super-Deep Continental Drilling and Deep Geophysical Sounding*; Fuchs, K., Kozlovsky, Y.A., Krivtsov, A.I., Zoback, M.D., Eds.; Springer: Berlin/Heidelberg, Germany, 1990; pp. 353–363.
19. Faber, E.; Gerling, P.; Dumke, I. Gaseous hydrocarbons of unknown origin found while drilling. *Org. Geochem.* **1988**, *13*, 875–879. [CrossRef]
20. Möller, P.; Weise, S.M.; Althaus, E.; Bach, W.; Behr, H.J.; Borchardt, R.; Bräuer, K.; Drescher, J.; Erzinger, J.; Faber, E.; et al. Paleofluids and recent fluids in the upper continental crust: Results from German continental drilling program (KTB). *J. Geophys. Res.* **1997**, *102*, 18233–18254. [CrossRef]

applied sciences

MDPI

Article

Lp-Norm Inversion of Gravity Data Using Adaptive Differential Evolution

Tao Song [1,*], Xing Hu [2], Wei Du [3,*], Lianzheng Cheng [3,4,5], Tiaojie Xiao [6] and Qian Li [7]

1 College of Mathematics and Information Science, Guiyang University, Guiyang 550005, China
2 School of Mathematics and Statistics, Yunnan University, Kunming 650500, China; huxing@mail.ynu.edu.cn
3 State Key Laboratory of Ore Deposit Geochemistry, Institute of Geochemistry, Chinese Academy of Sciences, Guiyang 550081, China; chenglianzh10@mails.ucas.ac.cn
4 School of Earth Sciences, Yunnan University, Kunming 650500, China
5 Yunnan Key Laboratory of Statistical Modelling and Data Analysis, School of Mathematics and Statistics, Yunnan University, Kunming 650091, China
6 Science and Technology on Parallel and Distributed Processing Laboratory, National University of Defense Technology, Changsha 410073, China; xiaotiaojie@nudt.edu.cn
7 Shanxi Institute of Geophysical and Geochemical Exploration, Yuncheng 044004, China; liqian-1025@163.com
* Correspondence: sx0051@gyu.edu.cn (T.S.); du-wei@mail.gyig.ac.cn (W.D.)

Abstract: As a popular population based heuristic evolutionary algorithm, differential evolution (DE) has been widely applied in various science and engineering problems. Similar to other global nonlinear algorithms, such as genetic algorithm, simulated annealing, particle swarm optimization, etc., the DE algorithm is mostly applied to resolve the parametric inverse problem, but has few applications in physical property inversion. According to our knowledge, this is the first time DE has been applied in obtaining the physical property distribution of gravity data due to causative sources embedded in the subsurface. In this work, the search direction of DE is guided by better vectors, enhancing the exploration efficiency of the mutation strategy. Besides, to reduce the over-stochastic of the DE algorithm, the perturbation directions in mutation operations are smoothed by using a weighted moving average smoothing technique, and the Lp-norm regularization term is implemented to sharpen the boundary of density distribution. Meanwhile, in the search process of DE, the effect of Lp-norm regularization term is controlled in an adaptive manner, which can always have an impact on the data misfit function. In the synthetic anomaly case, both noise-free and noisy data sets are considered. For the field case, gravity anomalies originating from the Shihe iron ore deposit in China were inverted and interpreted. The reconstructed density distribution is in good agreement with the one obtained by drill-hole information. Based on the tests in the present study, one can conclude that the Lp-norm inversion using DE is a useful tool for physical property distribution using gravity anomalies.

Keywords: gravity inversion; Lp-norm; adaptive differential evolution; weighted moving average smooth

1. Introduction

As an important geophysical approach, gravity surveying is an effective tool for explorations dealing with mineral deposits, oil and gas, and environmental problems [1–5]. Over the past decades, despite substantial progress having been made in the interpretation of gravity data, the identification and characterization of field sources remain a challenging task due to several limitations: the inherent non-uniqueness of the gravity inverse problem and the limited and inaccurate discrete data sets [6,7]. To deal with the non-uniqueness issue, some kinds of regularizations and/or constraint(s) integrating with other available geologic or geophysical information, such as subsurface structure, borehole data, and prior models, etc. are taken into account in the inversion process.

Generally, according to the kind of model parameters selected, there are two basic approaches to gravity inversion—parametric inversion (e.g., [7–10]) and physical property inversion (e.g., [2,3,11–13]). The methods of parametric inversion determine the geometrical properties of unknown density sources with a fixed density contrast, and offer a homogeneous body of the postulated density. Compared with parametric inversion, the physical property inversion methods allow the densities of the elements of a regular subsurface partition to vary. In this case, the solution can give great flexibility to recover the depths and complex shape of sources. Therefore, the inversion for physical properties is currently important and prevalent. As we know, inversion methods search for the possible solutions employing optimization techniques with linearly iterative approaches such as steepest descent method, conjugate gradients, etc. These optimization techniques have traditionally proven very difficult to solve the highly non-linear mathematical formulation since the iteration process can be prone to fall into the local minima. Moreover, considering the inherent ambiguity of inverse problem for the potential field, it is necessary to choose an appropriate starting model and add regularization constraints (see [2,14]). To avoid the weak points of the linearly iterative optimization procedures, global optimization techniques, such as genetic algorithm (GA), simulated annealing (SA), particle swarm optimization (PSO), differential evolution (DE), etc., are considered appropriate to solve the inverse problems [6,7,15–17]. Among them, genetic algorithms have been successfully used for physical property inversion of gravity data (e.g., [6,18]).

Differential evolution, which has been illustrated as a simple but efficient evolution algorithm, was first proposed by Storn and Price to solve Chebyshev polynomial problems [19]. In recent years, it has been verified that DE provides better accuracy, more robustness, and fast convergence speed than other evolution algorithms, such as GA and PSO [20–22]. Hence, it has attracted great attention in many scientific and engineering fields, such as neural network training [23–26], data mining [27,28], and image processing [29–32], etc. With regards to geophysical inverse problems, the applications of DE are mainly non-gravity methods. Balkaya [33] implemented an inversion program for self-potential (SP) and vertical electrical sounding (VES) by DE; the synthetic data and field data demonstrate the effectiveness of DE in geoelectrical inversion. Li et al. [34] proposed a modified Boltzmann Annealing Differential Evolution (BADE) algorithm, which uses an annealing strategy to avoid the local minima and solve the inversion problem in the directional resistivity logging-while-drilling (DRLWD) measurements; the results also show robustness and immunity to the non-uniqueness inversion problem of their method. Ekinci et al. [17,35] applied DE to both inverting model parameters from residual gravity anomalies and amplitude of the 2D analytic signal of magnetic anomalies. Their works show the applicability and effectiveness of this algorithm on both synthetic and field anomalies, but they mainly focus on the basic parameters of the field source. Balkaya et al. [36] achieved a DE inversion for total field magnetic anomalies caused by vertical-sided prismatic bodies, and they mainly focus on geometric structure inversion. Recently, Du et al. [37] reconstructed magnetic susceptibility by using an improved adaptive DE. According to the inverted results, their inversion methods can help maintain the sharp boundary of magnetic sources.

Apart from the above-mentioned studies, DE has few applications in physical property inversion, although this method has become popular for geophysical inversion. The main reason is that in DE iteration, the smooth constraint is difficult to apply directly to physical property parameters. Here, we present a DE algorithm incorporated with the lp regularization technique to resolve the inversion problem by determining the 2-D distribution of the subsurface density distribution in the mineral exploration. Our gravity inversion method is tested with synthetic examples, and then applied to an iron ore deposit. Finally, we briefly discuss the application of lp norm gravity inversion using DE.

2. Differential Evolution

2.1. Standard Differential Evolution Algorithm

DE is a population-based heuristic search method suitable for solving numerical optimization problems. Its search process includes four parts: initialization, mutation, crossover, and selection. Without loss of generality, let the optimization problem in D-dimensional space be $f(x), x \in \mathbb{R}^D$. For the G-th iteration, any i-th vector in the population can be expressed as $m_i^G = \left(m_{i,1}^G, m_{i,2}^G, \cdots, m_{i,D}^G\right)$, in which $i = 1, 2, 3 \cdots, NP$, NP is the population size. The evolution of these vectors is accomplished by repeatedly performing mutation, crossover, and selection operations.

(1) Initialization

In order to establish a starting point for the optimization process, the DE randomly generates the decision parameter in every vector of initial population according to the given range. Therefore, the j-th variable of the i-th vector can be defined as:

$$m_{ij}^0 = L_j + rand(0,1) \cdot (U_j - L_j), j \in [1, D] \tag{1}$$

In which, $rand(0,1)$ represents a uniformly distributed random number in the range of $(0,1)$, the lower and upper bounds of the j-th variable is expressed as L_j and U_j respectively. Generally, the upper and lower bounds of variables can be expressed as vectors, such as $L = (L_1, L_2, \ldots, L_D)$, $U = (U_1, U_2, \ldots, U_D)$.

(2) Mutation

After initialization, the mutation vector $\boldsymbol{v}_i^G$ is generated for each target vector $\boldsymbol{m}_i^G$ in the current population with a suitable mutation strategy. In essence, the mutation strategy of differential evolution can be regarded as a linear combination of multiple vectors. The widely used strategies are as follows (see [38]):

"DE/rand/1":

$$\boldsymbol{v}_i^G = \boldsymbol{m}_{r1}^G + F(\boldsymbol{m}_{r2}^G - \boldsymbol{m}_{r3}^G) \tag{2}$$

"DE/rand/2":

$$\boldsymbol{v}_i^G = \boldsymbol{m}_{r1}^G + F(\boldsymbol{m}_{r2}^G - \boldsymbol{m}_{r3}^G) + F(\boldsymbol{m}_{r4}^G - \boldsymbol{m}_{r5}^G) \tag{3}$$

"DE/best/1":

$$\boldsymbol{v}_i^G = \boldsymbol{m}_{best}^G + F(\boldsymbol{m}_{r1}^G - \boldsymbol{m}_{r2}^G) \tag{4}$$

"DE/best/2":

$$\boldsymbol{v}_i^G = \boldsymbol{m}_{best}^G + F(\boldsymbol{m}_{r1}^G - \boldsymbol{m}_{r2}^G) + F(\boldsymbol{m}_{r3}^G - \boldsymbol{m}_{r4}^G) \tag{5}$$

"DE/current-to-best/1":

$$\boldsymbol{v}_i^G = \boldsymbol{m}_i^G + F(\boldsymbol{m}_{best}^G - \boldsymbol{m}_i^G) + F(\boldsymbol{m}_{r1}^G - \boldsymbol{m}_{r2}^G) \tag{6}$$

"DE/current-to-pbest/1":

$$\boldsymbol{v}_i^G = \boldsymbol{m}_{r1}^G + F(\boldsymbol{m}_{pbest}^G - \boldsymbol{m}_{r1}^G) + F(\boldsymbol{m}_{r2}^G - \boldsymbol{m}_{r3}^G) \tag{7}$$

In which, $r1$, $r2$, $r3$, $r4$, and $r5$ are random integers that are different from each other in the range of $[1, NP]$, and all of them are not equal to i. The coefficient $F(F > 0)$ is called the scaling factor, which is used to control the size of the difference direction. $\boldsymbol{m}_{best}^G$ represents the best vector in the population and also has the optimal function value in the G generation. For more details on the differential evolutionary mutation strategy, please refer to the work of Das et al. [39].

(3) Crossover

The purpose of the crossover is to replace some variables in the target vector $\boldsymbol{m}_i^G$ with variables in the mutation vector $\boldsymbol{v}_i^{G+1}$, thereby the trial vector $\boldsymbol{m}_i^{G+1}$ is obtained. Both binomial crossover and exponential crossover are commonly used crossover mechanisms. Here, the binomial crossover is explained, in which at least one variable in the trial vector is inherited from the mutation vector, and can be expressed as:

$$u_{ij}^G = \begin{cases} v_{ij}^G, \text{ if } rand(0,1) \leq CR \text{ } or \text{ } j = jrand \\ m_{ij}^G, \text{ otherwise} \end{cases} \tag{8}$$

Here, $CR \in (0,1)$ is the crossover probability, which is used to control the number of variables inherited from the mutation vector in the trial vector. $jrand$ is an integer randomly selected in the range of $[1, D]$.

(4) Selection

In DE, a greedy mechanism is employed to select a better one between the trial vector $\boldsymbol{u}_i^G$ and the target vector $\boldsymbol{m}_i^G$ according to their function values. The selection operation is as follows (for a minimization problem):

$$\boldsymbol{m}_i^{G+1} = \begin{cases} \boldsymbol{u}_i^G, \text{ if } f(\boldsymbol{u}_i^G) \leq f(\boldsymbol{x}_i^G) \\ \boldsymbol{m}_i^G, \text{ otherwise} \end{cases} \tag{9}$$

where $f(\cdot)$ is the fitness value of the target and the trial vector.

2.2. Improved Differential Evolution Algorithm

(1) Improvement of Mutation Strategy

When conducting gravity inversion, to obtain a smooth density model, it is necessary to use a proper smoothing technique to process the disturbance direction. For the disturbance direction $d_{i,j}$ of variable $m_{i,j}$, the adjacent variables and weights information are represented as shown in Figure 1. Assuming the weight of each adjacent point is $W_{s,i,j}$, then the smoothed $d_{s,\, i,j}$ can be calculated by:

$$d_{s,\, i,j} = \frac{\sum_{l=-1}^{1} \sum_{m=-1}^{1} W_{s,i+l,j+m} d_{i+l,j+m}}{\sum_{l=-1}^{1} \sum_{m=-1}^{1} W_{s,i+l,j+m}} \tag{10}$$

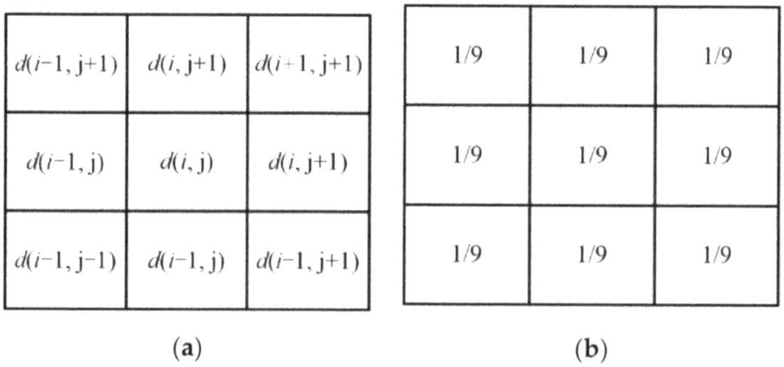

Figure 1. Square-shaped window of perturbation direction $d_{i,j}$. (**a**) $d_{i,j}$'s neighborhood; (**b**) The weight information for $d_{i,j}$ and its adjacent position.

In order to use the above formula repeatedly, the coefficients associated with all d_{ij} can be converted into a sparse matrix **S**, in which the k-th row stores the weight information of direction k and its adjacent elements. Finally, the smooth matrix is expressed as:

$$S = S^{f_s} \tag{11}$$

where the parameter f_s is the times of applying the smooth filter. Usually, we use $f_s = 2$, implying that the density of each k-element is calculated taking into account the values of the adjacent 18 elements. Obviously, A greater value of f_s increases the scope used for the smoothing around the k-element.

Here, the adaptive differential evolution algorithm (JADE) proposed by Zhang and Sanderson [40] is used, in which the mutation strategy is "current-to-pbest/1", showed in:

$$\boldsymbol{v}_i^G = \boldsymbol{m}_i^G + F_i\left(\boldsymbol{m}_{pbest}^G - \boldsymbol{m}_i^G\right) + F_i\left(\boldsymbol{m}_{r1}^G - \boldsymbol{m}_{r2}^G\right) \tag{12}$$

where $\boldsymbol{m}_{pbest}^G$ is randomly selected from the top $p_b * 100\%$ vectors in the G-th generation ($p_b \in [0,1]$). F_i is the scale factor related to the vector $\boldsymbol{x}_i$. $r1, r2$ used to construct the disturbance direction d_i for each mutant vector are mutually different random integers in the range of $[1, NP]$. Applying the smooth matrix to the perturbation direction formed by vectors r1 and r2, the new mutation strategy is:

$$\boldsymbol{v}_i^G = \boldsymbol{m}_i^G + F_i\left(\boldsymbol{m}_{pbest}^G - \boldsymbol{m}_i^G\right) + F_i S\left(\boldsymbol{m}_{r1}^G - \boldsymbol{m}_{r2}^G\right) \tag{13}$$

In which, the difference vector part formed by the *pbest* vector $\boldsymbol{m}_{pbest}^G$ and the target vector $\boldsymbol{m}_i^G$ is not smoothed, because smoothing will destroy the convex combination of $\boldsymbol{m}_{pbest}^G$ and $\boldsymbol{m}_i^G$.

(2) Adaptive Guided Evolution

In JADE, the parameter p_b in the "current-to-*p*best" mutation strategy is fixed throughout the evolution process and needs to be manually set according to the problem. Then, in order to improve the exploration ability of mutation mechanism, Tamabe and Fukunaga [41] assigned an independent $p_{b,i}$ to each target vector $\boldsymbol{m}_i$ by applying a uniform distribution. Here, the value of $p_{b,i}$ is adaptively and independently generated for each vector $\boldsymbol{x}_i$ based on the gaussian perturbation scheme. And $p_{b,i}$ is defined as:

$$p_{b,i} = randn_i(\mu_{p_b}, 0.1) \tag{14}$$

The average value μ_{p_b} is initialized to 0.5. If $p_{b,i}$ is not in the interval $[2/NP, 0.5]$, it will be truncated. At the end of each search, μ_{p_b} is updated as follows:

$$\mu_{p_b} = (1 - c_p)\mu_{p_b} + c_p \cdot \text{mean}_A(S_p) \tag{15}$$

Here, S_p is the set of successful p_b values in each iteration, $c_p \in (0, 1)$ is the learning rate of μ_{p_b}, mean_A represents the arithmetic mean. According to the above formula, it can be seen that the greedy here is controlled by μ_{p_b}, unlike in JADE, which is controlled by p_b. Therefore, μ_p can be set to a larger value at the beginning of the evolution process, such as $\mu_p = 0.5$, to maintain the diversity of the population.

The update method of μ_{p_b} in the above equation is an arithmetic average, which makes μ_p converge to a smaller value. In other words, as the search progresses, the parameter μ_p will decrease, thereby improving local exploitation capabilities.

2.3. Control Parameters Adaptation of DE Algorithm

Experimental results conducted by Wu et al. [38] have shown that parameter adaptation schemes in JADE [40] is the most effective. As a result, its procedures are extended here to adapt the parameters of gravity inversion.

In JADE, for each vector, the parameters F_i and CR_i are independently generated according to the successful F and CR values in last generation. The formula of generating CR_i can be expressed as follows:

$$CR_i = randn_i(\mu_{CR}, 0.1) \tag{16}$$

where $randn_i(\mu_{CR}, 0.1)$ returns a random value with a gaussian distribution. 0.1 is standard deviation; μ_{CR} is initialized to 0.5 and calculated using the following formula:

$$\mu_{CR} = (1 - c_{CR})\mu_{CR} + c_{CR}\ mean_A(S_{CR}) \tag{17}$$

where $c_{CR} \in (0, 1)$ is a constant, S_{CR} is the set of all successful CR values in last generation; $mean_A(\cdot)$ is the usual arithmetic mean.

Similarly, the adaptation of F_i is as follows:

$$\begin{cases} F_i = rand_{ci}(\mu_F, 0.1) \\ \mu_F = (1 - c)\mu_F + c\ mean_L(S_F) \\ mean_L = \frac{\sum_{F \in S_F} F^2}{\sum_{F \in S_F} F} \end{cases} \tag{18}$$

where $rand_{ci}(\mu_F, 0.1)$ means Cauchy distribution with location parameter μ_F and scale parameter 0; $mean_L(\cdot)$ is the Lehmer mean; and S_F is the set of successful scale factors.

3. Lp-Norm Gravity Inversion Based on Adaptive Differential Evolution

3.1. Forward Modeling

Assuming that the observed gravity data is a vector written as $\boldsymbol{d}$, whose length is N. The underground model is represented by $\boldsymbol{m}$, whose length is M. Then the forward modeling problem can be expressed as:

$$\boldsymbol{d} = \boldsymbol{G}(\boldsymbol{m}) \tag{19}$$

where $\boldsymbol{G}$ is the forward operator. In the traditional gravity inversion method, $\boldsymbol{G}$ is the linear forward matrix, but here $\boldsymbol{G}$ is no longer a linear matrix, while the finite volume method is adopted in the forward modeling [42].

3.2. Inversion Method

The objective function of Lp-norm gravity inversion can be expressed as:

$$\begin{gathered} \min\Phi(\boldsymbol{m}) = \Phi_d(\boldsymbol{m}) + \lambda\Phi_m(\boldsymbol{m}), \\ s.t.\ \boldsymbol{m}_a \le \boldsymbol{m} \le \boldsymbol{m}_b \end{gathered} \tag{20}$$

In which λ denotes the regularization factor, $\boldsymbol{m}_a, \boldsymbol{m}_b$ are the search range of the parameters, while there are M parameters need to be inverted, then the upper and lower bounds of the model parameters are constrained as $\boldsymbol{m}_a = (m_{a,1}, m_{a,2}, \cdots, m_{a,D})^T$, $\boldsymbol{m}_b = (m_{b,1}, m_{b,2}, \cdots, m_{b,D})^T$. $\Phi_d(\boldsymbol{m})$ is the misfit function of gravity data, which is defined as:

$$\Phi_d(\boldsymbol{m}) = \frac{\left\| \boldsymbol{W}_d\left(\boldsymbol{d}^{obs} - G(\boldsymbol{m})\right) \right\|_2^2}{\left\| \boldsymbol{W}_d \boldsymbol{d}^{obs} \right\|_2^2} \tag{21}$$

where $\boldsymbol{W}_d$ is the weight to the observed data, and determined by the followed equation [43,44]:

$$W_{d,i} = \frac{1}{\left|d_i^{obs}\right| + 0.5\left(d_{max}^{obs} - d_{min}^{obs}\right)},\ i \in [1, N] \tag{22}$$

In which, d_{max}^{obs} and d_{min}^{obs} denote the maximum and minimum data among the observed data, respectively. In Equation (21), the normalization process is carried out, and the purpose is to weaken the influence of the observation data error on the misfit function [45]. $\Phi_m(\boldsymbol{m})$ is regularization term of Lp $(1 \le p < \infty)$, and expressed as:

$$\Phi_m(\boldsymbol{m}) = \sum W_{m,}|m_i - m_{0,i}|^p \tag{23}$$

$W_{m,i}$ is the weight coefficient of parameter m_i, and it is the empirical information of the model, such as depth weighting value (see [2]). The notation $m_{0,i}$ denotes the reference model or 0 model. While the model distribution is known, m_0 can be used to strengthen the constraints on the inversion. Combining (21) and (23), the objective function of Lp-norm gravity inversion can be written as:

$$\min \Phi(\boldsymbol{m}) = \frac{\|\boldsymbol{W}_d(\boldsymbol{d}^{obs} - G(\boldsymbol{m}))\|_2^2}{\|\boldsymbol{W}_d\boldsymbol{d}^{obs}\|_2^2} + \lambda \sum W_{m,i}|m_i - m_{0,i}|^p, \quad s.t.\boldsymbol{m}_a \le \boldsymbol{m} \le \boldsymbol{m}_b \tag{24}$$

3.3. Adaptive Adjustment of Regularization Factor

The regularization factor λ determines the main fitting object in the inversion process. While $\lambda \to 0$, the observation data is mainly fitted; on the contrary, the model regularization term is mainly fitted. Chen et al. [45] and Zhdanov [46] have presented some adaptive techniques to adjust the regularization factor, but these methods are only suitable for conventional gradient-based iterative search algorithm, and are not suitable for global nonlinear DE algorithms. In order to fit DE algorithms, Du et al. [37] combined the ideas of Zhdanov [46] and Lu [47], then proposed the following update method:

$$\lambda^{G+1} = \begin{cases} \xi\lambda^G \ if\ \Phi_{d,\ mean}^G > \Phi_{d,\ mean}^{G-1} \\ \lambda^G \quad otherwise \end{cases}, \quad \Phi_{d,\ mean}^G = \frac{1}{NP}\sum \Phi_{d,\ i}^G, \tag{25}$$

where $\Phi_{d,\ mean}^G$ denotes the average misfit function value, $\xi \in (0,1)$ is attenuation coefficient, and equal to 0.65. At the 0-th iteration, the regularization factor is defined as:

$$\lambda^0 = 10\frac{\sum \Phi_{d,i}^0}{\sum \Phi_{m,i}^0}, i \in [1, NP] \tag{26}$$

The above method is successfully used for the inversion of the magnetic method, but this updated method ignores the phenomenon that the ratio of Φ_d and Φ_m is constantly changing during the search process. Therefore, in order to enable λ to balance the effects of Φ_d and Φ_m throughout the inversion process, the following updated method is proposed:

$$\lambda^{G+1} = \begin{cases} \xi\lambda^G \ if\ \Phi_{d,\ mean}^G \ge \Phi_{d,\ mean}^{G-1} \\ (1-c_\lambda)\lambda^G + c_\lambda \max(\lambda^G, \lambda^t),\ elseif\ \Phi_{d,\ mean}^G \le \delta_\lambda \\ \lambda^G \quad otherwise \end{cases}, \quad \lambda_t = \frac{\sum \Phi_{d,i}^G}{\sum \Phi_{m,i}^G}, \tag{27}$$

In which, $c_\lambda \in (0,1)$ denotes the learning rate, and is set at 0.8 based on the experimental results. λ_t is the ratio of $\Phi_{d,\ mean}^G$ and $\Phi_{m,\ mean}^G$ in the current population, δ_λ is a given threshold and is defined as follows:

$$\delta_\lambda = \frac{1}{2NP}\sum \Phi_{d,i}^0 \tag{28}$$

Compared with Equation (25), it can be found that the new update method of regularization factor is more complicated. At the same time, the new adjustment method is more reasonable due to the influence of λ_t.

3.4. Implement of Inversion Algorithm

The detailed implementation process of the inversion algorithm is shown in Algorithm 1.

Algorithm 1. Lp-norm inversion of gravity based on an improved adaptive differential evolution.

$G = 1$
Set $NP = 100$, $\mu_{CR} = 0.5$, $\mu_F = 0.5$, $\mu_{p_b} = 0.5$, $c = 0.1$, $c_p = 0.05$, initialize population $\boldsymbol{m} = (\boldsymbol{m}_{0,1}, \cdots, \boldsymbol{m}_{0,NP})$, initialize the regularity coefficient λ^0, evaluation the population $\boldsymbol{m}$.
While termination conditions are not met **do**
    $G = G + 1$
    For $i = 1$ to NP **do**
        $CR_i^G = randn_i(\mu_{CR}, 0.1)$;
        $F_i^G = randc_i(\mu_F, 0.1)$;
        $p_{b_i}^G = randn_i(\mu_{p_b}, 0.1)$
    End For
    For $i = 1$ *to* NP **do**
        Generate mutation vector according to mutation strategy (12)
        Generate trial vector $\boldsymbol{u}_i^G$ from (8)
    End For
    Evaluation test vector $\boldsymbol{u}^G$
    Form a new population $\boldsymbol{m}^{G+1}$ according to Equation (9)
    Update μ_{CR} and μ_F according to Equations (16)–(18), update μ_{p_b} according to (14), update λ by (27).
End While

In Algorithm 1, the termination condition is that the data fitting error reaches 5% or the maximum number of iterations is reached. And the maximum number of iterations is set to 100 M, M denotes the number of parameters to be inverted.

4. Simulation Tests

To illustrate the effectiveness and flexibility of the proposed inversion approach, four types of 2D single or combined prism models are set to test the DE algorithm used in this section, including (a) rectangular prism, (b) dipping prism, (c) parallel prism, and (d) U prism. All test models are list in Figure 2. The conclusion of the test results show that the method provides us with better resolution and can be used for interpreting the real geological case.

4.1. Parameter Setting

As mentioned before, three main control parameters (NP, F and CR) are used to adjust the search process of DE. Actually, when implementing the improved JADE for gravity inversion, corresponding parameters are set as: $NP = 100$; $\mu_{CR} = 0.9$; $\mu_F = 0.9$; $\mu_{pb} = 0.5$. And the density bound constraints for synthetic models are $0 \leq m \leq 1.1\ \mathrm{g/cm^3}$ for all variables in the DE population. When inverting the gravity data of the synthetic models, the initial model of the density distribution is created randomly from 0 to 10^{-2}. Besides, crossover rate sorting proposed by Zhou et al. is used to maintain the exploration ability of algorithm further [48]. The key idea of CR sorting mechanism is to assign a smaller CR to a vector with better fitness.

4.2. The Impact of Weighted Moving Average Smoothing

Here, the rectangular (Figure 2a) and dipping (Figure 2b) prism are adopted to verify the effectiveness of smoothing. To simplify the discussion, the norm of regularization term is set to 2 for two type of prism models. Additionally, considering that the smoothness of obtained models is controlled by the values of s_f factor, the different s_f values will be selected. In this part, s_f is set as 0, 1, 2 and 3, while $s_f = 0$ means no smoothness is applied in the inversion process. The inverted results for these two models are shown in Figures 3 and 4, respectively. Obviously, according to the models obtained with different

s_f values, incorporating the smoothness technique with mutation strategy of DE can help strengthen the continuity among the variables.

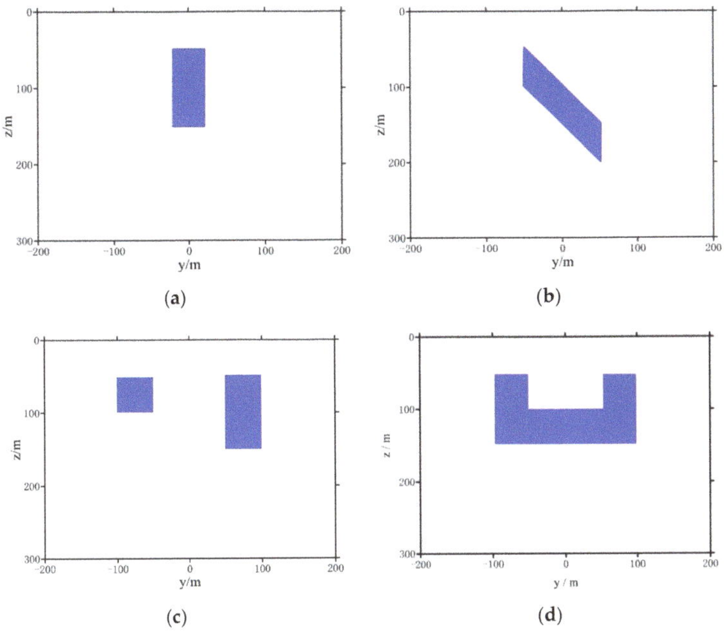

Figure 2. The cross−section through the x-axis of the 2D models with density 1 g/cm^3. (**a**) rectangular prism, (**b**) dipping prism, (**c**) parallel prism, (**d**) U prism.

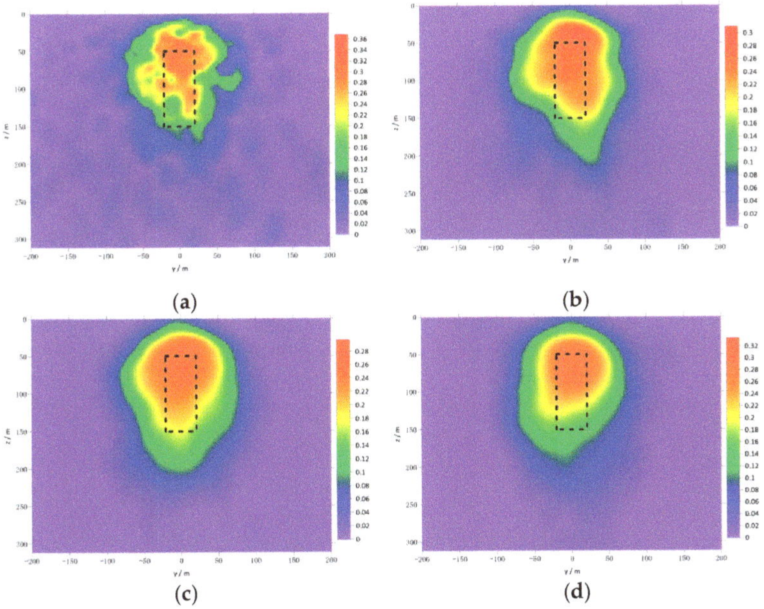

Figure 3. The inversion results of the model shown in Figure 2a with different s_f. (**a**) $s_f = 0$, (**b**) $s_f = 1$, (**c**) $s_f = 2$, (**d**) $s_f = 3$.

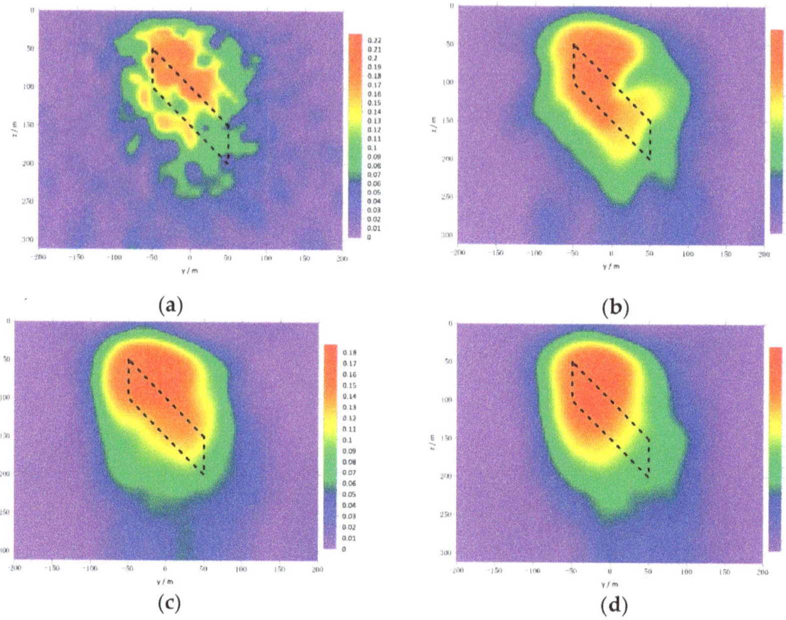

Figure 4. The inversion results of the model shown in Figure 2b with different s_f. (**a**) $s_f = 0$, (**b**) $s_f = 1$, (**c**) $s_f = 2$, (**d**) $s_f = 3$.

The fitting process to each s_f for the rectangular and dipping prism is shown in Figure 5.

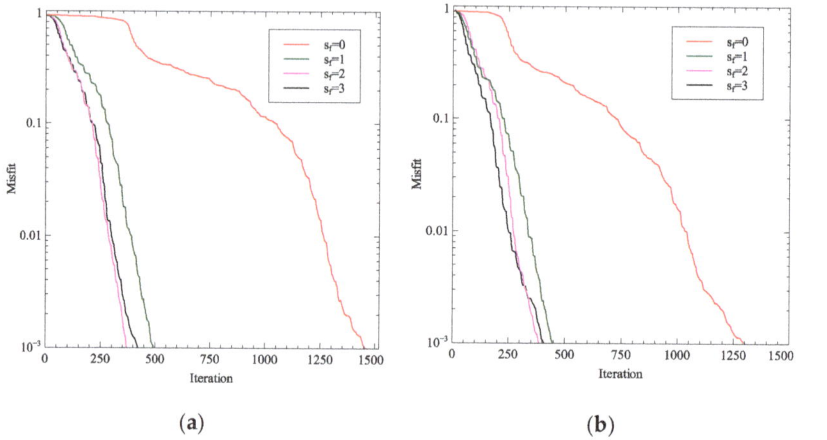

Figure 5. The iterative process corresponding to each s_f. (**a**) the fitting process for the rectangular prism, (**b**) the fitting process for the dipping prism.

Obviously from Figure 5 when $s_f \neq 0$, which mean smoothness is applied in the inversion process, the inversion can be significantly accelerated.

4.3. The Effect of p Value

For the Lp-norm based gravity inversion, the value of p actually plays a key role in determining the density distribution and shape of density source. To get a proper p value, the U prism model is selected to conduct the inversion with different p values. In this paper, the candidate values of p are 1, 1.1, 1.3, 1.5, 1.7, and 2, respectively. The inverted results are shown in Figure 6. We can observe that the density obtained by a larger p value is lower than the one with a smaller p. Besides, by comparing with the designed model, the large p values, especially when the p is close to 2, will increase the range of the predicted density model, and make the boundaries of model blur. On the contrary, a smaller p is helpful in describing the shape of the density model accurately, but it may weaken the continuity of the obtained model. Overall, in our opinion, it is appropriate to set p less than 1.5.

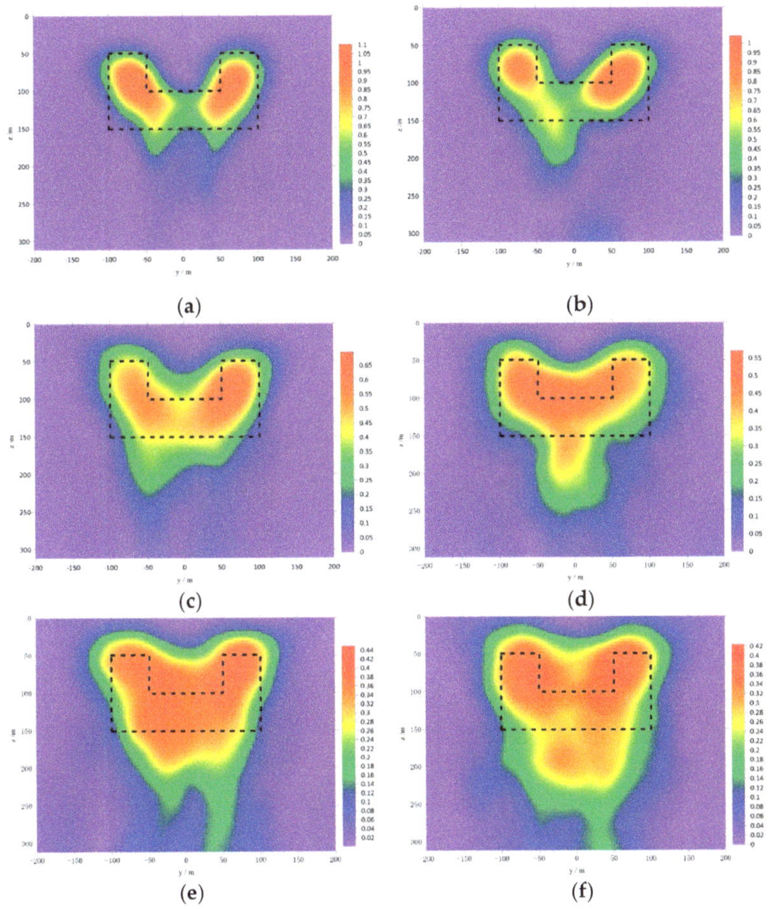

Figure 6. The inversion results of the model shown in Figure 2d with different p values. (**a**) $p = 1$, (**b**) $p = 1.1$, (**c**) $p = 1.3$, (**d**) $p = 1.5$, (**e**) $p = 1.7$, (**f**) $p = 2$.

4.4. Gravity Data with Gaussian Noise

To prove that a new method works correctly, it is essential to test it on data with noise since actual gravity data may be contaminated by noise. Here, Gaussian white noise with a mean value 0 and a standard deviation equal to 10% of the maximum value of synthetic data are added. By using the proposed inversion method, fittings between the observed and model response for these noisy data are shown in Figure 7. The inversion results of noise data are shown in Figure 8. According to the results, one can conclude that the Lp-norm inversion of gravity is robust, and capable of reflecting the real shape of the field source from the noisy data.

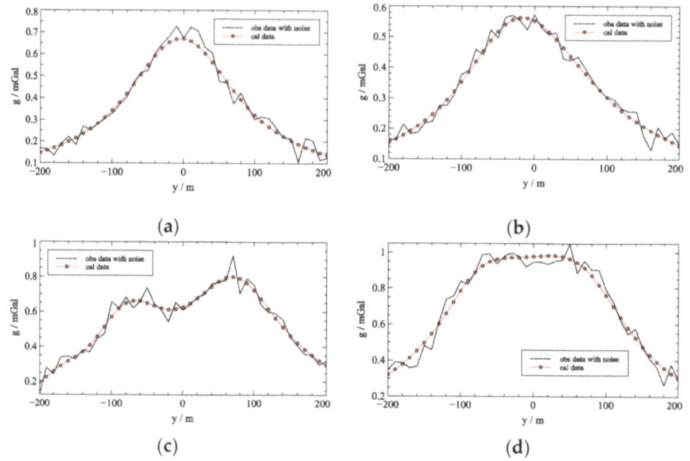

Figure 7. The observed data with noise and the model response for each model shown in Figure 2. (**a**) response of rectangular prism, (**b**) response of dipping prism, (**c**) response of parallel prism, (**d**) response of U prism.

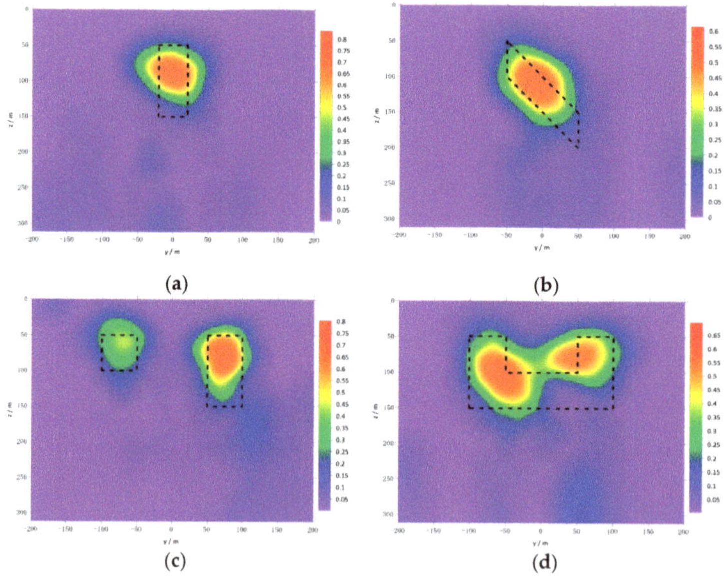

Figure 8. The inversion results of the model shown in Figure 2. (**a**) response of rectangular prism, (**b**) response of dipping prism, (**c**) response of parallel prism, (**d**) response of U prism.

5. Field Example

The Shihe iron ore deposit is located in the northern part of the Hengshan-Wutaishan dome in the Lvliang-Taihang fault block, North China. The main exposed strata are the Paleoproterozoic Jingangku Formation, the Mesoproterozoic Changcheng System, the Cambrian, and Quaternary. The Jingangku Formation is the ore-bearing unit that comprises metamorphic amphibolite-like rocks. The density of rock and mineral in this area is shown in Table 1. The garnet magnetite quartzite has the highest average density, and the difference between its density and the one of surrounding rock amphibolite has reached 0.38 g/cm^3.

Table 1. Density parameters of minerals and rocks in the Shihe area.

Mineral	Number of Specimens	Arithmetic Mean (g/cm^3)
PlagioclAse amphibolite	41	2.78
Garnet Magnetite Quartzite	30	3.16
Biotite Granulite	27	2.88
Amphibole Plagioclase	7	2.65
Layer	Number of specimens	Arithmetic mean (g/cm^3)
Quaternary system	6	1.53
Jingangku formation	106	2.78

Since there are negative residual anomalies in the original data, in order to fit this part of the data, the search range of Lp-norm gravity inversion in this area is set to [−0.5, 0.5] g/cm^3. The horizontal cell size for subsurface space is set to 20 m during the inversion process, and the vertical cell size changes proportionally. The results with different p values are shown in Figure 9; it can be concluded that the top buried depth of the iron ore is about 400 m, and the depth of the ore body extends more than 400 m.

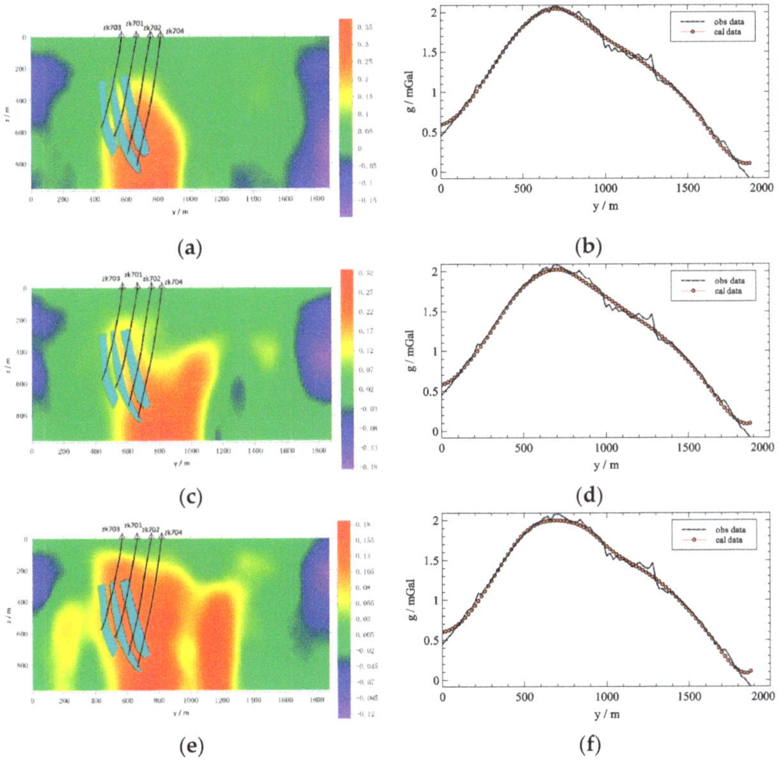

Figure 9. *Cont.*

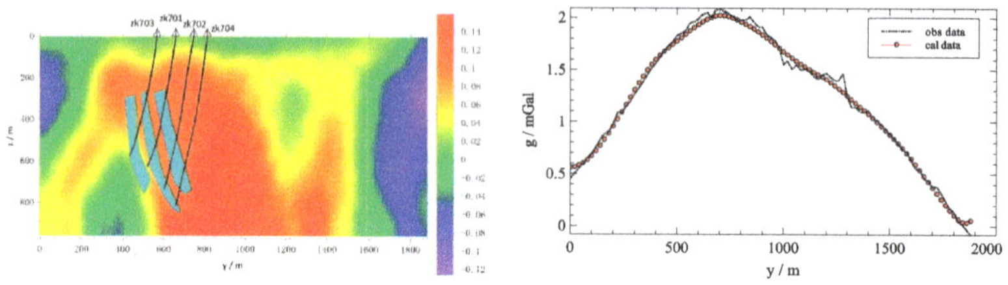

(g) (h)

Figure 9. The Lp-norm inversion results of abnormal gravity field in the Shihe area. (**a**,**c**,**e**,**g**) are the results while $p = 1$, $p = 1.1$, $p = 1.5$ and $p = 2$, respectively; the line denotes the drill-hole, the cyan block denotes the ore body, and (**b**,**d**,**f**,**h**) are the corresponding data fit lines.

Observed from Figure 9, the density distribution results obtained by Lp-norm regularization inversion with different p values are similar, and are in good agreement with the one obtained by drill-hole information. In Figure 9, the relation between the three inversion results and the geological interpretation of the ore body is also shown. Obviously, the inversion results Lp-norm with $p \leq 1.5$ are reliable, which shows that the Lp-norm inversion algorithm in this paper can improve the quality of inversion and bring the restored model closer to the real situation. Additionally, by comparing the L2-norm regularization results shown in Figure 9 with L1, L1.1, and L1.5, it is easy to see that the L2-norm inversion technique failed to draw the real shape of the ore deposit.

6. Conclusions

In this work, an attempt was made to test the applicability and effectiveness of adaptive DE on Lp-norm gravity inversion. As far as this work is concerned, this is the first attempt at applying DE for physical property inversion using gravity data. In the application of this stochastic search algorithm, to further improve the exploration ability, the evolved direction of the DE population in this paper is guided adaptively. Besides, in order to maintain the physical property to distribute in a continuous space, the perturbation direction in the mutation operation is smoothed by using the weighted moving average smooth technique. In addition, adaptive adjustment of the regularization factor is proposed to balance the effect of data misfit and Lp-norm regularization term. In the present algorithm, synthetic data experiments are performed using noise free and noisy data sets calculated from simple-shaped causative bodies. From the example of synthetic models, the adaptive DE algorithm with improved mechanisms shows favorable results and is capable of achieving a density distribution with sharp boundary by setting a proper p value. In addition, the proposed algorithm was also tested on a field data from the Shihe iron deposit; the predicted density distribution was verified by drill-hole information. Finally, in this work, we discussed the application of an improved JADE version in Lp-norm gravity inversion. Some recent DE variants demonstrated excellent performance for single objective optimization problems. Their application in geophysical inversion will be done in future research.

Author Contributions: Conceptualization, T.S. and X.H.; methodology, T.S. and L.C.; software, L.C. and T.X.; Formal analysis, T.S., X.H., and W.D.; Resources, Q.L.; writing—original draft preparation, T.S.; writing—review and editing, X.H. and W.D.; visualization, T.X.; funding acquisition, T.S. All authors have read and agreed to the published version of the manuscript.

Funding: This work was financially supported by the Discipline and Master's Site Construction Project of Guiyang University by Guiyang City Financial Support Guiyang University, grant number SX-2020; Natural Science Research Program of Guizhou Provincial Department of Education under grant number KY[2013]110; Youth Science and Technology Talent Growth Project from Guizhou Provincial Department of Education, grant number QJHKYZ[2020]075.

Institutional Review Board Statement: Not applicable.

Informed Consent Statement: Not applicable.

Data Availability Statement: The data of this study is available from the authors upon request.

Conflicts of Interest: The authors declare no conflict of interest.

References

1. Paterson, N.R.; Reeves, C.V. Applications of gravity and magnetic surveys: The state-of-the-art in 1985. *Geophysics* **1985**, *50*, 2558–2594. [CrossRef]
2. Li, Y.; Oldenburg, D.W. 3-D inversion of gravity data. *Geophysics* **1998**, *63*, 109–119. [CrossRef]
3. Fullagar, P.K.; Pears, G.; Hutton, D.; Thompson, A. 3D gravity and aeromagnetic inversion for MVT lead-zinc exploration at Pillara, Western Australia. *Explor. Geophys.* **2004**, *35*, 142–146. [CrossRef]
4. Hinze, W.J.; Von Frese, R.R.; Saad, A.H. *Gravity and Magnetic Exploration: Principles, Practices, and Applications*; Cambridge University Press: Cambridge, UK, 2013.
5. Afshar, A.; Norouzi, G.-H.; Moradzadeh, A.; Riahi, M.-A. Application of magnetic and gravity methods to the exploration of sodium sulfate deposits, case study: Garmab mine, Semnan, Iran. *J. Appl. Geophys.* **2018**, *159*, 586–596. [CrossRef]
6. Montesinos, F.G.; Arnoso, J.; Vieira, R. Using a genetic algorithm for 3-D inversion of gravity data in Fuerteventura (Canary Islands). *Int. J. Earth Sci.* **2005**, *94*, 301–316. [CrossRef]
7. Pallero, J.L.G.; Fernández-Martínez, J.L.; Fernández-Muñiz, Z.; Bonvalot, S.; Gabalda, G.; Nalpas, T. GravPSO2D: A Matlab package for 2D gravity inversion in sedimentary basins using the Particle Swarm Optimization algorithm. *Comput. Geosci.* **2021**, *146*, 104653. [CrossRef]
8. Zhou, X. Gravity inversion of 2D bedrock topography for heterogeneous sedimentary basins based on line integral and maximum difference reduction methods. *Geophys. Prospect.* **2013**, *61*, 220–234. [CrossRef]
9. Feng, J.; Meng, X.; Chen, Z.; Zhang, S. Three-dimensional density interface inversion of gravity anomalies in the spectral domain. *J. Geophys. Eng.* **2014**, *11*, 035001. [CrossRef]
10. Feng, J.; Zhang, S.; Meng, X. Constraint 3D density interface inversion from gravity anomalies. *Arab. J. Geosci.* **2015**, *9*, 56. [CrossRef]
11. Last, B.; Kubik, K.J.G. Compact gravity inversion. *Geophysics* **1983**, *48*, 713–721. [CrossRef]
12. Shamsipour, P.; Marcotte, D.; Chouteau, M. 3D stochastic joint inversion of gravity and magnetic data. *J. Appl. Geophys.* **2012**, *79*, 27–37. [CrossRef]
13. Meng, Z.-H.; Xu, X.-C.; Huang, D.-N. Three-dimensional gravity inversion based on sparse recovery iteration using approximate zero norm. *Appl. Geophys.* **2018**, *15*, 524–535. [CrossRef]
14. Silva, J.B.C.; Teixeira, W.A.; Barbosa, V.C.F. Gravity data as a tool for landfill study. *Environ. Geol.* **2008**, *57*, 749. [CrossRef]
15. Tarantola, A. *Inverse Problem Theory: Methods for Data Fitting and Model Parameter Estimation*; Elsevier: New York, NY, USA, 1987.
16. Yu, P.; Wang, J.L.; Wu, J.S. An Inversion of Gravity Anomalies by Using a 2.5 Dimensional Rectangle Gridded Model and the Simulated Annealing Algorithm. *Chin. J. Geophys.* **2007**, *50*, 756–764. [CrossRef]
17. Ekinci, Y.L.; Balkaya, Ç.; Göktürkler, G.; Turan, S. Model parameter estimations from residual gravity anomalies due to simple-shaped sources using Differential Evolution Algorithm. *J. Appl. Geophys.* **2016**, *129*, 133–147. [CrossRef]
18. Yao, C.; Hao, T.; Guan, Z.; Zhang, Y. High-Speed Computation and Efficient Storage in 3-D Gravity and Magnetic Inversion. *Chin. J. Geophys.* **2003**, *46*, 351–361. [CrossRef]
19. Storn, R.; Price, K. Differential Evolution–A Simple and Efficient Heuristic for global Optimization over Continuous Spaces. *J. Glob. Optim.* **1997**, *11*, 341–359. [CrossRef]
20. Voratas, K. Comparison of Three Evolutionary Algorithms: GA, PSO, and DE. *Ind. Eng. Manag. Syst.* **2012**, *11*, 215–223.
21. Kurban, T.; Civicioglu, P.; Kurban, R.; Besdok, E. Comparison of evolutionary and swarm based computational techniques for multilevel color image thresholding. *Appl. Soft Comput.* **2014**, *23*, 128–143. [CrossRef]
22. Mohamed, A.W.; Hadi, A.A.; Mohamed, A.K. Gaining-sharing knowledge based algorithm for solving optimization problems: A novel nature-inspired algorithm. *Int. J. Mach. Learn. Cybern.* **2020**, *11*, 1501–1529. [CrossRef]
23. Ilonen, J.; Kamarainen, J.-K.; Lampinen, J. Differential Evolution Training Algorithm for Feed-Forward Neural Networks. *Neural Process. Lett.* **2003**, *17*, 93–105. [CrossRef]
24. Piotrowski, A.P. Differential Evolution algorithms applied to Neural Network training suffer from stagnation. *Appl. Soft Comput.* **2014**, *21*, 382–406. [CrossRef]
25. Tang, Y.; Ji, J.; Zhu, Y.; Gao, S.; Tang, Z.; Todo, Y. A Differential Evolution-Oriented Pruning Neural Network Model for Bankruptcy Prediction. *Complexity* **2019**, *2019*, 8682124. [CrossRef]
26. Yılmaz, O.; Bas, E.; Egrioglu, E. The Training of Pi-Sigma Artificial Neural Networks with Differential Evolution Algorithm for Forecasting. *Comput. Econ.* **2021**. [CrossRef]
27. Sheng, W.; Wang, X.; Wang, Z.; Li, Q.; Zheng, Y.; Chen, S. A Differential Evolution Algorithm with Adaptive Niching and K-Means Operation for Data Clustering. *IEEE Trans. Cybern.* **2020**, *12*, 1–15. [CrossRef] [PubMed]
28. He, X.; Zhang, Q.; Sun, N.; Dong, Y. Feature Selection with Discrete Binary Differential Evolution. In Proceedings of the 2009 International Conference on Artificial Intelligence and Computational Intelligence, Shanghai, China, 7–8 November 2009; pp. 327–330.

29. Zhenkui, P.; Yanli, Z.; Zhen, L. Image segmentation based on Differential Evolution algorithm. In Proceedings of the 2009 International Conference on Image Analysis and Signal Processing, Linhai, China, 11–12 April 2009; pp. 48–51.
30. Ali, M.; Ahn, C.W.; Siarry, P. Differential evolution algorithm for the selection of optimal scaling factors in image watermarking. *Eng. Appl. Artif. Intell.* **2014**, *31*, 15–26. [CrossRef]
31. Cuevas, E.; Zaldívar, D.; Perez-Cisneros, M. Image Segmentation Based on Differential Evolution Optimization. In *Applications of Evolutionary Computation in Image Processing and Pattern Recognition*; Cuevas, E., Zaldívar, D., Perez-Cisneros, M., Eds.; Springer International Publishing: Berlin/Heidelberg, Germany, 2016; pp. 9–22. [CrossRef]
32. Bhandari, A.K.; Kumar, A.; Chaudhary, S.; Singh, G.K. A new beta differential evolution algorithm for edge preserved colored satellite image enhancement. *Multidimens. Syst. Signal. Process.* **2017**, *28*, 495–527. [CrossRef]
33. Balkaya, Ç. An implementation of differential evolution algorithm for inversion of geoelectrical data. *J. Appl. Geophys.* **2013**, *98*, 160–175. [CrossRef]
34. Li, H.; Wang, H.; Wang, L.; Zhou, X. A modified Boltzmann Annealing Differential Evolution algorithm for inversion of directional resistivity logging-while-drilling measurements. *J. Pet. Sci. Eng.* **2020**, *188*, 106916. [CrossRef]
35. Ekinci, Y.L.; Özyalın, Ş.; Sındırgı, P.; Balkaya, Ç.; Göktürkler, G. Amplitude inversion of the 2D analytic signal of magnetic anomalies through the differential evolution algorithm. *J. Geophys. Eng.* **2017**, *14*, 1492–1508. [CrossRef]
36. Balkaya, Ç.; Ekinci, Y.L.; Göktürkler, G.; Turan, S. 3D non-linear inversion of magnetic anomalies caused by prismatic bodies using differential evolution algorithm. *J. Appl. Geophys.* **2017**, *136*, 372–386. [CrossRef]
37. Du, W.; Cheng, L.; Li, Y. lp Norm Smooth Inversion of Magnetic Anomaly Based on Improved Adaptive Differential Evolution. *Appl. Sci.* **2021**, *11*, 1072. [CrossRef]
38. Wu, G.; Mallipeddi, R.; Suganthan, P.N.; Wang, R.; Chen, H. Differential evolution with multi-population based ensemble of mutation strategies. *Inf. Sci.* **2016**, *329*, 329–345. [CrossRef]
39. Das, S.; Mullick, S.S.; Suganthan, P.N. Recent advances in differential evolution—An updated survey. *Swarm Evol. Comput.* **2016**, *27*, 1–30. [CrossRef]
40. Zhang, J.; Sanderson, A.C. JADE: Adaptive Differential Evolution with Optional External Archive. *IEEE Trans. Evol. Comput.* **2009**, *13*, 945–958. [CrossRef]
41. Tanabe, R.; Fukunaga, A. Success-history based parameter adaptation for Differential Evolution. In Proceedings of the 2013 IEEE Congress on Evolutionary Computation, Cancun, Mexico, 20–23 June 2013; pp. 71–78.
42. Mazumder, S. *Numerical Methods for Partial Differential Equations*; Academic Press: New York, NY, USA, 2016.
43. Rao, K.; Jain, S.; Biswas, A. Global Optimization for Delineation of Self-potential Anomaly of a 2D Inclined Plate. *Nat. Resour. Res.* **2021**, *30*, 175–189. [CrossRef]
44. Sharma, S.P.; Biswas, A. Interpretation of self-potential anomaly over a 2D inclined structure using very fast simulated-annealing global optimization—An insight about ambiguity. *Geophysics* **2013**, *78*, WB3–WB15. [CrossRef]
45. Chen, X.B.; Zhao, G.-Z.; Tang, J.; Zhan, Y.; Wang, J.J. The Adaptive Regularized Inversion Algorithm (ARIA) for Magnetotelluric Data. *Chin. J. Geophys.* **2005**, *48*, 1005–1016. [CrossRef]
46. Zhdanov, M.S. Chapter 5-Nonlinear Inversion Technique. In *Inverse Theory and Applications in Geophysics*, 2nd ed.; Zhdanov, M.S., Ed.; Elsevier: Oxford, UK, 2015; pp. 129–177. [CrossRef]
47. Wu, Y.; Lu, J.; Sun, Y. Genetic Programming Based on an Adaptive Regularization Method. In Proceedings of the 2006 International Conference on Computational Intelligence and Security, Guangzhou, China, 3–6 November 2006; pp. 324–327.
48. Zhou, Y.; Yi, W.; Gao, L.; Li, X. Adaptive Differential Evolution with Sorting Crossover Rate for Continuous Optimization Problems. *IEEE Trans. Cybern.* **2017**, *47*, 2742–2753. [CrossRef]

applied sciences

MDPI

Article

Computation of Optimized Electrode Arrays for 3-D Electrical Resistivity Tomography Surveys

Kleanthis Simyrdanis, Nikos Papadopoulos * and Dimitrios Oikonomou

Laboratory of Geophysical-Satellite Remote Sensing and Archaeoenvironment (GeoSat ReSeArch Lab), Institute for Mediterranean Studies (IMS), Foundation for Research and Technology (FORTH), Nikiforou Foka 130, 74100 Rethymno, Greece; ksimirda@ims.forth.gr (K.S.); doikonomou@ims.forth.gr (D.O.)

* Correspondence: nikos@ims.forth.gr

Abstract: The present study explores the applicability and effectiveness of an optimization technique applied to electrical resistivity tomography data. The procedure is based on the Jacobian matrix, where the most sensitive measurements are selected from a comprehensive data set to enhance the least resolvable parameters of the reconstructed model. Two existing inversion programs in two and three dimensions are modified to incorporate this new approach. Both of them are selecting the optimum data from an initial comprehensive data set which is comprised of merged conventional arrays. With the two-dimensional (2-D) optimization approach, the most sensitive measurements are selected from a 2-D survey profile and then a clone of the resulting optimum profile reproduces a three-dimensional (3-D) optimum data set composed of equally spaced parallel lines. In a different approach, with the 3-D optimization technique, the optimum data are selected from a 3-D data set of equally spaced individual parallel lines. Both approaches are compared with Stummer's optimization technique which is based on the resolution matrix. The Jacobian optimization approach has the advantage of selecting the optimum data set without the need for the solution of the inversion problem since the Jacobian matrix is calculated as part of the forward resistivity problem, thus being faster from previous published approached based on the calculation of the sensitivity matrix. Synthetic 3-D data based on the extension of previous published works for the 2-D optimization case and field data from two case studies in Greece are tested, thus verifying the validity of the present study, where fewer measurements from the initial data set (about 15–50%) are able to reconstruct a model similar with the one produced from the original comprehensive data set.

Keywords: ERT; optimization; Jacobian; 3-D

1. Introduction

Over the last twenty years, electrical resistivity tomography (ERT) has been transformed into a widely used geophysical method for approaching several different near-surface applications. This was boosted by the improvement of the hardware's capabilities, the development of multichannel resistivity systems and the compilation of automated inversion and reconstruction algorithms to image and monitor the subsurface geoelectrical properties [1]. Two dimensional (2-D) ERT method is nowadays routinely applied in approaching diverse problems including the study of saline water intrusion in coastal aquifers [2], the tracing of water content dynamics in unsaturated zones [3], in geotechnical studies [4] and the monitoring of subsurface conductive pollutants [5].

However, 2-D approaches underperform in cases of complicated stratigraphy and intense inhomogeneity of the subsurface resistivity. Thus, significant research efforts have been placed on proposing and investigating the resolving capabilities of different measuring and processing strategies to map the subsurface resistivity structure within a three-dimensional (3-D) context [6–10]. Every surface or cross-hole ERT experiment involves the arrangement of a number of electrodes along a line for 2-D or on the nodes of a regular or irregular grid for 3-D resistivity surveying, respectively. In either case, the linearly independent apparent resistivity measurements are a function of the maximum number of electrodes that are simultaneously connected to any multiplexed/multichannel resistivity instrument. The number of measurements for an array using four-electrode configurations increases exponentially with the respective number of electrodes [11,12]. This renders the field collection of all the possible measurements practically impossible, even for an array composed of a relatively small number (~50) of electrodes. Thus, the routine practice relies on the collection of tomographic data using standard electrode arrays [13] and the integration of dense parallel or orthogonal 2-D lines to characterize the 3-D resistivity structure [14].

A step forward to exploit the full capability spectrum of the modern acquisition instruments was firstly introduced by the pioneering work of Stummer et al. [15], who presented an experimental optimization method to select electrode arrays that maximize the resolving capabilities of the inversion images. Their approach was based on the compilation of the comprehensive data set including all the standard and non-standard arrays, the associated resolution matrix and a goodness function that ranks the sensitivity of every possible measurement to changes in the subsurface parameters. Extensive comparisons among the reconstructed models resulting from the inversion of the comprehensive, standard, and optimized data sets revealed the significant potential of the proposed technique.

Wilkinson et al. [16] developed and tested some improved strategies for the automatic selection of the optimized ERT data sets using a linearized estimate of the model resolution matrix that quantifies the resolving capabilities of the observed data. One of the methods (Compare R) directly compared the model resolution matrices to enhance the spatial resolution while another one used a modified version of the goodness function (Modified GF) to ensure a high degree of orthogonality to the base data set. The effectiveness of the algorithms was validated through synthetic and real data, outlining their relative superiority in the optimization and resolving capabilities performance in relation to Stummer's method, though being more time consuming than Stummer's approach. Wilkinson et al. [17] presented an adaptive optimization method based on Compare R to compile optimum arrays for time-lapse monitoring surveys.

Loke et al. [18] addressed the time-consuming issues related to the automatic selection of the optimized data set using a fast computation strategy to reduce the traffic between the computer main memory and the CPU registers. They compared Compare R, Modified GF, hybrid Modified GF–Compare R and Stummer's method, using synthetic and field tests. The results verified the superiority of Compare R to all the other methods. Their tests outlined that the hybrid Modified GF–Compare R method produced results almost identical to Compare R with the advantage of being several times faster.

In another work by Loke et al. [19], it is emphasized that using a smaller step to add optimized measurements to the base data set benefits the final model resolution. Furthermore, the particular study referred to the advantages of replacing matrix–vector with matrix–matrix multiplications, the storing of temporary variables in the double-precision single instruction multiple data (SIMD) registers within the CPU and the use of a computer graphics processor unit (GPU) to decrease the calculation time for selecting the optimum configurations.

In a more recent study, Loke et al. [20] presented optimized arrays for 2-D resistivity surveys with buried probes, using implanted electrodes in the subsurface and on the ground surface. This geometry array was able to increase the depth resolution of resistivity surveys at sites where the profile length is limited. Apart from the fact that the resolution is better when optimal arrays are used, the specific study also manifested the increased resolution distances slightly beyond the lateral ends of the arrays and for depths below the buried array.

Coles and Morgan [21] followed a slightly different path for the sequential experimental design of optimum resistivity data sets by developing a determinant-based objective function that minimizes the posterior model uncertainty. The aspect of computational efficiency was addressed through formulae for updating determinants and matrix inverses without the need for direct calculation. The applicability of the method was demonstrated on single and adaptive borehole ERT surveys. Additionally, Wagner et al. [22] applied a new approach based on the resolution matrix to calculate the optimal electrode positions, wherein the results of the study confirmed that a sparse but well-conceived set of electrodes can provide a large part of the information content offered by comparably dense electrode distributions.

All the above optimization algorithms rely on the appraisals of the model resolution matrix that is directly related to the calculation of the generalized inverse, which can be extremely time-consuming in large 2-D [23] and 3-D problems [14]. Furman et al. [24] recommended a method to extract the optimal set of ERT arrays using genetic algorithms to identify the array with the optimal cumulative spatial sensitivity. Athanasiou et al. [25] proposed an alternative method to pick the optimal measurements solely on the examination of the sensitivity matrix entries. The extensive testing of the specific algorithm proposed by Athanasiou et al. [25] with surface, cross-hole and surface-to-hole electrode layouts demonstrated comparable results with previously published approaches.

The extension of the optimization procedure to generate a set of optimum measurements for 3-D survey grids poses new challenges to balance the time constraints and the need to increase the resolution analysis of the 3-D subsurface inversion images. Loke et al. [26] presented the first effort of computing optimized arrays for 3-D surface electrical surveys. In the specific work, they constructed the comprehensive data set with a mixture of in-line and offset alpha and beta configurations as well as respective equatorial arrays. The unstable measurements were filtered out by setting a maximum limit for the geometric factor and the sensitivity of the array geometric factor in the position of the electrodes. A similar approach was also followed in optimizing resistivity measurements collected from the layout of electrodes placed on nodes of non-regular grids [27]. Recently Uhlemann et al. [28] presented a novel algorithm combining the optimization of measurement configuration and electrode placement both for the 2-D and 3-D case.

The present study is focused on generating the optimum 3-D tomographic resistivity data sets based on the sensitivity matrix and employing two different formulations. At first, a 2-D optimization algorithm was the basis to extract optimum arrays along a single section and the optimized 3-D data set resulting from cloning these parallel 2-D optimized sections. The second approach involved the extension of an experimental algorithm in a 3-D context and the extraction of the optimized arrays from 3-D tomographic data sets composed of a dense network of parallel 2-D lines. The resolving capabilities of the optimized configurations produced by these two different strategies were compared using synthetic data generated from the extension of a known published 2-D resistivity model in the 3-D space. In either case, the comprehensive data set was composed of a mixture of commonly known electrode arrays. Both optimization techniques are compared with Stummer's approach. The efficiency of the 3-D optimized protocols in imaging the subsurface resistivity structure was also verified with two real field data examples from Greece.

2. Methodology

The experimental selection of the optimum measurements from an initial array based on the Jacobian matrix (see Appendix A) method was firstly introduced by Athanasiou et al. [25] for the case of the 2-D resistivity tomographic problem. The same technique expanded and incorporated within a 3-D resistivity modelling and inversion algorithm [14] to select the optimum tomographic data in the 3-D space. In both algorithms, the adjoint equation technique [29] is used to calculate the sensitivity matrix in a homogeneous 2-D or 3-D resistivity model, which associates variations in the resistivity of the parameters ρ_j with variations in the apparent resistivity data d_i. This was carried out to ensure the generality of the optimized protocols regardless of the target's or the background's property values.

Figure 1a illustrates a simplified model with 24 surface electrodes (indicated with black dots), on a 3-D rectangular grid with the associated parameterization of the subsurface composed of 60 blocks to describe the earth's resistivity. It is known that the parameters closer to the surface electrodes are more sensitive to any change since they have higher absolute Jacobian values (Figure 1a, green-colored blocks). Gradually, the parameters that are further away from the surface (Figure 1a, yellow-colored blocks) and to the edges (Figure 1a, red-colored blocks) are less sensitive (lower absolute Jacobian values) to any subsurface resistivity change. As an example, let us assume an experiment with 200,000 measurements collected from a network of 24 surface electrodes and a subsurface model discretized using these 60 parameters. The calculated Jacobian matrix will have 200,000 rows representing the measurements and 60 columns associated with the parameters (Figure 1b). The experimental workflow follows the next procedure after the calculation of the Jacobian matrix.

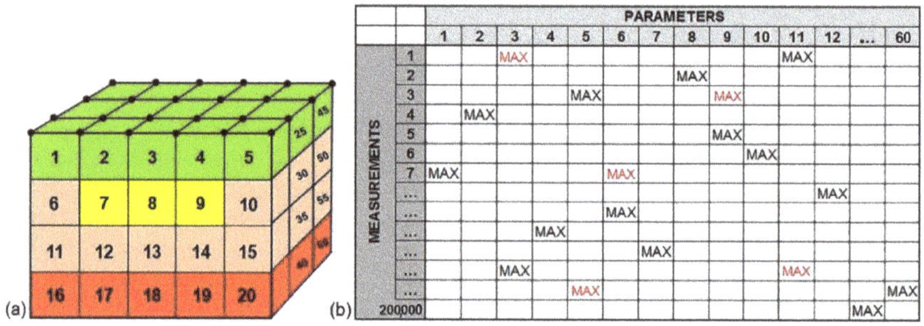

Figure 1. (**a**) Simplified 3-D model with 60 parameters, where the sensitivity value of each parameter is decreasing in larger distances from the surface electrodes (black dots). (**b**) Representation of the Jacobian matrix where the higher sensitivity absolute values (rows) are located for the 60 parameters of the model (columns) (detailed description in the text).

Firstly, the normalized cumulative Jacobian value is calculated for each parameter by summation of the absolute sensitivity values of all measurements for each parameter and the resultant value is divided by the number of measurements associated to a specific parameter. Then, all parameters are sorted in ascending order from the least to the most sensitive having the higher absolute sensitivity value. The specific sorting of the parameters from the weakest to the strongest sensitivity is performed to enforce the optimization algorithm to correlate the strongest array configuration to the weakest parameter and enhance the resolving capabilities of the deepest subsurface parameters.

For each parameter, the algorithm selects the electrode configuration with the maximum sensitivity value, which is stored as the optimum measurement for the specific parameter (Figure 1b, black-colored text 'MAX'). During the experimental procedure, if one measurement is already selected for a previous parameter (Figure 1b, red-colored text 'MAX'), the algorithm chooses the next best measurement for the specific parameter. In the end, the 'strongest' measurement (with the highest sensitivity value) is correlated with the weakest parameter. After the termination of the experimental procedure, the number of optimized measurements will be equal to the number of the parameters forming the model parameter space.

The specific optimization technique is repeated through iterative steps to add extra optimum measurements according to the operator's demand for increasing the subsurface resolution. At each iterative step, the extra added optimum measurements will be equal to the number of original model's parameters. In practice, in the above-mentioned example, after two steps, the protocol will have 120 optimum measurements (60 + 60), after three-step 180 optimum measurements (120 + 60), etc.

The comprehensive data set for a single 2-D profile was compiled from the combination of widely used common arrays: dipole–dipole (DD), pole–dipole (PD), pole–tripole (PT) and gradient (GR) (Figure 2). The 2-D and 3-D algorithms were used to produce the 2-D and 3-D optimum protocols, named 'Optim-2D' and 'Optim-3D', respectively. In both cases, the Jacobian matrix is calculated for a homogeneous model to ensure that the generated optimum protocols would apply to general resistivity surveys regardless of the target's or the background's property values. The algorithmic details of the procedure to extract the optimum protocolos with the Stummer's method [15] were followed in the specific work and the respective optimized configurations are coded as 'Res-STUM'.

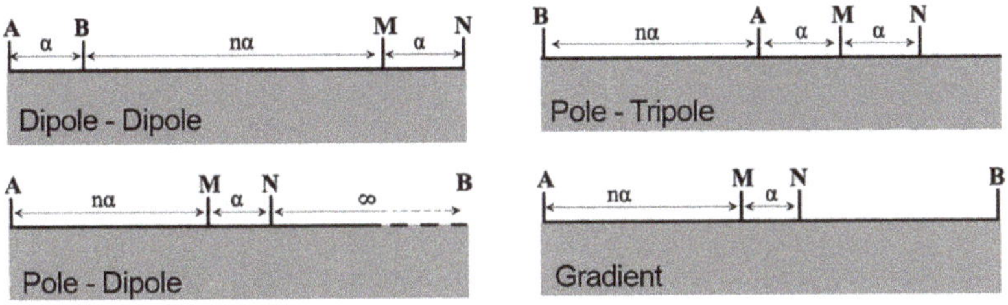

Figure 2. Representation of the basic arrays used to generate the comprehensive data set.

A demonstrative way of producing the 3-D optimum measurements from the comprehensive data set is shown in Figure 3. From the initial comprehensive (normal) 2-D tomographic data set (Figure 3, red-colored profile view), the 2-D algorithm is used to select the optimum data for a single 2-D profile (Figure 3, green-colored). Then, this optimum 2-D data set is cloned in multiple parallel profiles equally separated to create the 3-D optimum data set.

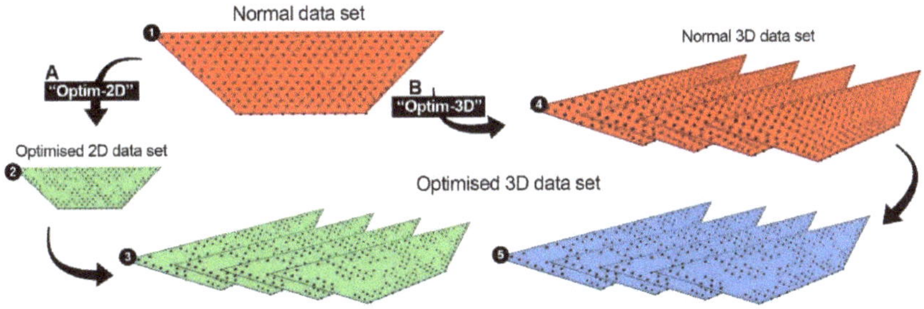

Figure 3. *Cont.*

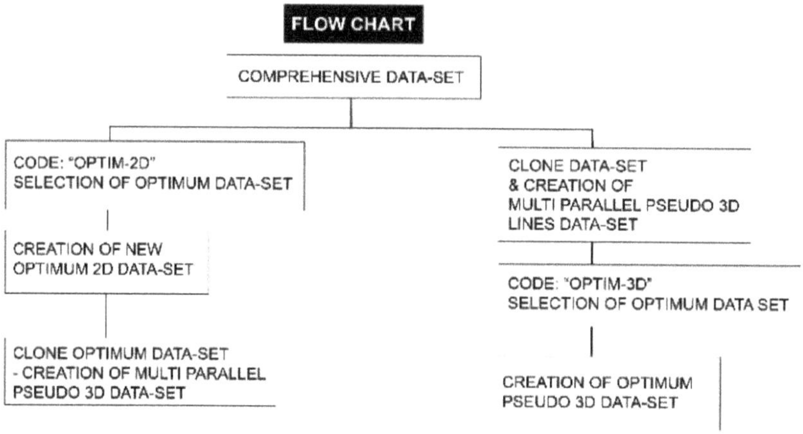

Figure 3. Optimization technique using: "Optim-2D" (2-D algorithm) data set (green-colored) and "Optim-3D" (3-D algorithm) data set (blue-colored). Comprehensive data set (red-colored).

On the other hand, for the 3-D optimization technique, the initial comprehensive 2-D data are firstly cloned in multiple equally separated parallel profiles (Figure 3, red-colored) and then the 3-D algorithm produces the 3-D optimum data (Figure 3, blue-colored). In the present study, both 2-D and 3-D optimization algorithms were allowed to run for specific steps to maintain equal measurements for the 'Optim-2D' and 'Optim-3D' protocols, respectively, for direct comparison purposes.

Table 1 shows the unit electrode spacing 'a' and the separation factor 'n' used to compile the 23,670 measurements of the comprehensive data set, comprised of dipole–dipole (DD), pole–dipole (PD), pole–tripole (PT) and gradient (GR) array, assuming 11 parallel profiles. The final resulted in "Optim-2D" and "Optim-3D" protocols each being only 14% of the initial comprehensive data set. The "Res-STUM" protocol used only 22% of the initial comprehensive data set. Figure 4 shows the percentage participation of each protocol that was used. A comparable quantitative analysis of the resolving capabilities of the different protocols based on the cumulative Jacobian values per parameter is shown in Figure 5. The cumulative values have been normalized according to the total number of the data to gain the flexibility of comparing protocols with uneven measurements. As is expected, higher sensitivity values are the ones closer to the surface. The normalized cumulative Jacobian of the "optim-3D' data set has higher values than the "optim-2D". Furthermore, the sensitivity pattern of the "optim-3D" is more intense in the deeper layers, thus indicating a higher capability in resolving deeper resistivity structures.

Table 1. Number of data for FULL protocol (percentage of initial array).

Protocol	a	na	Comprehensive	Optim-2D	Optim-3D	Stummer's
FULL	1–9	17	23,670 (100%)	3390 (14%)	3315 (14%)	5224 (22%)

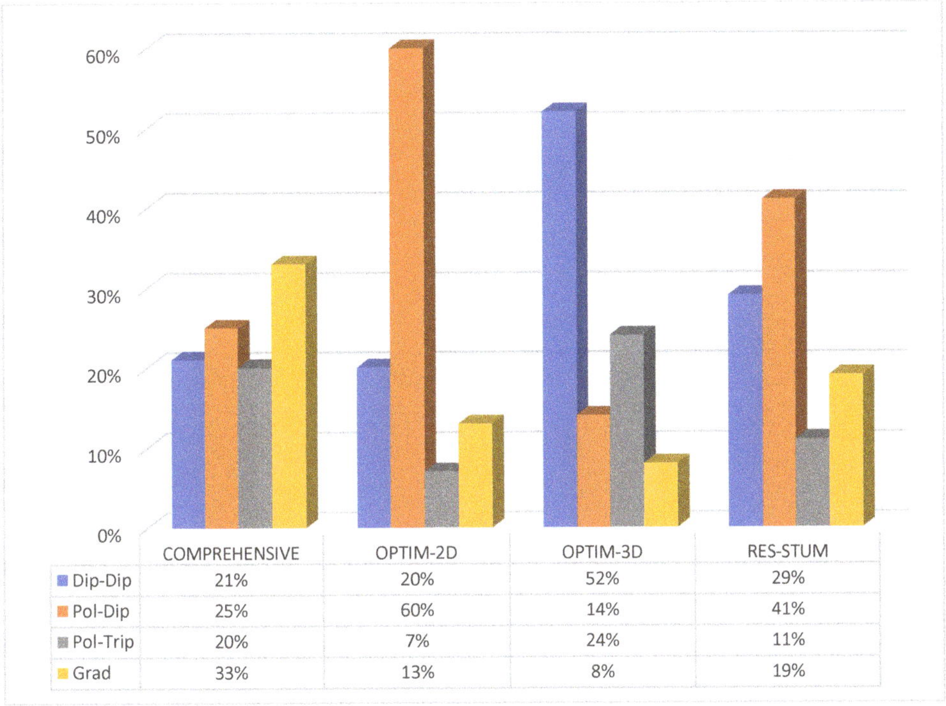

Figure 4. Participation of protocols used at data.

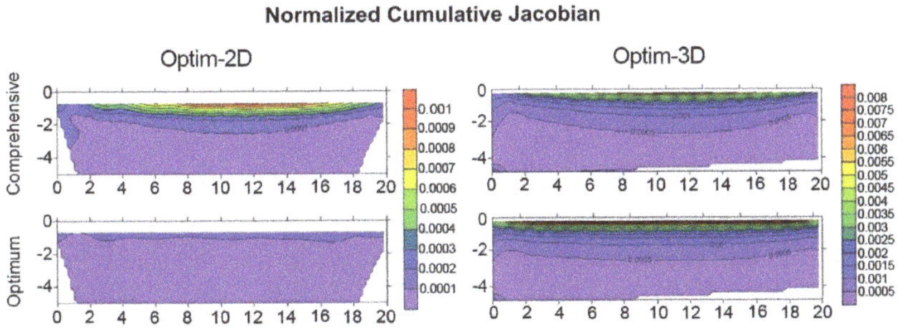

Figure 5. Normalized cumulative sensitivity results along a 2-D section for the comprehensive and optimum 3-D data sets resulted by the "optim-2D" and "optim-3D" strategy. Different scaling is used for presentation purposes since the sensitivity values differ.

3. Synthetic Model

The efficiency of the optimization algorithm was firstly tested with synthetic data in order to evaluate the resolving capabilities of the generated optimized protocols under constrained conditions, where the exact targets' position and resistivity values are predefined and well-known. The resistivity model used in the work of Stummer et al. [15] was extended in the 3-D space, composed of nine layers (D1–D9) with a constant thickness of 0.5 m (Figure 6). A homogeneous half-space of 1000 ohm.m lies below a less resistive superficial layer of 100 ohm.m one meter thick. A highly resistive target 'A' (3 × 2 × 1 m) with a resistivity of 10,000 ohm.m is buried at 0.5 m from the ground surface lying within the two horizontal layers D2–D3. A target 'B' with larger dimensions (5 × 4 × 1.5 m) is buried at 2 m depth from the surface within the layers D5 to D7. The resistivity of this block gradually decreases inwards from 100 to 10 ohm.m.

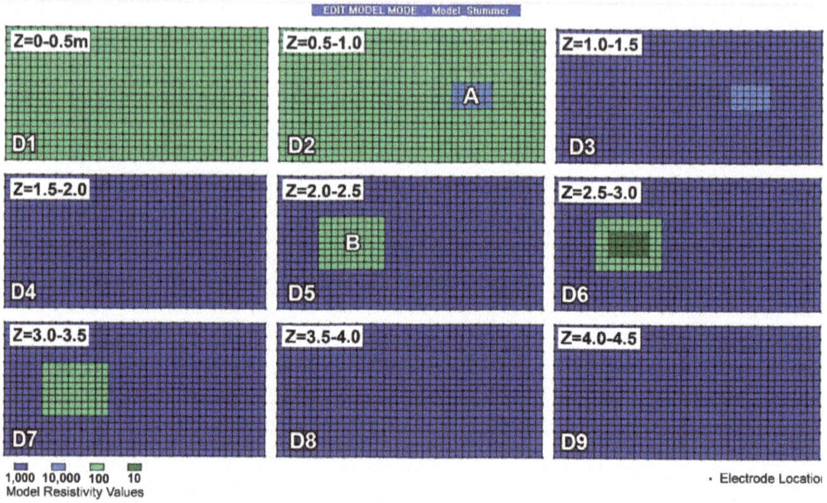

Figure 6. Synthetic Stummer's model.

The 3-D resistivity survey was composed of total 231 electrodes in total arranged in a rectangular grid of 20 m by 10 m along the X and Y axes, respectively. The inter-probe and inter-line spacing were set to 1 m on both axes. The tomographic data were collected along the 2-D profiles on the X-axis using the combination of the four traditional arrays (dd + pd + pt + gr) and the measuring parameters shown in Table 1. A 3-D finite element algorithm (RES3DMOD) was used to calculate the forward response of the specific model and extract the distribution of the apparent resistivity values for the comprehensive data in the 3-D subsurface space. The synthetic apparent resistivity data were then corrupted with 5% Gaussian noise on the modeled apparent resistivity values.

During the inversion procedure, the L1-norm method was used for data misfit and model roughness to account for the sharp boundaries of the targets [30]. A finite-element modelling subroutine was used to calculate the apparent resistivity values. The selected initial value for the damping factor was $\lambda = 0.1$, which was increasing with a depth factor of 1.01. The inversion procedure was terminated after five iterations. The common resistivity colored scale (10–10,000 ohm.m) and depth slices (D1–D9) are used in all inversion images for comparison purposes. Black squared rectangles indicate the expected exact position of targets 'A' and 'B'. The RMS error value varies between 3.7 and 4.1% among all the inversion images of the arrays that were used.

The comprehensive FULL protocol (Figure 7) reconstructs the targets 'A' and 'B' with comparable accuracy: the calculated resistivity value for target 'A' is approximately 1000 ohm.m. It is located inside the rectangular lines and the corresponding depth slices (D2 and D3). The target's 'B' resistivity value is close to 500 ohm.m. The target's location at the XY axis is well located but not on the Z-axis, where it is extended to layer depth D4.

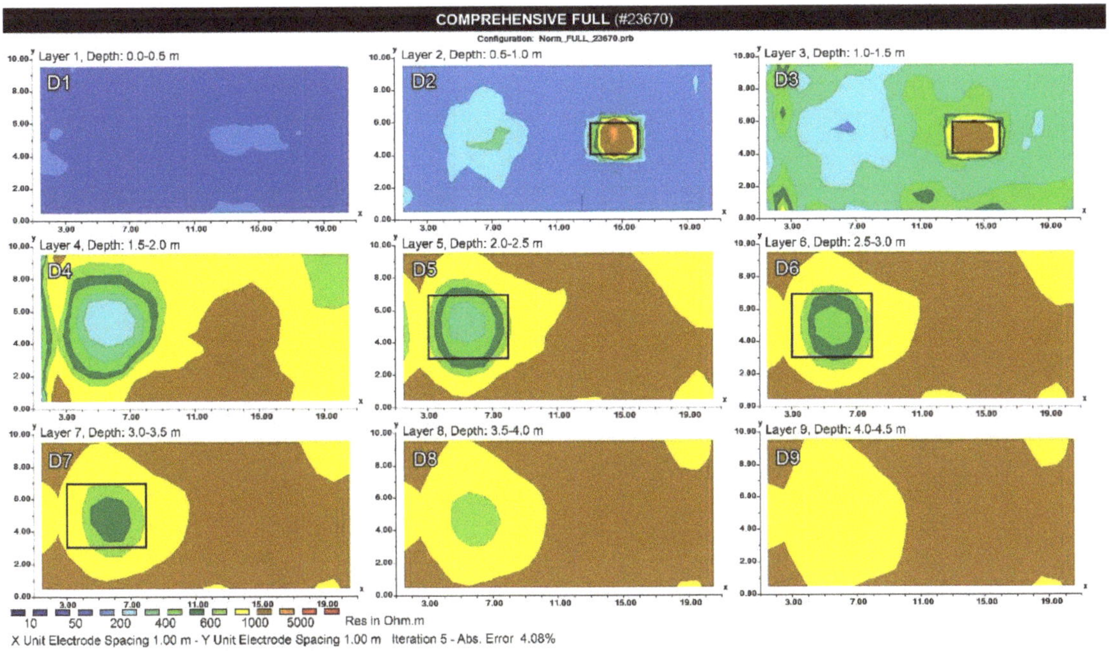

Figure 7. Inversion results from FULL comprehensive array.

The Optim-2D data set (Figure 8) is able to reconstruct the target's 'A' position inside the rectangular box but the resistivity value (500–1000 ohm.m) is lower than the comprehensive data set value. The target 'B' is also located inside the expected rectangular area with resistivity values that vary between 400–600 ohm.m and it is expanded at depth D4.

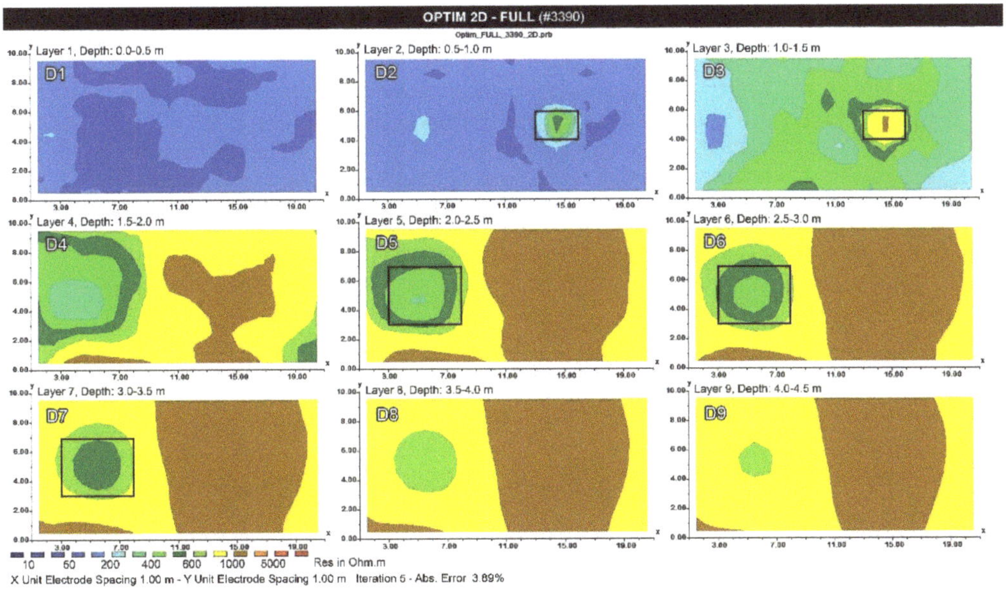

Figure 8. Inversion results from optimum array produced by 2D algorithm.

The higher and closer to true value resistivity value of 5000 ohm.m is presented with the Optim-3D data set regarding the reconstruction of target 'A' (Figure 9). Additionally, the position of target 'A' is inside the expected position. Target 'B' is reconstructed with a small shift downwards along the XY direction. The resistivity value varies between 400 and 500 ohm.m. Similarly, as noticed before, target 'B' is shifted downwards at depth level D4.

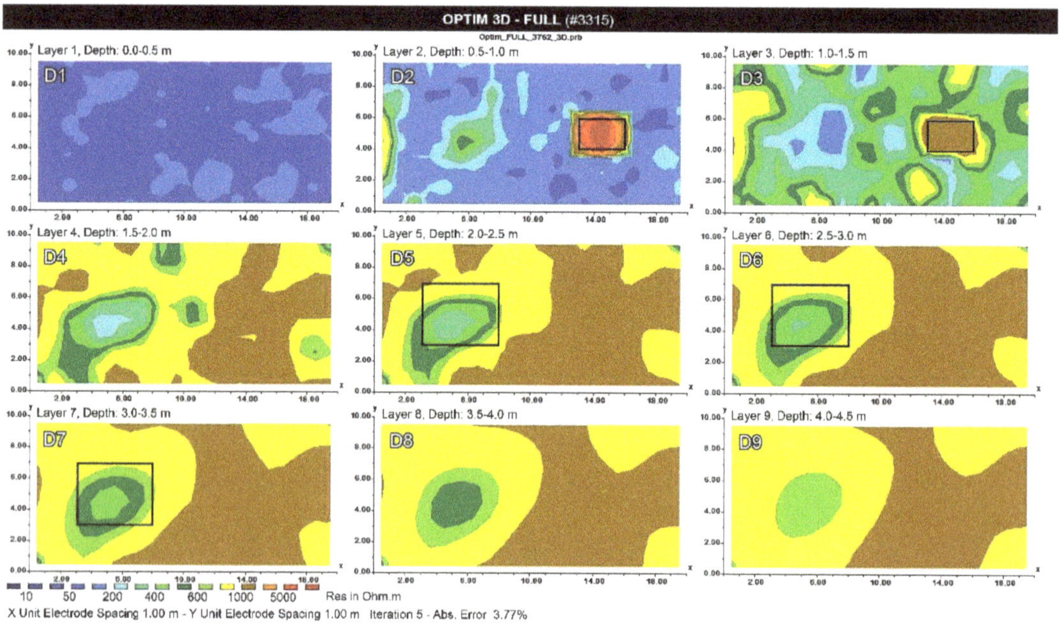

Figure 9. Inversion results from optimum array produced by 3D algorithm.

Stummer's optimum data reconstruct target 'A' only at layer D2 (not at D3) with higher resistivity values (>5000 ohm.m) and deformed along the Y direction (Figure 10). The layer D2 is reconstructed with values close to 1000 ohm.m, a value higher than the model (100 ohm.m). Target 'B' is also reconstructed at depth layers D3 and D4, where it should not be, but with resistivity values closer to the theoretical model (100 ohm.m).

It should be noted that both optimum data sets (Optim-2D, Optim-3D) are able to reconstruct the target's position while using only 14% of the initial comprehensive data set. Furthermore, the Optim-3D data set shows comparable accuracy regarding the target's resistivity value among all the other data set (comprehensive and Optim-2D) but the location of target 'B' was shifted slightly away from the expected location. All data sets reconstructed target 'B' at depth level D4 due to the smoothness constraints imposed in the inversion to stabilize the inversion.

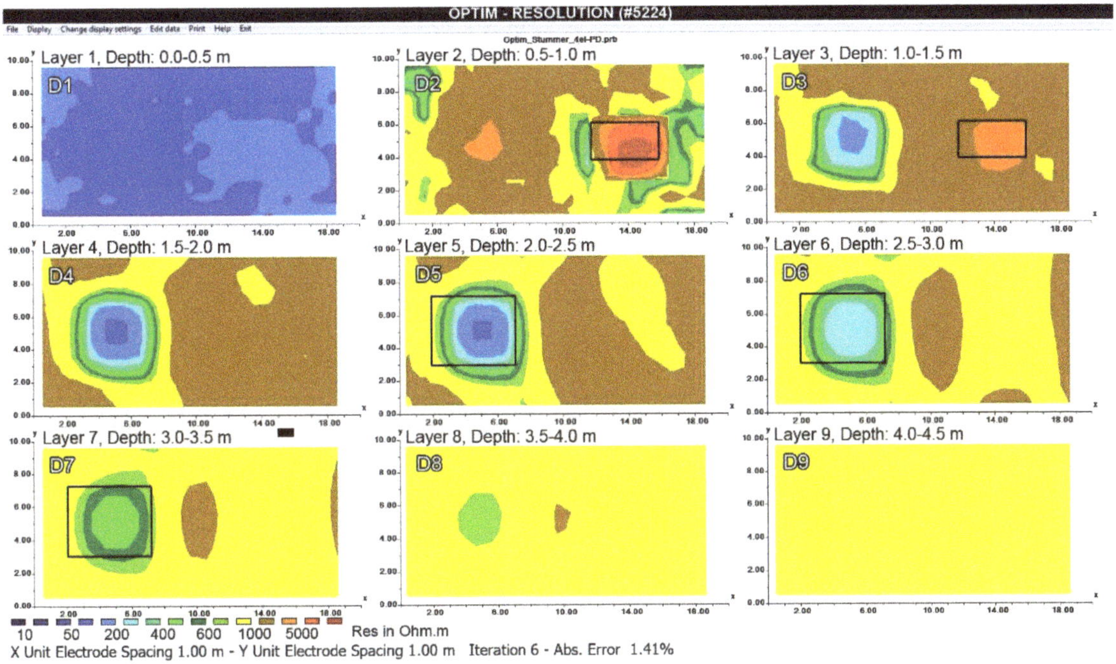

Figure 10. Inversion results from FULL optimum array produced by Stummer's algorithm.

4. Quantitative Analysis

The quantitative comparative analysis between the resistivity models resulting in the inversion of the comprehensive and optimum data sets was made feasible using the model misfit and the layer RMS. The above-mentioned techniques were used to validate the degree of model matching on the inverted model with respect to the true model as predefined by the initial user resistivity values. The model misfit is defined as the mean squared misfit between the true parameter model ρ_t and the inverted parameter model ρ_c resistivity value [27]:

$$Model\ Misfit = \left(\frac{1}{N} \sum_{i=1}^{N} [log(\rho_t(i)) - log(\rho_c(i))]^2 \right)^{0.5} \tag{1}$$

The layer RMS is the root mean square error for each layer defined as the percentage difference in the resistivity values between the true and the inverted model using the following equation [31]:

$$layer\ RMS = \sqrt{\frac{1}{N} \sum_{i=1}^{N} \left(\frac{\rho_c(i) - \rho_t(i)}{\rho_t(i)} \times 100 \right)^2} \tag{2}$$

The results of the calculations are shown in Figure 11. All values (model misfit and layer-RMS) are calculated for the FULL array only for five specific depth model layers (D2, D3, D5, D6 and D7) where the targets exist. The model misfit values of the Optim-3D data set (green-colored line) are profoundly decreased at the superficial depths (D2, D3) then equalize with the comprehensive and Optim-2D data sets at deeper levels (D5, D6, D7). The Optim-2D data set (yellow-colored line) has a higher value of model misfit at depth D2 from the Optim-3D data set and equal value with the comprehensive data set. At depth of D3, the Optim-2D data sets have lower model misfit values in comparison with the comprehensive data set and higher value in comparison with the Optim-3D data set. For the following depths (D5, D6, D7), the Optim-2D data set shares the same model misfit values with Optim-3D and comprehensive data set.

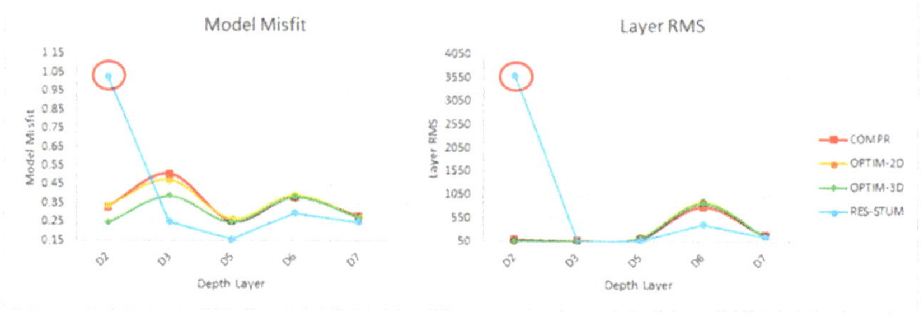

(**a**) (**b**)

Figure 11. Model misfit (**a**) and layer RMS (**b**) values for FULL array among comprehensive (red colored line), Optim-2D (yellow colored line), Optim-3D (green colored line) and Res-STUM (blue colored line) data set. Outliers indicated with red circles.

Regarding the layer-RMS values, the comprehensive data sets (red-colored line) share the same values with the optimum data sets (Optim-2D, Optim-3D) at superficial depth layers (D2, D3, D5). For deeper layer D6, the layer-RMS value of optimum data sets is higher than the comprehensive data set: Optim-2D data set has the highest value in comparison with the Optim-3D and the comprehensive data set. At the deepest layer D7, comprehensive, Optim-2D and Optim-3D share the same layer-RMS value.

The Res-STUM data (blue-colored line) presents a high value model misfit and layer-RMS at depth layer D2 showing a decreased efficiency of the optimized protocols to reconstruct the shallow depth layers. On the other hand, it seems to have lower values than the comprehensive data set, Optim-2D and Oprim-3D approaches at deeper depth layers.

The time to compute the 3-D optimized protocols with both the Jacobian and the Stummer's method will depend on the number of electrodes, the size of the comprehensive data set and the scheme to parametrize the subsurface in homogeneous 3-D resistivity blocks. Assuming a 3-D ERT survey where the number of electrodes and the parameters is fixed, the total time to extract the optimized data sets with the two algorithms will depend on the size of the comprehensive data set. Both algorithms seem to have a linear increasing trend relating the time and the number of measurements. However, the Jacobian method shows a relative reduction of more than 20% in the actual time needed to compile the optimized data sets, thus being significantly faster than Stummer's approach (Figure 12).

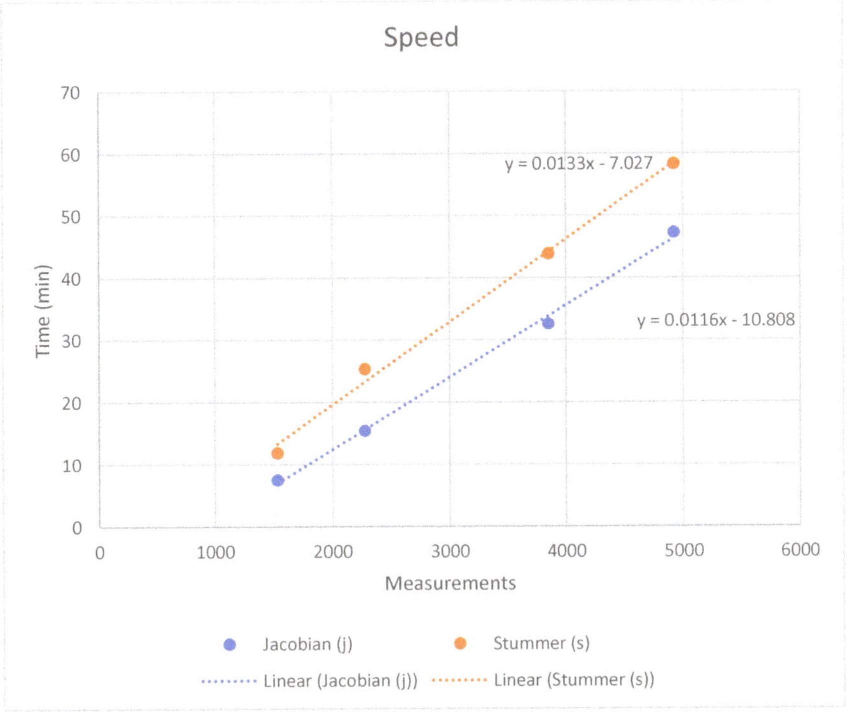

Figure 12. Speed comparison between Jacobian and Stummer optimization approach.

5. Field Data

5.1. Panormos (Rethymno)

The field data were collected for a geotechnical study at the city of Panormos, east of Rethymno (Crete, Greece) (Figure 13). The selected grid was 94 × 75 m, where 16 parallel survey lines (5 m apart) were used to collect the geoelectrical data. For each line, 48 electrodes were used with 2 m probe spacing. Each line collected 1216 (dipole–dipole) and 638 (gradient) data, producing an overall complete data set of 19,456 dipole–dipole and 10,208 gradient data in total from the 16 lines. All data were used to produce the comprehensive FULL data set (dd + gr), which consists of 29,664 measurements.

Figure 13. Map of Crete (Greece) indicating the location of the case studies discussed in the text.

The final Optim-2D optimized data set that was created by the 2D optimization procedure selected 15,214 measurements (10,035 dd and 5179 gr). The Optim-3D data set with 13,970 measurements were created from the 3-D optimization procedure which selected 8658 dipole–dipole and 5312 gradient data. The amount of optimum data that was selected from the 2-D and 3-D algorithms from the comprehensive data set is 51% and 47%, respectively. The Res-STUM optimum data set used 14,943 measurements (50%). All data were inverted with the same inversion parameters and are presented with a common resistivity scale from 3.4 to 205 ohm.m. The estimated RMS error for all inversion models is ranged from 3.0 to 3.87% after five or six iterations. The deepest depth layer reaches 15.0 m below the ground (Figures 14–17).

Based on the geoelectrical model in the research area, a surface layer of about 2.5 to 3.3 m thick is detected with resistivity values from 20 to 70 ohm.m. This layer coincides with the topsoil and the recent surface deposits. Within this layer, there are areas of increased resistivity values due to local concentrations of rock masses, resulting from the disintegration of the rocky background.

From the depth of 3.3 m up to the maximum depth of the geophysical survey (15 m), a formation with high resistivity values over than 70 ohm.m) appears, which is associated with the rocky background of the slates and shale in the wider area. The rocky background entirely covers the western area and extends east to a horizontal distance of 60 m. It also appears that the rocky background is not continuous and exhibits an intrinsic heterogeneity, as can be observed from changes in resistivity within the same formation. This means that the rock is fractured internally, mainly at sections where lower resistivity values are registered. Such an area is located on the northwest side of the area (X = 0–26 m and Y = 35–75 m) where it appears with resistivity values in the range of 30 to 40 ohm.m.

The most characteristic geophysical anomaly is the area to the eastern side, which appears with very low resistivity values below 20 ohm.m. This zone is superficial and appears to continue at the deeper slices. This area seems to lean westward and is bound west and east by the rocky background. To the east, this rocky background appears with high resistivity values mainly from the depth of 7 m.

The above range of low resistivity values can either be connected to a paleo-river due to the conductive signature of the sediments because of the water flowing in them. This area may also have originated from a local fragment of the rocky background so that, again, the water has the ability to move freely in this area and affect the values of resistivity.

The inversion results show a comparable accuracy among the resistivity models reconstructed by the inversion of the comprehensive, optim-2D, optim-3D and Res-STUM data sets: the mapping of the bedrock, which was the main interest of the study, is reconstructed at the same position and depth slices of both comprehensive and optimum inversion results.

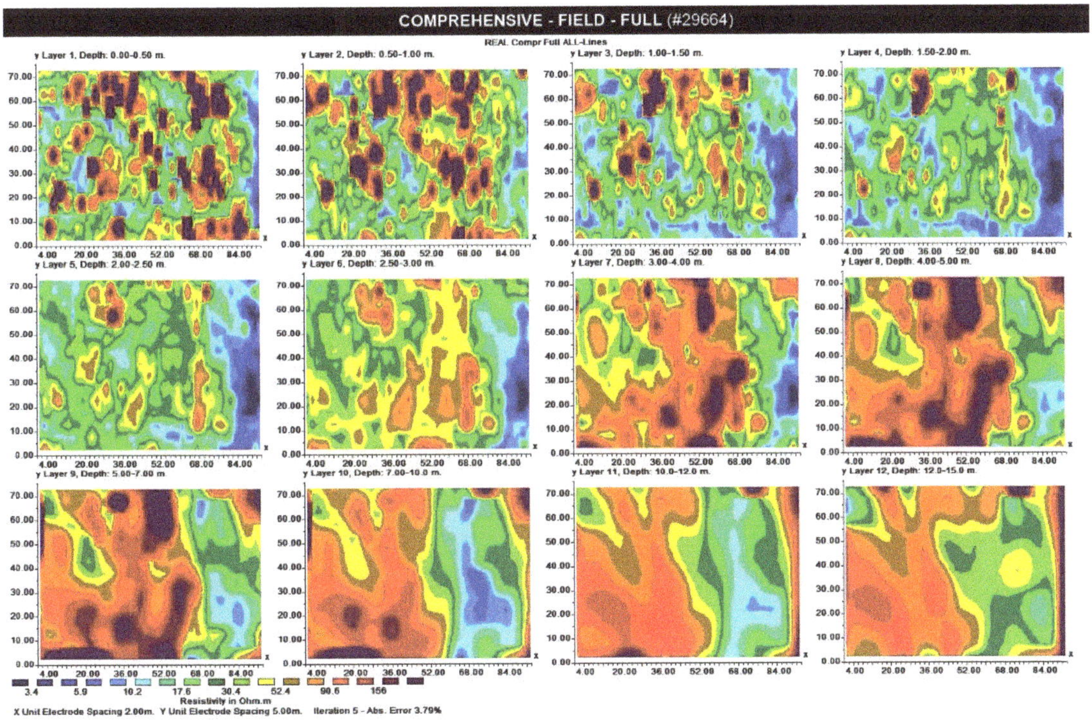

Figure 14. Inversion results from the comprehensive field data set.

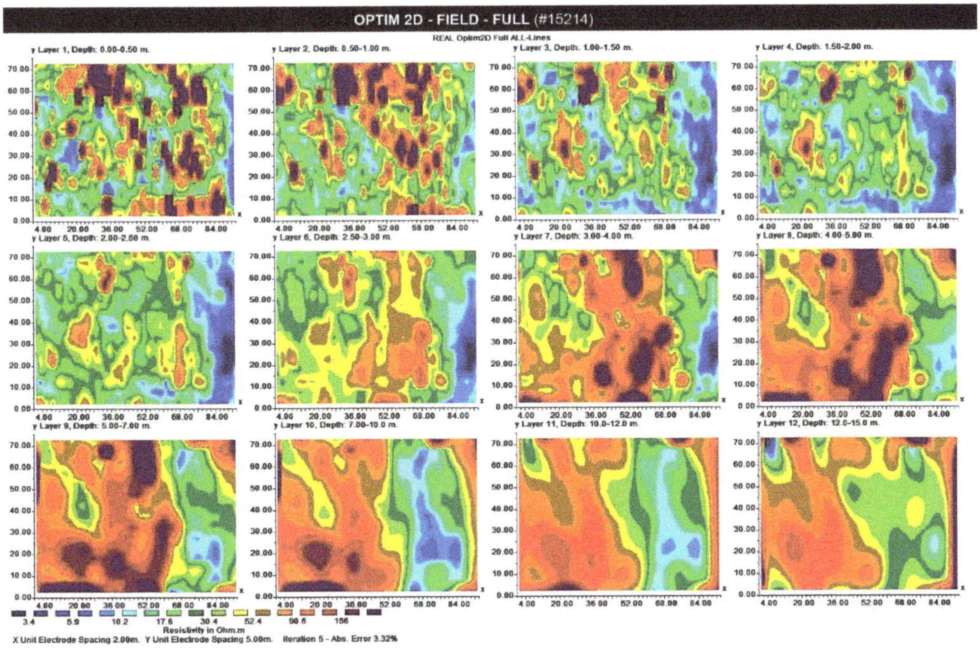

Figure 15. Inversion results from the Optim-2D field data set.

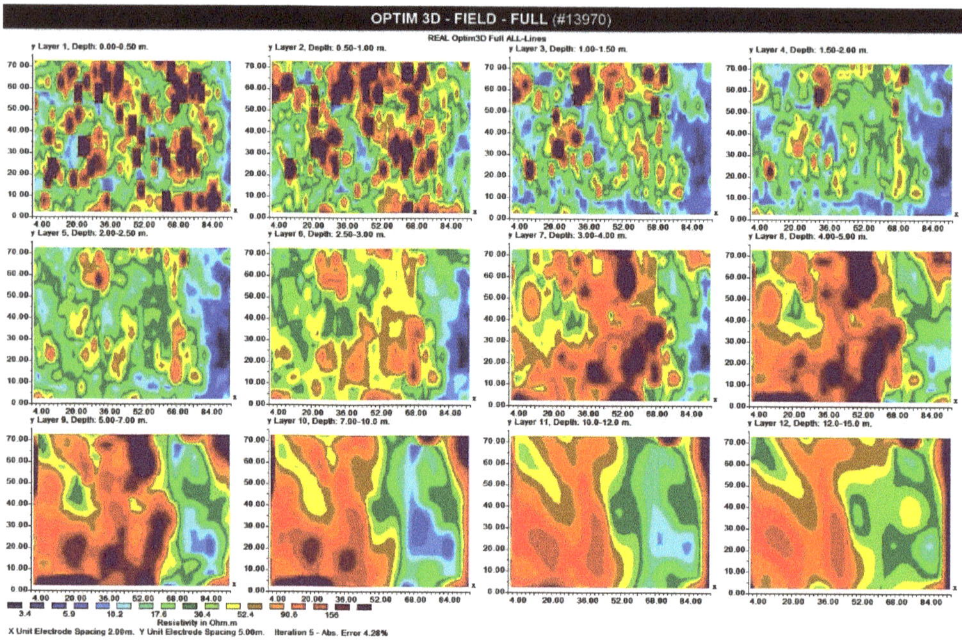

Figure 16. Inversion results from the Optim-3D field data set.

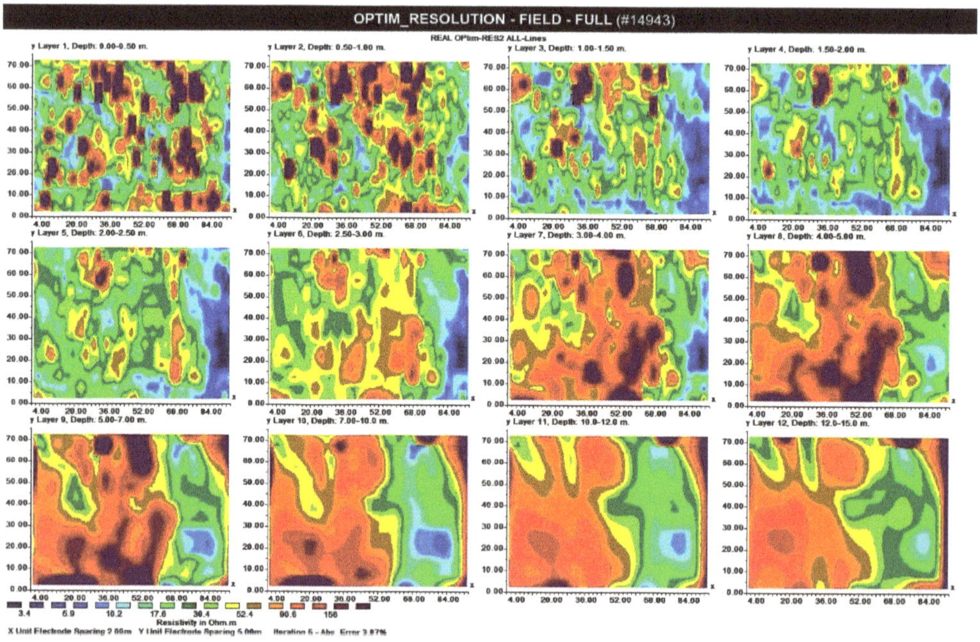

Figure 17. Inversion results from the Res-STUM field data set.

5.2. Elounda (Agios Nikolaos)

The field data were collected from the bay of Elounta (Crete, Greece). The selected grid was 30 × 47 m where 31 parallel survey lines (1 m apart) were used to collect the geoelectrical data in the marine environment. For each line, 48 submerged electrodes were used with 1 m probe spacing in depth varying from 0.07 m to 2.86 m and the resistivity of water was measured 0.2 ohm.m. Each line collected 1298 (pole–dipole forward and reverse), producing an overall complete data set of 40,204. All data were used to produce the comprehensive FULL data set.

The Optim-3D data set with 21,184 measurements were created from the 3-D Jacobian optimization procedure. The amount of optimum data that was selected from 3-D algorithm from the comprehensive data set was 50%. All data were inverted with the same inversion parameters and are presented with a common resistivity scale from 0.2 to 7 ohm.m. The estimated RMS error for all inversion models is from 1.56 to 2.21% after five iterations. The deepest depth layer reaches 3.0 m.

Both models (Figures 18 and 19) present similar results revealing a linear resistive area which extends from the depth of 0.5 m up to 2.5 m located between 23 and 38 m in the x direction and 18 to 21 m in the y direction. This area is correlated with the visible archaeological remains of a possible wall or tower. The relatively resistive area that extends from 29 to 41 m in the x direction across the y direction could be related to collapse parts of the aforementioned structure or with a poorly preserved wall. The area (0–25 m in x, 0–30 m in y) is dominated by the conductive sandy seafloor. A resistive target (1–7 m in x, 3–11 m in y) inside that area could be either beach rock, also visible in other parts of the bay, or some other archaeological relics.

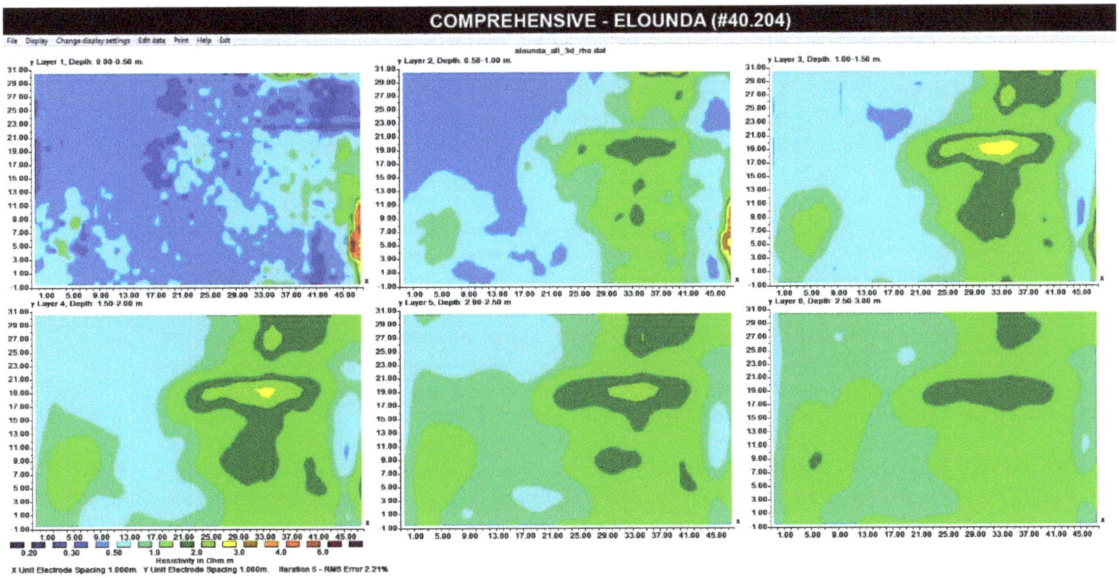

Figure 18. Inversion results from the comprehensive Elounda field data set.

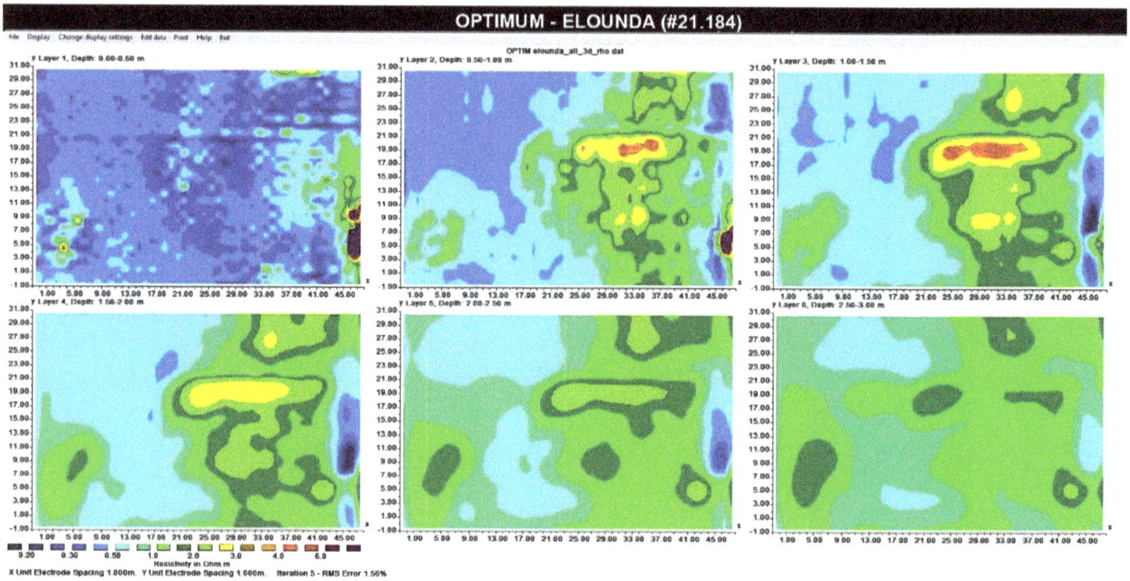

Figure 19. Inversion results from the Optimum Elounda field data set.

6. Conclusions

The work describes an experimental method to select electrode configurations that carry the maximum resolving capability in order to reconstruct the subsurface resistivity distribution. The technique is based on the sensitivity and the examination of Jacobian matrix entries. The basic differences of the specific method in relation to approaches utilizing the model resolution matrix to select the optimum array protocols have already been stressed for the 2-D resistivity problem. The main one is attributed to the fact that the optimum data set is created solely depending on each individual array. On the contrary, the initiation of any algorithm based on resolution matrix approaches needs the compilation of an initial base data set and the enrichment of this base set with added configurations coming from a comprehensive data set having every possible electrode combination based on the update of the resolution matrix. Thus, the sensitivity approach can be faster since the calculation of the Jacobian is part of the forward resistivity problem and there is no need to perform any inversion, as is the case for the resolution matrix approaches. The comparative tests performed in this work showed that the Jacobian method is about 20% faster than previous approaches using the resolution matrix. Furthermore, the Jacobian matrix method avoids the evaluation of the goodness function since the algorithm picks the most sensitive measurements. Therefore, the extra time gained in selecting optimum measurements is increased, which is amplified in cases of large 3-D resistivity tomographic problems and data.

The 3-D optimum protocols are produced from 3-D data sets compiled by the merging of multiple equally spaced parallel 2-D profiles assuming traditional arrays: dipole–dipole, pole–dipole, pole–tripole and gradient. The two different approaches involved the application of 2-D and 3-D optimization algorithms to select the optimum protocols ranged from 14% (synthetic data set) and ~50% (field data set) of the original full data sets. The cumulative Jacobian matrix (normalized sensitivity value of each parameter) showed that the optimum protocols have the same sensitivity values with the full data set, although fewer measurements are used than the traditional protocols.

The algorithms were validated with a synthetic model where the inversion results verified the comparable similarity of the reconstructed models between the comprehensive and the optimum (from 2-D or 3-D) data sets. The efficiency of optimization procedure was validated with field data from completely different environments using a comprehensive data set which is comprised of different traditional arrays (dipole–dipole and gradient, pole–dipole forward and reverse), where the inversion models of the optimum protocols showed similar results, although half of the initial complete data set was used for the inversion.

In general, the specific work shows the flexibility and effectiveness of an experimental procedure based on the sensitivity matrix in selecting optimum resistivity protocols in a 3-D context. The two different approaches in compiling the optimum 3-D tomographic protocols show comparable results in terms of the inverted resistivity models. In relatively more complicated subsurface geometry and resistivity distribution, the full 3-D optimization method can give slightly superior results in relation to the 2-D approach with minimum extra time. Finally, the optimized data sets can also be used as a basis for 3-D monitoring experiments to experience enhanced resolving capabilities of the time-lapse resistivity images for monitoring the subsurface contamination in environmental applications [32].

Author Contributions: K.S.: Software development, methodology, data processing/analysis/interpretation, writing—original draft preparation; N.P.: conceptualization, review and editing, funding acquisition; D.O.: data acquisition/interpretation, visualization, validation. All authors have read and agreed to the published version of the manuscript.

Funding: The compilation of this paper was supported by the project "DIATOPO-Cretan cultural landscapes over the time: highlighting the marine and mountainous environment of Mirabello" with MIS 5028190. DIATOPO is part of the Priority Axis "Strengthening the competitiveness, in-novation and entrepreneurship of Crete" of the Operational Program. "Crete" 2014–2020 and co-financed by the European Regional Development Fund.

Institutional Review Board Statement: Not applicable.

Informed Consent Statement: Not applicable.

Data Availability Statement: The spatial raw data collected in the different sites are freely available to the academic and research community from the corresponding author upon reasonable request.

Acknowledgments: The authors would like to thank the anonymous reviewers with their constructive criticism that improved the clarity of the paper.

Conflicts of Interest: The authors declare no conflict of interest.

Appendix A. Calculation of Jacobian Matrix

A fundamental key in solving the 3-D resistivity inversion problem comprises the determination of a relationship that links the variation of the potential (V_i) or the apparent resistivity data ($\rho_{\alpha i}$) in relation to a change in the subsurface resistivity (ρ_j) or conductivity (σ_j) [33]. The MxN two-dimensional matrix known as Jacobian (Sensitivity) matrix $\mathbf{J}_{ij}$ expresses this link:

$$\mathbf{J}_{ij} = \begin{bmatrix} \frac{\partial \log \rho_{a1}}{\partial \log \rho_1} & \cdots & \frac{\partial \log \rho_{a1}}{\partial \log \rho_N} \\ \vdots & \ddots & \vdots \\ \frac{\partial \log \rho_{aM}}{\partial \log \rho_1} & \cdots & \frac{\partial \log \rho_{aM}}{\partial \log \rho_N} \end{bmatrix} \tag{A1}$$

where M, N are the number of the apparent resistivity measurements and the resistivity model parameters, respectively. Notice that since the resistivities and apparent resistivities often span several orders of magnitude, their logarithms have been used in Equation (A1). In this work, the Jacobian matrix values were estimated using the adjoint equation approach.

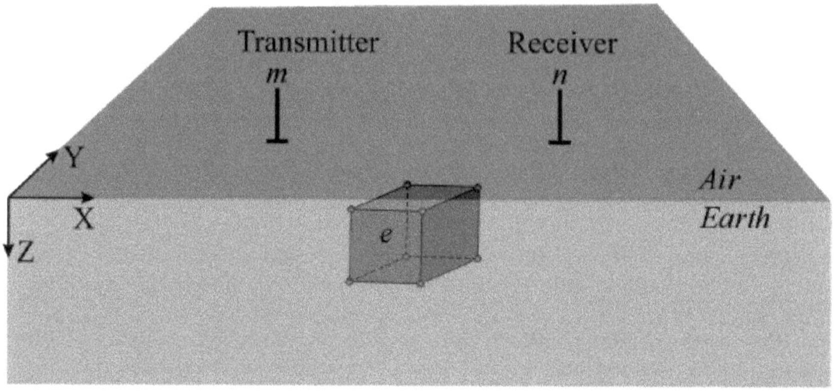

Figure A1. Relative location of a transmitter and a receiver located on nodes of the 3-D mesh with a hexahedral element in order to calculate the partial derivative of potential between the points m and n versus the element's conductivity $\frac{\partial V_{nm}}{\partial \sigma^{(e)}} = \iiint_e \left[\left(\frac{\partial V_n^{(e)}}{\partial x} \frac{\partial V_m^{(e)}}{\partial x} + \frac{\partial V_n^{(e)}}{\partial y} \frac{\partial V_m^{(e)}}{\partial y} + \frac{\partial V_n^{(e)}}{\partial z} \frac{\partial V_m^{(e)}}{\partial z} \right) dxdydz \right]$.

References

1. Loke, M.H.; Wilkinson, P.B.; Chambers, J.E. Computation of Optimized Arrays for 3-D electrical imaging surveys, Near Surface Geoscience. In Proceedings of the 19th European Meeting of Environmental and Engineering Geophysics, Bochum, Germany, 9–11 September 2013.
2. Kumar, V.S.; Dhakate, R.; Amarender, B.; Sankaran, S. Application of ERT and GPR for demarcating the saline water intrusion in coastal aquifers of Southern India. *Environ. Earth Sci.* **2016**, *75*, 1–17. [CrossRef]
3. De Carlo, L.; Berardi, M.; Vurro, M.; Caputo, M.C. Geophysical and hydrological data assimilation to monitor water content dynamics in the rocky unsaturated zone. *Environ. Monit. Assess.* **2018**, *190*, 310. [CrossRef]
4. Lucas, D.R.; Fankhauser, K.; Springman, S.M. Application of Geotechnical and Geophysical Field Measurements in an Active Alpine Environment. *Eng. Geology* **2017**, *219*, 32–51. [CrossRef]
5. Simyrdanis, K.; Papadopoulos, N.; Soupios, P.; Kirkou, S.; Tsourlos, P. Characterization and monitoring of subsurface contamination from Olive Oil Mills' waste waters using Electrical Resistivity Tomography. *Sci. Total Environ.* **2018**, *637*, 991–1003. [CrossRef]
6. Chambers, J.E.; Ogilvy, R.D.; Kuras, O.; Cripps, J.C.; Meldrum, P.I. 3D electrical imaging of known targets at a con-trolled environmental test site. *Environ. Geol.* **2002**, *41*, 690–704. [CrossRef]
7. Gharibi, M.; Bentley, L.R. Resolution of 3-D Electrical Resistivity Images from Inversions of 2-D Orthogonal Lines. *J. Environ. Eng. Geophys.* **2005**, *10*, 339–349. [CrossRef]
8. Papadopoulos, N.; Tsourlos, P.; Tsokas, G.; Sarris, A. Efficient ERT measuring and inversion strategies for 3D imaging of buried antiquities. *Near Surf. Geophys.* **2007**, *5*, 349–361. [CrossRef]
9. Tejero-Andrade, A.; Cifuentes-Nava, G.; Chávez, R.E.; López-González, A.E.; Delgado-Solórzano, C. L- and CORNER-arrays for 3D electric resistivity tomography: An alternative for geophysical surveys in urban zones. *Near Surf. Geophys.* **2015**, *13*, 355–368. [CrossRef]
10. Robbins, A.R.; Plattner, A. Offset-Electrode Profile Acquisition Strategy for Electrical Resistivity Tomography. *J. Appl. Geophys.* **2018**, *151*, 66–72. [CrossRef]
11. Noel, M.; Xu, B. Archaeological investigation by electrical resistivity tomography: A preliminary study. *Geophys. J. Int.* **1991**, *107*, 95–102. [CrossRef]
12. Xu, B.; Noel, M. On the completeness of data sets with multi-electrode systems for electrical resistivity survey. *Geophys. Prospect.* **1993**, *41*, 791–801. [CrossRef]
13. Dahlin, T.; Zhou, B. A numerical comparison of 2D resistivity imaging with 10 electrode arrays. *Geophys. Prospect.* **2004**, *52*, 379–398. [CrossRef]
14. Papadopoulos, N.G.; Tsourlos, P.; Papazachos, C.; Tsokas, G.N.; Sarris, A.; Kim, J.H. An Algorithm for the Fast 3-D Resistivity Inversion of Surface Electrical Resistivity Data: Application on Imaging Buried Antiquities. *Geophys. Prospect.* **2011**, *59*, 557–575. [CrossRef]
15. Stummer, P.; Maurer, H.; Green, A.G. Experimental design: Electrical resistivity data sets that provide optimum subsurface information. *Geophysics* **2004**, *69*, 120–139. [CrossRef]
16. Wilkinson, P.B.; Meldrum, P.I.; Chambers, J.E.; Kuras, O.; Ogilvy, R.D. Improved strategies for the automatic selection of optimized sets of ERT measurement configurations. *Geophys. J. Int.* **2006**, *167*, 1119–1126. [CrossRef]

17. Wilkinson, P.B.; Uhlemann, S.; Meldrum, P.I.; Chambers, J.E.; Carriere, S.; Oxby, L.S.; Loke, M.H. Adaptive time-lapse optimized survey design for electrical resistivity tomography monitoring. *Geophys. J. Int.* **2015**, *203*, 755–766. [CrossRef]
18. Loke, M.; Wilkinson, P.; Chambers, J. Fast computation of optimized electrode arrays for 2D resistivity surveys. *Comput. Geosci.* **2010**, *36*, 1414–1426. [CrossRef]
19. Loke, M.H.; Wilkinson, P.B.; Chambers, J. Parallel computation of optimized arrays for 2-D electrical imaging surveys. *Geophys. J. Int.* **2010**, *183*, 1302–1315. [CrossRef]
20. Loke, M.; Kiflu, H.; Wilkinson, P.; Harro, D.; Kruse, S. Optimized arrays for 2D resistivity surveys with combined surface and buried arrays. *Near Surf. Geophys.* **2015**, *13*, 505–518. [CrossRef]
21. Coles, D.A.; Morgan, F.D. A method of fast, sequential experimental design for linearized geophysical inverse problems. *Geophys. J. Int.* **2009**, *178*, 145–158. [CrossRef]
22. Wagner, F.M.; Günther, T.; Schmidt-Hattenberger, C.; Maurer, H. Constructive optimization of electrode locations for target-focused resistivity monitoring. *Geophysics* **2015**, *80*, E29–E40. [CrossRef]
23. Tsourlos, P.; Papapdopoulos, N.; Yi, M.-J.; Kim, J.-H.; Tsokas, G. Comparison of measuring strategies for the 3-D electrical resistivity imaging of tumuli. *J. Appl. Geophys.* **2014**, *101*, 77–85. [CrossRef]
24. Furman, A.; Ferré, T.P.A.; Warrick, A.W. Optimization of ERT surveys for monitoring transient hydrological events using perturbation sensitivity and genetic algorithms. *Vadose Zone J.* **2004**, *3*, 1230–1239. [CrossRef]
25. Athanasiou, E.; Tsourlos, P.; Papazachos, C.; Tsokas, G. Optimizing Electrical Resistivity Array Configurations by Using a Method Based on the Sensitivity Matrix. In Proceedings of the Near Surface 2009—15th EAGE European Meeting of Environmental and Engineering Geophysics, Dublin, Ireland, 7–9 September 2009.
26. Loke, M.H.; Wilkinson, P.B.; Uhlemann, S.S.; Chambers, J.E.; Oxby, L.S. Computation of optimized arrays for 3-D electrical imaging surveys. *Geophys. J. Int.* **2014**, *199*, 1751–1764. [CrossRef]
27. Loke, M.; Wilkinson, P.; Chambers, J.; Uhlemann, S.; Sorensen, J. Optimized arrays for 2-D resistivity survey lines with a large number of electrodes. *J. Appl. Geophys.* **2015**, *112*, 136–146. [CrossRef]
28. Uhlemann, S.; Wilkinson, P.B.; Maurer, H.; Wagner, F.M.; Johnson, T.C.; Chambers, J.E. Optimized survey design for electrical resistivity tomography: Combined optimization of measurement configuration and electrode placement. *Geophys. J. Int.* **2018**, *214*, 108–121. [CrossRef]
29. McGillivray, P.R.; Oldenburg, D.W. Methods for Calculating Frechet Derivatives and Sensitivities for the Non-Linear Inverse Problem: A Comparative Study. *Geophys. Prospect.* **1990**, *38*, 499–524. [CrossRef]
30. Loke, M.; Acworth, I.; Dahlin, T. A comparison of smooth and blocky inversion methods in 2D electrical imaging surveys. *Explor. Geophys.* **2003**, *34*, 182–187. [CrossRef]
31. Perren, L.J. Investigating the Performance of Electrical Resistivity Arrays. Doctoral Dissertation, Virginia Tech, Blacksburg, VA, USA, 2005.
32. Simyrdanis, K.; Tsourlos, P.; Papadopoulos, N.; Kirkou, S. Monitoring of contaminant flow using time-lapse optimized 3-D electrical resistivity tomography in a supervised experimental test site. In Proceedings of the Near Surface Geoscience 2016-22nd European Meeting of Environmental and Engineering Geophysics, Barcelona, Spain, 4–8 September 2016.
33. Tsourlos, P.; Ogilvy, R. An algorithm for the 3-D Inversion of Tomographic Resistivity and Induced Polarization data: Preliminary Results. *J. Balk. Geophys. Soc.* **1999**, *2*, 30–45.

applied sciences

MDPI

Article

Mean Dynamic Topography Modeling Based on Optimal Interpolation from Satellite Gravimetry and Altimetry Data

Yihao Wu *, Jia Huang *, Hongkai Shi and Xiufeng He

School of Earth Sciences and Engineering, Hohai University, Nanjing 211100, China; shkcehui@hotmail.com (H.S.); xfhe@hhu.edu.cn (X.H.)
* Correspondence: yihaowu@hhu.edu.cn (Y.W.); hj123@hhu.edu.cn (J.H.)

Abstract: Mean dynamic topography (MDT) is crucial for research in oceanography and climatology. The optimal interpolation method (OIM) is applied to MDT modeling, where the error variance–covariance information of the observations is established. The global geopotential model (GGM) derived from GOCE (Gravity Field and Steady-State Ocean Circulation Explorer) gravity data and the mean sea surface model derived from satellite altimetry data are combined to construct MDT. Numerical experiments in the Kuroshio over Japan show that the use of recently released GOCE-derived GGM derives a better MDT compared to the previous models. The MDT solution computed based on the sixth-generation model illustrates a lower level of root mean square error (77.0 mm) compared with the ocean reanalysis data, which is 2.4 mm (5.4 mm) smaller than that derived from the fifth-generation (fourth-generation) model. This illustrates that the accumulation of GOCE data and updated data preprocessing methods can be beneficial for MDT recovery. Moreover, the results show that the OIM outperforms the Gaussian filtering approach, where the geostrophic velocity derived from the OIM method has a smaller misfit against the buoy data, by a magnitude of 10 mm/s (17 mm/s) when the zonal (meridional) component is validated. This is mainly due to the error information of input data being used in the optimal interpolation method, which may obtain more reasonable weights of observations than the Gaussian filtering method.

Keywords: mean dynamic topography; optimal interpolation; global geopotential models; geostrophic velocity

1. Introduction

Mean dynamic topography (MDT) is the deviation between the mean sea surface (MSS) and the geoid, which plays an important role in disciplines like oceanography and climatology. In the field of geodesy, an accurate MDT is necessary for the unification of the land–sea vertical datum and research on the Earth's rotation [1,2]. In addition, in the field of oceanography, the parameters of MDT are of great importance in the study of ocean circulation, sea-level change, and energy transport mechanisms in the ocean. The ocean circulation derived from the MDT can be applied to monitor the movement of pollutants, which is of great significance for marine environmental monitoring [3]. Moreover, the ocean current system promotes the energy exchange between the high and low latitudes of the earth, which in turn affects global climate change. The stability and variability of ocean circulation and its related transfer of extreme heat and freshwater can have a significant impact on local and even global climate change. For example, the Kuroshio warm current and the Oyashio cold current have important impacts on the weather and climate of China and Japan; the Gulf Stream has an important impact on the climate of North America and Europe; and the changes in ocean currents in the equatorial Pacific and El Niño and La Niña are also closely related to the global climate characteristics.

Based on the geodetic approach, the MSS and the geoid can be obtained by satellite altimetry technology and the gravity geopotential model (GGM), respectively [4], while the difference between them is the MDT. However, limited by the observation technique, the GGM would contribute to a higher level of errors than the MSS in solving the MDT [5]. In recent years, with the development of gravity satellites, especially the successive launch of the Gravity Recovery and Climate Experiment mission (GRACE), the Gravity Field and Steady-State Ocean Circulation Explorer mission (GOCE), and the GRACE Follow-On mission, the accuracy and resolution of GGM have been significantly improved [6–8]. These dedicated satellite missions facilitate many applications in earth science, including understanding ocean circulations in oceanography and studying tectonic–geodynamic activities [9]. Currently, the highest degree and order (d/o) of the satellite-only GGM can be up to 300 d/o, corresponding to a spatial resolution of about 66 km [10].

However, the spatial resolution of MSS can currently reach $1'$ (~10 km), which is much higher than that of the GGM [11]. As a result, because of the spectral inconsistency, it will cause problems such as spectral aliasing and spectral leakage when removing the geoid signal directly from the MSS [5]. Accordingly, a filtering procedure is mandatory to unify the spectrum between the MSS and the geoid. Researchers now have investigated different filtering methods for reasonable MDT solutions, such as Gaussian filtering and wavelet filtering [12–14]. The weighting scheme of Gaussian filtering is usually only related to distance, which may not be the best method. In addition, Vianna et al. [15] developed an adaptive filtering method based on singular spectrum analysis to solve high-resolution MDT. This method is adequate for the finite domain MDT processing since it minimizes smoothing and does not lose boundary points. The signal and noise can effectively separate by this method. But these filtering methods still lead to signal leakage and distortion problems, and it is difficult to give the error distribution of the MDT model. Compared with the previous studies, Rio et al. [5,16] introduced an optimal interpolation method (OIM) in a combination of the MSS and GGM, which considered the covariance between observations. The estimated MDT can be smoothed while preserving more detailed signals, which mitigates the problems of signal leakage or distortion, and error information of the MDT is estimated. The key point of the OIM focuses on the construction of the covariance function, involving the estimation of the parameters like the variance of the MDT and the correlation radius.

The accurate information of the MSS and especially the GGM has a direct influence on the MDT reconstruction. In fact, the GOCE mission now can detect more accurate gravity signals thanks to the equipment of the gradient gravimeter in the GOCE gravity satellite. Specifically, the European Space Agency (ESA) has released six generations of GGM based on GOCE gravity data and direct approach; the fourth-generation model (GO_CONS_GCF_2_DIR_R4, DIRR4) assimilated data from 2009 to 2012, and the fifth-generation model (GO_CONS_GCF_2_DIR_R5, DIRR5) and the sixth-generation model (GO_CONS_GCF_2_DIR_R6, DIRR6) used all the data from 2009 to 2013. In addition, the DIRR6 recalibrated the satellite's orbits compared to the DIRR5 [17]. The influence of these differences between GGMs on the constructed MDTs can be further investigated. In addition, the recently released DTU18 MSS incorporates the latest satellite altimetry data, especially Jason, CyroSat-2, and SARAL/Altika, which is expected to further improve the accuracy of the MSS.

The aim of this study includes three aspects. First, we quantify the effects on MDT recovery based on the recently released GOCE-derived GGMs, which are expected to further improve the MDT modeling in terms of spatial resolution and accuracy. In addition, the choice of MSS is studied, in which the recently published MSS models are compared to see if the choice of MSS has nonignorable effects on MDT modeling. Then, we compare the performances of the traditionally used Gaussian filtering method with the optimal interpolation approach, providing an insight with respect to the choice of approach in MDT recovery.

The structure of this study is as follows: In Section 2, the main principle of the OIM in MDT modeling is briefly introduced. Then, the estimation of the key parameters in the OIM is introduced. Numerical experiments are shown in Section 3, in which the Kuroshio is selected as the research area, and the oceanographic and geodetic MDTs and the buoy-measured ocean geostrophic current-velocity profiles are used for comparison and validation. The summary of this study and a brief conclusion are given in Section 4.

2. Data and Methods

2.1. Study Area and Data

The study area was located in a part of the Kuroshio current (20–40° N, 125–145° E), which is shown in Figure 1 (image adapted from Shaded Relief, http://www.shadedrelief.com/ (accessed on 15 March 2021)). The Kuroshio current is the western boundary in the wind-driven, subtropical, and subarctic circulations of the North Pacific Ocean. The Kuroshio brings the high temperature and high salinity seawater of the equatorial Pacific Ocean to the vast offshore sea areas, which has a huge impact on the ocean, meteorology, and hydrology of these areas. Moreover, the Kuroshio also transports a large amount of heat from the low latitudes of the North Pacific to the middle and high latitudes, so its seasonal and interannual changes have an important impact on the climate and environment of neighboring countries and even the world [5,18]. There is important scientific significance in studying the MDT and geostrophic current in this area.

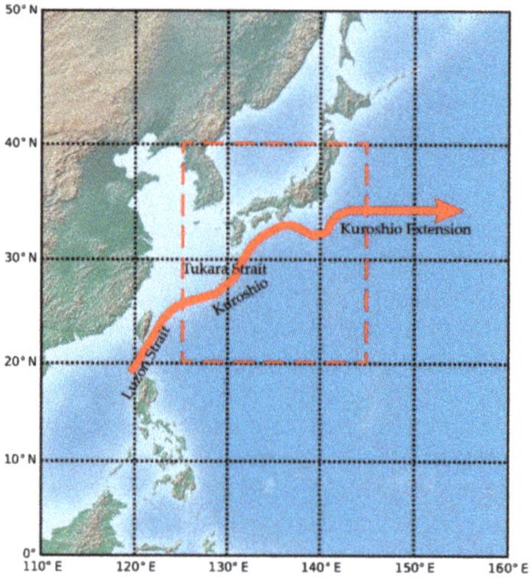

Figure 1. The study area is identified by a red dotted rectangle, and the solid red line represents the Kuroshio current. The Kuroshio passes through the Luzon Strait, and the majority flows northward along the east coast of Taiwan into the East China Sea. Then it flows out of the East China Sea from the Tukara Strait, and finally flows eastward offshore near 35° N, becoming the Kuroshio extension.

2.1.1. Mean Sea-Surface Model

The mean sea-surface model is a key element for modeling MDT. DTU15 MSS and DTU18 MSS, published by the Danish Technical University (DTU), were chosen to construct MDT in the paper. DTU15 MSS uses the last 4 years of Cryosat-2 data to improve the accuracy of the MSS model. The resolution of DTU15 MSS is $1' \times 1'$. DTU18 MSS is based on the previous generation model DTU15 MSS, incorporating 3 years of Sentinel-3A and 7 years of Cryosat-2 altimetry data from two satellites that provide data in SAR mode, and using the new tide model FES2014, which allows for a more accurate MSS [11]. DTU18 MSS has a spatial resolution of $1' \times 1'$ and a reference time of 1993–2012.

2.1.2. Global Geopotential Models

In order to facilitate the estimation of errors of the geoid, this study introduces multi-generation, GOCE-derived GGMs solved by the direct and time-wise approaches: the GO_CONS_GCF_2_DIR_R4 (DIRR4), GO_CONS_GCF_2_DIR_R5 (DIRR5), and

GO_CONS_GCF_2_DIR_R6 (DIRR6); and the GO_CONS_GCF_2_TIM_R4 (TIMR4), GO_CONS_GCF_2_TIM_R5 (TIMR5), and GO_CONS_GCF_2_TIM_R6 (TIMR6) [9,19]. Table 1 summarizes detailed information about these GOCE-derived GGMs. Figure 2 shows the error-degree variance in geoid height calculated by different GOCE-derived GGMs. It shows that the period of the assimilated data was extended with the development of the generation, and the supplement of data improved the modeling accuracy. The accuracy of the direct solution model was higher than that of the time domain solution model. With the update of the GOCE gravity data and the update of the solution method, the error of the time domain solution model showed a significant downward trend, while in the direct solution model, the accuracy of DIRR5 was higher than that of DIRR4, but the accuracy of DIRR6 was higher than DIRR5 before order 170, and lower than DIRR5 when the order was greater than 170. This may be due to the parameters set in the solution of these two models. While different from the data used, this result did not mean that the accuracy of DIRR6 was lower than DIRR5. Regarding the error of the GOCE satellite gravity field model, ESA has issued a related report that used externally measured GPS/leveling data to evaluate the accuracy of the multigeneration gravity field model. The results showed that the accuracy of the DIRR6 was better than that of the DIRR5 [20].

Table 1. Summary of the multigeneration GOCE-derived GGMs.

Geoid Model	Maximum d/o	Assimilated GOCE Data
DIRR4	260	Nov. 2009–Aug. 2012
DIRR5	300	Nov. 2009–Oct. 2013
DIRR6	300	Oct. 2009–Oct. 2013
TIMR4	250	Nov. 2009–Jun. 2012
TIMR5	280	Nov. 2009–Oct. 2013
TIMR6	300	Oct. 2009–Oct. 2013

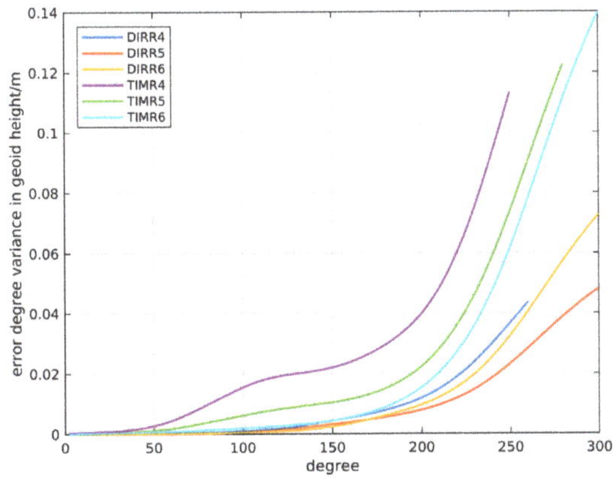

Figure 2. The error degree variance in geoid height calculated with different GOCE-derived GGMs.

2.1.3. Ocean Numerical Models and Reference MDT

In order to make the comparison between the MDT solutions and the empirical MDTs, several ocean numerical models; i.e., Simple Ocean Data Assimilation 3 (SODA3), Ocean and Sea-Ice Reanalysis System (ORAS5), Copernicus, and geodetic model CNES-CLS13MDT were used in this study. The spatial resolutions of the SODA3, ORAS5, and Copernicus ocean models were all 1/4°, but the reference time of these MDTs was different; detailed information is listed in Table 2. Based on Sea Level Anomalies (SLA) data provided by Aviso (ftp://ftp-access.aviso.altimetry.fr/climatology (accessed on 19 May 2020)), the reference time of these MDTs was unified to the same time period [21].

Table 2. Details of the ocean numerical models.

Ocean Numerical Model	Spatial Resolution	Time Scale	References
SODA3	1/4°	1980–2017	Carton et al. [22]
ORAS5	1/4°	1979–2018	Zuo et al. [23]
Copernicus	1/4°	1993–2019	Product User Guide [24]
CNES-CLS13MDT	1/4°	1993–2012	Rio et al. [25]

2.1.4. Drifting-Buoy Data

In situ ocean buoy data can be used to compare and verify MDT, because the geostrophic ocean current velocities can be extracted from buoy data and derived from MDT. The buoy data were obtained from the Global Drifter Program (GDP) provided by the U.S. National Oceanic and Atmospheric Administration (NOAA) and the Atlantic Ocean and Meteorological Laboratory (AOML), and they have been quality-controlled and kriged in order to provide 6-hour velocity measurements. The GDP uses satellite-tracked drifting buoys on the ocean surface to obtain current velocity, temperature, and salinity data. This study used buoy data from 1993–2016, and the buoys' observed current parameters included geostrophic, tidal, Ekman, inertial, and high-frequency nongeostrophic currents. Since only the geostrophic ocean current could be deduced by the MDT, the buoy observations needed to be processed to remove the nongeostrophic component. The Ekman component can be modeled based on the data of the wind speed and wind stress. Thus, the Ekman current model developed by Rio [16] was used to remove the Ekman current from the buoy data, and a 3-day low-pass filter was applied to the residual buoy current to remove tidal, inertial, and high-frequency nongeostrophic components [26,27]. The mean zonal and meridional geostrophic velocities were finally averaged into 1/4° grids. Moreover, to evaluate the accuracy of the geostrophic velocities, the errors of this data set was estimated, which can be given as the standard deviation of the observations divided by the square root of the number of observations [5]. The statistics of the errors are shown in Table 3. It shows that the RMS (root mean square) of the errors of the zonal (meridional) drifter geostrophic velocity were 14.3 mm/s (11.9 mm/s), respectively.

Table 3. Statistics of the errors of the drifter geostrophic velocity (units: mm/s).

Geostrophic Velocity	Max	Mean	RMS
Zonal component	823	4.2	14.3
Meridional component	445.8	6.9	11.9

2.2. *Optimal Interpolation Method*

The MDT is expressed by the difference between the MSS and geoid, which is given by:

$$h = \bar{\eta} - N \quad (1)$$

with the MDT h, the MSS $\bar{\eta}$, and the geoid N, which should be combined under the same coordinate system and tide system. Considering the GOCE-derived GGM and the MSS models used in this study, the models were transformed to the GRS80 and zero-tide system.

In order to mitigate the signal distortion and leakage problems induced by the traditional filtering methods, this study adopted the OIM (also known as multivariate objective analysis) for the MDT modeling, which was first applied to oceanographic parameter estimation by Bretherton [28]. The basic principle of this method can be understood as a linear weighted average method for the observations, thus it is essential for the determination of the variance/covariance information. Compared to the traditional Gaussian filtering method, the OIM retains more detailed signals in the combination and helps derive a more reasonable MDT solution. Supposing there are N points used for MDT modeling, the weights for each point are expressed as the following:

$$\eta_i = \sum_{j=1}^{N} A_{ij}^{-1} \varphi_j \tag{2}$$

where A_{ij} denotes the covariance between point i and j. For each estimated point r and the observation point i, the estimated MDT value $\langle h \rangle(r)$ could be expressed as:

$$\langle h \rangle(r) = \sum_{i=1}^{N} C_{ri} \eta_i \tag{3}$$

where C_{ri} denotes the covariance between r and i.

Considering a linear weighted estimator, the estimated MDT $\langle h \rangle(r)$ could be rewritten as:

$$\langle h \rangle(r) = \sum_{j=1}^{N} \alpha_{rj} \varphi_j \tag{4}$$

where α_{rj} denotes the weighting factors. Considering the difference between the estimated MDT $\langle h \rangle(r)$ and the true MDT $h(r)$, the error variance between them is expressed as follows:

$$\begin{aligned} \overline{(h(r) - \langle h \rangle(r))^2} &= \overline{\left(h(r) - \sum_{j=1}^{N} \alpha_{rj} \varphi_j\right)^2} \\ &= \mathbf{C}_{rr} - 2 \sum_{j=1}^{N} \alpha_{rj} \mathbf{C}_{rj} \\ &+ \sum_{i=1}^{N} \sum_{j=1}^{N} \alpha_{ri} \alpha_{rj} \mathbf{A}_{ij} \\ &= \sum_{i=1}^{N} \sum_{j=1}^{N} \left(\alpha_{ri} - \sum_{i\prime=1}^{N} \mathbf{C}_{ri\prime} \mathbf{A}_{i\prime i}^{-1}\right)\left(\alpha_{rj} - \sum_{j\prime=1}^{N} \mathbf{C}_{rj\prime} \mathbf{A}_{j\prime j}^{-1}\right) \mathbf{A}_{ij} + \mathbf{C}_{rr} \\ &- \sum_{i=1}^{N} \sum_{j=1}^{N} \mathbf{C}_{ri} \mathbf{C}_{rj} \mathbf{A}_{ij}^{-1} \end{aligned} \tag{5}$$

To minimize the error variance in Equation (5), let $\alpha_{ri} = \sum_{i=1}^{n} \mathbf{C}_{ri\prime} \mathbf{A}_{i\prime i}^{-1}$, and then Equation (4) can be rewritten as:

$$\langle h \rangle(r) = \sum_{i=1}^{N} \alpha_i O(r_i), \alpha_i = \sum_{j=1}^{N} \mathbf{A}_{i,j}^{-1} \mathbf{C}_{r,j} \tag{6}$$

with the estimated MDT value $\langle h \rangle(r)$ at point r; raw observation $O(r_i)$, which is the difference between MSS and geoid; the covariance matrix of the observation **A**; and covariance vector **C** between the observed and estimated MDT. Assuming that the observations are isotropic and homogeneous, then the covariance of the observations at points i and j would only depend on the distance d_{ij} between the two points:

$$\mathbf{A} = \left(\langle \sigma^2 \rangle C(d_{ij}) + \langle \varepsilon_i \varepsilon_j \rangle\right)_{i,j=1,N}, \mathbf{C}_r = \left(\langle \sigma^2 \rangle C(d_{rj})\right)_{j=1,N} \tag{7}$$

where σ^2 represents the a priori MDT variance, and ε_i denotes the error on the observation located at i. This study assumed that the errors at different locations were uncorrelated with each other, and that the errors of the observations were uncorrelated with the observations. $C(r)$ denotes the a priori covariance function of MDT, which was suggested by Arhan and Colin de Verdiere [29]:

$$C(r) = \left(1 + r + \frac{1}{6} r^2 - \frac{1}{6} r^3\right) e^{-r} \tag{8}$$

where $r = \sqrt{(\frac{x}{x_0})^2 + (\frac{y}{y_0})^2}$, and x_0 and y_0 denote the zonal and meridional correlation radii of the MDT, respectively. Consequently, the errors of the estimated MDT could be expressed as:

$$\varepsilon(r) = \sigma^2 - \sum_{i=1}^{N}\sum_{j=1}^{N} \mathbf{A}_{i,j}^{-1}\mathbf{C}_{r,i}\mathbf{C}_{r,j} \quad (9)$$

Theoretically, the mean of the values to be estimated should be zero when applying the OIM, but it is difficult to satisfy this condition in practice. Therefore, in this study, a large-scale prior MDT was removed from the observation, which could be obtained by applying a Gaussian filter on the raw MDT; that is, the difference between the selected MSS and GGM, so that the mean of the remaining residuals was close to zero, and the residuals were then treated as the observations in MDT modeling. The large-scale prior MDT depends on the filtering radius of Gaussian filtering. The appropriate filtering radius can be estimated by the trial and error method. In this study, the filtering radius was set to 400 km.

The MDT value of each grid point was estimated based on the procedure of its surrounding points using the OIM for each grid point, where the key parameters are the error of the observations, the a priori variance of the observations, and the covariance between the observations. There are two estimation methods for the error of the observations, one of which can estimate the error of the MDT by using the error information of the MSS and GGM according to the error propagation theory [30]. However, the calculations of this method will increase as the maximum d/o of the GGM increases. In this study, an informal error was introduced for the replacement of the error of the GGM, where several higher-order GGMs like GECO, EGM2008, SGG-UGM-1, and XME2019e_2159 [31–34] were used to calculate the geoid of the study area. Then, the mean of these geoids was calculated and the difference between the geoid calculated by the GGM used in this study and the mean value was taken as the error of the GGM [35]. Notably, this informal error did not consider factors such as the omission error of GGM, which may lead to a smaller result than the true values. Second, in order to derive an accurate estimation of the observation error, an a priori MDT (e.g., DTU13MDT) was used to compare with the raw MDT, and the RMS differences between the two models could be computed in a 1° grid as the error of the observation. However, this error information can be overestimated due to the variability of the mean circulation by the a priori MDT. Therefore, in this study, the two methods were combined, and for each grid point, the observation error was set to the larger one in the two methods.

The a priori variance and correlation function of the MDT can be calculated by using the a priori MDT (e.g., CNES-CLS18 MDT). Since the large-scale signal of the MDT was removed in this experiment, only the covariance of the residual part of the MDT needed to be estimated. First, the a priori MDT was Gaussian filtered to obtain the large-scale signals, with the same filter radius as the large-scale MDT. Then, this large-scale MDT was removed from the a priori MDT, and only the variance of the residual part of the MDT signal was calculated and used as the variance of the a priori MDT. In addition, the correlation radius parameter in the covariance function of Equation (8) was the key parameter in the OIM, which refers to the corresponding x_0 and y_0 when the result of the covariance function is zero value and is generally related to the latitude and the ocean basin [5]. Similarly, the residual part of the MDT could be used for the calculation of the correlation radius in the correlation function: the study area was divided into several subregions, in which the corresponding correlation radius was calculated. It was found that a better MDT solution could be modeled when the subregion was divided into 8° × 8° subregions. In each subregion, the corresponding empirical covariance was calculated based on the distance between the residual MDT grids:

$$C(\psi) = \frac{\sum A_j A_k f_i f_k}{\sum A_j A_k} \quad (10)$$

where ψ is the spherical distance between grids, f is the MDT value of the grid on the spherical surface, and A is the corresponding area. If the subregions are divided by equal areas, this can be simplified to:

$$C(\psi) = \frac{\sum f_i f_k}{n} \tag{11}$$

where n is the number of the groups of the grids with a spherical distance of ψ, which could be split by Equation (12):

$$\psi - \frac{\Delta\psi}{2} < \psi_{PQ} < \psi + \frac{\Delta\psi}{2} \tag{12}$$

and fit C (ψ) as an empirical covariance to a known covariance function (Equation (8)) to determine the correlation radii in zonal and meridional in each region, the results of which are shown in Figure 3. The sensitivity of the inputs in modeling should be considered [36]. The sensitivity tests to the covariance of the observations have been analyzed by Rio [37], and indicated that the correlation scales obtained with different numerical models showed similar patterns. The recently released CNES-CLS18 MDT was used to estimate the variance and correlation function of the MDT in this paper.

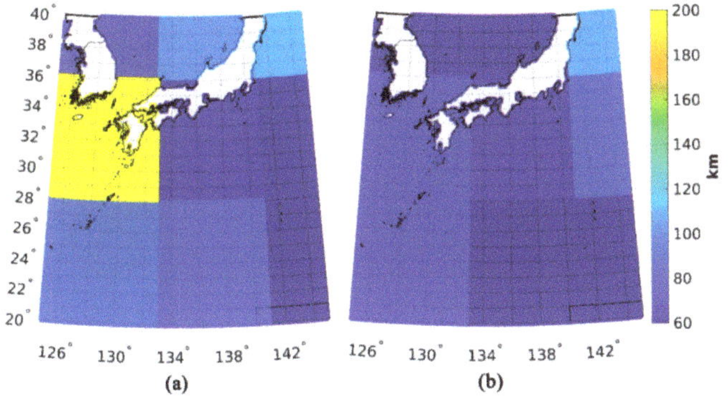

Figure 3. The correlation radii in (**a**) zonal and (**b**) meridional of the study area.

In order to evaluate the performance of the detailed signal recovery of the OIM for the MDT, the geostrophic ocean current velocity could be deduced from the MDT [38]:

$$\begin{cases} u_g = -\frac{\gamma}{fR}\frac{\partial h}{\partial \phi} \\ v_g = \frac{\gamma}{fR\cos\phi}\frac{\partial h}{\partial \lambda} \end{cases} \tag{13}$$

where u_g and v_g represent the geostrophic ocean current velocity in the zonal and meridional direction, respectively; ϕ and λ represent the latitude and longitude of the point, respectively; γ represents the normal gravity, $f = 2\omega \cdot \sin\phi$ is the Coriolis force, ω is the Earth's rotation rate, and h is the MDT.

3. Results and Discussion

3.1. MDT Solutions from Different GOCE-Derived GGM

In order to evaluate the performance of the MDT solution derived from different GOCE-derived GGM, the MDT with a spatial resolution of 10′ and a reference time of 1993–2012 was solved, which combined the DTU18 MSS with the GOCE-derived GGMs, including the DIRR4, DIRR5, DIRR6, TIMR4, TIMR5, and TIMR6. The method for constructing MDT was the optimal interpolation method, which is described in detail above. The large-scale MDT was estimated by using the Gaussian filtering method with a 400 km filtering radius. The CNES-CLS18 MDT was used for estimating the variance and covariance of the MDT.

The MDTs solved by the OIM through the combination of the DTU18 MSS and GOCE-derived GGMs by means of the direct approach are shown in Figure 4, where Figure 4a–c are the MDTs based on the DIRR4, DIRR5, and DIRR6, denoted as DTU18DIRR4-MDT, DTU18DIRR5-MDT, and DTU18DIRR6-MDT, respectively; while Figure 4d–f are the time-wise solutions, which are denoted as DTU18TIMR4-MDT, DTU18TIMR5-MDT, and DTU18TIMR6-MDT, respectively. In order to verify the reliability of the MDT solutions, the reference MDT derived based on the average of the oceanographic and geodetic MDTs models; i.e., SODA, ORAS5, CNES-CLS13MDT, and Copernicus, were used for comparison. However, the evaluations of the error levels of the four MDTs, in addition to the reference MDT, were needed. Thus, these models were first referenced to 1993–2016; that is, the same as the buoy data, and the corresponding geostrophic ocean current was deduced and then compared with the in situ data, as shown in Table 4. The results showed that the RMS of the difference between geostrophic current velocity derived from reference MDT and the buoy data was relatively smaller in comparison: 112 mm/s (91 mm/s) in the zonal (meridional) directions, respectively, which may be due to the smoothness of the noise. As a result, the reference MDT could be used for the comparative analysis in this study.

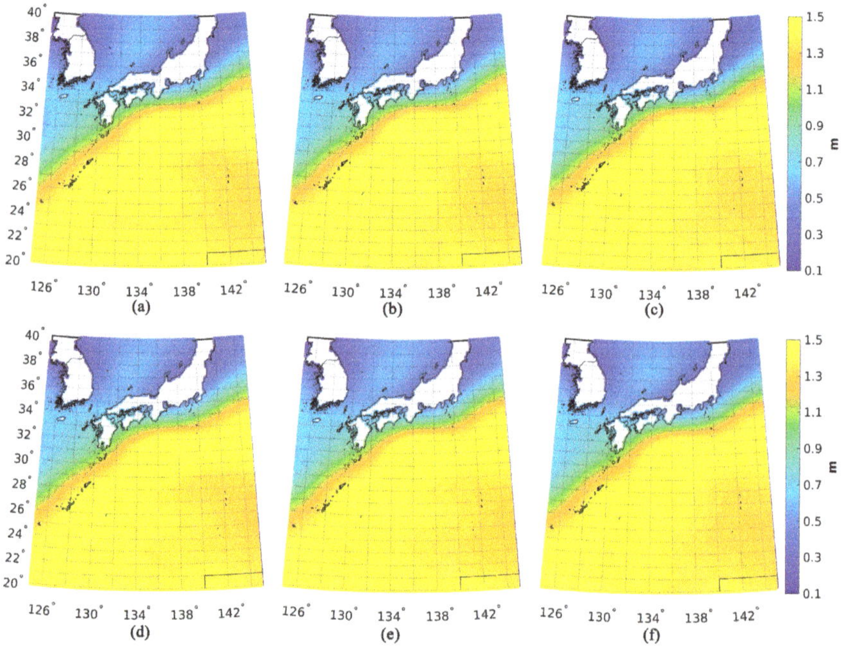

Figure 4. The estimated MDT based on (**a**) DIRR4, (**b**) DIRR5, (**c**) DIRR6, (**d**) TIMR4, (**e**) TIMR5, and (**f**) TIMR6.

The differences between the reference MDT and the estimated MDTs are shown in Figure 5. It can be seen that the differences between the MDT solutions by OIM and the reference MDT were mainly concentrated in the current region in Figure 1, while there were fewer differences between the filtered MDT and reference MDT in most regions. This may be because the MDT models also were calculated through the filtering procedure, which screened out some of the detailed signals and is therefore not evident in Figure 5. Table 5 shows the statistical results of these comparisons (the mean differences are removed). The RMS of the difference between the DTU18DIRR6-MDT solution and reference MDT was 77.0 mm, which was 2.4 mm lower than that of the DTU18DIRR5-MDT solution and 5.4 mm lower than that of the DTU18DIRR4-MDT solution. In addition, the RMS of the difference between the DTU18TIMR6-MDT and reference MDT was 0.5 mm (5.7 mm) lower than that of DTU18TIMR5-MDT (DTU18TIMR4-MDT), suggesting that the development of the GGM could contribute to the MDT modeling. The recently satellite-only GGM was derived from all of the available GOCE gravity gradiometry data and recalibrated satellite orbit data. This illustrated that the accumulation of GOCE data and updated data preprocessing methods can be beneficial for MDT recovery.

Table 4. Statistics of the differences in the geostrophic velocities between the buoy data and the MDT models (units: mm/s).

MDT	COMPONENTS	MIN	MAX	RMS
SODA3	u	−758.2	497.0	113.0
	v	−619.8	541.7	94.5
ORAS5	u	−766.1	784.9	132.6
	v	−652.4	419.7	97.3
CNES-CLS13MDT	u	−724.2	539.8	113.4
	v	−647.6	478.0	93.4
Copernicus	u	−728.1	548.3	115.6
	v	−619.3	523.4	93.2
Reference MDT	u	−707.8	530.6	112.3
	v	−625.0	459.8	90.7

Note: the mean values of the misfit of the geostrophic velocity between the buoy data and the ocean models are removed.

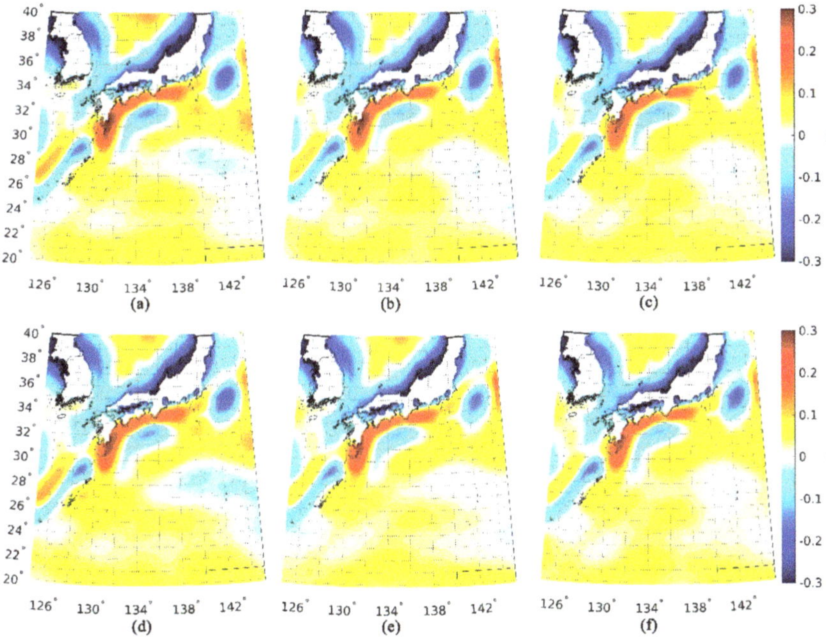

Figure 5. The differences between (**a**) DTU18DIRR4-MDT, (**b**) DTU18DIRR5-MDT, (**c**) DTU18DIRR6-MDT, (**d**) DTU18TIMR4-MDT, (**e**) DTU18 TIMR5MDT, (**f**) DTU18TIM R6-MDT and the reference MDT.

Table 5. Statistics of the differences between the reference MDTs and the estimated MDT (units: mm).

MDT Solutions	Min	Max	RMS
DTU18DIRR4-MDT	−434.6	292.6	82.4
DTU18DIRR5-MDT	−389.1	284.8	79.4
DTU18DIRR6-MDT	−383.3	270.2	77.0
DTU18TIMR4-MDT	−446.6	281.1	83.2
DTU18TIMR5-MDT	−402.5	272.7	78.0
DTU18TIMR6-MDT	−376.6	269.2	77.5

Note: the mean values of the misfit between the estimated MDTs and the reference MDT model are removed.

3.2. MDT Solutions from Different DTU-Based MSS

The selection of the MSS is of great importance to MDT modeling. In this case, the difference of the inputs of the OIM was the selection of MSS; therefore, the parameters such as the filtering radius of calculating the large-scale MDT and the prior MDT to estimate the variance and correlation function of the MDT were not considered. Although the accuracy of MSS in the open sea can reach the centimeter level [6], the accuracy of the MSS reduces rapidly near the coastal zone due to the defects of satellite altimetry. In this study, the DTU15 MSS and DTU18 MSS were used for MDT modeling, and the error distribution of the two MSSs is shown in Figure 6a,b, in which it can be seen that the error of DTU18 MSS was significantly smaller than that of DTU15 MSS, especially in the region around the coast. Table 6 gives the statistical information of the error of the MSS in the study area, and the accuracy of the DTU18 MSS improved by about 1 mm relative to the DTU15 MSS. The MDT solutions derived from the combination of the DIRR6 GGM and the DTU15 MSS, DTU18 MSS are shown in Figure 7a,b, and are denoted as DTU15DIRR6-MDT and DTU18DIRR6-MDT, respectively. It is shown that the two MDTs were similar to each other; thus, the reference MDT was used to perform the comparison, and the differences between the MDTs and the reference MDT are shown in Figure 7c,d. The statistical results of the differences are shown in Table 7 (with the mean differences removed), from which we can see that the RMS of the difference between DTU18DIRR6-MDT and reference MDT was 0.7 mm smaller than that of DTU15DIRR6-MDT, which indicated that the improvement of MSS could not significantly improve the quality of the final MDT. The reason is that the improvement of the DTU18 MSS focused on the coastal area with respect to DTU15 MSS, but the improvement was limited in the other area.

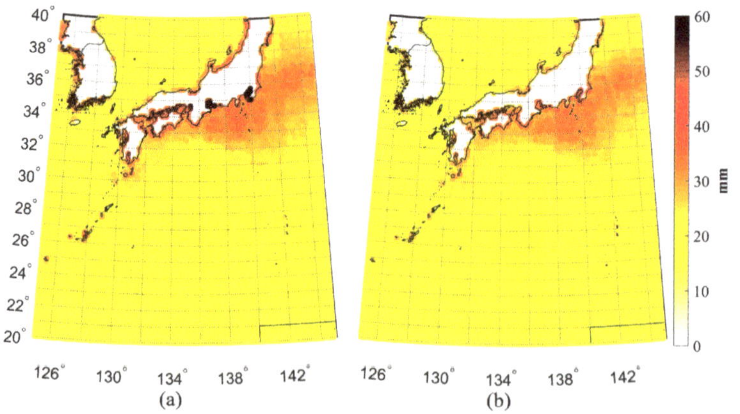

Figure 6. The error distribution of (**a**) DTU15 MSS; (**b**) DTU18 MSS.

Table 6. Statistics of the errors of the MSS models (units: mm).

MSS Models	Min	Max	Mean	RMS
DTU15 MSS	14.7	61.0	21.9	22.2
DTU18 MSS	13.5	48.0	20.7	21.1

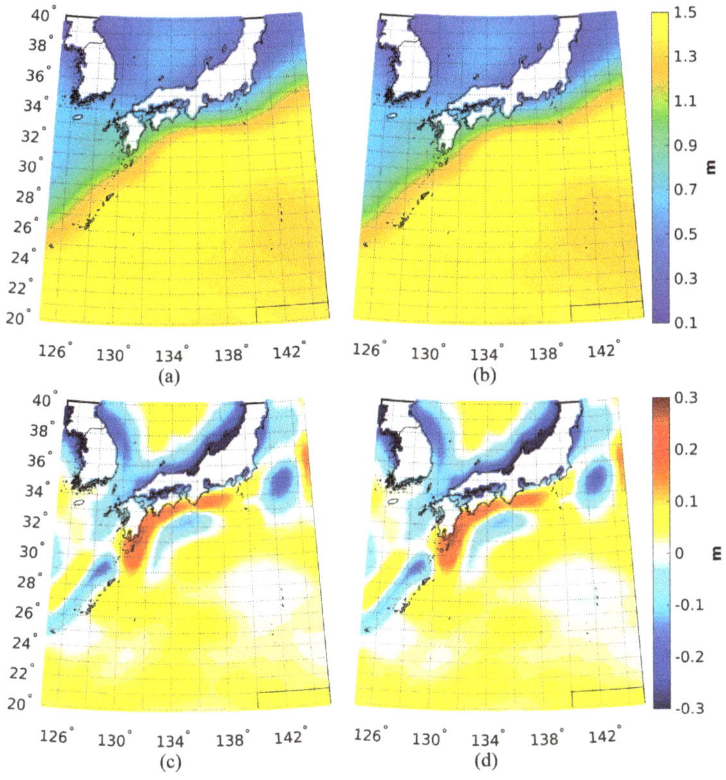

Figure 7. The MDTs of the (**a**) DTU15DIRR6-MDT and (**b**) DTU18DIRR6-MDT, and the corresponding differences between them and the reference MDT (**c**,**d**).

Table 7. Statistics of the differences between the estimated MDTs and the reference MDT (Units: mm).

MDT Solutions	Min	Max	RMS
DTU15DIRR6-MDT	−381.5	276.2	77.7
DTU18DIRR6-MDT	−383.3	270.2	77.0

Note: the mean values of the misfit between the estimated MDTs and the reference MDT model are removed.

3.3. Validation

Compared to the Gaussian filtering method, one of the advantages is that the OIM can give the error distribution of the MDT based on the observations. Figure 8 shows the estimated error of the MDT calculated by the OIM, and the error was larger near the Kuroshio current. The RMS of the errors of DTU18DIRR4-MDT, DTU18DIRR5-MDT, and DTU18DIRR6-MDT was 6.1 mm, 6.0 mm, and 5.8 mm, respectively, which reflects the contribution of the improved GGM. Furthermore, in order to compare the OIM and Gaussian filter method, the MDTs based on DTU18 MSS and GOCE-derived GGMs were derived using a Gaussian filter method with a filter radius of 70 km [39], and the results were denoted as GF-R4-MDT, GF-R5-MDT, and GF-R6-MDT, respectively. The in situ geostrophic ocean current velocity profiles were used for validation, in which the reference time of the MDT solution; i.e., the DTU18DIRR4-MDT, DTU18DIRR5-MDT, DTU18DIRR6-MDT, GF-R4-MDT, GF-R5-MDT, and GF-R6-MDT, was unified to the same time period with the buoy data (1993–2016) using SLA provided by Aviso. Then, the associated geostrophic ocean current velocity was calculated using Equation (13). It was notable that the exportation of the geostrophic ocean current velocity from the MDT led to an error amplification [38], so a spatial filter was used to reduce the noise; the filter radius was selected as 60 km through experiments. In addition, the geostrophic ocean current velocity exported from the MDT was interpolated to the point of the buoy data, and the differences between them are shown in Figures 9 and 10, where the differences in the zonal and meridional directions are shown, respectively. The geostrophic velocities derived from the OIM method had smaller misfits against the buoy data than those of the Gaussian filter method, especially in coastal areas. The statistical results of the difference are shown in Table 8, in which it can be seen that the RMS of the differences between the DTU18DIRR6-MDT-derived zonal (meridional) components and the in situ data were 122.6 mm/s (93.7 mm/s), which were 1.2 mm/s (1.8 mm/s) lower than that of the DTU18DIRR5-MDT, and 2.2 mm/s (3.5 mm/s) lower than that of the DTU18DIRR4-MDT. In addition, when making the comparison between the OIM solution and the Gaussian filtered solution, the RMS of the difference between the zonal (meridional) geostrophic ocean current velocity derived from the Gaussian filtered solution against the in situ data were about 10 mm/s (17 mm/s) larger than those of the OIM solution, which indicated that the OIM was better than the Gaussian filtering method. The comparison between Tables 4 and 8 shows that the RMS of the difference between the geostrophic ocean current velocity of the MDT models and the in situ data was about 10 mm/s lower than that of the estimated MDT, which indicated that the accuracy of the MDT models was similar to the estimated MDT.

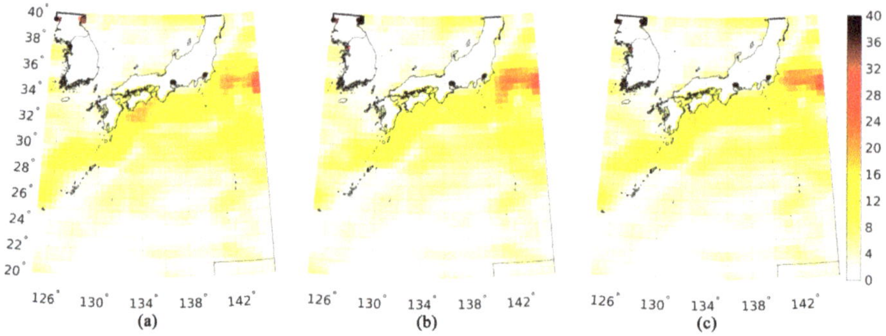

Figure 8. The estimated errors the MDT solutions (**a**) DTU18DIRR4-MDT; (**b**) DTU18DIRR5-MDT; (**c**) DTU18DIRR6-MDT.

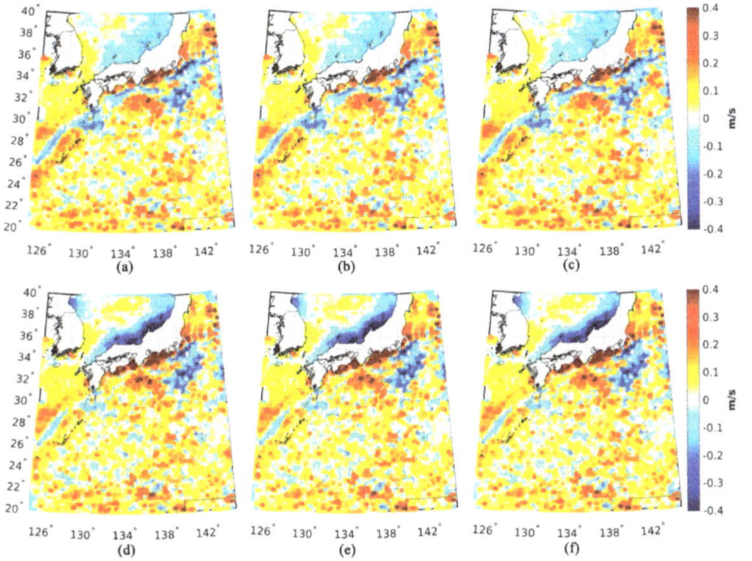

Figure 9. The differences in zonal geostrophic velocities between buoy data and information deduced from the MDT solutions: (**a**) DTU18DIRR4-MDT; (**b**) DTU18DIRR5-MDT; (**c**) DTU18DIRR6-MDT; (**d**) GF-R4-MDT; (**e**) GF-R5-MDT; (**f**) GF-R6-MDT.

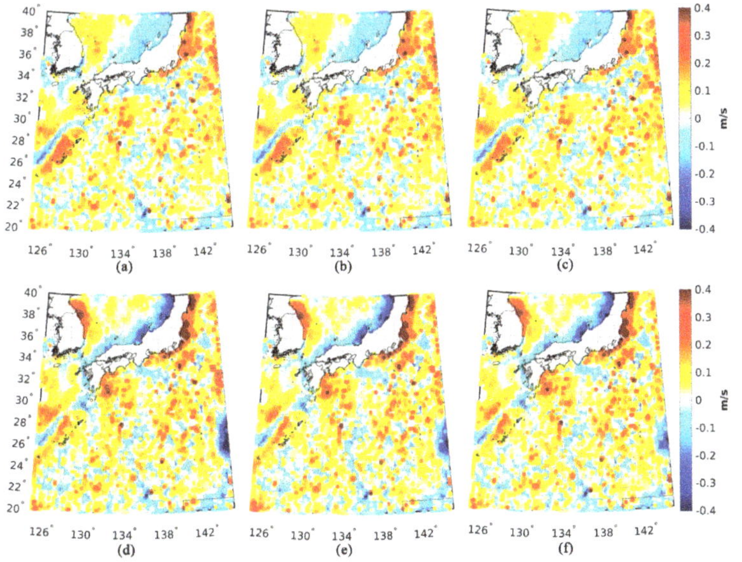

Figure 10. The differences in meridional geostrophic velocities between buoy data and information deduced from the MDT solutions: (**a**) DTU18DIRR4-MDT; (**b**) DTU18DIRR5-MDT; (**c**) DTU18DIRR6-MDT; (**d**) GF-R4-MDT; (**e**) GF-R5-MDT; (**f**) GF-R6-MDT.

Table 8. Statistics of the differences in the geostrophic velocities between the buoy data and the information deduced from the MDT solutions (units: mm/s).

MDT Solutions	Geostrophic Velocity	Min	Max	RMS
DTU18DIRR4-MDT	u	−692.3	654.7	124.8
	v	−629.7	511.7	97.2
DTU18DIRR5-MDT	u	−698.5	646.8	123.8
	v	−629.3	510.4	95.5
DTU18DIRR6-MDT	u	−695.6	648.3	122.6
	v	−620.6	510.9	93.7
GF-R4-MDT	u	−707.7	766.6	133.4
	v	−675.9	651.6	119.3
GF-R5-MDT	u	−727.4	765.0	133.0
	v	−608.8	641.3	113.7
GF-R6-MDT	u	−722.6	763.9	132.9
	v	−622.9	637.2	110.8

Note: the mean values of the misfit of the geostrophic velocity between the buoy data and the MDT models are removed.

4. Conclusions

We studied the optimal interpolation method for regional MDT recovery based on GOCE-derived GGMs and mean sea surface. In particular, the performances of different GGMs in MDT recovery were studied, and the contributions introduced from the recently released GOCE-derived GGMs were quantified and assessed. Moreover, the choice of MSS model was investigated, and the effects on MDT modeling introduced from different MSS models were quantified. In addition, we compared the traditionally used Gaussian filtering method with the optimal interpolation method in regional MDT modeling. The numerical results over the Kuroshio showed that:

(1) The use of the recently released GOCE-derived GGM improved the accuracy of the computed MDT. The RMS of the difference between the MDT derived from sixth-generation GOCE-derived GGM modeled by the direct approach and independent ocean data was reduced by a magnitude of 5.4 mm (2.4 mm), compared to the result derived from the fourth-generation (fifth-generation) model.

(2) The choice of recently published MSS models may have had limited effects on MDT recovery. The RMS of the difference between the MDT modeled from DTU18MSS and ocean data was only 0.7 mm lower than that computed from DTU15MSS.

(3) For further comparison, the buoy data was introduced, and the geostrophic velocities derived from DTU18MSS and different GOCE-derived GGMs were compared with the buoy data. The results revealed that the RMS of the differences between the zonal (meridional) velocities derived from the sixth-generation GGM and buoy data were 1.2 mm/s (1.8 mm/s) less than that derived from the fifth-generation model, and 2.2 mm/s (3.5 mm/s) less than the fourth-generation GGM, respectively. This also suggested that the improvements in the GGM could contribute to the accuracy improvement of the MDT modeling.

(4) Moreover, the numerical experiments showed that the optimal interpolation method outperformed the Gaussian filtering approach, in which the geostrophic velocity derived from the optimal interpolation method had a smaller misfit against the buoy data by a magnitude of 10 mm/s (17 mm/s) when the zonal (meridian) component was validated. This was mainly due to the optimal interpolation method using the error information of the input data, which may have obtained more reasonable weights of observations than the Gaussian filtering method.

Author Contributions: Conceptualization, J.H. and Y.W.; methodology, J.H. and Y.W.; software, J.H. and Y.W.; validation, J.H., Y.W. and H.S.; formal analysis, J.H. and Y.W.; investigation, J.H.; resources, Y.W.; data curation, J.H.; writing—original draft preparation, J.H. and H.S.; writing—review and editing, Y.W., H.S. and X.H.; visualization, J.H.; supervision, X.H.; project administration, Y.W.; funding acquisition, Y.W. All authors have read and agreed to the published version of the manuscript.

Funding: This research was funded by the National Natural Science Foundation of China, grant number 42004008; the Natural Science Foundation of Jiangsu Province, China, grant number BK20190498; and the Fundamental Research Funds for the Central Universities, grant number B200202019.

Institutional Review Board Statement: Not applicable.

Informed Consent Statement: Not applicable.

Data Availability Statement: The global geopotential models can be publicly accessed at http://icgem.gfz-potsdam.de/tom_longtime (accessed on 16 May 2020). DTU15MSS, DTU18MSS, and their associated error grids are available at https://ftp.space.dtu.dk/pub/ (accessed on 16 May 2020). CNES-CLS13MDT was accessed at ftp://ftp-access.aviso.altimetry.fr/auxiliary/mdt/mdt_cnes_cls2013_global (accessed on 16 May 2020). SODA3 was accessed at http://www.atmos.umd.edu/~ocean/index_files/soda3.12.2_mn_download.htm (accessed on 18 May 2020). ORAS5 was accessed at https://icdc.cen.uni-hamburg.de/thredds/catalog/ftpthredds/EASYInit/oras5/ORCA025/catalog.html (accessed on 18 May 2020). Copernicus was accessed at https://cds.climate.copernicus.eu/cdsapp#!/dataset/satellite-sea-level-global?tab=overview (accessed on 20 May 2020). Drifter buoy data were accessed at ftp://ftp.aoml.noaa.gov/phod/pub/buoydata (accessed on 25 May 2020).

Acknowledgments: We gratefully acknowledge the funders of this study.

Conflicts of Interest: The authors declare no conflict of interest.

References

1. Wu, Y.; Abulaitijiang, A.; Featherstone, W.E.; McCubbine, J.C.; Andersen, O.B. Coastal gravity field refinement by combining airborne and ground-based data. *J. Geod.* **2019**, *93*, 2569–2584. [CrossRef]
2. Hwang, C.; Hsu, H.J.; Featherstone, W.E.; Cheng, C.C.; Yang, M.; Huang, W.; Wang, C.Y.; Huang, J.F.; Chen, K.H.; Huang, C.H.; et al. New gravimetric-only and hybrid geoid models of Taiwan for height modernisation, cross-island datum connection and airborne LiDAR mapping. *J. Geod.* **2020**, *94*, 1–22. [CrossRef]
3. He, Y.C.; Gao, Y.Q.; Wang, H.J.; Ola, J.M.; Yu, L. Transport of nuclear leakage from Fukushima nuclear power plant in the north pacific. *Acta Oceanol. Sin.* **2012**, *34*, 12–20. [CrossRef]
4. Wu, Y.; Luo, Z. The approach of regional geoid refinement based on combining multi-satellite altimetry observations and heterogeneous gravity data sets. *Chin. J. Geophys.* **2016**, *59*, 1596–1607. [CrossRef]
5. Rio, M.H.; Hernandez, F. A mean dynamic topography computed over the world ocean from altimetry, in situ measurements, and a geoid model. *J. Geophys. Res. Oceans* **2004**, *109*. [CrossRef]
6. Chambers, D.P. Evaluation of new GRACE time-variable gravity data over the ocean. *Geophys. Res. Lett.* **2006**, *33*. [CrossRef]
7. Wu, Y.; Zhou, H.; Zhong, B.; Luo, Z. Regional gravity field recovery using the GOCE gravity gradient tensor and heterogeneous gravimetry and altimetry data. *J. Geophys. Res. Solid Earth* **2017**, *122*, 6928–6952. [CrossRef]
8. Kornfeld, R.P.; Arnold, B.W.; Gross, M.A.; Dahya, N.T.; Klipstein, W.M.; Gath, P.F.; Bettadpur, S. GRACE-FO: The gravity recovery and climate experiment follow-on mission. *J. Spacecr Rockets* **2019**, *56*, 931–951. [CrossRef]
9. Eppelbaum, L.; Katz, Y.; Klokochnik, J.; Kosteletsky, J.; Zheludev, V.; Ben-Avraham, Z. Tectonic Insights into the Arabian-African Region inferred from a Comprehensive Examination of Satellite Gravity Big Data. *Glob. Planet. Chang.* **2018**, *171*, 65–87. [CrossRef]
10. Bruinsma, S.L.; Förste, C.; Abrikosov, O.; Lemoine, J.M.; Marty, J.C.; Mulet, S.; Rio, M.H.; Bonvalot, S. ESA's satellite-only gravity field model via the direct approach based on all GOCE data. *Geophys Res. Lett.* **2014**, *41*, 7508–7514. [CrossRef]
11. Andersen, O.; Knudsen, P.; Stenseng, L. A New DTU18 MSS Mean Sea Surface—Improvement from SAR Altimetry. In Proceedings of the 25 Years of Progress in Radar Altimetry Symposium, Ponta Delgada, Azores, Portugal, 24–29 September 2018.
12. Knudsen, P.; Bingham, R.; Andersen, O.; Rio, M.H. A global mean dynamic topography and ocean circulation estimation using a preliminary GOCE gravity model. *J. Geod.* **2011**, *85*, 861–879. [CrossRef]
13. Wang, W.; Wen, H.; Liu, H.; Chen, B.; Jia, X. A method for determining high-precision geostrophic currents in the eastern seas of China. *Sci. Surv. Mapp.* **2019**, *44*, 37–43. [CrossRef]
14. Zhang, Z.; Lu, Y.; Hsu, H. Detecting surface geostrophic currents using wavelet filter from satellite geodesy. *Sci. China Ser. D Earth Sci.* **2007**, *50*, 918–926. [CrossRef]

15. Vianna, M.L.; Menezes, V.V.; Chambers, D.P. A high resolution satellite-only GRACE-based mean dynamic topography of the South Atlantic Ocean. *Geophys. Res. Lett.* **2007**, *34*, L24604. [CrossRef]
16. Rio, M.H.; Guinehut, S.; Larnicol, G. New CNES-CLS09 global mean dynamic topography computed from the combination of GRACE data, altimetry, and in situ measurements. *J. Geophys. Res. Oceans* **2011**, *116*. [CrossRef]
17. Förste, C.; Abrykosov, O.; Bruinsma, S.; Dahle, C.; König, R.; Lemoine, J.M. *ESA's Release 6 GOCE Gravity Field Model by Means of the Direct Approach Based on Improved Filtering of the Reprocessed Gradients of the Entire Mission (GO_CONS_GCF_2_DIR_R6)*; Data Publication; GFZ Data Services: Potsdam, Germany, 2019. [CrossRef]
18. Joh, Y.; Di Lorenzo, E. Interactions between Kuroshio Extension and Central Tropical Pacific lead to preferred decadal-timescale oscillations in Pacific climate. *Sci. Rep.* **2019**, *9*, 13558. [CrossRef]
19. Brockmann, J.M.; Schubert, T.; Schuh, W.D. An Improved Model of the Earth's Static Gravity Field Solely Derived from Reprocessed GOCE Data. *Surv. Geophys.* **2021**, *42*, 277–316. [CrossRef]
20. Gruber, T. GOCE Release 6 Products and Performance. In Proceedings of the 27th IUGG General Assembly, Presentation, Montreal, QC, Canada, 8–18 July 2019.
21. Bingham, R.J.; Haines, K. Mean dynamic topography: Intercomparisons and errors. *Philos. Trans. R. Soc. A* **2006**, *364*, 903–916. [CrossRef]
22. Carton, J.A.; Chepurin, G.A.; Chen, L. SODA3: A new ocean climate reanalysis. *J. Clim.* **2018**, *31*, 6967–6983. [CrossRef]
23. Zuo, H.; Balmaseda, M.A.; Tietsche, S.; Mogensen, K.; Mayer, M. The ECMWF operational ensemble reanalysis–analysis system for ocean and sea ice: A description of the system and assessment. *Ocean. Sci.* **2019**, *15*. [CrossRef]
24. The Climate Data Store. Available online: https://cds.climate.copernicus.eu/cdsapp#!/dataset/satellite-sea-level-global?tab=doc (accessed on 20 May 2020).
25. Rio, M.H.; Mulet, S.; Picot, N. Beyond GOCE for the ocean circulation estimate: Synergetic use of altimetry, gravimetry, and in situ data provides new insight into geostrophic and Ekman currents. *Geophys. Res. Lett.* **2014**, *41*, 8918–8925. [CrossRef]
26. Lumpkin, R.; Johnson, G.C. Global ocean surface velocities from drifters: Mean, variance, El Niño–Southern Oscillation response, and seasonal cycle. *J. Geophys. Res. Oceans* **2013**, *118*, 2992–3006. [CrossRef]
27. Laurindo, L.C.; Mariano, A.J.; Lumpkin, R. An improved near-surface velocity climatology for the global ocean from drifter observations. *Deep Sea Res. 1 Oceanogr. Res. Pap.* **2017**, *124*, 73–92. [CrossRef]
28. Bretherton, F.P.; Davis, R.E.; Fandry, C.B. A technique for objective analysis and design of oceanographic experiments applied to MODE-73. *Deep Sea Res. Oceanogr. Abstr.* **1976**, *23*, 559–582. [CrossRef]
29. Arhan, M.; De Verdiére, A.C. Dynamics of eddy motions in the eastern North Atlantic. *J. Phys. Oceanogr.* **1985**, *15*, 153–170. [CrossRef]
30. Balmino, G. Efficient propagation of error covariance matrices of gravitational models: Application to GRACE and GOCE. *J. Geod.* **2009**, *83*, 989–995. [CrossRef]
31. Gilardoni, M.; Reguzzoni, M.; Sampietro, D. GECO: A global gravity model by locally combining GOCE data and EGM2008. *Stud. Geophys. Geodaet.* **2016**, *60*, 228–247. [CrossRef]
32. Pavlis, N.K. An Earth Gravitational model to degree 2160: EGM08. In Proceedings of the 2008 General Assembly of the European Geosciences Union, Vienna, Austria, 13–18 April 2008.
33. Liang, W.; Xu, X.; Li, J.; Zhu, G. The determination of an ultra high gravity field model SGG-UGM-1 by combining EGM2008 gravity anomaly and GOCE observation data. *Acta Geodaet. Cartogr. Sin.* **2018**, *47*, 425–434. [CrossRef]
34. Pail, R.; Fecher, T.; Barnes, D.; Factor, J.F.; Holmes, S.A.; Gruber, T.; Zingerle, P. Short note: The experimental geopotential model XGM2016. *J. Geod.* **2018**, *92*, 443–451. [CrossRef]
35. Bingham, R.J.; Haines, K.; Lea, D.J. How well can we measure the ocean's mean dynamic topography from space? *J. Geophys. Res. Oceans* **2014**, *119*, 3336–3356. [CrossRef]
36. Asheghi, R.; Hosseini, S.A.; Saneie, M.; Shahri, A.A. Updating the neural network sediment load models using different sensitivity analysis methods: A regional application. *J. Hydroinform.* **2020**, *22*, 562–577. [CrossRef]
37. Rio, M.H.; Pascual, A.; Poulain, P.M.; Menna, M.; Barceló, B.; Tintoré, J. Computation of a new mean dynamic topography for the Mediterranean Sea from model outputs, altimeter measurements and oceanographic in situ data. *Ocean. Sci.* **2014**, *10*, 731–744. [CrossRef]
38. Bai, X.; Yan, H.; Zhu, Y. GOCE steady-state sea surface dynamic terrain and geostrophic circulation based on regional filtering. *Chin. J. Geophys.* **2015**, *58*, 1535–1546. [CrossRef]
39. Andersen, O.; Knudsen, P.; Stenseng, L. The DTU13 MSS (mean sea surface) and MDT (mean dynamic topography) from 20 years of satellite altimetry. In *IGFS2014*; Springer: Cham, Switzerland, 2015; pp. 111–121. [CrossRef]

applied sciences

MDPI

Article

Multiscale Post-Seismic Deformation Based on cGNSS Time Series Following the 2015 Lefkas (W. Greece) Mw6.5 Earthquake

Filippos Vallianatos [1,2,*] and Vassilis Sakkas [1,2]

1 Section of Geophysics—Geothermics, Department of Geology and Geoenvironment, National and Kapodistrian University of Athens, Panepistimiopolis, GR15784 Athens, Greece; vsakkas@geol.uoa.gr
2 Institute of Physics of Earth's Interior and Geohazards, UNESCO Chair on Solid Earth Physics and Geohazards Risk Reduction, Hellenic Mediterranean University Research Center, GR73100 Grete, Greece
* Correspondence: fvallian@geol.uoa.gr

Abstract: In the present work, a multiscale post-seismic relaxation mechanism, based on the existence of a distribution in relaxation time, is presented. Assuming an Arrhenius dependence of the relaxation time with uniform distributed activation energy in a mesoscopic scale, a generic logarithmic-type relaxation in a macroscopic scale results. The model was applied in the case of the strong 2015 Lefkas Mw6.5 (W. Greece) earthquake, where continuous GNSS (cGNSS) time series were recorded in a station located in the near vicinity of the epicentral area. The application of the present approach to the Lefkas event fits the observed displacements implied by a distribution of relaxation times in the range $\tau_{min} \approx 3.5$ days to $\tau_{max} \approx 350$ days.

Keywords: post seismic relaxation; cGNSS; hierarchically constrained dynamics; Lefkas earthquake

1. Introduction

Earthquake rupture creates static stress changes within the lithosphere that are large enough to produce transient deformation that can be observed on the Earth's surface. Satellite geodesy using GNSS techniques for observing Earth's surface displacements have advanced in such a level that the post-seismic deformation can be monitored with high accuracy [1].

To study post-seismic relaxation, various mechanisms have been proposed, such as those of poroelasticity of the rock medium [2,3], shallow after-slip on the main faulting plane or nearby faults [4,5], and viscoelastic relaxation [6]. The after-slip mechanisms are localized and follow logarithmic decay [4,7], while the viscoelastic one is supported by a bulk deformation process and exhibits exponential decay [8–10].

In [11], it was demonstrated that, during the first 6 months following the 1999 Chi-Chi earthquake (M = 7.6, Taiwan), both the post-seismic relaxation and the cumulative number of aftershocks followed the same temporal dependence, which reduced to a logarithmic one after a long time from the main event. In addition, logarithmic deformation could be explained by an exhaustion process, which is based on the assumption that the number of sources of strain-producing events evolve according to a thermally activated process, in which the activation energy is stress dependent [12].

The recent strong earthquakes present examples where the macroscopic post-seismic deformation is interpreted in view of the above relaxation mechanisms, which are based on the assumption of a single relaxation time estimated for each post seismic relaxation observed [13–18].

Experimental results and theoretical models suggest that post seismic relaxation is thermally activated and enhanced by the presence of fluids [19]. Deformation is controlled by a variety of temperature-dependent healing processes that are related to the crack size [20,21]. Since different crack sizes have different activation energy, the associated relaxation processes operate at different timescales [22], indicating an intrinsic organization [23] with a distribution of relaxation times, instead of the use of a single one. The latter implies that post-seismic relaxation could be viewed as the superposition of several different relaxation times, each one activated in different mesoscopic domains of the macroscopic relaxed seismic volume. Thus, the macroscopic relaxation observed results from the superposition of different relaxation times on the microscopic–mesoscopic scale, which obviously operates on different spatiotemporal scales. The recovery process that returns the deformed volume to equilibrium after a strong earthquake could be characterized as a slow dynamics process [24], because an instantaneous perturbation causes a transient response, which moves the system to a state of equilibrium [25–30]. It is, thus, possible that relaxation, which varies with the logarithm of time, can be interpreted as the result of the superposition of relaxation mechanisms with a spectrum of relaxation times.

The purpose of this work is to give a description of the post seismic deformation relaxation. It is based on the thermodynamic Arrhenius model for the distribution of the relaxation times [31] and describes the post-seismic recovery as a slow thermodynamic process taking place in a mesoscopic scale. The model predicts a physically accepted finite value just after the co-seismic perturbation, a logarithmic time dependence for the time period between the minimum and maximum relaxation time, and an exponential decay while approaching an equilibrium state a long time after the main event. Based on that, the explanation that relaxation occurs as a logarithm of time does not depend on the details of Earth's physical and chemical properties, but is based on a generic, or even a universal, multiscale property of the relaxation mechanism, as resulted from the Arrhenius expression, which connects relaxation time with the activation energy of the relaxation process.

To test the proposed approach for the post-seismic relaxation, positioning data from the continuous GNSS station PONT [32] and www.geodesy.gein.noa.gr (accessed on 30 December 2020) located in the southern part of Lefkas Island were used. This cGNSS station is located in the close vicinity of the earthquake epicenter of the 17 November 2015 Mw6.5, shallow event (~10 km depth), that occurred on the island, causing severe damages and human losses.

2. A Thermodynamic Model of Post Seismic Relaxation

A post-seismic transient deformation stage, where strain varies with time at a decreasing rate, occurs following the instantaneous deformation during an earthquake. Transient deformation is followed by the return to steady-state deformation, where the strain rate is a constant after some time following the main event.

Following [33], we define a relaxation function $R(t)$, where t is the time after the main seismic event, as a perturbation of the displacement given by the generic expression:

$$u(t) = u_\infty + S\,R(t) \quad (1)$$

where u_∞ is the equilibrium value of displacement after a long time and S is a scaling factor. Notice that $R(t \to \infty) \approx 0$, while $u_\infty = u(t \to \infty)$ is the remaining distortion at the end of the relaxation process. The initial value u_o of the post-seismic deformation immediately after the co-seismic perturbation δu_c is given from Equation (1) as $u_o = u(t = 0^+) = u_\infty + S\,R_O$, where $R_0 = R(t = 0)$. The scaling factor S of the relaxation constrains the rheological properties of the medium [34,35] and depends on the rheology of the crust and upper mantle, as well as on the local conditions around the site (i.e., crustal heterogeneities). In addition, S depends on the stress released in the medium and reflects the response of the medium (i.e., of the magnitude of the earthquake event) after the co-seismic displacement due to earthquake.

In most of the cases a simple relaxation mechanism with a single relaxation time τ that corresponds to a relaxation function $\exp(-t/\tau)$ (i.e., not varying logarithmically with time) is the simplest rheological model introduced to explain post-seismic deformation, known as the linear Maxwell model [6]. In this case:

$$R_o = 1,\ u_o = u_\infty + S \text{ and } u(t) = u_\infty + S\, e^{\frac{-t}{\tau}} \tag{2}$$

while the normalized displacement is $D(t) = \frac{u(t)}{u_o} = \frac{1+\frac{S}{u_\infty}e^{\frac{-t}{\tau}}}{1+\frac{S}{u_\infty}}$. In Figure 1, we plot the displacement as a function of normalized time $\frac{t}{\tau}$. In the very early stage, expanding the exponential in Equation (2) leads us to $u(t) = u_\infty + S\left(1 - \frac{t}{\tau}\right)orD(t) = 1 - \frac{\frac{S}{u_\infty}}{1+\frac{S}{u_\infty}}\frac{t}{\tau}$, a linear expression of normalized displacement with time, which for $\frac{S}{u_\infty} \gg 1$, leads to D = 1 − (t/τ) with the slope equal to $1/\tau$, and independent of S.

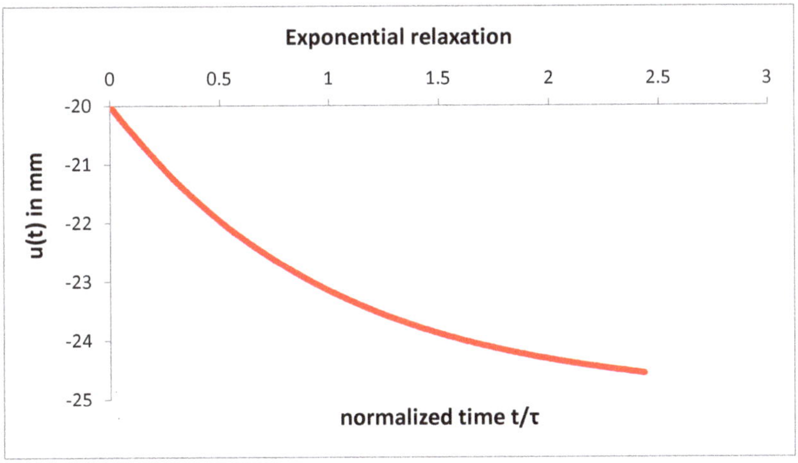

Figure 1. Example of exponential relaxation (see Equation (1)) with S = 5 mm, u_∞ = −25 mm and τ = 300 days.

We proceeded to study a post-seismic relaxation process with a spectrum of relaxation times within the range defined by the minimum and maximum values τ_{min}, τ_{max} of relaxation time, respectively. If $f(\tau)$ is the distribution function of the relaxation time, then:

$$R(t) = \int_{\tau_{min}}^{\tau_{max}} f(\tau) e^{\frac{-t}{\tau}} d\tau \tag{3}$$

The expression (3) is based on the assumption that the total macroscopic relaxation observed is a result of the superposition of relaxation processes operating with different relaxation times τ. This approach could be viewed as a generalization of the Burgers body rheological model [8,36,37], expanding the two discrete relaxation times to a continuum spectrum in the range values τ_{min}, τ_{max}.

Assuming a relaxation process with activation energy E, then following Arrhenius' law [31], the relaxation time is given by:

$$\tau = \tau_0 e^{\frac{E}{k_B T}} \tag{4}$$

with τ_0 a constant proexponential factor, T the temperature, and k_B the Boltzmann constant. If the activation energy is distributed following a distribution function $N(E)$, then the distribution function $f(\tau)$ for the relaxation times satisfies the expression $f(\tau) = N(E)(dE/d\tau)$. Using Expression (4), $d\tau/dE = (\tau_0/k_B T)\exp(E/k_B T) = \tau/k_B T$, hence:

$$f(\tau) = \frac{k_B T N(E)}{\tau} \tag{5}$$

In the case of a uniform distribution of the activation energy, bounded by a minimum E_{min} and maximum E_{max} activation energy, $N(E)$ has a constant value N_o, which, according to Equation (5) for a temperature T, leads to a distribution of relaxation times $f(\tau) = (k_B T N_0)/\tau$, which is proportional to $1/\tau$. It is evident that the minimum τ_{min} and maximum τ_{max} relaxation times are related to the corresponding minimum and maximum activation energies by the expressions $\tau_{min} = \tau_0 e^{\frac{E_{min}}{k_B T}}$ and $\tau_{max} = \tau_0 e^{\frac{E_{max}}{k_B T}}$ and, thus, $\Delta E = E_{max} - E_{min} = kT \ \ln(\frac{\tau_{max}}{\tau_{min}})$.

The macroscopically observed relaxation is given by the superposition of relaxation processes, expressed as:

$$R(t) = C \int_{\tau_{min}}^{\tau_{max}} \frac{1}{\tau} e^{-t/\tau} d\tau \tag{6}$$

where $C = T\ N_o$ is a constant for a given T and N_o, which is omitted in the calculation merging it with the scaling parameter S. We note that N_o is a constant value for a uniform distribution of energy $N(E)$ between its minimum E_{min} and maximum E_{max} value. The weight factor $1/\tau$ in Equation (6) is a result of the application of Arrhenius' law due to mesoscale relaxation processes operating on different timescales. In Figure 2, we present examples of the relaxation function with time on a logarithmic scale along the horizontal axis with a constant τ_{min} and various values of τ_{max}. In a logarithmic scale, the curve in Figure 2 decreases almost linearly, up to a time $t \approx \tau_{max}$, and at the end, it flattens out. Thus, the relaxation function $R(t)$ is proportional to $\ln(t)$ for $t << \tau_{max}$. Based on that, the maximum relaxation time τ_{max} could be estimated as the time where the relaxation curve flattens out.

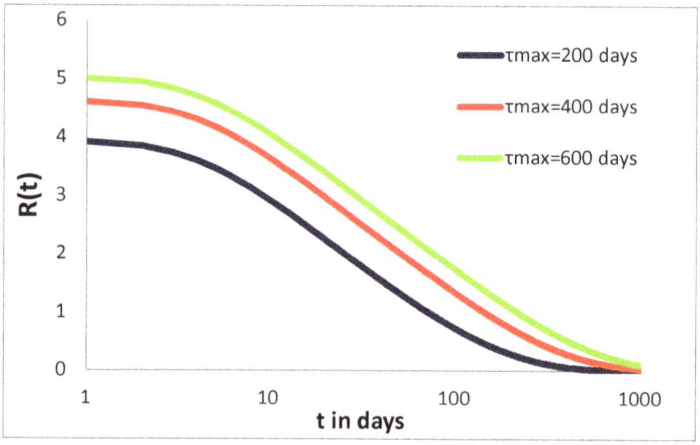

Figure 2. The relaxation function $R(t)$ (see Equation (6) in the text) for τ_{min} = 5 days and τ_{max} = 200, 400, 600 days.

The integral in Equation (6) is the exponential integral $E_1(x)$ [38], which can not be solved analytically but behaves like a negative exponential for very large values of the argument and decays on a logarithmic mode in the intermediate range of the relaxation curve $R(t)$ while deviating from a logarithmic time dependence for $t < \tau_{min}$. Furthermore, as results from Equation (6) when $t = 0$, $R(t)$ can be calculated analytically:

$$R_0 = R(t = 0) = \ln\left(\frac{\tau_{max}}{\tau_{min}}\right) \tag{7}$$

supporting the physical condition to have a finite relaxation function for $t = 0$. Since $\Delta E = E_{max} - E_{min} = kT\ln\left(\frac{\tau_{max}}{\tau_{min}}\right) = kTR_0$, the finite value of the relaxation function for $t = 0$ is required in order for the energy range ΔE to be a finite quantity. Taking the derivative of $R(t)$ with respect to t, we form the following equation:

$$\frac{dR}{dt} = \frac{1}{t}\left(e^{\frac{-t}{\tau_{min}}} - e^{\frac{-t}{\tau_{max}}}\right) \tag{8}$$

which, for $\tau_{min} \approx t \ll \tau_{max}$, expands exponentially and leads to $\left[\frac{dR(t)}{dt}\right]_{t=0} = -\left(\frac{1}{\tau_{min}} - \frac{1}{\tau_{max}}\right)$ and

$$R(t) \approx \ln\left(\frac{\tau_{max}}{\tau_{min}}\right) - \left(\frac{1}{\tau_{min}} - \frac{1}{\tau_{max}}\right)t = R_0 - \left(\frac{1}{\tau_{min}} - \frac{1}{\tau_{max}}\right)t \tag{9}$$

In the observed cases where $\tau_{max} \gg \tau_{min}$, the slope of $R(t)$ vs. t at $t = 0$ is well approximated as $\left[\frac{dR(t)}{dt}\right]_{t=0} = -\frac{1}{\tau_{min}}$, while the ratio of the slope to $R(0)$ follows the expression: $\frac{1}{R(t=0)}\left[\frac{dR(t)}{dt}\right]_{t=0} = \frac{1}{\tau_{min}} \ln\left(\frac{\tau_{max}}{\tau_{min}}\right) = \frac{R_o}{\tau_{min}}$.

Introducing expression (9) in (1), the very early stage displacement immediately following the co-seismic one is given by the expression:

$$u(t) = u_\infty + SR_0 - S\left(\frac{1}{\tau_{min}} - \frac{1}{\tau_{max}}\right)t = u_o - F_1 t \tag{10}$$

where the slope of Equation (10) is $F_1 = S\left(\frac{1}{\tau_{min}} - \frac{1}{\tau_{max}}\right) \approx S/\tau_{min}$ for $\tau_{min} \ll \tau_{max}$ and $u_o = u_\infty + SR_0$.

In the case when $\tau_{min} \ll \tau_{max}$ and the time t is such that $\tau_{min} \ll t \ll \tau_{max}$, we could approximate exponentials such as $e^{\frac{-t}{\tau_{min}}} \approx 0$ and $e^{\frac{-t}{\tau_{max}}} \approx 1$ and Equation (8) results to:

$$dR/dt \approx -1/t$$

which, after integration, gives:

$$R(t) \approx R_1 - \ln(t) \text{ for } \tau_{min} \ll \tau \ll \tau_{max} \tag{11}$$

with R_1 an integration constant. Equation (12) reveals that the relaxation function $R(t)$ exhibits a logarithmic behavior in an intermediate time interval bounded by the relaxation times τ_{min} and τ_{max} with a slope -1 in a graph of $R(t)$ versus ln (t). It becomes evident that when multiplying $R(t)$ with a scale factor S in Equation (1), there is a modification of the slope of the postseismic relaxation versus ln (t) related to the particular characteristics of earthquake event and the seismotectonic settings and the rheology of the deformed zone as well. Substituting Equation (11) into (1), displacement obtains logarithmic behavior:

$$u(t) = u_\infty + SR_1 - S\ln t \tag{12}$$

for $\tau_{min} \ll \tau \ll \tau_{max}$ with a slope $F_2 = S$.

Defining τ^* as the point where the transition from the linear to logarithmic behavior starts, there is a continuation of displacement (as given by Equations (10) and (12), respectively), that permits the estimation of R_1 as:

$$R_1 = \ln\left(\frac{\tau_{max}}{\tau_{min}}\right) + \ln \tau^* - \left(\frac{1}{\tau_{min}} - \frac{1}{\tau_{max}}\right)\tau^* \quad (13)$$

In order to estimate the order of magnitude of R_1, we calculate its value for two reasonable values of τ^*, as that of $\tau^* = \tau_{min}$ and $\tau^* = 2\tau_{min}$. For $\tau^* = \tau_{min}$, (i.e., the transition in the logarithmic range starts in the very early stage of deformation), we have $R_1 = \ln \tau_{max} - 1 + \frac{\tau_{min}}{\tau_{max}}$, which for $\frac{\tau_{min}}{\tau_{max}} \ll 1$, leads to:

$$R_1 = \ln \tau_{max} - 1$$

giving $R = -\ln\left(\frac{t}{\tau_{max}}\right) - 1$. Since $R(t)$ is positively defined, the above expression is valid for $t < \tau_{max}/e \approx 0.37\tau_{max}$. For a more reasonable estimation, we set $\tau^* = 2\tau_{min}$, leading to $R_1 = \ln \tau_{max} + \ln 2 - 2(1 - \frac{\tau_{min}}{\tau_{max}})$, which for $\frac{\tau_{min}}{\tau_{max}} \ll 1$, leads to:

$$R_1 = \ln \tau_{max} + \ln 2 - 2,$$

with $R = -\ln\left(\frac{t}{\tau_{max}}\right) + \ln 2 - 2$. From the positive definition of $R(t)$, we conclude that the above expression is valid for $t < \frac{\tau_{max}}{2-\ln 2} \approx 0.76\tau_{max}$. A comparison between Equations (10) and (13) indicates that the slope of the logarithmic part $F_2 = S$ is scaled with that of early stage deformation behavior $F_1 = S\left(\frac{1}{\tau_{min}} - \frac{1}{\tau_{max}}\right)$. For $\tau_{min} \ll \tau_{max}$, this scaling can provide an estimation of τ_{min}, since $\frac{F_2}{F_1} = \tau_{min}$.

Analyzing the relaxation function (3) for the case $\tau_{min} \ll \tau_{max} \ll t$, the long time behavior of $R(t)$ can be estimated, replacing the upper limit τ_{max} with infinity so that [33]

$$R(t) \approx \frac{\tau_{max}}{t} e^{-\frac{t}{\tau_{max}}} \quad (14)$$

indicating that, for $t \gg \tau_{min}$, the relaxation function is controlled only by τ_{max} and not τ_{min}. In that case, the relaxation time τ_{max} is predominant, and $R(t)$ leads to $R(t) = e^{\left(\frac{-t}{\tau_{max}}\right)}$, implying that the term τ_{max}/t is associated to the relaxation processes with relaxation time $\tau_{min} < \tau < \tau_{max}$. We note that, since the time dependence of $1/t$ is slow compared to $e^{\left(\frac{-t}{\tau_{max}}\right)}$, the relaxation function could be well approximated as an exponential a long time after the main event. Furthermore, the exponential relaxation, as presented in Equation (2), is a special case of the general expression of the relaxation function (3) and results when τ_{min} and τ_{max} have similar values, i.e., when $\tau_{min} = \tau_0 - \Delta$ and $\tau_{max} = \tau_0 + \Delta$, which leads to $R(t) \approx \frac{2\Delta}{\tau} e^{-\frac{t}{\tau_0}}$ presenting an exponential decay over time.

3. Seismotectonic Setting, Data Selection, and Analysis

Intense seismic activity in the Central Ionian Islands of Cephalonia and Lefkas during the last few years, along with several large events (M > 6) over the last century [39,40], clearly highlight the key role of this area in the kinematic process in the Eastern Mediterranean. The area is dominated by a major regional seismogenic zone known as the Cephalonia Transform Fault (CTF) (Figure 3), as well as several other faulting features, resulting in generation of the highest seismic energy release in Europe [41,42]. The broader area is characterized by a complex kinematic and geodynamic regime. The area links the oceanic subduction boundary in the south to the collision between the Apulian microplate and Eurasia in the north [43–45]. This complicated tectonic regime creates a stress field that generates intense seismic activity and ground deformation on both a local and regional scale.

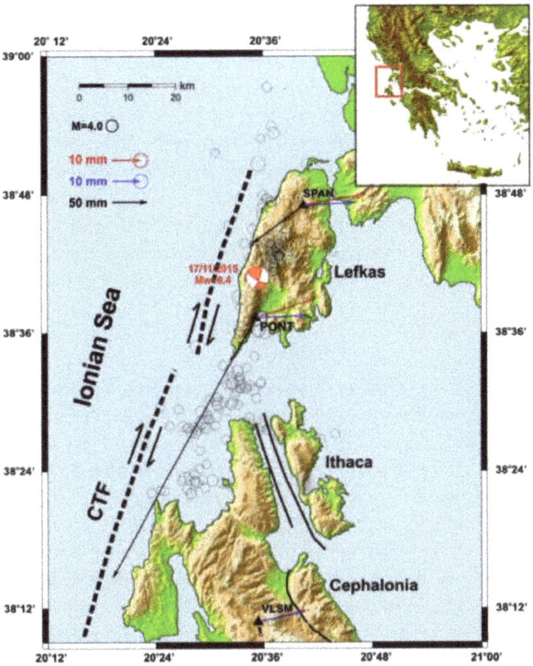

Figure 3. Map of the central Ionian islands of Lefkas, Cephalonia, and Ithaca, showing the main faulting zones and marking the continuous GNSS stations (triangles) PONT and SPAN in Lefkas and VLSM in Cephalonia islands. Circles represent the earthquake epicenters (M > 3) after the 17 November 2015 main event. Red and blue vectors represent the cGNSS horizontal velocities before and after the 2015 earthquake, respectively. Black arrows show the co-seismic deformation. CTF: Cephalonia Transform Fault.

Based on seismological data and geodetic measurements, the CTF zone is described as a dextral strike-slip faulting type system. This system is composed of two separate segments that run along the western shores of the Cephalonia and Lefkas Islands. The southern segment, offshore of Cephalonia, extends for approximately 90 km on a SW–NE direction and ends on the northern part of the island. The northern segment, offshore of western Lefkas Island, strikes a NE–SW direction for ~50 km, ending close to north-western Cephalonia. This segment dips to the ESE and exhibits dextral strike-slip motion combined with a small thrust component [45].

Over the last twenty years, several strong earthquakes have devastated the islands of Lefkas and Cephalonia. In August 2003, a strong event (M_W6.3) occurred off-shore of the north-western coast of Lefkas Island [46–50]. In late January–early February 2014, two large-magnitude earthquakes (M_W5.9 and M_W6.1) were located in the western part of Cephalonia [51,52]. However, these events were not linked directly with the CTF zone.

This study focuses on a strong seismic event that took place in near proximity to the PONT continuous GNSS station. On 17 November 2015, a strong earthquake (M_W6.4) occurred in the southern part of Lefkas Island [53,54] just north of Cephalonia. The main event was followed by a significant post-seismic activity that extended southwards towards Cephalonia. The hypocenter of this earthquake was located at a depth of ~10 km, and the fault plane solutions demonstrated a NNE–SSW striking dextral strike-slip seismic fault with a reverse component that dipped east at a high angle of about $70° \pm 5°$. The after-shock seismic activity extended along three distinctive clusters north and south of the main event [54]. The northern cluster seemed to extends towards the epicentral area of the 2003 strong earthquake, the central one was located in the vicinity of the main event, while the southern one extended towards Cephalonia Island and about 15 km southwest of the epicenter. The spatial coverage of the aftershocks indicated less dense distribution close to the epicentral area, while away from the main shock, there was higher concentration of post-seismic activity. The latter may indicate that, during the main event, higher energy was released in the main ruptured fault, which comprised part of the major CTF zone, while the post-seismic activity was triggered in nearby secondary and adjacent local faults, away from the main rupture plain [54]. The post seismic events extended in a NE–SW direction for approximately 50 km, extending north towards the 2003 earthquake and south towards the northern Cephalonia. The southern seismic cluster occurred in the same area that was also activated during all the previous intense seismic periods in 2003 and in 2014 [48,49].

3.1. GNSS Data Analysis

The epicentral area of the 2015 Lefkas earthquake, located to the southern part of the island, is remarkably close to the local continuous GNSS station PONT [32]. This station has been operating in the area since 2007. The GNSS data are processed daily, using 30 s RINEX files, in order to calculate the daily coordinates of the station (Figure 4). From the time series of PONT station coordinates, the station velocity was estimated for the period 2009–2015. Prior to the 2015 strong event, the station exhibited the anticipated regional motion of the area, and the velocity was calculated as V_{East} = (20.45 ± 0.03) mm/yr, V_{North} = (6.81 ± 0.02) mm/yr, and V_{Up} = (−0.71 ± 0.08) mm/yr with respect to IGb08 reference frame (http://igscb.jpl.nasa.gov/network/refframe.html) (accessed on 30 December 2020). The earthquake of 17 November 2015 caused a co-seismic displacement of the PONT station of 360 mm to south, 201 mm to the west, and a subsidence of 73 mm (Figure 4).

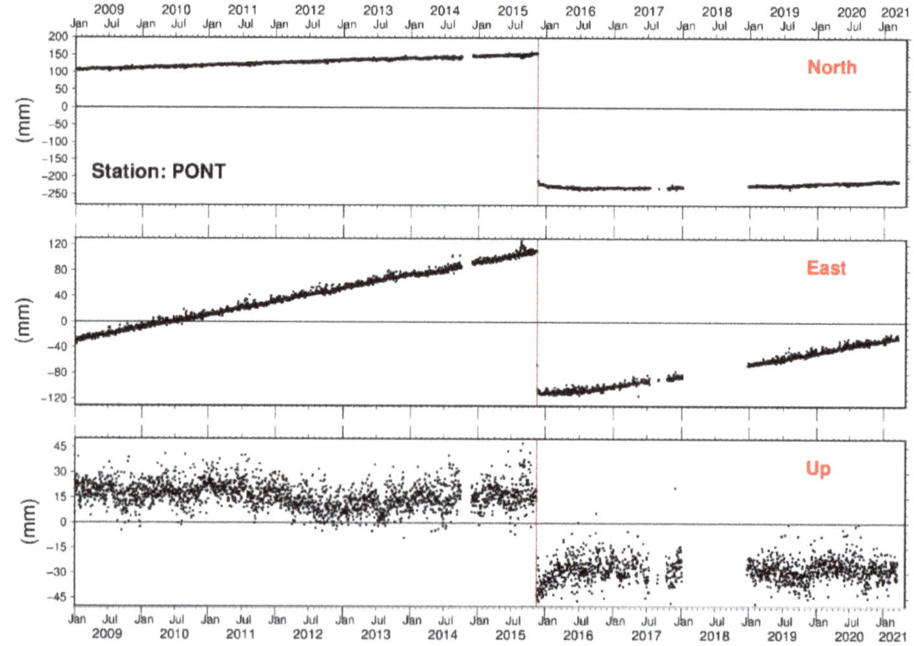

Figure 4. Time series of PONT continuous GNSS stations located in Lefkas island. Red line indicates the date of the 2015 seismic event.

The files from PONT station were processed using the Bernese Software version 5.2 [55]. The software allows the estimation of epoch wise receiver coordinates in precise point positioning mode (PPP), as well as in the double difference mode. For the present research, coordinates were calculated after the earthquake of 2015 on a static mode. To get the best results, the PPP technique was used as a first step to get a priori coordinate values, which were, consequently, introduced in the more precise double difference method. In the processing procedure, GNSS data from 25 IGS and EUREF cGNSS stations from Europe, Greece, Cyprus, and Turkey were used to define the local reference frame. Ambiguities were solved using several resolution strategies depending on the baseline length between the stations: the wide-lane and narrow-lane for the medium baseline (<10–200 km) and the quasi-ionosphere-free (QIF) strategy for long baselines (<1000–2000 km). For the troposphere modelling, the dry and wet Vienna mapping function (VMF) of Bohm et al. (2006), the dry and wet Niell model [56], as well as the Saastamoinen model [57] were the models used to analyze the GNSS microwave data in different stages of the processing. Effects of the solid earth tides were taken into account and corrections of the ocean tide loading were applied based on the FESS2004 model. In analogy to site displacements caused by ocean tide loading, the loading effect of the atmospheric tides were taken into account in the processing of the GNSS data. The latter was done by introducing a specific file in the processing containing atmospheric tidal loading coefficients from a global grid-based model [58]. Moreover, the absolute antenna phase center corrections were used in the processing and to better improve the final solutions files containing the precise orbital parameters; the Earth orientation parameters and the satellite clock corrections were taken from the Center for Orbit Determination in Europe (CODE) (DOI:10.7892/boris.75876.4). On a weekly basis, the calculated coordinates were evaluated for the repeatability error, and in case of large deviations from the weekly solution, the values were excluded. The latter resulted in an unweighted RMS error, in mm, for each component that varied from 0.5 to 1.01 in the east component, 0.8 to 2.1 in the north one, and 1.8 to 4.6 for the up component. However, that was not applied for the first ten days after the main event. The daily coordinates were estimated with respect to IGb08 reference frame.

3.2. Post Seismic Time Series Analysis

The post seismic displacement evolution is made up of two contributions: the displacement $u(t)$ due to the relaxation one on the fault zone and the long term $V_{pst}\,t$, where V_{pst} is the long term loading velocity a long time after the earthquake when the system has reached a new equilibrium state. In our case, $V_{pstEast}$ = (17.48 $\pm$ 0.07) mm/yr, $V_{pstNorth}$ = (3.72 $\pm$ 0.07) mm/yr, and V_{pstUp} = ($-$0.44 $\pm$ 0.18) mm/yr, which was removed from the observed postseismic data. To apply our approach, we analyzed the post seismic relaxation observed after the 17 November 2015 Mw6.5, Lefkas island earthquake. In Figure 5, the North–South displacement, which was the event's predominant post seismic displacement, presented as a function of time for a period 1000 days after the event. For energy reasons, the minimum relaxation time τ_{min} cannot be equal to zero in order to have a converged relaxation function. From Figure 5, the transition from the linear to logarithmic part is observed, permitting the estimation of $\tau_{min} \approx$ 3–5 days. Since the longest relaxation time τ_{max} is the time where the relaxation stops, from Figure 5, we estimate $\tau_{max} \approx 350$ days. The aforementioned estimations of τ_{min} and τ_{max} result in a value $R_o \approx 4.2$ to 4.7. Using the observed values of $u(t)$, we estimate that an average value of displacement in the range from 300 to 600 days is about -26 mm, a value that could be used to approximate u_∞. In Figure 5, the observed displacement, along with the theoretical one, calculated using Equation (1), with S = 4.8 mm, $u_\infty = -26$ mm, $\tau_{min} \approx 3.5$ days, and $\tau_{max} \approx 350$ days (i.e., $R_o \approx 4.6$), is presented as a very good agreement. Furthermore, the time evolution of displacement suggests a slope $F_2 \approx -4.21$ for the logarithmic part and an approximate value $F_1 \approx -1.09$ for the short linear early part of deformation. Comparing the above estimation with Equations (10) and (13), we conclude that, since $\tau_{min} << \tau_{max}$, the scaling of the slopes leads to $\tau_{min} \approx 3.8$ days, in agreement with our estimation from Figure 5.

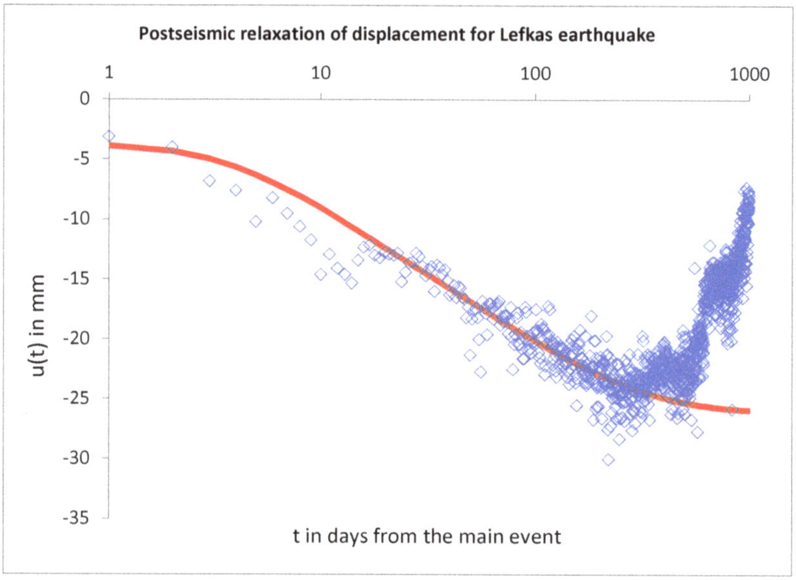

Figure 5. The observed displacement, along with the theoretical calculated using Equation (1) (see text), with S = 4.8 mm, $u_\infty = -26$ mm, $\tau_{min} \approx 3.5$ days, and $\tau_{max} \approx 350$ days (i.e., $R_o \approx 4.6$) is presented as a very good agreement.

A function to present the slow dynamics of postseismic relaxation could be defined in terms of the asymptotic long-term value u_∞ as [59]:

$$\Omega(t) = \frac{u(t) - u_\infty}{u_o - u_\infty} \tag{15}$$

Substituting Equation (10) in (15) leads us to:

$$\Omega(t) = \frac{R(t)}{R_o} \tag{16}$$

In Figure 6, the experimentally estimated $\Omega(t)$ function for the first 400 days of the post-seismic relaxation after the Lefkas main event is given, along with the calculated one using Equation (16) for $\tau_{min} \approx 3.5$ days and $\tau_{max} \approx 350$ days, respectively. A comparison of the observed and calculated values is given in Figure 7, along with a dichotomous line.

Furthermore, we note that $\int_0^\infty \Omega(t)dt = \frac{\tau_{max} - \tau_{min}}{R_o}$, which, after using the experimental values of Ω, leads to $\tau_{max} \approx 360$ days in quite good agreement with the value used in Figures 5 and 6. We note that the law values of $\frac{R(t)}{R_o}$ in Figure 7 correspond to the logarithmic part of relaxation, while the high values are related with the part just after the rupture where the uncertainty in the estimated displacement is higher.

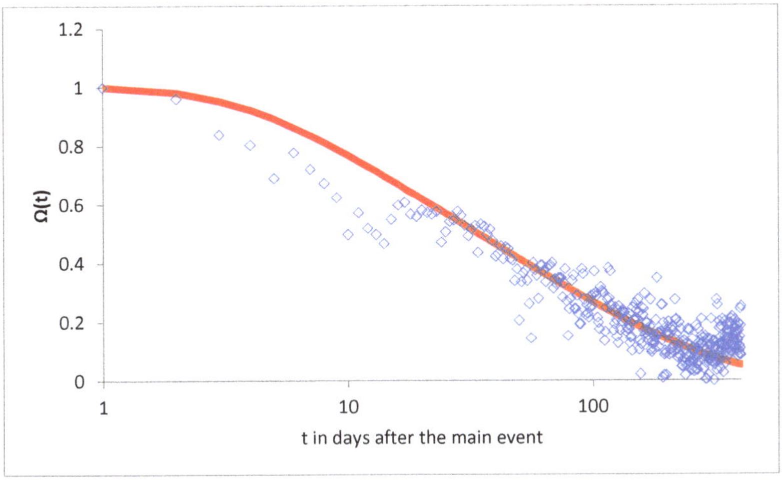

Figure 6. The experimentally estimated $\Omega(t)$ function for the first 400 days of the post-seismic relaxation after the Lefkas main event, along with the calculated one using Equation (16) for $\tau_{min} \approx 3.5$ days and $\tau_{max} \approx 350$ days, respectively.

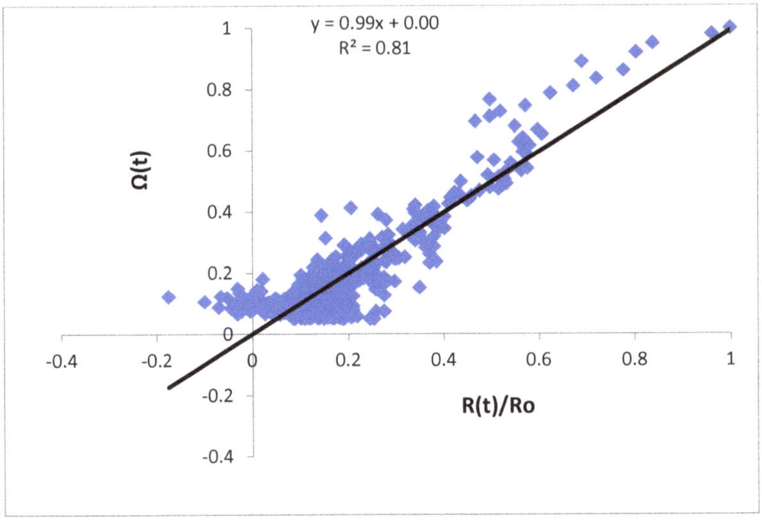

Figure 7. A comparison of the experimentally estimated $\Omega(t)$ function with the calculated values $R(t)/\text{Ro}$ for the post-seismic relaxation period after the Lefkas main event for $\tau_{min} \approx 3.5$ days and $\tau_{max} \approx 350$ days, respectively (see text).

4. Discussion

Most of the models used to describe post-seismic relaxation are based on the assumption of a single relaxation time estimated for each relaxation mechanism proposed [4,11,14,60,61]. A co-seismic deformation leaves the system in a metastable state taking place in the broad earthquake zone. In the present work, the post-seismic relaxation is viewed as a superposition of several different relaxation times, each one activated in different mesoscopic domains of the macroscopic relaxed seismic volume. The macroscopic relaxation observed is a result of the superposition of different relaxation times on the microscopic–mesoscopic scale. This slow thermodynamic recovery could be explained by a physical model based on the Arrhenius Equation [62].

The lithosphere as a dynamical system under a post-seismic relaxation is a complex systems with strong interactions between its mesoscopic "subvolumes" [63] and, thus, could be viewed as a macroscopic system driven by hierarchically constrained dynamics [28] involved in the macroscopically observed relaxation process as a sequential series of correlated mesoscopic processes.

The view of hierarchical constrained dynamics could be a physical pathway to describe many of the complex systems that present logarithmic relaxation [64–66]. In the case of post-seismic phase, the applied tectonic load drives the system to fracture, and its relaxation creates a reorganization of the stress field in some selected mesoscopic regions of the system. The deformed subvolumes nearest to the fractured region are rearranged and the system reorganized. These new stress changes will affect the deformed subvolumes outside the area nearest to the fracture, permitting their relaxation, and so on. The above picture suggests that the relaxation in the fractured fault system is slow because it consists of a large number of correlated and strongly interacting reorganization steps with an increasing relaxation time. It seems to be appropriate to describe macroscopical post-seismic relaxation in terms of a lithosphere that presents a hierarchical pattern, with the faster relaxation to be associated with the subvolume close to earthquake focus regions and, due to strong correlations and long-range interactions, the slower relaxation time to be related with the "outer" deformed, but in the broad vicinity of epicenter, regions. The latter is in agreement with that proposed for complex relaxed systems with cluster interactions, where the relaxation time τ scales with the relaxation volume as $\tau \sim V^{\alpha}$, where α is an exponent related macroscopically with the fractal geometry of the subvolume distribution and the dynamical nature of relaxation [67–69]. This type of hierarchically constrained dynamics is able to derive the logarithmic relaxation frequently observed and associate it with thermally activated processes. We note that this behavior can result either from the parallel relaxation of subvolumes or by a sequential relaxation series of strongly correlated complex events [28].

Our results are in agreement with that presented in [70], where a Burridge-Knopoff (BK) model is used to describe the time dependence of the mean stress, assuming that a uniform statistical distribution of stress values is acting on each block (i.e., deformation in subvolumes) in the relaxing lithospheric system.

In [70], the post-seismic normalized stress relaxation of the strongly interacted elements of the BK model presents a stress relaxation that could be viewed as hierarchically constrained since blocks increase the stress on their nearby elements when they slip, and as a consequence, the probability that their neighbors will slip increases as well. They conclude that the normalized mean stress $<\sigma>$ during the post-seismic relaxation in a BK model with a uniform distribution of stresses [71] is:

$$< \sigma > = \int_0^1 \sigma e^{-\nu t \exp\left(-\beta \varepsilon_m (1-\sigma^2)\right)} d\sigma \tag{17}$$

where ε_m is the maximum size of energy barrier, ν is an attempt frequency to jump over the energy barrier, and $\beta = 1/k_B T$. Approximation of (17) leads to the following behavior in short, intermediate, and long time periods, respectively:

$$\lim_{t \to 0} < \sigma > = \frac{1}{2} - \frac{1}{2\beta\varepsilon_m}(1 - \exp(-\beta\varepsilon_m))\nu t \tag{18}$$

$$< \sigma >_{int} \approx \frac{1}{2} - \frac{1}{2\beta\varepsilon_m}(\gamma + \ln(\nu t)) \tag{19}$$

where γ is the Euler–Mascheroni constant ($\gamma \approx 0.577$).

$$\lim_{t \to \infty} < \sigma > = \frac{\exp(\beta\varepsilon_m)}{2\beta\varepsilon_m} \frac{\exp(-\nu t \exp(-\beta\varepsilon_m)}{\nu t} \tag{20}$$

which, for very long time periods, t is proportional to $\exp(-t/\tau)$ with $\tau = \exp(\beta\varepsilon_m)/\nu$.

A comparison of Equations (18)–(20) with Equation (10), (12), and (14), respectively, implies that $S \sim \frac{1}{2\beta\varepsilon_m}$, supporting the hypothesis that the scaling factor depends on temperature and the energy barrier that must be overcome before the relaxed subvolume moves to a new stress stage. In addition, $\tau_{max} \sim \frac{1}{\nu}\exp(\beta\varepsilon_m)$ and $\tau_{min} \sim \frac{1}{\nu}$ with $\ln\left(\frac{\tau_{max}}{\tau_{min}}\right) \sim \beta\varepsilon_m$ being an expression similar to that defined in the present approach using the Arrhenius expression, if $\varepsilon_m = E_{max} - E_{min}$ is the maximum size of energy barrier.

The relaxation function, as presented in Equation (6), explains the logarithmic relaxation observed. In addition, it results in finite values, both just after the perturbation (i.e., $t = 0$) and for a long time after the earthquake event ($t \to \infty$). The relaxation function $R(t)$ can be explained by Arrhenius' law, presenting a universal behavior of the relaxation process. Following Equation (6) and the ideas of hierarchically constrained dynamics, the faster relaxation mechanisms, related to small relaxed volumes (small τ), offer a more significant contribution to the relaxation process than the slower relaxation mechanisms (large τ related with the broader relaxed area), as indicated by the presence of term $1/\tau$ in the expression of $R(t)$, which permits greater weight to fast relaxation processes. We note that the superposition of relaxation mechanisms, as expressed in Equation (6), implies that, for early time periods, the fast relaxation mechanism associated with small relaxed volumes that have a relaxation time close to τ_{min} dominates, while for late time periods, the slow relaxation mechanism, related to the relaxation of volumes farther from the epicenter and present relaxation time close to τ_{max}, contributes most. It is reasonable to assume that the maximum relaxation time depends on the perturbation that triggers the relaxation and, thus, on the deformed volume related with the main event, along with the geotectonic conditions in the deformed region, such as stress, temperature, and fluids.

In addition, measurements of τ_{min} could give information on the fastest relaxation mechanism, which is scaled with the smallest deformed subvolume, even if it is sometimes not clearly observed due to the immediate transition of the deformation process into the logarithmic range, resulting in an unreliable estimation of τ_{min}.

Applying the generic hierarchically constrained dynamics view to the North–South displacement of the post-seismic relaxation of Lefkas 17 November 2015, Mw6.5 earthquake, we conclude that this approach describes with good agreement the displacement as estimated using cGNSS recordings during the post-seismic relaxation period, leading to the estimation of $\tau_{min} \approx 3.5$ days and $\tau_{max} \approx 350$ days, respectively.

This is a rather new approach that is used to describe post-seismic relaxation based on the assumption of a spectrum of relaxation times that are superposed, each of them activated in different mesoscopic domains of the macroscopic relaxed seismic volume. This slow thermodynamic recovery (i.e., a logarithmic time dependence) could be explained by a physical model based on the Arrhenius equation. With such an approach, we are led to finite values for $t = 0$, without assuming any empirical constitutional equation.

5. Concluding Remarks

In the present work the post-seismic relaxation is viewed as a superposition of several different relaxation times, each one activated in different mesoscopic domains of the macroscopic relaxed seismic volume. The macroscopic relaxation that is observed is a result of the superposition of different relaxation times on the microscopic–mesoscopic scale. This approach is a new one since most of the models used to describe post-seismic relaxation are based on the assumption of a single relaxation time estimated for each relaxation mechanism proposed. The slow thermodynamic recovery of the post-seismic relaxation could be explained by a physical model based on the Arrhenius equation and an appropriate distribution of relaxation times.

In summary, we can state that, in the present work:

a. A multiscale post-seismic relaxation mechanism based on the existence of a distribution in relaxation time is presented.

b. Assuming an Arrhenius dependence of the relaxation time with the uniform distributed activation energy in a mesoscopic scale, we conclude there is a generic logarithmic type relaxation on a macroscopic scale.
c. The hierarchically constrained dynamics model could be used to understand the evolution of post seismic relaxation.
d. The model was applied in the case of 2015 Lefkas (Greece) Mw6.5 earthquake, where cGNSS time series were recorded in a station located in the vicinity of the epicentral area. The application of the present approach to the Lefkas event fits the observed displacements, implying a distribution of relaxation times in the range $\tau_{min} \approx 3.5$ days to $\tau_{max} \approx 350$ days.

These results offer a stimulating finding for further studies that will investigate post-seismic deformation timeseries for other strong earthquakes, which may provide new insights into the underlying dynamics of earthquake relaxation processes, along with the possible parameters that influence this behaviour (i.e., fault type, local conditions, regional stress field, etc.).

Author Contributions: Conceptualization, F.V.; methodology, F.V.; software, F.V. and V.S.; validation, F.V.; data curation, V.S.; writing—original draft preparation, F.V.; writing—review and editing, F.V. and V.S. All authors have read and agreed to the published version of the manuscript.

Funding: This research was funded by the Operational Program "Competitiveness, Entrepreneurship and Innovation" (NSRF 2014–2020) and co-financed by Greece and the European Union (European Regional Development Fund).

Data Availability Statement: Data are available in www.geodesy.gein.noa.gr (accessed on 30 December 2020).

Acknowledgments: We acknowledge support of this study by the project "HELPOS—Hellenic Plate Observing System" (MIS 5002697), which is implemented under the action "Reinforcement of the Research and Innovation Infrastructure", funded by the Operational Program "Competitiveness, Entrepreneurship and Innovation" (NSRF 2014–2020) and co-financed by Greece and the European Union (European Regional Development Fund).

Conflicts of Interest: The authors declare no conflict of interest.

References

1. Hearn, E.H. What can GPS data tell us about the dynamics of post-seismic deformation? *Geophys. J. Int.* **2003**, *155*, 753–777. [CrossRef]
2. Nur, A.; Byerlee, J.D. An effective stress law for elastic deformation of rocks with fluids. *J. Geophys. Res.* **1971**, *76*, 414–419. [CrossRef]
3. Wang, H.F. *Theory of Linear Poroelasticity with Applications to Geomechanics and Hydrogeology;* Princeton University Press: Princeton, NJ, USA, 2000; Volume 2, p. 276.
4. Marone, C.J.; Scholtz, C.H.; Bilham, R. On the mechanics of earthquake afterslip. *J. Geophys. Res.* **1991**, *96*, 8441–8452. [CrossRef]
5. Pollitz, F.F.; Bürgmann, R.; Segall, P. Joint estimation of afterslip rate and postseismic relaxation following the 1989 Loma Prieta earthquake. *J. Geophys. Res.* **1998**, *103*, 26975–26992. [CrossRef]
6. Montési, L.G.J. Controls of shear zone rheology and tectonic loading on postseismic creep. *J. Geophys. Res.* **2004**, *109*, B10404. [CrossRef]
7. Marone, C.; Raleigh, C.B.; Scholz, C.H. Frictional behavior and constitutive modeling of simulated fault gouge. *J. Geophys. Res.* **1990**, *95*, 7007–7025. [CrossRef]
8. Hetland, E.A.; Hager, B.H. The effects of rheological layering on post-seismic deformation. *Geophys. J. Int.* **2006**, *166*, 277–292. [CrossRef]
9. Wang, L.; Hainzl, S.; Sinan Özeren, M.; Ben-Zion, Y. Postseismic deformation induced by brittle rock damage of aftershocks. *J. Geophys. Res.* **2010**, *115*, B10422. [CrossRef]
10. Ingleby, T.; Wright, T.J. Omori-like decay of postseismic velocities following continental earthquakes. *Geophys. Res. Lett.* **2017**, *44*, 3119–3130. [CrossRef]
11. Perfettini, H.; Avouac, J.-P. Postseismic relaxation driven by brittle creep: A possible mechanism to reconcile geodetic measurements and the decay rate of aftershocks, application to the Chi-Chi earthquake, Taiwan. *J. Geophys. Res.* **2004**, *109*. [CrossRef]

12. Mallick, N.; Ciliberto, S.; Roux, S.G.; Di Stephano, P.; Vanel, L. Aftershocks in thermally activated rupture of indented glass. In Proceedings of the 12th International Conference on Fracture, ICF-12, Ottawa, ON, Canada, 12–17 July 2009; Volume 6, pp. 4564–4572.
13. Bürgmann, R.; Segall, P.; Lisowski, M.; Svarc, J. Postseismic strain following the 1989 Loma Prieta earthquake from GPS and leveling measurements. *J. Geophys. Res.* **1997**, *102*, 4933–4955. [CrossRef]
14. Yu, S.B.; Kuo, L.C.; Hsu, Y.J.; Su, H.H.; Liu, C.C.; Hou, C.; Lee, J.F.; Shin, T.C. Preseismic deformation and coseismic displacements associated with the 1999 Chi-Chi, Taiwan, earthquake. *Bull. Seismol. Soc. Am.* **2001**, *91*, 995–1012. [CrossRef]
15. Hearn, E.H.; Bürgmann, R.; Reilinger, R.E. Dynamics of İzmit earthquake postseismic deformation and loading of the Düzce earthquake hypocenter. *Bull. Seismol. Soc. Am.* **2002**, *92*, 172–193. [CrossRef]
16. Takahashi, H.; Nakao, S.; Okazaki, N.; Koyama, J.; Sagiya, T.; Ito, T.; Ohya, F.; Sato, K.; Fujita, Y.; Hashimoto, M.; et al. GPS observation of the first month of postseismic crustal deformation associated with the 2003 Tokachi-oki earthquake (M_{JMA} 8.0), off southeastern Hokkaido. *Earth Planets Space* **2004**, *56*, 377–382. [CrossRef]
17. Anugrah, B.; Meilano, I.; Gunawan, E.; Efendi, J. Estimation of postseismic deformation parameters from continuous GPS data in northern Sumatra after the 2004 Sumatra-Andaman earthquake. *Earthq. Sci.* **2015**, *28*, 347–352. [CrossRef]
18. Tobita, M. Combined logarithmic and exponential function model for fitting postseismic GNSS time series after 2011 Tohoku-Oki earthquake. *Earth Planets Space* **2011**, *68*, 1–12. [CrossRef]
19. Marone, C. Laboratory-derived friction laws and their application to seismic faulting. *Annu. Rev. Earth Planet. Sci.* **1998**, *26*, 643–696. [CrossRef]
20. Kawada, Y.; Nagahama, H.; Hara, H. Irreversible thermodynamic and viscoelastic model for power-law relaxation and attenuation of rocks. *Tectonophysics* **2006**, *427*, 255–263. [CrossRef]
21. Brantut, N.; Heap, M.; Baud, P.; Meredith, P. Mechanisms of time-dependent deformation in porous limestone. *J. Geophys. Res.* **2014**, *119*, 5444–5463. [CrossRef]
22. Bocquet, L.; Charlaix, E.; Ciliberto, S.; Crassous, A. Moisture-induced ageing in granular media and the kinetics of capillary condensation. *Nature* **1998**, *396*, 735–737. [CrossRef]
23. Renard, F.; Dysthe, D.; Feder, J.; Meakin, P.; Morris, S.; Jamtveit, B. Pattern formation during healing of fluid-filled cracks: An analog experiment. *Geofluids* **2009**, *9*, 365–372. [CrossRef]
24. Ten Cate, J.A.; Shankland, T.J. Slow dynamics in the nonlinear elastic response of Berea sandstone. *Geophys. Res. Lett.* **1996**, *21*, 3019–3022. [CrossRef]
25. Gibbs, M.; Evetts, J.; Leake, J. Activation energy spectra and relaxation in amorphous materials. *J. Mater. Sci.* **1983**, *18*, 278–288. [CrossRef]
26. Shaknhovich, E.; Gutin, A. Relaxation to equilibrium in the random energy model. *Eurphys. Lett.* **1989**, *9*, 569–574. [CrossRef]
27. Ten Cate, J.; Smith, E.; Guyer, R. Universal slow dynamics in granular solids. *Phys. Rev. Lett.* **2000**, *85*, 1020–1023. [CrossRef] [PubMed]
28. Brey, J.; Prados, A. Slow logarithmic relaxation in models with hierarchically constrained dynamics. *Phys. Rev. E* **2001**, *63*, 21108. [CrossRef] [PubMed]
29. Amir, A.; Oreg, Y.; Imry, Y. On relaxations and aging of various glasses. *Proc. Natl. Acad. Sci. USA* **2012**, *109*, 1850–1855. [CrossRef] [PubMed]
30. Zaitsev, V.; Gusev, V.; Tournat, V.; Richard, P. Slow relaxation and aging phenomena at the nanoscale in granular materials. *Phys. Rev. Lett.* **2014**, *112*, 108302. [CrossRef]
31. Varotsos, P.; Alexopoulos, C. *Thermodynamics of Point Defects and Their Relation with the Bulk Properties*; Amelinckx, S., Gevers, R., Nihoul, J., Eds.; Elsevier: Amsterdam, The Netherlands, 1986.
32. Ganas, A.; Drakatos, G.; Rontogianni, S.; Tsimi, C.; Petrou, P.; Papanikolaou, M.; Argyrakis, P.; Boukouras, K.; Melis, N.; Stavrakakis, G. NOANET: The new permanent GPS network for Geodynamics in Greece. *Geophys. Res. Abstr.* **2008**, *10*, EGU2008-A-04380.
33. Snieder, R.; Sens-Schönfelder, C.; Wu, R. The time dependence of rock healing as a universal relaxation process, a tutorial. *Geophys. J. Int.* **2017**, *208*, 1–9. [CrossRef]
34. Ranalli, G. *Rheology of the Earth*; Allen and Unwin Inc.: Winchester, MA, USA, 1987.
35. Ergintav, S.; Bürgmann, R.; McClusky, S.; Çakmak, R.; Reilinger, R.E.; Lenk, O.; Barka, A.; Özener, H. Postseismic deformation near the İzmit earthquake (17 August 1999, M 7.5) rupture zone. *Bull. Seismol. Soc. Am.* **2002**, *92*, 194–207. [CrossRef]
36. Pollitz, F.F. Transient rheology of the uppermost mantle beneath the Mojave Desert, California. *Earth Planet. Sci. Lett.* **2003**, *215*, 89–104. [CrossRef]
37. Pollitz, F.F. Transient rheology of the upper mantle beneath central Alaska inferred from the crustal velocity field following the 2002 Denali earthquake. *J. Geophys. Res.* **2005**, *110*. [CrossRef]
38. Abramowitz, M.; Stegun, I.A. *Handbook of Mathematical Functions: With Formulas, Graphs, and Mathematical Tables*; Dover Books on Mathematics: Dover, NY, USA, 1965.
39. Papazachos, B.C. Seismicity of the Aegean and surrounding area. *Tectonophysics* **1009**, *178*, 287–308.
40. Louvari, E.; Kiratzi, A.A.; Papazachos, B.C. The Cephalonia transform fault and its extension to western Lefkada island (Greece). *Tectonophysics* **1999**, *308*, 223–236. [CrossRef]

41. Le Pichon, X.; Chamot Rooke, N.; Lallemant, S.; Noomen, R.; Veis, G. Geodetic determination of the kinematics of central Greece with respect to Europe: Implications for eastern Mediterranean tectonics. *J. Geophys. Res.* **1995**, *100*, 12675–12690. [CrossRef]
42. Kokinou, E.; Papadimitriou, E.; Karakostas, V.; Kamberis, E.; Vallianatos, F. The Kefalonia Transform Zone (offshore Western Greece) with special emphasis to its prolongation towards the Ionian Abyssal Plain. *Mar. Geophys. Res.* **2006**, *27*, 241–252. [CrossRef]
43. Van Hinsbergen, D.J.J.; van der Meer, D.G.; Zachariasse, W.J.; Meulenkamp, J.E. Deformation of western Greece during Neogene clockwise rotation and collision with Apulia. *Int. J. Earth Sci.* **2006**, *95*, 463–490. [CrossRef]
44. Vassilakis, E.; Royden, L.; Papanikolaou, D. Kinematic links between subduction along the Hellenic trench and extension in the Gulf of Corinth, Greece: A multidisciplinary analysis. *Earth Planet. Sci. Lett.* **2011**, *303*, 108–120. [CrossRef]
45. Lekkas, E.; Mavroulis, S.; Carydis, P.; Alexoudi, V. The 17 November 2015 Mw 6.4 Lefkas (Ionian Sea, Western Greece) Earthquake: Impact on Environment and Buildings. *Geotech. Geol. Eng.* **2018**, *36*, 2109–2142. [CrossRef]
46. Karastathis, V.; Mouzakiotis, E.; Ganas, A.; Papadopoulos, G.A. High-precision relocation of seismic sequences above a dipping Moho: The case of the January–February 2014 seismic sequence on Cephalonia island (Greece). *Solid Earth* **2015**, *6*, 173–184. [CrossRef]
47. Karakostas, V.; Papadimitriou, E.; Papazachos, C.B. Properties of the 2003 Lefkada, Ionian Islands, Greece, earthquake seismic sequence and seismicity triggering. *Bull. Seismol. Soc. Am.* **2004**, *94*, 1976–1981. [CrossRef]
48. Papadimitriou, P.; Kaviris, G.; Makropoulos, K. The Mw = 6.3 2003 Lefkada Earthquake (Greece) and induced stress transfer changes. *Tectonophysics* **2006**, *423*, 73–82. [CrossRef]
49. Lagios, E.; Sakkas, V.; Papadimitriou, P.; Damiata, B.N.; Parcharidis, I.; Chousianitis, K.; Vassilopoulou, S. Crustal deformation in the Central Ionian Islands (Greece): Results from DGPS and DInSAR analyses (1995–2006). *Tectonophysics* **2007**, *444*, 119–145. [CrossRef]
50. Saltogianni, V.; Stiros, S. A two-fault model of the 2003 Leucas (Aegean Arc) earthquake based on topological inversion of GPS data. *Bull. Seismol. Soc. Am.* **2015**, *105*, 2510–2520. [CrossRef]
51. Papadopoulos, G.A.; Karastathis, V.; Koukouvelas, I. The Cephalonia, Ionian Sea (Greece), sequence of strong earthquakes of January-February (2014): A first report. *Res. Geophys.* **2015**, *4*, 5441. [CrossRef]
52. Karakostas, V.; Papadimitriou, E.; Mesimeri, M.; Gkarlaouni, C.; Paradisopoulou, P. The 2014 Kefalonia doublet (Mw6.1 and Mw6.0) central Ionian Islands, Greece: Seismotectonic implications along the Kefalonia transform fault zone. *Acta Geophys.* **2015**, *63*, 1–16. [CrossRef]
53. Chousianitis, K.; Konca, A.O.; Tselentis, G.-A.; Papadopoulos, G.A.; Gianniou, M. Slip model of the 17 November 2015 M = 6.5 Lefkada earthquake from the joint inversion of geodetic and seismic data. *Geophys. Res. Lett.* **2016**, *43*, 7973–7981. [CrossRef]
54. Papadimitriou, E.; Karakostas, V.; Mesimeri, M.; Chouliaras, G.; Kourouklas, C. The Mw6.5 17 November 2015 Lefkada (Greece) Earthquake: Structural Interpretation by Means of the Aftershock Analysis. *Pure Appl. Geophys.* **2017**, *174*, 3869–3888. [CrossRef]
55. *Bernese GNSS Software Version 5.2. User Manual*; Dach, R.; Lutz, S.; Walser, P.; Fridez, P. (Eds.) Astronomical Institute, University of Bern; Bern Open Publishing: Bern, Switzerland, 2015; ISBN 978-3-906813-05-9. [CrossRef]
56. Niell, A.E. Global mapping functions for the atmosphere delay at radio wavelengths. *J. Geophys. Res.* **1996**, *101*, 3227–3246. [CrossRef]
57. Saastamoinen, I.I. Contribution to the theory of atmospheric refraction. *Bull. Géodésique* **1972**, *105*, 279–298. [CrossRef]
58. Ray, R.D.; Ponte, R.M. Barometric tides from ECMWF operational analyses. *Ann. Geophys.* **2003**, *21*, 1897–1910. [CrossRef]
59. Tsallis, C. Some open points in nonextensive statistical mechanics. *Int. J. Bifurc. Chaos* **2012**, *22*, 1230030. [CrossRef]
60. Savage, J.C.; Svarc, J.L. Post seismic deformation associated with the 1992 M_W = 7.3 Landers earthquake, southern California. *J. Geophys. Res.* **1997**, *102*, 7565–7577. [CrossRef]
61. Savage, J.; Svarc, J.; Yu, S.-B. Postseismic relaxation and transient creep. *J. Geophys. Res.* **2005**, *110*, B11402. [CrossRef]
62. Ten Cate, J. Slow dynamics of earth materials: An experimental overview. *Pure Appl. Geophys.* **2011**, *168*, 2211–2219. [CrossRef]
63. Vallianatos, F.; Chatzopoulos, G. A Complexity View into the Physics of the Accelerating Seismic Release Hypothesis: Theoretical Principles. *Entropy* **2018**, *20*, 754. [CrossRef] [PubMed]
64. Tarasov, A.; Titov, K. Relaxation time distribution from time domain induced polarization measurements. *Geophys. J. Int.* **2007**, *170*, 31–43. [CrossRef]
65. Zaitsev, V.; Gusev, V.; Castagnede, B. Thermoelastic mechanism for logarithmic slow dynamics and memory in elastic wave interactions with individual cracks. *Phys. Rev. Lett.* **2003**, *90*, 75501. [CrossRef] [PubMed]
66. Ostrovsky, L.; Lebedev, A.; Riviere, J.; Shokouhi, P.; Wu, C.; Stuber Geesey, M.A.; Johnson, P.A. Long-Time Relaxation Induced by Dynamic Forcing in Geomaterials. *J. Geophys. Res.* **2019**, *124*, 5003–5013. [CrossRef]
67. Brouers, F.; Sotolongo-Costa, O. Universal relaxation in nonextensive systems. *Eurphys. Lett.* **2003**, *62*, 808–814. [CrossRef]
68. Brouers, F.; Sotolongo-Costa, O. Relaxation in heterogeneous systems: A rare events dominated phenomenon. *Phys. A Stat. Mech. Appl.* **2005**, *356*, 359–374. [CrossRef]
69. Vallianatos, F. On the non-extensive nature of the Isothermal Depolarization relaxation Currents in cement mortars. *J. Phys. Chem. Solid* **2012**, *73*, 550–553. [CrossRef]
70. Huisman, B.; Fasolino, A. Logarithmic relaxation due to minimization of interactions in the Burridge-Knopoff model. *Phys. Rev. E* **2006**, *74*, 26110. [CrossRef] [PubMed]
71. Persson, B.N.J. Theory of friction: Stress domains, relaxation and creep. *Phys. Rev. B* **1995**, *51*, 13568–13585. [CrossRef] [PubMed]

applied sciences

MDPI

Article

Detection of Preferential Water Flow by Electrical Resistivity Tomography and Self-Potential Method

Karlis Kukemilks * and Jean-Frank Wagner

Department of Geology, University of Trier, Behringstraße 21, D-54296 Trier, Germany; wagnerf@uni-trier.de
* Correspondence: kukemilks.karlis@inbox.lv; Tel.: +49-651-201-4644

Abstract: This study explores the hydrogeological conditions of a landslide-prone hillslope in the Upper Mosel valley, Luxembourg. The investigation program included the monitoring of piezometer wells, hydrogeological field tests, analysis of drillcore records, and geophysical surveys. Monitoring and field testing in some of the observation wells indicated very pronounced preferential flow. Electrical resistivity tomography (ERT) and self-potential geophysical methods were employed in the study area for exploration of the morphology of preferential flowpaths. Possible signals associated with flowing groundwater in the subsurface were detected; however, they were diffusively spread over a relatively large zone, which did not allow for the determination of an exact morphology of the conduit. Analysis of drillcore records indicated that flowpaths are caused by the dissolution of thin gypsum interlayers in marls. For better understanding of the site's hydrogeological settings, a 3D hydrogeological model was compiled. By applying different subsurface flow mechanisms, a hydrogeological model with thin, laterally extending flowpaths embedded in a porous media matrix showed the best correspondence with field observations. Simulated groundwater heads in a preferential flow conduit exactly corresponded with the observed heads in the piezometer wells. This study illustrates how hydrogeological monitoring and geophysical surveys in conjunction with the newest hydrogeological models allow for better conceptualization and parametrization of preferential flow.

Keywords: preferential flow; resistivity tomography; self-potential mapping; hydrogeological modeling

1. Introduction

The aim of this study is the identification and characterization of preferential flow in a study site by using hydrogeological and geophysical investigation techniques. Field investigation delivers boundary conditions for groundwater simulation, which allows for better characterization of preferential flow in the subsurface.

Preferential groundwater flow has been extensively studied in recent decades because it has a growing importance in several research fields such as in contaminant transport [1], the construction of nuclear waste repositories [2,3], the prevention of dam leakages [4,5], and in slope failures [6–9].

Various types of discontinuities in rock or soil such as surface or subsurface layer topography, tree root decay, animal caves, or soil/rock cracks can result in preferential flow [10,11]. Preferential flow routes can form complex networks [12], and moreover, they can evolve over time [13].

Recent advances in numerical modeling allow for the consideration of preferential flow in hydrogeological simulations [14,15]. Beven and Germann [16] identified the following preferential flow modeling approaches: single continuum, dual continuum, dual permeability, and dual porosity. In the case of single continuum, dual continuum, and dual permeability modeling approaches, the morphology of separated fractures is not represented in the model. Instead, modified hydraulic conductivity functions for single continuum models or regularly distributed permeable inclusions are applied to the model [16]. Such an approach allows for the consideration of the impact of preferential flowpaths in a generalized way if the impact of individual conduits can be neglected.

The dual-porosity approach considers the morphology of individual flowpaths, and the water exchange between the conduit and the encompassing matrix is simulated [16]. However, this approach is numerically challenging and demands advanced investigation techniques for the identification and parametrization of the preferential flow conduits in the subsurface. Consequently, depending on the demanded model's resolution, an appropriate approach of preferential flow conceptualization in the model should be selected. Data characterizing preferential flow in the field site can be attained by selecting appropriate field investigation techniques, and they are a prerequisite for the building of a representative groundwater model.

Various geophysical and hydrogeological approaches have been used recently for the identification of preferential flow pathways [17–22]. The main hydrogeological approaches are single-borehole tests, hydraulic tests, cross-borehole hydraulic tests, and tracer tests [23]. However, these methods have limited applicability if the groundwater-bearing fracture is not in the proximity of the borehole [17].

Non-invasive geophysical techniques such as electrical resistivity tomography (ERT) have been widely applied for the investigation of lithological settings [18,19] and groundwater bodies [20,21,24,25]. ERT determines the resistivity distribution in the subsurface, which allows for the identification of saturated zones of soil and rock. Both the flow of groundwater and of electrical current in saturated rock or soil rely on similar physical principles, which can be conceptualized through the flow in porous and fractured media [24]. The conduits within the rock matrix can radically increase conductivity both in terms of water and electrical current [6,24,26]. By using the discrete-dual-porosity modeling approach, the impact of anisotropy on the electrical resistivity of rock massif has been illustrated [24]. It should be added that anisotropy of electrical conductivity in subsurface is affected by other factors such as solute concentration variability. This effect is particularly emphasized under partially saturated conditions [27].

Preferential flowpaths can be caused by different types of heterogeneities in the subsurface. Without discrete fracture networks, as discussed before, highly permeable interlayers, subsurface layer topography, or animal burrows can serve as preferential flowpaths as well [10,11]. The integrated use of geophysical and subsurface monitoring techniques increases the accuracy of geophysical surveys for application in different scales and geological settings. The combined use of ERT and active seismics allowed for monitoring of water flow in the vadose zone with very high spatial and temporal resolution [28].

The effect of groundwater flow in fractured media on electrical resistivity distribution and streaming potential signals has been modeled numerically [24,26]. For this research, we have selected ERT and self-potential techniques that provide a physical link both with water content and water flow in the subsurface [24,26,28].

A self-potential geophysical method using non-polarizing electrodes has been employed for measuring the streaming potential associated with preferential flow paths [18,29–33]. Contrary to ERT, when an external source of current is applied, the self-potential method measures electrical charges developing naturally in the subsurface [34–36]. There are various sources of self-potentials. Redox potentials cause the most pronounced self-potential anomalies up to 1500 mV [37]. Anomalies caused by streaming potentials have lower magnitude, and they are generated by electrolytic groundwater flowing in porous or fractured media [35,36]. Water in the space between mineral grains is never neutral, as there is an excess of charge in pore water to balance the charge on the surface of silicate and aluminosilicate mineral particles [35]. The presence of the excess charges in pore water causes the phenomenon of streaming potential when the charges are dragged by flowing groundwater. As a consequence, in unconfined aquifers, positive self-potential anomalies can be observed in the direction of the preferential flow [18,35].

The literature review shows that preferential flow has an importance in many fields because it can radically alter the velocity and rate of groundwater flow. The newest hydrogeological models are able to consider various kinds of preferential flow mechanisms in a groundwater simulation. However, advanced investigation and testing in the field site is a prerequisite for the adequate conceptualization and parametrization of the preferential flowpaths in numerical models.

This study was conducted in the study site of Deisermillen (Luxembourg), Upper Mosel region (Figure 1a), where in the autumn of 1964, a dangerous landslide occurred [38,39]. The typical morphological features of the landslide such as hummocky topography can be seen in a high-resolution digital elevation model (Figure 1b).

In 1965, several drillings were done under the supervision of the State Geological Survey [39] in order to investigate the possible causes of the Deisermillen landslide. In 1974, additional fieldworks were done to conduct remediation measures in the landslide's area [40]. From previous research campaigns, 11 borehole records in all are available in the Archive of Geological Survey of Luxembourg [39,40]. In 2018–2019, an extensive field research program was conducted including 8 new geological drillings, hydrogeological monitoring and testing, and geophysical exploration to investigate the lithological and hydrogeological setting of the landslide's area. All of the borehole records from previous and actual studies were used for the geological characterization of the study area and the building of a 3D geological model (Figure 1c,d).

The uppermost lithological sequence in the study area is the dolomites of the Upper Muschelkalk (mo). This formation appears as a steeply inclined cliff in the western part of the site (Figure 1b–d).

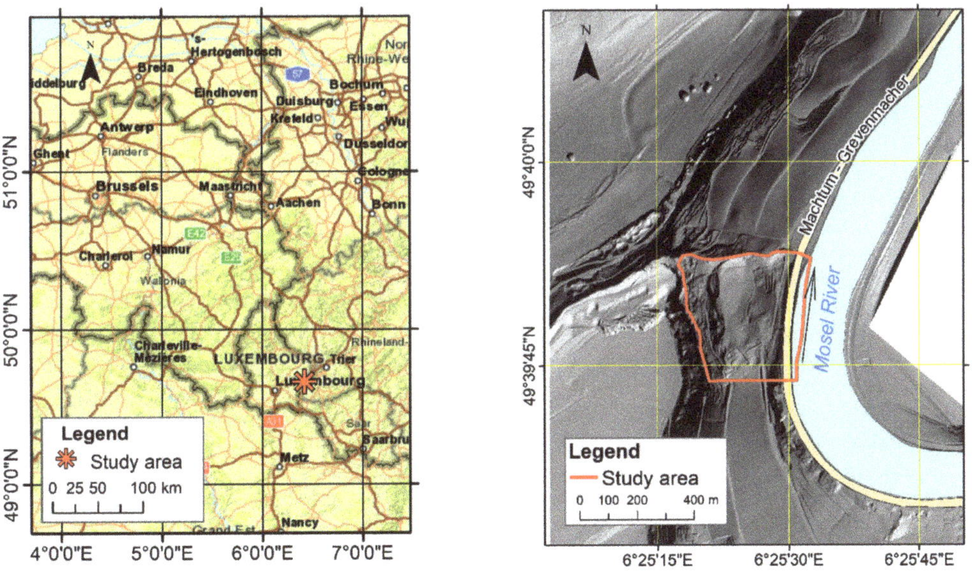

(a) (b)

Figure 1. *Cont.*

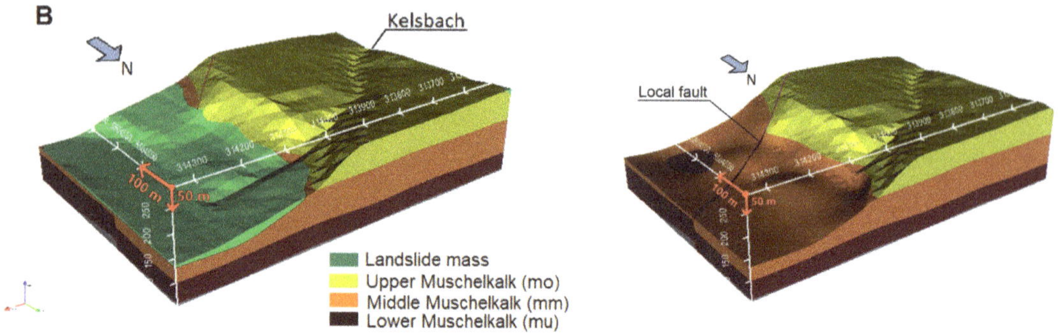

(c) (d)

Figure 1. (**a**) Regional location of the study area (modified after [41]), (**b**) hillshade map showing location of the study area affected by the landslide in the Mosel River valley, a 3D geological model of the study area (**c**) with and (**d**) without landslide mass cover (processed from [42,43]).

This formation is underlain by a 70 m thick Middle Muschelkalk (mm) formation composed of gypsiferous marls. These marls are covered by the 6 to 14 m thick, less consolidated, landslide mass extending from the Mosel River to the Upper Muschelkalk cliff (Figure 1c). The landslide mass cover consists of dolomite blocks embedded in a silty matrix. The landslide mass, through which the landslide occurred, was formed during a long-term morphological evolution of the hillslope by alteration and erosion of the Middle Muschelkalk marls and rockfall of dolomite blocks from the Upper Muschelkalk cliff. There is a local fault trending in a SW-NE direction along the southern border of the study area (Figure 1d). It can be assumed that the fault marks the maximal extension of the landslide to the south as more resistant rocks of the Lower Muschelkalk are located at shallow depth southward of the fault.The top of the Lower Muschelkalk (mu) formation is located at 30.7–42.5 m depth below the surface northward of the fault in the study area.

The lithological units of Lower and Middle Muschelkalk correspond to the international chronostratigraphic stage of Anisian and Upper Muschelkalk to Ladinian, respectively.

The hydrogeological settings of the study area are characterized by fracture flow in a carstic network of the dolomitic Upper Muschelkalk layers. It is supposed that groundwater flows through the spaces in the dolomites, accumulates on the underlying less conductive Middle Muschelkalk, and is drained in the landslide mass where it occasionally reappears as springs. In the study area, several groundwater springs can be observed discharging from the landslide mass deposits (Figure 2a). Additional water supply for the springs is provided by the Kelsbach Creek that flows on the plateau at the rear of the dolomite cliff (Figure 2a). The creek completely disappears in the dolomite massif, and only during periods of intensive rainfall is a surplus of water drained through the creek bed to the Mosel River. Tracer tests carried out in 1968 confirmed the presence of the karstic network in the dolomitic rock massif, resulting in a groundwater flow from Kelsbach Creek to the springs in the landslide's area [44].

Previous studies describe preferential flow in the dolomites of Upper Muschelkalk, whereas the marls of Middle Muschelkalk are supposed to be less conductive material [44]. Therefore, the initial conceptual hydrogeological model of the study site assumed intense groundwater flow discharging from the dolomites and following the way through the landslide mass cover toward the Mosel River. However, the field studies described later below indicated that an intense preferential flow in the marls can occur as well, as it is shown in Figure 2b.

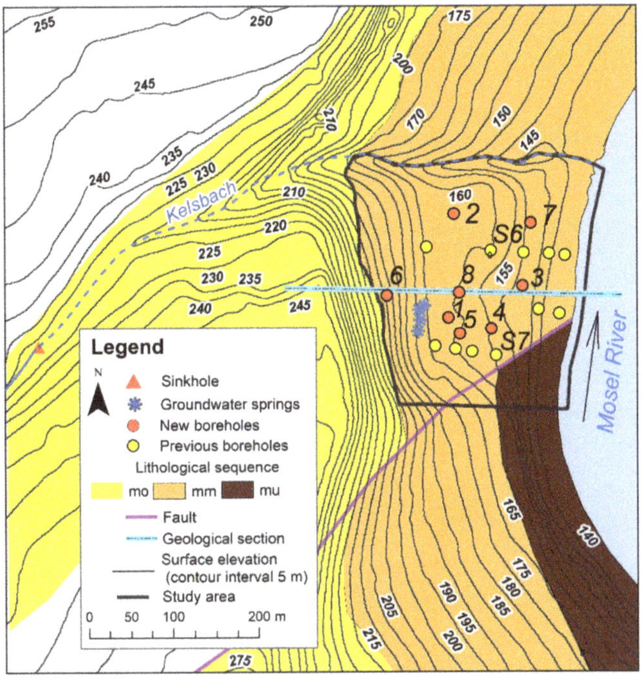

(**a**)

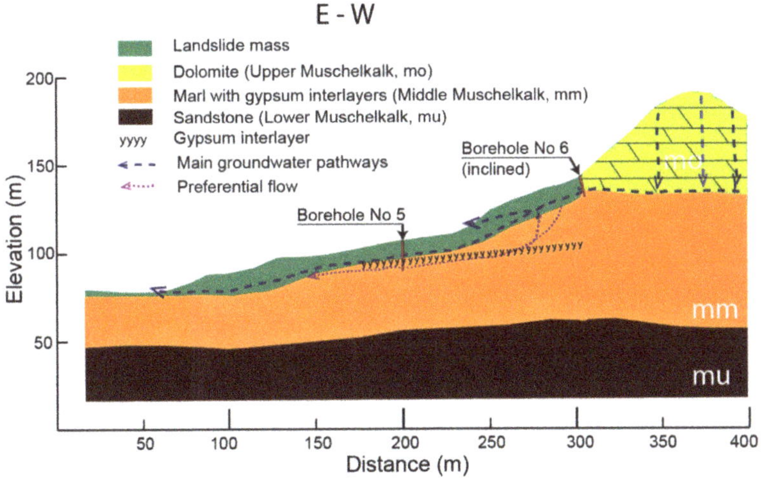

(**b**)

Figure 2. (**a**) Geological sketch of the landslide area showing distribution of boreholes and groundwater springs (**b**) geological section of the area showing conceptual hydrogeological settings (processed from [42,43]).

2. Materials and Methods

In this study, geological and hydrogeological survey techniques were applied in conjunction with geophysical methods to investigate the hydrogeological conditions of the study site. The geophysical techniques were applied with particular focus on preferential flow, with a scope to identify and parametrize the flow pathways in the subsurface. Subsequently a hydrogeological model considering preferential pathways was compiled and validated through observed groundwater heads in piezometer wells.

In addition to 11 boreholes conducted during previous research campaigns [39,40] in the study site, new boreholes 1–8 were drilled (Figure 2a) with a core driller, and samples were stored in boxes. The depth of the drillings was set two meters below the landslide mass–marl interface. The new boreholes Nos 1–8 and the 11 previous borehole records were used for the building of a 3D geological model applying GeoModeller software (Figure 1c,d). Further piezometers were installed in the new boreholes 1–7 (Figure 2a) equipped with barometric sounds measuring groundwater heads at a frequency of 15 min.

The measurements of the groundwater heads are available for the time period from 7 March 2018 to 1 April 2019. The measurements were extended over a period of one year to consider seasonal fluctuation of the groundwater table.

Injection tests were conducted in boreholes No 3 and 5 (Figure 2). These boreholes were selected because preferential flow was expected there due to a loud noise of flowing water and almost constant observed groundwater heads. Water was pumped into the boreholes at an average rate of 0.8 L/sec; however, no increase of the groundwater table was attained.

Two geophysical techniques, ERT and self-potential mapping, were used for exploration of the hydrogeological settings of the study site. The ERT was used for the location of groundwater bodies, while self-potential mapping was selected for identification of groundwater flow in the subsurface. These two geoelecterical methods are complimentary and have been successfully applied for hydrogeophysical exploration [35].

The basic principle of ERT can be explained through four electrode measurements; namely, an electrical current is injected into the ground through two electrodes and voltage is measured between the second two [34]. An apparent resistivity of the ground can be calculated from applied current and measured voltage between the electrodes. When changing the arrangement of the electrodes, characteristics such as penetration depth, resolution, accuracy etc. can be varied [37]. An inversion procedure is applied to convert the pseudosection into a true resistivity section for a heterogeneous subsurface [34,37].

As the charge carrier in most soils is electrolytic groundwater and mineral–grain interfaces, ERT can be effectively applied for the identification of groundwater in the subsurface [24,34,35]. In the study site of Deisermillen, an ERT survey with 13 profiles was conducted using a LIPPMANN 4-point light 10 W device. Using a maximum number of 50 electrodes with 2 m electrode spacing and a maximum profile length of 100 m allowed the attainment of approximately 15 m penetration depth. However, the length and number of electrodes varied because blocks of rock or intense vegetation crossed the way. The position of the first profile's start and end points was georeferenced using a GNSS receiver (HiPer HR Topcon Deutschland Positioning GmbH, Hamburg, Germany). Horizontal accuracy of the GNSS measurement did not exceed 1 m during the fieldwork. A further 12 additional profiles were aligned parallel to the first profile using measuring tape, wooden sticks, and wires. The ERT profiles were orientated perpendicularly to the expected direction of groundwater flow, and 4 m spacing was set between the profiles. The surveys were conducted using Wenner electrode allocation due to a high signal-to-noise ratio [37]. The ZondRes3D software was used for creating a joint 3D ERT model from 13 profiles. Before inversion data were processed with the ZondRes Quality control module, unrealistic data points were removed. Ten inversion iterations were applied to maximally reduce the difference between observed and calculated apparent resistivity.

Streaming potential is one of the sources causing self-potential signals, which are triggered by the movement of electrolytic groundwater in porous media [34]. The magnitude of the streaming potential can range up to 100 mV [37]. Moreover, in unconfined aquifers along the preferential flow path, a negative anomaly can be observed in the upstream, and a positive anomaly can be observed in the downstream part of the flow path [18,34]. When conducting the self-potential measurements, the base electrode is commonly kept in a fixed location as a reference, while the removable electrode is replaced along a specified raster [34]. Initially, self-potential mapping with a coarse raster of 2 × 5 m was conducted for the identification of locations with self-potential anomalies in the study area. Further in the location of anomalies, measurements were repeated with a refined raster of 2 × 2 m. A regular measurement raster composed of several separated plots was marked in the field using measuring tape, wires, and wooden sticks. Furthermore, the corner points of the each plot were georeferenced using a GNSS receiver. Two Ag/AgCl non-polarizing electrodes and a high impedance voltmeter with a sensitivity of 0.1 mV were used. The repeatability of the self-potential measurements in the locations of the anomalies was validated over a period of one year, and regular measurements showed that the anomalies remain stable.

A 3D geological model was developed to provide a geological context for the interpretation of geophysical measurements. GeoModeller software was employed for the building of a 3D geological model applying 19 boreholes. Layer interfaces from the 3D geological model were further used for the building of a hydrogeological model. The 3D hydrogeological model with a focus on preferential flow was developed (Figure 3). The aim of the model was to find a physically-based explanation for observed groundwater heads by applying different subsurface flow mechanisms. When only flow in porous media was simulated, we were not able to explain the observed groundwater flow in the strata of less conductive marl; therefore, preferential flow was considered in the model's simulations. A discrete fracture modeling approach applying the HydroGeoSphere software was used. The HydroGeoSphere (HGS) is a 3D fully-integrated surface and subsurface flow simulator [45,46], and with this, surface flow and subsurface flow in porous, fractured, and dual media for both saturated and unsaturated conditions can be modeled [45]. This allows for the application of the HGS code for advanced simulations in the vadose zone [47–49]. A detailed explanation of the governing equations used in the HGS code are available in the HGS user manual [46] and in a publication by Brunner and Simmons [45].

In the hydrogeological simulation, the recharging of groundwater in the model's domain occurs through the first-type boundary condition of the rainfall rate, which is applied to the model's surface domain (Figure 3a). Initially, the average precipitation amounts were applied for a period of several years, which allowed for the saturation of the model's domain. Furthermore, daily precipitation amounts observed by the meteorological observation station of Nittel from 7 March 2018 to 1 April 2019 were used in the model [50]. The first-type (Dirichlet) boundary condition of the prescribed hydraulic head was applied in the eastern part of the model's domain to consider the infiltration of groundwater from the Mosel River, and also in the western part, which was equal with the groundwater table in borehole No 6 to consider the afflux of groundwater from the carstic network in the eastern part. We assume that the discharge of groundwater from the carstic network in the dolomite cliff occurs in a narrow zone where springs occur (Figure 2a). To the outer boundary of the model's domain, a critical depth boundary condition was applied, which allowed water to leave the model's domain without influencing the upstream groundwater heads or surface water depths [46]. The morphology of the preferential flowpath in the model was assumed to be a 7 mm thin, lateral conduit dissecting borehole No 5 where intense preferential flow was located (Figure 3c). The width of the conduit was arbitrarily assumed. The conduit was laterally embedded in the marl strata until it reached the marl–landslide mass interface. Furthermore, the conduit followed the interface allowing the drainage of water toward Mosel River. The model's structural sublayering was adopted to the above explained conceptual model of the flowpath (Figure 3b). A discrete fracture modeling approach was employed for the simulation of groundwater flow in the preferential flowpath. An exchange between the fracture and the encompassing matrix was considered in the HydroGeoSphere simulation, meaning that the saturation of the fracture can vary depending on afflux from the encompassing porous media [46].

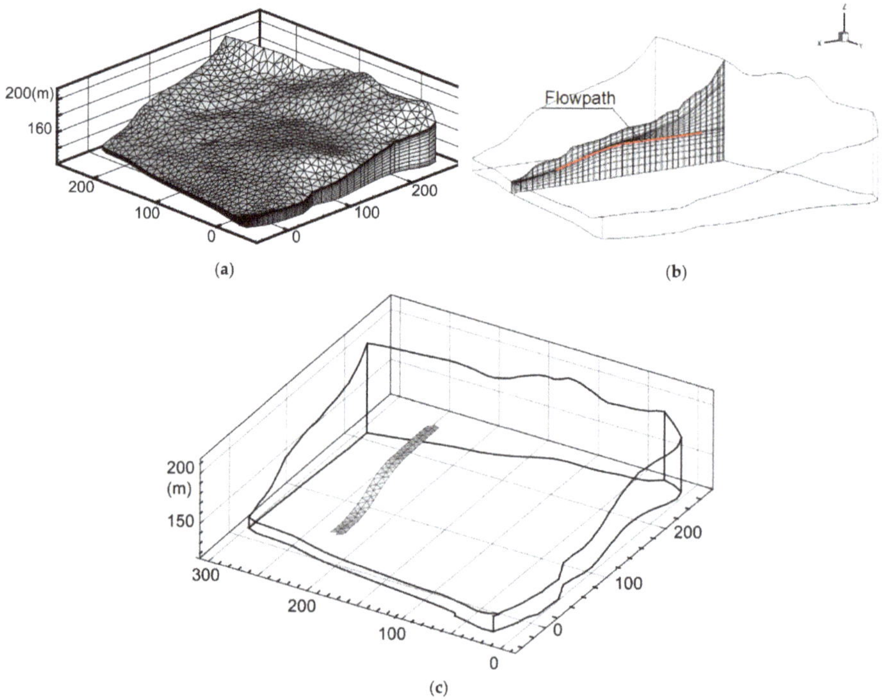

Figure 3. (**a**) The structure of the hydrogeological model, (**b**) the model's sub layering, (**c**) morphology of the lateral flowpath.

3. Results

From a hydrogeological viewpoint, the two most conductive geological units in the study site are the carstic dolomites and the overlaying conductive landslide mass. However, field investigation revealed that in the strata of less conductive marls highly permeable conduits exist. Figure 4 illustrates that two systems of groundwater flow are present, namely one above the less conductive marls and another in the marl sediments, which itself is represented by boreholes 3 and 5.

Boreholes 3, 5, and partially also 4 and 6 have an untypical temporal variability of their hydraulic heads, because independently of precipitation, their hydraulic heads remain constant. This is especially emphasized in borehole No 5. In boreholes No 3 and 5 (Figure 4), a loud noise of rapidly flowing water can be heard. Further pumping tests were conducted in these boreholes to get detailed insights about the possible causes of such an untypical behavior of groundwater in the boreholes. An injection test with a flowrate of 0.8 L/s was conducted in boreholes No. 3 and 5; however, no increase of the groundwater head could be observed there. This indicates that very highly conductive conduits in the low permeable deposits of the marls exist. This phenomenon is also confirmed by the previous borehole records available in the archive of the Geological Survey of Luxembourg [40]. Namely, in borehole S7 at a depth of 24.5 to 25.10 m and in borehole S6 (Figure 2a) at a depth of 18.6 to 19.25 m, a cavity was identified [40].

Unfortunately, no detailed description of these cavities is present. Drawings of the cores indicate that the cavities are located in the proximity of gypsiferous interlayers. Therefore, it can be speculated that the preferential flowpaths in the new boreholes No 3 and 5 are also caused by dissolution of the gypsiferous interlayers. After a detailed analysis of the new drillcore records in borehole No 5, a thin gypsum interlayer was discovered at the same depth where preferential groundwater flow occurs (Figure 5). For detailed insight into the geological settings, a geological section through borehole No 5 was drawn (Figure 2b).

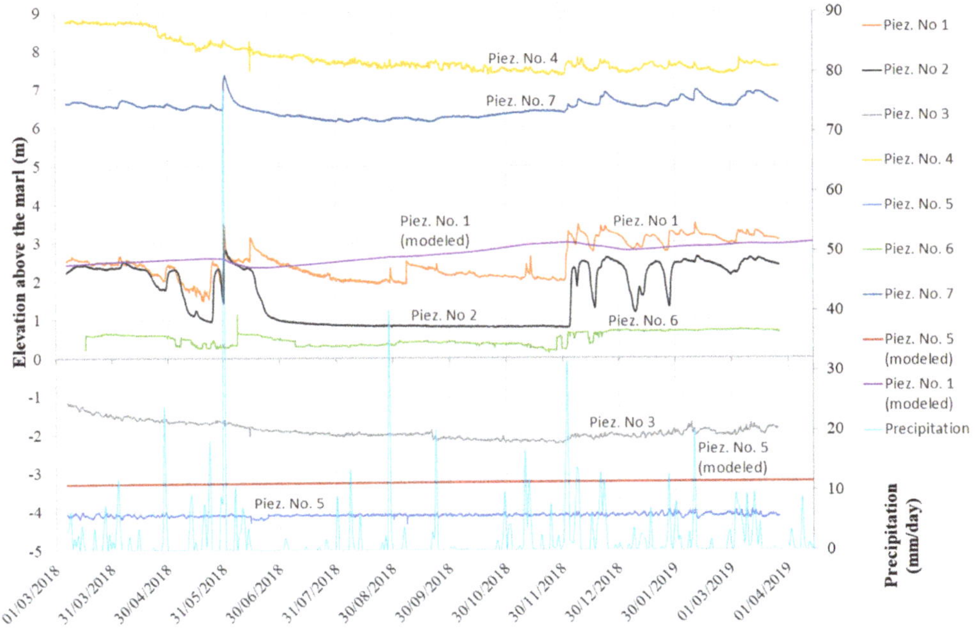

Figure 4. Temporal changes of the groundwater heads in piezometer wells Nos 1–7 from 7 March 2018 to 1 April 2019. 0-elevation corresponds with the surface of the marl.

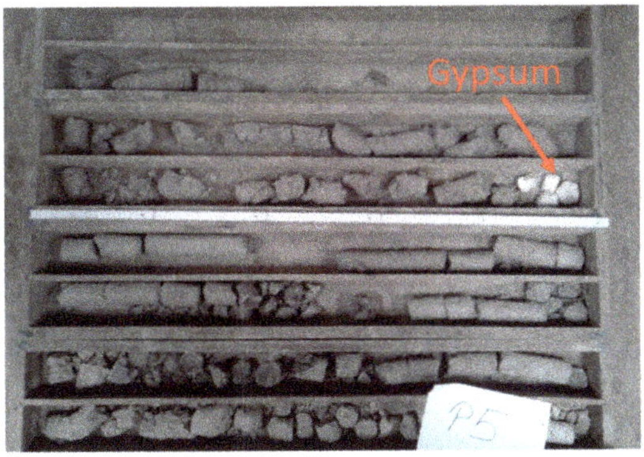

Figure 5. Gypsum interlayer in the core records of borehole No 5.

Both ERT and self-potential mapping were used to obtain detailed information about the preferential flowpaths in the study area. The ERT was employed for the identification of groundwater bodies caused by groundwater discharging from the carstic network and flowing downslope through the landslide mass to the Mosel River. The self-potential mapping was used for the identification of streaming potential caused by preferential groundwater flow.

The results of the self-potential mapping are explicitly illustrated in Figure 6. Relatively homogenous self-potential values can be observed in the northern part of the study area; however, downslope from the discharge area of groundwater springs, emphasized self-potential anomalies are present. Several couples of positive and negative self-potential anomalies can be observed in the study site (Figure 6). The maximum magnitude of the self-potential signal exceeds approximately −103/+98 mV. The polarity, magnitude, and orientation of the self-potential anomalies indicate that they might be caused by the streaming potential. In Figure 6, the main self-potential anomalies are marked by blue arrows. They are orientated in the direction of the expected groundwater flow with a positive polarity downslope. An exception is the anomaly in the southern part, which is probably associated with a groundwater flowpath adjoining the main pathway. A closer look at the subsurface morphology of the less conductive marl indicates a local depression that corresponds with the direction of the supposed streaming potentials. It can be speculated that the surface of the marl constrains groundwater flow, which corresponds with the direction of the measured streaming potentials.

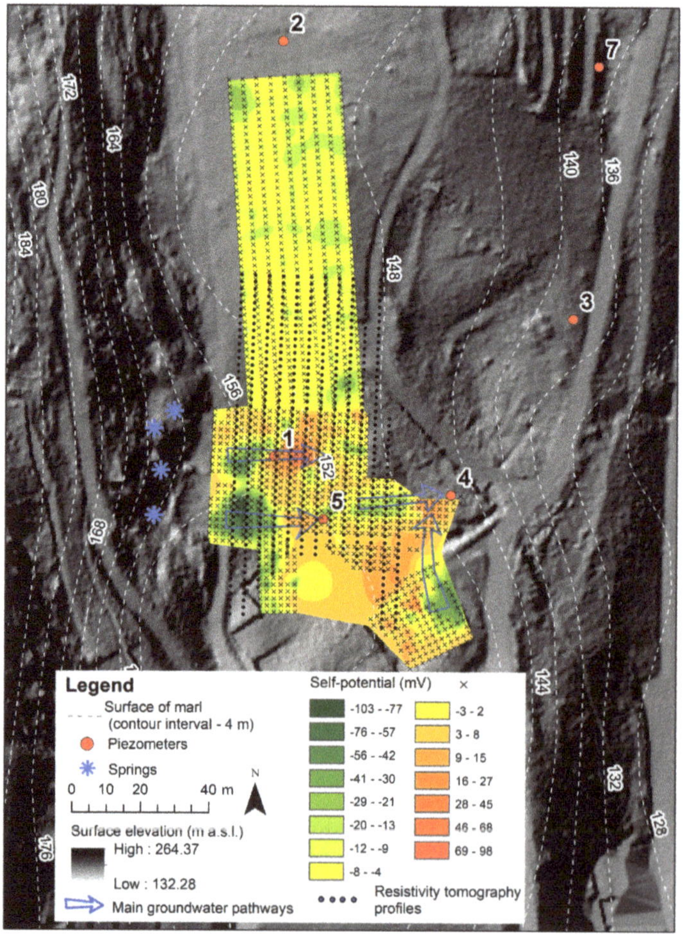

Figure 6. Location self-potential mapping in the study area.

The intense preferential flow within the marls, as revealed in boreholes No 3 and 5, may not cause the streaming potential anomalies. Possibly self-potential signals occur due to groundwater flow in the conductive landslide mass directed by the topography of the underlain marl surface.

When considering the ERT model, highly conductive zones can be observed in deeper sections of the ERT survey while low conductivities can be observed in the uppermost part (Figure 7). Moreover, the less conductive zones are more fragmented, as they may perhaps represent landslide mass in the unsaturated zone with large spaces between the dolomite blocks resulting in increased electrical resistivity. Highly conductive zones are better connected as they may be related through groundwater pathways (Figure 7a,c).

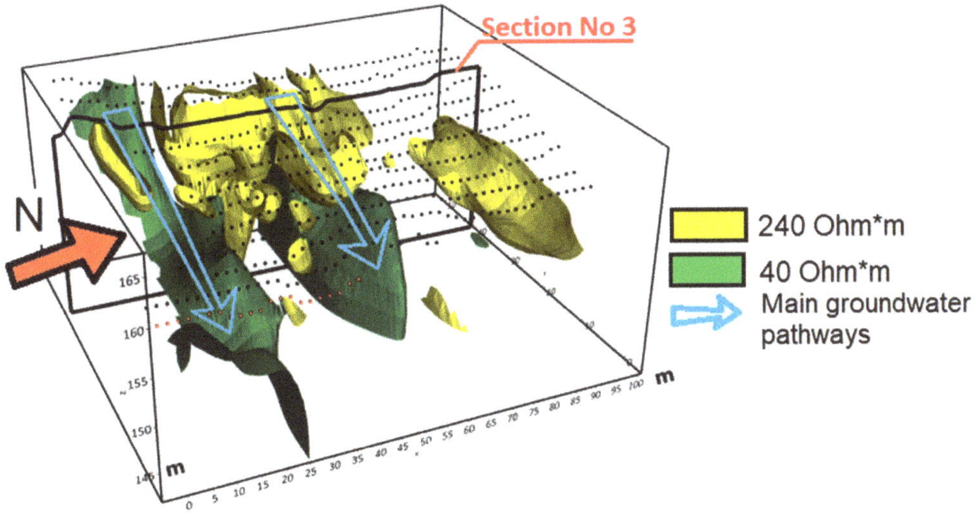

(a)

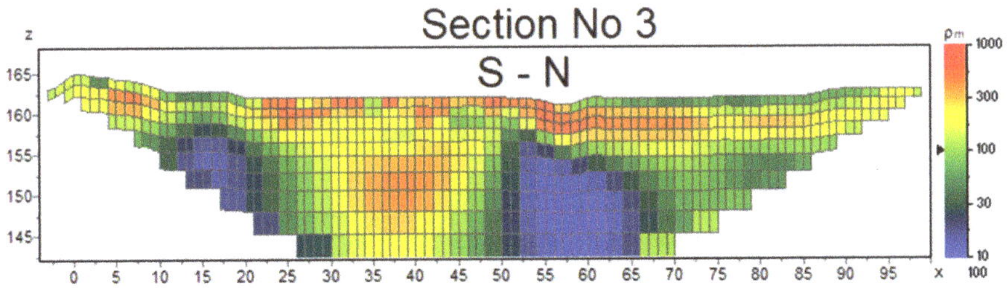

(b)

Figure 7. *Cont.*

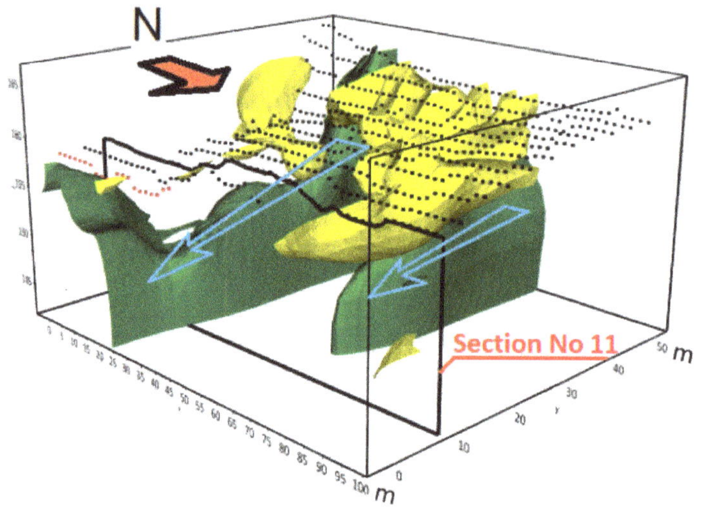

(**c**)

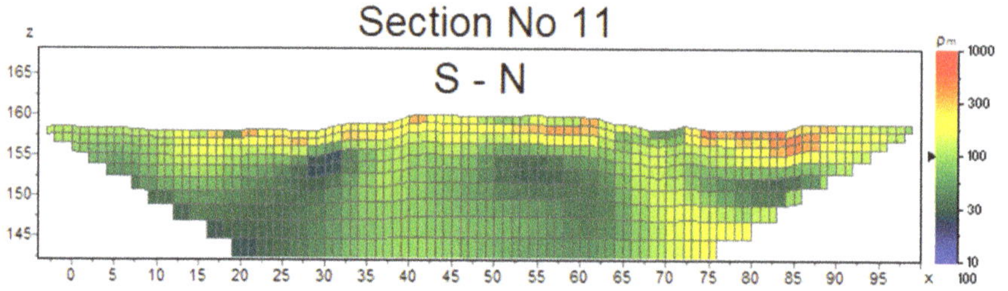

(**d**)

Figure 7. Three-dimensional (3D) high and low resistivity zones in the subsurface created through the inversion of 13 ERT profiles. The 3D model in view from (**a**) SE and (**c**) NE (the black dots represent the location of each electrode) and two characteristic sections (**b**) No 3 and (**d**) No 11 of the 3D model.

A 3D hydrogeological model of the study site was compiled for the time period from 7 March 2018 to 1 April 2019 when the results of groundwater monitoring were available. This 3D model considering preferential flow was compiled for the simulation of groundwater flow in the conductive marl. The hydrogeological model considered daily precipitation rates and saturated and unsaturated flow both in porous and fractured media.

In the hydrogeological model, a 7 mm thin lateral conduit was embedded in the layer of marls intersecting borehole No 5 where the most pronounced preferential groundwater flow was observed (Figure 2b).

Modeled groundwater heads in two observation points corresponding with boreholes No 1 and 5 are illustrated in Figure 4. In borehole No 1, flow in porous media can be observed; both observed and simulated heads indicate response to precipitation events, which is not the case in borehole No 5 where preferential flow occurs. The groundwater simulation with the thin laterally extending conduit allowed for the attainment of an exact correspondence between the modeled and observed groundwater heads in piezometer well No 5, where preferential flow occurs.

4. Discussion

This study was conducted in a study site in the valley of Upper Mosel, Luxembourg, where mass movements are frequently present. The aim of the study was to explore the hydrogeological conditions of a slope affected by a recent landslide.

Observations of the groundwater heads in piezometer wells indicated very diverse responses to changing precipitation in the study site. The almost constant hydraulic heads with high flow rates in two boreholes could be explained through preferential flow in the marl strata. The dissolution of gypsum could have caused thin, lateral conduits in the marl deposits. In boreholes No 3 and 5, constant groundwater heads are sustained because water discharges into these conduits.

In the inclined borehole No 6, which was installed in the dolomite cliff (Figure 2b), alternation between the two governing groundwater heads can be observed (Figure 4). This phenomenon could be explained through presence of two fracture planes dissecting the piezometer's screen. Possibly if the conductivity of the lowest conduit is relatively low and high inflow of groundwater occurs, the level of the next lateral conduit is reached where it remains stable until the water inflow decreases.

Hydrogeological field testing and observations in boreholes at the field site also revealed that preferential flowpaths have high flow capacity. However, the piezometer observations delivered no information about the morphology and spatial extension of preferential flowpaths.

The newest 3D geophysical surveys are promising techniques for the investigation of the morphology of subsurface flowpaths. The 3D ERT indicated possible locations of saturated groundwater bodies associated with preferential flow heading toward the Mosel River; however, the resistivity method was not able to detect increased flowrate in the subsurface, which is characteristic of preferential flow. The self-potential method was employed for the measurement of streaming potentials associated with preferential flow into the subsurface. When integrating both geophysical methods—ERT and self-potential—conducted in the study area, it can be observed that the main groundwater pathways toward the Mosel River start from spring discharges (Figures 6 and 7). Both geophysical methods reveal self-potential anomalies and highly conductive zones in the same locations. Two main groundwater pathways starting from spring discharge could be recognized both in the ERT and self-potential surveys. The pathway in the northern part disappears from geophysical surveys; perhaps it diverges or moves toward deeper layers in the marls. The pathway in the southern part continues in both surveys through the study area toward the Mosel River.

The aforementioned field research allowed the measurement of signals associated with preferential flow such as steaming potentials and the increase of electrical conductivity along the flowpath; however, possible signals of flowing groundwater are diffusely spread over a relatively large area. Despite intensive testing, the morphology of the flowpath remained unclear. A detailed analysis of the drillcores indicated that a possible cause of intense preferential flow may be the dissolution of gypsum interlayers in the marls of Middle Muschelkalk. Thin, laterally extended conduits, caused by the dissolution of the gypsum, are proposed as the most likely morphological shapes of preferential flow. To validate this hypothesis, a 3D hydrogeological model considering flow both in porous and fractured media was developed. This model with a lateral preferential flow path embedded in a porous matrix (Figure 3c) provides an excellent correspondence with observations in piezometers where preferential flow is present. Nevertheless, many unknowns still remain such as lateral extension, aperture, or the discharge location of the flowpath. Furthermore, the impact of the hydrogeological settings on the stability of the slope is unclear and will be investigated in a separate article.

In the future, advanced field testing with a higher injected water amount could allow for the determination of flow capacity in the flowpath. Field tests integrated with a numerical groundwater model would allow for the estimation of the aperture of the flowpath. However, an injection test should be conducted with particular caution because the injection of high water amounts in a landslide prone slope could cause the reactivation of the landslide. Recent advances in the modeling of the self-potential signals could allow estimating the groundwater seepage velocity, which can be used for validation of the hydrogeological model.

This study shows that an integrated research program including hydrogeological monitoring, hydrogeological field testing, and geological and geophysical exploration could be necessary for the exploration of preferential groundwater flow. Despite recent advances in geophysical exploration, the determination of an exact morphology for preferential flowpaths is not straightforward. The newest hydrogeological models are able to conduct fully integrated simulations of preferential groundwater flow. By adopting the hydrogeological model to observations, the site's hydrogeological settings can be explored in a very detailed scale. Thus, numerical models allow for better conceptualization and parametrization of preferential flowpaths in the subsurface.

5. Conclusions

Hydrogeological investigation methods in combination with geophysical prospecting were used for the investigation of the hydrogeological settings of a landslide-prone slope in the Upper Mosel valley.

Observations in piezometer wells show that in the Deisermillen study site, two groundwater flow systems are present: the first being groundwater flow above the less permeable marls and the second located within the marl sediments where preferential flow occurs. Hydrogeological observations in piezometer wells and hydrogeological testing revealed an intense preferential flow in the less conductive marls, which was not expected to occur there. Nevertheless, observations in boreholes are 2D measurements which are not suitable for the complete exploration of heterogeneous groundwater flow conditions.

A geophysical investigation was also conducted for the investigation of the spatial extension of flowpaths in the subsurface. By applying ERT and self-potential mapping techniques, preferential flow could have been identified in a zone of the hillslope where groundwater from the carstic network enters the landslide mass cover and flows toward the Mosel River. However, it is not entirely clear which preferential flow system is the source of the measured self-potential anomalies because the signals are widely spread over a large area. As the interpretation of the geophysical measurements was not straightforward, assumptions based on the geological data were necessary for the conceptualization of the preferential flowpaths in the numerical model.

The hydrogeological model allowed for the validation of the assumed morphology of a thin, laterally extending preferential flowpath through field observations in piezometer wells.

The results of this study show that the identification and parametrization of the preferential flow using conventional techniques such as hydrogeological testing and monitoring can be ambiguous. Other data sources, such as knowledge about the local geological settings and numerical modeling could be necessary for better conceptualization and parametrization of preferential flow in the subsurface.

Integrated application of models, for instance geophysical techniques used for the validation of the hydrogeological simulations, can help to explore the preferential flow. Applying inverted streaming potential signals that the permeability and seepage velocity of the preferential flowpaths can be derived [33]. In the study area of Deisermillen, modeling of the streaming potential signals can be used for validation of the hydrogeological model during the subsequent research.

Author Contributions: Conceptualization, K.K. and J.-F.W.; methodology, K.K.; software, K.K.; validation, K.K. and J.-F.W.; formal analysis, K.K. and J.-F.W.; investigation, K.K.; resources, J.-F.W.; data curation, K.K. and J.-F.W.; writing—original draft preparation, K.K.; writing—review and editing, J.-F.W.; visualization, K.K.; supervision, J.-F.W.; project administration, K.K.; funding acquisition, J.-F.W. All authors have read and agreed to the published version of the manuscript.

Funding: This research received no external funding.

Institutional Review Board Statement: Not applicable.

Informed Consent Statement: Not applicable.

Data Availability Statement: Publicly available datasets were analyzed in this study. This data can be found here: https://map.geoportail.lu/theme/main?lang=en&version=3&zoom=14&X=716922&Y=6389557&rotation=0&layers=1788-314-206&opacities=0.75-1-1&bgLayer=blank (accessed on 15 April 2021), https://www.agrimeteo.lu/__c12577b4003e3132.nsf/0/d6d6c28de01e95c7c1257dc600436335?OpenDocument&TableRow=10.1.1#10.1 (accessed on 15 October 2020). Other data sources presented in this study are available on request from the corresponding author. The data are not publicly available because they were provided by a private company.

Acknowledgments: We would like to thank the staff members of the company EURASOL S.A. for their engagement and patience during the field work. Especially, we would like to acknowledge the contribution of director Robert Heintz who sponsored drilling works of eight boreholes in the study area and provided us with groundwater monitoring data over a period more than one year long. We would also like to thank geomorphology students of University of Trier, especially Mathias Jackel, Judith Koschorke, Stefan Habetz, Pascal Wintrich, and technical assistant Yannick Hausener, who helped during geophysical surveys and Robert Colbach and Birgit Kausch for supporting this study with archive materials. Finally, we would like to acknowledge researchers of the Centre for Hydrogeology and Geothermics of University of Neuchâtel, especially Professor Philip Brunner for the help with the modeling software.

Conflicts of Interest: The authors declare no conflict of interest. The funders had no role in the design of the study; in the collection, analyses, or interpretation of data; in the writing of the manuscript, or in the decision to publish the results.

References

1. Clothier, B.E.; Green, S.R.; Deurer, M. Preferential flow and transport in soil: Progress and prognosis. *Eur. J. Soil Sci.* **2008**, *59*, 2–13. [CrossRef]
2. Tsang, C.-F.; Neretnieks, I.; Tsang, Y. Hydrologic issues associated with nuclear waste repositories. *Water Resour. Res.* **2015**, *51*, 6923–6972. [CrossRef]
3. Birkholzer, J.T.; Mukhopadhyay, S.; Tsang, Y.Y.W. The Impact of Preferential Flow on the Vaporization Barrier Above Waste Emplacement Drifts at Yucca Mountain, Nevada. *Nucl. Technol.* **2004**, *148*, 138–150. [CrossRef]
4. Bolève, A.; Janod, F.; Revil, A.; Lafon, A.; Fry, J.-J. Localization and quantification of leakages in dams using time-lapse self-potential measurements associated with salt tracer injection. *J. Hydrol.* **2011**, *403*, 242–252. [CrossRef]
5. Ikard, S.J.; Rittgers, J.; Revil, A.; Mooney, M.A. Geophysical Investigation of Seepage Beneath an Earthen Dam. *Groundwater* **2015**, *53*, 238–250. [CrossRef]
6. Kukemilks, K.; Wagner, J.-F.; Saks, T.; Brunner, P. Conceptualization of preferential flow for hillslope stability assessment. *Hydrogeol. J.* **2018**, *26*, 439–450. [CrossRef]
7. Shao, W.; Bogaard, T.A.; Bakker, M.; Greco, R. Quantification of the influence of preferential flow on slope stability using a numerical modelling approach. *Hydrol. Earth Syst. Sci.* **2015**, *19*, 2197–2212. [CrossRef]
8. Shao, W.; Bogaard, T.; Bakker, M.; Berti, M. The influence of preferential flow on pressure propagation and landslide triggering of the Rocca Pitigliana landslide. *J. Hydrol.* **2016**, *543*, 360–372. [CrossRef]
9. Krzeminska, D.M.; Bogaard, T.A.; van Asch, T.W.J.; van Beek, L.P.H. A conceptual model of the hydrological influence of fissures on landslide activity. *Hydrol. Earth Syst. Sci.* **2012**, *16*, 1561–1576. [CrossRef]
10. Bogaard, T.A.; Greco, R. Landslide hydrology: From hydrology to pore pressure. *WIREs Water* **2016**, *3*, 439–459. [CrossRef]
11. Lu, N.; Godt, J.W. *Hillslope Hydrology and Stability*; Cambridge University Press: Cambridge, UK, 2013.
12. Noguchi, S.; Tsuboyama, Y.; Sidle, R.C.; Hosoda, I. Morphological Characteristics of Macropores and the Distribution of Preferential Flow Pathways in a Forested Slope Segment. *Soil Sci. Soc. Am. J.* **1999**, *63*, 1413–1423. [CrossRef]
13. Harp, E.L.; Wells, W.G., II; Sarmiento, J.G. Pore pressure response during failure in soils. *GSA Bull.* **1990**, *102*, 428–438. [CrossRef]
14. Nieber, J.L.; Warner, G.S. Soil pipe contribution to steady subsurface stormflow. *Hydrol. Processes* **1991**, *5*, 329–344. [CrossRef]
15. Stadler, L.; Hinkelmann, R.; Helmig, R. Modeling Macroporous Soils with a Two-Phase Dual-Permeability Model. *Transp. Porous Media* **2012**, *95*, 585–601. [CrossRef]
16. Beven, K.; Germann, P. Macropores and water flow in soils revisited. *Water Resour. Res.* **2013**, *49*, 3071–3092. [CrossRef]
17. Ellefsen, K.J.; Hsieh, P.A.; Shapiro, A.M. Crosswell seismic investigation of hydraulically conductive, fractured bedrock near Mirror Lake, New Hampshire. *J. Appl. Geophys.* **2002**, *50*, 299–317. [CrossRef]
18. Revil, A.; Cary, L.; Fan, Q.; Finizola, A.; Trolard, F. Self-potential signals associated with preferential ground water flow pathways in a buried paleo-channel. *Geophys. Res. Lett.* **2005**, *32*. [CrossRef]

19. Uhlemann, S.; Kuras, O.; Richards, L.A.; Naden, E.; Polya, D.A. Electrical resistivity tomography determines the spatial distribution of clay layer thickness and aquifer vulnerability, Kandal Province, Cambodia. *J. Asian Earth Sci.* **2017**, *147*, 402–414. [CrossRef]
20. Yeh, H.-F.; Lin, H.-I.; Wu, C.-S.; Hsu, K.-C.; Lee, J.-W.; Lee, C.-H. Electrical resistivity tomography applied to groundwater aquifer at downstream of Chih-Ben Creek basin, Taiwan. *Environ. Earth Sci.* **2015**, *73*, 4681–4687. [CrossRef]
21. Pidlisecky, A.; Moran, T.; Hansen, B.; Knight, R. Electrical Resistivity Imaging of Seawater Intrusion into the Monterey Bay Aquifer System. *Groundwater* **2016**, *54*, 255–261. [CrossRef]
22. Colombero, C.; Baillet, L.; Comina, C.; Jongmans, D.; Vinciguerra, S. Characterization of the 3-D fracture setting of an unstable rock mass: From surface and seismic investigations to numerical modeling. *J. Geophys. Res. Solid Earth* **2017**, *122*, 6346–6366. [CrossRef]
23. National Research Council. *Rock Fractures and Fluid Flow: Contemporary Understanding and Applications*; The National Academies Press: Washington, DC, USA, 1996; p. 568.
24. Roubinet, D.; Irving, J. Discrete-dual-porosity model for electric current flow in fractured rock. *J. Geophys. Res. Solid Earth* **2014**, *119*, 767–786. [CrossRef]
25. Hubbard, S.S.; Linde, N. 2.15—Hydrogeophysics. In *Treatise on Water Science*; Wilderer, P., Ed.; Elsevier: Oxford, UK, 2011; pp. 401–434.
26. Roubinet, D.; Linde, N.; Jougnot, D.; Irving, J. Streaming potential modeling in fractured rock: Insights into the identification of hydraulically active fractures. *Geophys. Res. Lett.* **2016**, *43*, 4937–4944. [CrossRef]
27. Jougnot, D.; Jiménez-Martínez, J.; Legendre, R.; Le Borgne, T.; Méheust, Y.; Linde, N. Impact of small-scale saline tracer heterogeneity on electrical resistivity monitoring in fully and partially saturated porous media: Insights from geoelectrical milli-fluidic experiments. *Adv. Water Resour.* **2018**, *113*, 295–309. [CrossRef]
28. Blazevic, L.A.; Bodet, L.; Pasquet, S.; Linde, N.; Jougnot, D.; Longuevergne, L. Time-Lapse Seismic and Electrical Monitoring of the Vadose Zone during a Controlled Infiltration Experiment at the Ploemeur Hydrological Observatory, France. *Water* **2020**, *12*, 1230. [CrossRef]
29. Fournier, C. Spontaneous potentials and resistivity surveys applied to hydrogeology in a volcanic area: Case history of the Chaîne des Puys (Puy-de-Dôme, France). *Geophys. Prospect.* **1989**, *37*, 647–668. [CrossRef]
30. Birch, F.S. Imaging the Water Table by Filtering Self-Potential Profiles. *Groundwater* **1998**, *36*, 779–782. [CrossRef]
31. Perrier, F.E.; Petiau, G.; Clerc, G.; Bogorodsky, V.; Erkul, E.; Jouniaux, L.; Lesmes, D.; Macnae, J.; Meunier, J.M.; Morgan, D.; et al. A One-Year Systematic Study of Electrodes for Long Period Measurements of the Electric Field in Geophysical Environments. *J. Geomagn. Geoelectr.* **1997**, *49*, 1677–1696. [CrossRef]
32. Soueid Ahmed, A.; Revil, A.; Steck, B.; Vergniault, C.; Jardani, A.; Vinceslas, G. Self-potential signals associated with localized leaks in embankment dams and dikes. *Eng. Geol.* **2019**, *253*, 229–239. [CrossRef]
33. Ahmed, A.S.; Revil, A.; Bolève, A.; Steck, B.; Vergniault, C.; Courivaud, J.R.; Jougnot, D.; Abbas, M. Determination of the permeability of seepage flow paths in dams from self-potential measurements. *Eng. Geol.* **2020**, *268*, 105514. [CrossRef]
34. Everett, M.E. (Ed.) *Near-Surface Applied Geophysics*; Cambridge University Press: Cambridge, UK, 2013.
35. Revil, A.; Jardani, A. *The Self-Potential Method: Theory and Applications in Environmental Geosciences*; Cambridge University Press: Cambridge, UK, 2013.
36. Jouniaux, L.; Maineult, A.; Naudet, V.; Pessel, M.; Sailhac, P. Review of self-potential methods in hydrogeophysics. *Comptes Rendus Geosci.* **2009**, *341*, 928–936. [CrossRef]
37. Knödel, K.; Krummel, H.; Lange, G. (Eds.) *Geophysik*; Springer: Berlin/Heidelberg, Germany, 1997.
38. Heyl, K.-E. *Ein Hangrutsch bei Nittel (Obermosel). Notizenblatt des Hessischen Landesamtes für Bodenforschung zu Wiesbaden*; Hessian Agency for the Environment and Geology: Wiesbaden, Germany, 1971; Volume 99, pp. 347–360.
39. Bintz, J. *Raport Géologique Concerne: Situation Geologiques au Glissement "Deysermühle"*; Service Géologique, Administration des Ponts et Chaussées: Luxembourg, 1965.
40. Johan Keller GmbH. *Sanierung Rutschung Deysermühle. Aufschluß und Pegelbohrungen*; Service Géologique, Administration des Ponts et Chaussées: Luxembourg, 1974.
41. *OpenStreetMap Contributors, and the GIS User Community*; World Street Map, Esri: Redlands, CA, USA, 2017.
42. Service Géologique, Administration des Ponts et Chaussées. *Droits Réservés à l'Etat du Grand-Duché de Luxembourg*; Service Géologique, Administration des Ponts et Chaussées: Luxembourg, 2020.
43. Administration du Cadastre et de la Topographie. Droits Réservés à l'Etat du Grand-Duché de Luxembourg. Available online: https://map.geoportail.lu/theme/main?lang=en&version=3&zoom=14&X=716922&Y=6389557&rotation=0&layers=1788-314-206&opacities=0.75-1-1&bgLayer=blank (accessed on 15 April 2021).
44. Neumann-Redlin, C. *Hydrogeologische und Hydrochemische Untersuchungen im Oberen Muschelkalk und Keuper Luxemburgs. Eberhard-Karls-Universität zu Tübingen*; Service Géologique du Luxembourg: Luxembourg, 1971; Volume 22, pp. 1–87.
45. Brunner, P.; Simmons, C.T. HydroGeoSphere: A Fully Integrated, Physically Based Hydrological Model. *Groundwater* **2012**, *50*, 170–176. [CrossRef]
46. Aquanty Inc. *HydroGeoSphere: User Manual, Release 1.0*; Aquanty Inc.: Waterloo, ON, Canada, 2013.
47. Banks, E.W.; Brunner, P.; Simmons, C.T. Vegetation controls on variably saturated processes between surface water and groundwater and their impact on the state of connection. *Water Resour. Res.* **2011**, *47*. [CrossRef]

48. Schilling, O.S.; Doherty, J.; Kinzelbach, W.; Wang, H.; Yang, P.N.; Brunner, P. Using tree ring data as a proxy for transpiration to reduce predictive uncertainty of a model simulating groundwater–surface water–vegetation interactions. *J. Hydrol.* **2014**, *519*, 2258–2271. [CrossRef]
49. Xie, Y.; Cook, P.G.; Brunner, P.; Irvine, D.J.; Simmons, C.T. When Can Inverted Water Tables Occur Beneath Streams? *Groundwater* **2014**, *52*, 769–774. [CrossRef] [PubMed]
50. Agrarmeteorologie Luxemburg. Wetterstation Nittel. Available online: https://www.agrimeteo.lu/__c12577b4003e3132.nsf/0/d6d6c28de01e95c7c1257dc600436335?OpenDocument&TableRow=10.1.1#10.1 (accessed on 15 October 2020).

applied sciences

MDPI

Article

Uncertainty of the 2D Resistivity Survey on the Subsurface Cavities

Yonatan Garkebo Doyoro [1,2], Ping-Yu Chang [2,3,*] and Jordi Mahardika Puntu [2]

[1] Earth System Science, Taiwan International Graduate Program (TIGP), Academia Sinica, Taipei 115, Taiwan; yonatangrkb978@gmail.com
[2] Department of Earth Science, National Central University, Taoyuan 320, Taiwan; jordipuntu@gmail.com
[3] Earthquake-Disaster, Risk Evaluation and Management Centre, National Central University, Taoyuan 320, Taiwan
* Correspondence: pingyuc@gmail.com; Tel.: +866-3-4227151

Abstract: We examined the uncertainty of the two-dimensional (2D) resistivity method using conceptual cavity models. The experimental cavity study was conducted to validate numerical model results. Spatial resolution and sensitivity to resistivity perturbations were also assessed using checkerboard tests. Conceptual models were simulated to generate synthetic resistivity data for dipole-dipole (DD), pole-dipole (PD), Wenner–Schlumberger (WS), and pole-pole (PP) arrays. The synthetically measured resistivity data were inverted to obtain the geoelectric models. The highest anomaly effect (1.46) and variance (24,400 Ω·m) in resistivity data were recovered by the DD array, whereas the PP array obtained the lowest anomaly effect (0.60) and variance (2401 Ω·m) for the shallowest target cavity set at 2.2 m depth. The anomaly effect and variance showed direct dependency on the quality of the inverted models. The DD array provided the highest model resolution that shows relatively distinct anomaly geometries. In contrast, the PD and WS arrays recovered good resolutions, but it is challenging to determine the correct anomaly geometries with them. The PP array reproduced the lowest resolution with less precise anomaly geometries. Moreover, all the tested arrays showed high sensitivity to the resistivity contrasts at shallow depth. The DD and WS arrays displayed the higher sensitivity to the resistivity perturbations compared to the PD and PP arrays. The inverted models showed a reduction in sensitivity, model resolution, and accuracy at deeper depths, creating ambiguity in resistivity model interpretations. Despite these uncertainties, our modeling specified that two-dimensional resistivity imaging is a potential technique to study subsurface cavities. We inferred that the DD array is the most appropriate for cavity surveys. The PD and WS arrays are adequate, while the PP array is the least suitable for cavity studies.

Keywords: cavity; depth of investigation; modeling; resistivity imaging; resolution; sensitivity; uncertainty

1. Introduction

Resistivity imaging is a geophysical method, which has become a powerful tool to investigate shallow subsurface features. The technique has been widely used in several fields of geosciences, such as structural geological [1–5], hydrogeological [6,7], geohazard [8,9], and environmental [10–13] studies. Subsurface cavity structures most commonly occur in carbonate terrain through the dissolution of carbonate rocks. Isostatic subsidence of the developed cavity can impose substantial damage on structural, environmental, and human values. Engineering structures require subsurface investigation to map and assess the presence of cavities, which reduces construction risks. Several geophysical methods are applied in order to determine the size, position, and depth of the subsurface cavities. Among the geophysical methods, resistivity imaging is a commonly used technique to identify cavity structures [12,14–16].

Despite the applicability success, resistivity imaging often leads to non-unique models of the subsurface. It is well-known that the resistivity method produces uncertainties in final model interpretation because of the inherited ambiguity in data inversion. This is because the inversion of electrical resistivity data is a non-linear problem, while the solutions are obtained using a linearized forward model [17–19]. The resistivity survey measures a limited number of data points at a discrete electrode spacing, which also constrained to obtain a unique resistivity response of the subsurface structures [20]. Thus, the delineation and mapping of subsurface cavities remain challenging in environmental studies using resistivity imaging [11,21].

Different factors can be considered in cavity studies, such as target size, depth, orientation, geometry, filling materials, and resistivity contrast with the host medium. The measured apparent resistivity data gives an average value representing a volume of geological material. The volume-averaging property of the resistivity method may obscure the small-scale cavity. The sparse sampling and less sensitivity characteristics of the resistivity method with increased survey depth may also obscure cavity information [15, 22]. These factors may delimit the technique and the type of array to be used in resistivity imaging [23,24]. Even though the cavity depth is small and situated at a relatively deeper depth, substantial resistivity contrast between the cavity and the host medium can reveal suitable imaging results [12,21,25].

Moreover, selecting the best electrode configuration to recover the material variations over a cavity zone is the main problem in resistivity study. Among the commonly used electrode configurations, the dipole-dipole (DD) array is widely used to detect subsurface lateral changes, yet it is more vulnerable to measurement noises and often produces a low signal-to-noise ratio. The pole-dipole (PD) configuration has relatively higher signal strength compared to the DD array. However, the PD array provides asymmetrical resistivity data that adds bias to the inverted models [21,25], whereas the Wenner–Schlumberger (WS) configuration is moderately sensitive to both lateral and vertical variations. The WS array also has a higher signal-to-noise ratio, yet it recovers a relatively lower resistivity contrast compared to the DD arrays [26,27]. On the other hand, the pole-pole (PP) array can produce comprehensive horizontal data coverage and survey greater depths as compared to other types of arrays. However, its remote electrode is highly sensitive to ambient (telluric) noises [21,25]. Therefore, each array has its advantages and limits in relation to cavity imaging, and a comprehensive study is needed to ascertain which configuration is the most suitable one.

Many studies have been carried out to compare the pros and cons of the most commonly used resistivity imaging arrays. The study in Reference [22] stressed on the measurement sensitivities of the arrays to compare the Wenner, dipole-dipole, pole-dipole, pole-pole, Wenner–Schlumberger, and gradient arrays. The dipole-dipole, pole-dipole, and pole-pole arrays were compared based on the depth of investigation [28]. Another study [25] performed numerical simulation for five synthetic models using different arrays: pole-pole, pole-dipole, half-Wenner, Wenner, Schlumberger, dipole-dipole, Wenner-β, γ-array, moving gradient, and midpoint-potential-referred measurement arrays. They suggested the moving gradient array, pole-dipole, dipole-dipole, and Schlumberger arrays. However, the final choice of the arrays should be decided based on the type of expected geology as the subsurface geological distributions have a significant impact on the arrays' imaging ability. Various numerical simulations for different noise perturbations were conducted by emphasizing the data-acquiring ability of the arrays: Wenner, dipole-dipole, Wenner–Schlumberger, and linear grid [29]. Comparison of arrays (Wenner, Wenner–Schlumberger, dipole-dipole, pole-dipole, and pole-pole) for imaging buried tunnels showed better results using the Wenner–Schlumberger and dipole-dipole arrays [30]. The Wenner-Alph and Wenner–Schlumberger showed a better result than the dipole-dipole array for reconstructing the landslide model [31]. Among the Wenner, dipole-dipole, pole-dipole, and pole-pole arrays used for detecting the cleared-wall foundations, the Wenner and dipole-dipole arrays were selected as the best methods [32]. From the numerical modeling using the dipole-dipole, pole-dipole, dipole-pole, and pole-pole arrays, the pole-dipole array showed the best result for modeling sinkhole [33]. The dipole-dipole array imaged the sinkhole better than the Wenner and Schlumerger arrays [21]; in contrast, the Wenner array showed better model resolution than the dipole-dipole array for surveying the underground three-dimensional cavity [34]. Even though these studies have evaluated the effectiveness of resistivity imaging for different geological conditions, there are no studies that assess the constraints of resistivity imaging as well as the configuration best-suited for imaging subsurface cavities.

In this study, we present the resistivity imaging uncertainties using synthetic and experimental cavity models. We simulated a cavity model with different depths, sizes, orientations, geometries, and filling materials, as these features substantially change the model result of the resistivity imaging [25]. Additionally, arrays' spatial resolution and sensitivity were examined. The modeling was performed for the DD, PD, WS, and PP arrays since the resolution of data inversion is also constrained by the type of arrays. A model with cells of a given resistivity value was constructed, and the apparent resistivity data were formulated using the finite-element method. Inversion of measured resistivity data was provided with the two-dimensional (2D) geoelectric models. We examined the constraints of the resistivity method on recovering cavity information for different cavity models. Moreover, the potential capability of arrays that can reveal the cavity structures was tested and identified.

2. Methodology

Resistivity imaging is usually applied in unknown geologic conditions, which cannot fully recover the true subsurface structure for the measured data [35]. Using a synthetic model under the controlled circumstance is a fundamental approach to assessing imaging method accuracy. A synthetic modeling so-called forward-inverse cycle consists of standard geophysical inversion processes, which inverts the synthetically measured data by the mathematical method to get specific resistivity distribution. It also implements a forward calculation that predicts a set of resistivity data based on a given synthetic model [36]. We used forward-inverse modeling to examine resistivity imaging uncertainties and identify appropriate arrays for cavity probing based on model accuracy, resolution, and sensitivity. In the following section, we systematically present modeling procedures and data analysis.

2.1. Forward-Inverse Modeling

We developed resistivity conceptual models to represent karstic subsurface structures in limestone terrain. In nature, cavities have different depth of burial, size, geometry, and orientation that can be air-filled, water-filled, or sediment-filled [37,38]. To study the effect of these cavity properties on the recovering ability of the 2D resistivity methods (DD, PD, WS, and PP arrays), we generate different synthetic models. The cavity depths, sizes, orientations, and geometries were considered in our numerical modeling. Since the survey depth has a significant effect on the resistivity model resolution, we focused on the arrays' performance with cavity depths. We set an air-filled cavity model at six different depths, naming cavity targets as T_1, T_2, T_3, T_4, T_5, and T_6, with the depths of 2.2, 4.2, 6.2, 8.2, 10.2, and 12.2 m, respectively (as shown in Figure 1). The target cavity T_2 was used for further analysis as its depth is more common in cavernous limestone terrain. We numerically modelled the conductive cavities by using a filling of dry and saturated clay; however, the resistivity of the cavity filling material is a function of the particle size, saturation, and resistivity of the pore fluid [39].

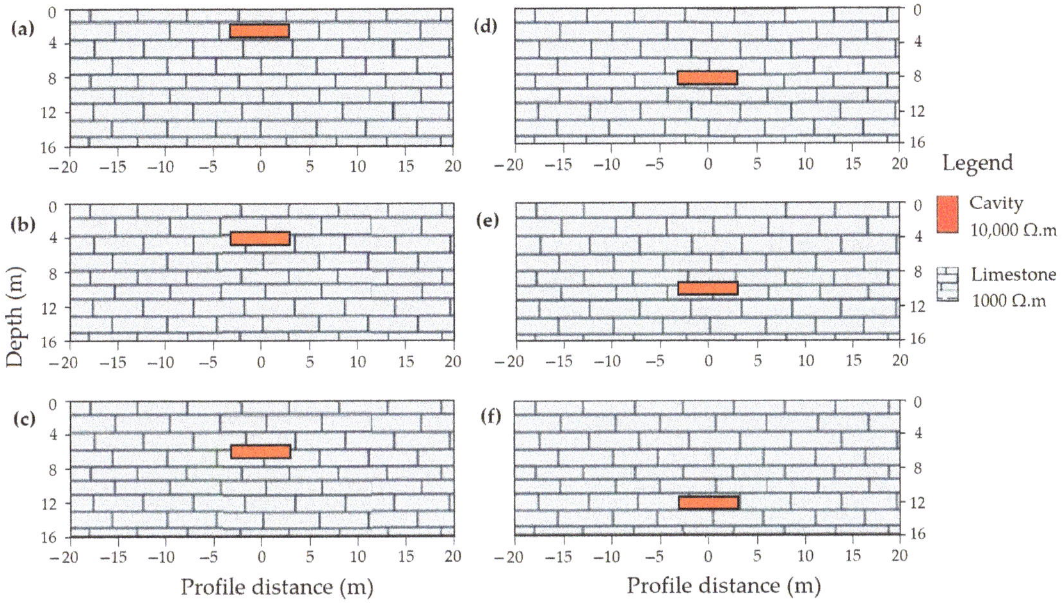

Figure 1. Resistivity synthetic models that represent a target cavity set in limestone unit at six different depths: (**a**) Target cavity T_1 at 2.2 m, (**b**) target cavity T_2 at 4.2 m, (**c**) target cavity T_3 at 6.2 m, (**d**) target cavity T_4 at 8.2 m, (**e**) target cavity T_5 at 10.2 m, (**f**) target cavity T_6 at 12.2 m depth. The depth indicates the mid-point of the cavity. For better illustration, we presented only 40 m horizontal section of the models.

As this study specifically examines the case of cavities in karstic terrain, the conceptual models' host medium was considered as uniform dry limestone rock with a resistivity value of 1000 Ω·m. Only in the case of conductive cavity modeling, when the water table is above the cavity, did we use a resistivity value of 450 Ω·m for the host limestone [38,40]. We used the resistivity value of 10,000 Ω·m for the air-filled cavity [11], 90 Ω·m for the cavity filled with dry clay [41], and 50 Ω·m for the cavity filled with saturated clay [33]. The arrays such as DD, PD, WS, and PP were used for surface probing, which apply 89 electrodes with 1 m spacing.

Furthermore, the synthetic reconstruction test, known as the checkerboard resolution test, was implemented in this study. The checkerboard test is often applied in seismic studies to examine the lateral resolution of the tomographic image. The alternating pattern of positive and negative anomalies along a spatial dimension is used to develop the checkerboard model. Inversion was performed for synthetically measured data to evaluate the recovered model resolution [42–44].

We applied the checkerboard resolution test to assess the resolution of resistivity tomogram in-depth cross-section along the survey profiles. Three different size scenarios of checkerboard cells were used to examine the anomaly size effect on the spatial resolution of the recovered models. We used a checkerboard cell size with 1 (vertical) × 2 m (horizontal) dimensions in scenario 1. In scenario 2, the size increased by the factor of two, i.e., 2 × 4 m cell size, while the factor of four was used in scenario 3, i.e., 4 × 8 m cell size. The background resistivity (1000 Ω·m) was perturbed by ±20% resistivity anomaly; thus, the checkerboard cells were assigned by alternating resistivity values of 800 and 1200 Ω·m.

Sensitivity in resistivity study can be used in different ways [22,45–48]. In this study, sensitivity was used to assess how the arrays are sensitive to different resistivity perturbations (contrasts). In addition to evaluating the spatial resolution of the inverted tomogram, we applied the checkerboard test to examine the sensitivities of the array (DD,

PD, WS, and PP). We used a checkerboard cell size of 2 m in the vertical direction and 4 m in the horizontal direction with background resistivity of 1000 Ω·m, which was perturbed by ±20%, 40%, 60%, and 80% anomalies.

The conceptual models were simulated to generate synthetic resistivity data. We implemented the AGI EarthImager^{TM} 2D software [49] for the model simulations. The finite-element modeling calculates the resistivity data using a partial differential equation [50]. Throughout the entire lateral and vertical mesh formulations, 1.0 m thickness and depth increment factor were used. Finer meshes of 0.5 × 0.5 m were applied to increase the forward modeling accuracy. The simulation was perturbed using Gaussian noise (a random value generated with zero mean and 3% standard deviation) in order to integrate the field data-acquisition noise effects.

We applied a robust inversion scheme to recover the 2D geoelectric model. It optimizes [51] the least-square method [52] by integrating the weighting matrices, giving approximately equal weight for data misfit and model roughness vector in the inversion processes. Robust inversion also provides improved inversion results for sharper resistivity models. It obtains more compacted anomalies, and it is a sound method to reduce data outliers [53]. Consistently, in our modeling, the robust inversion provided minor inversion artefacts, especially for large resistivity contrasts over a short profile distance.

The use of a larger grid size makes the inversion process more efficient; however, it leads to low model resolutions. Further discretization of the model parameters, on the contrary, does not show substantial changes in the model structure rather than smoothening the transition zones [53]. Hence, we applied an optimal grid of 1 × 1 m for the inversion of apparent resistivity data. The damping and smoothing factors were set to 10 initially, and these factors were reduced iteratively.

2.2. Data Analysis

We analyzed the numerically measured apparent resistivity data using a statistical variance, which describes how far the data are spread out from the average value. Variance was also used to examine the association of measured resistivity data with the cavities since the resistivity method explicitly depends on the subsurface resistivity variations. A higher variance may indicate larger resistivity contrast and vice versa. Hence, the variance (σ_ρ) for the measured apparent resistivities (ρ_i) with the mean value (μ_ρ) was calculated as follows:

$$\sigma_\rho{}^2 = \frac{1}{N}\sum_{i=1}^{N}(\rho_i - \mu_\rho)^2 \tag{1}$$

The anomaly effect (AE) was first introduced by Militzer [54], commonly used to evaluate arrays' measuring abilities. The anomaly effect was also used to examine arrays' applicability for different subsurface features, and it is calculated using the expression [25,55]:

$$AE = \left|\max\left(\frac{\rho_a}{\rho_{av}}\right) - \min\left(\frac{\rho_a}{\rho_{av}}\right)\right| \tag{2}$$

where ρ_{max}, ρ_{min}, and ρ_{av} represent the maximum, minimum, and average apparent resistivities, respectively. We applied the anomaly effect to assess the recovering abilities of the arrays and inspect the corresponding image quality as a function of cavity depth.

Data misfit between the measured and predicted apparent resistivity using relative root mean square (RMS) error [56] was used to assess arrays' inversion error with cavity depth.

The depth of investigation (DOI) index [28,57] calculated the depth below, for which the obtained model influenced by the reference model rather than the measured data, by inverting the same dataset twice with different reference models (m_{01} and m_{02}) using 0.1

and 10 times of the average apparent resistivity. DOI value of the bottom model (R_b) was used to normalize the DOI index, and it is presented as:

$$\mathrm{DOI}\ (x, z) = \frac{m_1(x, z) - m_2\ (x, z)}{R_b(m_{01} - m_{02})} \quad (3)$$

where m_1 and m_2 indicate the inverted resistivity values of the same cell for the two inversions.

We also examined the arrays' model accuracy based on the image correlation between the actual and inverted cavity models. The similarities were quantified using a statistical correlation coefficient that is expressed by:

$$\mathrm{Correl}\ (\rho_i, \rho_a) = \frac{\sum(\rho_i - \mu_i)(\rho_a - \mu_a)}{\sqrt{\sum(\rho_i - \mu_i)^2(\rho_a - \mu_a)^2}} \quad (4)$$

where ρ_i is the inverted resistivity data points, ρ_a is the actual model resistivity data points, μ_i is the mean value for the inverted resistivity data, and μ_a is the mean value for the actual resistivity data.

3. Results and Discussion

3.1. Array Detection Ability

The data recovering ability of different arrays may vary when applied to a specific subsurface structure [47]. In this study, we used the statistical variance to examine the measuring ability of arrays with regard to cavity depth. As depicted in Figure 2, when modeling the target cavity T_1 (at 2.2 m depth), resistivity data exhibited the variances of 24,400, 9058, 6341, and 2401 Ω·m for the DD, PD, WS, and PP arrays, respectively. In contrast, for the target cavity T_6 (at 12.2 m depth), resistivity data showed a statistical variance of 1100, 1044, 936, and 900 Ω·m for the DD, PD, WS, and PP arrays, respectively. The DD array depicted the maximum variance in resistivity data, whereas the PD and WS arrays produced moderate variances. The resistivity data from the PP array indicated the least variance. The results also show a linear dependence between the variance of the resistivity data and the accuracy of the cavity model recovered. Figure 2 shows that the decrease in resistivity data variance with increasing cavity depth is likely due to decreasing measurement sensitivity with survey depth [22].

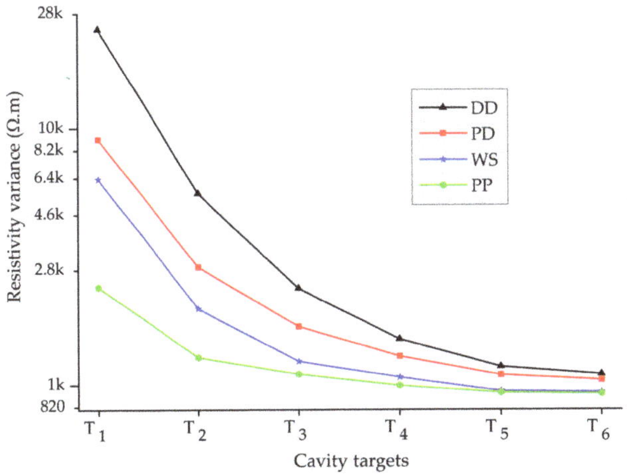

Figure 2. The variance of the numerically measured resistivity data for air-filled cavity situated at different survey depths.

To obtain reliable and high-resolution resistivity tomograms, the arrays used in the measurements should provide data with maximum anomaly information [47]. Thus, we calculated the anomaly effect of the numerically measured resistivity data, which shows the recovering ability of the arrays for different air-filled cavity depths. As shown in Figure 3, the highest anomaly effects of 1.46, 1.16, 0.74, and 0.6 for the corresponding arrays of DD, PD, WS, and PP were shown for the simulation of the shallowest cavity T_1. In contrast, the arrays DD, PD, WS, and PP exhibited the lowest anomaly effects of 0.245, 0.24, 0.19, and 0.18 respectively, for the deepest cavity T_6. The dielectric nature of the cavity causes the arrays to acquire resistivity data with higher anomaly effects, mainly for the shallow cavity depth [11]. The DD array provided maximum anomaly effect, while the PP array exhibited minimum anomaly effect, similar to other studies [55]. In comparison, the PD and WS arrays depicted moderate anomaly effects. The anomaly effect shows a linear relationship with the measurement sensitivity of the array [47]. This could be the reason behind the decline of anomaly effects with an increasing cavity depth (Figure 3). The higher sensitivity near the electrode location may enable the arrays to get a more pronounced anomaly effect. On the other hand, the low sensitive zone of the far electrode position might limit the arrays to recover the higher anomaly effect.

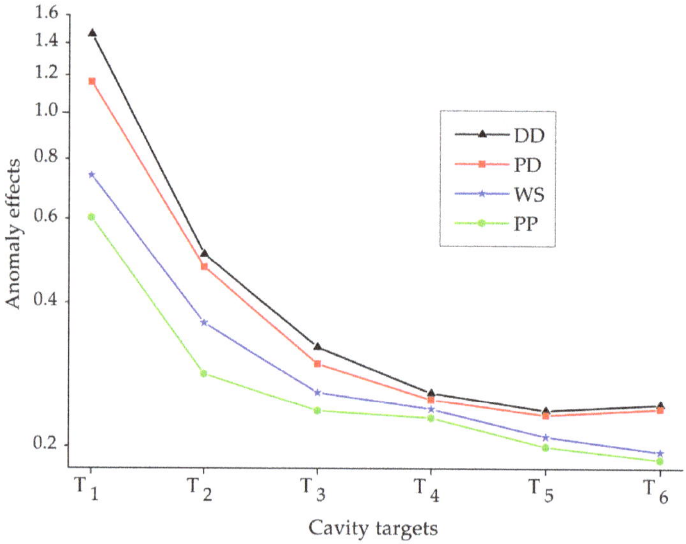

Figure 3. The anomaly effect of the arrays with regard to the air-filled cavity survey depths.

3.2. Inversion Error

The presence of noisy data points can create a large discrepancy between the inverted and the resistivity model, producing artefacts in the inverted resistivity models [58]. In this study, we used the statistical 95% confidence interval to identify those noisy data points. Figure 4 exhibits the model predicted versus synthetically measured apparent resistivity datasets for the target cavity T_2. About 5% of the data points situated on the upper and lower limit of the scatter plots were specified as outliers. The predicted resistivity was significantly overestimated for smaller values of measured resistivity data points that may be related to the background medium (host limestone unit). On the other hand, the predicted resistivity was considerably underestimated for the extreme values of measured resistivity, probably due to the smoothing of the resistivity contrast between the background medium and the cavity by the inversion method [59].

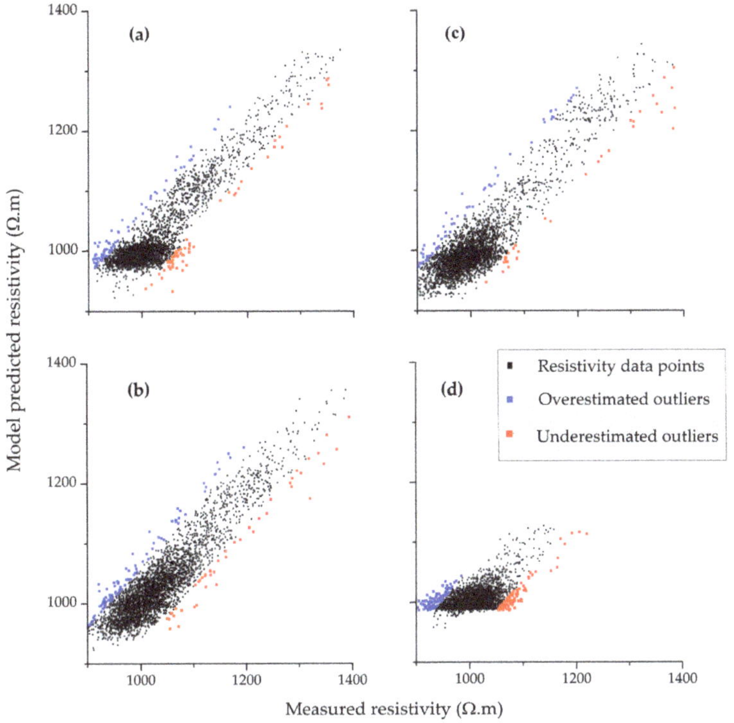

Figure 4. The scatter plot of the model predicted and synthetically measured apparent resistivity data for the arrays: (**a**) dipole-dipole (DD) array, (**b**) pole-dipole (PD) array, (**c**) Wenner–Schlumberger (WS) array, and (**d**) pole-pole (PP) array. Overestimated and underestimated data outliers are displayed on the upper and lower boundaries.

The inversion model with a smaller error usually provides a more accurate image; however, it does not necessarily mean that an inverted model with a smaller RMS error can give a realistic final model. This is because the inversion algorithm will sometimes tend to overfit the measured data for a larger number of iterations. For large resistivity contrasts, the inversion algorithm can progressively increase the inverted resistivity value from one iteration to the next without a considerable change in RMS error [60]. Consequently, a best-inverted model is often attained at a relatively low iteration number [61]. In cavity modeling, we used the model outputs with a maximum number of 4 iterations to reduce data overfit as further iterations did not improve the results. The result also depicted a considerable RMS error for inversion of high-contrast resistivity data. It is well-known that the surface resistivity method can acquire high-contrast resistivity data for shallower structures (e.g., cavity) than the deeper ones. As a result, the inversion of shallower cavity models exhibited higher RMS error than the deeper models (Figure 5), yet the inversion processes were stopped as long as the absolute data misfit value closes perturbation noise level (3%). Resistivity modeling may also reveal high model resolution, although the inversion provides large RMS error and vice versa. For example, the DD array showed a higher model resolution than the PP array; however, the DD array produced a larger RMS error than the PP array (Figure 5).

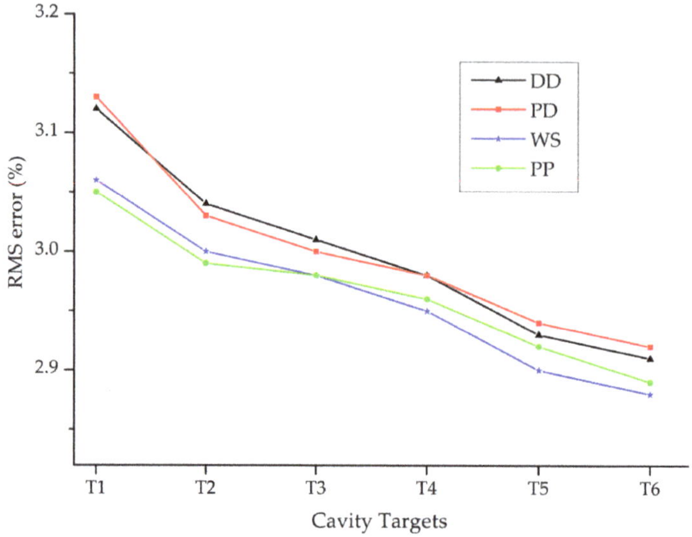

Figure 5. Arrays' inversion error for varying air-filled cavity depths.

3.3. Model Accuracy, Resolution, and Sensitivity

3.3.1. Model Accuracy

Depth of Cavity

We evaluated the model accuracy of arrays numerically using the image correlation between the actual and inverted resistivity models, as shown in Figure 6. The DD model correlation values were highest, varying from 0.29 (T_6) to 0.82 (T_1). On the other hand, the PD and WS models depicted moderate image correlation, ranging from 0.28 (T_6) to 0.75 (T_1) for the PD models and 0.24 (T_6) to 0.81 (T_1) for the WS models. The lowest correlation values were exhibited by the PP array, varying from 0.19 (T_6) to 0.71 (T_1). The correlation coefficients were directly associated with the accuracy of the obtained models.

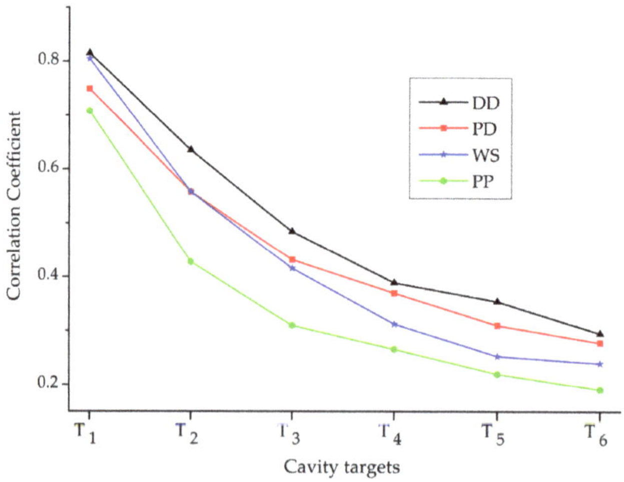

Figure 6. The statistical correlation between the inverted resistivity models and actual air-filled cavity models.

Moreover, the cavity region showed significant changes in model accuracy and resolution based on cavity depth. For the convenience of description, we categorized the air-filled cavity targets as shallower (T_1 and T_2), intermediate (T_3 and T_4), and deeper (T_5 and T_6) based on the cavity situated depths.

The shallower cavity models situated at 2.2 m (T_1) and 4.2 m (T_2) depths were accurately reconstructed by the DD, PD, and WS arrays, as depicted in Figures 7, 8 and 9a,b. The obtained images have also shown a strong statistical correlation with actual models (Figure 6). The PP array yielded a relatively lower resolution than the tested arrays (Figure 10a,b). The resistivity imaging displayed relatively wider anomaly sizes regarding the actual cavity models (Figures 7, 8, 9 and 10a,b). The obtained geometry accuracy was considerably varied for the upper and lower anomaly boundaries. For instance, the upper boundary of the T_1 anomaly coincided with an actual cavity model, while the bottom boundary extended beyond that of the actual models (Figure 7a). Generally, inversion artefact increases with the survey depth; however, the inversion for high resistivity contrast may also provide a noticeable artefact, especially nearby the anomalous body [62]. Even though we applied the robust inversion algorithm, inversion for the shallow depth cavities has generated considerable artefacts, as depicted in Figures 7, 8, 9 and 10a,b. The significant artefacts in inversions performed in all the air-filled cavity modeling had the resistivity value below 950 Ω·m, that is often misinterpreted as subsurface physical features.

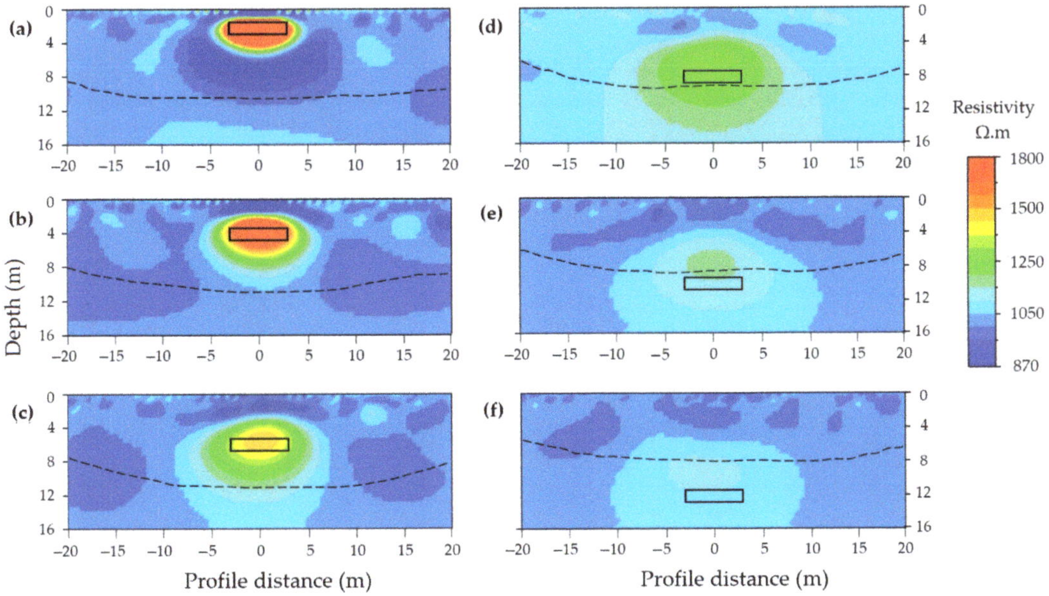

Figure 7. The inverted resistivity model using dipole-dipole (DD) array for the cavity targets: (**a**) T_1, (**b**) T_2, (**c**) T_3, (**d**) T_4, (**e**) T_5, (**f**) T_6. The rectangular boxes indicate the actual air-filled cavities. The broken lines represent the depth of investigation (DOI) threshold depths. For better anomaly illustration, we presented only a 40 m lateral model section.

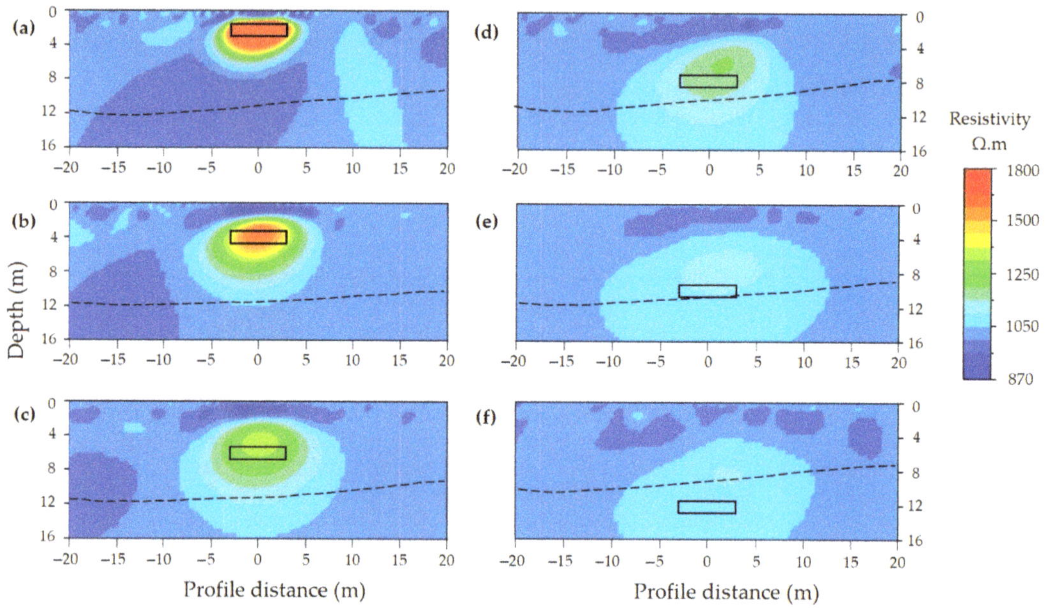

Figure 8. The inverted resistivity model using pole-dipole (PD) array for the cavity targets: (**a**) T_1, (**b**) T_2, (**c**) T_3, (**d**) T_4, (**e**) T_5, (**f**) T_6. The rectangular boxes indicate the actual air-filled cavities. The broken lines represent the DOI threshold depths.

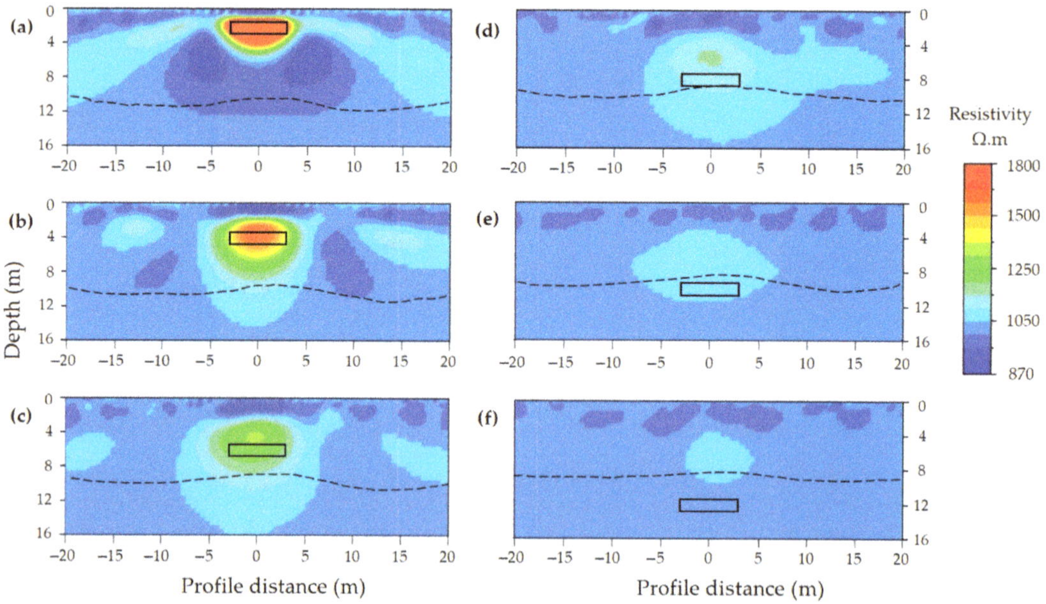

Figure 9. Inverted resistivity model using Wenner–Schlumberger (WS) array for the cavity targets: (**a**) T_1, (**b**) T_2, (**c**) T_3, (**d**) T_4, (**e**) T_5, (**f**) T_6. The rectangular boxes indicate the actual air-filled cavities. The broken lines represent the DOI threshold depths.

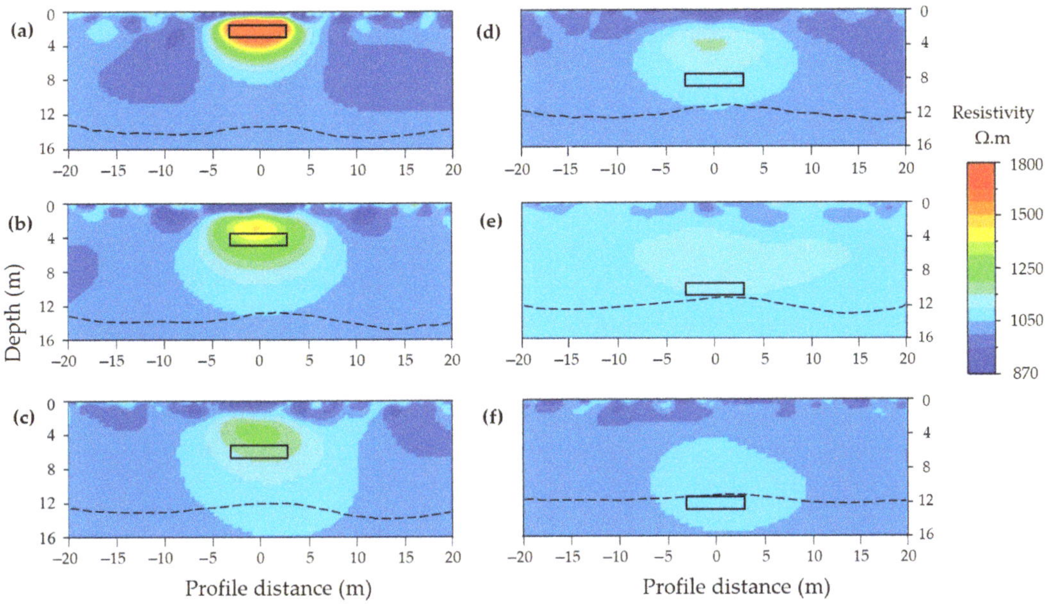

Figure 10. Inverted resistivity model using pole-pole (PP) array for the cavity targets: (**a**) T_1, (**b**) T_2, (**c**) T_3, (**d**) T_4, (**e**) T_5, (**f**) T_6. The rectangular boxes indicate the actual air-filled cavities. The broken lines represent the DOI threshold depths.

Modeling for the intermediate cavities located at 6.2 m (T_3) and 8.2 m (T_4) depths showed reasonable model resolution in the tested arrays (Figures 7, 8, 9 and 10c,d). The DD array displayed slightly better resolution than the other arrays. Figure 6 shows the moderate correlation between the inverted images and the actual models. The obtained anomaly sizes were noticeably overestimated, whereas the anomaly depths were underestimated by about 0.4–1.3 m with regard to actual cavity models. In Figures 7, 8, 9 and 10c,d, the artefacts were clearly observed on the shallow part of the inverted models, particularly on the DD models. However, the resistivity imaging showed the intermediate cavity anomalies, the size overestimation and depth underestimation were limited the proper inference of the cavity information.

In contrast, the deeper cavity models set at 10.2 m (T_5) and 12.2 m (T_6) depths were not recovered adequately by any of the arrays (Figures 7, 8, 9 and 10e,f). The obtained image correlations to the real model were weak (Figure 6). As shown in Figures 7, 8, 9 and 10e,f, the anomaly sizes were highly overestimated, making it very challenging to interpret the cavity geometries without prior subsurface information. The result also showed about 1.6–3.0 m anomaly depth underestimation relative to the actual cavity locations. As the cavity depth increases, significant resistivity underestimation occurred in anomaly zones. Furthermore, the shallow parts of the final models were contaminated by patched inversion artefacts. These poor model accuracies can lead to inappropriate interpretation in resistivity imaging.

The anomaly responses of different arrays for a particular air-filled cavity were also examined graphically. We produced a one-dimensional (1D) horizontal resistivity profile from the inverted model of the cavity T_2, which crossed the mid-point of the cavity at 4.2 m depth (as shown in Figure 11). We extracted the resistivity values along the 1D profile for the DD, PD, WS, and PP arrays. The DD array showed the highest (steepest) anomaly gradient over the cavity zone, with the peak resistivity value of 2220 Ω·m. In contrast, the PP array exhibited the lowest (gentlest) anomaly gradient, with the peak resistivity value of 1400 Ω·m. The 1D profile of the PD and WS arrays showed moderate anomaly gradients that have the peak resistivity value of 1850 and 1700 Ω·m, respectively.

The arrays that provided the steep anomaly gradient can recover cavity boundaries more accurately, while the arrays that yield the gentle anomaly gradient prevent the inference of the cavity boundaries.

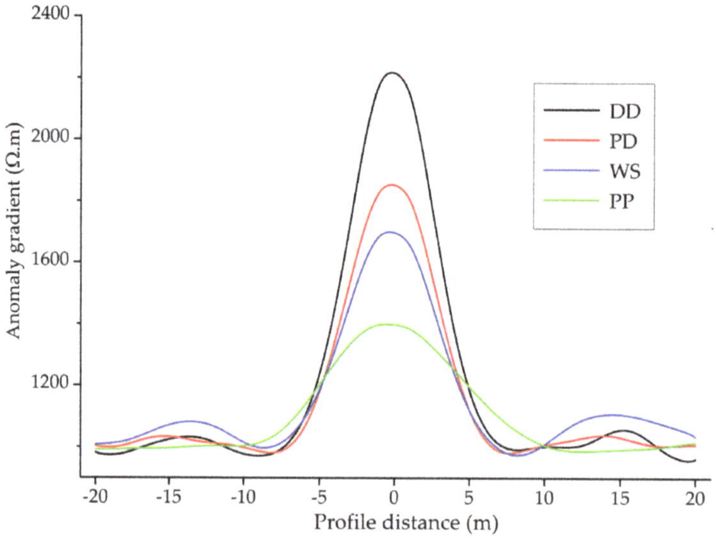

Figure 11. Resistivity anomaly gradient of the arrays for the target cavity T_2.

The 1D resistivity profiles were also used to assess the effect of air-filled cavity depth on the anomaly response of the arrays. Since the DD model provides the highest anomaly gradient, we used its models to generate 1D horizontal profiles for different depth cavity cases (Figure 12). The resistivity profile data were extracted along the profile lines that pass via the mid-points of the cavities T_1, T_2, T_3, T_4, T_5, and T_6 at depths of 2.2, 4.2, 6.2, 8.2, 10.2, and 12.2 m, respectively. The shallower cavity models showed a more robust resistivity amplitude than the cavity set at the relatively deeper position. For instance, the upper cavity T_1 indicates the maximum anomaly amplitude with the peak resistivity value of 6470 Ω·m. In comparison, the bottom cavity T_6 exhibits a minimum anomaly amplitude with a peak resistivity value of 1080 Ω·m. The substantial decrease in resistivity amplitude (gradient) as an increase in cavity depth can limit obtaining the anomaly geometries (Figure 12).

Furthermore, the depth of investigation (DOI) index may show variation based on the type of arrays and the resistivity distributions of the survey domain. We delineated the more reliable zones of the inverted models since higher credibility is given for the DOI patterns with values below 0.1 [28]. The zone below the threshold depth (DOI $\leq$ 0.1) is considered a low sensitive zone and assumed to be poorly imaged [63]. Air-filled cavity model simulation showed a small change of DOI index with cavity position, which is slightly shifted up as the cavity depth increases (Figures 7–10). Consistent with our results, other studies [57] depicted the variation of DOI indexes based on the resistivity distribution of the survey domain. In our study, a significant variation of DOI index was also observed for different types of arrays. The inverted models have displayed the average DOI threshold depth of 8 m for DD, 9.8 m for WS, 10.5 m for PD, and 12.7 m for PP models (Figures 7–10). For the same number of electrodes, equal spacing, and profile length, the largest DOI threshold depth shown by the PP array may be due to its greater penetration depth [22,25]. The PP array depicted the largest DOI threshold depth; however, it obtained the lowest model resolution with blurred cavity anomalies (Figure

10). This is likely related to the least sensitivity of the PP array to the subsurface resistivity variations [22,47].

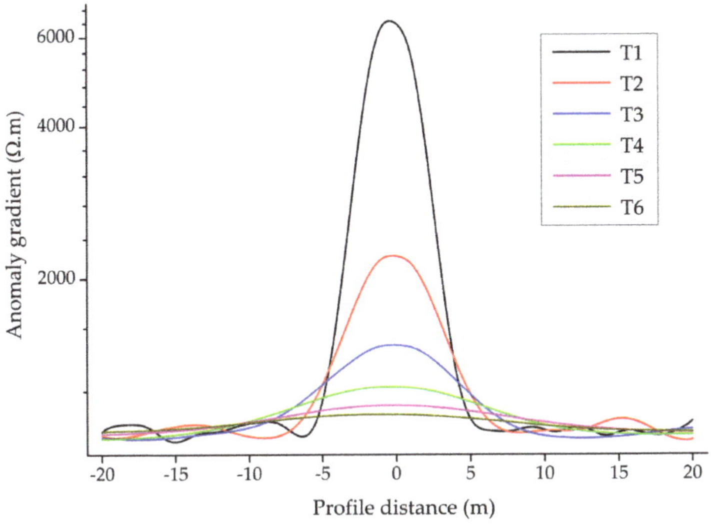

Figure 12. Resistivity anomaly gradient for the air-filled cavity set at different survey depths using the dipole-dipole (DD) array.

Size of Cavity

The air-filled cavity size scenarios were used since the cavity size is an important factor in the array-model resolution. We used three cavity sizes as three scenarios: 0.5 × 6 m, 1.5 × 6 m, and 3 × 6 m respectively, all set at 4.2 m depth.

The cavity size significantly affects the resolution of the resistivity tomogram. The small cavity size in scenario 1 was not correctly resolved in any tested arrays; however, the DD array showed a relatively better result (Figure 13). The small size cavity can be highly smoothened by the inversion process, in addition to the low-resistivity measurement on the recording instrument that nearly approaches the host resistivity value (Figure 13). We show the results of scenario 2 with the results of the previous section for the similar-sized cavity T_2 for the sake of comparison (Figure 14). Scenario 2 was adequately resolved in the DD, PD, and WS arrays; however, the PD and WS arrays' resolution were slightly lower than the DD array, whereas the PP array poorly recovered the cavity in scenario 2. Moreover, all the tested arrays resolved scenario 3 cavities (Figure 15), but the PP array showed a relatively lower resolution than the other arrays. For increasing cavity size, the model resolution significantly improved, while the inversion artefacts considerably increased. The result showed that the overestimation in anomaly size was less pronounced as cavity size increases. By considering the homogenous host medium, cavity targets larger than the anomaly size in scenario 1 could be recovered effectively by the DD array. The DD and WS arrays could be effective for the cavity targets greater than or equal to the anomaly size in scenario 2. The PP array could be effective for the cavity size having a size greater than or equal to the anomaly in scenario 3.

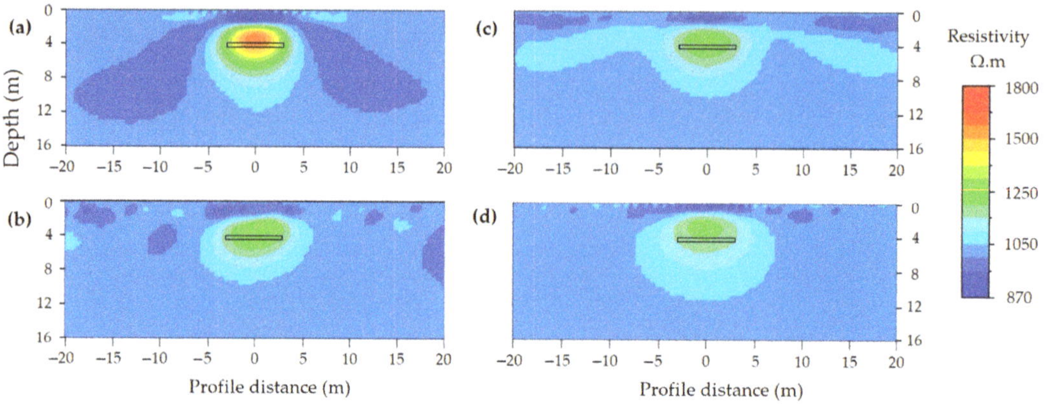

Figure 13. The inverted resistivity model for scenario 1 with 0.5 × 6 m cavity size using: (**a**) DD array, (**b**) PD array, (**c**) WS array, (**d**) PP array. The rectangular box indicates the actual cavity target.

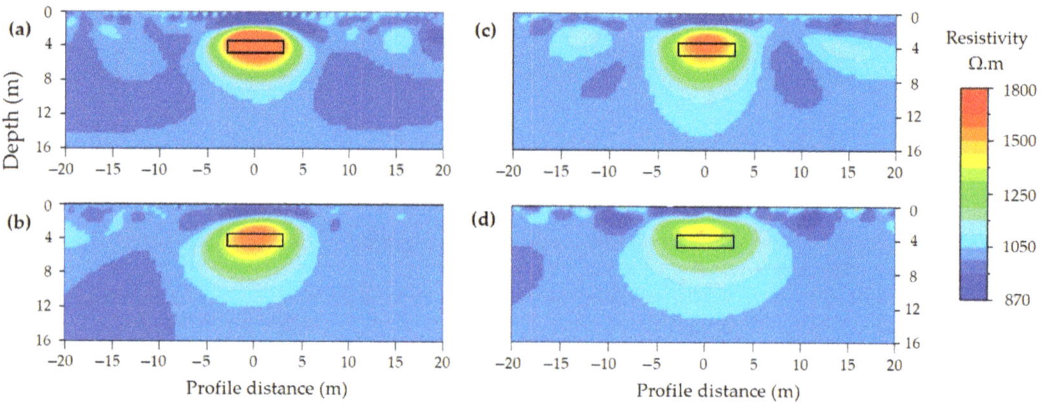

Figure 14. The inverted resistivity model for scenario 2 with 1.5 × 6 m cavity size using: (**a**) DD array, (**b**) PD array, (**c**) WS array, (**d**) PP array. The rectangular box indicates the actual cavity target.

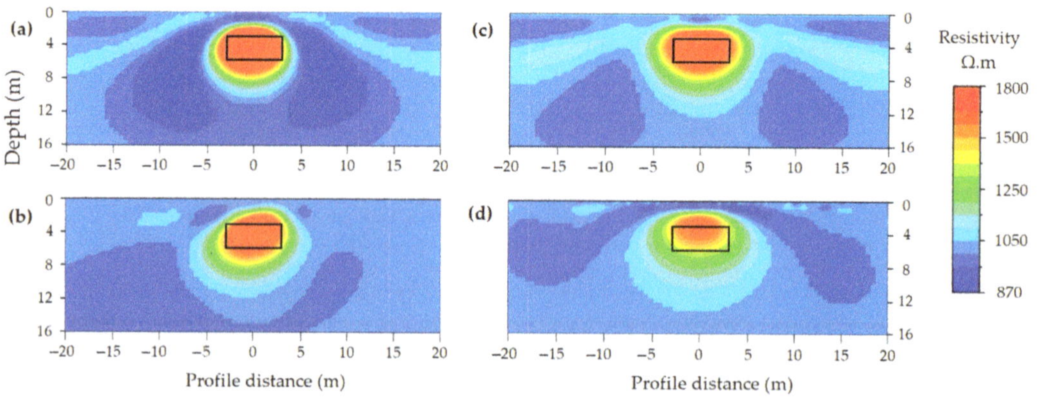

Figure 15. The inverted resistivity model for scenario 3 with 3 × 6 m cavity size using: (**a**) DD array, (**b**) PD array, (**c**) WS array, (**d**) PP array. The rectangular box indicates the actual cavity target.

Orientation of Cavity

In karstic terrain, cavities are not usually aligned in the horizontal direction; thus, we examined the surface resistivity methods using a cavity target (0.5 × 6 m) set at a different inclination (I), such as the horizontal (I = 0°), inclined (I = 45°), and vertical (I = 90°). As the horizontal cavity models are discussed in the earlier sections (Figure 14), we present the inclined (Figure 16) and vertical (Figure 17) cavity models. The result showed that the orientation and geometry of the inclined cavity were correctly resolved in the DD and PD arrays. The WS and PP arrays recovered the upper part of the inclined cavity, but they did not properly detect the deeper parts (Figure 16). As shown in Figure 17, the PD and DD arrays also accurately recovered the orientation and geometry of the vertical cavity, whereas these were poorly reproduced in the WS and PP arrays. All the arrays recovered the geometry of the top cavity boundary since the resistivity method detected a high signal near the electrode position. For increasing cavity depth, the inverted resistivity value decreased, and the cavity anomaly width considerably increased. Pronounced inversion artefacts were observed in the inclined and vertical cavity models. It is well-known that the DD array is highly sensitive to the lateral variation of the subsurface structures [21,64]. However, our results show that the DD array can also resolve cavities oriented in the horizontal and inclined directions. The PD array could be more effective for vertical and inclined cavity structures compared to the horizontal structures. The WS array is known to be moderately sensitive to both lateral and vertical variations [22,27]. However, in our results, it showed relatively better resolution for the horizontal cavity than the inclined and vertical ones. The PP array depicted low resolution for all the tested orientations of the cavity.

Geometry of Cavity

Obtaining the exact geometry of the subsurface structure using the resistivity method is a challenging task [22,59]. As shown in Figure 18, we simulated two air-filled cavity geometries: the L-shaped cavity (left side) and the triangular-shaped cavity (right side). The L-shaped cavity geometry was slightly recovered in the DD and PD arrays. The WS and PP array did not reproduce the geometry of the L-shaped cavity. The shape of the triangular cavity was not reproduced in all the tested arrays. Even though the resistivity methods show a high anomaly over the buried cavity zone, its geometry cannot be adequately reconstructed for moderate to complex cavity structures. This is possibly associated with the smoothness effect of the inversion method [59].

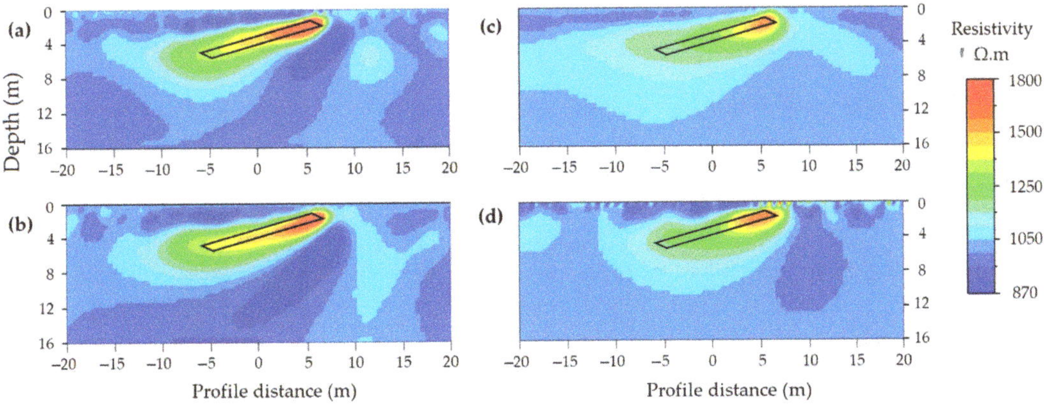

Figure 16. The inverted resistivity model for the inclined (45°) cavity using: (**a**) DD array, (**b**) PD array, (**c**) WS array, (**d**) PP array. The box indicates the actual cavity target.

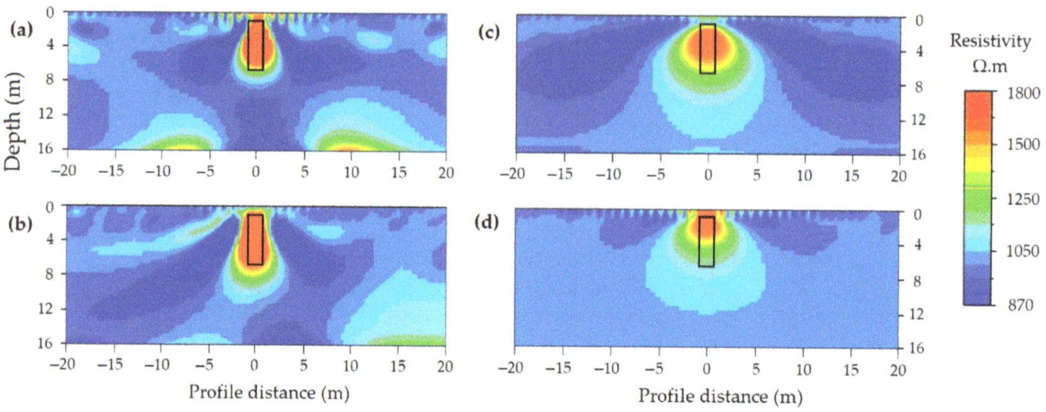

Figure 17. The inverted resistivity model for the vertical (90°) cavity size using: (**a**) DD array, (**b**) PD array, (**c**) WS array, (**d**) PP array. The rectangular box indicates the actual cavity target.

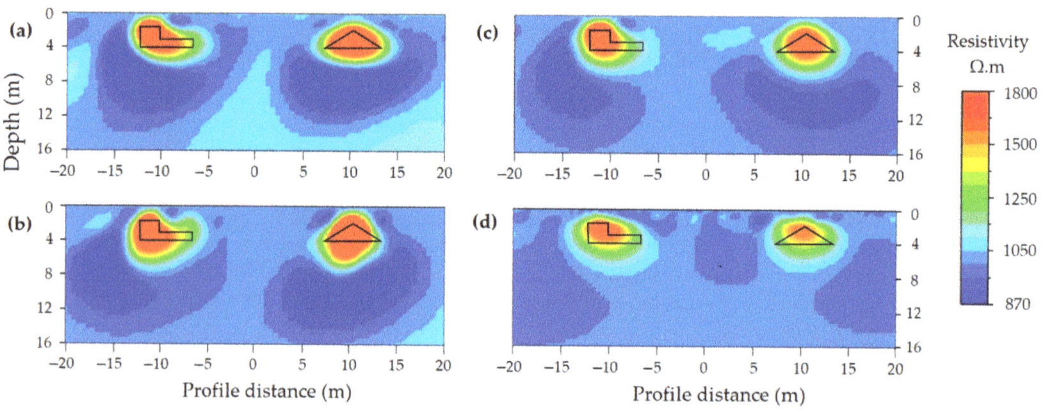

Figure 18. The inverted resistivity models for the cavity geometries using: (**a**) DD array, (**b**) PD array, (**c**) WS array, (**d**) PP array. The solid line indicates the actual cavity targets.

Conductive Cavity Modeling

The air-filled cavity showed a high anomaly in resistivity imaging, while a conductive cavity appeared as a low anomaly. For conductive cavity modeling, a cavity with a 1.5×6 m dimension set at 4.2 m depth was filled with dry clay (90 Ω·m) and saturated clay (50 Ω·m).

Figure 19 shows the inverted model results for a cavity filled with dry clay. The DD array accurately resolved the cavity; however, the anomaly size was higher than the actual cavity size. The PD and WS arrays also showed relatively high resolution, and their anomaly sizes were wider compared to the actual cavity. The recovered cavity orientation was tilted in the case of the PD array. The PP array showed the lowest resolution, and its anomaly geometry was unclear. The noticeable inversion artefact had a resistivity value above 1160 Ω·m that can lead to inappropriate model interpretation in resistivity studies. These artefacts were pronounced at shallow depth.

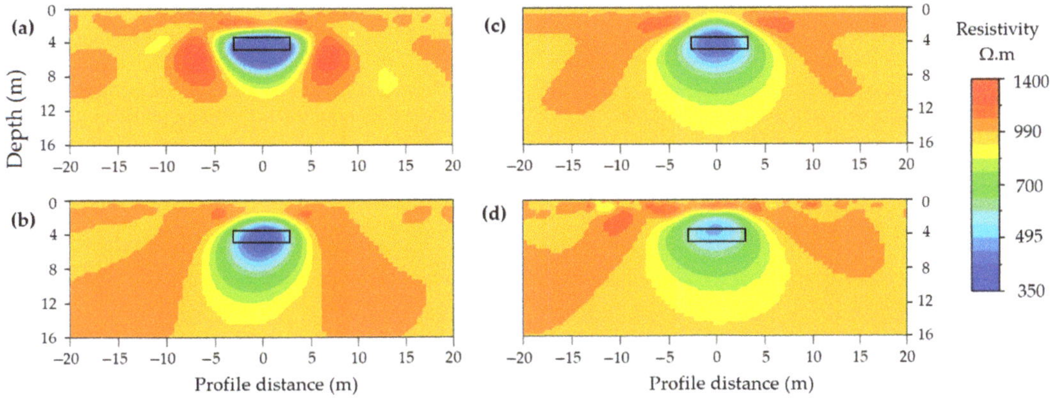

Figure 19. The inverted resistivity model for a cavity filled with dry clay using: (**a**) DD array, (**b**) PD array, (**c**) WS array, (**d**) PP array. The rectangular box indicates the actual cavity target.

Significantly low resistivity may be detected in resistivity imaging over a cavity filled with clay and water [65]. We consider the groundwater level (2 m) above the cavity depth for the saturated clay-filled condition, as shown in Figure 20. The model inversion result showed that the saturated clay-filled cavity was accurately resolved in the DD array. The PD and WS arrays obtained high resolution; however, the WS array showed a slightly lower resolution than the PD array. The PP array identified the cavity zone, but its resolution is lower than the other arrays. All the arrays exhibited noticeably wider anomaly sizes with regard to the actual cavity size. The boundary between unsaturated and saturated limestone was precisely resolved in the DD, WS, and PP arrays, while the recovered boundary was slightly deeper in the PD array. The cavity filled with conductive materials showed a slight improvement in the resolution of cavity anomaly compared to the air-filled cavity model, particularly in the PP array (Figure 20). In line with our study, the DD array field study shows the highest resolution for the conductive cavity [21,66]. The numerical study for sinkhole modeling showed lower resolution in the DD array than the PD array, contradicting our study [33]. The field study for cavity detection displayed relatively high resolution in the PD array, while the PP array showed low resolution [67], consistent with our results.

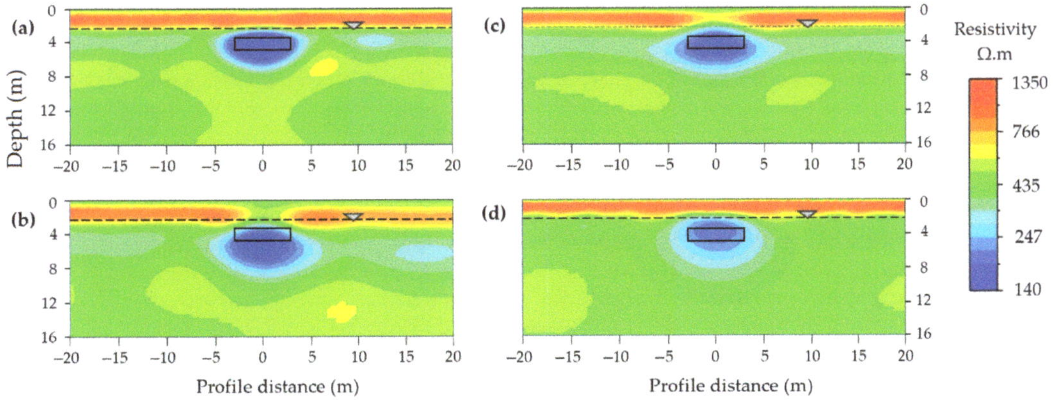

Figure 20. The inverted resistivity model for the saturated clay-filled cavity using: (**a**) DD array, (**b**) PD array, (**c**) WS array, (**d**) PP array. The rectangular box indicates the actual cavity target. The broken line represents the actual groundwater level.

Experimental Cavity Study

A field experiment was also conducted to validate the numerical modeling results. Our numerical model cavities were embedded in a limestone unit; however, in the field experiment, we used moderately compacted sandy clay as the host unit around an empty plastic box which acted as the cavity (Figure 21). In our experiment, the one-tenth scale of the cavity target T_2 was used. Thus, the plastic box's center with $0.15 \times 0.15 \times 0.6$ m size was situated at 0.42 m depth. The experimental resistivity datasets were measured using 89 electrodes of a 4-point light 10W (LGM Lippmann) resistivity meter with 0.1 m spacing (Figure 21).

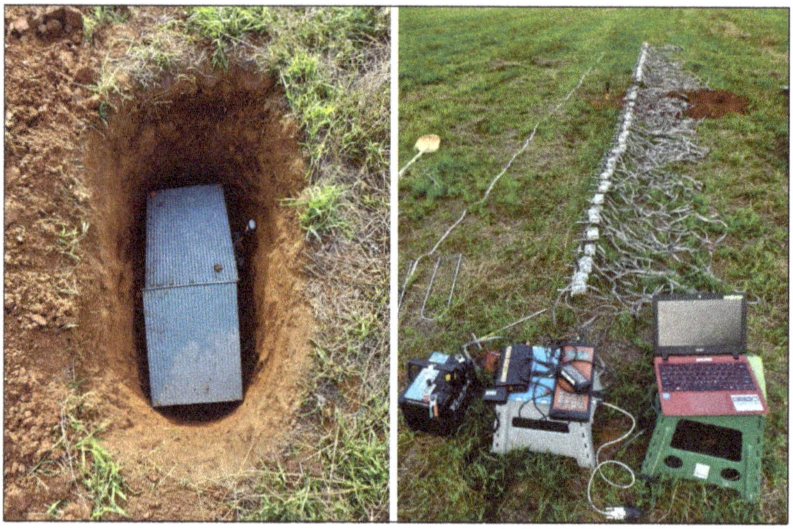

Figure 21. The experimental data collection procedure for resistivity imaging.

Figure 22 shows the experimental resistivity survey results for the DD, PD, WS, and PP arrays. The DD and PD arrays showed a high resistivity anomaly at the cavity zone, whereas the lower anomaly amplitude was obtained by WS and PP arrays. The experimental cavity geometry was more adequately reconstructed in the DD and WS arrays; however, they obtained wider and shallower anomalies. The geometry of the cavity in the PD array was not properly restored—its left side anomaly was elongated to the greater depth. The PP array exhibited a triangular anomaly instead of providing the rectangular cavity shape. Compared to the numerical results, the experimental cavity anomaly in all the tested arrays showed considerable underestimation of anomaly depth. The depth underestimation in cavity studies can lead to inappropriate interpretation, similar to other field studies [26,68]. Since our study was conducted in shallow sandy clay, the patchily distributed high and low resistive features were inferred as the inversion artefacts. The DD, PD, and PP arrays artefact agreed with synthetic model results, whereas a relatively greater artefact was observed in the WS arrays. As the experimental cavity was set in low-resistive sandy clay, its obtained resistivity values were considerably underestimated compared to the numerical cavity, which was embedded in limestone rock. Based on the experimental result, the WS and DD arrays are more appropriate for the subsurface cavity study, whereas the PD array is reasonable. The PP array is the least suitable, particularly in obtaining the cavity geometry and position.

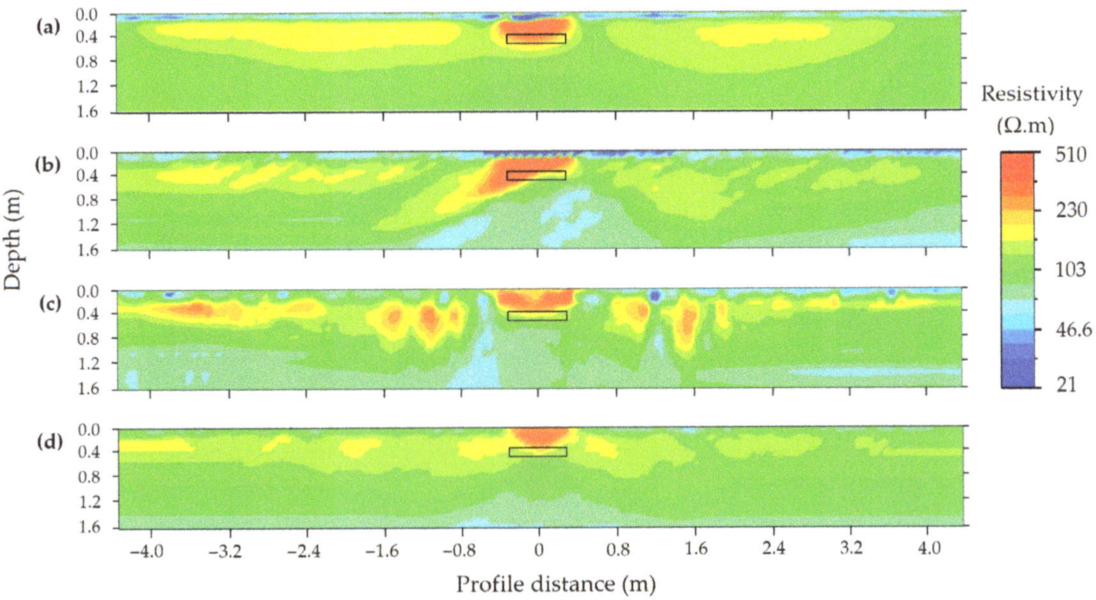

Figure 22. The inverted resistivity model of the experimental data for the: (**a**) DD array, (**b**) PD array, (**c**) WS array, (**d**) PP array. The rectangular box indicates the buried empty plastic box.

Our numerical and experimental study can help to decide proper arrays for subsurface target (e.g., cavity) studies. In line with our results, the field study and numerical modeling have shown that the DD array can accurately resolve the subsurface structures such as cavity [38,67], thin dike [25], buried wall [29,32], and sinkhole [21]. These structures were moderately recovered by the WS and PD arrays [24,26,69]. The pole-pole array is generally the poorest in terms of model accuracy and image quality [25,32], consistent with our study. The array effectiveness is also constrained by the target depth-to-dimension ratio [32]. In our results, the shallowest cavities with considerable size were recovered effectively (Figures 7, 8, 9 and 10a). For example, the DD, WS, and PD arrays well recovered a cavity diameter larger than one-fourth times its depth, similar to other studies [24]. If the anomaly geometry is not considered, those arrays can detect relatively deeper cavities. Thus, the DD array could be the most effective, PD and WS arrays could be moderate, and the PP array could be the least effective for surveying subsurface targets such as environmental or archaeological features (e.g., buried cavity, tombs, and wall), and geological structures (e.g., dike and thin channels).

3.3.2. Spatial Resolution

The spatial resolution of resistivity imaging is a function of many factors, including subsurface material distribution, data accuracy, array sampling scheme, and the parameters used in the model (e.g., smoothing factors) [70,71]. In this study, we used three different anomaly sizes as input models with the same inversion parameters to assess the spatial resolution of the arrays (DD, PD, WS, and PP).

Figure 23 shows the resolution test for scenario 1 with a checkerboard cell size of 1×2 m. In this scenario, the DD array can resolve the anomalies in both rows; however, the resolution is significantly reduced for the second row. Other arrays can only resolve the top row of anomalies. The WS and PP arrays depicted smearing at selected locations, whereas the PD array showed vertical smearing across the profile. The anomaly geometries were not well resolved for the PD array.

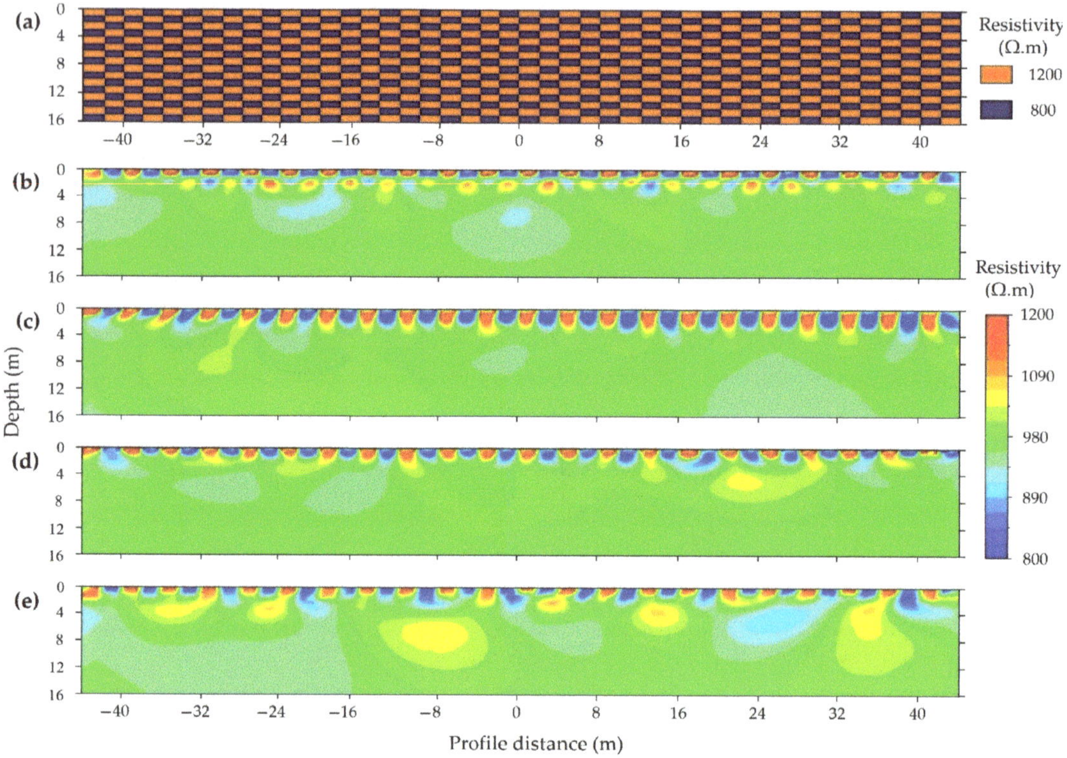

Figure 23. Scenario 1 spatial resolution test for the 1 m (vertical) × 2 m (horizontal) checkerboard cell sizes. (**a**) Checkerboard model, (**b**) DD array, (**c**) PD array, (**d**) WS array, (**e**) PP array.

The second scenario with a checkerboard cell size of 2 × 4 m was better resolved in general compared to scenario 1, as shown in Figure 24. The DD array was able to recover the first and second rows, but the resolution was slightly reduced for the second row. The PD and WS arrays restored the entire checkerboard anomalies for the first row; however, the resolution of the second row was significantly reduced. The PP array recovered only the first row, and its anomaly was considerably smeared. The geometries of the checkerboard patterns were more accurately recovered by the DD, PD, and WS arrays, while the geometries could not be reproduced correctly for the PP array.

The larger size of 4 × 8 m was considered in the third scenario, as shown in Figure 25. The DD array resolved the first and the second rows; however, the resolution was slightly reduced for the second row. The PD, WS, and PP arrays consistently imaged the first row input checkerboard structures, but these were not resolved correctly in the second row. The second row checkerboards' geometry was not accurately recovered for all the arrays and displayed elongated checker anomalies instead of rectangular cells.

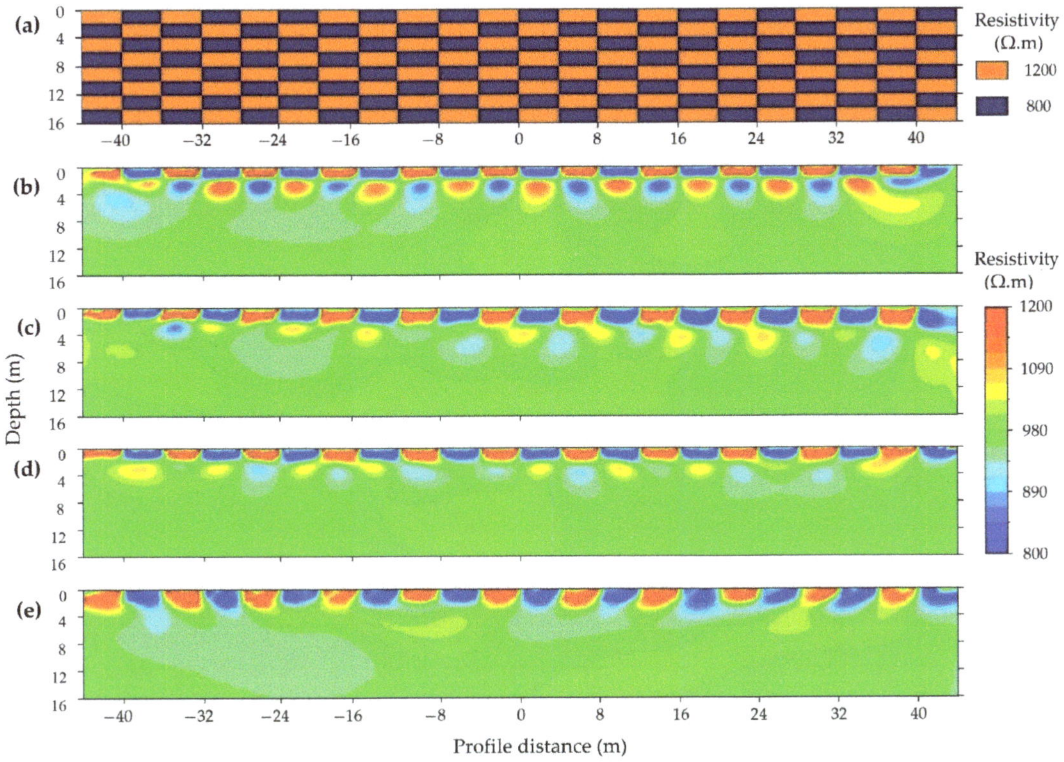

Figure 24. Scenario 2 spatial resolution test for the 2 m (vertical) × 4 m (horizontal) checkerboard cell sizes. (**a**) Checkerboard model, (**b**) DD array, (**c**) PD array, (**d**) WS array, (**e**) PP array.

The checkerboard test results depicted that even though we used the same arrays, the resolution of the recovered anomalies can vary based on their sizes. In general, the checkerboard resolution test of the first scenario reproduced about 12.5% checkerboard patterns (Figure 23), while the second scenario (twice the size of scenario 1) restored about 25% of the checkerboard (Figure 24). The third scenario (twice the size of scenario 2) reproduced about 50% of the checkerboard patterns (Figure 25). Thus, the increase in target size can noticeably increase the obtained model resolution and recovery depth. If the uniform background resistivity anomaly has a low amplitude compared to the targeted anomaly, then the depth of recovery is higher (Figure 7). However, all the alternating anomalies have similar amplitudes in the checkerboard test (Figure 24), and the recovery depth is low. This is likely related to the concentration of current density at the first-row of the checkerboard, which significantly decreases with depth, as current density is concentrated at the top layer in stratified formation [72,73]. As the resistivity data points are not uniformly distributed in the profile section, the inverted model resolution can considerably be biased. As shown in Figures 23 and 24, the center of the survey profiles is well recovered, but smearing is observed at the end of the profiles.

We can choose the best array type for surveying shallow subsurface features supporting the checkerboard resolution test with different studies if prior information is available. Among the tested arrays, the DD array could be the most appropriate method for studying concealed small-scale features with less than or equal to the checkerboard size in scenario 1, such as void, tombs, and thin dikes [26,33,68]. The geological or environmental structures (e.g., thin buried channels) with the checkerboard size in scenario 2 could be recovered effectively by the DD, PD, and WS arrays [25,30]. All the tested arrays could resolve the subsurface features (e.g., tunnels, sinkholes) having a size greater than or equal to the checkerboard size in scenario 3 [21,30]. These arrays could be more effective if the targets-embedded media are homogenous.

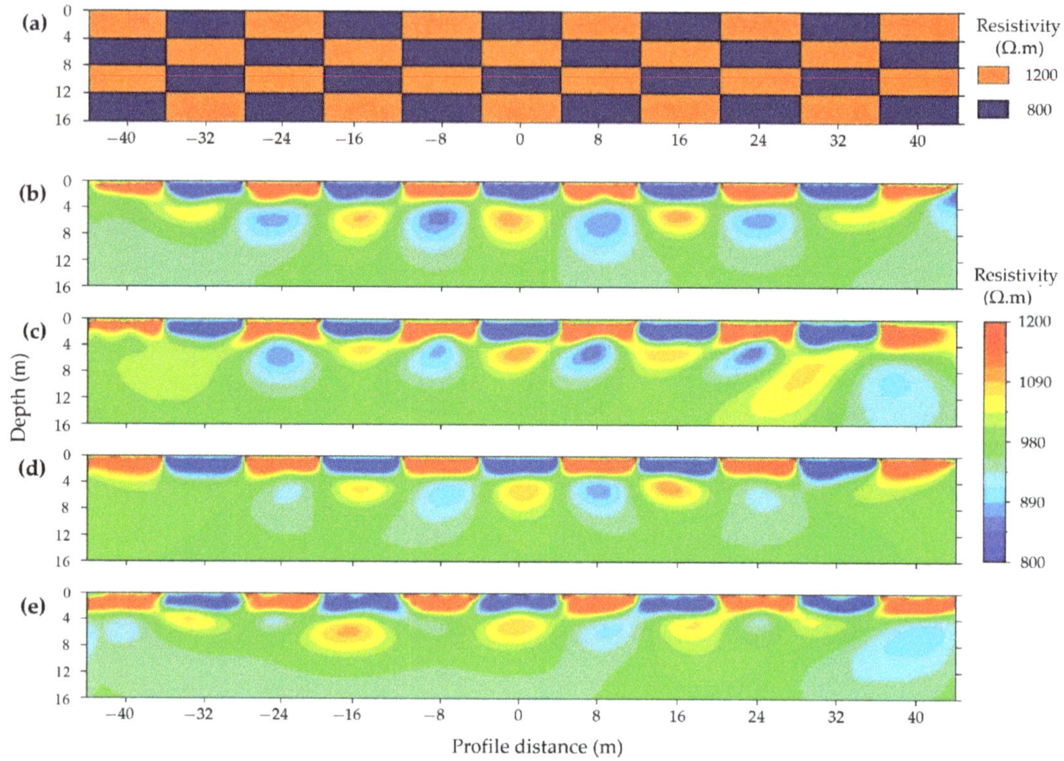

Figure 25. Scenario 3 spatial resolution test for the 4 m (vertical) × 8 m (horizontal) checkerboard cell sizes. (**a**) Checkerboard model, (**b**) DD array, (**c**) PD array, (**d**) WS array, (**e**) PP array.

3.3.3. Sensitivity to Resistivity Contrasts

The detection of subsurface structure using the resistivity method is mainly dependent on resistivity contrast, which may constrain the inverted model resolution [16]. We used different resistivity perturbations (±20%, 40%, 60%, and 80%) to examine array (DD, PD, WS, and PP) sensitivities. As shown in Figure 26, the DD array can resolve the first two rows of the checkerboards for the tested perturbations. However, the model resolution of the DD array was considerably enhanced for increased resistivity perturbation, reflecting the high sensitivity of the DD array. The high sensitivity of the DD array helped to study various subsurface structures with low [25,74] to high [32,75] resistivity contrasts. In contrasts, the PD array recovered the first row of the checkerboard patterns (Figure 27); however, the anomaly sizes were significantly overestimated for the higher values of negative perturbations (−60% and −80%). It can slightly detect the positive 20% perturbation of the second-row checkerboards. The PD array resolution was slightly improved as the resistivity contrast increased, indicating the less sensitivity of the PD array to the resistivity perturbations. The WS array can recover the first row of the checkerboard anomalies in all the perturbations as well as high perturbations (60% and 80%) of the second row (Figure 28). However, the resolution of the WS array is relatively lower than the DD array, and its resolution considerably improved for the increased resistivity contrasts, indicating high sensitivity to resistivity perturbations. The PP array adequately reproduced the first-row checkerboards for all the perturbations. Its resolution slightly increased, but the improvements are negligible. This is likely related to the low sensitivity of the PP array for both low- and high-resistivity contrasts (Figure 29). As shown in Figures 26–29, all the arrays recovered the first row of the checkerboard patterns for the tested resistivity perturbations. This may be associated with the high sensitivity of the resistivity methods near the vicinity of the electrode positions.

Furthermore, the checkerboard tests were used to assess the inversion artefacts. The extensive anomaly features observed below 2 m depth in Figure 23, below 4 m depth in Figures 24 and 26–29, and below 8 m depth in Figure 25 were not in the input model and interpreted as inversion artefacts. The pronounced artefacts at the edge of the profiles (e.g., Figure 26) are likely related to limited data points or no data coverages [76]. As shown in Figures 26–29, the artefact contaminations were slightly increased as the degree of resistivity perturbation increases, as the cross-borehole resistivity experiment for analyzing subsurface heterogeneity between two boreholes has also shown [70]. Even though the DD array has high resolving power, the recovered models were highly contaminated by the inversion artefacts (Figure 26). The PD and PP arrays produced moderate artefacts (Figures 27 and 29), whereas the WS array showed smaller artefacts (Figure 28). The inversion artefacts show a linear association with signal-to-noise ratio; for instance, the low signal-to-noise ratio of the DD array correlates with greater artefacts and the high signal-to-noise ratio of the WS array correlates with lesser inversion artefacts [25, 77].

We present the summary of the results from the experimental model and all the synthetic modeling performed in this study in Table 1. It can help readers to decide appropriate array types for the planned subsurface cavity studies.

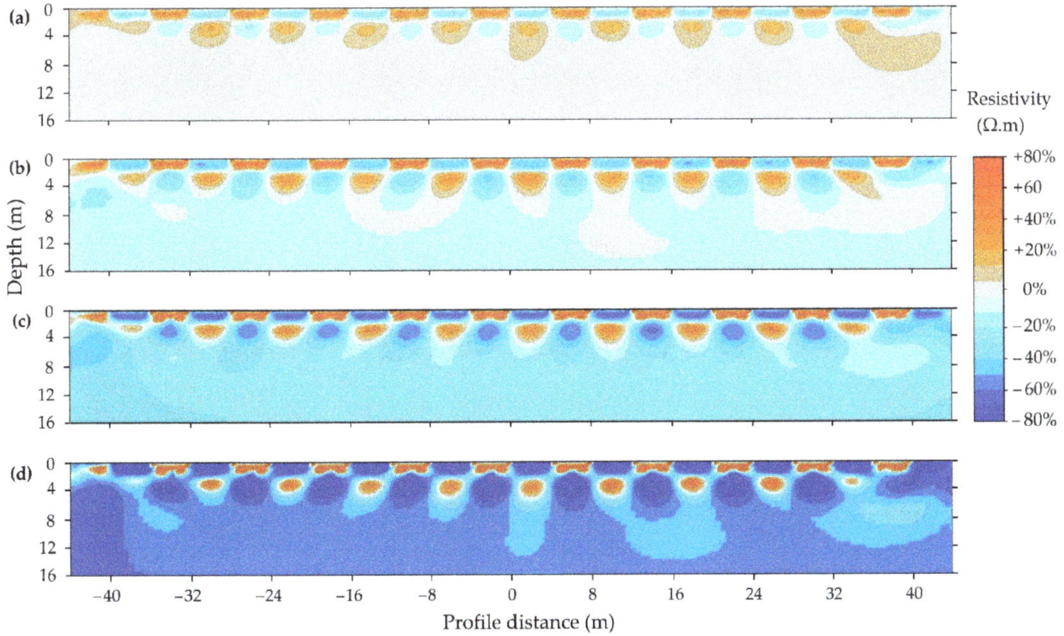

Figure 26. Dipole-dipole (DD) array sensitivity to resistivity perturbations. The inverted model for: (**a**) ±20% perturbation, (**b**) ±40% perturbation, (**c**) ±60% perturbation, (**d**) ±80% perturbation.

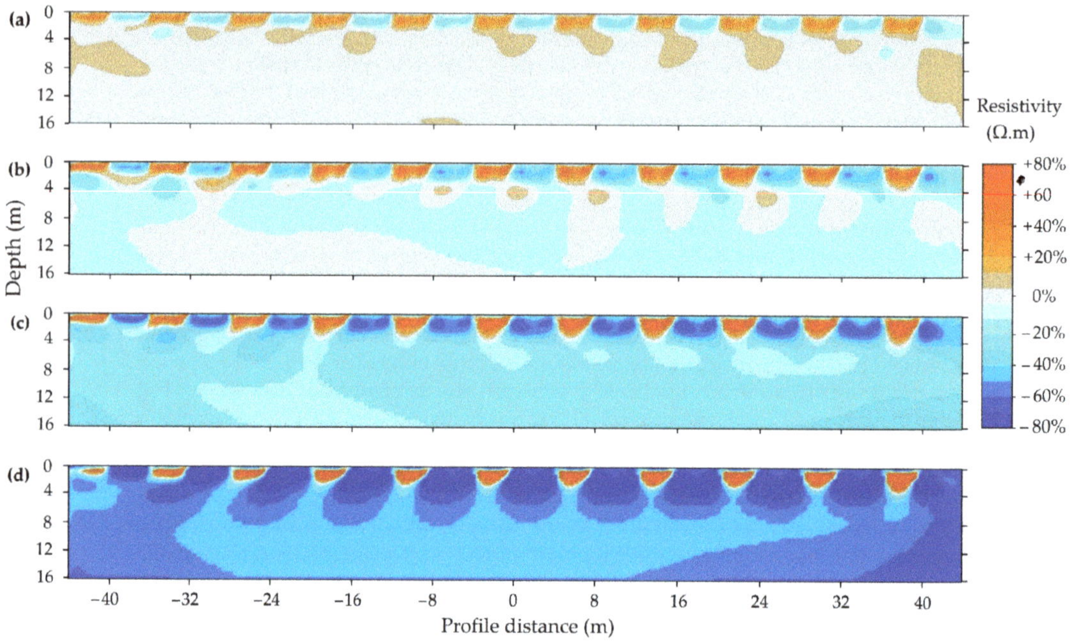

Figure 27. Pole-dipole (PD) array sensitivity to resistivity perturbations. The inverted model for: (**a**) ±20% perturbation, (**b**) ±40% perturbation, (**c**) ±60% perturbation, (**d**) ±80% perturbation.

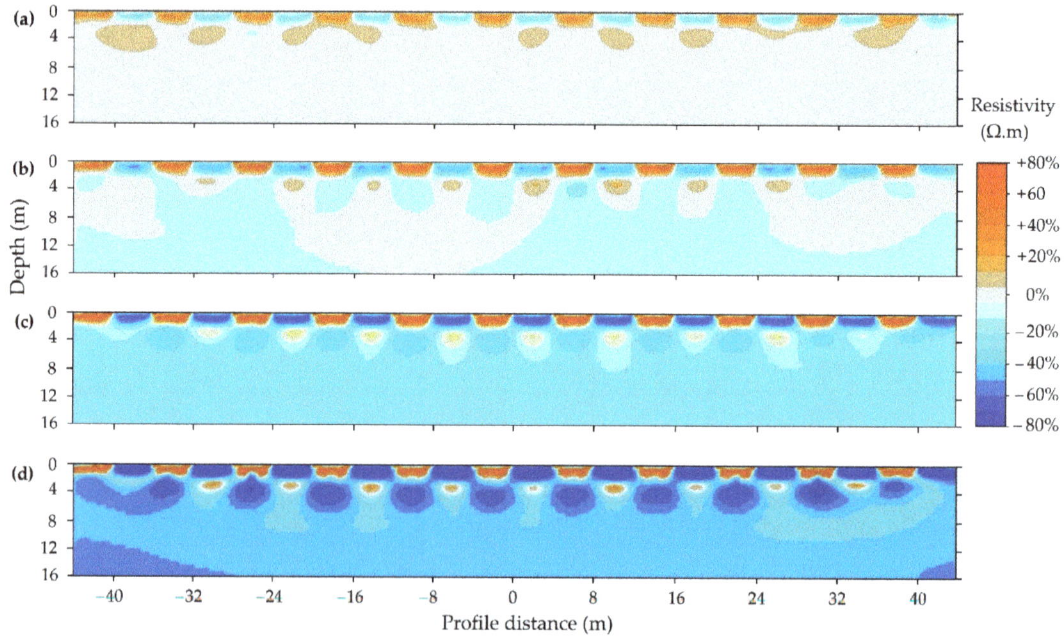

Figure 28. Wenner–Schlumberger (WS) array sensitivity to resistivity perturbations. The inverted model for: (**a**) ±20% perturbation, (**b**) ±40% perturbation, (**c**) ±60% perturbation, (**d**) ±80% perturbation.

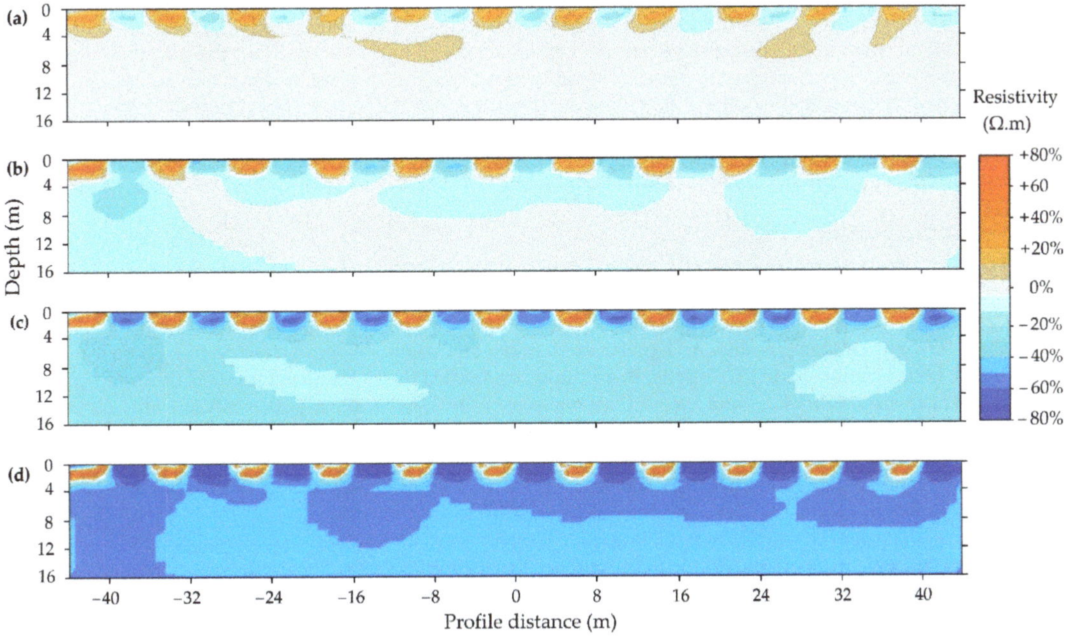

Figure 29. Pole-pole (PP) array sensitivity to resistivity perturbations. The inverted model for: (**a**) ±20% perturbation, (**b**) ±40% perturbation, (**c**) ±60% perturbation, (**d**) ±80% perturbation.

Table 1. Adequacy of the resistivity imaging types for cavity study: highly effective = xxx, moderately effective = xx, and less effective = x.

Adequacy of Arrays for Cavity	DD	PD	WS	PP
Shallow depth (≲4 m) [1]	xxx	xxx	xxx	xx
Intermediate depth (≲8 m and >4 m) [1]	xx	x	x	x
Deeper depth (>8 m) [1]	x	x	x	-
Small size (≲0.5 m) [2]	x	-	-	-
Intermediate size (≲3 m and >0.5 m) [2]	xxx	xx	xx	x
Large size (>3 m) [2]	xxx	xxx	xxx	xx
Horizontal cavity [1,2]	xxx	xx	xx	x
Vertical cavity [1]	xx	xx	x	x
Inclined cavity [1]	xxx	xxx	x	x
Dry clay-filled cavity [1,2]	xxx	x	xx	x
Saturated clay-filled cavity [1,2]	xxx	xxx	xxx	xx

[1] Considering a 1.5 × 6 m cavity size. [2] Considering 4.2 m depth. Except for dry and saturated clay, all the scenarios considered an air-filled cavity. The host medium is considered as homogenous limestone. For the heterogeneous background, the checkerboard cell size scenarios could be used.

In this study, we observed that a specific resistivity imaging method might provide varying model accuracy, resolution, and anomaly responses for inversion of a single cavity model situated at different survey depths. Such ambiguity in subsurface cavity studies can substantially be reduced by applying two or more geophysical methods. Various methods have different capabilities to resolve the subsurface structures; thus, they complement each other and obtain detailed subsurface information. The joint inversion of two separate datasets may improve the overall resolution and match the inverted and the actual model [78]. Moreover, the joint interpretation of different outputs from two or more geophysical methods can minimize model ambiguity [77]. For example, a joint interpretation of results from resistivity tomography and ground penetration radar survey can reduce the non-uniqueness of the inverted resistivity model. Particularly in cavity study, the upper and lower boundaries of the cavity will be clearly determined by high-resolution ground penetration radar; however, it usually fails to provide information about the host medium and the cavity-filling materials. Instead, the resistivity imaging quite well detects the cavity-filling materials and the surrounding formation, but it cannot resolve the accurate geometry of the cavity [19]. Hence, using a joint interpretation of the model results and performing a joint inversion of different methods can reduce those inherited uncertainties of the resistivity imaging.

4. Conclusions

This study examined the uncertainties of resistivity imaging and identified the efficient arrays for subsurface cavity studies. Different numerical examples were demonstrated to examine the depth, size, orientation, geometry, and filling material effects on the array's imaging ability. An experimental cavity study was also performed to compare with the numerical model results. In addition, we studied the spatial resolution for different checkerboard sizes and assessed arrays' sensitivities to resistivity perturbations. The conceptual models were simulated to generate apparent resistivity data. The detection abilities of the arrays were statistically analyzed using variances and anomaly effects of the synthetically measured resistivity data. The apparent resistivity data were inverted to recover geoelectric models, and the model accuracies were evaluated.

Based on cavity depth modeling, the DD array exhibited the highest anomaly effect (1.46) and variance (24,400 $\Omega \cdot m$) in measured resistivity data for the shallowest cavity (T_1) set at 2.2 m depth. In contrast, the PP array showed the lowest anomaly effect (0.18) and variance (900 $\Omega \cdot m$) for the deeper cavity (T_6) set at 12.2 m depth. The anomaly effect and the variance of resistivity data decreased for increasing cavity depths, limiting the recovery of the cavity information. The DOI threshold depths were delineated, and the inverted models were inspected. As cavity depth increases, the statistical image correlation between the inverted and actual models was significantly decreased. The anomaly sizes were considerably overestimated as cavity depth increases, which can highly restrict the recovery of accurate geometries.

The arrays' spatial resolution was significantly reduced as the size of the checkerboard decreased. All the tested arrays accurately resolved the first row of the checkerboard cells, but the second-row cells were considerably smeared, especially in the WS, PD, and PP arrays. Based on sensitivity tests, all the arrays were highly sensitive to the resistivity contrasts at the vicinity to electrode location. The DD and WS arrays showed higher sensitivity to resistivity perturbations than the PD and PP arrays. The PD displayed slightly better sensitivity compared to the PP array. Considerable artefact contaminations occurred for the DD, PD, and PP models, whereas minor contamination was observed on the WS models, which correlates with the signal-to-noise ratios of these arrays.

The study identified the ideal subsurface target probing arrays based on model accuracy, resolution, and sensitivity. We infer that the DD array is the most effective for studying subsurface targets (e.g., cavity), the PD and WS arrays are adequate, whereas the PP array is the least suitable for target probing. The sensitivity and resolution were substantially reduced as the depth increased; thus, a single cavity set at varying depth resulted in different solutions. This leads to ambiguity in model interpretations of the resistivity method. Under these circumstances, integrating prior information, performing a joint inversion of different parameters, and jointly interpreting the model outputs of two or more methods can significantly reduce the inherent uncertainties of the resistivity method.

Author Contributions: Y.G.D. was responsible for data analysis, simulations, and writing the draft. P.-Y.C. supervised and instructed results analysis and modified the manuscript. J.M.P. provided comments on results discussions with the other two authors. All authors have read and agreed to the published version of the manuscript.

Funding: This research was funded by the Minister of Science and Technology (MOST) of Taiwan under the project number of MOST 108-2638-E-008-001-MY2. This research was also partly supported by Earthquake-Disaster, Risk Evaluation and Management Centre, E-DREaM, from the Featured Areas Research Centre Program within the framework of the Higher Education Sprout Project by the Ministry of Education (MOE) in Taiwan.

Institutional Review Board Statement: Not applicable.

Informed Consent Statement: Not applicable.

Data Availability Statement: Data will be available upon request to the authors.

Acknowledgments: We would like to thank Partul Ranjan for his valuable revision and idea. We are grateful to Yitea Seneshaw for proofreading and comments. We also thank Shie Meng-Shiun, Ding-Jiun Lin, Yoseph Arba, and Lingerew Nebere, who contributed to experimental data collection. The comments and suggestions from anonymous reviewers have significantly improved the paper.

Conflicts of Interest: The authors declare no conflict of interest.

References

1. Chang, P.-Y.; Chang, L.-C.; LEE, T.-Q.; Chan, Y.-C.; Chen, H.-F. Examining Lake-bottom structures with the resistivity imaging method in Ilan's Da-Hu Lake in Northeastern Taiwan. *J. Appl. Geophy.* **2015**, *119*, 170–177.
2. Fazzito, S.Y.; Rapalini, A.E.; Cortés, J.M.; Terrizzano, C.M. Characterization of Quaternary faults by electric resistivity tomography in the Andean Precordillera of Western Argentina. *J. S. Am. Earth Sci.* **2009**, *28*, 217–228. [CrossRef]
3. Ho, G.-R.; Chang, P.Y.; Lo, W.; Liu, C.-M.; Song, S.-R. New evidence of regional geological structures inferred from reprocessing and resistivity data interpretation in the Chingshui-Sanshing-Hanchi area of Southwestern Ilan County, NE Taiwan. *Terr. Atoms. Ocean Sci.* **2014**, *25*, 491. [CrossRef]
4. Tsai, J.P.; Chang, P.Y.; Yeh, T.C.J.; Chang, L.C.; Hsiao, C.T. Constructing the Apparent Geological Model by Fusing Surface Resistivity Survey and Borehole Records. *J. Groundw.* **2019**, *57*, 590–601. [CrossRef]
5. Caputo, R.; Piscitelli, S.; Oliveto, A.; Rizzo, E.; Lapenna, V. The use of electrical resistivity tomographies in active tectonics: Examples from the Tyrnavos Basin, Greece. *J. Geodyn.* **2003**, *36*, 19–35. [CrossRef]
6. Chang, P.-Y.; Chang, L.-C.; Hsu, S.-Y.; Tsai, J.-P.; Chen, W.-F. Estimating the hydrogeological parameters of an unconfined aquifer with the time-lapse resistivity-imaging method during pumping tests: Case studies at the Pengtsuo and Dajou sites, Taiwan. *J. Appl. Geophy.* **2017**, *144*, 134–143. [CrossRef]
7. Zhang, G.; Zhang, G.-B.; Chen, C.-C.; Chang, P.-Y.; Wang, T.-P.; Yen, H.-Y.; Dong, J.-J.; Ni, C.-F.; Chen, S.-C.; Chen, C.-W. Imaging rainfall infiltration processes with the time-lapse electrical resistivity imaging method. *Pure Appl. Geophys.* **2016**, *173*, 2227–2239. [CrossRef]
8. Chang, P.-Y.; Chen, C.-c.; Chang, S.-K.; Wang, T.-B.; Wang, C.-Y.; Hsu, S.-K. An investigation into the debris flow induced by Typhoon Morakot in the Siaolin Area, Southern Taiwan, using the electrical resistivity imaging method. *Geophys. J. Int.* **2012**, *188*, 1012–1024. [CrossRef]
9. Drahor, M.G.; Göktürkler, G.; Berge, M.A.; Kurtulmuş, T.Ö. Application of electrical resistivity tomography technique for investigation of landslides: A case from Turkey. *J. Environ. Geol.* **2006**, *50*, 147–155. [CrossRef]
10. Gómez-Ortiz, D.; Martín-Velázquez, S.; Martín-Crespo, T.; De Ignacio-San José, C.; Lillo, J. Application of electrical resistivity tomography to the environmental characterization of abandoned massive sulphide mine ponds (Iberian Pyrite Belt, SW Spain). *Near Surf. Geophys.* **2010**, *8*, 65–74. [CrossRef]
11. Cardarelli, E.; Cercato, M.; Cerreto, A.; Di Filippo, G. Electrical resistivity and seismic refraction tomography to detect buried cavities. *Geophys. Prospect.* **2010**, *58*, 685–695. [CrossRef]
12. Van Schoor, M. Detection of sinkholes using 2D electrical resistivity imaging. *J. Appl. Geophy.* **2002**, *50*, 393–399. [CrossRef]
13. Gabarrón, M.; Martínez-Pagán, P.; Martínez-Segura, M.A.; Bueso, M.C.; Martínez-Martínez, S.; Faz, Á.; Acosta, J.A.J.M. Electrical resistivity tomography as a support tool for physicochemical properties assessment of near-surface waste materials in a mining tailing pond (El Gorguel, SE Spain). *Minerals* **2020**, *10*, 559. [CrossRef]
14. Batayneh, A.T. Archaeogeophysics–archaeological prospection–A mini review. *J. King Saud Univ. Sci.* **2011**, *23*, 83–89. [CrossRef]
15. Mochales, T.; Casas, A.; Pueyo, E.; Pueyo, O.; Román, M.; Pocoví, A.; Soriano, M.; Ansón, D. Detection of underground cavities by combining gravity, magnetic and ground penetrating radar surveys: A case study from the Zaragoza area, NE Spain. *Environ. Geol.* **2008**, *53*, 1067–1077. [CrossRef]
16. Das, P.; Mohanty, P.R. Resistivity imaging technique to delineate shallow subsurface cavities associated with old coal working: A numerical study. *Environ. Earth Sci.* **2016**, *75*, 661. [CrossRef]
17. Loke, M.; Barker, R. Practical techniques for 3D resistivity surveys and data inversion. *Geophys. Prospect.* **1996**, *44*, 499–523. [CrossRef]

18. Sharma, S.; Verma, S. Solutions of the inherent problem of the equivalence in direct current resistivity and electromagnetic methods through global optimization and joint inversion by successive refinement of model space. *Geophys. Prospect.* **2011**, *59*, 760–776. [CrossRef]
19. Orlando, L. GPR to constrain ERT data inversion in cavity searching: Theoretical and practical applications in archeology. *J. Appl. Geophy.* **2013**, *89*, 35–47. [CrossRef]
20. Narayan, S.; Dusseault, M.B.; Nobes, D.C. Inversion techniques applied to resistivity inverse problems. *Inverse Probl.* **1994**, *10*, 669. [CrossRef]
21. Zhou, W.; Beck, B.F.; Adams, A.L. Effective electrode array in mapping karst hazards in electrical resistivity tomography. *J. Environ. Geol.* **2002**, *42*, 922–928. [CrossRef]
22. Loke, M. *Tutorial: 2-D and 3-D Electrical Imaging Surveys*; Geotomo Software: Penang, Malaysia, 2004.
23. Yang, X.; Lagmanson, M.B. Planning resistivity surveys using numerical simulations. In Proceedings of the 16th EEGS Symposium on the Application of Geophysics to Engineering and Environmental Problems, San Antonio, TX, USA, 6–10 April 2003; p. cp-190-00047.
24. Santos, F.A.M.; Afonso, A.R.A. Detection and 2D modelling of cavities using pole–dipole array. *J. Environ. Geol.* **2005**, *48*, 108–116. [CrossRef]
25. Dahlin, T.; Zhou, B. A numerical comparison of 2D resistivity imaging with 10 electrode arrays. *Geophys. Prospect.* **2004**, *52*, 379–398. [CrossRef]
26. Amini, A.; Ramazi, H. CRSP, numerical results for an electrical resistivity array to detect underground cavities. *Open Geosci.* **2017**, *9*, 13–23. [CrossRef]
27. Martinez-Lopez, J.; Rey, J.; Duenas, J.; Hidalgo, C.; Benavente, J. Electrical Tomography Applied to the Detection of Subsurface Cavites. *J. Caves Karst Stud.* **2013**, *75*, 28–37. [CrossRef]
28. Oldenburg, D.W.; Li, Y. Estimating depth of investigation in dc resistivity and IP surveys. *Geophysics* **1999**, *64*, 403–416. [CrossRef]
29. Martorana, R.; Fiandaca, G.; Casas Ponsati, A.; Cosentino, P. Comparative tests on different multi-electrode arrays using models in near-surface geophysics. *J. App. Geophy.* **2009**, *6*, 1–20. [CrossRef]
30. Hassan, A.A.; Kadhim, E.H.; Ahmed, M.T. Performance of Various Electrical Resistivity Configurations for Detecting Buried Tunnels Using 2D Electrical Resistivity Tomography Modelling. *Diyala J. Eng. Sc.* **2018**, *11*, 14–21. [CrossRef]
31. Rezaei, S.; Shooshpasha, I.; Rezaei, H. Reconstruction of landslide model from ERT, geotechnical, and field data, Nargeschal landslide, Iran. *Bull. Eng. Geol. Environ.* **2019**, *78*, 3223–3237. [CrossRef]
32. Eissa, R.; Cassidy, N.; Pringle, J.; Stimpson, I. Electrical resistivity tomography array comparisons to detect cleared-wall foundations in brownfield sites. *Q. J. Eng. Geol.* **2020**, *53*, 137–144. [CrossRef]
33. Park, M.K.; Park, S.; Yi, M.-J.; Kim, C.; Son, J.-S.; Kim, J.-H.; Abraham, A.A. Application of electrical resistivity tomography (ERT) technique to detect underground cavities in a karst area of South Korea. *Environ. Earth Sci.* **2014**, *71*, 2797–2806. [CrossRef]
34. Neyamadpour, A.; Wan Abdullah, W.; Taib, S.; Neyamadpour, B. Comparison of Wenner and dipole–dipole arrays in the study of an underground three-dimensional cavity. *J. App. Geophy.* **2010**, *7*, 30–40. [CrossRef]
35. Boschi, L. Measures of resolution in global body wave tomography. *Geophys. Res. Lett.* **2003**, *30*. [CrossRef]
36. Mewes, B.; Hilbich, C.; Delaloye, R.; Hauck, C. Resolution capacity of geophysical monitoring regarding permafrost degradation induced by hydrological processes. *Cryosphere* **2017**, *11*, 2957–2974. [CrossRef]
37. Butler, D.J.N.-S.G.; Activity, H. Detection and characterization of subsurface cavities, tunnels and abandoned mines. *Near Surf. Geophys. Hum. Act.* **2008**, 578–584.
38. Satitpittakul, A.; Vachiratienchai, C.; Siripunvaraporn, W. Factors influencing cavity detection in Karst terrain on two-dimensional (2-D) direct current (DC) resistivity survey: A case study from the western part of Thailand. *Eng. Geol.* **2013**, *152*, 162–171. [CrossRef]
39. Smith, R.C.; Sjogren, D.B. An evaluation of electrical resistivity imaging (ERI) in Quaternary sediments, southern Alberta, Canada. *Geosphere* **2006**, *2*, 287–298. [CrossRef]
40. Keller, G.V.; Carmichael, R. Electrical properties of rocks and minerals. *CRC Handb. Phys. Prop. Rocks* **1982**, *1*, 217–293.
41. Lowrie, W.; Fichtner, A. *Fundamentals of Geophysics*; Cambridge University Press: Cambridge, UK, 2020.
42. Zhao, D.; Hasegawa, A.; Kanamori, H. Deep structure of Japan subduction zone as derived from local, regional, and teleseismic events. *J. Geophys. Res. Solid Earth* **1994**, *99*, 22313–22329. [CrossRef]
43. Zhao, D.; Yanada, T.; Hasegawa, A.; Umino, N.; Wei, W. Imaging the subducting slabs and mantle upwelling under the Japan Islands. *Geophys. J. Int.* **2012**, *190*, 816–828. [CrossRef]
44. Zhang, Y.; Wang, B.; Lin, G.; Ouyang, Y.; Wang, T.; Xu, S.; Song, L.; Wang, R. Three-Dimensional P-wave Velocity Structure of the Zhuxi Ore Deposit, South China Revealed by Control-Source First-Arrival Tomography. *Minerals* **2020**, *10*, 148. [CrossRef]
45. Athanasiou, E.; Tsourlos, P.; Papazachos, C.; Tsokas, G. Combined weighted inversion of electrical resistivity data arising from different array types. *J. App. Geophy.* **2007**, *62*, 124–140. [CrossRef]
46. Günther, T. *Inversion Methods and Resolution Analysis for the 2D/3D reconstRuction of Resistivity Structures from DC Measurements*; Universitätsbibliothek der TU BAF: Freiberg, Germany, 2005.
47. Okpoli, C.C. Sensitivity and resolution capacity of electrode configurations. *Geophys. J. Int.* **2013**, *2013*, 608037. [CrossRef]
48. Oldenborger, G.A.; Routh, P.S.; Knoll, M.D. Sensitivity of electrical resistivity tomography data to electrode position errors. *Geophys. J. Int.* **2005**, *163*, 1–9. [CrossRef]

49. AGI. *Instruction Manual for EarthImager™ 2D Version 2.4. 0.*; Advanced Geosciences, Inc.: Austin, TX, USA, 2009.
50. Rubin, Y.; Hubbard, S.S. *Hydrogeophysics*; Springer Science & Business Media: Berlin/Heidelberg, Germany, 2006; Volume 50.
51. Wolke, R.; Schwetlick, H. Iteratively reweighted least squares: Algorithms, convergence analysis, and numerical comparisons. *SIAM J. Sci. Comput.* **1988**, *9*, 907–921. [CrossRef]
52. Degroot-Hedlin, C.; Constable, S. Occam's inversion to generate smooth, two-dimensional models from magnetotelluric data. *Geophysics* **1990**, *55*, 1613–1624. [CrossRef]
53. Ren, Z.; Kalscheuer, T. Uncertainty and resolution analysis of 2D and 3D inversion models computed from geophysical electromagnetic data. *Surv. Geophys.* **2020**, *41*, 47–112. [CrossRef]
54. Militzer, H.; Rösler, R.; Lösch, W. Theoretical and experimental investigations for cavity research with geoelectrical resistivity methods. *Geophys. Prospect.* **1979**, *27*, 640–652. [CrossRef]
55. Aizebeokhai, A.; Olayinka, A. Anomaly effects of arrays for 3d geoelectrical resistivity imaging using orthogonal or parallel 2d profiles. *Afr. J. Environ. Sci. Technol.* **2010**, *4*, 446–454.
56. Chai, T.; Draxler, R.R. Root mean square error (RMSE) or mean absolute error (MAE)–Arguments against avoiding RMSE in the literature. *Geosci. Model Dev.* **2014**, *7*, 1247–1250. [CrossRef]
57. Hilbich, C.; Marescot, L.; Hauck, C.; Loke, M.; Mäusbacher, R. Applicability of electrical resistivity tomography monitoring to coarse blocky and ice-rich permafrost landforms. *Permafr. Periglac. Process.* **2009**, *20*, 269–284. [CrossRef]
58. Yang, X.; Chen, X.; Carrigan, C.R.; Ramirez, A.L. Uncertainty quantification of CO_2 saturation estimated from electrical resistance tomography data at the Cranfield site. *Int. J. Greenh. Gas Control.* **2014**, *27*, 59–68. [CrossRef]
59. Portniaguine, O.; Zhdanov, M.S. Focusing geophysical inversion images. *Geophysics* **1999**, *64*, 874–887. [CrossRef]
60. Hauck, C.; Mühll, D.V. Inversion and interpretation of two-dimensional geoelectrical measurements for detecting permafrost in mountainous regions. *Permafr. Periglac. Process.* **2003**, *14*, 305–318. [CrossRef]
61. Olayinka, A.; Yaramanci, U.; Geophysics, E. Choice of the best model in 2-D geoelectrical imaging: Case study from a waste dump site. *Eur. J. Environ. Sci.* **1999**, *3*, 221–244.
62. Carey, A.M.; Paige, G.B.; Carr, B.J.; Dogan, M.J. Forward modeling to investigate inversion artifacts resulting from time-lapse electrical resistivity tomography during rainfall simulations. *J. Appl. Geophy.* **2017**, *145*, 39–49. [CrossRef]
63. Caterina, D.; Beaujean, J.; Robert, T.; Nguyen, F. A comparison study of different image appraisal tools for electrical resistivity tomography. *Near Surf. Geophys.* **2013**, *11*, 639–658. [CrossRef]
64. Seaton, W.J.; Burbey, T.J. Evaluation of two-dimensional resistivity methods in a fractured crystalline-rock terrane. *J. Appl. Geophy.* **2002**, *51*, 21–41. [CrossRef]
65. Kim, J.-H.; Yi, M.-J.; Hwang, S.-H.; Song, Y.; Cho, S.-J.; Synn, J.-H. Integrated geophysical surveys for the safety evaluation of a ground subsidence zone in a small city. *J. Geophys. Eng.* **2007**, *4*, 332–347. [CrossRef]
66. Festa, V.; Fiore, A.; Parise, M.; Siniscalchi, A.; Studies, K. Sinkhole evolution in the Apulian karst of southern Italy: A case study, with some considerations on sinkhole hazards. *J Caves Karst Stud.* **2012**, *74*, 137–147. [CrossRef]
67. Muhammad, F.; Samgyu, P.; Young, S.S.; Ho Kim, J.; Mohammad, T.; Adepelumi, A.A. Subsurface cavity detection in a karst environment using electrical resistivity (er): A case study from yongweol-ri, South Korea. *Earth Sci. Res. J.* **2012**, *16*, 75–82.
68. Saribudak, M.; Hawkins, A.; Stoker, K. Do air-filled caves cause high resistivity anomalies? A six-case study from the Edwards Aquifer Recharge Zone in San Antonio, Texas. *Houst. Geol. Soc. Bull.* **2012**, *54*, 41–49.
69. Gourdol, L.; Clément, R.; Juilleret, J.; Pfister, L.; Hissler, C. Large-scale ERT surveys for investigating shallow regolith properties and architecture. *Hydrol. Earth Syst. Sci. Discuss.* **2018**, *2018*, 1–39.
70. Nimmer, R.E.; Osiensky, J.L.; Binley, A.M.; Williams, B.C. Three-dimensional effects causing artifacts in two-dimensional, cross-borehole, electrical imaging. *J. Hydrol.* **2008**, *359*, 59–70. [CrossRef]
71. Daily, W.; Ramirez, A. Electrical resistance tomography during in-situ trichloroethylene remediation at the Savannah River Site. *J. Appl. Geophy.* **1995**, *33*, 239–249. [CrossRef]
72. Verdet, C.; Anguy, Y.; Sirieix, C.; Clément, R.; Gaborieau, C. On the effect of electrode finiteness in small-scale electrical resistivity imaging. *Geophysics* **2018**, *83*, EN39–EN52. [CrossRef]
73. Rucker, D.F.; Loke, M.H.; Levitt, M.T.; Noonan, G.E. Electrical-resistivity characterization of an industrial site using long electrodes. *Geophysics* **2010**, *75*, WA95–WA104. [CrossRef]
74. Hermawan, O.R.; Putra, D.P.E. the effectiveness of wenner-schlumberger and dipole-dipole array of 2d geoelectrical survey to detect the occurring of groundwater in the gunung kidul karst aquifer system, Yogyakarta, Indonesia. *J. Appl. Geol.* **2016**, *1*, 71–81. [CrossRef]
75. Andrej, M.; Uros, S.; Studies, K. Electrical Resistivity Imaging of Cave Divaska Jama, Solvania. *J. Caves Karst. Stud.* **2012**, *74*, 235–242. [CrossRef]
76. Dahlin, T.; Loke, M.H. Resolution of 2D Wenner resistivity imaging as assessed by numerical modelling. *J. Appl. Geophy.* **1998**, *38*, 237–249. [CrossRef]
77. Adepelumi, A.; Fayemi, O. Joint application of ground penetrating radar and electrical resistivity measurements for characterization of subsurface stratigraphy in Southwestern Nigeria. *J. Geophys. Eng.* **2012**, *9*, 397–412. [CrossRef]
78. Gallardo, L.A.; Meju, M.A. Joint two-dimensional DC resistivity and seismic travel time inversion with cross-gradients constraints. *J. Geophys. Res. Solid Earth* **2004**, *109*. [CrossRef]

applied sciences

MDPI

Article

Seismic Data Interpretation and Petrophysical Analysis of Kabirwala Area Tola (01) Well, Central Indus Basin, Pakistan

Naveed Ahmad [1], Sikandar Khan [2,*] and Abdullatif Al-Shuhail [3]

1 Department of Marine Geophysics, Ocean University of China, Qingdao 266100, China; naveedahmadgeophy@gmail.com
2 Department of Mechanical Engineering, King Fahd University of Petroleum and Minerals, Dhahran 31261, Saudi Arabia
3 Department of Geosciences, King Fahd University of Petroleum and Minerals, Dhahran 31261, Saudi Arabia; ashuhail@kfupm.edu.sa
* Correspondence: sikandarkhan@kfupm.edu.sa

Abstract: Well logging is a significant procedure that assists geophysicists and geologists with making predictions regarding boreholes and efficiently utilizing and optimizing the drilling process. The current study area is positioned in the Punjab Territory of Pakistan, and the geographic coordinates are 30020′10 N and 70043′30 E. The objective of the current research work was to interpret the subsurface structure and reservoir characteristics of the Kabirwala area Tola (01) well, which is located in the Punjab platform, Central Indus Basin, utilizing 2D seismic and well log data. Formation evaluation for hydrocarbon potential using the reservoir properties is performed in this study. For the marked zone of interest, the study also focuses on evaluating the average water saturation, average total porosity, average effective porosity, and net pay thickness. The results of the study show a spotted horizon stone with respect to time and depth as follows: Dunghan formation, 0.9 s and 1080.46 m; Cretaceous Samana Suk formation, 0.96 s and 1174.05 m; Datta formation, 1.08 s and 1400 m; and Warcha formation, 1.24 s and 1810 m. Based on the interpretation of well logs, the purpose of petrophysical analysis was to identify hydrocarbon-bearing zones in the study area. Gamma ray, spontaneous potential, resistivity, neutron, and density log data were utilized. The high zone present in the east–west part of the contour maps may be a possible location of hydrocarbon entrapment, which is further confirmed by the presence of the Tola-01 well.

Keywords: 2D seismic interpretation; contour mapping; petrophysical analysis; porosity

1. Introduction

Well logging is a significant tool that assists geophysicists and geologists with making predictions regarding hydrocarbon reservoirs and provides important information such as rock composition, characteristics, and integrity as well as the presence of fluids and hydrocarbon. The collected data from well logs can be utilized to perform feasibility studies relevant to the physical properties of the area of interest under geophysical work. The oil and gas industries utilize estimated petrophysical parameters from well log data, which is a crucial part of the exploration and production processes [1,2]. These petrophysical parameters also assist with interpreting the subsurface for detection and estimation of hydrocarbon reservoirs. Furthermore, the oil and gas industries mostly use wireline logging to achieve continuous records regarding formation rock properties [3–5].

Seismic data interpretation is commonly used for hydrocarbon exploration and development studies [6–9]. The interpretation of seismic data requires an understanding of the subsurface formations and wave propagation effects. Seismic data consist of reflections that represent the geological structures as well as the reservoir stratigraphy, fabric, and fluid content. Al-Shuhail et al. [3,5] presented 2D viscoelastic models of the Ghawar and central fields of Saudi Arabia and created corresponding multi-component synthetic seismic datasets for hydrocarbon seismology exploration. Similarly, petrophysics plays a significant part in the chemical and physical properties of rocks and their internal fluids. Petrophysical data are generally achieved from well logs and can be used for qualitative and quantitative analysis. The evaluation of petrophysical analysis of well logging provides core properties such as lithology, grain size, porosity, water saturation, clay volume, and permeability [10]. Petrophysics plays an important role in seismic interpretation and has experienced a great leap forward in the past 10 years from significant advances in seismic data processing practices, including full waveform inversion, attribute analysis, and amplitude versus offset methods used in the estimation of reservoir properties from pre-stack and post-stack seismic data [7]. The findings of the literature review are summarized in Table 1.

The main objective of the current research is to interpret the subsurface structure and reservoir characterization of Kabirwala area Tola (01) located in Punjab platform, Central Indus Basin, utilizing 2D seismic and well log data. The study area (Tola (01) well) is in the Central Indus Basin (CIB). The Central Indus Basin (CIB) consists of Punjab Platform, Suleiman Depression (Zindapir Inner Fold Zone, Mari Bugti Inner Fold Zone), along with the Suleiman Fold Belt. As shown in the below Figure 1a, the Tola (01) well is situated in the Kabirwala area near district Multan. The Kabirwala extent is positioned in the Punjab Territory, the geographical coordinates are 30020′10″ North and 70043′30″ East. Kabirwala is one of the four tehsils in Khanewal district Multan. It lies in Central Indus Basin Punjab Platform, Pakistan. Five seismic lines—875-KBR-230, 875-KBR-229, 875-KBR-231, 875-KBR221, and 875-KBR-220—and well logs of Tola-01 were used to explain the subsurface structures and demarcation of zones that have fair potential for hydrocarbon accumulation. The main aim is formation evaluation for hydrocarbon potential using the reservoir properties. In addition, time and depth contour maps are generated to delineate lateral extension and closure of the reservoir. Based on well log interpretation, the main aim of petrophysical analysis is to recognize hydrocarbon-bearing zones in the projected study area. The logging data of gamma ray, spontaneous potential, resistivity, neutron, and density were utilized, as shown in Figure 1a. The exploration license of Tola (01) was granted to Oil and Gas Development Company Limited (OGDCL) in June 1992. The acquisition and processing of 2D seismic data took place on Fort Abbas field in 1994. Exploratory wells Tola-01 and Nandpur-01 were drilled in 1994 and 1996, respectively, on the Punjab platform to discover the petroleum potential of Infra Cambrian reservoir rocks. The input data used in the current research study were taken from OGDCL.

Table 1. Findings of literature review.

Reference	Parameters Used	Simulation Method (2D, 3D, 4D)	Petrophysical Analysis Used (Yes/No)	Stratigraphy Interpretation Used (Yes/No)	Objective of Research Study
[11]	Volume of shale, porosity, hydrocarbon saturation, water saturation	3D	Yes	Yes	Discovery of hydrocarbon reserve resource accumulations in Pennay field
[12]	Effective porosity, clay content, and water saturation	–	Yes	Yes	Proposal of joint estimation of petrophysical properties combining statistical rock physics and Bayesian seismic inversion

Table 1. *Cont.*

Reference	Parameters Used	Simulation Method (2D, 3D, 4D)	Petrophysical Analysis Used (Yes/No)	Stratigraphy Interpretation Used (Yes/No)	Objective of Research Study
[13]	Volume	3D	Yes	Yes	Estimation of brittleness of resource plays in 3D by integrating petrophysics and seismic data analysis
[14]	Joint probability function, porosity, thickness	3D	Yes	Yes	Estimation procedure for extrapolating wireline data from existing wells using geostatistical inversion of post-stack 3D seismic data
[15]	Porosity, water saturation and hydrocarbon saturation	3D	Yes	Yes	Determination of reservoir properties and estimated volume of hydrocarbons within reservoirs
[16]	Porosity, net-to-gross, water saturation, hydrocarbon saturation	3D	Yes	Yes	Reserve evaluation of hydrocarbon-bearing sands
[17]	Volume of shale, porosity, water saturation, formation water resistivity, and hydrocarbon saturation	2D	Yes	Yes	Structural interpretation and hydrocarbon potential of Balkassar oil field
[18]	Density, neutron, self-potential, and resistivity	2D	Yes	Yes	Structural interpretation of Joya Mair oil field using 2D seismic data and petrophysical analysis
[19]	Porosity, permeability, moderate net to gross, and low water saturation	3D	Yes	Yes	Application of 3D static model using 3D seismic and well log data for optimization and development of hydrocarbon potential in KN field of Niger Delta Province

The current research work is organized as follows: Section 2 describes the tectonics of the proposed area, Section 3 describes the generalized stratigraphy of the area, and Section 4 presents a comprehensive analysis of the seismic interpretation and the well log analysis, Section 5 consists of a detailed petrophysical analysis, Section 6 presents the discussion, and finally, Section 7 presents the conclusions of the current research work.

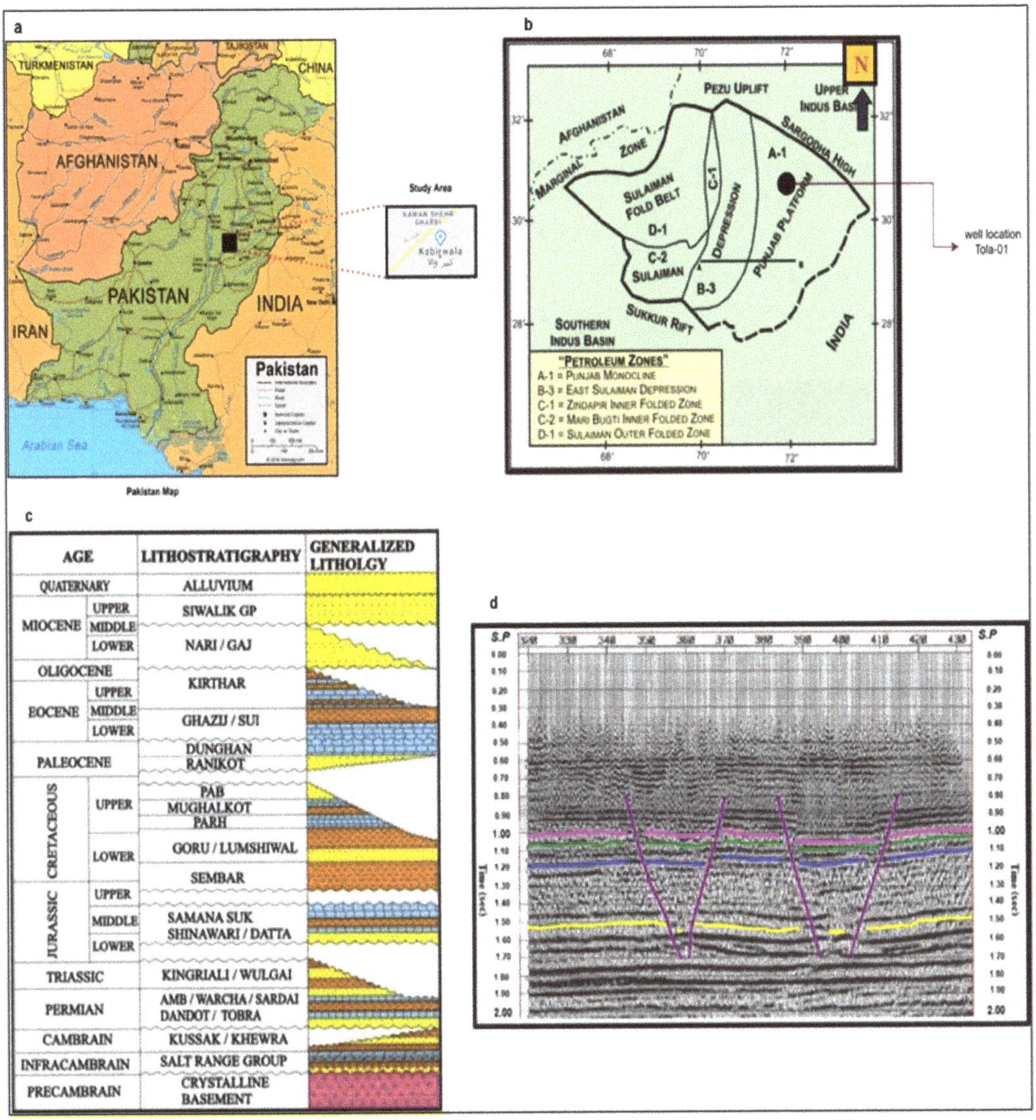

Figure 1. (**a**) Location of the study area. (**b**) Tectonic map of Central Indus Basin and subdivisions of petroleum zones (after [20]). (**c**). Generalized stratigraphic chart of the study area [21]. (**d**). Seismic section with marked horizon.

2. Tectonics of the Area

There are three subdivisions of the Central Indus Basin (CIB): the Punjab platform, Sulaiman fold belt, and Sulaiman depression. The Punjab platform is located at the east–west of the Sulaiman depression and fold belt, as shown in the structural map in Figure 1b [8]. The Punjab platform is a westward-dipping monocline covered with unconsolidated Quaternary deposits with a maximum thickness of about 500 m [9], and the area is almost all desert and neglected surface outcrops [10]. As shown in Figure 1b, Sargodha high is at the north of the CIB, Indian Shield is at the east, an axial belt is at the west, and finally Sukker Rift is located at the south [10]. The Punjab platform was the least affected during Himalayan orogeny; therefore, it possesses non-tectonic origin structures [11]. It also has no exposure to bedrocks. The primogenital rock existing in the basin is counted in the Triassic age. Owing to the absence of active tectonic elements in the CIB, it is almost stable [12].

3. Generalized Stratigraphy of the Area

Stratigraphy is mainly concerned with the study of rock layers (strata) and layering (stratification). Stratigraphy has two related subfields: lithostratigraphy and biostratigraphy. The generalized stratigraphy of the study area is discussed in the following subsections.

3.1. Pliocene/Miocene Stratigraphy

The Pliocene/Miocene stratigraphy of the area includes the following formations:

3.1.1. Nagri Formation

Its lithology is constituted of sandstone with dependent clay as well as conglomerate. Its topmost association is with Dhok Pathan formation, which is transitionary. The age is Late Miocene.

3.1.2. Chinji Formation

Its lithology constitutes of red clay with dependent ash gray or brownish-gray sandstone. It is exclusively limited to the southern bisection of the eastern Suleiman range and is not elaborated in the repose of the Lower Indus Basin. It is substantially superimposed beside the Nagri formation. The age is Late Miocene [21].

3.2. Oligocene Stratigraphy

The Oligocene stratigraphy of the area includes the following formation:

Nari Formation

The name "Nari series", derived from the Nari River in the Kirthar range, was introduced by Khokhar et al. [22]. The Nari formation consists of sandstone, shale, and subordinate limestone. The lower part of the formation, the Nal member, is predominantly composed of crystalline limestone, which is white to gray, brown or yellow shelly, nodular or rough surfaced, thin to thick bedded, and even massive. Thin stringers of dark shale and of thin bedded and fine-grained brown sandstone are also present.

3.3. Eocene Stratigraphy

The Eocene stratigraphy of the area includes the following formations:

3.3.1. Habib Rahi Formation

The Habib Rahi formation was formalized by the Stratigraphic Committee from Habib Rahi limestone of Tainsh et al. [23]. It can be described as a grayish-brown and buff hard limestone that is weathering white. It is fine-grained, thin-bedded, dominantly argillaceous, and in a few places grades into marl.

3.3.2. Pirkoh Formation

The Pirkoh limestone member was introduced by Hemphill and Kidwai [24]. Rehman et al. [25] defined the formation as follows: lithologically, the Pirkoh formation consists of limestone that is light gray to chalky white, buff to brown, fine-grained, mostly thin and regularly bedded, and often argillaceous. The formation commonly contains subordinate beds of soft, shaly limestone, and dark gray calcareous claystone.

3.4. Paleocene Stratigraphy

The Paleocene stratigraphy of the area includes the following formations:

3.4.1. Ghazij Sui Member

The overall lithology in the early Eocene episode is shale, as well as marl in the Lower Indus Basin, along with adjacent areas of the axial belt. The Ghazij Sui member consists of Kirthar formation. The Kirthar formation is separable into four components in sections of eastern Sulaiman province as Habib Rahi limestone, Sirki, Pir koh, and Drazinda components [26].

3.4.2. Dunghan Formation

The Dunghan formation was familiarized by Kazmi [27]. Malkani et al. [28] nominated the order portion to be adjacent to Harnai (latitude 300 08′38″ N, longitude 670 59′33″ E) and renewed the close rank Dunghan formation. It is constituted of limestone, shale, and marl. The limestone is gray to yellowish, fine to intermediate bedded with conglomerates. The shale is gray and khaki as well as calcareous. The marl is brown to gray, fine to intermediate bedded, and fine-grained. Sideward, these formational facies are dissimilar at places with dense limestone precipitates, while at some positions, there is insignificant limestone evidence. The Sui main is the topmost portion of the Dunghan limestone because of its fluctuating manner. It is dense in the Zindapir, Duki, Sanjawi, Harand, and Mughal Kot portions, but it is minute in the Rakhi Gaj and in Mekhtar regions. Petroleum evidence is common in this formation, notably in the Khatan portion [27]. Adjacent to the axial belt, it has disconformity at the base, while the topmost association with the Shaheed Ghat formation is transitionary. It has many mega forms. Its age is deliberated as Late Paleocene, notably excessively to Early Eocene. Even though it is prolonged, all the Paleocene in the Ziarat region, as well as the axial belt area where the Sangiali and Rakhi Gaj formations, i.e., the lower, along with the middle Sangiali group, is not promoted. For instance, the Ziarat Laterite exhibited K-T boundary is contacted through the Parh along the Dunghan formation [28].

3.4.3. Ranikot Formation

The Ranikot formation was subsequently nominated as the Ranikot fortress in the Laki range adjacent to Sindh. Furthermore, the Ranikot group is recognized as infra-Nummulitic. The Ranikot formation has been segregated into three formations, recognized as the Khadro, Bara (lower Ranikot sandstone), and Lakhra (upper Ranikot limestone) formations. The lower part has sandstone of brownish yellow color and shale along with limestone. Correspondingly, the lower Ranikot has diversified sandstone and shale along with gray to brown limestone, and the shale is also present in the Lakhra formation [29].

3.5. Cretaceous Stratigraphy

The Cretaceous stratigraphy of the area includes the following formations:

3.5.1. Sembar Formation

The name "Sembar formation" was introduced by Williams [30], after Sembar Pass in the Mari hills, to include the lower part of the "Belemnite beds" and "Belemnite shales". The "Belemnite beds" of Oldham also include the overlying Goru formation and the Parh limestone. According to Fatmi [31], the Sembar formation consists of black silty shale with interbeds of black siltstone and nodular rusty weathering argillaceous limestone beds or concretions. The upper contact is generally gradational with the Goru formation.

3.5.2. Goru Formation

The name "Goru formation" was introduced by Williams [30]. As described by Fatmi [31], the Goru formation consists of inter bedded limestone, shale, and siltstone.

The limestone is fine-grained, thin-bedded, and light to medium gray and olive gray. The interbedded shale and siltstone are gray, greenish gray, and locally maroon in color.

3.5.3. Parh Limestone

The term "Parh" was introduced by Blanford [32]. The Parh limestone, as described by Fatmi [31], is a lithologically very distinct unit. It is a hard, light gray, white, cream, olive green, thin-to medium-bedded, lithographic to porcellaneous, argillaceous and occasionally platy to slabby limestone with subordinate calcareous shale and marl intercalations. The formation is correlated with the Kawagarh formation of Kohat-Potwar province. The lower contact is with the Goru formation, while the upper contact is with the Moghal Kot formation.

3.5.4. Moghal Kot Formation

The term "Mughal Kot formation" was applied by Williams [30]. Lithologically, as stated by Fatmi [31], the formation comprises dark gray calcareous mudstone and calcareous shale with intercalations of quartos sandstone and light gray argillaceous limestone.

3.5.5. Pab Sandstone

The term "Pab sandstone" was introduced by Eschard et al. [33]. The formation typically consists of quartos sandstone, which is white, cream, or brown, weathering yellow brown, medium to coarse-grained thick bedded to massive, and shows cross-stratification. Some marl and argillaceous limestone, similar to that of the Parh limestone, are found intercalated.

3.6. *Jurassic Stratigraphy*

The Jurassic stratigraphy of the area includes the following formations:

3.6.1. Samana Suk Formation

The name came from a mountain peak in the Samana range, which is revealed. It is constituted of limestone that is absolutely fine-grained and clay, along with sand. The lower association is with the Shinawri and topmost with the Chichali formation. The environment of precipitation is slightly nautical [34].

3.6.2. Shinawri Formation

It lies in Shinawri village in the western part of the Samana range. It has sparse to well-stratified limestone nodular marl, calcareous along with non-calcareous shale, and calcareous sandstone. Its lower association is conformable with the Datta and topmost with the Samana Suk formation. Its environment of precipitation is nautical [35].

3.6.3. Datta Formation

It is positioned in Datta Nala of the Surghar range. The lithology is constituted of arenaceous, siltstone, and shale. It has no peculiar petrified report; however, some carbonaceous remnants are present. Based on the law of superposition, its age is Early Jurassic. Datta shale acted as the superior originator rock and sandstone as the superior reservoir rock. Its environment of precipitation is deltaic [36].

3.7. *Triassic Stratigraphy*

The Triassic stratigraphy of the area includes the following formations:

3.7.1. Kingriali Formation

It is positioned in Zaluch Nala, Tapan Wahan, at the Khisor range. It has two components, nominated as the Doya component, which is constituted of sandstone, dolomite, dolomitic sandstone, and shale, and the Vanjari component, which is mainly dolomitic. Its environment of precipitation is tidal flats [37].

3.7.2. Tredian Formation

It has also two components, Landa and Khatkiari. The Landa component's lithology is sandstone along with shale, while the Khatkiari component is constituted of compacted white-appearing sandstone. Its environment of precipitation is fluvial (non-marine) [38].

3.8. Permian Stratigraphy

The Permian stratigraphy of the area includes the following formations:

3.8.1. Amb Formation

The Amb formation is positioned in Amb village. It has three units: the lower unit is constituted of shale, the intermediate is constituted of limestone, and the topmost is constituted of shale. As a result of the known index of petrified fusulinids, its age is nominated as Late Permian. Its lower association is conformable with the Sardhai and topmost with the Wargal formation. Its environment of precipitation is nautical [39].

3.8.2. Sardhai Formation

The Sardhai formation is positioned in Sardhai Gorge. It is constituted of clays of various appearances with a minute extent of sandstone and siltstone. The topmost part of Sardhai is calcareous. The limestone strata have well-improved petrification. Its environment of precipitation is fluvial delta. The age is Early-Middle Permian [40].

3.8.3. Warcha Formation

Its lithology is constituted of intermediate to coarse-grained zigzag stratified sandstone, conglomerate in places, and interbeds of shale are also present. The pebbles of the unit are particularly of granite of pink appearance along with quartzite. Based on the law of superposition, its age is Early Permian [41].

All of these formations overlay the Salt Range formation, Khewra sandstone, and Kussak formation (composed of sandstone and glauconitic shale), respectively, as shown in Figure 1c.

4. Seismic Data Interpretation

Seismic interpretation is the description of the subsurface geological features from seismic data collected during the extraction process. The drilling team can perform the interpretation of seismic data to determine the structural models beneath the earth's surface. Amplitude, frequency, and phase are the main seismic attributes that are generally used in the interpretation process.

The main aim of interpretation is to identify the different reflectors or horizons as interfaces between different geological formations. Interpretation is mainly required in order to obtain structural and stratigraphic information. During interpretation, the horizons and faults are marked for better representation. In the beginning, interpretation was done manually on paper sections, but nowadays, it is done with the help of powerful computer systems. With graphic support, computer-aided interpretation systems are used in the industry, as shown in Figure 1d. Seismic sections can forecast configurations of up to tens of kilometers. A fault with throw that is less than one-quarter of the wavelength of the seismic wave will be very complicated to mark out in the seismic section [42]. The current study area is situated in an extensional regime, so the general structure is normal related, i.e., horst and graben structure. A normal fault will progress due to the extensional regime. Overall, four faults are apparent on the seismic section, which increases the complexity of the research area. These are marked upon the detection of abrupt variations in the condition of reflectors and misrepresentation or disappearance of the reflection underneath the faults.

The purpose of interpretation is to understand how these structures are formed, and it is necessary for mapping and marking the horizons above and below the target zone. The strata in an exposure or outcropping of sedimentary rock can range from layers as thin as paper, which are known as lamina (plural: laminae or laminations), to beds tens of feet thick. Generally, the more stable and consistent the environmental conditions during deposition, the thicker the strata will be.

The seismic interpretation procedure consists of various tracks reserved to find an explanation of the seismic sector on the subject of all stratigraphic structures and sequences [13]. In the oil and gas industries, the critical objectives of seismic interpretation are to detect hydrocarbon pathways and explore its flow [14]. In the case of the Kabirwala lines, the reflectors are constituted to be fundamentally flat; thus, a conduit cannot be assembled for hydrocarbon migration along with maturation. Seismic interpretation is also necessary for selecting the appropriate geological reservoir for long-term CO_2 sequestration [43–49].

The basic interpretation stages are based on seismic well log data of the Tola-01 well in the Kabirwala area and on seismic lines. As given in Tables 2 and 3, five migrated seismic lines, navigation data of seismic lines, well log data, and formation tops for the Tola-01 well were accessible for this project. Seismic data analysis consists of stratigraphic and structural analysis. The structure and stratigraphy of the subsurface is evaluated from the seismic reflection data [50]. Seismic interpretation involves converting velocity and time into the depth of subsurface reflecting interfaces to translate seismic data into geological images [5].

Table 2. Seismic line information.

Sr. No.	Seismic Line	Line Style	Direction
1	875-KBR-230	Strike	N-S
2	875-KBR-229	Strike	N-S
3	875-KBR-231	Dip	E-W
4	875-KBR-221	Dip	E-W
5	875-KBR-220	Dip	E-W

Table 3. Well log data of the study area.

Sr. No.	Tola-01	Unit
1	Sonic (DT)	μsec/ft
2	Neutron porosity (NPHI)	Fraction
3	Formation density (RHOB)	gm/cm^3
4	Resistivity (MSFL) (LLD) (LLS)	Ω.m
5	Gamma ray (GR)	API
6	Spontaneous potential (SP)	mV
7	Caliper (CALI)	Inches

Figure 2a points out the "Base Map" of the study area (Tola-01 well), which is located in the Punjab platform of Central Indus Basin (CIB). The key purpose of the "Base Map" is to point out the particular direction of the seismic lines, shot points, as well as the well position on the seismic line. In the current study, the "Base Map" points out two strike lines and three dip lines. The two strike lines have the shooting trend from the north to south direction. However, the three dip lines have the shooting trend from the east to west direction. The location of the Tola (1) well is added in the base map. The Tola (1) well is located at 875-KBR-230 line in the Base Map. Owing to the seismic data interpretation, four reflectors are defined in the seismic sections. The well tops of Tola-01 are utilized for the correlation. The horizons are the Dunghan, Samana Suk, Datta, and Warcha formations, as shown in Table 4. Similarly, the horizons of seismic lines are given in detail in Figures 3–7. The Tola (01) well is located on the 875-KBR-230 line in the base map. The 875-KBR-230 line is used as a reference for marking our seismic sections. The well tops of the Tola-01 well were used to locate the required reflectors on the seismic section. The well tops data give us the formation tops with a required depth. As far as the well tops of the Tola-01 well are concerned, the tops of the required formations, which are Dunghan, Samana suk, Datta and Warcha, are given in Tables 4 and 5.

The fit line curve created from the velocity windows calculation and the graph help us find the time against the depth on the time section. The time values of the prescribed formations have been obtained using the time depth chart. Then, these time values obtained from the time-depth chart of well Tola-01 were projected on the seismic line 844-KBR-230. Furthermore, a line is drawn from the prescribed shot point crossways on the seismic traces toward the second time vertically, and the values of the time are also pinpointed on the well line over the seismic line. The time picked on the seismic line is shown in Tables 4 and 5. For labeling the seismic horizons and faults, the time of each reflector will be recognized at various shot points, and the travel time of each reflector will be marked along with faults on the y-axis, in contrast to the shot points on the x-axis.

In the present study, there are four marked reflectors with normal faults on each line. Gridding of these lines was assembled for a contouring plan as well as to interpret the configuration model of these lines for the arrangement of contouring. The horst along with the graben configurations that existed on the seismic sections (Figures 3–7) may be a proper vicinity for the gathering of hydrocarbons. Thus, the labeled seismic sections are the promoted sections of the reflectors, which point out the subsurface configurations in the time domain. The values of the time depth are taken on the seismic line 845-KBR-45. Therefore, the trace on a specific time, using the referred time scaling, is used to locate the time on the well spot. In addition, after four traces are marked, four of them are therefore spread, using the common line joining the trace. The trace is observed by watching it parallel to the paper orientation rather looking from the top. Then, the horizons are marked on the seismic line. The time contour maps of four horizons are organized and demonstrated in Figures 8a, 9a, 10a and 11a. The average stacking velocities are used to convert the two-way travel time (TWT) of these four horizons into depth at Tola-01, and the resulting depth structure maps are shown in Figures 12a, 13a, 14a and 15a. The seismic line 844-KBR-5, which is a strike line, was reserved as a control line for the correlation.

4.1. Seismic Time and Depth Sections

The first step is to solve the velocity windows, which are given on the top of the seismic lines. As shown in the Base Map in Figure 2a, the Tola-01 well lies on the seismic line 875-KBR-230; therefore, the velocity window is solved for the KBR-230 seismic section. On the velocity window, one column is for the time in milliseconds, and the other column is for the RMS (root-mean-square) velocity. Then, the RMS velocity is multiplied with the one-way travel time in milliseconds to give the depth by using the formula S = VT. The one-way travel time is calculated by dividing the time by 2. This column gives the depth of each time during the recording of a seismic wave traveling under the surface. For more insight into the velocity window, the depth needs to be plotted against the time window [51,52]. The velocity windows are first solved to obtain the time values, and then, the depth is plotted on the horizontal axis against the time on the vertical axis. By using the reasoning that the subsurface structures can be predicted by transforming the time into depth, the TVT chart is constructed. The perception of transforming time to depth is accurate to point out the subsurface structures because the structure is flat and the map is in terms of the two-way travel time (TWT) of sound waves. Thus, a map of the seismic time section is proposed to point out the structure in the subsurface. The depth is plotted with time, as shown in Figure 2b.

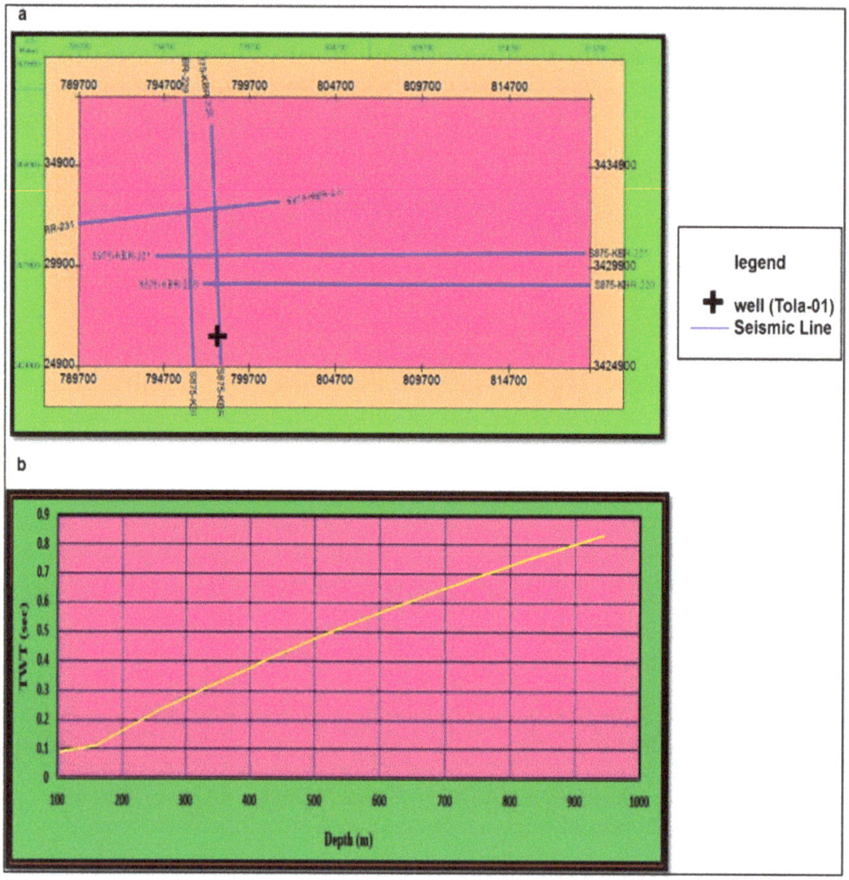

Figure 2. (**a**) Base map of the study area. (**b**) Time-depth (TD) chart of the study area.

Table 4. Formation top values of marked reflectors.

Formation Name	Depth
Dunghan	1080.46 m
Samana Suk	1174.05 m
Datta	1400 m
Warcha sandstone	1810 m

Table 5. Marked horizons obtained from time-depth chart estimated in Tola-01 well.

Formation Name	Time
Dunghan	0.9 s
Samana Suk	0.96 s
Datta	1.08 s
Warcha sandstone	1.24 s

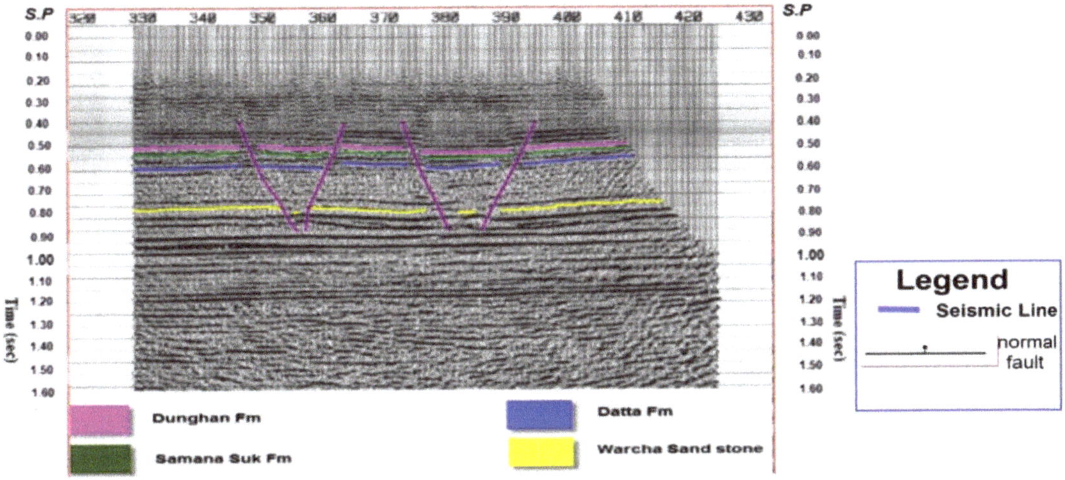

Figure 3. Marked seismic horizons of 875-KBR-220 line.

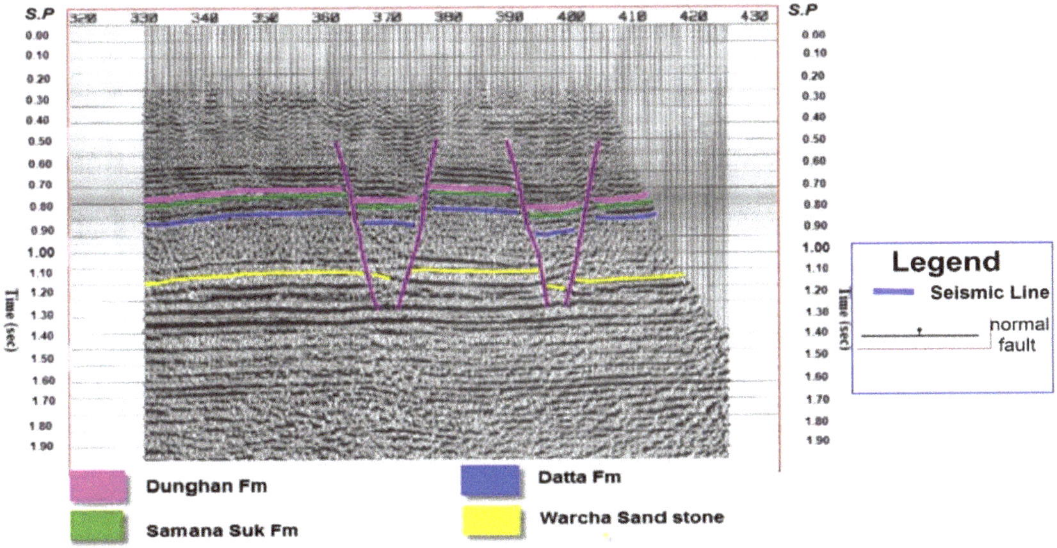

Figure 4. Marked seismic horizons of 875-KBR-221 line.

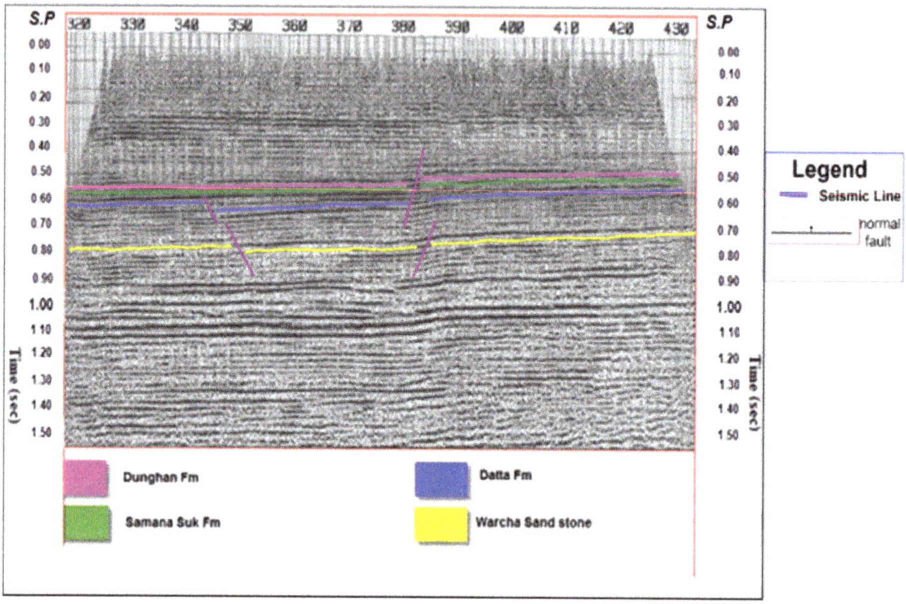

Figure 5. Marked seismic horizons of 875-KBR-229 line.

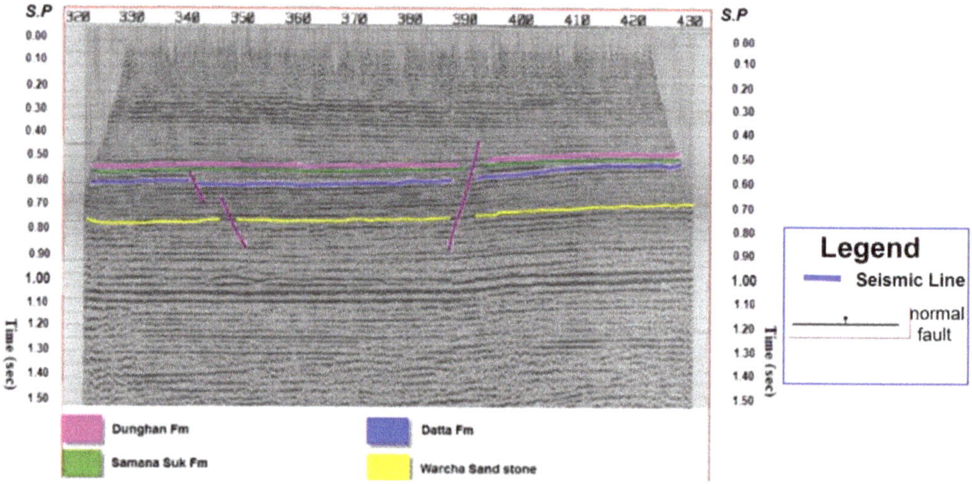

Figure 6. Marked seismic horizons of 875-KBR-230 line.

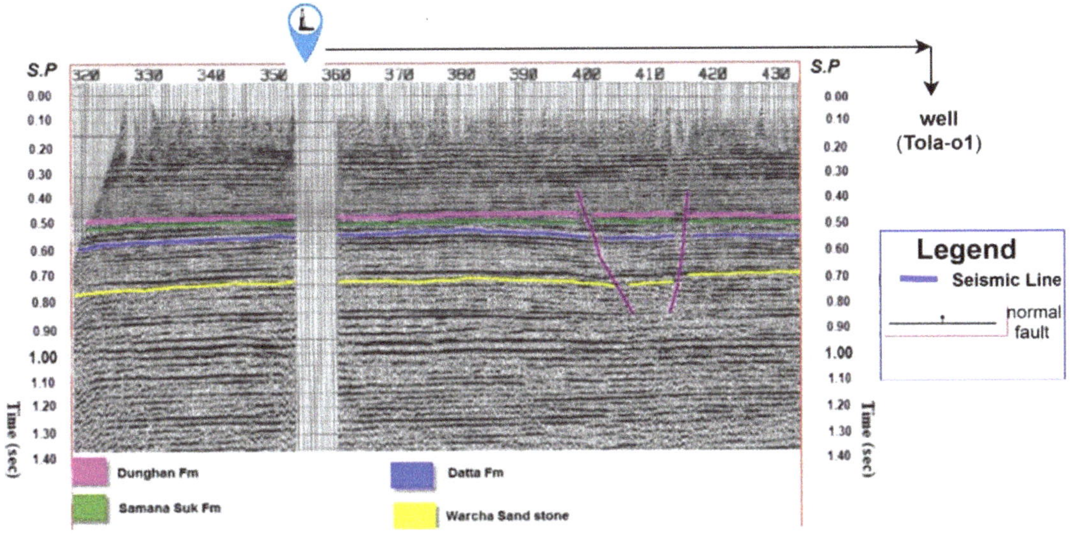

Figure 7. Marked seismic horizons of 875-KBR-231 line.

4.2. Marking Reflectors

To interpret the seismic sections, it is important to point out a particular well on a specific line to be used as a reference [53]. In this study, the seismic line and the well tops of the Tola-01 well were used to pinpoint the prescribed reflectors on that particular seismic section. Then, the nature of the marked reflectors is traced toward the seismic section used in this study. The well tops of Tola-01 of the prescribed formation are Dunghan, Samana Suk, Datta, and Warcha formations. The Punjab platform is tectonically less influenced area because of its greater outdistance from the belt, as shown in Tables 4 and 5.

4.3. Time and Depth Contour Maps and Surfaces Maps

A mark that joins the track of equivalent values is called a contour mark or line. Such drawings display the sharpness of angles (slope), the promotion top of the subsurface of the deposited rock stratum (elevation), and the corresponding two-way travel time of the prospect in milliseconds [54].

4.3.1. Time Contour Maps

A contour map is simply defined as a line that exhibits parameters on a 2D map, such as depth or elevation [20]. With the help of time rates, the contours are created to show the time of the waves that reflect back from pronounced condensed reflectors or strata due to the density/velocity contrast or simply acoustic impedance. The time contour is initiated on a map for every reflector. In our research, the reflectors of relevance are the Dunghan, Samana Suk, Datta, and Warcha formations. The time contour maps of the four essential reflectors on five seismic sections are projected on the Base Map. Justified to the depth along with other diverging capacities in the path of the traveling waves, the reflection time is further computed in order to give an idea about the structures and the time delay of the seismic wave feedback.

4.3.2. Depth Contour Maps

A depth contour map points out the structures of the area and contributes to the overall information about the depth along with its range of precise formation [21]. The given depth contour maps are constructed with the help of the reflection time rates of the seismic lines of the Kabirwala area, which contains 875-KBR 231, 875-KBR-230, 875-KBR-229, 875-KBR-221, and 875-KBR-220. On the seismic section, there are four reflectors: Dunghan, Samana Suk, Datta, and Warcha formations. It is assured by the contour maps that the reflectors are laterally flat, which can be seen in their patterns. These reflectors are marked as T1, T2, T3, and T4 on the seismic sections along with the depth contour graphs of these reflectors. For a single depth value, the depth contour map points out the same depth contour line, and the contour contributes to help us build a three-dimensional structure of a horizon showing its depth as a surface.

4.4. Time Contour and Time Surfaces Maps of Dunghan Formation

The time contour map of the Dunghan formation points out a monocline structure dipping in the eastward orientation. The reflection time increases from 0.93 to 1.14 s, as shown in Figure 8a. The time contour surface of the Dunghan formation was generated by joining the points of equal time. These time values were assigned a color bar, which can be used as a guide for interpretation. On the color bar, red shows the shallowest point, and purple-blue shows the deepest point (Figure 8b). The 3D view of the particular formation surface helps to comprehend the monocline structure in the Punjab platform region northward.

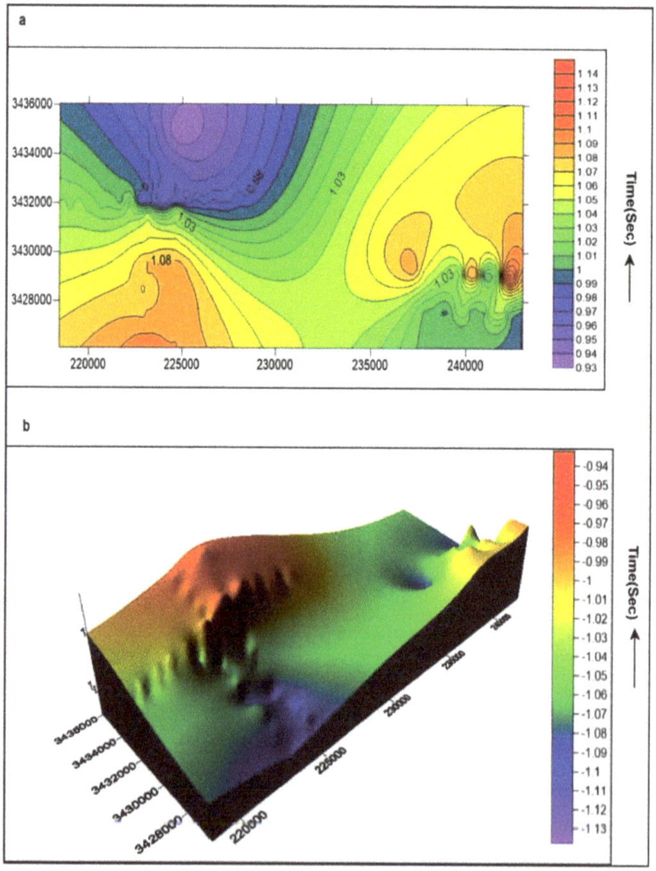

Figure 8. (**a**) Time contour map of the Dunghan formation, (**b**) Time surface map of the Dunghan formation.

The counter map is generally defined as the line showing the same parameter (depth elevation) on a 2D map. The contours are made with the help of the time values, which represent the time of the wave to reflect from a dense point reflector such as R1, R2, R3, and R4. The time counters are generated on the map, taking the value of reflector time filling a line. In the current research, the reflectors of interest are Dubghan, Samana Suk, Datta, and Warcha formations, and these are plotted on the counter section referring to their reflection time values. The reflection time is plotted in order to give an idea about the structure and time delay of the seismic wave response due to the depth and other deflections in the path of the wave.

4.5. Time Contour and Time Surfaces Maps of Samana Suk Formation

The time contour map of the Samana Suk formation also points toward the monocline structure dipping moderately in the eastward direction. The increase in reflection time is from 0.97 to 1.19 s, which is shown in Figure 9a. The time contour surface of the Samana Suk formation was generated by joining the points of equal time. These time values were assigned a color bar, which can be used as a guide for interpretation. On the color bar, red shows the shallowest point and purple-blue shows the deepest point (Figure 9b).

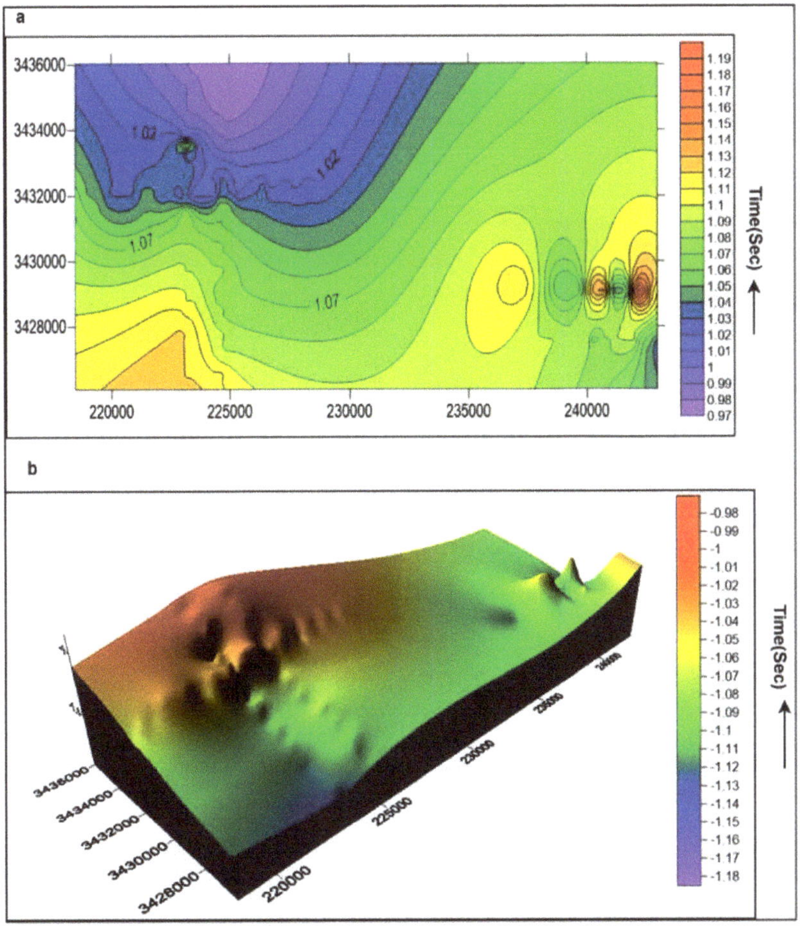

Figure 9. (**a**) Time contour map of the Samana Suk formation, (**b**) Time surface map of Samana Suk formation.

4.6. Time Contour and Time Surfaces Maps of Datta Formation

The time contour map of the Datta formation also has a monocline structure, similar to the other two formations, describing the characteristics of the Punjab platform. The increase in reflection time is from 1.02 to 1.28 sec, which is shown in Figure 10a. The time contour surface of the Datta formation was generated by joining the points of equal time. These time values were assigned a color bar, which can be used as a guide for interpretation. On the color bar, red shows the shallowest point and purple-blue shows the deepest point (Figure 10b).

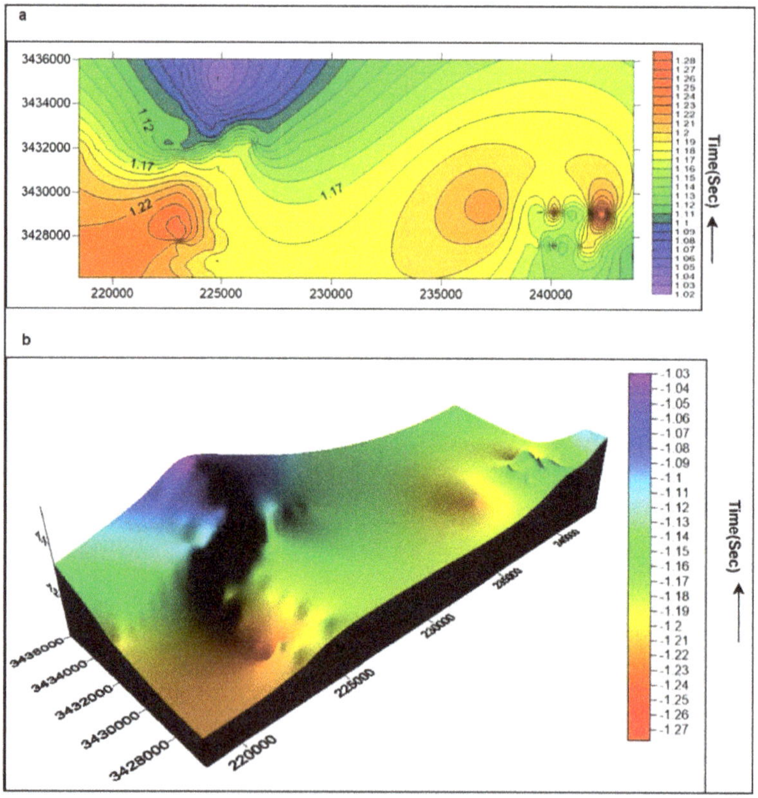

Figure 10. (**a**) Time contour map of the Datta formation. (**b**) Time surface map of Datta formation.

4.7. Time Contour and Time Surfaces Maps of Warcha Formation

The time contour map of the Warcha formation also points toward the monocline structure, dipping moderately in the eastward direction. The increase in reflection time is from 1.34 to 1.64 s, as shown in Figure 11a. The time contour surface of the Warcha formation was generated by joining the points of equal time. These time values were assigned a color bar, which can be used as a guide for interpretation. On the color bar, red shows the shallowest point and purple-blue shows the deepest point (Figure 11b).

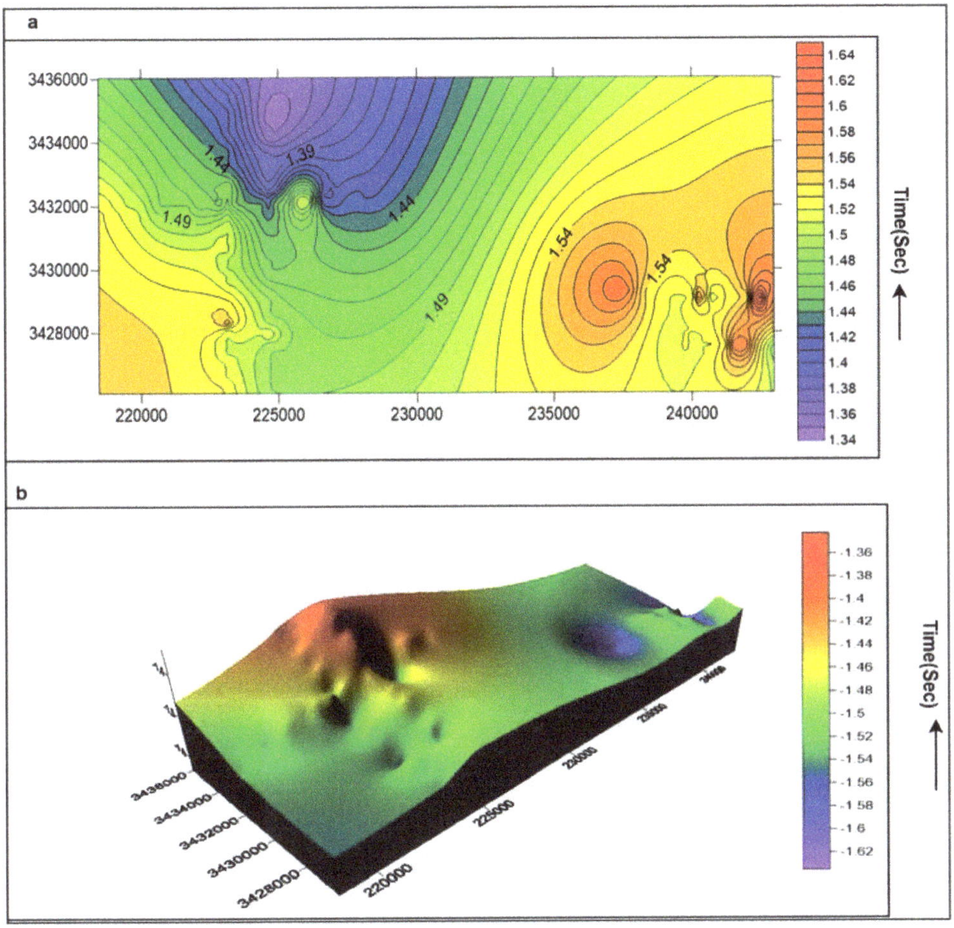

Figure 11. (**a**) Time contour map of the Warcha formation. (**b**) Time surface map of the Warcha formation.

4.8. Depth Contour and Depth Surfaces Maps of Dunghan Formation

In comparison to the time contour map, the depth contour map of the Dunghan formation contributes to the same diagram of the dipping monocline structure of the Punjab platform area of the CIB. The depth value of the Dunghan formation plunges eastward and the decrease in depth value is from 1370 to 1110 m, as shown in Figure 12a. The depth contour surface of the Dunghan formation was generated by joining the points of equal depth. These depth values were assigned a color bar, which can be used as a guide for interpretation. On the color bar, red shows the shallowest point and purple-blue shows the deepest point (Figure 12b).

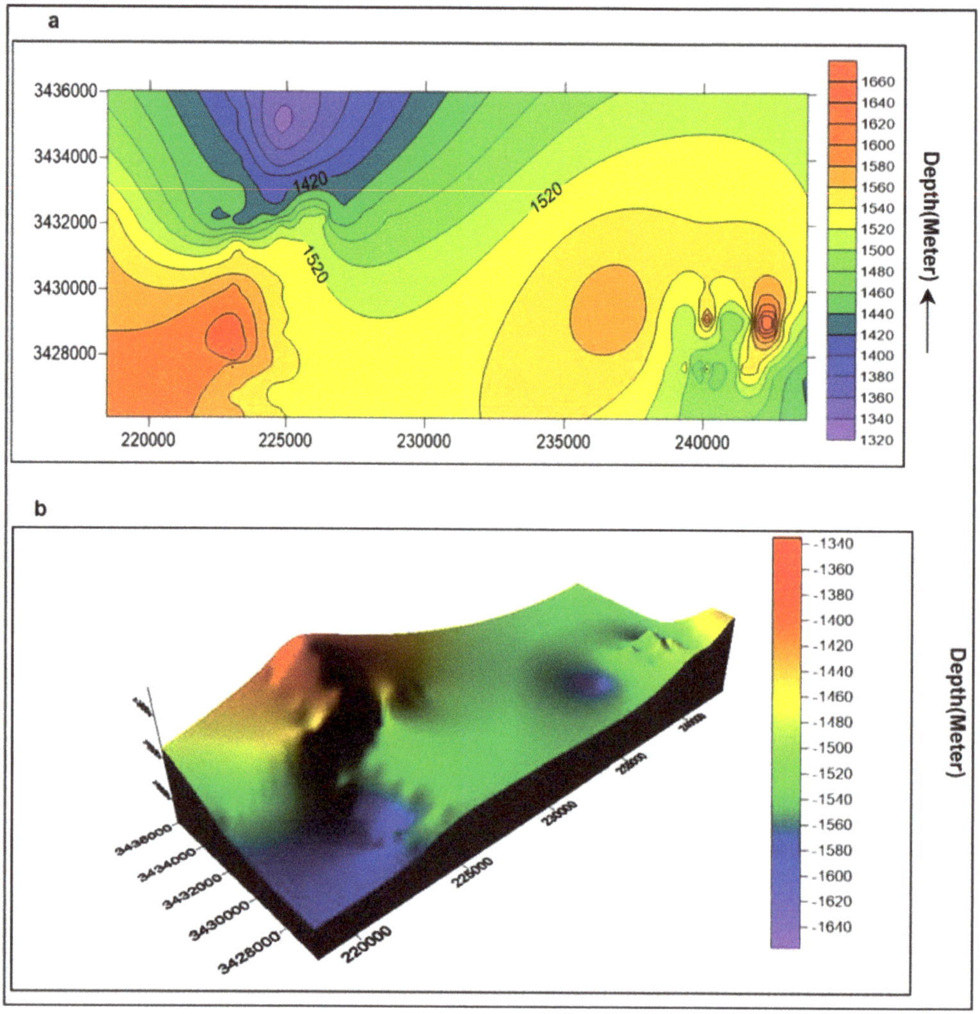

Figure 12. (**a**) Depth contour map of the Dunghan formation. (**b**) Depth surface map of the Dunghan formation.

4.9. Depth Contour and Depth Surfaces Maps of Samana Suk Formation

The depth contour map of the Samana Suk formation points out descending depth northward from 1450 to 1180 m, as shown in Figure 13a. The depth contour surface of the Samana Suk formation was generated by joining the points of equal depth. These depth values were assigned a color bar, which can be used as a guide for interpretation. On the color bar, red shows the shallowest point and purple-blue shows the deepest point (Figure 13b).

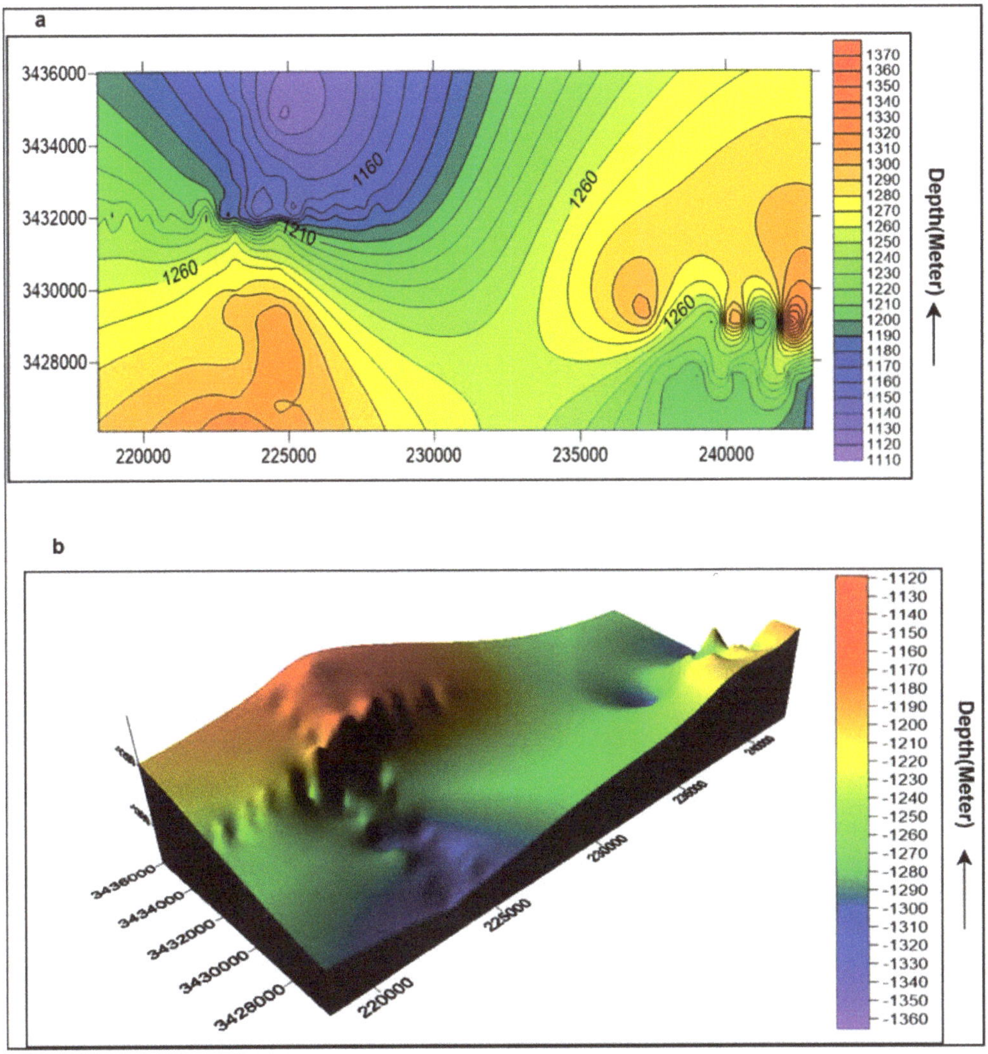

Figure 13. (**a**) Depth contour map of the Samana Suk formation. (**b**) Depth surface map of the Samana Suk formation.

4.10. Depth Contour and Depth Surfaces Maps of Datta Formation

The depth contour map of the Datta formation uses a similar template as the Samana Suk formation. The inclusive structure is the monocline structure of the Punjab platform, even though fluctuating variations in depth are shown. The decrease in depth value is from 1660 to 1320 m, as shown in Figure 14a. The depth contour surface of the Datta formation was generated by joining the points of equal depth. These depth values were assigned a color bar, which can be used as a guide for interpretation. On the color bar, red shows the shallowest point and purple-blue shows the deepest point (Figure 14b).

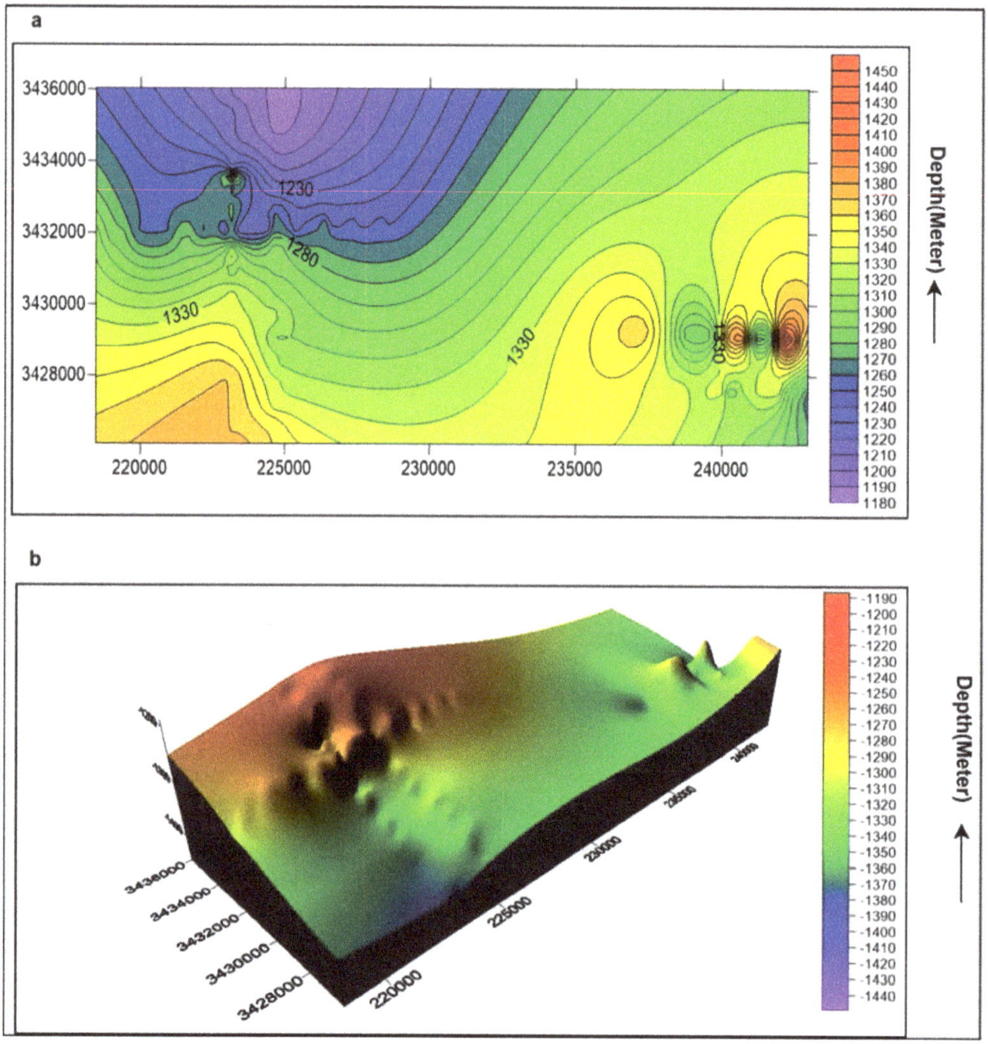

Figure 14. (**a**) Depth contour map of the Datta formation. (**b**) Depth surface map of the Datta formation.

4.11. Depth Contour and Depth Surfaces Maps of Warcha Formation

The depth contour map of the Warcha formation uses a similar template as the Datta formation. The inclusive structure is the monocline structure of the Punjab platform, even though fluctuating variations in depth are shown. The decrease in depth value is from 2400 to 1940 m, as shown in Figure 15a. The depth contour surface of the Warcha formation was generated by joining the points of equal depth. These depth values were assigned a color bar, which can be used as a guide for interpretation. On the color bar, red shows the shallowest point and purple-blue shows the deepest point (Figure 15b).

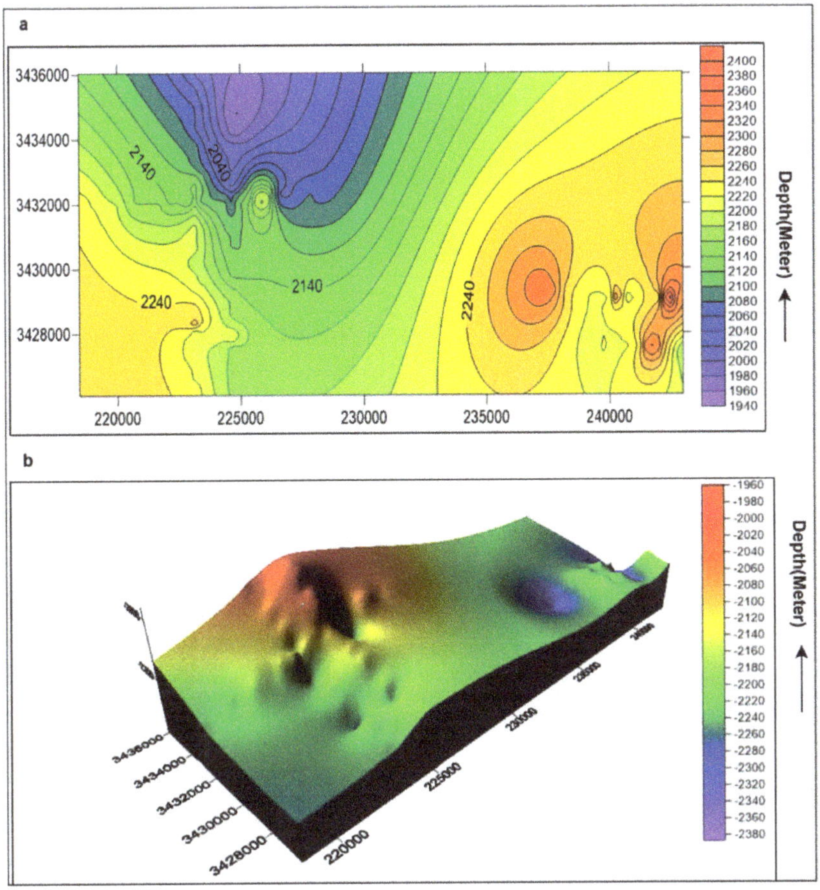

Figure 15. (**a**) Depth contour map of the Warcha formation. (**b**) Depth surface map of the Warcha formation.

5. Petrophysical Analysis

The main purpose of the petrophysical analysis is to retrieve information and data from the well logs [7,55,56]. Data utilized in the current research study include log curves of gamma ray, neutron, density, and resistivity logs. The clean zone is spotted by means of the gamma ray log. The separation in the resistivity curves is LLS (Letro-Log Shallow) MSFL (Micro Spherically Focused Log) and LLD (Letro-Log Deep) and the highest values of resistivity indicate the occurrence of hydrocarbons. The crossover between the neutron and density logs indicates the existence of hydrocarbons in the chosen clean zone. The subsequent zone of concentration is marked in Table 6, and the petrophysical parameters are shown in Table 7.

Table 6. Marked zones of interest encountered in Tola-01 well.

Formation Name	Depth
Dunghan	1080.46 m
Samana Suk	1174.05 m
Datta	1400 m
Warcha	1810 m

Table 7. Petrophysical properties and their average values.

Petrophysical Property	Average Value
Volume of shale	69%
Average porosity (PHIT)	27.8%
Effective porosity (PHIE)	8.2%
Water saturation (SW)	18%
Density porosity	27%
Hydrocarbon saturation (SH)	82%

5.1. Calculation of Volume of Shale

The volume of shale (Vsh) is also called the dirtiness of the reservoir. Figure 16 shows the Vsh considered in the zone of interest. The following equation was used for Vsh calculation:

$$\text{Vsh} = (\text{GR log} - \text{GR min})/(\text{GR max} - \text{GR min}) \quad (1)$$

where GR log is the gamma ray log reading, GR max is the maximum gamma ray value, and GR min is the minimum gamma ray value.

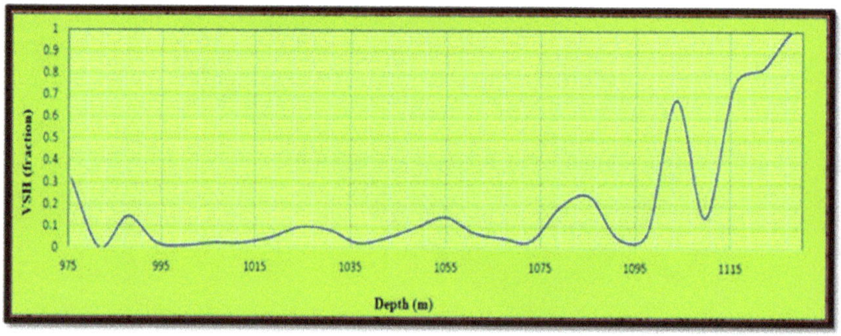

Figure 16. Volume of shale for selected zone.

5.2. Total Porosity

Basically, porosity is described as the quantity of void places (pores) relative to the overall volume of a rock [57]. These pores were formed in the rock during its deposition, which is termed as primary porosity, in contrast to the disintegration of particles through water or cracking, which is termed as secondary porosity. It is signified by ϕ and is often articulated in percentage. Porosity can be calculated using dissimilar techniques. In the current study, porosity was found using two methods [58]: density and neutron logs.

$$\text{Total Porosity} = (\text{Density Porosity} + \text{Neutron Porosity})/2 \quad (2)$$

Total porosity (PHIT) is shown in Figure 17.

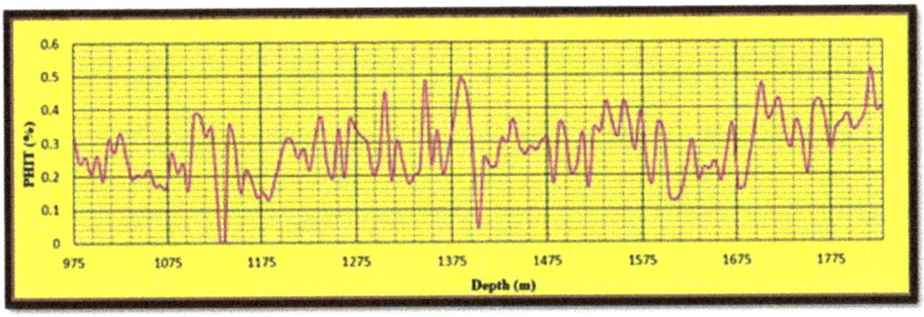

Figure 17. Total porosity variation along the depth.

5.3. Neutron Porosity

Neutron porosity records porosity, based on the influence of the accumulation on fast neutrons emanated by means of a source [59]. Hydrogen has an extreme impact on decelerating and catching neutrons. Hydrogen often resides in pore liquid, and the neutron porosity record reacts predominantly to porosity. The neutron porosity log is shown in Figure 18.

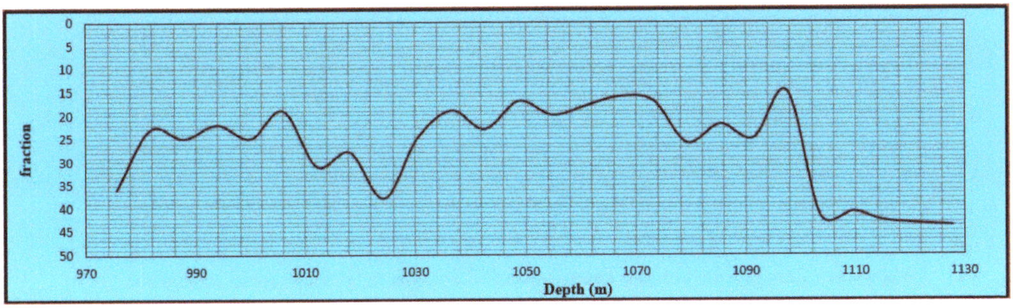

Figure 18. Neutron porosity variation along the depth.

Density Porosity

The density record is converted to porosity through the following relation:

$$\Phi_D = \rho_{ma} - \rho_b/\rho_{ma} - \rho_{fl} \quad (3)$$

where Φ_D is density porosity, ρ_{ma} is matrix density (g/cm^3), ρ_b is formation bulk density (log value) (g/cm^3), and ρ_{fl} is the density of the fluid saturating the rock adjacent to the drill-hole, which is frequently mud filtrate (g/cm^3).

Incorrect values may perhaps result when conducting the analysis in evaporates or gas accumulations. They generate lower porosity values than expected authentic values. The density porosity log is shown in Figure 19.

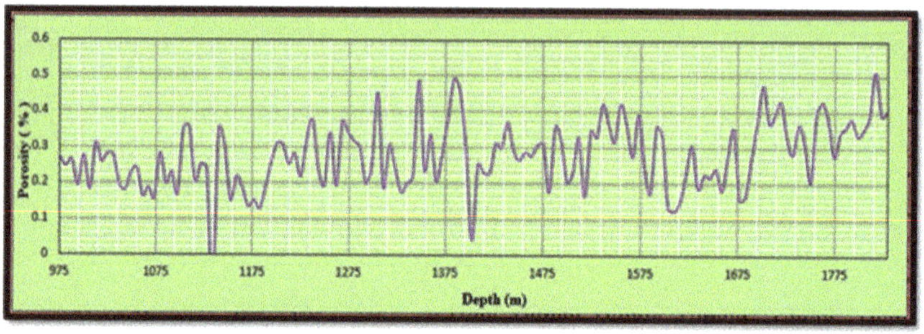

Figure 19. Density porosity variation along the depth.

5.4. Effective Porosity

Effective porosity (PHIE) is the summation of totally connected pores [7]. It is calculated by the following formula:

$$\text{Effective Porosity} = \text{Total Porosity} \times (1 - \text{Vsh}). \quad (4)$$

The effective porosity log is shown in Figure 20.

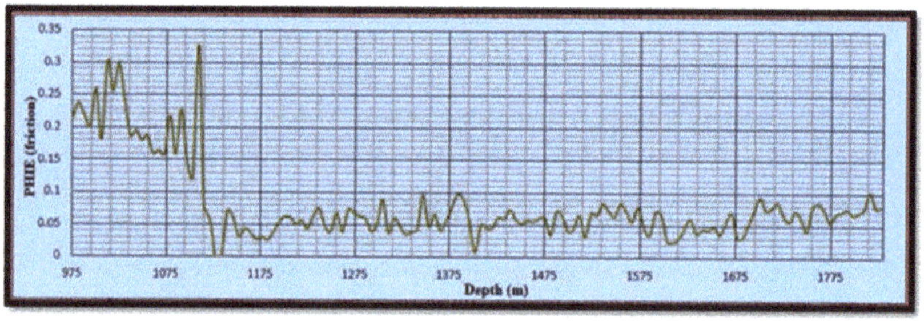

Figure 20. Effective porosity variation along the depth.

5.5. Saturation of Water

Saturation of water (S_w) is calculated by using Archie's equation [26]:

$$S_W = \sqrt{(R_W/(R_t * \varphi e^2))} \quad (5)$$

where S_w is saturation of water, R_w is resistivity of water, R_t is resistivity of the true zone, and φ_e is effective porosity; the log is shown in Figure 21.

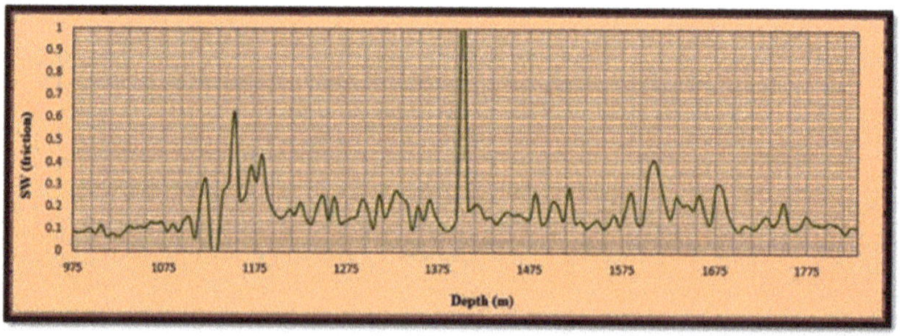

Figure 21. Water saturation along the depth.

5.6. Saturation of Hydrocarbon

Hydrocarbon saturation (S_h) is the most crucial feature due to the potential of the reservoir to generate hydrocarbons [27]. The result is shown in Figure 22, and the following formula was used for its calculation:

$$S_h = 1 - S_w. \quad (6)$$

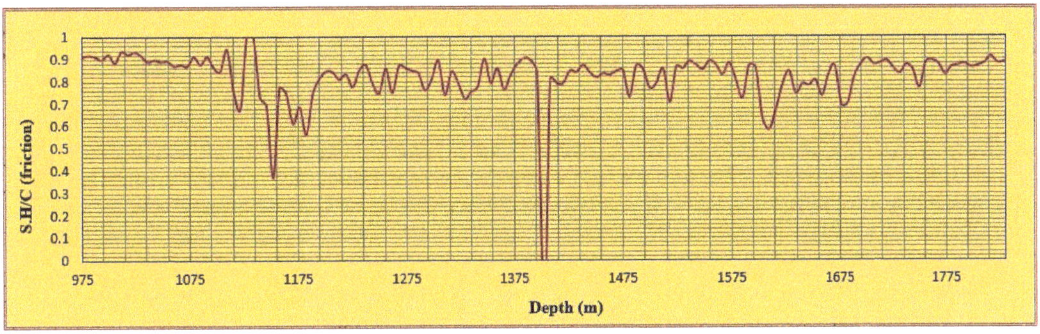

Figure 22. Hydrocarbon saturation along the depth.

In order to find out any hydrocarbon indicator (HI), all the well logs are shown together in Figure 23 in order to find out the hydrocarbon-bearing sand/carbonate and shale zones.

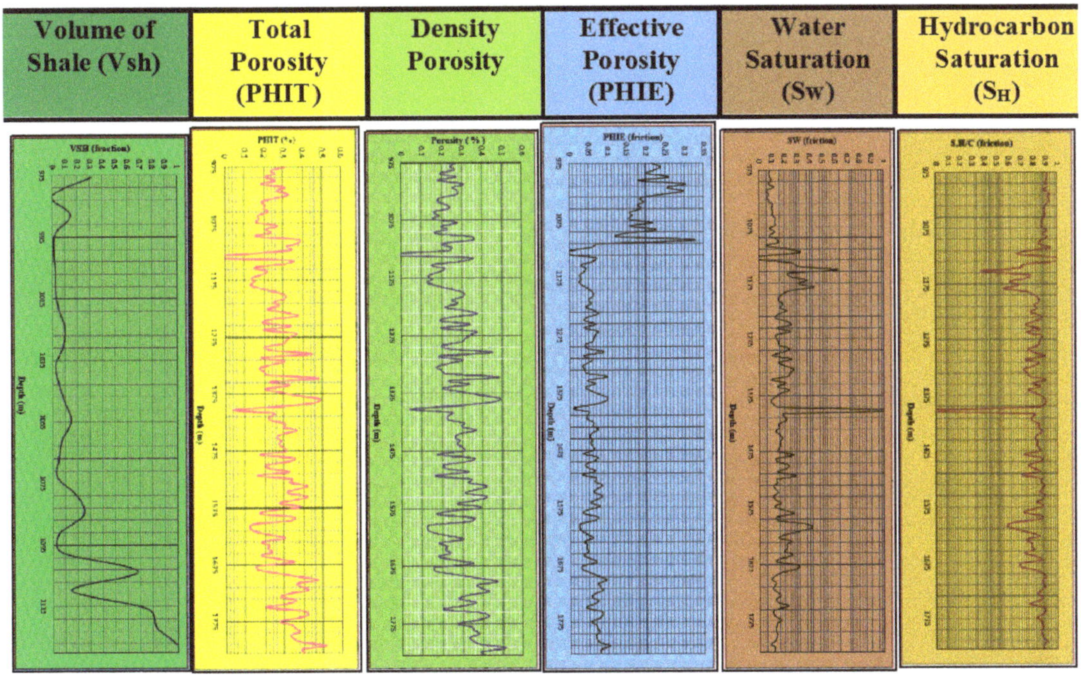

Figure 23. Combined well logs.

6. Discussion

In Figure 16, the volume of shale mostly varies from 40 to 80% up to the depth of 1115 m. This percentage shows that the lithology is partially clean and partially dirty and presents some presence of sand; therefore, it indicates a reservoir zone.

In Figure 20, effective porosity varies from 3 to 15% up to depths of 975 to 1115 m then decreases up to 1175 m, which shows that after 1115 m, compaction starts, which damages the interconnectivity of pores.

In Figure 21, the saturation of water lies almost in the range of 0–10%. Water saturation continuously changes with depth in this variation due to the fracture or compaction present in the formations. This is represented by the zone of less water saturation and high hydrocarbon saturation. Comparing Figures 21 and 22, hydrocarbon saturation and water saturation, we can clearly identify that these two graphs are inverse images of each other. This indicates that with increased hydrocarbon saturation, the water saturation decreases, and vice versa. This shows the zones of less water saturation and high hydrocarbon saturation.

7. Conclusions

The objective of the current research work was to interpret the subsurface structure and reservoir characterization of Kabirwala area Tola (01), located in Punjab platform, Central Indus Basin, utilizing 2D seismic and well log data. Formation evaluation for hydrocarbon potential using the reservoir properties was performed in this study. The outcomes of the current study are summarized as follows:

- The seismic data interpretation confirms that normal faults with horst and graben structures are favorable for the accumulation of hydrocarbons.
- The high zone present in the northwestern part of the contour maps could be a possible location of hydrocarbon entrapment, which is further confirmed by the presence of the Tola01 well.
- For the marked zone of interest, the average water saturation is 18%, average total porosity is 27.8%, average effective porosity is 8.2%, and net pay thickness is 84.14 m.
- The time section confirms that the formations are shallower in the northwestern direction and deeper along the southeastern direction. The stratigraphic conditions of these Jurassic age formations led to a complete petroleum system in the area.
- As a limitation, the current study needs further and suitable calibration with more actual well data to prove reliability and efficiency.
- In the future, we will implement 3D seismic data interpretation in the proposed field of our research. Similarly, we will also attempt to use the inversion algorithm for further exploration to obtain rich information.

Author Contributions: Conceptualization, N.A., S.K. and A.A.-S.; methodology, N.A., S.K. and A.A.-S.; software, N.A.; validation, N.A., S.K. and A.A.-S.; formal analysis, N.A. and S.K.; investigation, N.A., S.K. and A.A.-S.; resources, S.K. and A.A.-S.; data curation, N.A.; writing—original draft preparation, N.A., S.K. and A.A.-S.; visualization, N.A. and S.K.; supervision, S.K. and A.A.-S.; project administration, S.K. and A.A.-S.; funding acquisition, A.A.-S. All authors have read and agreed to the published version of the manuscript.

Funding: This research received funding from the College of Petroleum Engineering and Geosciences of KFUPM.

Institutional Review Board Statement: Not applicable.

Informed Consent Statement: Not applicable.

Data Availability Statement: The data are available by contacting the corresponding author.

Acknowledgments: The authors appreciate and acknowledge the support provided by King Fahd University of Petroleum and Minerals (KFUPM), KSA.

Conflicts of Interest: The authors declare no conflict of interest.

References

1. Li, J.-Y.; Chen, X.; Hao, Z.-J.; Rui, Z.-H. A Study on Multiple Time-Lapse Seismic AVO Inversion. *Chin. J. Geophys.* **2005**, *48*, 974–981. [CrossRef]
2. Ou, C.; Li, C.; Rui, Z.; Ma, Q. Lithofacies distribution and gas-controlling characteristics of the Wufeng–Longmaxi black shales in the southeastern region of the Sichuan Basin, China. *J. Pet. Sci. Eng.* **2018**, *165*, 269–283. [CrossRef]
3. Al-Shuhail, A.A.; Alshuhail, A.A.; Khulief, Y.A.; Salam, S.A.; Chan, S.A.; Ashadi, A.L.; Al-Lehyani, A.F.; Almubarak, A.M.; Khan, M.Z.U.; Khan, S.; et al. KFUPM Ghawar digital viscoelastic seismic model. *Arab. J. Geosci.* **2019**, *12*, 245. [CrossRef]
4. Liu, Y.; Chen, S.; Guan, B.; Xu, P. Layout optimization of large-scale oil–gas gathering system based on combined optimization strategy. *Neurocomputing* **2019**, *332*, 159–183. [CrossRef]
5. Khan, A.M.; Al-Juhani, S.G.; Abdul, S. Digital viscoelastic seismic models and data sets of central Saudi Arabia in the presence of near-surface karst features. *J. Seism. Explor.* **2020**, *29*, 15–28.
6. Khan, S.; Khulief, Y.A.; Al-Shuhail, A. Mitigating climate change via CO_2 sequestration into Biyadh reservoir: Geomechanical modeling and caprock integrity. *Mitig. Adapt. Strat. Glob. Chang.* **2018**, *24*, 23–52. [CrossRef]
7. Khan, S.; Al-Shuhail, A.A.; Khulief, Y.A.; Al-Shuhail, A. Numerical modeling of the geomechanical behavior of Ghawar Arab-D carbonate petroleum reservoir undergoing CO_2 injection. *Environ. Earth Sci.* **2016**, *75*, 1499. [CrossRef]
8. Khan, S.; Khulief, Y.; Al Shuhail, A. The effect of injection well arrangement on CO_2 injection into carbonate petroleum reservoir. *Int. J. Glob. Warm.* **2018**, *14*, 462. [CrossRef]
9. Khan, S.; Khulief, Y.A.; Al-Shuhail, A.A.; Al-Shuhail, A. Numerical Modeling of the Geomechanical Behavior of Biyadh Reservoir Undergoing CO_2 Injection. *Int. J. Géomeéch.* **2017**, *17*, 04017039. [CrossRef]
10. Das, B.; Chatterjee, R. Well log data analysis for lithology and fluid identification in Krishna-Godavari Basin, India. *Arab. J. Geosci.* **2018**, *11*, 231. [CrossRef]
11. Fajana, A.O.; Ayuk, M.A.; Enikanselu, P.A.; Oyebamiji, A.R. Seismic interpretation and petrophysical analysis for hydrocarbon resource evaluation of 'Pennay' field, Niger Delta. *J. Pet. Explor. Prod. Technol.* **2018**, *9*, 1025–1040. [CrossRef]
12. Grana, D.; Della Rossa, E. Probabilistic petrophysical-properties estimation integrating statistical rock physics with seismic inversion. *Geophysics* **2010**, *75*, O21–O37. [CrossRef]
13. Zhang, B.; Zhao, T.; Jin, X.; Marfurt, K.J. Brittleness evaluation of resource plays by integrating petrophysical and seismic data analysis. *Interpretation* **2015**, *3*, T81–T92. [CrossRef]
14. Grijalba-Cuenca, A.; Torres-Verdin, C.; van der Made, P. Geostatistical inversion of 3D seismic data to extrapolate wireline petrophysical variables laterally away from the well. In Proceedings of the SPE Annual Technical Conference and Exhibition, Society of Petroleum Engineers, Dallas, TX, USA, 1–4 October 2000.
15. Sanuade, O.A.; Akanji, A.O.; Olaojo, A.A.; Oyeyemi, K.D. Seismic interpretation and petrophysical evaluation of SH field, Niger Delta. *J. Pet. Explor. Prod. Technol.* **2018**, *8*, 51–60. [CrossRef]
16. Nwaezeapu, V.C.; Ezenwaka, K.C.; Ede, T.A. Evaluation of hydrocarbon reserves using integrated petrophysical analysis and seismic interpretation: A case study of TIM field at southwestern offshore Niger Delta oil Province, Nigeria. *Egypt. J. Pet.* **2019**, *28*, 273–280. [CrossRef]
17. Shah, S.B.A.; Abdullah, W.H. Structural interpretation and hydrocarbon potential of Balkassar oil field, eastern Potwar, Pakistan, using seismic 2D data and petrophysical analysis. *J. Geol. Soc. India* **2017**, *90*, 323–328. [CrossRef]
18. Zahid, M.; Khan, A.; ur Rashid, M.; Saboor, A.; Ahmad, S. Structural interpretation of Joya Mair oil field, south Potwar, Upper Indus Basin, Pakistan, using 2D seismic data and petrophysical analysis. *J. Himal. Earth Sci.* **2014**, *47*, 73.
19. Adeoti, L.; Onyekachi, N.; Olatinsu, O.; Fatoba, J.; Bello, M. Static Reservoir Modeling Using Well Log and 3-D Seismic Data in a KN Field, Offshore Niger Delta, Nigeria. *Int. J. Geosci.* **2014**, *5*, 93–106. [CrossRef]
20. Aadil, N.; Sohail, G.M. 3D geological modeling of Punjab platform, Middle Indus Basin Pakistan through integration of Wireline logs and seismic data. *J. Geol. Soc. India* **2014**, *83*, 211–217. [CrossRef]
21. Khalid, P.; Naeem, M.; Afzal, M.H.; Din, Z.U.; Yasin, Q. Petroleum play analysis, structural and stratigraphic interpretation of cretaceous sequence, Punjab platform, central indus basin, Pakistan. *Sci. Int.* **2014**, *26*, 727–778.
22. Khokhar, Q.D.; Hakro, A.; Solangi, S.H.; Siddiqui, I.; Abbasi, S.A. Textural Evaluation of Nari Formation, Laki Range, Southern Indus Basin, Pakistan. *Sindh Univ. Res. J.* **2016**, *48*. Available online: https://sujo-old.usindh.edu.pk/index.php/SURJ/article/view/2557/2093 (accessed on 25 January 2021).
23. Tainsh, H.R.; Stringer, K.V.; Azad, J. Major gas fields of West Pakistan. *AAPG Bull.* **1959**, *43*, 2675–2700.
24. Hemphill, W.R.; Kidwai, A.H. *Stratigraphy of the Bannu and Dera Ismail Khan areas, Pakistan*; USGS: Reston, VA, USA, 1973. [CrossRef]
25. Rehman, S.U.; Riaz, M.A.; Ahmed, M.; Ullah, M.F.; Kashif, M.; Rehman, F. Sedimentology of pir koh formation exposed at Dholi and Rakhi Gaj, Central Sulaiman Range, Pakistan. *J. Biol. Environ. Sci.* **2017**, *11*, 224–234.
26. Wandrey, C.J.; Law, B.E.; Shah, H.A. *Sembar Goru/Ghazij Composite Total Petroleum System, Indus and Sulaiman-Kirthar Geologic Provinces, Pakistan and India*; US Department of the Interior, US Geological Survey: Reston, VA, USA, 2004.
27. Kazmi, A.H. Stratigraphy Of The Dungan Group In Kach-Ziarat Area, NE Balochistan. *J. Himal. Earth Sci.* **1988**, *21*, 117–130.
28. Malkani, M.S. Updated stratigraphy and mineral potential of Sulaiman Basin, Pakistan. *Sindh Univ. Res. J.-SURJ* **2010**, *42*.
29. Hakro, A.D.; Baig, M.A.A. Depositional environment of the Bara formation, fort ranikot area, Sindh. *Sindh Univ. Res. J.* **2013**, *45*. Available online: https://sujo-old.usindh.edu.pk/index.php/SURJ/article/view/1640/0 (accessed on 25 January 2021).

30. Williams, M.D. Stratigraphy of the Lower Indus Basin, West Pakistan. In Proceedings of the 5th World Petroleum Congress, New York, NY, USA, 30 May–5 June 1959.
31. Fatmi, A.; Rawson, P. The first Early Cretaceous ammonite faunas from Baluchistan. *Cretac. Res.* **1993**, *14*, 91–100. [CrossRef]
32. Blanford, W.T. 5. On some of the Specific Identifications in Dr. Güther's Second Report of Collection of Indian Reptiles obtained by the British Museum. *J. Zool.* **2009**, *44*, 635–637. [CrossRef]
33. Eschard, R.; Albouy, E.; Gaumet, F.; Ayub, A. Comparing the depositional architecture of basin floor fans and slope fans in the Pab Sandstone, Maastrichtian, Pakistan. *Geol. Soc. Lond. Spéc. Publ.* **2004**, *222*, 159–185. [CrossRef]
34. Malkani, M.S. Trackways evidence of sauropod dinosaurs confronted by a theropod found from Middle Jurassic Samana Suk Limestone of Pakistan. *Sindh Univ. Res. J.* **2007**, *39*, 1–14.
35. Shah, S.B.A.; Ahmed, A. Hydrocarbon source rock potential of Paleocene and Jurassic deposits in the Panjpir oilfield subsurface, Punjab Platform, Pakistan. *Arab. J. Geosci.* **2018**, *11*, 607. [CrossRef]
36. Iqbal, S.; Jan, I.U.; Akhter, M.G.; Bibi, M. Palaeoenvironmental and sequence stratigraphic analyses of the Jurassic Datta Formation, Salt Range, Pakistan. *J. Earth Syst. Sci.* **2015**, *124*, 747–766. [CrossRef]
37. Abdulghani, A.; Ghazi, S.; Riaz, M.; Zafar, T. *Sedimentary Fabrics and Diagenetic Features of the Late Triassic Kingriali Formation, Khisor-Marwat Ranges, Pakistan*; NISCAIR-CSIR: New Delhi, India, 2020.
38. Sarjeant, W.A.S. *Acritarchs and Tasmanitids from the Mianwali and Tredian Formations (Triassic) of the Salt and Surghar Ranges, West Pakistan*; AAPG: Tulsa, OK, USA, 1973.
39. Smith, J.J.; Hasiotis, S.T.; Kraus, M.J.; Woody, D.T. Naktodemasis bowni: New ichnogenus and ichnospecies for adhesive meniscate burrows (AMB), and paleoenvironmental implications, Paleogene Willwood Formation, Bighorn Basin, Wyoming. *J. Paléontol.* **2008**, *82*, 267–278. [CrossRef]
40. Jan, I.U.; Stephenson, M.H.; Khan, F.R. Palynostratigraphic correlation of the Sardhai Formation (Permian) of Pakistan. *Rev. Palaeobot. Palynol.* **2009**, *158*, 72–82. [CrossRef]
41. Ullah, K.; Khattak, N.; Qureshi, A.; Akram, M.; Khan, H.; Nisar, A. Search for Uranium Source in Warcha Sandstone, Salt Range, Pakistan, using SSNTD Technique. *Radiat. Meas.* **2005**, *40*, 491–495. [CrossRef]
42. Badley, M.E. Practical Seismic Interpretation by Michael E. Badley. *J. Acoust. Soc. Am.* **1987**, *82*, 1100. [CrossRef]
43. Chopra, S.; Marfurt, K.J. Seismic attributes—A historical perspective. *Geophysics* **2005**, *70*, 3SO–28SO. [CrossRef]
44. Atlas, R.M.; Bartha, R. Hydrocarbon biodegradation and oil spill bioremediation. In *Advances in Microbial Ecology*; Springer: Berlin/Heidelberg, Germany, 1992; pp. 287–338.
45. Khan, S.; Khulief, Y.A.; Al-Shuhail, A.A. Alleviation of pore pressure buildup and ground uplift during carbon dioxide injection into Ghawar Arab-D carbonate naturally fractured reservoir. *Environ. Earth Sci.* **2018**, *77*, 449. [CrossRef]
46. Khan, S.; Khulief, Y.A.; Al-Shuhail, A.A. Effects of reservoir size and boundary conditions on pore-pressure buildup and fault reactivation during CO_2 injection in deep geological reservoirs. *Environ. Earth Sci.* **2020**, *79*, 1–23. [CrossRef]
47. Khan, S.; Khulief, Y.A.; Al-Shuhail, A.A. Coupled Reservoir-Geomechanical Modeling and Stability Analysis during CO_2 Injection into Minjur Sandstone Reservoir. In *World Environmental and Water Resources Congress 2018*; ASCE: Reston, VA, USA, 2018; pp. 1–13.
48. Khan, S.; Khulief, Y.; Al-Shuhail, A. Reservoir Geomechanical Modeling and Ground Uplift During CO2 Injection Into Khuff Reservoir. In *Proceedings of the Computational Fluid Dynamics*; ASME International: New York, NY, USA, 2019; Volume 2, pp. 2–46.
49. Khan, S.; Khulief, Y.A.; Al-Shuhail, A.A. Reservoir Geomechanical Modeling during CO_2 Injection into Deep Qasim Reservoir: A Study Focused on Mitigating Climate Change. In *World Environmental and Water Resources Congress 2020: Groundwater, Sustainability, Hydro Climate/Climate Change, and Environmental Engineering*; ASCE: Reston, VA, USA, 2020; pp. 29–40.
50. Quadri, V.-N.; Shuaib, S.M. Hydrocarbon prospects of southern Indus basin, Pakistan. *AAPG Bull.* **1986**, *70*, 730–747.
51. Raza, H.A.; Ahmed, W.; Ali, S.M.; Mujtaba, M.; Alam, S.; Shafeeq, M.; Iqbal, M.; Noor, I.; Riaz, N. Hydrocarbon prospects of Punjab platform Pakistan, with special reference to Bikaner-Nagaur Basin of India. *Pak. J. Hydrocarb. Res.* **2008**, *18*, 1–33.
52. Kadri, I. *Petroleum Geology of Pakistan*; Pakistan Petroleum Limited: Karachi City, Pakistan, 1995.
53. Hasany, T.; Aftab, M.; Siddiqui, R.A. Refound exploration opportunities in Infracambrian and Cambrian sediments of Punjab Platform, Pakistan. In Proceedings of the Annual Technical Conference, Islamabad, Pakistan, 27–28 March 2007; pp. 31–62.
54. Kazmi, A.H.; Jan, M.Q. *Geology and Tectonics of Pakistan*; Graphic Publishers: Portland, OR, USA, 1997.
55. Well Logging. Available online: https://gardaerlangga.wordpress.com/2014/07/06/well-logging-definisi-dansejarahnya/ (accessed on 17 January 2021).
56. Amosu, A.; Sun, Y. MinInversion: A Program for Petrophysical Composition Analysis of Geophysical Well Log Data. *Geosciences* **2018**, *8*, 65. [CrossRef]
57. Schön, J.H. *Physical Properties of Rocks: Fundamentals and Principles of Petrophysics*; Elsevier: Amsterdam, The Netherlands, 2015.
58. De Beer, F.C.; Middleton, M.F.; Hilson, J. Neutron radiography of porous rocks and iron ore. *Appl. Radiat. Isot.* **2004**, *61*, 487–495. [CrossRef] [PubMed]
59. Elkatatny, S.; Mahmoud, M.; Tariq, Z.; Abdulraheem, A. New insights into the prediction of heterogeneous carbonate reservoir permeability from well logs using artificial intelligence network. *Neural Comput. Appl.* **2017**, *30*, 2673–2683. [CrossRef]

applied sciences

Article

Application of the Segmented Correlation Technology in Seismic Communication with Morse Code

Yuanjie Jiang [1], Yuda Chen [1], Ruyun Tian [1], Longxu Wang [1], Shixue Lv [1], Jun Lin [1,2] and Xuefeng Xing [1,2,*]

[1] College of Instrumentation and Electrical Engineering, Jilin University, Changchun 130061, China; jiangyj19@mails.jlu.edu.cn (Y.J.); chenyd19@mails.jlu.edu.cn (Y.C.); tianry18@mails.jlu.edu.cn (R.T.); wanglx17@mails.jlu.edu.cn (L.W.); lvsx15@mails.jlu.edu.cn (S.L.); lin_jun@jlu.edu.cn (J.L.)

[2] Key Laboratory of Geophysical Exploration Equipment, Ministry of Education, Jilin University, Changchun 130061, China

* Correspondence: xing@jlu.edu.cn; Tel.: +86-138-4414-5155

Abstract: Seismic communication might promise to revolutionize the theory of seismic waves. However, one of the greatest challenges to its widespread adoption is the difficulty of signal extraction because the seismic waves in the vibration environments, such as seas, streets, city centers and subways, are very complex. Here, we employ segmented correlation technology with Morse code (SCTMC), which extracts the target signal by cutting the collected data into a series of segments and makes these segments cross-correlate with the decoded signal to process the collected data. To test the effectiveness of the technology, a seismic communication system composed of vibroseis sources and geophones was built in an environment full of other vibration signals. Most notably, it improves the signal-to-noise ratio (SNR), extending the relay distance and suppressing other vibration signals by using technology to deal with seismic data generated by the system.

Keywords: signal extraction; correlation; SNR; vibroseis sources; seismic communication

1. Introduction

As early as 1899, scientists started to discuss how to implement a system of communication which could transmit signals through the earth [1]. During World War I, French physicists invented a geophone to detect underground activities in the battlefield, such as underground mining and tunnel excavation [2]. From the middle of the 1980s to the early 1990s, frequent seismic activities threatened the lives and property of miners, which drew the attention of the Canadian federal government, the Ontario provincial government and major mining companies. As a result, the technology of geophones gradually became a significant monitoring method for mine safety and ground control [3]. During the gulf war, the American military took full advantage of the electronic-jamming aircraft to destroy the air defense systems of Iraqi army, which paralyzed the radar systems of the Iraqi army so that the American military quickly won a victory on the battlefield [4]. During a mine disaster in Chile in August 2010, victims hammered a drill and attached a slip of paper to the drill to attract the attention of rescuers [5]. Over the last three decades, many methods have been proposed for communication in special circumstances where radio cannot operate, such as during mine disasters or wars, as mentioned above. To overcome this problem, Hanafy conducted a seismic communication test, where hammering signals are encoded and filtered by a band-pass filter with a hammer and geophones [6,7]. Hao delineated the framework of the rake reception system by analyzing the multipath fading, which adopted time diversity and maximum ratio combination to reduce the bit error rate of propagation about seismic waves [8,9]. Zhang discussed the transmission of seismic signals with direct sequence spread spectrum technology (DSSS), orthogonal frequency division multiplexing technology (OFDM) and code division multiple access technology (CDMA) [10]. He pointed out that a RAKE receiver could be used to solve the problem of multipath fading of seismic waves by MATLAB simulation [11].

With the continuous development of urban underground space, many cities also face similar problems [12–15]. Some countries would like to only develop shallow underground space to ensure the safety of people and quickly evacuate the population in case of accidents. For example, most underground spaces of cities in China are less than 30 m. However, the increasing cases of urban disaster are seriously endangering the health of the public and the safety of property. For example, on 2 December 2019, more than 1100 rescue workers and 192 emergency rescue vehicles were deployed in Guangzhou to search for underground victims due to the sudden collapse of the ground. On 7 January 2020, a worker was trapped for nine hours in a 30 m deep bomb shelter when an elevator suddenly failed, and his cell phone had no radio signal.

Maintaining the connection between the underground and the ground when the communication of radio or cable is not available is an urgent task [16]. This is a feasible project, as Hanafy mentioned, but there are many shock signals which might interfere with target signals. This study will discuss the influence of other vibration signals (especially the footstep signal) in urban environments and explore how to make use of the SCTMC to extract target signals. For the sake of illustration, this paper will be divided into the following five parts: the principles and methods in seismic communication, the experimental process, the analysis of experimental data, discussion of experimental results and conclusion about the segmented correlation technology.

2. Principles and Methods

Seismic communication systems mainly include four parts: the mode of seismic communication, facilities employed in the experiment, modulation of transmitted signal and the method of signal extraction, as described in the following sections.

2.1. Communication Mode

The establishment of a communication system first needs to determine the method of signal encoding. In different applications, Morse code is one of the most commonly used methods to realize information transmission about short signal [17–19]. Compared with other methods, Morse code relies on a stable and unchanging wireless signal so that its wireless communication equipment is simpler and can be used in high noise and weak signal environments [20]. These features will contribute to the realization of seismic communication, so Morse code is adopted in this paper. By encoding information into dot, dash, symbol spaces, character spaces and word spaces, the information can be designed in various orders to deliver different punctuation marks, numbers and letters. Based on the principle of Morse code, a communication network can be set up with seismic transceivers, which can encode, decode, receive and emit seismic signals [21–23], as described in Figure 1.

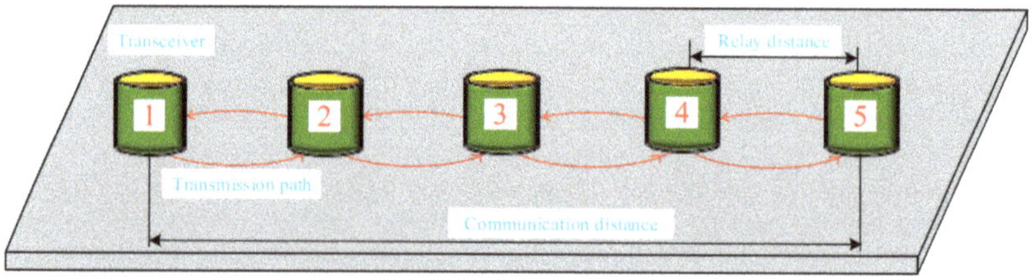

Figure 1. A long-distance transmission scheme about seismic signals.

A complete signal, which is compiled by transceivers, consists of a beginning signal, a Morse signal and an end signal. When transceiver 2 receives the beginning signal from transceiver 1, transceiver 2 is activated and decodes the seismic signal. When transceiver 2 receives the end signal from transceiver 1, transceiver 2 interrupts the communication with transceiver 1 and transmits the decoded signal in the form of a complete signal.

Then the compiled signals received by other transceivers can be transmitted again so that seismic signals travel a long distance. In this system, the communication distance is the length between the first transceiver and the last transceiver, and the distance between the two adjacent transceivers is the relay distance. This paper will discuss how to take advantage of the SCTMC to extend them. Although the integrated seismic transceiver has not been manufactured yet, it can be taken over by some instruments like vibroseis sources, hammer sources and geophones.

2.2. Experimental Facilities

When a disaster occurs, it is better for victims to send out a distress signal with a hammer or small vibroseis source [24–26]. Hammer sources, whose signals are actually shock signals, send out seismic signals by hitting the ground with gravity [27,28]. However, the components of vibration signals in urban environments are very complex, such as a lot of footstep signals (a shock signal) in crowded places, and it is unclear whether the hammer sources are suitable for urban communication. For instance, the amplitudes of the hammering signals obtained at the acquisition may be close to that of the footstep signals after hammering signals pass through the earth, as shown in Figure 2a. The footstep signals and the hammer signals are all weakened when the collected data are filtered by a 5–100 Hz band-pass filter accepted in the experiment of Hanafy, as shown in Figure 2b. The phenomenon occurs because the bandwidths of their frequencies are overlapped, which may increase the bit error rate of the system, as shown in Figure 2c,d.

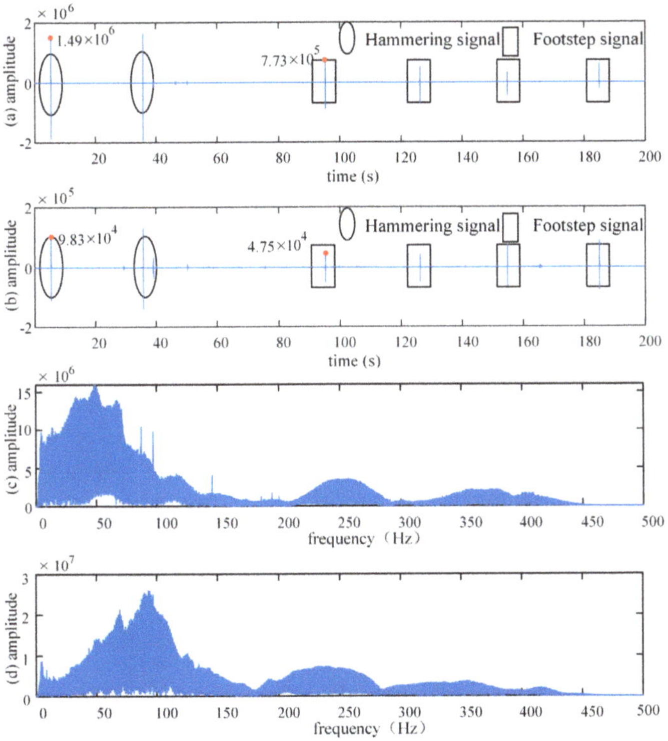

Figure 2. The comparison of hammering signals and footstep signals; (**a**) Hammer signals and footstep signals in raw data; (**b**) Hammering signals and footstep signals filtered by a 5–100 Hz band-pass filter; (**c**) The spectrum of footstep signals in raw data; (**d**) The spectrum of hammer signals in raw data.

Unlike hammer sources, the vibroseis source is a continuous vibration source whose duration and frequency can be controlled [29–31]. To test the effect of vibroseis signals on the results, footstep signals are synthetically generated, as shown in Figure 3a. After filtering 200 Hz single-frequency signals generated by vibroseis sources with a 180–220 Hz band-pass filter, the footstep signals are significantly weakened while the vibroseis signals are hardly affected, as shown in Figure 3b. The vibroseis sources accepted in this paper have great advantages in signal modulation compared with hammer sources.

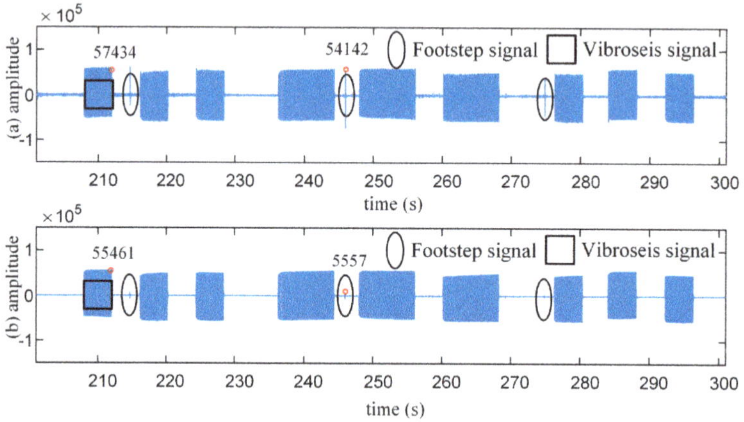

Figure 3. The amplitude changes of signals before and after filtering; (**a**) Vibroseis signals and footstep signals before filtering; (**b**) Vibroseis signals and footstep signals after filtering.

The vibroseis source sends seismic signals to the underground in the form of radiant energy, which takes several seconds to release energy. Therefore, the relative durations can be defined: marks—dot = 4 s and dash = 8 s, spaces—symbol space = 4 s, word space = 8 s and sentence space = 12 s.

2.3. Signal Modulation

The modulation of transmitted signals is performed with a 500 N electromagnetic driven vibroseis source developed by Jilin university, which mainly transmits a linear sweep signal [32,33], whose function is defined as follows:

$$s(t) = A(t)\sin 2\pi(F_1 + \frac{F_2 - F_1}{2T_D}t)t\ , \quad 0 \le t \le T_D \tag{1}$$

where $s(t)$ stands for the linear sweep signal; $A(t)$ represents the cosine amplitude envelope function of the linear scanning signal; F_1 shows the initial frequency; F_2 identifies the termination frequency; T_D reveals the duration of the scanning signals. It is straightforward to adjust the frequency and duration of signals so that vibroseis sources can output the desired signals. When the frequencies (F_2 and F_1) are set to 200 and the T_D is set to 4, the output of the vibroseis source is a 200 Hz single-frequency signal with length 4 s.

2.4. Signal Extraction

The cross-correlation adopted in this study is the digital characteristic that describes the statistical relation between two signals [34–36]. In other words, when the collected signal and the decoded signal are very similar, the target signal can be extracted. The decoded signal is designed to have the same frequency and duration as the target signal without noise [37]. The direct cross-correlation function can be expressed as:

$$R_{XH}(t_1, t_2) = E[X(t_1)H(t_2)] \tag{2}$$

where $R_{XH}(t_1, t_2)$ stands for the direct cross-correlation signal; $X(t_1)$ represents the collected signal at time t_1; $H(t_2)$ identifies the decoded signal at time t_2. Although cross-correlation technology can effectively extract the target signal and suppress partial surface waves in seismic exploration, it is difficult to get the correct information from the direct cross-correlation. When the save our souls (SOS) signal is correlated directly, the cross-correlation signal will be different from our target signal (SOS), as shown in Figure 4. The vibroseis sources transmit the seismic signal intermittently, so the final signal we get is a discrete signal.

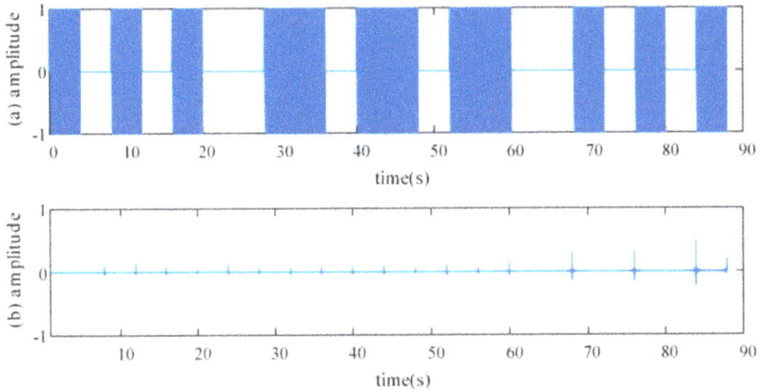

Figure 4. The direct cross-correlation in signal simulation; (**a**) The 200 Hz SOS signal in the simulation diagram; (**b**) The direct cross-correlation about SOS signal in the simulation diagram.

Based on this reality, the segmented correlation technology is proposed to deal with this problem. According to the rules of Morse code, the collected data are divided into a certain number of segments, and then these fragments are correlated with the decoded signal respectively, and the segmented correlation function is as follows.

$$\Phi_{XH}(u) = \sum_{n=1}^{N} E[X_n(u)H_n(u)] \quad (3)$$

where $\Phi_{XH}(u)$ is the segmented correlation signal; $X(u)$ shows the nth fragment of the collected signal with length u; $H(u)$ the nth fragment of the decoded signal with length u.

If the amplitude of a fragment exceeds a pre-selected threshold value, then that fragment is interpreted as "1", otherwise the fragment is interpreted as "0". So, we can interpret the complete signal according to the following definition:

- "1010" for the beginning of the message;
- "1110" for end of message;
- The signal between the beginning signal and the end signal is considered to be the intermediate signal, which is the message.

3. Experiment

In the course of experiment, a seismic communication system is first set up with a 500 N vibroseis source, 5 CDJS2C-2 geophones and 5 GEIWS RII seismometers in the Geological Palace of Changchun, Jilin, China. A 500 N vibroseis source sends out seismic signals encoded in Morse code, whose volume is no more than 8000 cm^3 so that it is easy to carry. The vibroseis, whose power supply system can be adjusted according to the actual situation, is powered by alternating current (AC) in the urban power grid or the lithium battery system with voltage of 220 V and frequency of 50 Hz. The CDJS2C-2 geophones are used to collect signals through the earth, whose sensitivity is 2 + 10% v/cm/s. The GEIWSR-II seismometers are employed to record and store seismic data from the geophones at a sampling rate of 1000 Hz. In this system, the distance between the first geophone and the source is 20 m, the distance between geophones is 10 m and the distribution of their locations as shown in Figure 5. Finally, a series of instructions (including a 200 Hz SOS signal) are sent at the appointed time in the environment with various electromagnetic equipment, people and elevators. According to the Morse table, the binary code of the SOS signal should be 1010100110110110010101.

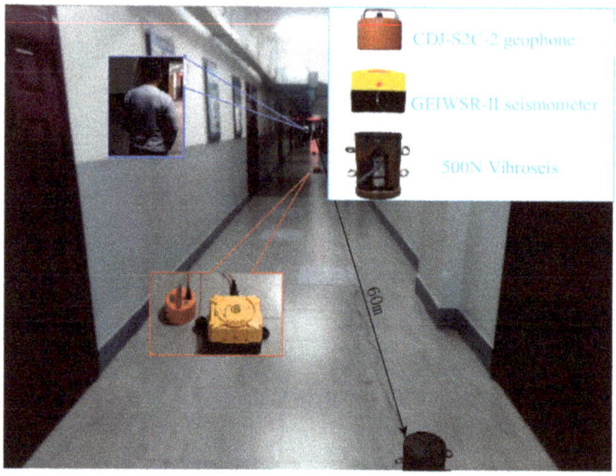

Figure 5. The distribution of 5 CDJ-S2C-2 geophones, 5 GEIWSR-II seismometers and a 500 N vibroseis source.

4. Data Analysis

The energy of the seismic signal decreases as the distance increases, and the target signal at 60 m is certainly the weakest [38–40]. Through reading the collected signal, whose SNR is 3.83 dB at 60 m, the target signal is completely drowned in noise as shown in the Figure 6a.

The signal in Figure 6a is too different from that in Figure 4a to distinguish the target signal, so the existence of the target signal needs to be identified. Spectrum analysis is first adopted to process the collected data, which can detect the frequency of the target signal [41,42]. According to results of the spectral analysis in Figure 6b, the frequencies of collected signals are mainly concentrated below 100 Hz while the 200 Hz frequency of the target signal is not detected. Moreover, even if a 180–220 Hz band-pass filter is employed to filter the collected data, the target signal is still not found, while a lot of other vibration signals are obtained, such as footstep signals and elevator signals in Figure 6c. That is to say, the target signal is submerged by other vibration signals so that it is impossible to realize seismic communication within 60 m with a band-pass filter at the site under these conditions.

To further extract the target signal, the SCTMC is employed to process the collected data obtained at 60 m. During the seismic communication experiment, the time of the seismic signal transmitted by the source is designed so that the rough position of the target signals can be predicted in collected data at 60 m. Then the SCTMC cuts the collected data into 4 s length segments, which are cross correlated with the decoded signal. By fine-tuning the position of cutting in collected data according to the cross-correlation results, the target signal (its SNR is 7.5 dB) is extracted while other vibration signals (including footstep signals) are suppressed. When the threshold of the correlation signal is set to 8.5, the binary code 10101010100110110110010101110 of the collected signal can be obtained, as show in Figure 6d. Then, we take advantage of the source to transmit the SOS signal again in this location, and similar results are obtained 60 m away from the source. At this time, the communication distance is 120 m and the relay distance is 60 m.

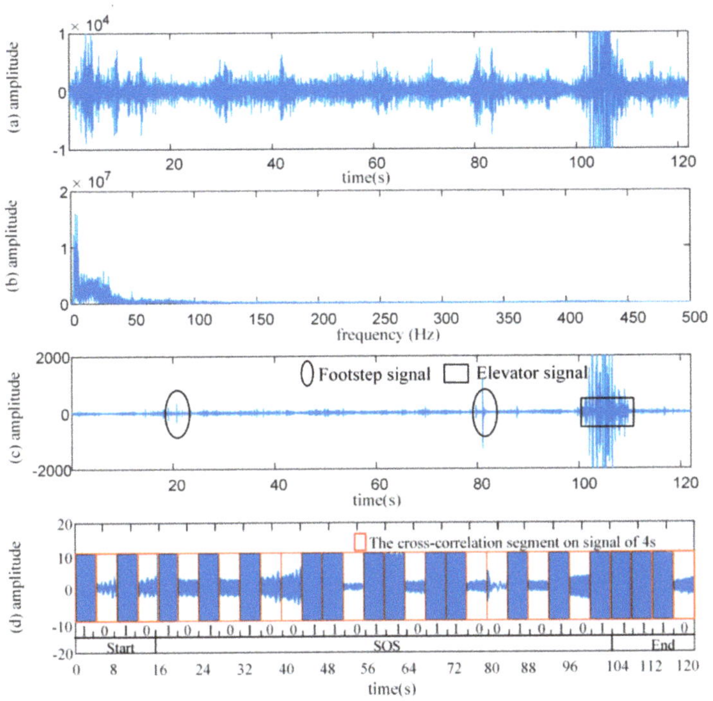

Figure 6. The collected signal at 60 m; (**a**) The original data at 60 m; (**b**)The spectral analysis of the original data; (**c**) The target signal extraction with a band-pass filter; (**d**) The target signal extraction with the segmented correlation technology.

5. Discussion

In this study, we have presented the SCTMC to extract the target signal (SOS) and built a seismic communication system. By comparing the results of footstep signals, vibroseis signals and hammer signals, the signal transmitter suitable for complex environments is selected. According to Morse code, a vibroseis source is employed to transmit a 200 Hz SOS signal, which is received by geophones and recorded by seismometers at 60 m. However, the energy of the target signal is so weak that the band-pass filter cannot extract the target signal. Hence, the SCTMC is put forward to handle the seismic data, which extracts the target signal and suppresses other signals. Although its amplitude is only 0.2% of the amplitude about the original signal (in Figure 6a), the extracted signal can be converted into the binary code 10101010100110110110010101110 by setting a reasonable threshold. In this study, the communication distance can also be extended by improving the filtering algorithm or increasing the number of vibroseis and geophones. This kind of information recognition is at the expense of information transmission, which is relatively low at present. However, with the development of seismic communication technology, the relay distance will be extended or the communication speed will be improved.

6. Conclusions

In this paper, we first discuss the influence of other vibration signals on the target signal using different transmitting devices, such as hammers and vibroseis sources. By contrast, vibroseis sources accepted in this study have great advantages in signal modulation compared with hammer sources. However, it is difficult to extract the target signal because there are a large number of strong vibration signals in a complex environment. In order to solve the problem, the SCTMC is proposed to extract the target signal in the collected data. The experimental results show that the SCTMC is well combined with the advantages of vibroseis sources to extract the target signal. At its core, it effectively extracts the similar signal, suppresses other signals, improves the SNR of seismic signal and extends the relay distance of the communication system, which greatly reduces the cost of the seismic communication system. Implementing the transmission of information recognition with SCTMC is a breakthrough which offers a better application of the vibroseis sources and geophones, constitutes an innovative form of wireless communication and lays the foundation for the long-distance communication for the seismic communication system. Maybe in the near future, this technology might be widely employed in our production and life.

Author Contributions: Y.J.: writing—original draft preparation; Y.C., R.T., L.W., S.L., J.L. and X.X.: offering some suggestions and guidance for the original draft preparation; Y.J., and Y.C.: conducting the experiments or partial experiments. All authors have read and agreed to the published version of the manuscript.

Funding: This research was funded by National Natural Science Foundation of China, grant number 41827803.

Institutional Review Board Statement: Not applicable.

Informed Consent Statement: Not applicable.

Data Availability Statement: The data generated or analyzed in this study are included in the submitted article.

Acknowledgments: Authors acknowledge Jilin University who provides instruments, funding and experimental sites. Authors would also like to thank anonymous reviewers who have provided many helpful comments on the manuscript.

Conflicts of Interest: The authors declare no conflict of interest.

References

1. Forooshani, A.E.; Bashir, S.; Michelson, D.G.; Noghanian, S. A survey of wireless communications and propagation modeling in underground mines. *IEEE Commun. Surv. Tutor.* **2013**, *15*, 1524–1545. [CrossRef]
2. Ralchenkoa, M.; Svilansb, M.; Samsona, C.; Roperbl, M. Finite-difference time-domain modelling of through-the-Earth radio signal propagation. *Comput. Geosci.* **2015**, *85*, 184–195. [CrossRef]
3. Xu, N.W.; Li, T.B.; Dai, F.; Li, B.; Zhu, Y.G.; Yang, D.S. Microseismic monitoring and stability evaluation for the large scale underground caverns at the Houziyan hydropower station in Southwest China. *Eng. Geol.* **2015**, *188*, 48–67. [CrossRef]
4. Allagui, I.; Akdenizli, B. The Gulf Information War and the role of media and communication technologies. *Int. J. Commun.* **2019**, *13*, 1287–1300.
5. Rashid, F.; Edmondson, A.C.; Leonard, H.B. Leadership lessons from the chilean mine rescue. *Harv. Bus. Rev.* **2013**, *91*, 113–134.
6. Hanafy, S.M.; Cao, W.; McCarter, K.; Schuster, G.T. Using super-stacking and super-resolution properties of time-reversal mirrors to locate trapped miners. *Lead. Edge* **2009**, *28*, 302–307. [CrossRef]
7. Hanafy, S.M.; Schuster, G.T. Two applications of time reversal mirrors: Seismic radio and seismic radar. *J. Acoust. Soc. Am.* **2011**, *130*, 1985–1994. [CrossRef]
8. Hao, J.J.; Wang, F.Y. Multipath characteristics of TTE stratified medium channel of elastic wave signal communication. *J. China Coal. Soc.* **2012**, *37*, 695–699.
9. Hao, J.J.; Sun, X.C. Analysis and comparison of several through-the-earth communication technologies for mining. *J. Hunan Univ. Sci. Tech.* **2014**, *1*, 59–63.
10. Zhang, B. Study of Applicability for Modulation Method in Elastic Wave Stratum Communication System. Master's Thesis, Shandong University of Science and Technology, Qingdao, China, 2016.
11. He, Q.H.; Chen, X.Z.; Zhang, B.; Wu, Y.R. Design of Elastic Wave Receiving Scheme in Stratum Communication System. *Electron. Qual.* **2017**, *1*, 97–102.
12. Qiao, Y.K.; Peng, F.L.; Wang, Y. Monetary valuation of urban underground space: A critical issue for the decision-making of urban underground space development. *Land Use Policy* **2017**, *69*, 12–24. [CrossRef]
13. Zhang, J.; Liu, S.; Yang, C. Detection of urban underground cavities using seismic scattered waves: A case study along the Xuzhou Metro Line 1 in China. *Near Surf. Geophys.* **2020**. [CrossRef]
14. Wang, Y.; Shi, G.; Tian, X. Experimental Study on City Road Collapse under Vibrating Load. *Shock Vib.* **2020**, *2020*, 6074658. [CrossRef]

15. Abidi, A.; Demehati, A.; Banouni, H. The Importance of Underground Cavities Detection in the Choice of Constructible Areas: Case of the Agglomeration of Fez (Morocco). *Geot. Geol. Eng.* **2017**, *36*, 1919–1932. [CrossRef]
16. Zhao, S.; Zhou, C.; Shen, X. Investigation of VLF Transmitter Signals in the Ionosphere by ZH-1 Observations and Full-Wave Simulation. *J. Geophys. Res.-Space Phys.* **2019**, *124*. [CrossRef]
17. Yuan, Y.; Sun, Z.H.; Wei, Z.H.; Jia, K.B. Deep Morse: A Deep Convolutional Learning Method for Blind Morse Signal Detection in Wideband Wireless Spectrum. *IEEE Access* **2019**, *7*, 80577–80587. [CrossRef]
18. Juneja, S.; Joshi, P. Design and development of a low cost and reliable writing aid for visually impaired based on Morse code communication. *Technol. Disabil.* **2020**, *32*, 59–67. [CrossRef]
19. Wang, X.; Zhang, M.; Zhou, H. A Robust Real-Time Automatic Recognition Prototype for Maritime Optical Morse-Based Communication Employing Modified Clustering Algorithm. *Appl. Sci.* **2020**, *10*, 1227. [CrossRef]
20. Junker, F.B.; Schlaffke, L.; Axmacher, N. Impact of multisensory learning on perceptual and lexical processing of unisensory Morse code. *Brain Res.* **2021**, *1775*, 147259. [CrossRef]
21. Dani, V.; Jalihal, D.; Ramaiyan, V. Variable Length Coding for Asynchronous Communication. *IEEE Commun. Lett.* **2020**, *24*, 2157–2161. [CrossRef]
22. Cao, Q.; Yu, C.; Cheng, X.F. Polysquaramides: Rapid and Stable Humidity Sensing for Breath Monitoring and Morse Code Communication. *Sens. Actuator B Chem.* **2020**, *320*, 128390. [CrossRef]
23. Cho, J. Prefix-Free Code Distribution Matching for Probabilistic Constellation Shaping. *IEEE Trans. Commun.* **2019**, *68*, 670–682. [CrossRef]
24. Zheng, X.; Wang, H.; Guo, J. Method for multi-information drilling detection after mining disasters. *Comput. Electr. Eng.* **2020**, *86*, 106726. [CrossRef]
25. He, J.F.; Zhao, Z.Q.; Yin, Q.L. Design and optimisation on rapid rescue well-drilling technology with large-diameter pneumatic hammers. *Int. J. Min. Reclam. Environ.* **2020**, *34*, 19–33. [CrossRef]
26. Gaurav, D.; Abdullah, A.T.; Ahmad, T. Extracting 220 Hz information from 55 Hz field data by near-field superresolution imaging. *Geophys. J. Int.* **2016**, *206*, 197.
27. Allstadt, K.E.; Farin, M.; Iverson, R.M. Measuring Basal Force Fluctuations of Debris Flows Using Seismic Recordings and Empirical Green's Functions. *J. Geophys. Res.-Earth Surf.* **2020**, *125*. [CrossRef]
28. Yordkayhun, S.; Suwan, J.N. A university-developed seismic source for shallow seismic surveys. *J. Appl. Geophys.* **2012**, *82*, 110–118. [CrossRef]
29. Jia, Z.; Lu, W. CNN-Based Ringing Effect Attenuation of Vibroseis Data for First-Break Picking. *IEEE Geosci. Remote Sens. Lett.* **2019**, *16*, 1319–1323. [CrossRef]
30. Butler, K.E.; Bernd, K.; Pugin André, J.-M. Multimode seismoelectric phenomena generated using explosive and vibroseis sources. *Geophys. J. Int.* **2018**, *213*, 836–850. [CrossRef]
31. Marcello, S.; Gaetano, F.; Philippe, R. A strongly heterogeneous hydrothermal area imaged by surface waves: The case of Solfatara, Campi Flegrei, Italy. *Geophys. J. Int.* **2016**, *205*, 1813–1822.
32. Parker, L.; Thurber, C.; Zeng, X. Active-Source Seismic Tomography at the Brady Geothermal Field, Nevada, with Dense Nodal and Fiber-Optic Seismic Arrays. *Seismol. Res. Lett.* **2018**, *89*, 1629–1640. [CrossRef]
33. Naghizadeh, M.; Snyder, D.; Cheraghi, S. Acquisition and Processing of Wider Bandwidth Seismic Data in Crystalline Crust: Progress with the Metal Earth Project. *Minerals* **2019**, *9*, 145. [CrossRef]
34. Lin, M.; Qin, J.; Wang, G. Multi-scale cross-correlation analysis of temporal and spatial seismic data. *Eur. Phys. J. B* **2020**, *93*. [CrossRef]
35. Li, G.; Chen, J.; Peng, D. Short communication: The lag response of daily milk yield to heat stress in dairy cows. *J. Dairy Sci.* **2020**, *104*, 981–988. [CrossRef] [PubMed]
36. Cao, C.; Yang, H.; Zhang, H. Passive Detection of Low-Altitude Signal Sources Using an Improved Cross-Correlation Algorithm. *Appl. Sci.* **2018**, *8*, 2348. [CrossRef]
37. Borio, D. Cross-correlation Codeless Processing of BOC Modulated Signals. *IET Radar Sonar Navig.* **2019**, *13*, 1998–2007. [CrossRef]
38. Littel, G.F.; Thomas, A.M.; Baltay, A.S. Using tectonic tremor to constrain seismic wave attenuation in cascadia. *Geophys. Res. Lett.* **2018**, *45*, 9579–9587. [CrossRef]
39. Mukhopadhyay, S.; Sharma, J.; Del-Pezzo, E. Study of attenuation mechanism for Garwhal–Kumaun Himalayas from analysis of coda of local earthquakes. *Phys. Earth Planet. Int.* **2010**, *180*, 7–15. [CrossRef]
40. Dobrynina, A.A.; Predein, P.A.; Sankov, V.A. Spatial variations of seismic wave attenuation in the South Baikal Basin and adjacent areas (Baikal rift). *Geod. Tect.* **2019**, *10*, 147–166. [CrossRef]
41. Guo, L.; Jiang, J.; Tan, F. Channelized broadband signal spectrum analysis based on weighted overlap-add structure. *Rev. Sci. Instrum.* **2016**, *87*, 105123. [CrossRef]
42. Ciuonzo, D.; Rossi, P.S. Non-colocated Time-Reversal MUSIC: High-SNR Distribution of Null Spectrum. *IEEE Signal Process. Lett.* **2017**, *24*, 397–401. [CrossRef]

applied sciences

MDPI

Article

Ambient Noise Measurements to Constrain the Geological Structure of the Güevéjar Landslide (S Spain)

José Delgado [1,*], Juan José Galiana-Merino [2], Francisco J. García-Tortosa [3], Jesús Garrido [4], Luca Lenti [5], Salvatore Martino [6], José A. Peláez [7], Martín J. Rodríguez-Peces [8], Carlos Sanz de Galdeano [9] and Juan L. Soler-Llorens [1]

1 Departamento Ciencias de la Tierra y Medio Ambiente, Universidad de Alicante, 03080 Alicante, Spain; jl.soler@ua.es
2 Departamento Física, Ingeniería de Sistemas y Teoría de la Señal, Universidad de Alicante, 03080 Alicante, Spain; jj.galiana@ua.es
3 Departamento de Geología, Campus Las Lagunillas, Universidad de Jaén, 23071 Jaén, Spain; gtortosa@ujaes.es
4 Departamento Ingeniería Civil, Campus Fuentenueva, Universidad de Granada, 18071 Granada, Spain; jega@ugr.es
5 IFSTTAR (GERS/SRO), Université Gustave Eiffel, 14-20 Boulevard Newton, Champs-sur-Marne, F-77447 Marne la Vallée, France; luca.lenti@ifsttar.fr
6 Departamento Earth Science and Research Centre for Geological Risk (CERI), Università di Roma "La Sapienza", Piazzale Aldo Moro, 5, 00185 Roma, Italy; salvatore.martino@uniroma1.it
7 Departamento de Física, Campus Las Lagunillas, Universidad de Jaén, 23071 Jaén, Spain; japelaez@ujaen.es
8 Departamento Geodinámica, Estratigrafía y Paleontología, Facultad de Geología, Universidad Complutense de Madrid, C/José Antonio Novais 12, 28040 Madrid, Spain; martinjr@ucm.es
9 Instituto Andaluz de Ciencias de la Tierra, CSIC-Universidad de Granada, Av. Las Palmeras 4, 18100 Armilla, Spain; csanz@ugr.es
* Correspondence: jose.delgado@ua.es

Abstract: The reactivation of very large landslides may cause severe damage to society. Its prevention and management requires detailed information on the geometry and structure of these landslides, but the use of standard techniques (boreholes) may be prohibitive from an economic point of view. To overcome these difficulties, geophysical techniques are of special interest because they allow for studying very large areas at a reasonable cost. In this paper, we present a case study wherein the analysis of ambient noise allowed us to produce a model of a large landslide near Granada (southern Spain). The geometry and location of the failure zone, as well as the assessment of the state of involved materials, were estimated by combining two available boreholes and different geophysical techniques (downhole tests and the spectral analysis of ambient noise, horizontal to vertical spectral ratios (HVSR) and the frequency-wavenumber (f-k) methods). The results have allowed us to differentiate between values within the landslide mass with respect to those of stable materials, and to perform for the first time a comprehensive geological model of this unstable mass. Differences were also observed within the landslide mass (earth flow vs. slide zones), which are attributed to differences in the degree of alteration and the disturbance of the internal structure of materials constituting the landslide mass. These results show that techniques based on the measurement of ambient noise are of special interest for studying very large, highly remolded landslide masses.

Keywords: landslide; geophysical prospecting; ambient noise; f-k technique; HVSR

1. Introduction

The occurrence of landslides is a geohazard that affects many places in the world. Landslides may be induced by natural events (rain, earthquakes, volcanoes) and by human activities. When landslides are large, the damage toll may be very high for the society, including casualties [1]. The possible reactivation of such large landslides also represents a severe threat to society [2]. Land-use planning and civil works may mitigate the associated hazard, but for the correct implementation of countermeasures and the management of this kind of hazard, knowing the characteristics of landslides (typology, volume, etc.) is of great interest. Thus, the detailed study of existing large landslides is of prime importance.

Evaluating the possible reactivation of existing large landslides is a complex task. This evaluation requires detailed geological and geomechanical data to assess the stability of a landslide mass in its change when the triggering factors affect the mass. However, this type of study has to face the difficulty of obtaining data for defining the geometry of the landslide mass and characterizing the mechanical and physical properties of the involved materials. The use of traditional in situ methods, such as drilling boreholes and taking undisturbed soil samples, may become economically prohibitive due to their cost, the size of the area to study and the difficulty of drilling in unstable slopes. An economical alternative is needed for the study of very large landslides. For this purpose, the use of geophysical methods is the best solution [3]. Jongmans and Garambois [4] summarized the advantages of these methods for this purpose: (1) these methods are flexible, quick and deployable on slopes, (2) they are not invasive and do not disturb the medium, providing information about the internal structure of the landslide mass, and (3) they allow for investigating large volumes of material in a fast way.

Among the available geophysical techniques, those based on the analysis of seismic noise have become the preferred [3,5]. This is due to their low cost and the type of results they offer (thickness and shear-wave velocity, Vs, of the materials studied), which are useful for defining the inner structure of landslide masses and for studies of their seismic response. The horizontal to vertical spectral ratio (H/V or HVSR) method [6] and different array techniques [7–10] are the most used procedures for a quick and easy characterization of the soil. The analysis of the HVSR results provides an estimation of the fundamental resonance frequency of the SH waves [11]. The analysis of the array data provides an estimation of the dispersion curve. After that, the resulting curves obtained from the HVSR and array techniques can be inverted in order to estimate the corresponding Vs profile at the site under study. When only one of the HVSR or dispersion curves is available at one site, they can be inverted separately. In other cases when both results are accessible at the same site, a joint inversion of the HVSR and dispersion curves is carried out for the soil characterization [12]. In any case, there is no single solution for the inversion issue. The best solutions can be estimated in terms of a misfit function. Two of the most used inversion methods are the neighborhood [13,14] and the genetic algorithms [15].

The Güevéjar landslide (Granada, southern Spain, Figure 1) is an outstanding example of very large landslide that occurred in the past, likely related to climate conditions different to the present ones. It has had multiple earthquake-induced reactivations, but no activity has been observed due to other triggering factors [16]. The seismic-reactivation history of this landslide started on 1 November 1755, when the Lisbon earthquake (moment magnitude Mw 8.7) struck and reactivated this pre-existing landslide [17]. Because of landslide mass movement, the town of Güevéjar, located on the central part of the landslide mass, was severely damaged [18], but the town was reconstructed when landslide movement stopped. A century later, on 24 December 1884, a Mw 6.5 earthquake struck the provinces of Granada and Málaga in Spain (Figure 1), inducing a new reactivation of this landslide [19]. Historical documents describe that due to the downslope movement of the earth flow existing at the toe of the landslide, a temporary damming of the river existing at the base of the slope occurred [20]. Because of the renewal of movement, the town was so severely damaged that it was finally abandoned. A new town was built approximately 700 m west of its original location, outside of the landslide limits (Figure 2). During the last few decades, old roads and trails have been asphalted, and new houses have been built within the landslide mass.

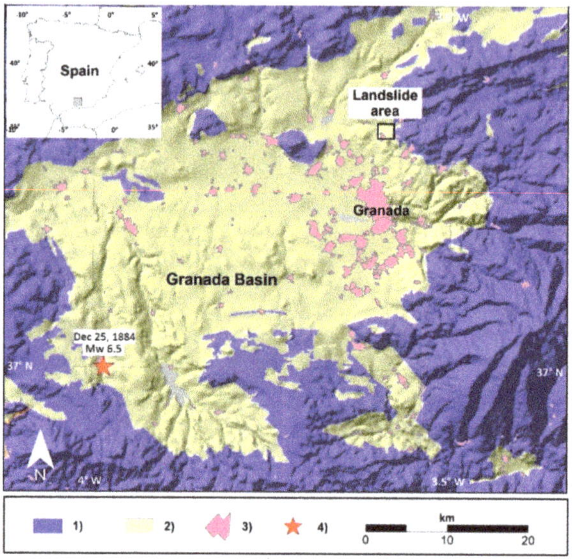

Figure 1. Location map of the study area. (**1**) Basin basement, (**2**) basin fill (Neogene sedimentary rocks), (**3**) urban areas and (**4**) location of the 1884 earthquake (Mw 6.5) that reactivated the landslide.

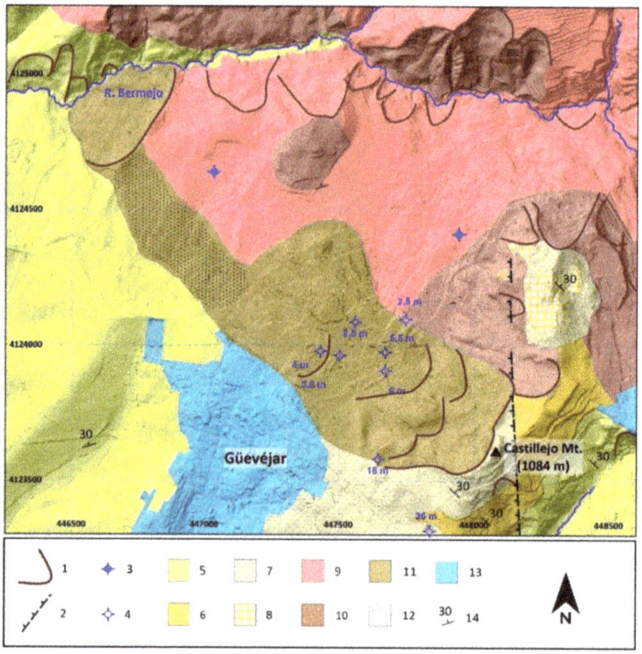

Figure 2. Geological map of the landslide area. (**1**) Landslide scarps, (**2**) faults, (**3**) springs, (**4**) well points and depth to water table (in m), (**5**) marls unit (MU unit), (**6**) red silts unit (RSU unit), (**7**) marls of the Marl and Limestone unit (MLU unit), (**8**) limestone of the Marl and Limestone unit (MLU unit), (**9**) red clays of the Recent Red Deposits unit (RRD unit), (**10**) other landslides, (**11**) Güevéjar landslide, (**12**) earth-flow in the Güevéjar landslide, (**13**) urban areas and (**14**) strike and dip of bedding.

Despite the size and the seismic history of this landslide, it has been poorly studied. In addition to the studies that describe the reactivation of the landslide during past earthquakes [18], a geological model for this landslide was proposed based on ground surface geological data [21]. In this last study, several landslides existing in the area were grouped into a single one. Based on the proposed model, a study into earthquake-induced reactivation criteria, estimating the magnitude of earthquakes that could reactivate this landslide and the maximum epicentral distance of the occurrence of such events, has been performed [22]. More recently, the mobility of the upper part of this landslide, based on a model created from a limited amount of geophysical data and ground surface geological data, was analyzed through quantitative numerical approaches [23].

In this paper, we perform a comprehensive study of this landslide from a geophysical point of view, with the use of several different seismic techniques based on the measurement of ambient noise. Specifically, single-point measurements and array data are analyzed using the HVSR method and the frequency-wavenumber (f-k) technique [8,24], respectively. Subsequently, the inversion of the HVSR and dispersion curves is carried out by applying the neighborhood algorithm. The results obtained with these methods are constrained by data from boreholes and ground surface geology. The idea is to create the most complete possible model of the considered landslide, useful for later studies about seismic scenarios that could reactivate this landslide in the future. The estimated Vs profiles might be subsequently used for assessing site effects [25].

2. Geological Setting

The study zone is located approximately 6 km north of the city of Granada (southern Spain). The landslide affects materials of the sedimentary fill of the Neogene basin of Granada, in the central sector of the Betic Cordillera (Figure 1). This area likely has the highest tectonic activity and seismicity rate in Spain.

The sedimentary record of the basin is up to 2 km in thickness [26]. The basin fill is divided into two different sets: a lower group constituted by marine sedimentary rocks, Upper Miocene in age, and an upper one, of continental sedimentary rocks, Upper Miocene (Turolian) to Quaternary in age [27].

Focusing on the landslide area, all geological formations cropping out in the area belong to the continental group (Figure 2). A first unit consisting of silts, sands and marls (MU unit) is found at the bottom of the stratigraphic series. This unit is more than 100 m thick and shows some transitions between silts, sands and yellowish marls. Marls are more frequent towards the western part of the landslide area, while silts are predominant to the south. This formation contains some thin layers of gypsum and/or limestone. Its age has been identified as late Miocene [28,29]. A second unit consists of red silts, including some layers of clays, sands and highly cemented conglomerates (RSU unit). Both the MU and RSU units are parts of the same depositional environment. The MU unit constitutes the distal deposits of the alluvial fan system developed during the late Miocene in the area, while the RSU unit represents the proximal deposits of this system. Above these units, marls and lacustrine limestone (MLU unit) deposits exist (late Miocene in age). They crop out at the top of Mount Castillejo and in nearby areas (Figure 2). Finally, red clays, sands and conglomerates (recent red deposits, RRD unit) crop out widely north of the landslide area (Pliocene-Pleistocene in age).

This landslide is approximately 2000 m long (from top to toe) with a maximum width of 600 m in its central part. Down the slope, the width of instability decreases and varies between 300 and 400 m near its toe. The landslide may be described as a complex one *sensu* Cruden and Varnes [30], consisting of a roto-translational earth slide in the upper and middle parts of the slope, and an earth flow downslope (toe of the landslide). The slide affects most of the geological units previously described. At the head of the slide, RSU and MLU units are clearly recognizable in the scarps existing in this area. They show vertical displacement of 2 to 5 m. Several tilted blocks consisting of limestone (MLU unit) resting on red silts (RSU unit) may be found in the upper half of the slide. Due to agricultural activity, the structure of the materials is hardly recognizable in the remaining parts of the landslide. In the lower part, at the base of the slide and in the zone of earth flow, the landslide also involves the RRD unit. In some places, where small ravines have eroded the landslide mass, in situ marls (MU unit) are recognizable.

Historical data of the landslide after the 1884 reactivation describe that large and deep ground cracks were recognizable in the middle part of the landslide, defining large, undisturbed blocks of materials that slid down slope [31]. From the base of the slide body, historical data also describe large debris (earth flow) that flowed downslope, damming the river Bermejo [20].

3. Methodology

3.1. Horizontal to Vertical Spectral Ratios (HVSR) Method

The application of the HVSR technique to ambient vibration signals or seismic noise provides an estimation of the fundamental resonance frequency associated with the site under study. This relation is the fundamental basis of the HVSR analysis [6,32].

Many experiments and theoretical 1D investigations have demonstrated that a strong impedance contrast between the upper layers and the underlying ones (stiffer layers) provides a clear HVSR peak around the fundamental resonance frequency [33–39]. The HVSR method has become accepted in microzonation studies, since its application and interpretation is straightforward and suitable [40–43].

The practical implementation of the HVSR method requires only one three-component sensor for the recordings, which makes this technique relatively cheap and easy to implement. In this study, seismic noise measurements were recorded using a broadband seismic station (Guralp CMG-6TD). Approximately 45 min to one hour of seismic noise was captured using a sampling frequency of 100 Hz, recorded in Guralp Compressed Format (GCF). A total of 26 measurements were taken in March 2019, in days of little or no wind in order to avoid disturbances at low frequencies (Figure 3).

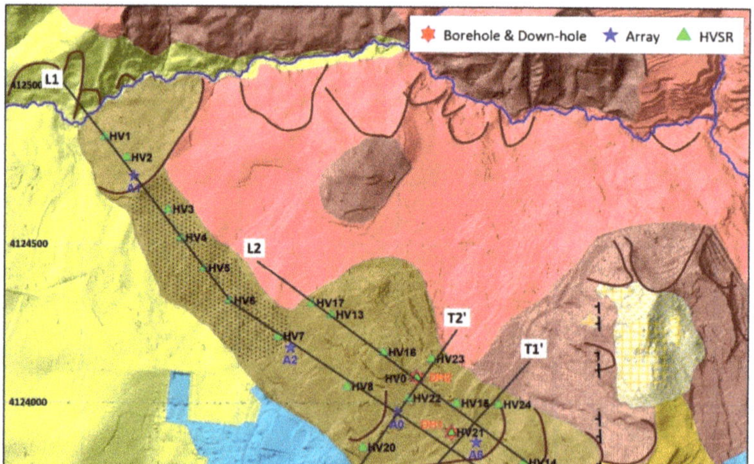

Figure 3. Location of sites of measurement in the Guëvéjar landslide area. The colors and signatures are the same as in Figure 2.

The HVSR curves have been estimated using the GEOPSY software from the Sesarray package [44]. The recorded signals have passed through an anti-triggering filter, and then they have been divided into non-overlapping windows of frequency-dependent length. Each window was tapered by a 5% cosine function and its length was 50 periods of the frequency analyzed at each moment [45]. After that, the frequency response was obtained for each window and component, and smoothed applying a Konno and Omachi window (coefficient b = 40) [46]. Subsequently, for each window, the squared average of the horizontal components was computed and then the horizontal-to-vertical spectral ratio was calculated. Finally, the HVSR curves from the selected windows have been averaged to provide the mean HVSR curve and its corresponding standard deviation.

3.2. Downhole Measurements

Geotechnical exploration was carried out in September 2014 to develop a first approximation of the soil characteristics. More specifically, two boreholes were drilled along the main direction of the landslide (Figure 3) just where the historical town was located, to the northeast of the present Güevéjar town. The maximum depths were 60 (borehole S2, Figure 4) and 70 m (S1, Figure 4), respectively.

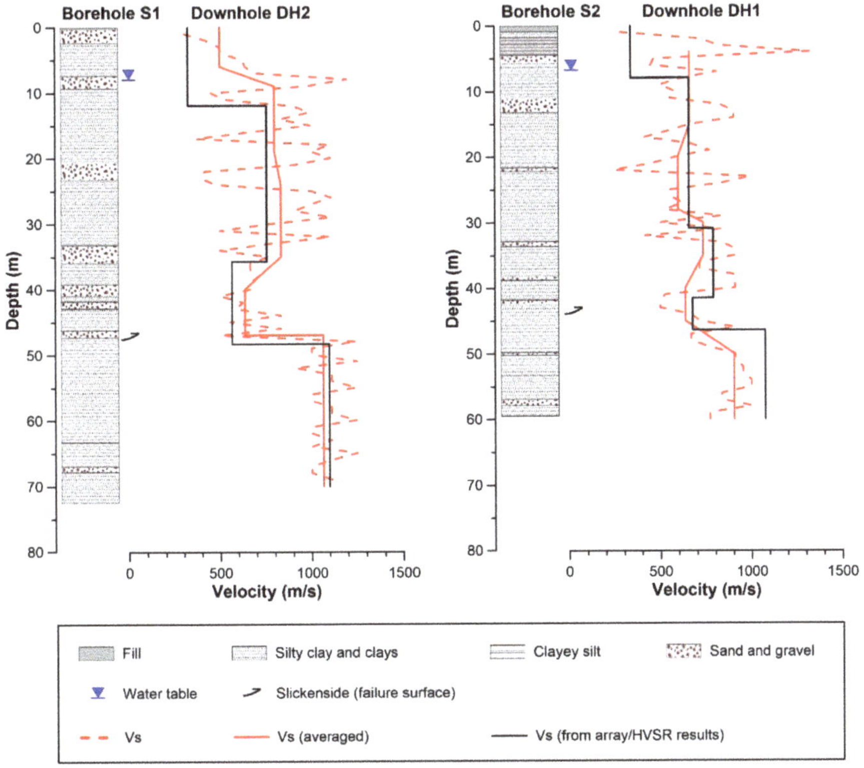

Figure 4. Soil columns and downhole results obtained at boreholes drilled in the landslide. The plot also depicts results obtained from the inversion of noise measurements at the same locations (see Figure 3 for location of the boreholes).

The tests were performed by Geytex SL company. They operated an exploration seismograph (model GEODE 12, 24 bits, Geometrics, San Jose, CA, USA) with triaxial geophones and a heavy hammer as the seismic source. The hammer hits a steel rod that had been previously fixed to the ground surface. Measurements were taken every meter along the borehole until reaching its base.

3.3. Frequency-Wavenumber (f-k) Analysis

The f-k method is an array technique based on the analysis of seismic noise. Assuming stationary noise in both time and space, the incoming waves can be represented as a function of the frequency and the vector velocities through the frequency-wavenumber spectral density function [8,24]. Thus, the wave velocities and their respective directions of approach are derived as functions of the frequency.

The power spectrum can be obtained through the maximum likelihood method (MLM) [8] or the beam-forming method (BFM) [24], which is the one used in this work. Additional information can be found elsewhere [47–49].

The seismic noise array measurements were taken around the Güevéjar landslide at eight different sites (Figure 3). The implementation of the arrays was achieved with seismic refraction/reflection equipment consisting of twelve 4.5 Hz vertical geophones connected to a multichannel seismic recorder (SmartSeis ST, Geometrics, San Jose, CA, USA). A circular layout without a central station was deployed for the array layout (Figure 5a). For all the sites, the diameter of the arrays was configured to 20 m, which is the maximum possible aperture allowed by the cable length of the reflection/refraction equipment used. Taking into account that the wavelengths obtained by the f-k method can be several times greater than the array diameter [50,51], this configuration would be enough for the soil classification purposes [52].

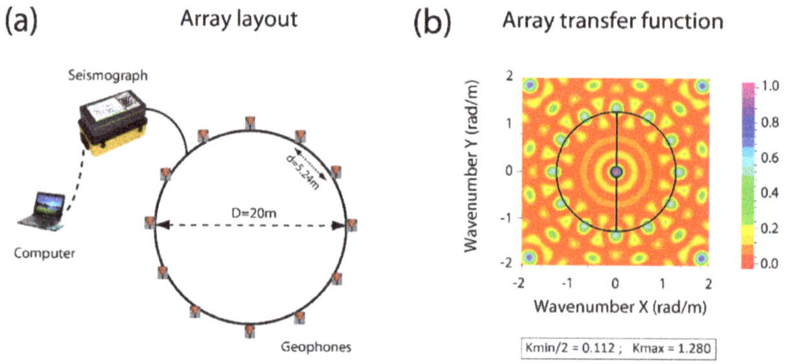

Figure 5. (**a**) Array layout; (**b**) corresponding theoretical response. The k limits corresponding to the maximum resolution (kmin/2) and the aliasing limit (kmax) are also indicated.

The data acquisition consisted of approximately fifty records of 32 s without overlapping, sampled at 500 Hz and subsequently resampled at 100 Hz. Although the natural frequency of the geophones is 4.5 Hz, this equipment allows for estimating dispersion curves above approximately 2 Hz [53,54].

In Figure 5b, we show the theoretical response of the array, providing estimates of the aliasing (kmax) and the maximum resolution (kmin/2) limits, in terms of the wavenumber.

The f-k analysis has also been accomplished using the GEOPSY software. The initial configuration requires the establishing of the grid of wavenumber (k) points to be analyzed, indicating the maximum value (grid size) as well as the minimum step (grid step). These values determine the aliasing limit and the maximum resolution, respectively. For the arrangement carried out in this work, the analysis of the theoretical array response provides a maximum resolution of 0.06 rad/m (kmin/4) and an aliasing limit of 0.64 rad/m (kmax/2) (see Figure 5b). The real array response depends on the number of sensors and the sensor geometry, but also on the slowness and the wavenumbers of the seismic phases observed within the array.

With these considerations, a dispersion curve is first estimated for each analyzed 32 s window, which corresponds with the length of the recordings. Finally, a mean dispersion curve is computed by averaging the curves obtained from all the windows.

3.4. Inversion of the Dispersion and HVSR Curves

The HVSR and dispersion curves are inverted using the neighborhood algorithm to obtain an optimal estimation of the Vs profiles [14]. The inversion process explores a defined parameter space (number of layers, S-wave velocity, P-wave velocity, density and layer thickness) in search of possible solutions. The results previously obtained from borehole drills and downhole experiments have helped to constrain the number of layers and their corresponding thickness and S-wave velocity ranges. Poisson's ratio and density values have been set up for all the layers in the range of 0.2–0.5 and 1600–2200 kg/m^3, respectively, which correspond to the characteristics of the materials in the studied landslide.

The ellipticity of the fundamental mode Rayleigh wave or the dispersion curve is estimated for each computed model (forward problem), and then contrasted with the empirical curves using a misfit function [55], which reflects quantitatively the difference between the generated models and the observed results.

For the calculation of the Vs profiles, the Dinver software from the Sesarray package has been used [44]. This software allows the inversion of dispersion and ellipticity curves. In the case of the dispersion curves, they are obtained directly from the f-k analysis. In the case of the ellipticty curves, they were previously extracted from the experimental HVSR curves [42,56] and then used as input data of the Dinver software. Alternatively, HVSR curves can also be used as input data if some considerations are taken into account [41,57]. For large velocity contrast, Fäh et al. [58,59] showed that HVSR curves are dominated by the ellipticity of the fundamental mode Rayleigh wave in the frequency band between the fundamental resonant frequency and the next minimum, with its shape being dominated by the specific layering of the sediments. With regards to these indications, in this work the frequency band around the main resonant peak has been considered for the inversion process. Only in the cases where a secondary peak or hump has also been identified have we broadened the frequency interval to include also this peak, as this could provide information about other velocity contrasts or interfaces.

4. Results and Discussion

4.1. Downhole Results

We have used the information provided from two boreholes drilled in the area (Figure 3) to identify the main characteristics of the soil columns. In Figure 4, the stratigraphic information from the boreholes and the Vs profiles deduced from the downhole test are depicted. The soil columns observed at both boreholes show a series of reddish silty clay and clayey silt levels with thin sand or conglomerate levels. At depths of 43 m (at borehole S2) and 48 m (S1), the soil samples showed polished surfaces, which are interpreted as slickensides formed by the slip along a failure surface of the landslide. The water table is found at less than 10 m in both boreholes.

The first borehole (S1) was drilled to a depth of 70 m (Figure 4). From the downhole study, we deduced a three-layer Vs profile. The first layer has a thickness of 5–10 m and a Vs value less than 500 m/s. The second layer reaches a depth of 40–43 m with a Vs value of approximately 800–900 m/s. Finally, the depth of the third layer is between 43 and 50 m with a Vs of 600–700 m/s. Below 43–50 m, hard materials are found with a Vs value of approximately 1000 m/s.

The second borehole (S2) was drilled to a depth of 60 m (Figure 4). From the downhole study, we deduced a four-layer Vs profile. The first layer of fills and clayey silts presents a maximum thickness of 5 m and a Vs value less than 500 m/s, which is similar to that observed at S1. The depth of the second layer lies between 25 and 30 m with an average Vs value of approximately 700 m/s. The third layer reaches a depth of

35–40 m and presents a Vs value of 800–900 m/s. Finally, the depth of the fourth layer is found at approximately 45–50 m, with a Vs value of approximately 700 m/s. Below 45–50 m, hard materials are found, presenting a Vs value close to 1000 m/s.

The deeper layers on both profiles are located below the surface failures identified in the soil columns, and represent the in situ materials. They show a certain contrast in the measured velocities with respect to shallower layers (Figure 4). Sudden increases in Vs velocities in S1 just below the failure surface are evident. In S2, this contrast is not as clear, but even in this case, it is recognizable that the Vs is lower at the zone of failure than the values observed in materials above and below this zone.

4.2. HVSR Analysis Resonance Frequencies

In Figure 6, some representative cases of the HVSR curves obtained along the landslide are shown. The locations of some of these sites, close to the landslide flanks, may explain the presence of some broad peaks with rather low amplitudes. At these sites, the wavefields associated with local noise sources are more complex, since the wavefields also include additional waves diffracted from the lateral heterogeneities. Experimental tests and numerical simulations have shown that, for local surface sources, the HVSR curves obtained at laterally varying structures present broader and lower maxima, which may even be hard to identify [60].

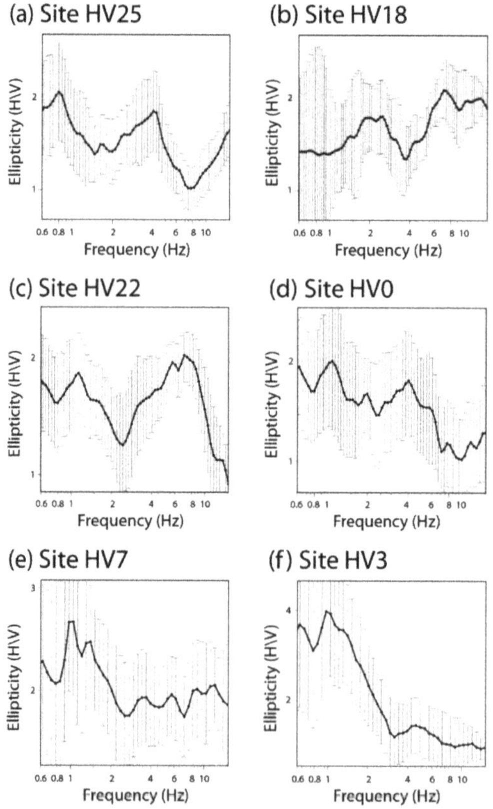

Figure 6. Examples of HVSR curves estimated in the central area of the landslide. Check Figure 3 for the location of the sites.

In our study case, the resonance frequency, f, associated with the contrast between the sediments and the stiffer materials below the landslide, can be related to the average shear-wave velocity of this one-dimensional stratigraphy, Vs, and the corresponding thickness, h, through the equation [33]:

$$f\,(\mathrm{Hz}) = Vs/(4\,h) \quad (1)$$

Therefore, attending to the borehole information, it can be estimated that the resonance frequencies of interest for the landslide characterization present values higher than 2 Hz. In this case, peaks below 2 Hz might be associated with deeper contrasts from the contacts between materials of the sedimentary fill of the Granada basin (or between this fill and the basement).

4.3. S-Wave Velocity Profiles

HVSR and dispersion curves have been inverted separately, as they were obtained at different sites within the landslide. The number and the thickness of layers, as well as the range of Vs values obtained from the downhole results (see Section 4.1), have been used to constrain the parameter space in the sites around the central area of the landslide. In the case of the thickness and Vs values, an additional margin around the value ranges deduced from the boreholes has been included because the measurement points do not match exactly at the borehole sites (see Figure 3). For the sites closest to the first borehole (S1), the following model has been used: a first layer with maximum thickness of 12 m and shear-wave velocity between 150 and 550 m/s; a second layer that reaches a maximum depth between 35 and 43 m and has a Vs value between 750 and 950 m/s; and a third layer whose maximum depth reaches 45–53 m, with a Vs value of 550–750 m/s. Below this, a hard layer is configured with a minimum depth of 48 m and velocities between 900 and 1100 m/s.

In the cases of the sites closest to the second borehole (S2), a four-layer model has been applied with the following characteristics: a first layer with maximum thickness of 8 m and shear-wave velocity between 150 and 550 m/s; a second layer whose maximum depth reaches 22–33 m, with Vs value of 650–750 m/s; a third layer with maximum depth and shear-wave velocity in the ranges 33–43 m. and 750–800 m/s, respectively; and a fourth layer that reaches a limit depth of 45–53 m., and a Vs of 650–700 m/s. Below, a hard layer is configured with a minimum depth of 48 m and velocities between 900 and 1000 m/s.

In all cases, an additional harder layer has been added to these models in order to take into account possible deeper velocity contrasts. The profiles obtained at these two sites are shown in Figure 4, showing a remarkable similarity with the results of the downhole tests.

For those sites closer to the expected border of the landslide, a simpler model has been used with two sedimentary layers over two harder layers, representing the in situ stable materials. The superficial layer is included to take into account the shallower materials, disturbed by agricultural activities. The inclusion of two harder layers allows also for adjusting the estimated model to the HVSR peaks found below 2 Hz, which might be associated with deeper contrasts.

The fitting carried out by the software is based on a multimode surface-wave modeling through a vertically heterogeneous, elastic and isotropic medium. The model generation process implies the computation of the theoretical ellipticity, or dispersion curves, of the fundamental mode Rayleigh wave (forward model), and a comparison with the respective experimental curves. The similarity between the theoretical and the empirical HVSR or dispersion curves is expressed quantitatively by a misfit value.

Figures 7 and 8 depict some examples of the Vs profiles estimated using the HVSR curves, and the dispersion curves obtained by the f-k method, respectively. From the estimated profiles, we can observe how the obtained results with both techniques are coherent with each other.

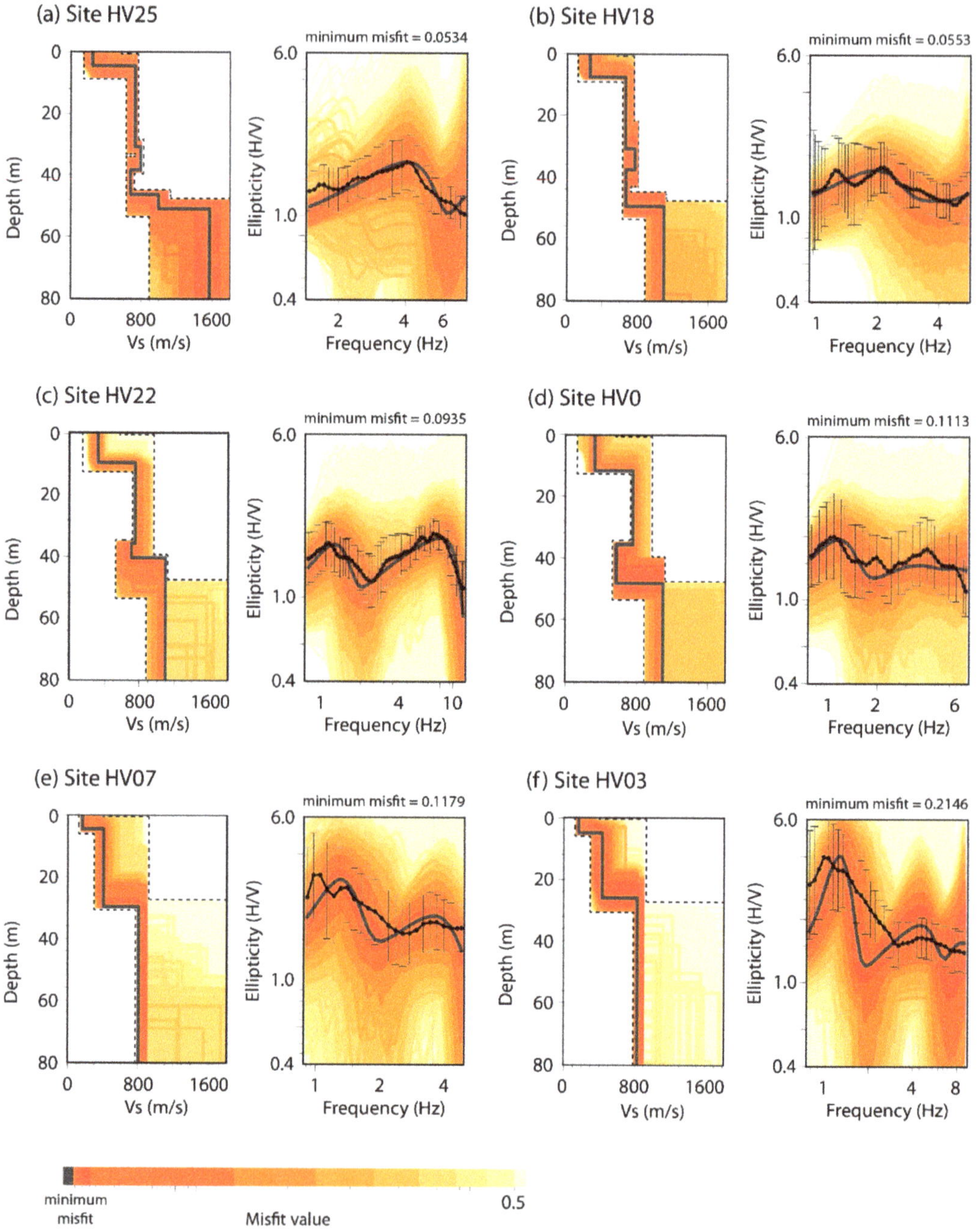

Figure 7. HVSR inversion for six sites (see Figure 3 for location of sites). The best ellipticity and velocity models (dark grey lines), together with models lying inside the minimum misfit + 10% (dark orange) and all the tested models, are shown. The black curve and bars in the ellipticity plots are the average and standard deviation, respectively, of the experimental HVSR curve. The dashed lines in the velocity profiles indicate the search bounds.

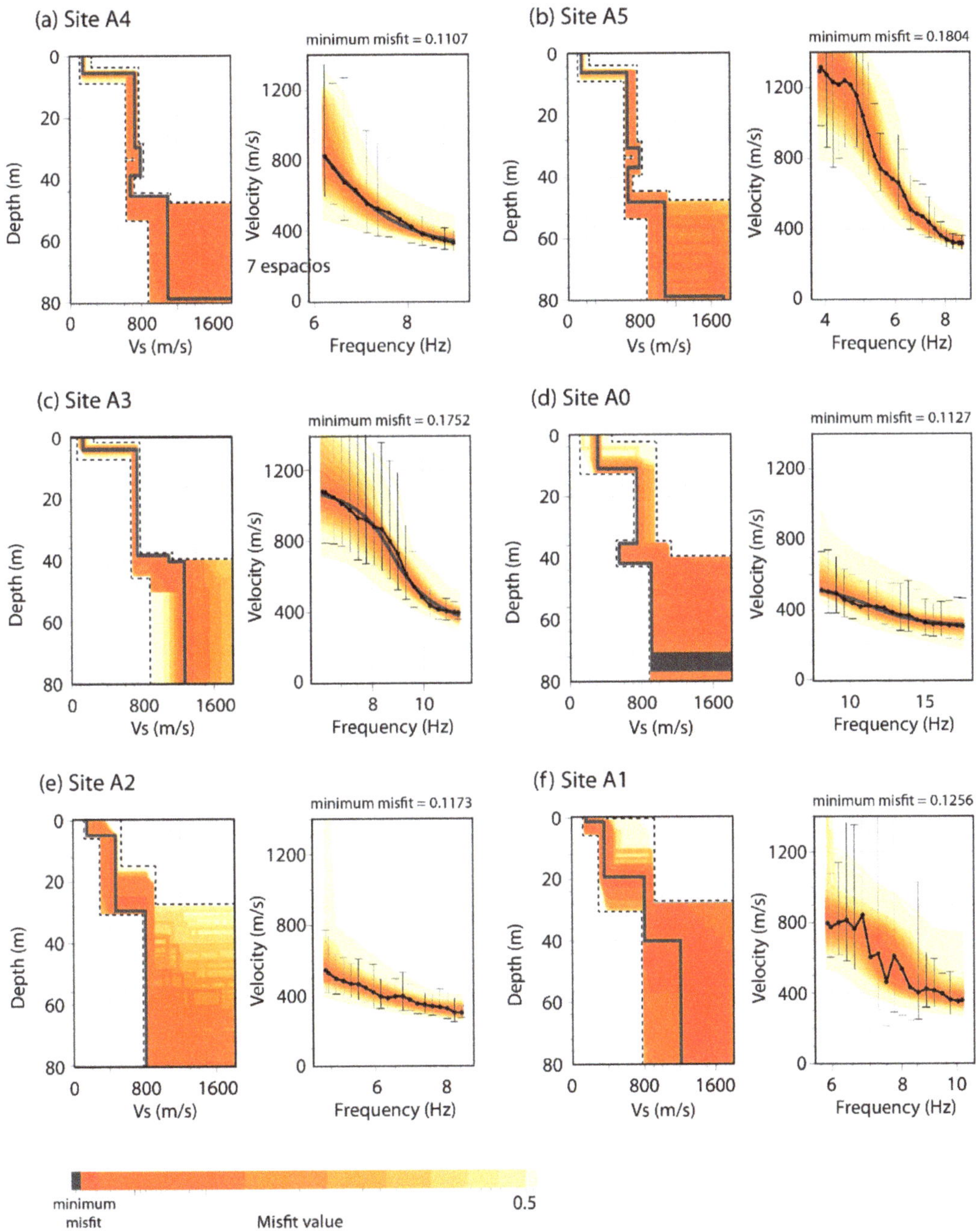

Figure 8. Dispersion curve inversion for six different sites (see Figure 3 for location of sites). The best dispersion and velocity models (dark gray lines), together with models lying inside the minimum misfit + 10% (dark orange) and all the tested models, are shown. The black curve and bars in the velocity-frequency plots are the average and standard deviation, respectively, of the experimental dispersion curve. The dashed lines in the velocity profiles indicate the search bounds.

4.4. Landslide Geological Structure

The results obtained from the geophysical studies on the Güevéjar landslide are integrated and synthesized with ground surface geology in order to build, for the first time, a model of the whole unstable mass (Figures 9 and 10). The landslide limits are based on data from boreholes and on the contrast in shear wave velocities found with depth. Superficial limits are based on field observations and on observations supplied by land owners.

The results show that the whole area is characterized by a shallow layer with very low Vs velocity (average value of 251 m/s, ranging from 139 to 422 m/s). Its thickness varies irregularly (6.6 m on average, ranging from 1.1 up to 11.9 m), although the greater values are found in the central part of the slide area (thickness > 5 m), where a gentle slope exists (Figures 9 and 10). The obtained values are in agreement with those observed in the available borehole soil columns. This layer is found elsewhere, in both landslide and stable terrain areas. It is interpreted as a shallow layer of alteration due to several not exclusive possible causes (weathering, agricultural activities, fills or the deposition of debris in low areas).

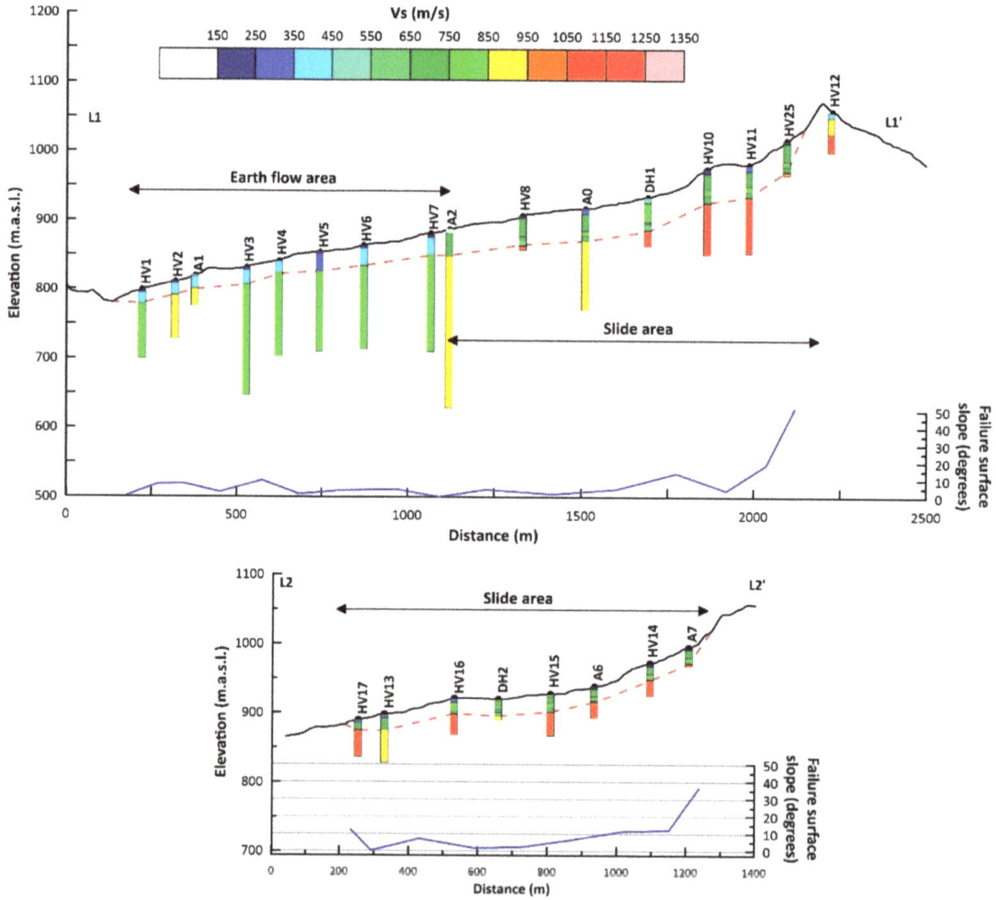

Figure 9. Longitudinal cross sections of the landslide. See Figure 3 for location of sections. The dashed red line shows the location of the failure surface.

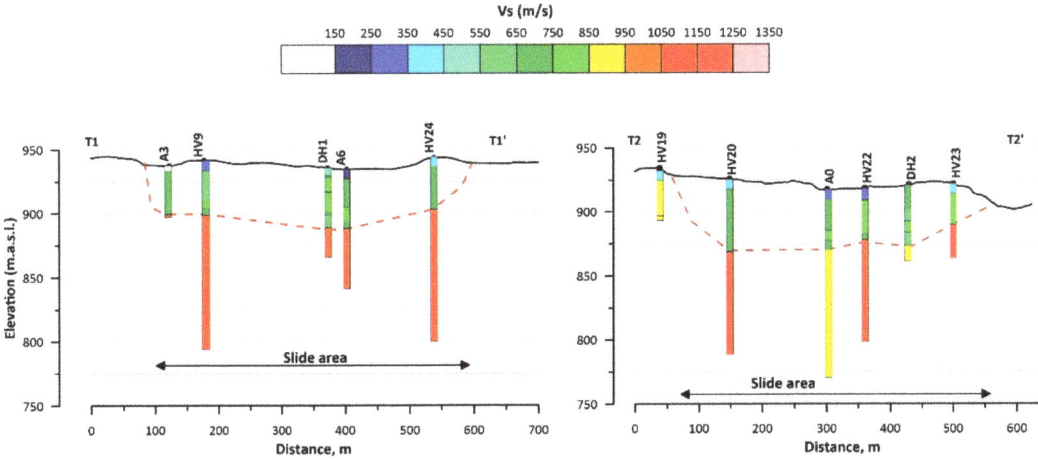

Figure 10. Transversal cross sections of the landslide. See Figure 3 for location of sections.

In the landslide area, below the shallow layer, a second layer exists exhibiting a complex pattern of velocities, depending on whether it is found in the earth flow or in the slide deposits. At the lower parts of the slope, corresponding with the area where the earth flow is, a layer with moderate Vs exists, averaging 375 m/s (ranging from 340 to 427 m/s), with thicknesses varying in the range of 16–25 m (Figure 9). Such thicknesses reduce towards the flanks of the instability, as is seen in some ravines that have partially eroded landslide mass. The observed Vs velocities may be related to the degree of disturbance of the materials involved in the landslide. The materials involved in the earth flow are characterized by a high degree of disturbance, and they are altered to a soil state. A significant reduction in Vs is expected with respect to the same undisturbed materials.

In the zone occupied by slide deposits, this second layer may comprise several sublayers. Downhole tests show moderate Vs values (500–800 m/s; Figure 4) for these sublayers. Similar results are obtained at most of the sites where HVSR (Figure 7) or arrays (Figure 8) measurements were accomplished. The average Vs value of all sublayers is 710 m/s, markedly higher than the Vs of earth flow deposits. This may be explained by the lower degree of disturbance of the materials involved. Between all measurement sites, the sum of the thicknesses of all sublayers varies from 24 m up to 48 m (Figures 9 and 10).

The lower limit of landslide, or failure surface, is attributed to a sudden increase in Vs with depth observed below the layer previously described, in accordance with results observed in the boreholes drilled and downhole tests. Below the earth flow area, Vs > 800 m/s, resulting in a high contrast with earth flow deposits (average increase in Vs > 100%). The resulting surface shows a gentle dip towards the base of the slope (Figure 9). Below the slide area, materials show an average Vs of 1060 m/s (ranging from 910 to 1100 m/s; average increase > 50%), and the failure surface is more irregular than in the earth flow area (Figures 9 and 10), with changes in relation to the secondary failures recognizable in the field (Figure 2). It is interesting to note that, for the in situ materials, lower values of Vs coincide with the spatial distribution of marls of the MU unit, while higher Vs values were found in areas where RSU and MLU materials crop out (Figures 3, 9 and 10).

The resulting model of the landslide shows that its maximum thickness is approximately 56 m in the western part (site HV20; Figures 3 and 10). This value results from the sum of thicknesses of surficial layer and that below it. This thickness decreases slowly to 45–50 m towards the crown area of the landslide (sites A5, HV10, HV14 and HV18). The thickness of earth flow ranges between 20 and 30 m at the center of the lobe, where the measurements were performed. Towards the flanks, thickness decreases progressively. The reconstructed geometry is consistent with the results obtained from the numerical modeling performed in previous studies [23], which justifies the main local resonance as related to the landslide thickness, while landslide mobility seems to be more strictly linked to its longitudinal dimension (length). This justifies the unexpected reactivation of the Güevéjar landslide in the past, such as when it was triggered by the 1755 Lisbon earthquake.

5. Conclusions

The high-resolution engineering–geological modeling and characterization of large landslides is complex when the application of standard geotechnical techniques is unfeasible. This is mainly due to the large number of required investigations and the difficulty of drilling in unstable slopes. As an alternative, seismic methods based on ambient noise measurements are used to characterize the geometry and the underground distribution of materials. Specifically, HVSR and array measurements (f-k analysis) have been combined properly along the studied landslide area. In such a large and complex geological structure, the data analysis is carried out from the central area of the landslide towards its lateral limits. In addition, geotechnical information, when available, has to be used to constrain the range of the expected Vs models estimated from the inversion of HVSR and/or dispersion curves. The combination of different techniques provides more robustness to the estimated Vs profiles.

Therefore, the results obtained here from the use of ambient noise and their interpretation with several geophysical techniques are useful for building a comprehensive model for the Güevéjar landslide based on the contrast in velocities between in situ materials and the landslide mass. This model allows for distinguishing two zones characterized by different landslide mechanisms: the first one, located at the upper slope, is related to a sliding slope failure, and the second one, located on the downslope, is related to an earth flow. It is also possible to highlight differences in Vs values within the landslide mass: minimum Vs values are found in the earth flow that exists at the toe of the landslide (average Vs = 375 m/s), while higher Vs values appear in the middle and upper parts of the landslide (average Vs = 710 m/s). These changes may be interpreted as a function of the degree of disturbance/alteration of the geological materials involved in the landslide: the earth flow implies the total loss of the internal structure of the materials, whereas the slide dislodges the materials into blocks that move as a whole, preserving most of their internal structure.

These results show the usefulness of the set of geophysical methods implemented in this study, based on the measurement and inversion of ambient noise data, for analyzing very large landslides at a reasonable cost. The relation between the degree of material damage and the resulting Vs values makes these techniques of special interest for studying multiple landslide mechanisms, such as sliding, or disorganized ones, such as flows.

Author Contributions: Conceptualization, J.D., J.G., L.L. and S.M.; methodology, J.J.G.-M., J.A.P., M.J.R.-P. and J.L.S.-L.; software, J.J.G.-M., L.L. and S.M.; validation, J.D., F.J.G.-T., J.G. and C.S.d.G.; formal analysis, J.J.G.-M. and S.M.; investigation, all authors; data curation, J.D. and J.J.G.-M.; writing—original draft preparation, J.D. and J.J.G.-M.; writing—review and editing, all authors; supervision, J.D.; project administration, J.D. and J.J.G.-M.; funding acquisition, J.D. and J.J.G.-M. All authors have read and agreed to the published version of the manuscript.

Funding: This study was funded by the EU (FEDER), by the Secretaría de Estado de Investigación, Desarrollo e Innovación of the Spanish government (projects CGL2015-65602-R and CGL2016-77688-R), by the Junta de Andalucía (project GGI3002IDIN), the Programa Operativo FEDER Andalucía 2014–2020—call made by the University of Jaén 2018, and by Research Groups VIGROB-184 and VIGROB-116 (University of Alicante).

Data Availability Statement: Data will be available under request to authors.

Acknowledgments: The authors greatly appreciate the support, comments and suggestions made by C. López Casado, formerly at University of Granada, that greatly improved the work.

Conflicts of Interest: The authors declare no conflict of interest. The funders had no role in the design of the study; in the collection, analyses, or interpretation of data; in the writing of the manuscript, or in the decision to publish the results.

References

1. Sassa, K. The ISDR-ICL Sendai Partnerships 2015–2025: Background and content. In *Advance Culture of Living with Landslides, ISDR-ICL Sendai Pertnerships 2015–2025*; Sassa, K., Mikos, M., Yin, Y., Eds.; Springer: Berlin/Heidelberg, Germany, 2017; Volume 1, pp. 3–21. [CrossRef]
2. Lari, S.; Frattini, P.; Crosta, G.B. A probabilistic approach for landslide hazard analysis. *Eng. Geol.* **2014**, *182*, 3–14. [CrossRef]
3. Pazzi, V.; Morelli, S.; Fanti, R. A review of the advantages and limitations of geophysical investigations in landslide studies. *Int. J. Geophys.* **2019**, *27*, 2983087. [CrossRef]
4. Jongmans, D.; Garambois, S. Geophysical investigation of landslides: A review. *Bull. Soc. Géol. France* **2007**, *178*, 101–112. [CrossRef]
5. D'Amico, S.; Panzera, F.; Martino, S.; Iannucci, R.; Paciello, A.; Lombardo, G.; Galea, P.; Farrugia, D. Chapter 12-Ambient noise techniques to study near-surface in particular geological conditions: A brief review. In *Innovation in Near-Surface Geophysics*; Persico, R., Piro, S., Linford, N., Eds.; Elsevier: Amsterdam, The Netherlands, 2019; pp. 419–460. [CrossRef]
6. Nakamura, Y. A method for dynamic characteristics estimation of subsurface using microtremors on the ground surface. *Quart. Rep. Railw. Tech. Res. Inst. Jpn.* **1989**, *30*, 25–33.
7. Aki, K. Space and time spectra of stationary stochastic waves, with special reference to microtremors. *Bull. Earthq. Res. Inst. Tokyo Univ.* **1957**, *25*, 415–457.
8. Capon, J. High-resolution frequency-wavenumber spectral analysis. *Proc. IEEE* **1969**, *57*, 1408–1419. [CrossRef]
9. Ling, S.; Okada, H. An extended use of the spatial autocorrelation method for the estimation of geological structure using microtremors. In Proceedings of the 89th SEGJ Conference, Nagoya, Japan, 12–14 October 1993; pp. 44–48.
10. Park, C.B.; Miller, R.D.; Xia, J.; Ivanov, J. Multichannel analysis of surface waves (MASW)-active and passive methods. *Lead. Edge* **2007**, *26*, 60–64. [CrossRef]
11. Bonnefoy-Claudet, S.; Cornou, C.; Bard, P.Y.; Cotton, F.; Moczo, P.; Kristek, J.; Fah, D. H/V ratio: A tool for site evaluation. Results from 1-D noise simulations. *Geophys. J. Int.* **2006**, *67*, 827–837. [CrossRef]
12. Picozzi, M.; Parolai, S.; Richwalski, S.M. Joint inversion of H/V ratios and dispersion curves from seismic noise: Estimating the S-wave velocity of bedrock. *Geophys. Res. Lett.* **2005**, *32*, L11308. [CrossRef]
13. Sambridge, M. Geophysical inversion with a neighbourhood algorithm I. Searching a parameter space. *Geophys. J. Int.* **1999**, *138*, 479–494. [CrossRef]
14. Wathelet, M. An improved neighborhood algorithm: Parameter conditions and dynamic scaling. *Geophys. Res. Lett.* **2008**, *35*. [CrossRef]
15. Yamanaka, H.; Ishida, H. Application of genetic algorithms to an inversion of surface wave dispersion data. *Bull. Seismol. Soc. Am.* **1996**, *86*, 436–444.
16. Delgado, J.; Peláez, J.A.; Tomás, R.; García-Tortosa, F.J.; Alfaro, P.; López Casado, C. Seismically-induced landslides in the Betic Cordillera (S Spain). *Soil Dyn. Earthq. Eng.* **2011**, *31*, 1203–1211. [CrossRef]
17. Martínez Solares, J.M.; López Arroyo, A. The great historical 1755 earthquake. Effects and damage in Spain. *J. Seismol.* **2004**, *8*, 275–294. [CrossRef]
18. Sanz, E. Le mouvement de versant de Güevejar (Grenade) au cours des tremblements de terre de Lisbonne (1755) et d'Andalousie (1884). *Bull. Interintl. Assoc. Eng. Geol.* **1997**, *56*, 83–87.
19. Instituto Geográfico Nacional. *El Terremoto de Andalucía del 25 de Diciembre de 1884*; Publicaciones IGN: Madrid, Spain, 1980; 139p.
20. Fernández Castro, M.; Lasala, J.P.; Cortázar, D.; Gonzalo Tarín, J. Terremoto de Andalucía: Informe de la Comisión nombrada para su estudio dando cuenta del estado de los trabajos en 7 de marzo de 1985. *Boll. Com. Mapa Geol. Esp.* **1885**, *12*, 1–105.
21. Jiménez-Pintor, J.; Azor, A. El Deslizamiento de Güevéjar (provincia de Granada): Un caso de inestabilidad de laderas inducida por sismos. *Geogaceta* **2006**, *40*, 287–290.
22. Rodríguez-Peces, M.J.; García-Mayordomo, J.; Azañón, J.M.; Insúa, J.M.; Jiménez Pintor, J. Constraining pre-instrumental earthquake parameters from slope stability back-analysis: Palaeoseismic reconstruction of the Güevéjar landslide during the 1st November 1755 Lisbon and 25th December 1884 Arenas del Rey earthquakes. *Q. Intern.* **2011**, *242*, 76–89. [CrossRef]
23. Martino, S.; Lenti, L.; Delgado, J.; Garrido, J.; López-Casado, C. Application of a Characteristic Periods-Based (CPB) approach to estimate earthquake-induced displacements of landslides through dynamic numerical modelling. *Geophys. J. Int.* **2016**, *206*, 85–102. [CrossRef]
24. Lacoss, R.T.; Kelly, E.J.; Toksöz, M.N. Estimation of seismic noise structure using array. *Geophysics* **1969**, *29*, 21–38. [CrossRef]
25. Ohrnberger, M.; Scherbaum, F.; Krüger, F.; Pelzing, R.; Reamer, S.K. How good are shear wave velocity models obtained from inversion of ambient vibrations in the lower Rhine embayment (N.W. Germany). *Boll. Geofis. Teor. Appl.* **2004**, *45*, 215–232.
26. Rodríguez-Fernández, J.; Sanz de Galdeano, C. Late orogenic intramontane basin development: The Granada basin, Betics (Southern Spain). *Basin Res.* **2006**, *18*, 85–102. [CrossRef]

27. Fernández, J.; Soria, J.M.; Viseras, C. Stratigraphic architecture of the Neogene basins in the central sector of the Betic Cordillera (Spain): Tectonic control and base level changes. In *Tertiary Basins of Spain: The Stratigraphic Record of Crustal Kinematics*; Friend, P.F., Dabrio, C.J., Eds.; Cambridge University Press: Cambridge, UK, 1996; pp. 353–365.
28. Dabrio, C.J.; Fernández, J.; Peña, J.A.; Ruiz-Bustos, A.; Sanz de Galdeano, C. Rasgos sedimentarios de los conglomerados miocenos del borde Noreste de la Depresión de Granada. *Estud. Geol.* **1978**, *343*, 89–97.
29. Dabrio, C.J.; Fernández, J.; Peña, J.A.; Ruiz-Bustos, A.; Sanz de Galdeano, C. Interpretation sedimentaire des materiaux néogènes du bord nord-est du Bassin Grenade (Espagne). *C.R. Somm. Seanc. Soc. Geol. France* **1978**, *3*, 121–123.
30. Cruden, D.M.; Varnes, D.J. Landslide type and processes. In *Landslides. Investigation and Mitigation*; Special Report 247; Turner, A.K., Schuster, R.L., Eds.; Transportation Research Board, National Academy Press: Washington, DC, USA, 1996; pp. 36–75.
31. Meléndez, B.; Fuster, J.M. *Geología*; Paraninfo: Madrid, Spain, 1966; p. 687.
32. Nogoshi, M.; Igarashi, T. On the propagation characteristics estimations of subsurface using microtremors on the ground surface. *J. Seism. Soc. Japan* **1971**, *23*, 264–280.
33. Lermo, J.; Chávez-García, F.J. Site effect evolution using spectral ratios with only one station. *Bull. Seismol. Soc. Am.* **1993**, *83*, 1574–1594.
34. Lachet, C.; Bard, P.Y. Numerical and theoretical investigations on the possibilities and limitations of the Nakamura's technique. *J. Phys. Earth.* **1994**, *42*, 377–397. [CrossRef]
35. Field, E.H.; Jacob, K.H. A comparison and test of various site response estimation techniques, including three that are non reference–site dependent. *Bull. Seismol. Soc. Am.* **1995**, *85*, 1127–1143.
36. Gitterman, Y.; Zaslavsky, Y.; Shapira, A.; Shtivelman, V. Empirical site response evaluations: Case studies in Israel. *Soil Dyn. Earthq. Eng.* **1996**, *15*, 447–463. [CrossRef]
37. Seekins, L.C.; Wennerberg, L.; Marghereti, L.; Liu, H.P. Site amplification at five locations in San Fransico, California: A comparison of S waves, codas, and microtremors. *Bull. Seismol. Soc. Am.* **1996**, *86*, 627–635.
38. Albarello, D.; Lunedei, E. Alternative interpretations of horizontal to vertical spectral ratios of ambient vibrations: New insights from theoretical modeling. *Bull. Earthq. Eng.* **2010**, *8*, 519–534. [CrossRef]
39. Schwellenbach, I.; Hinzen, K.-G.; Petersen, G.M.; Bottari, C. Combined use of refraction seismic, MASW, and ambient noise array measurements to determine the near-surface velocity structure in the Selinunte Archaeological Park, SW Sicily. *J. Seismol.* **2020**, *24*, 753–776. [CrossRef]
40. D'Amico, V.; Picozzi, M.; Baliva, F.; Albarello, D. Ambient noise measurements for preliminary site effects characterization in the urban area of Florence, Italy. *Bull. Seismol. Soc. Am.* **2008**, *98*, 1373–1388. [CrossRef]
41. Mundepi, A.K.; Galiana-Merino, J.J.; Asthana, A.K.L.; Rosa-Cintas, S. Soil characteristics in Doon Valley (northwest Himalaya, India) by inversion of H/V spectral ratios from ambient noise measurements. *Soil Dyn. Earthq. Eng.* **2015**, *77*, 309–320. [CrossRef]
42. Issaadi, A.; Semmane, F.; Yelles-Chaouche, A.K.; Galiana-Merino, J.J.; Layadi, K. A shear-wave velocity model in the city of Oued-Fodda (Northern Algeria) from Rayleigh wave ellipticity inversion. *Appl. Sci.* **2020**, *10*, 1717. [CrossRef]
43. Panzera, F.; Lombardo, G.; Imposa, S.; Grassi, S.; Gresta, S.; Catalano, S.; Romagnoli, G.; Tortorici, G.; Patti, F.; Di Maio, E. Correlation between earthquake damage and seismic site effects: The study case of Lentini and Carlentini, Italy. *Eng. Geol.* **2018**, *240*, 149–162. [CrossRef]
44. Wathelet, M.; Chatelain, J.-L.; Cornou, C.; Di Giulio, G.; Guillier, B.; Ohrnberger, M.; Savvaidis, A. Geopsy: A User-Friendly Open-Source Tool Set for Ambient Vibration Processing. *Seismol. Res. Lett.* **2020**, *91*, 1878–1889. [CrossRef]
45. Wathelet, M. Array recordings of ambient vibrations: Surface-Wave Inversion. PhD. Thesis, Université de Liège, Liège, Belgium, 2005.
46. Konno, K.; Ohmachi, T. Ground motion characteristics estimated from spectral ratio between horizontal and vertical components of microtremors. *Bull. Seismol. Soc. Am.* **1998**, *88*, 228–241.
47. Scherbaum, F.; Hinzen, K.-G.; Ohrnberger, M. Determination of shallow shear wave velocity profiles in the Cologne/Germany area using ambient vibrations. *Geophys. J. Int.* **2003**, *152*, 597–612. [CrossRef]
48. Horike, M. Inversion of phase velocity of long period microtremors to the S-wave velocity structure down to the basement in urbanized areas. *J. Phys. Earth* **1985**, *33*, 59–96. [CrossRef]
49. Okada, H. *The Microtremor Survey Method, Geophysical Monograph Series 12*; Society of Exploration Geophysicists: Tulsa, OK, USA, 2003.
50. Asten, M.W.; Henstridge, J.D. Array estimators and use of microseisms for reconnaissance of sedimentary basins. *Geophysics* **1984**, *49*, 1828–1837. [CrossRef]
51. Arai, H.; Tokimatsu, K. S-wave velocity profiling by inversion of microtremor H/V spectrum. *Bull. Seismol. Soc. Am.* **2004**, *94*, 53–63. [CrossRef]
52. NEHRP. *Recommended Seismic Provisions for New Buildings and Other Structures*; FEMA P-750. 2009 Edition; NIST: Gaithersburg, MA, USA, 2009.
53. Galiana-Merino, J.J.; Mahajan, A.K.; Lindholm, C.; Rosa-Herranz, J.; Mundepi, A.K.; Rai, N. Seismic noise array measurements using broadband stations and vertical geophones: Preliminary outcomes for the suitability on f-k analysis. *Bull. Earthq. Eng.* **2011**, *9*, 1309–1325. [CrossRef]

54. Rosa-Cintas, S.; Galiana-Merino, J.J.; Rosa-Herranz, J.; Molina, S.; Giner-Caturla, J. Suitability of 10 Hz vertical geophones for seismic noise array measurements based on frequency-wavenumber and extended spatial autocorrelation analyses. *Geophys. Prospect.* **2012**. [CrossRef]
55. Wathelet, M.; Jongmans, D.; Ohrnberger, M. Surface-wave inversion using a direct search algorithm and its application to ambient vibration measurements. *Near Surf. Geophys.* **2004**, *2*, 211–221. [CrossRef]
56. Panzera, F.; Tortorici, G.; Romagnoli, G.; Marletta, G.; Catalano, S. Empirical evidence of orthogonal relationship between directional site effects and fracture azimuths in an active fault zone: The case of the Mt. Etna lower eastern flank. *Eng. Geol.* **2020**, *279*, 105900. [CrossRef]
57. Panzera, F.; Romagnoli, G.; Tortorici, G.; D'Amico, S.; Rizza, M.; Catalano, S. Integrated use of ambient vibrations and geological methods for seismic microzonation. *J. Appl. Geophys.* **2019**, *170*, 103820. [CrossRef]
58. Fäh, D.; Kind, F.; Giardini, D. A theoretical investigation of average H/V ratios. *Geophys. J. Int.* **2001**, *145*, 535–549. [CrossRef]
59. Fäh, D.; Kind, F.; Giardini, D. Inversion of local S-wave velocity structures from average H/V ratios, and their use for the estimation of site-effects. *J. Seismol.* **2003**, *7*, 449–467. [CrossRef]
60. SESAME. *Guidelines for the Implementation of the H/V Spectral Ration Technique on Ambient Vibrations Measurements, Processing and Interpretation*; SESAME European research project WP12_ Deliverable D23.12; Euratom: Rome, Italy, 2004.

applied sciences

MDPI

Article

Viscoelastic Model and Synthetic Seismic Data of Eastern Rub'Al-Khali

Septriandi A. Chan [1,†], Paul Edigbue [1,†], Sikandar Khan [2,†], Abdul L. Ashadi [1,†] and Abdullatif A. Al-Shuhail [1,*,†]

1 Geosciences Department, King Fahd University of Petroleum & Minerals, Dhahran 34464, Saudi Arabia; g201305750@kfupm.edu.sa (S.A.C.); g201305630@kfupm.edu.sa (P.E.); latif.demak@gmail.com (A.L.A.)
2 Mechanical Engineering Department, King Fahd University of Petroleum & Minerals, Dhahran 34464, Saudi Arabia; sikandarkhan@kfupm.edu.sa
* Correspondence: ashuhail@kfupm.edu.sa
† These authors contributed equally to this work.

Abstract: The Rub' Al-Khali basin in Saudi Arabia remains unexplored and lacks data availability due to its remoteness and the challenging nature of its terrain. Thus far, there are neither digital geologic models nor synthetic seismic data from this specific area accessible for testing research techniques and analysis. In this study, we build a 2D viscoelastic model of the eastern part of the Rub' Al-Khali basin and generate a corresponding dual-component seismic data set. We compile high-resolution depth models of compressional- and shear-wave velocities, density, as well as compressional- and shear-wave quality factors from published data. The compiled models span Neoproterozoic basement up to Quaternary sand dunes. We then use the finite-difference technique to model the propagation of seismic waves in the compiled viscoelastic medium of eastern Rub' Al-Khali desert. In particular, we generate vertical and horizontal components of the shot gathers with accuracy to the fourth and second orders in space and time, respectively. The viscoelastic models and synthetic seismic datasets are made available in an open-source site for prospective researchers who desire to use them for their research. Users of these datasets are urged to make their findings also accessible to the geoscience community as a way of keeping track of developments related to the Rub' Al-Khali desert.

Keywords: viscoelastic model; synthetic seismic data; eastern Rub-Al-Khali; Saudi Arabia

1. Introduction

Numerical modeling of seismic wave propagation is a powerful way to understand real seismic data acquired in an area, because these methods closely simulate seismic waves from their generation at the source until recording at the receiver. Seismic numerical modeling is best done in a 3D sense in order to closely simulate wave propagation in 3D earth models. However, it can also be done in 1D, 2D, and 2.5D models in some applications to reduce the simulation time. Various methods are used to simulate 3D seismic wave propagation including finite-difference, e.g., [1–4], finite-volume, e.g., [5–7], and Lattice–Boltzmann, e.g., [8–10]. In particular, the finite-difference method covers an intuitive and practical modeling approach because it can be used for single-and multi-dimensional spaces and wave equations of any complexity. In the literature, the viscoelastic models and synthetic seismic data, calculated using finite-difference method, have been used as input in various studies [11–14]. Regardless of the simulation method, increasing levels of wave equation complexity can be used to model seismic wave propagation depending on how realistic the depth model is. The simplest but least realistic model uses the acoustic wave equation where only P-wave and density are considered. On the other hand, the most complex and realistic model uses an anisotropic poro-viscoelastic wave equation, which addresses almost all types of seismic waves and phenomena [15].

The highest level of modeling is not practical because of the need to know many generally unknown parameters (i.e., anisotropy) about subsurface layers, in addition to being very computationally intensive. The viscoelastic wave equation offers a practical compromise, especially in large subsurface models spanning a wide range of velocities. In addition to modeling P-waves, S-waves, and surface waves, viscoelasticity also addresses an elastic wave behavior. The Rub′ Al-Khali basin is situated below the Quaternary sand dunes and occupies most of the southern part of Saudi Arabia. It covers a large area that extends approximately 1300 km in the east–west direction and 300 km in the north–south direction extending from southwest Saudi Arabia toward the United Arab Emirates (UAE) [16] (Figure 1A). This basin is divided by the Qatar arch into two sub-basins: East and West Rub′ Al-Khali sub-basins (Figure 1B). In comparison with the UAE part, Rub′ Al-Khali basin in Saudi Arabia remains unexplored and lacks openly available data due to its remoteness and the challenging desert conditions. Most of the hydrocarbon fields in Saudi Arabia in the Rub′ Al-Khali basin are mainly located on its northeastern portion, including Shaybah, Ramlah, North, and South Kidan fields [17]. Alfara et al. [18] described the main challenges and uncertainties in the interpretation of 3D seismic data in the Shaybah field due to the complex and rugged topography. They successfully improved the quality of the data and resolved the uncertainties by combining 3D seismic data and borehole seismic measurements including vertical seismic pro ling (VSP) surveys. Stewart et al. [19,20] carried out a study on the Mesozoic petroleum system of the Rub′ Al-Khali basin, particularly the Triassic to Jurassic siliciclastic reservoirs. They also described the structural evolution of the Rub′ Al-Khali basin from the Precambrian to Neogene using a combination of regional seismic, gravity, and magnetic data.

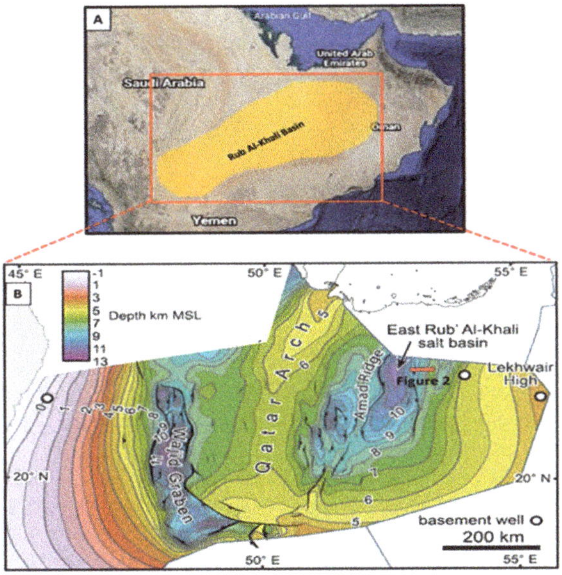

Figure 1. (**A**) Satellite image showing the location of the Rub′ Al-Khali basin. (**B**) Top basement depth structure map [20].

Thus far, there are no publicly-available synthetic seismic data nor digital geological models of the Rub′ Al-Khali basin. In this study, we build a 2D viscoelastic model of the eastern Rub′ Al-Khali, which is close to the border of the UAE (Figure 1B) and generate the associated dual-component seismic data sets. Similar models and synthetic seismic data sets from the northwest, eastern, and central parts of Saudi Arabia have been made publicly available by the authors [21–23]. Models of P-wave velocity (V_p),

S-wave velocity (V_s), Density (ρ), P-wave quality factor (Q_P), and S-wave quality factor (Q_s) that cover Neoproterozoic basement to Quaternary sand are first compiled from available literature [24]. These models are subsequently utilized in generating the 2D synthetic viscoelastic dual-component seismic data sets that include vertical component data and horizontal component data recorded by surface geophones using a conventional 2D acquisition geometry. The horizons, models, and synthetic seismic data sets are made publicly available at the following link: (https://www.dropbox.com/sh/9t099hhn1h8nea8/AAChpqVg3R_km2MlEMJ0SuxLa?dl=0).

The synthetic seismic data sets are saved in SEGY files, while horizons and faults are saved in a CSV text file.

2. Geology of the Rub' Al-Khali Basin

Due to lack of outcrop exposure (except in the west margin), the Rub' Al-Khali basin evolution is known only from geophysical data. Stewart [20] reported the process of development of structures in the Rub' Al-Khali basin from the Late Precambrian to Neogene observed on seismic, magnetic, gravity, and well data. This data shows a relatively complete Phanerozoic succession with north–south and northwest–southeast trending basement structure (Figure 1B) that are formed during East African Antarctic Orogeny. The structural trend in the Rub' Al-Khali basin has been affected and reactivated during several periods including Ediacaran—Cambrian Ara/Hormuz salt, Devonian to Late Carboniferous Hercynian orogeny, early Mesozoic Gondwana break up, and finally in the Late Cretaceous and Tertiary time due to effects of the Arabian plate colliding with the Eurasian plate, resulting in the emplacement of ophiolites in the Oman Mountains.

3. Development of Digital Depth Models

There are almost no published data about the eastern parts of the Rub' Al-Khali basin in Saudi Arabia. Only limited published data on the geological or geophysical properties were available for the Rub' Al-Khali basin in Saudi Arabia. In this study, we performed an extensive literature review to compile one geophysical and geological model in the eastern Rub' Al-Khali area using references specific to Rub' Al-Khali. In cases where data are not available in these references, we obtain the missing data from references studying the same formations in nearby areas of Saudi Arabia, UAE, or Oman. In terms of geologic column in the Rub' Al-Khali basin, we found the following twenty-three formations: Quaternary (F-1), Hofuf Dam Hadrukh (F-2), Dammam (F-3), Rus (F-4), Umm Er Radhuma (F-5), Aruma (F-6), Wasia (F-7), Shuaiba (F-8), Biyadh (F-9), Hith (F-10), Arab (F-11), Hanifa & Tuwaiq Mountain (F-12), Dhruma (F-13), Marrat (F-14), Minjur (F-15), Jilh (F-16), Sudair (F-17), Khuff (F-18), Unayzah (F-19), Qusaiba (F-20), Qasim (F-21), Saq (F-22), and Basement (F-23) (Figure 2). Eventually, twenty-three layers and two major faults were defined and digitized at variable intervals (Figure 2). The Phanerozoic stratigraphy column of the eastern Arabian Peninsula is given in Figure 3 [25]. The stratigraphy of layers and faults of the model were built mostly based on the geologic information provided by [19]. The elastic properties such as densities and velocities were determined using median values from digitized well-log data provided by several published articles as reported in Table 1 [18,22,23,26–34], and available Vp and density data were plotted in Figure 4. If logs are not available, values for these parameters from nearby areas are used. Due to the lack of availability of S-wave velocities for some formations, we use Vp-Vs relations to complete the whole stratigraphic column. We used the relation developed by [26] for carbonate lithologies, and relations presented by [22] for sandstones and shales. The remaining properties, which are P- and S-wave quality factors (Qp and Qs), were calculated using a method proposed by [27], which involves taking the square root of the corresponding value of P- and S-wave, respectively. The complete geophysical properties including P-wave velocity, S-wave velocity, density, lithology, P- and S- wave quality factors of each formation are listed in Table 1. Moreover, Figure 5 shows plots of the compiled Vp, Vs, ρ, Qp, and Qs models.

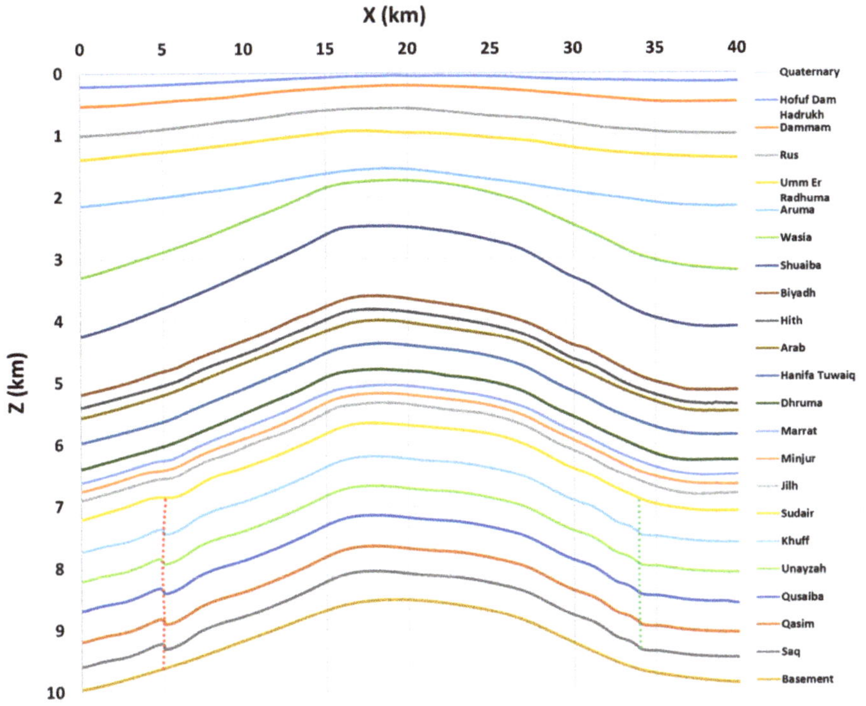

Figure 2. Digitized 2D model of the Rub' Al-Khali basin. Solid curves indicate horizons while dotted lines indicate faults.

Table 1. Lithology, densities, compressional and shear-wave speeds, compressional and shear wave quality factors of layers of the eastern Rub' Al-Khali basin model.

Layer	Formation	Lithology	ρ (kg/m^3)	V_p (m/s)	V_s (m/s)	Q_p	Q_s
F-1	Quaternary	Sand	1500 [a]	850 [a]	501 [a]	29.15 [i]	22.36 [i]
F-2	Hofuf Dam Hadrukh	Mixed limestone and siliciclastic	1877 [b]	1835 [b]	1180 [c]	42.84 [i]	34.35 [i]
F-3	Dammam	Limestone	2289 [b]	3110 [b]	1870 [d]	55.77 [i]	43.24 [i]
F-4	Rus	Anhydrite	2397 [b]	5252 [b]	2470 [d]	65.32 [i]	49.70 [i]
F-5	Umm Er Radhuma	Limestone	2028 [b]	3318 [b]	1978 [d]	57.60 [i]	44.47 [i]
F-6	Aruma	Limestone	2037 [b]	3558 [b]	1672 [d]	52.25 [i]	40.89 [i]
F-7	Wasia	Sandstone	2277 [b]	3233 [b]	2142 [c]	56.86 [i]	46.28 [i]
F-8	Shuaiba	Limestone	2037 [b]	3010 [b]	1818 [d]	54.86 [i]	42.63 [i]
F-9	Biyadh	Sandstone	2364 [b]	4045 [b]	2701 [c]	63.60 [i]	51.97 [i]
F-10	Hith	Anhydrite	2874 [e]	4483 [e]	2328 [e]	66.96 [i]	48.24 [i]
F-11	Arab	Limestone	2400 [e]	5399 [e]	2748 [e]	73.47 [i]	52.42 [i]
F-12	Hanifa & Tuwaiq Mountain	Limestone	2550	5698 [e,k]	2903 [e]	75.48 [i]	53.88 [i]
F-13	Dhruma	Limestone	2458 [f,k]	5033 [f,k]	2870 [d]	70.94 [i]	53.57 [i]
F-14	Marrat	Shale	2410 [f]	3272 [f]	1436 [c]	57.20 [i]	37.89 [i]
F-15	Minjur	Sandstone	2394 [f]	3930 [f]	2499 [c]	62.69 [i]	49.99 [i]
F-16	Jilh	Dolomite	2400 [f,k]	4823 [f,k]	2761 [d]	69.45 [i]	52.54 [i]
F-17	Sudair	Shale	2372 [g,k]	5182 [g,k]	2674 [g]	71.99 [i]	51.71 [i]
F-18	Khuff	Dolomite	2639 [g,k]	4953 [g,k]	2530 [g]	70.38 [i]	50.3 [i]
F-19	Unayzah	Sandstone	2405 [g]	3752 [g]	2085 [g]	61.25 [i]	45.66 [i]
F-20	Qusaiba	Shale	2486 [g]	3898 [g]	2143 [g]	62.43 [i]	46.29 [i]
F-21	Qasim	Sandstone	2380 [h]	3685 [h]	2453 [c]	60.70 [i]	49.53 [i]
F-22	Saq	Sandstone	2350 [h]	3765 [h]	2508 [c]	61.36 [i]	50.08 [i]
F-23	Basement	Igneous and metamorphic	2800 [j]	6380 [j]	3580 [j]	79.87 [i]	59.83 [i]

[a] [28], [b] [18], [c] [22,23], [d] [26], [e] [29], [f] [30], [g] [31], [h] [32], [i] [27], [j] [33], [k] [34].

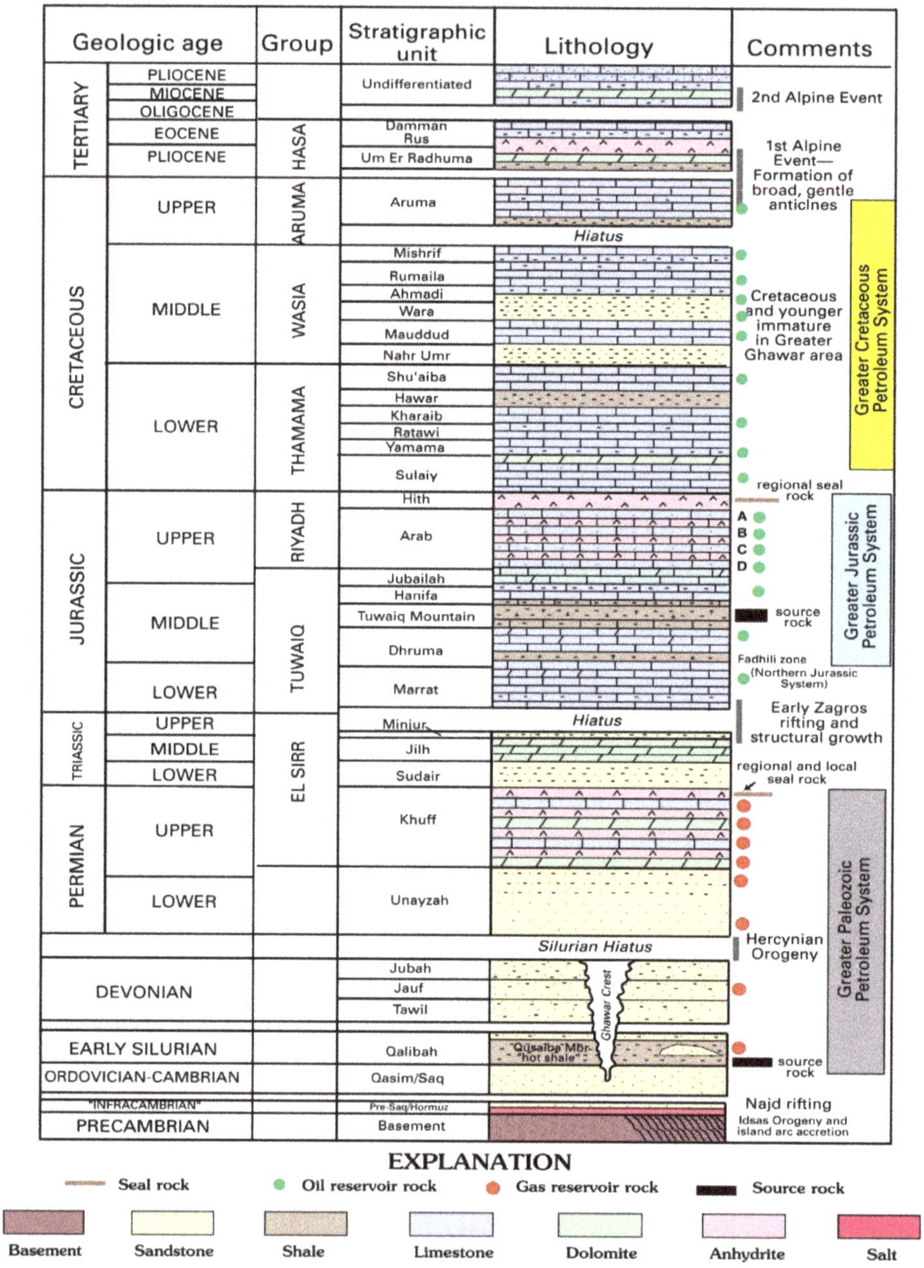

Figure 3. Phanerozoic stratigraphy column of the eastern Arabian Peninsula [25].

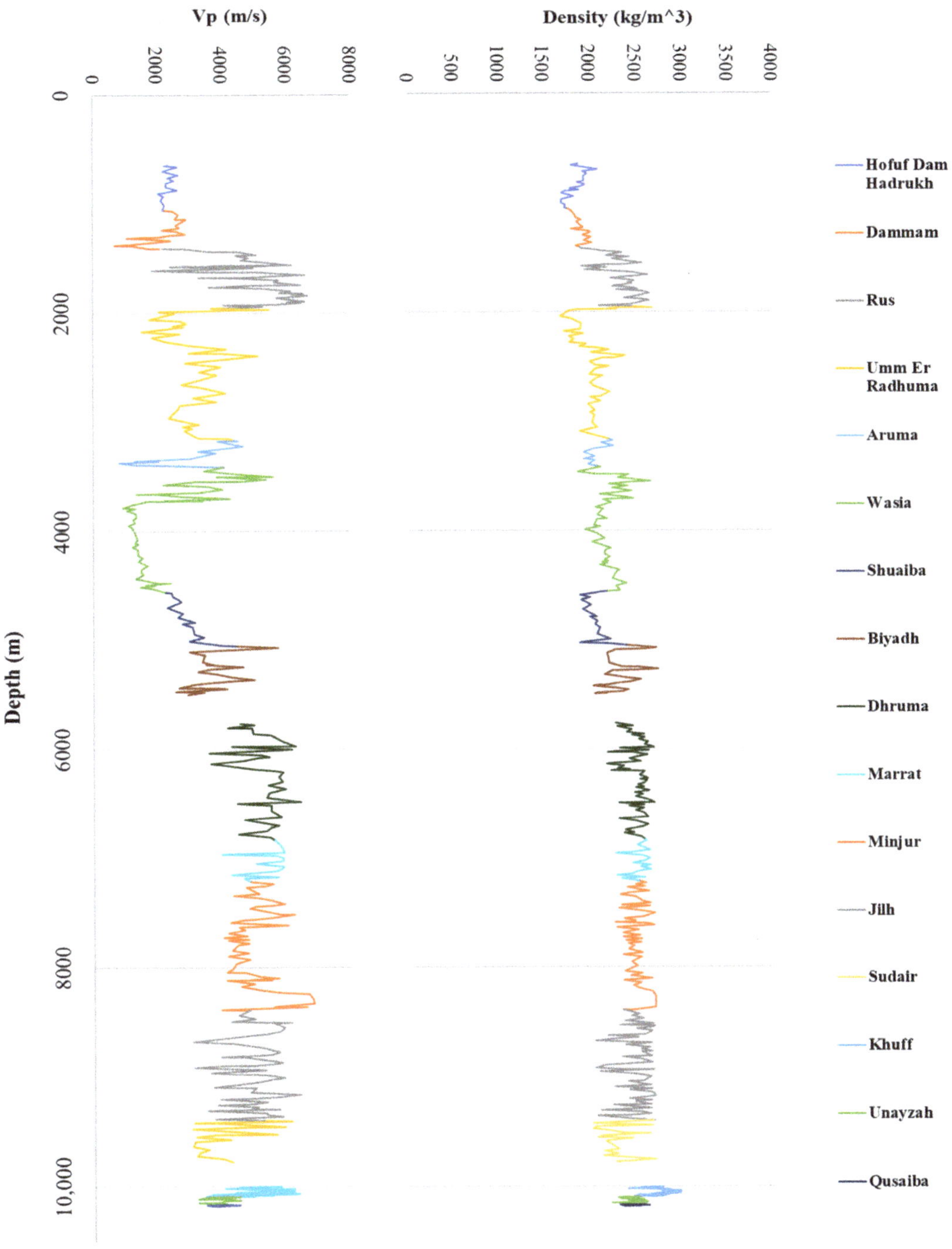

Figure 4. V_p and ρ plot of selected formations [18,30,31,34].

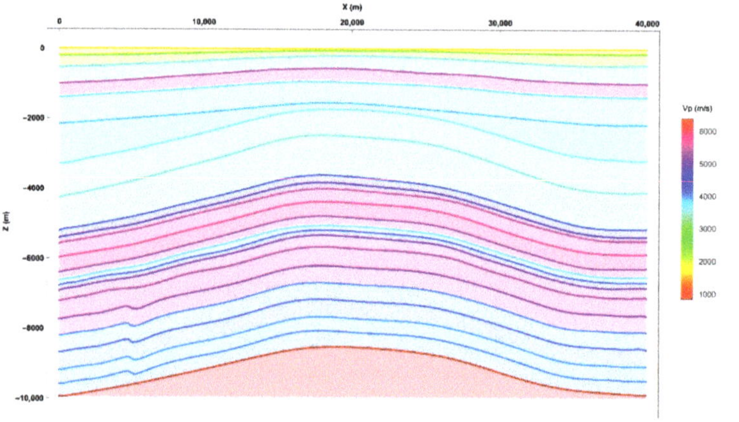

(a)

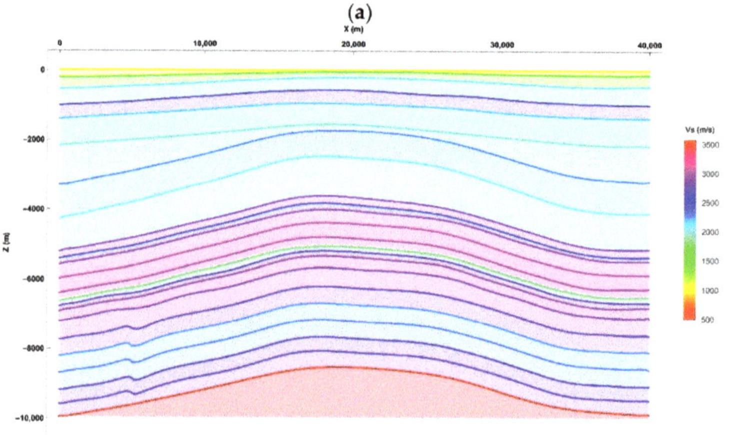

(b)

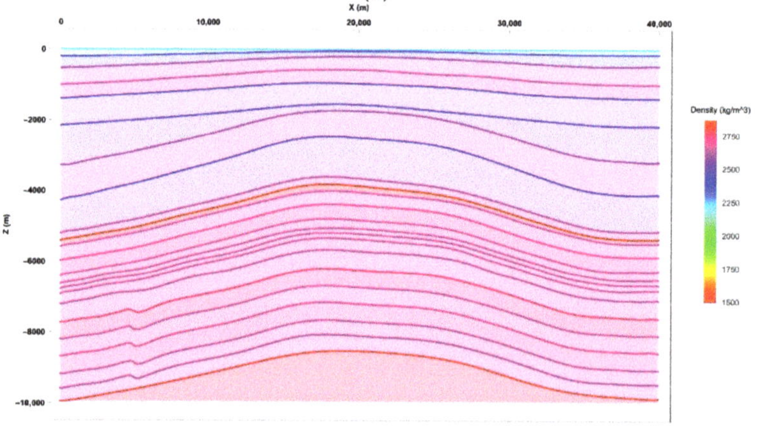

(c)

Figure 5. *Cont.*

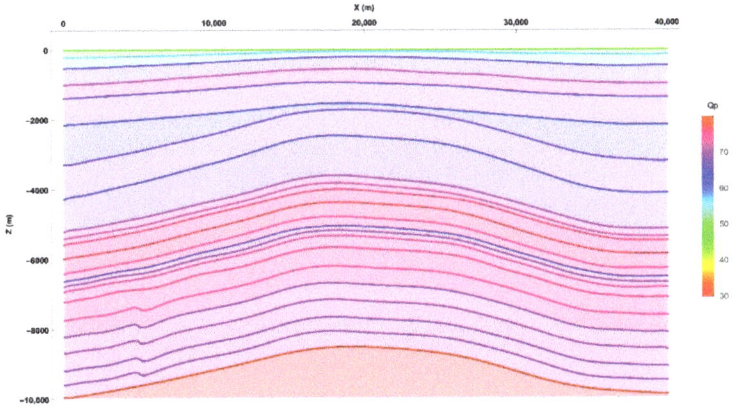

(d)

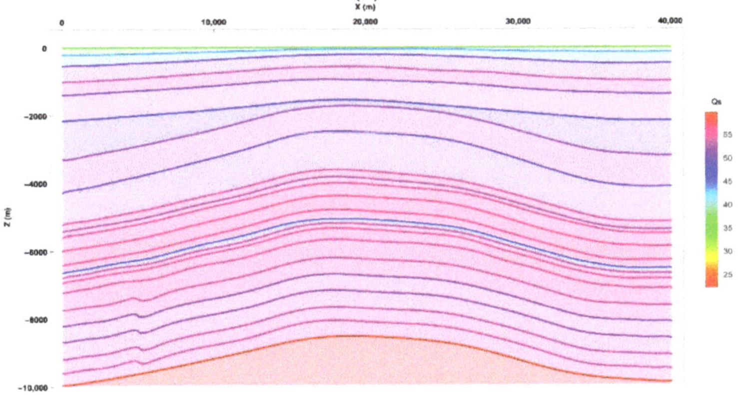

(e)

Figure 5. (**a**) Vp model; (**b**) Vs model; (**c**) density model; (**d**) Qp model; (**e**) Qs model.

4. The Synthetic Seismic Data Generation Procedure

We apply 2D finite-difference approximation methodology to calculate the synthetic seismic data, using the available viscoelastic model inputs. This method involves the application of 2D wave equation to simulate the propagation of seismic waves in viscoelastic media that is determined by a modified Newton's equation of motion and Hooke's law, as described in [35,36]. We propagate the waves in a 2D viscoelastic medium by computing the 3 strain tensor components along the x-axis and z-axis, and also compute their associated stress tensors. Then, we calculate the divergence of the stresses which generate accelerations that are further extrapolated to produce displacements to succeeding time steps. The finite-difference method used here utilizes staggered grids to compute the spatial derivatives using the centered finite-difference approximations. The finite-difference algorithm inputs the viscoelastic properties (i.e., Vp, Vs, Qp, and Qs) from the compiled models to compute Lame's coefficients and the wave's relaxation parameters.

The finite-difference viscoelastic scheme begins with generation and propagation of the source at an initial time t = 0. Then it computes the wave-speeds using the finite-difference scheme by increasing time in little time steps while modifying the wave-speed in both x and z-directions, respectively. Subsequently, this process updates the stress field and makes it staggered in time relative to particle wave-speed fields. The process is repeated with another time step, updating the particle wave-speeds in this case until it covers the entire grid. At the boundaries of the grid, we utilize absorbing boundary conditions to minimize the effects of reflections. Usually, the computation of the finite-difference solutions relies on the frequency band of the source wavelet. In this case, we apply a typical Ricker wavelet having a zero-phase and a 20-Hz peak frequency. The finite-difference solution of the wave equation is usually associated with numerical dispersion if appropriate grid spacing is not used. To avoid this problem, we apply the general formula for preventing dispersion in the finite-difference scheme with accuracy close to spatial fourth-order and close to second order in time. This means that we must have a minimum of five grid points for every wavelength within a viscoelastic finite-difference scheme [37]:

$$\Delta x = \Delta z \leq \frac{V_{min}}{5f_{max}} = 1.5\ \mathrm{m} \tag{1}$$

The grid sizes Δx (along x-axis) and Δz (along z-axis) are assumed to be equal, Vmin is the minimum wave speed, and fmax is the maximum frequency. In our model, Vmin = 501 m/s corresponding to the minimum shear-wave speed and fmax = 60 Hz, which is the frequency at which the wavelet amplitude spectrum drops below 1% of its maximum value. Consequently, we calculate a value of x = z = 1.5625 m, which is approximated to 1.5 m. Besides the dispersion criteria, the time-marching step of the finite-difference also needs to fulfill the Courant-Friedrichs-Lewy (CFL) stability criteria for convergence in 2D media [37], which is given as:

$$\Delta t \leq \Delta z \frac{0.606\Delta x}{V_{max}} = 0.14\ \mathrm{msec} \tag{2}$$

where Vmax is the maximum wave speed in the model. We use the basement compressional-wave speed as the maximum velocity (i.e., Vmax = 6380 m/s), then we compute the time step as t = 0.142853 ms, which is approximated to 0.14 ms. All these parameters are embedded in the 2D finite-difference wave field modelling open-source code [37], which we utilized for generating the viscoelastic synthetic seismic data. To satisfy the boundary conditions, we used absorbing criteria with a taper of 375 grid points at the bottom, right, and left edges of the model and a free-surface boundary condition for the top margin of the model. We compute the taper using the formula described in [37]:

$$\mathrm{Taper} = \frac{4V_{max}}{\Delta x \times f_{max}} = 375 \tag{3}$$

Regarding the data acquisition, we apply a conventional 2D seismic land survey geometry. Starting from point x = 0 m, sources cover the whole horizontal length to point x = 40,000 m with a source interval of 50 m. Similarly, receivers cover a horizontal length from point x = 0 m to point x = 40,000 m, but with a receiver separation of 25 m. Receivers and shots are placed at 15 m depth below the top interface [37] to avoid the free-surface effect. The offset ranges from −40,000 to 40,000 m, which is desirable for most seismic data analysis and processing such as; velocity analysis, anisotropy, and amplitude variation with offset (AVO) analysis. Parameters used for the generation of synthetic seismic data sets are tabulated and presented in Table 2. We generated both horizontal and vertical components of the synthetic viscoelastic seismic data sets. Samples of the vertical component and horizontal component of displacements recorded at the beginning, center, and end of the model are presented in Figures 6 and 7, respectively.

For generating the synthetic seismic data, we used a PC workstation running a Linux operating system with an Intel® Xeon® Processor E5-2687 3.0 GHz 10C and a memory of 256 GB 2133 MHz DDR4 ECC RDIMM. A single shot gather consumed approximately 36 h to generate on this workstation. We attribute this length to the relatively low minimum velocity and high maximum frequency of our model.

Table 2. List of parameters used to construct synthetic seismic data.

Parameter	Value
Source wavelet	20 Hz zero-phase Ricker
Time sampling interval for finite-difference calculation	0.14 ms
Square-grid size for finite-difference calculation	1.5 m
Receiver spacing	25 m
Shot spacing	50 m
Recording time sampling interval	4.0 ms
Total recording time	6.0 s
Total number of receivers	1535
Receivers x-axis	0 to 40,000 m
Receivers z-axis	15 m
Shots x-axis	0 to 40,000 m
Shots z-axis	15 m
Offset range	−40,000 to 40,000 m
Total number of shots	801 shots

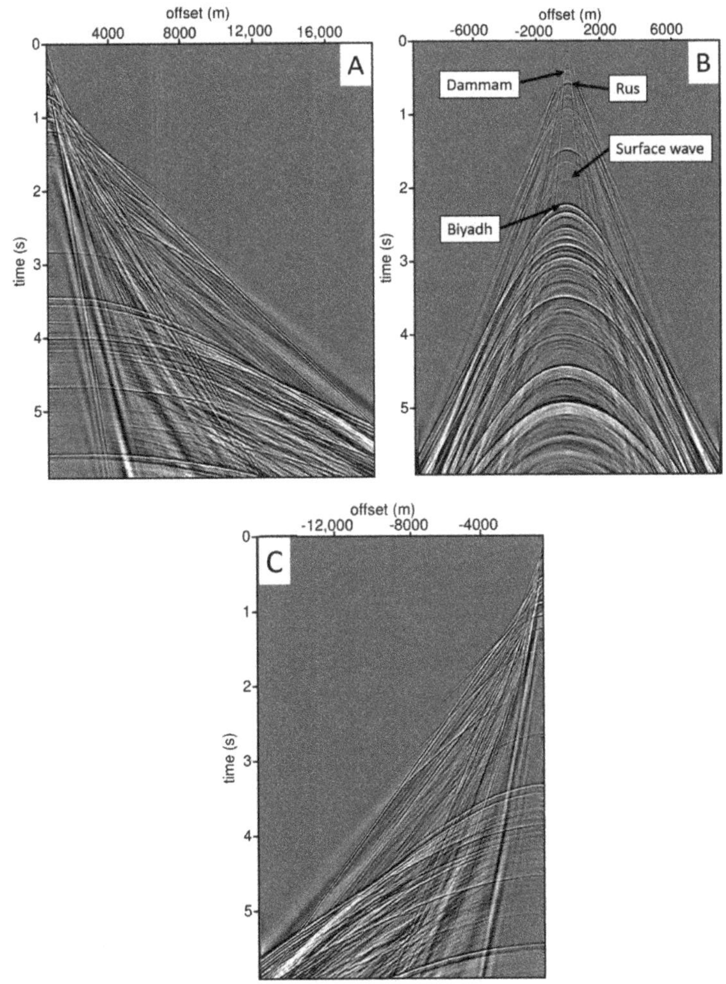

Figure 6. Vertical component of synthetic seismic data for shot located at (**A**) x = 1 m, (**B**) x = 2000 m, and (**C**) x = 40,000 m. Some prominent seismic events are indicated on (**B**).

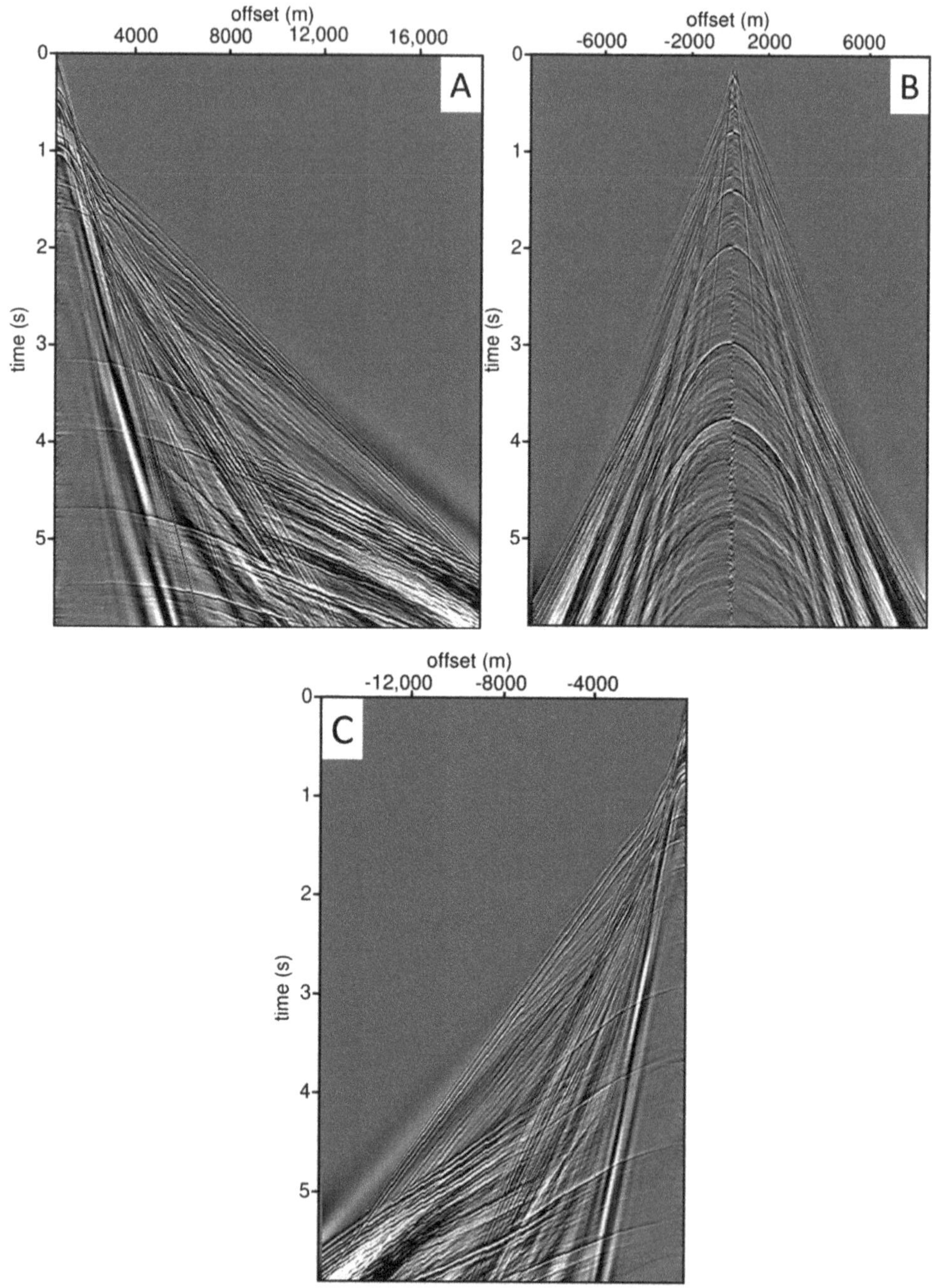

Figure 7. Horizontal component of synthetic seismic data for shot located at (**A**) x = 1 m, (**B**) x = 2000 m, and (**C**) x = 40,000 m.

5. Conclusions

We have put together a 2D viscoelastic model from the eastern Rub' Al-Khali basin and produced the associated dual-component synthetic seismic data sets. Regardless of the large number of layers involved in the model, the data appears to have good quality in capturing most important reservoirs including the relatively deep Khuff Formation. We provided the geologic model and associated seismic data sets, available online publicly for those who want to test their methods, and we encourage them to make their output open publicly too.

Author Contributions: Conceptualization, S.A.C., P.E., S.K., A.L.A. and A.A.A.-S.; methodology, S.A.C., P.E., S.K., A.L.A. and A.A.A.-S.; software, S.A.C., P.E., S.K., A.L.A. and A.A.A.-S.; validation, S.A.C., P.E., S.K., A.L.A. and A.A.A.-S.; formal analysis, S.A.C., P.E., S.K., A.L.A. and A.A.A.-S.; investigation, S.A.C., P.E., S.K., A.L.A. and A.A.A.-S.; resources, S.A.C., P.E., S.K., A.L.A. and A.A.A.-S.; data curation, S.A.C., P.E., S.K., A.L.A. and A.A.A.-S.; writing—original draft preparation, S.A.C., P.E., S.K., A.L.A. and A.A.A.-S.; writing—review and editing, S.A.C., P.E., S.K., A.L.A. and A.A.A.-S.; visualization, S.A.C., P.E., S.K., A.L.A. and A.A.A.-S. All authors have read and agreed to the published version of the manuscript.

Funding: This research received funding from the College of Petroleum Engineering and Geosciences of KFUPM through Startup Fund Grant number SF 18060.

Institutional Review Board Statement: Not applicable.

Informed Consent Statement: Not applicable.

Data Availability Statement: Publicly available datasets were analyzed in this study. This data can be found here: [https://www.dropbox.com/sh/9t099hhn1h8nea8/AAChpqVg3R_km2MlEMJ0SuxLa?dl=0].

Acknowledgments: The authors appreciate and acknowledge the support provided by King Fahd University of Petroleum and Minerals (KFUPM) by providing all the essential resources to conduct this study.

Conflicts of Interest: The authors declare no conflict of interest.

References

1. Li, Q.; Wu, G.; Wu, J.; Duan, P. Finite difference seismic forward modeling method for fluid–solid coupled media with irregular seabed interface. *J. Geophys. Eng.* **2019**, *16*, 198–214. [CrossRef]
2. Wang, J.; Hong, L. Stable optimization of finite-difference operators for seismic wave modeling. *Studia Geophys. Geod.* **2020**, *64*, 452–464. [CrossRef]
3. Liu, J.; Yong, W.A.; Liu, J.; Guo, Z. Stable Finite-Difference Methods for Elastic Wave Modeling with Characteristic Boundary Conditions. *Mathematics* **2020**, *8*, 1039. [CrossRef]
4. Wang, Y.; Ning, Y.; Wang, Y. Fractional Time Derivative Seismic Wave Equation Modeling for Natural Gas Hydrate. *Energies* **2020**, *13*, 5901. [CrossRef]
5. Maranzoni, A.; Mignosa, P. Seismic-generated unsteady motions in shallow basins and channels. Part II: Numerical modelling. *Appl. Math. Model.* **2019**, *68*, 712–731. [CrossRef]
6. Xu, Y.; Chen, X.; Han, D.; Zhang, W. Complex topography and media implementation using hp-adaptive DG-FEM for seismic wave modeling. EGU General Assembly 2020, Lund, Sweden, Online, 4–8 May 2020, EGU2020-21047. Available online: https://meetingorganizer.copernicus.org/EGU2020/EGU2020-21047.html (accessed on 1 June 2020).
7. Asfour, K.; Martin, R.; El Baz, D. Numerical Modeling of Attenuation and Nonlinear Effects in Computational Seismology. EGU General Assembly Conference Abstracts. 2020, p. 91. Available online: https://ui.adsabs.harvard.edu/abs/2020EGUGA..22...91A/abstract (accessed on 1 June 2020).
8. Jiang, C.; Zhou, H.; Xia, M.; Chen, H.; Zhang, Y.; Jiang, S. Acoustic Wave Simulation by Lattice Boltzmann Method with D2Q5 and D2Q9 of Different Relaxation Times. In Proceedings of the 81st EAGE Conference and Exhibition, London, UK, 3–6 June 2019; European Association of Geoscientists & Engineers: Houten, The Netherlands, 2019; pp. 1–5. [CrossRef]
9. Sakemi, Y.; Hitoshi Mikada, J.T. A quantitative analysis of the passive seismic emission tomography using the lattice Boltzmann method. In Proceedings of the Japan Geoscience Union Meeting, Chiba, Japan, 26–30 May 2019; p. STT42–P04.
10. Mora, P.; Morra, G.; Yuen, D.A. A concise python implementation of the lattice Boltzmann method on HPC for geo-fluid flow. *Geophys. J. Int.* **2020**, *220*, 682–702.
11. Khan, S.; Khulief, Y.A.; Al-Shuhail, A. Mitigating climate change via CO_2 sequestration into Biyadh reservoir: Geomechanical modeling and caprock integrity. *Mitig. Adapt. Strateg. Glob. Chang.* **2019**, *24*, 23–52. [CrossRef]

12. Khan, S.; Al-Shuhail, A.; Khulief, Y.A. Numerical modeling of the geomechanical behavior of Ghawar Arab-D carbonate petroleum reservoir undergoing CO_2 injection. *Environ. Earth Sci.* **2016**, *75*, 1499. [CrossRef]
13. Khan, S.; Khulief, Y.A.; Al-Shuhail, A. The effect of injection well arrangement on CO_2 injection into carbonate petroleum reservoir. *Int. J. Glob. Warm.* **2018**, *14*, 462–487. [CrossRef]
14. Khan, S.; Khulief, Y.A.; Al-Shuhail, A. Numerical modeling of the geomechanical behavior of Biyadh reservoir undergoing CO_2 injection. *Int. J. Geomech.* **2017**, *17*, 04017039. [CrossRef]
15. Carcione, J.; Poletto, F.; Farina, B.; Craglietto, A. Simulation of seismic waves at the earth's crust (brittle-ductile transition) based on the Burgers model. *Solid Earth* **2014**, *5*, 1001. [CrossRef]
16. Soliman, F.; Al Shamlan, A. Review on the geology of the Cretaceous sediments of the Rub'al-Khali, Saudi Arabia. *Cretac. Res.* **1982**, *3*, 187–194. [CrossRef]
17. Nairn, A.; Alsharhan, A. *Sedimentary Basins and Petroleum Geology of the Middle East*; Elsevier: Amsterdam, The Netherlands, 1997.
18. Alfara, M.N.; Nebrija, E.L.; Ferguson, M.D. The Challenge of Interpreting 3-D Seismic in Shaybah Field, Saudi Arabia. *GeoArabia* **1998**, *3*, 209–226.
19. Stewart, S.; Reid, C.; Hooker, N.; Kharouf, O. Mesozoic siliciclastic reservoirs and petroleum system in the Rub'Al-Khali basin, Saudi Arabia. *Aapg Bull.* **2016**, *100*, 819–841.
20. Stewart, S. Structural geology of the Rub'Al-Khali Basin, Saudi Arabia. *Tectonics* **2016**, *35*, 2417–2438. [CrossRef]
21. Al-Shuhail, A.A.; Mousa, W.A.; Alkhalifah, T. KFUPM-KAUST Red Sea model: Digital viscoelastic depth model and synthetic seismic data set. *Lead. Edge* **2017**, *36*, 507–511. [CrossRef]
22. Al-Shuhail, A.A.; Alshuhail, A.A.; Khulief, Y.A.; Salam, S.A.; Chan, S.A.; Ashadi, A.L.; Al-Lehyani, A.F.; Almubarak, A.M.; Khan, M.Z.U.; Khan, S.; et al. KFUPM Ghawar digital viscoelastic seismic model. *Arab. J. Geosci.* **2019**, *12*, 245. [CrossRef]
23. Khan, A.M.; Al-Juhani, S.G.; Abdul, S. Digital Viscoelastic Seismic Models and Data Sets Of Central Saudi Arabia in the Presence of Near-Surface Karst Features. *J. Seism. Explor.* **2020**, *29*, 15–28.
24. Zhou, H.W. Multi-scale tomography for crustal P and S velocities in southern California. *Pure Appl. Geophys.* **2004**, *161*, 283–302. [CrossRef]
25. Pollastro, R.M. *Total Petroleum Systems of the Paleozoic and Jurassic, Greater Ghawar Uplift and Adjoining Provinces of Central Saudi Arabia and Northern Arabian-Persian Gulf*; US Geological Survey; US Department of the Interior: Washington, DC, USA, 2003; p. 2202-H.
26. Ameen, M.S.; Smart, B.G.; Somerville, J.M.; Hammilton, S.; Naji, N.A. Predicting rock mechanical properties of carbonates from wireline logs (A case study: Arab-D reservoir, Ghawar field, Saudi Arabia). *Mar. Pet. Geol.* **2009**, *26*, 430–444. [CrossRef]
27. Mittet, R. A simple design procedure for depth extrapolation operators that compensate for absorption and dispersion. *Geophysics* **2007**, *72*, S105–S112. [CrossRef]
28. Bridle, R.; Ley, R.; Al-Mustafa, A. Near-surface models in Saudi Arabia. *Geophys. Prospect.* **2007**, *55*, 779–792. [CrossRef]
29. Liu, L.; Kharji, M.; Chatterjee, H.S. Seismic attributes as exploration tools: A case study from the Jurassic in Saudi Arabia. In Proceedings of the IPTC 2013: International Petroleum Technology Conference, Beijing, China, 26–28 March 2013; European Association of Geoscientists & Engineers: Houten, The Netherlands, 2013; p. cp-350.
30. Dasgupta, S.N.; Hong, M.R.; La Croix, P.; Al-Mana, L.; Robinson, G. Prediction of reservoir properties by integration of seismic stochastic inversion and coherency attributes in super Giant Ghawar field. In *SEG Technical Program Expanded Abstracts 2000*; Society of Exploration Geophysicists: Tulsa, OK, USA, 2000; pp. 1501–1505.
31. Macrides, C.G.; Kelamis, P.G. A 9C-2D land seismic experiment for lithology estimation of a Permian clastic reservoir. *GeoArabia* **2000**, *5*, 427–440.
32. Al-Ahmadi, M.E. Hydrogeology of the Saq aquifer northwest of Tabuk, northern Saudi Arabia. *Earth Sci.* **2009**, *20*, 15. [CrossRef]
33. Mooney, W.D.; Gettings, M.E.; Blank, H.R.; Healy, J.H. Saudi Arabian seismic-refraction profile: A traveltime interpretation of crustal and upper mantle structure. *Tectonophysics* **1985**, *111*, 173–246. [CrossRef]
34. Sharland, P.; Archer, R.; Davies, R.B.; Simmons, M.D.; Sutcliffe, O.E. *Arabian Plate Sequence Stratigraphy*; GeoArabia Special Publication: Tysons, VA, USA, 2001; Volume 2, p. 371.
35. Robertsson, J.O.; Blanch, J.O.; Symes, W.W. Viscoelastic finite-difference modeling. *Geophysics* **1994**, *59*, 1444–1456. [CrossRef]
36. Talukdar, K.; Behera, L. Sub-basalt Imaging of Hydrocarbon-Bearing Mesozoic Sediments Using Ray-Trace Inversion of First-Arrival Seismic Data and Elastic Finite-Difference Full-Wave Modeling Along Sinor–Valod Profile of Deccan Syneclise, India. *Pure Appl. Geophys.* **2018**, *175*, 2931–2954. [CrossRef]
37. Thorbecke, J. *Finite-Difference Wavefield Modeling*. 2016.

applied sciences

MDPI

Article

l_p Norm Smooth Inversion of Magnetic Anomaly Based on Improved Adaptive Differential Evolution

Wei Du [1,*], Lianzheng Cheng [1,2,*] and Yuanfang Li [1,2]

[1] State Key Laboratory of Ore Deposit Geochemistry, Institute of Geochemistry, Chinese Academy of Sciences, Guiyang 550081, China; liyuanfang@mail.gyig.ac.cn
[2] College of Earth and Planetary Sciences, University of Chinese Academy of Sciences, Beijing 100049, China
* Correspondence: du-wei@mail.gyig.ac.cn (W.D.); chenglianzheng@mail.gyig.ac.cn (L.C.)

Abstract: Due to the approved applicability of differential evolution (DE) in geophysical problems, the algorithm has been widely concerned. The DE algorithms are mostly applied to solve the geophysical parametric estimation based on specific models, but they are rarely used in solving the physical property inverse problem of geophysical data. In this paper, an improved adaptive differential evolution is proposed to solve the l_p norm magnetic inversion of 2D data, in which the perturbation direction in the mutation strategy is smoothed by using the moving average technique. Besides, a new way of updating the regularization coefficient is introduced to balance the effect of the model constraint adaptively. The inversion results of synthetic models demonstrate that the presented method can obtain a smoother solution and delineate the distributions of abnormal bodies better. In the field example of Zaohuoxi iron ore deposits in China, the reconstructed magnetic source distribution is in good agreement with the one inferred from drilling information. The result shows that the proposed method offers a valuable tool for magnetic anomaly inversion.

Keywords: magnetic inversion; l_p norm; adaptive differential evolution; regularization coefficient

1. Introduction

The magnetic method is widely used in mineral resource exploration and structure investigation. To determine the depth, position, and shape of the magnetic anomaly body, the magnetic data inversion is usually needed, which mainly includes parameter inversion [1,2], imaging inversion [3–5], and physical property inversion [6]. Physical property inversion can recover the shape and depth of complex sources without depending on a specific model. Therefore, the inversion of physical properties has become one of the most important and commonly used methods. Linear iterative methods, such as the steepest descent method, Newton's method, and conjugate gradient method, are usually used in physical property inversion [7–9]. However, the gradient-based algorithms are independent of the initial guess to start the optimization process and are easy to trap into local minima for nonlinear problems [10]. Contrary to the conventional approaches, the metaheuristic methods like differential evolution (DE) do not require good initial solutions when searching the global minimum.

DE [11] is a population-based metaheuristic global optimization algorithm. Because of its fast convergence and easy implementation, it has been widely used to solve the optimization of practical problems, such as electromagnetic optimization [12,13], pattern recognition [14], signal processing [15–20], engineering application [21–23], and other inversion problems [24–27]. Considering the dependency of control parameters [28], some scholars tried to adjust them by using adaptive or self-adaptive manners.

Liu and Lampinen proposed an adaptive DE based on fuzzy logic (FADE) [29]. Brest optimized the scaling factor (F) and crossover rate (CR) by encoding them into individuals as parameters [30]. Zhang and Sanderson automatically adjust F and CR based on historical experience information, then proposed an adaptive DE with an optional external archive (JADE) [31]. In JADE, F, and CR are generated based on Cauchy distribution and Gaussian distribution, respectively. Besides, JADE introduces a new mutation strategy named "current-to-pbest", which can balance the global exploration and local exploitation capability of the algorithm better than the traditional ones. Then, researchers have proposed success-history based DE (SHADE) by improving the technique of the parameter adjustment in JADE [32]. In recent years, a series of DE algorithms with superior performance have been presented based on SHADE, such as using linear population size reduction to improve the search performance of SHADE (LSHADE) [33], improved LSHADE algorithm (iLSHADE) [34], ensemble sinusoidal differential covariance matrix adaptation with Euclidean neighborhood (LSHADE-EpSin) [35], LSHADE with semi-parameter adaptation hybrid with CMA-ES (LSHADESPACMA) [36], single objective real-parameter optimization algorithm (jSO) [37], etc.

DE has few applications in magnetic data inversion. For example, some scholars first applied DE into 3D inversion of the magnetic anomaly to estimate the source parameters such as the horizontal extension, the depth of top and bottom interface, the magnetic dip, and declination [38]. Other scholars have applied DE to analytic signal amplitude (ASA) inversion of two-dimensional (2D) magnetic data [39]. As far as magnetic inversion is concerned, there is a huge difference between the observed data and the inverted parameters, which means that one needs to recover a large number of model parameters by using a small amount of physical field information. Hence, the magnetic inversion is ill-conditioned and exists in multiple solutions. To address the mentioned issues, it is necessary to add the prior information as a constraint to reduce non-uniqueness and make the calculated model as close to the real one as possible. In addition, since the forward modeling calculation in this paper is based on the finite volume method, the relation of magnetic source parameters and magnetic data is nonlinear. The l_p norm based objective function aggravates the non-linearity further.

This article has systematically studied the $l_p(1 \leq p \leq 2)$ norm inversion of the magnetic data based on the proposed adaptive DE. The proposed algorithm can improve the global search ability and solution precision. Besides, considering that the original DE failed to obtain a smooth model, we designed a new mutation strategy to solve this problem. Although the inversion regularized with l_p norm has been widely solved by applying the gradient-based methods [40–44], the application of adaptive DE in the physical property of magnetic data is studied for the first time.

2. Method

2.1. Conventional Differential Evolution Algorithm

(1) Initialization

Like other evolution algorithms (EAs), DE initializes the population within a given search range in a certain way, such as uniform random initialization, chaos initialization, opposition learning initialization, clustering initialization, etc. [45–49]. In the conventional DE algorithm, the population size (NP) with D dimension is initialized as follows:

$$m_{ij}^{G} = L_j + rand(0,1) \cdot (U_j - L_j), G = 0,\ i \in [1, NP],\ j \in [1,\ D]\ , \quad (1)$$

where U_j and L_j are the upper and lower bounds of the j-th variable, respectively, $rand(0,1)$ generates a random number between [0, 1] according to a uniform distribution, and G represents the generation number of the evolutionary search.

(2) Mutation

After the initialization, a mutant vector $\boldsymbol{v}_i^G$ will be created for each target vector $\boldsymbol{m}_i^G$ according to the mutation strategy. The most widely used mutation strategies are listed as follows [50,51],

"DE/rand/1":

$$\boldsymbol{v}_i^G = \boldsymbol{m}_{r1}^G + F\left(\boldsymbol{m}_{r2}^G - \boldsymbol{m}_{r3}^G\right), \quad (2)$$

"DE/rand/2":

$$v_i^G = m_{r1}^G + F\left(m_{r2}^G - m_{r3}^G\right) + F\left(m_{r4}^G - m_{r5}^G\right), \tag{3}$$

"DE/best/1":

$$v_i^G = m_{best}^G + F\left(m_{r1}^G - m_{r2}^G\right), \tag{4}$$

"DE/best/2":

$$v_i^G = m_{best}^G + F\left(m_{r1}^G - m_{r2}^G\right) + F\left(m_{r3}^G - m_{r4}^G\right), \tag{5}$$

"DE/current-to-best/1":

$$v_i^G = m_i^G + F\left(m_{best}^G - m_i^G\right) + F\left(m_{r1}^G - m_{r2}^G\right), \tag{6}$$

"DE/current-to-pbest/1":

$$v_i^G = m_{r1}^G + F\left(m_{pbest}^G - m_{r1}^G\right) + F\left(m_{r2}^G - m_{r3}^G\right), \tag{7}$$

where m_{best}^G is the individual vector, which has the best fitness value at the current generation. $r1$, $r2$, $r3$, $r4$, $r5$ were mutually different random integers between 1 and NP. F is a scale factor within the range of (0, 1].

(3) Crossover

In DE, to obtain the trial vector u_i^G by replacing certain variables of the target vector m_i^G with corresponding mutant vector v_i^G, a crossover operation was needed. There were two types of crossover schemes: Exponential crossover and binominal crossover [33]. In the binominal crossover, the following formula is used to generate the trial vector,

$$u_{ij}^G = \begin{cases} v_{ij}^G, \text{ if } rand(0,1) \le CR \text{ or } j = jrand \\ m_{ij}^G, \text{ otherwise} \end{cases}, \tag{8}$$

where $CR \in [0, 1]$ is the crossover rate that controls which or how many components were inherited from the mutant vector. $jrand$ was a randomly selected integer in the range [1, D] to ensure that one of the variables in the trial vector must come from the mutant vector.

(4) Selection

DE adopts a one-to-one selection method. The fitness value of the target and the trial vectors determine which one will be selected into the population of the next generation. The selection scheme is given by:

$$m_i^{G+1} = \begin{cases} u_i^G, \text{ if } f(u_i^G) \le f(x_i^G) \\ m_i^G, \text{ otherwise} \end{cases}, \tag{9}$$

where $f(\cdot)$ is the fitness value of the target and the trial vectors.

2.2. Adaptive Adjustment of Control Parameters

In the conventional DE algorithm, F and CR are fixed values. Different problems usually need to set different parameters to obtain optimal performance of the algorithm. To avoid manually adjusting these parameters, it was an ideal strategy to adopt an adaptive method to update F and CR based on successful experience continuously. Considering that the improved DE algorithm shared the same architecture with JADE [31], in this part, a detailed description of parameters adjustment will be instructed.

Firstly, $F_i \in (0,1]$ is defined as,

$$F_i = randc_i(\mu_F,\ 0.1), \tag{10}$$

where $randc_i(\mu_F,\ 0.1)$ denotes the Cauchy distribution with location parameter μ_F and scale parameter 0.1. If $F_i \leq 0$, it was regenerated by the execution of Equation (11) until an effective value was obtained; if $F_i > 1$, it was truncated to 1.0. The initial value of μ_F was 0.5. In the subsequent evolution process, it was updated according to the scale factor F_i fed back from the successful individuals. At the end of each iteration, these successful F_i are stored in the set S_F, and μ_F is updated as follows,

$$\mu_F = (1-c)\mu_F + c \cdot mean_L(S_F), \tag{11}$$

where the learning rate c is a positive constant in the range $(0,1)$; $mean_L(\cdot)$ is the Lehmer mean and calculated as follows,

$$mean_L(S_F) = \frac{\sum_{F \in S_F} F^2}{\sum_{F \in S_F} F}. \tag{12}$$

In addition, each individual in JADE has an independent crossover rate CR_i in the range $(0,1)$. Unlike F_i, the generation of CR_i is controlled by the normal distribution,

$$CR_i = randn_i(\mu_{CR},\ 0.1), \tag{13}$$

where $randn_i(\mu_{CR},\ 0.1)$ returns a random value with a normal distribution $(\mu_{CR},\ 0.1)$. If $CR_i < 0$ or $CR_i > 1$, it is truncated to 0 or 1. μ_{CR} is updated as follows [31],

$$\mu_{CR} = (1-c)\mu_{CR} + c \cdot mean_A(S_{CR}), \tag{14}$$

where $mean_A(\cdot)$ is the arithmetic mean; S_{CR} is the set of successful CR values in the current generation.

2.3. *Improved Adaptive Differential Evolution Algorithm*

The evolution algorithm has its advantages in solving non-smooth and non-differentiable problems. However, the inversion of magnetic data requires that the restored model is smooth and continuous in space. Therefore, it was necessary to improve the DE algorithm suitable for the optimization of magnetic inversion.

2.3.1. Initialization

Generally, the global search algorithm does not depend on the population initialization. Unfortunately, if the random initialization was carried out according to Equation (1), it will bring difficulties in determining the regularization coefficient. Therefore, the population in this paper was initialized randomly in a small range according to Algorithm 1.

Algorithm 1 Population Initialization

01: **For** $i = 1$ to NP Do
02: **For** $j = 1$ to $nele$
03: $m(i,j) = 1^{-3} \cdot rand(0,1)$
04: **End For**
05: **End For**

2.3.2. Construction of the Smooth Matrix

There were many ways to establish smooth matrix **S**, such as moving average, Gaussian smoothing, k-nearest smoothing, median smoothing, etc. [52–55]. In this paper, the moving average method was adopted. For the model parameters $m_{i,j}$, the adjacent elements in space are shown in Figure 1. Set the weight of each adjacent point $W_{s,i,j}$, then the smoothed $m_{s,\ i,j}$ is

$$m_{s,\ i,j} = \frac{\sum_{l=-1}^{1}\sum_{m=-1}^{1} W_{s,i-l,j-m} m_{i-l,j-m}}{\sum_{l=-1}^{1}\sum_{m=-1}^{1} W_{s,i-l,j-m}}. \quad (15)$$

$m_{i-1,j+1}$	$m_{i,j+1}$	$m_{i+1,j+1}$
$m_{i-1,j}$	$m_{i,j}$	$m_{i+1,j}$
$m_{i-1,j-1}$	$m_{i-1,j-1}$	$m_{i+1,j-1}$

(**a**)

$\frac{1}{9}$	$\frac{1}{9}$	$\frac{1}{9}$
$\frac{1}{9}$	$\frac{1}{9}$	$\frac{1}{9}$
$\frac{1}{9}$	$\frac{1}{9}$	$\frac{1}{9}$

(**b**)

Figure 1. The square-shaped window of $m_{i,j}$. (**a**) The adjacent elements of $m_{i,j}$, (**b**) the weight of adjacent position parameters of $m_{i,j}$.

In order to reuse the above formula, the coefficients corresponding to all elements can be converted into sparse matrix **S**. The *k*-th row of matrix **S** is used to store the weight information of neighborhood elements of *k*-th cell. Finally, the smooth matrix is,

$$S = S^{p_f}, \quad (16)$$

where p_f is the time of smoothness, and the specific value should be determined by trial-and-error. In general, $p_f = 1$ or 2.

2.3.3. Mutation Strategy Based on the Smooth Search Direction

Due to the randomness of the nonlinear global search algorithm, the parameters obtained by inversion are usually discontinuous. Šešum and Tošić assumed that combining the smoothing filters with a genetic algorithm can improve the quality of obtained results [56]. However, Wu et al. believed that smoothing the individuals will reduce the population diversity [57]. Then, another smoothing strategy was proposed, which only smooths the model before the forward modeling calculation, and the smooth model did not replace the parameters in the original population. In addition, some scholars used the particle swarm optimization method to make the obtained solution closer to the designed model by smoothing the velocity direction of the particle [10]. In DE, we smoothed the individuals, which produced the differential direction. Let the smoothing matrix be **S**, then the smooth vector $\boldsymbol{m}_{s,i}$ of individual $\boldsymbol{m}_i$ can be defined as,

$$\boldsymbol{m}_{s,i} = \boldsymbol{S} \cdot \boldsymbol{m}_i. \quad (17)$$

Obviously, smooth difference vectors can be constructed by selecting different smooth vectors for all differential evolution mutation strategies. For simplicity, we only considered the "current-to-pbest/1" mutation strategy of Equation (12). When the smooth vector of an individual is used, it evolves into:

$$\boldsymbol{v}_i^G = \boldsymbol{m}_i^G + F_i\left(\boldsymbol{m}_{pbest}^G - \boldsymbol{m}_i^G\right) + F_i\left(\boldsymbol{m}_{s,\ r1}^G - \boldsymbol{m}_{s,\ r2}^G\right), \quad (18)$$

In the above formula, the difference vector formed by the pbest individual and the target vector *i* does not use the smooth vector of the related individual because the convex combination property of $\boldsymbol{m}_{pbest}^G$ and $\boldsymbol{m}_i^G$ will be destroyed by the smooth process. At the same time, the mutant vector formed entirely based on the smooth vector will reduce the diversity of the population.

2.3.4. Improvement of the Selection Strategy

The disturbance direction of DE was generated by the smooth individual vector. When the target vector $\boldsymbol{m}_i$ was replaced by its trial vector $\boldsymbol{u}_i$, the smooth vector also needs to be recalculated and stored, and Equation (14) is changed as,

$$\boldsymbol{m}_i^{G+1} = \begin{cases} \boldsymbol{u}_i^G, \text{ if } f(\boldsymbol{u}_i^G) \leq f(\boldsymbol{m}_i^G) \\ \boldsymbol{m}_i^G, \text{ otherwise} \end{cases} \boldsymbol{m}_{s,i}^{G+1} = \begin{cases} S\boldsymbol{u}_i^G, \text{ if } f(\boldsymbol{u}_i^G) \leq f(\boldsymbol{m}_i^G) \\ \boldsymbol{m}_{s,i}^G, \text{ otherwise} \end{cases}. \tag{19}$$

From Equation (19), the selection operation of the improved DE algorithm should keep the trial vectors and store their corresponding smooth vectors.

2.3.5. Sorting Crossover Rate

The main function of sorting crossover rate was to prevent the rapid loss of excellent individuals. It also plays a role in preventing the average value of the crossover rate from rapidly decreasing. For more details, please refer to [58].

2.3.6. Other Related Adjustments

In this paper, $NP = 100$, $\mu_{CR} = 0.9$, $\mu_F = 0.9$, the maximum number of estimates $maxNFES = 1000 \times D$, D was the number of the elements. Note that the μ_{CR} and μ_F were inconsistent with those of the JADE algorithm. The existing study shows that the separable problems required a small crossover probability [59], while the inseparable problem required a larger crossover probability. Considering that the magnetic anomaly field was the comprehensive reflection of the subspace magnetic sources, it indicated that the objective function of magnetic data was inseparable. Therefore, it was more appropriate to maintain a larger crossover probability. In addition, maintaining a larger μ_F was helpful to improve the exploration of DE when using smooth vectors to generate the differential vector.

2.4. Magnetic Anomaly Inversion

2.4.1. Forward modeling of Magnetic Anomaly

In this paper, the finite volume method (FVM) was used to synthesize the magnetic anomaly data [60]. Let the forward modeling operator be $\mathcal{F}$, then the relationship between the observed data $\boldsymbol{d}$ and the model parameter $\boldsymbol{m}$ is defined as follows:

$$\boldsymbol{d} = \mathcal{F}(\boldsymbol{m}). \tag{20}$$

In the forward modeling process, the demagnetization effect was considered, although this effect was negligible for the sources of weak susceptibility.

2.4.2. The Construction of the Inverse Problem

The objective function of l_p norm magnetic anomaly inversion can be expressed as follows,

$$\begin{aligned} \min \Phi(\boldsymbol{m}) &= \Phi_d(\boldsymbol{m}) + \lambda\Phi_m(\boldsymbol{m}) \\ s.t.\ \boldsymbol{m}_a &\leq \boldsymbol{m} \leq \boldsymbol{m}_b, \end{aligned} \tag{21}$$

λ is the regularization coefficient; $\Phi_d(\boldsymbol{m})$ is the data fitting objective function, which is generally l_2 norm; $\Phi_m(\boldsymbol{m})$ is the model objective function. In most inversion problems, $\Phi_m(\boldsymbol{m})$ is generally l_2 norm. Here, we consider the more general $l_p(1 \leq p \leq 2)$ norm. $\boldsymbol{m}_a$ and $\boldsymbol{m}_b$ are the range of the model parameters. If there are D optimized parameters, the upper and lower bounds of the model parameters are set as $\boldsymbol{m}_a = (m_{a,1}, m_{a,2}, \cdots, m_{a,D})^T$, $\boldsymbol{m}_b = (m_{b,1}, m_{b,2}, \cdots, m_{b,D})^T$. The data fitting objective function can be defined as,

$$\Phi_d(\boldsymbol{m}) = \frac{\left|\left|\sigma_d\left(\boldsymbol{d}_{obs} - \mathcal{F}(\boldsymbol{m})\right)\right|\right|_2^2}{\left|\left|\sigma_d \boldsymbol{d}_{obs}\right|\right|_2^2}, \tag{22}$$

where σ_d is the weighted matrix formed by the reciprocals of the noise in the observation data. In the above formula, the data fitting objective function used normalization processing to weaken the data errors' influence on data fitting objective function [61]. Correspondingly, we define the model objective function as,

$$\Phi_m(\boldsymbol{m}) = \sum W_{m,i}|m_i - m_{0,i}|^p, \tag{23}$$

where $W_{m,i}$ is the weight coefficient of element i, and used to store the prior information of the model, such as depth weighting [6]; m_0 is an initial model (the reference model or 0). Combining Equations (22) and (23), the objective function of magnetic inversion can be written as follows,

$$\begin{aligned} \min \Phi(\boldsymbol{m}) &= \frac{\left|\left|\sigma_d\left(\boldsymbol{d}_{obs} - \mathcal{F}(\boldsymbol{m})\right)\right|\right|_2^2}{\left|\left|\sigma_d \boldsymbol{d}_{obs}\right|\right|_2^2} + \lambda \sum W_{m,i}|m_i - m_{0,i}|^p \\ &s.t.\ \boldsymbol{m}_a \le \boldsymbol{m} \le \boldsymbol{m}_b. \end{aligned} \tag{24}$$

2.4.3. Adaptive Regularization Coefficient

The regularization coefficient λ controls the weight of the model objective function. Since the model objective function was all non-negative, its value range was $\lambda \in [0, +\infty)$. When λ = 0, the objective function only contains the data fitting term; when $\lambda \to +\infty$, the influence of the data fitting term can be ignored. Scholars proposed an adaptive regularization method based on the average fitness of the population [62]. Some scholars set the initial value of the regularization coefficient in a certain way at the beginning of the iteration [63] and then decreased the initial value continuously until λ less than a certain threshold. DE was a population-based search algorithm, which needs to combine the above two methods to constrain the inversion process effectively. Therefore, a new adaptive regularization coefficient scheme is proposed:

$$\lambda^{G+1} = \begin{cases} q_1\lambda^G & if\ \Phi_{d,\ mean}^G > \Phi_{d,\ mean}^{G-1} \\ q_2\lambda^G & elseif\ \Phi_{d,\ mean}^G = \Phi_{d,\ mean}^{G-1} \\ \lambda^G & otherwise \end{cases}, \tag{25}$$

where $\Phi_{d,\ mean}^G$ is the average value of the data objective function of the Gth-generation in the population; $q_1, q_2 \in (0,1)$; when $\Phi_{d,\ mean}^G$ increases or does not change, it is necessary to reduce the regularization coefficient λ to weaken the influence of the model constraint term. The reason is that the EAs cannot ensure each model is updated. At the first iteration, the regularization coefficient is defined as,

$$\lambda^0 = 10\frac{\Phi_{d,mean}^0}{\Phi_{m,mean}^0}, \tag{26}$$

where $\Phi_{m,mean}^0$ is the average value of the model constraint term.

2.4.4. Construction of the Model Weight Vector W_m

(1) The depth weighting function

There are many depth weighting methods, and the method proposed in [6] is more commonly used. The depth weighting function can be expressed as,

$$W_z = (z + z_0)^{-\beta}, \tag{27}$$

where β is the attenuation index, for 2D magnetic inversion, $\beta = 2$ [10]; z_0 is a constant, which is related to the observation height of the measuring point; z is the depth of the element's center. When p-norm is considered, the weighting coefficient of element i can be obtained by applying Equation (27):

$$W_{z,i} = (z_i + z_0)^{-\frac{\beta}{p}}, \tag{28}$$

(2) Weighting of element area

The area weighting matrix is a diagonal sparse matrix formed by the area of the element. For any element i, if the size of it is known to be $\Delta y_i, \Delta z_i$, then,

$$W_{V,i} = \Delta y_i \cdot \Delta z_i \tag{29}$$

When the element size of the inversion space is uniform, the area of each element is the same, which makes no sense to employ area weighting. However, when the inversion element was non-uniform, an area weighting matrix was necessary.

In summary, we have considered two kinds of model weight information, the weight coefficient of element i is,

$$\begin{gathered} W_{m,i} = \frac{W_{V,i} W_{z,i}}{\sum W_{V,i} W_{z,i}} \\ W_m = [W_{m,1}, W_{m,2}, \cdots, W_{m,D}]^T \end{gathered} \tag{30}$$

where W_m is the weight vector of the model. For example, the effect of depth weighting was to suppress the shallow elements and avoid model concentration near the surface.

2.4.5. Implementation of the Inversion Algorithm

In the above, we have discussed the objective function, adaptive regularization coefficient, mutation strategy based on a smooth vector of individual, and control parameters of adaptive DE. In this section, the adaptive DE algorithm for 2-D l_p norm magnetic anomaly inversion was systematically presented. The flowchart of the DE algorithm is shown in Figure 2. Its main content includes the following parts:

(1) Load the observation data, and generate the inversion grid according to the inversion area. Store the coefficient sparse matrix related to the control equation for the forward modeling calculation. Create model weighted vectors and store them. The geomagnetic parameters related to magnetic anomaly are known.
(2) Population initialization of DE: Given the population size and the initial μ_{CR} and μ_F, and the population vector is initialized by Algorithm 1. Smooth the vectors of the population and store them, and initialize the Regularization coefficient by Equation (26).
(3) Mutation and crossover: The variation vector $\boldsymbol{v}$ is generated by Equation (18), and the trial vector $\boldsymbol{u}$ is generated by Equation (8).
(4) Selection and update of DE control parameters: The smooth trial vector $\boldsymbol{u}_s$ is evaluated. According to Equation (19), to determine whether to update the population $\boldsymbol{m}$ and the individual vector in the population $\boldsymbol{m}_s$. Update μ_{CR} and μ_F according to Equations (14) and (11).
(5) Termination condition: Judge whether the termination condition is satisfied. If it is true, the current inversion result will be output; otherwise, turn to step (6).

(6) Update the regularization coefficient. If the regularization coefficient changes, it is necessary to update the objective function value of the population under the new regularization coefficient and turn to step (3).

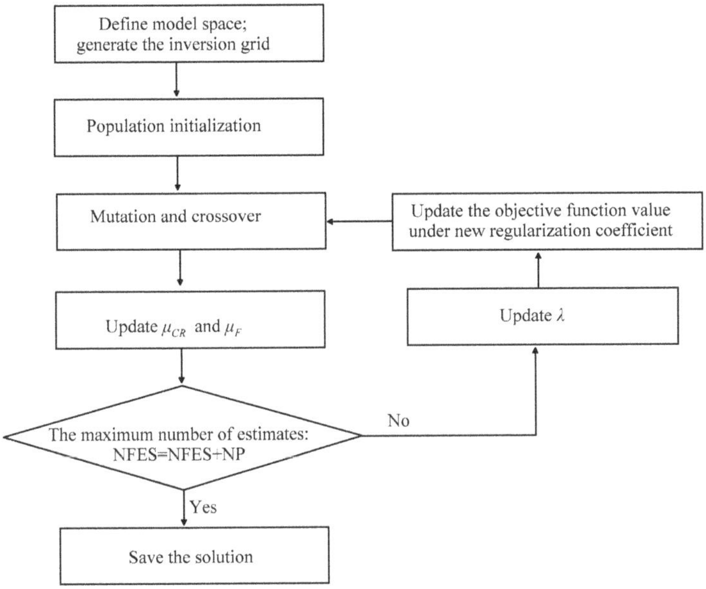

Figure 2. Flowchart of the differential evolution (DE) algorithm for the inversion of magnetic data.

The pseudo-code of the inversion Algorithm 2 is as follows:

Algorithm 2 l_p Norm Inversion Algorithm of Magnetic Anomaly Based on Improved DE

01: Data preprocessing. Read in the observation data, and generate the inversion grid and forward modeling sparse matrix, and then generate the weight vector W_m of the model
02: $G = 1$; $NFES = 0$
03: Set $NP = 100$, $\mu_{CR} = 0.9$, $\mu_F = 0.9$, $c = 0.1$. The population initialization based on Algorithm 1: $\boldsymbol{m} = (\boldsymbol{m}_{0,1}, \cdots, \boldsymbol{m}_{0,NP})$. Generate smooth population $\boldsymbol{m}_s$. Initialize the regularization coefficient λ^0, $NFES = NFES + NP$
04: **While** termination conditions are not satisfied do
05: $G = G + 1$
06: **For** $i = 1$ to NP do
07: $CR_i^G = randn_i(\mu_{CR}, 0.1)$
08: $F_i^G = randc_i(\mu_F, 0.1)$
09: **End For**
10: **For** $i = 1$ *to* NP do
11: Using the difference vector of smooth individuals, the mutation vector is generated according
to the mutation strategy in Equation (18)
12: According to Equation (8), the trial vector $\boldsymbol{u}_i^{G+1}$ is generated
13: **End For**
14: Evaluate the trial vector $\boldsymbol{u}_i^{G+1}$
15: According to Equation (19), the new populations $\boldsymbol{m}^{G+1}$ and $\boldsymbol{m}_s^{G+1}$ are formed.
16: Update μ_{CR} and μ_F by Equations (14) and (11)
17: The maximum number of estimates $NFES = NFES + NP$
18. **End While**

3. Model Inversion and Discussion

This section focuses on the influence of the mutation strategy, norm value, and the upper and lower bounds of magnetic parameters on the inversion results. For all theoretical models, without specified explanation, the magnetic inclination $I_0 = 60^{\circ}$, line direction $A_0 = 0^{\circ}$, geomagnetic field amplitude $|\boldsymbol{B}_0| = 5 \times 10^4 nT$.

3.1. The Influence of Smoothness on Inversion Results

Two simple models were used to test the influence of smooth individual vectors on the inversion results. Figure 3a is the cross-section through the x-axis of a 2-D magnetic dyke model. The dyke was buried at a depth of 50 m and extended to 150 m. Figure 3b is the cross-section through the x-axis of a 2-D magnetic dyke model at a dip angle of 135°.

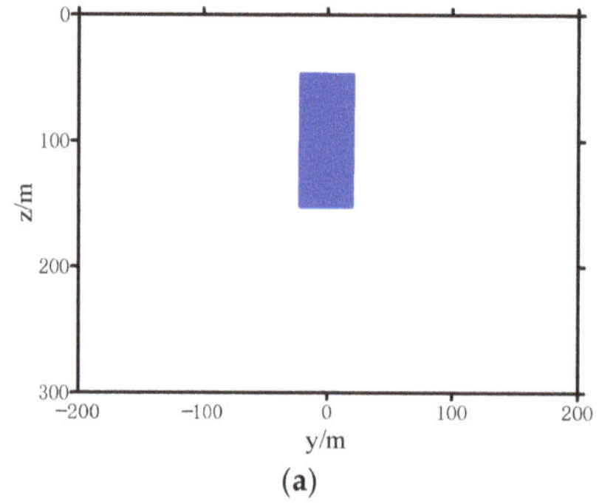

Figure 3. The cross-section through the x-axis of the 2D models. (**a**) Model 1, (**b**) Model 2.

The norm p of the model objective function is 2, and only the positive constraint ($m \geq 0\ SI$) is imposed on the model, without the upper limit of magnetic susceptibility. The magnetic anomaly distribution of the model is shown in Figure 4.

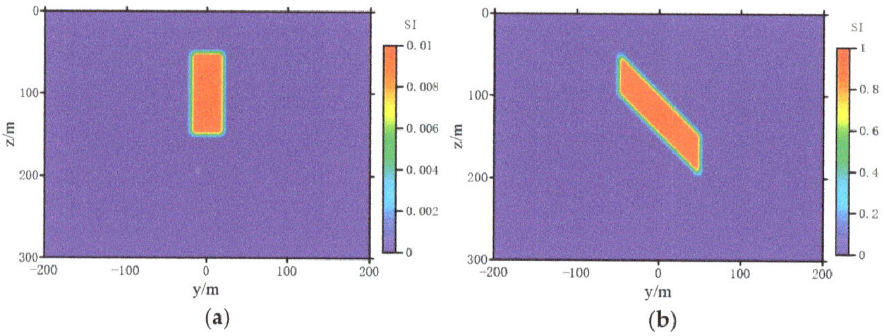

Figure 4. The tested models used for studying the effect of smooth vector strategy. (**a**) Model 1, (**b**) Model 2.

Figure 5 is the magnetic anomaly curve when the magnetic susceptibility χ was 0.01 SI. In order to ensure the accuracy of the simulation, the size of the forward model element was 5 × 5 m. During inversion, too many mesh sections will aggravate the ill-posedness of inversion. Therefore, the horizontal grid spacing was set as 10 m, and the grid spacing in the depth direction was a sparse grid that increased proportionally, and Algorithm 2 was used for inversion.

In order to compare the conventional adaptive differential evolution and the improved one, the smooth and non-smooth inversion results are shown in Figure 6. It is not difficult to observe that the proposed algorithm was capable of obtaining better results than the original algorithm. Figure 6 shows that the continuity of the model was poor, the inversion results were divergent, and the background noise was strong without considering the smoothness. When $p_f = 2$, the recovery of the model was better than $p_f = 1$. It can also be found from the above two figures that the magnetic parameters recovered by the inversion were 0.01 SI lower than the theoretical values without adding upper bound constraint. The main reason for this phenomenon is that l_2 norm inversion tends to obtain the smoothest model. For the abnormal body with a simple shape, the result can roughly reflect the dip information of the geologies. However, when the shape of the abnormal body tends to be complex, it can seldom depict the edge information of the model. In the following study, we set $p_f = 2$ to smooth the individuals in the population by default.

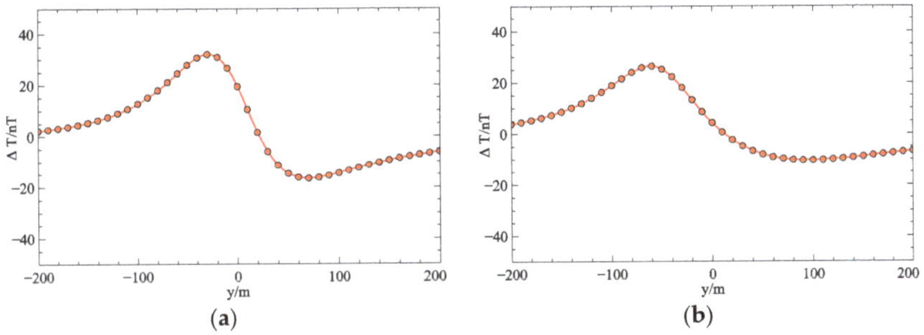

Figure 5. Anormal magnetic field ΔT of Model 1 and 2. (**a**) Model 1, (**b**) Model 2.

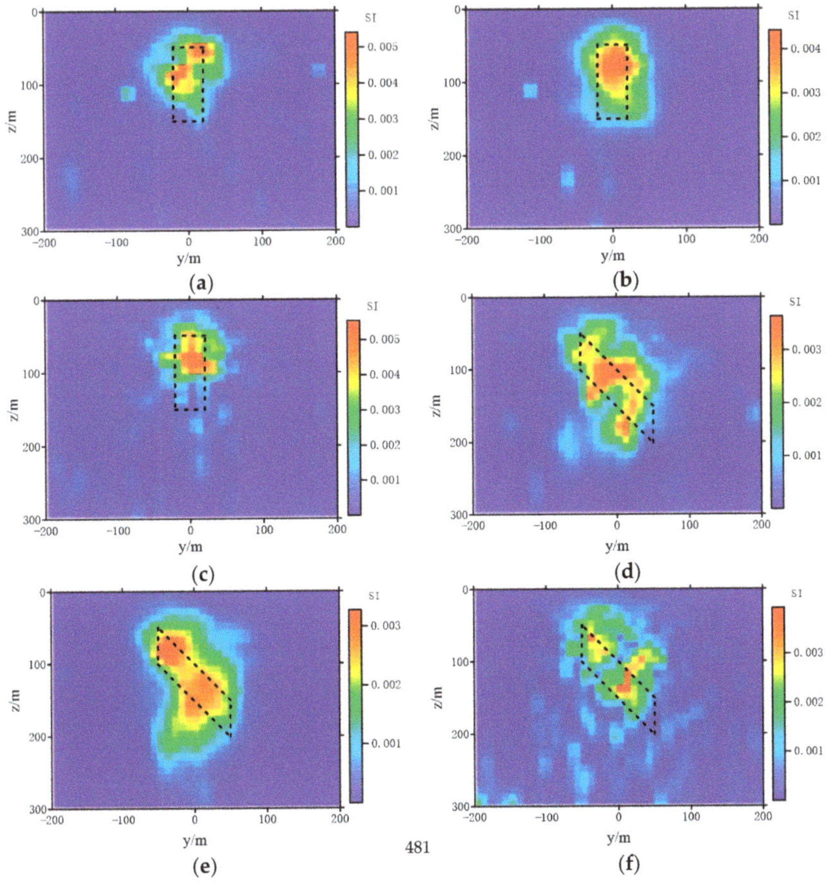

3.2. The Influence of P in Model Objective Function on Inversion Results

Figure 7 is the cross-section through the x-axis composed of two 2-D magnetic dyke models.

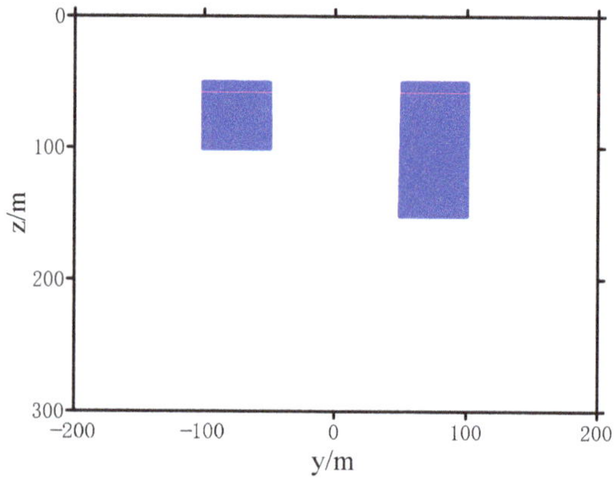

Figure 7. The cross-section through the x-axis of the model.

Set the magnetic susceptibility $\chi = 0.01$ SI and 0.02 SI. When $p = 1.0, 1.1, 1.3, 1.5, 1.7, 2$, we used Algorithm 2 to perform smooth inversion for Model 3 (Figure 8). The inversion results are shown in Figure 9. From the inversion results, we can observe that when p was large, the restored magnetic susceptibility was lower than the real value. The smaller p was, the closer the inversion result was to the real value. However, on the other hand, the larger p was, the larger the range of abnormal bodies distribution will be compared with the real model. Moreover, there will be interference between different anomalies to produce false anomalies. Generally speaking, $p \in [1, 1.5]$ was a suitable choice without the constraint of the upper bound of magnetic susceptibility. In Figure 10, due to the setting of the termination conditions, the field values obtained by inversion with each p-norm can fit the observed data. In summary, to balance the smoothness of the model and maintain the edge of the restored model, we set $p = 1.2$ in the subsequent work.

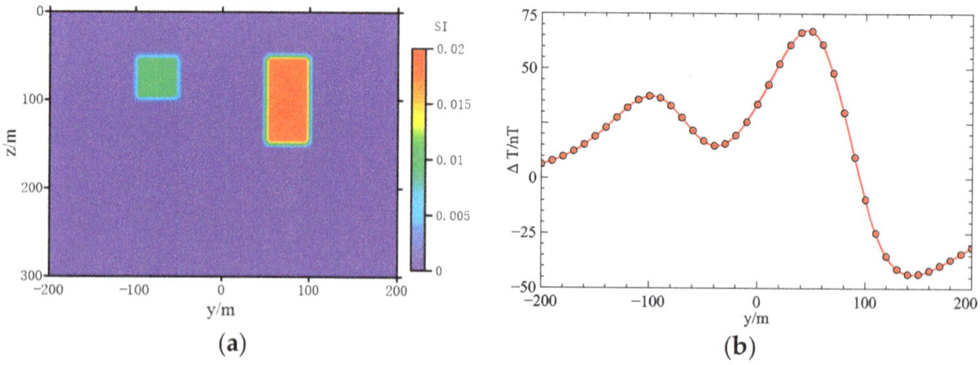

Figure 8. Susceptibility model 3 and its ΔT field. (**a**) Spatial distribution of magnetic susceptibility of model 3, (**b**) the ΔT of model 3.

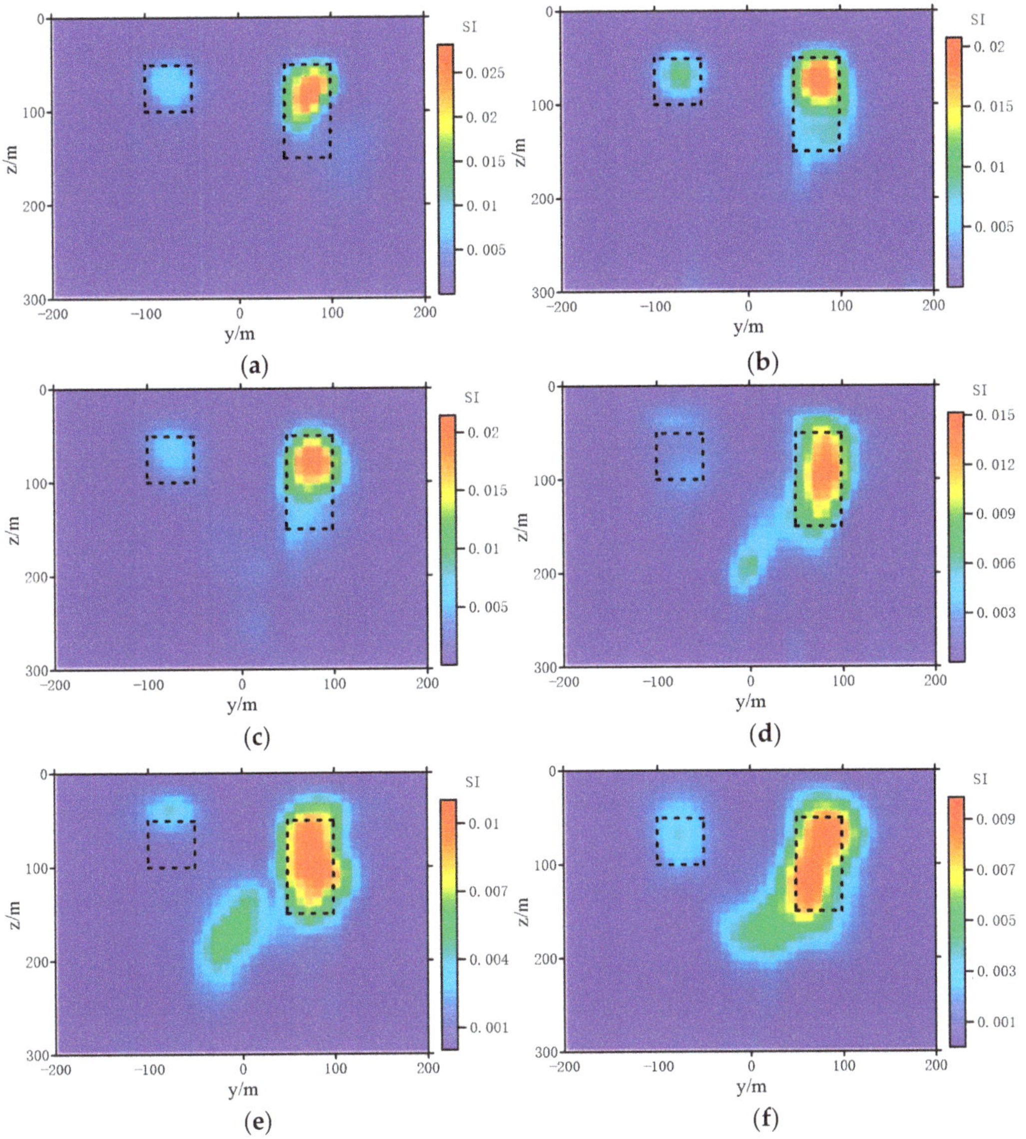

Figure 9. The inversion results of model 3 by using different p value. (**a**) $p = 1.0$, (**b**) $p = 1.1$, (**c**) $p = 1.3$, (**d**) $p = 1.5$, (**e**) $p = 1.7$, (**f**) $p = 2$.

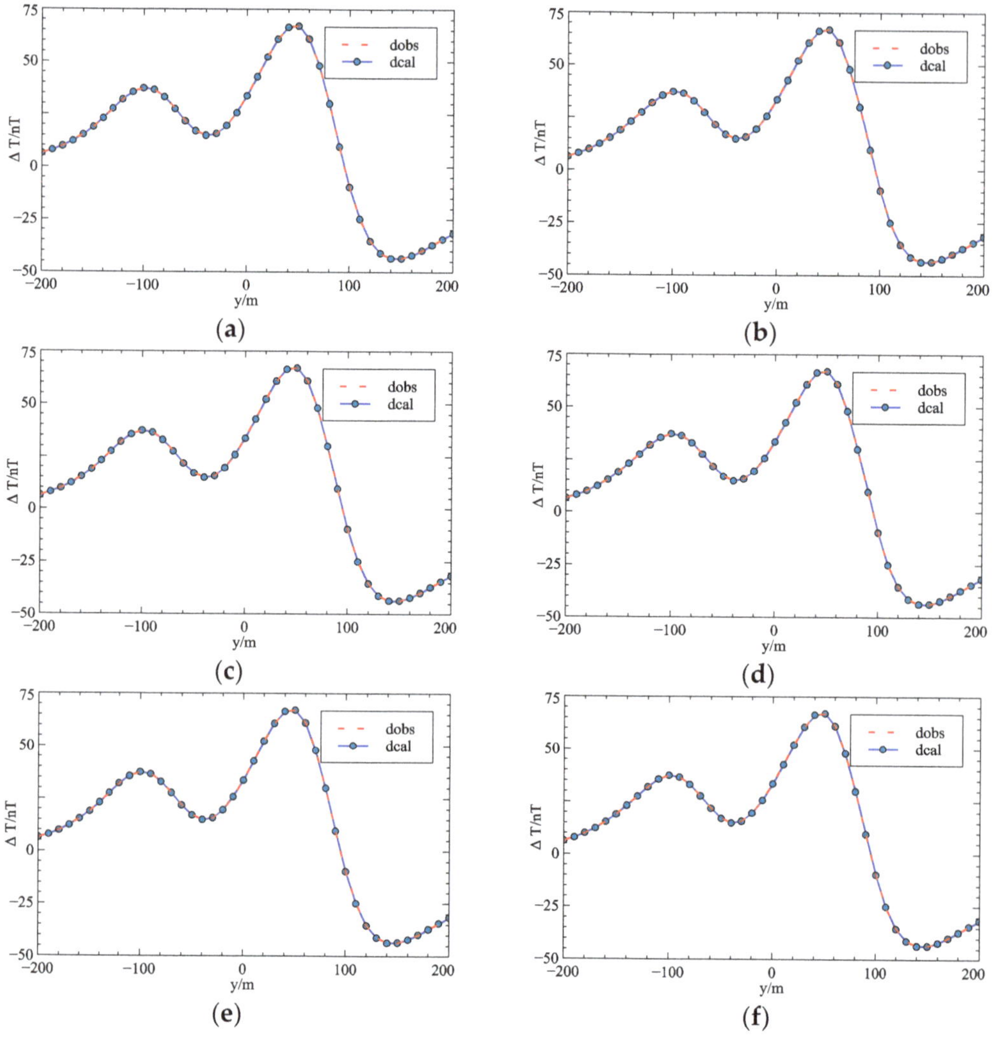

Figure 10. The data reconstruction results of different p values. (**a**) $p = 1$, (**b**) $p = 1.1$, (**c**) $p = 1.3$, (**d**) $p = 1.5$, (**e**) $p = 1.7$, (**f**) $p = 2$.

3.3. Influence of Noise on the Inversion Results

In the objective function, noise suppression was realized by the data misfit term. In most cases, the norm value of the data objective function q equals 2. We add Gaussian white noise with the mean value of 0 and the standard deviation of the percentage of the abnormal amplitude. The forward modeling data of Model 3 are superimposed with different degrees of noise, and the data after adding noise is shown in Figure 11. The set maximum number of iterations is 100D. The inversion results are shown in Figure 12. With the increase of noise, the abnormal shape deviates from the real model, and false anomalies appear in the deep. In Figure 13, the observed data and predicted inversion results with different noises are compared. The inversion results can reflect the original shape of the observed data.

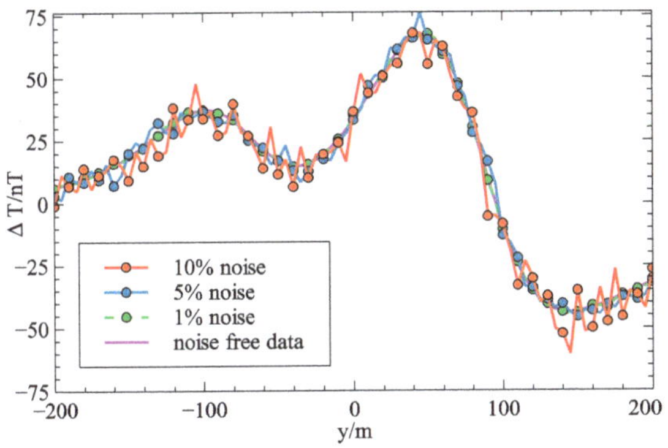

Figure 11. The abnormal magnetic data of Model 3 with different noise level.

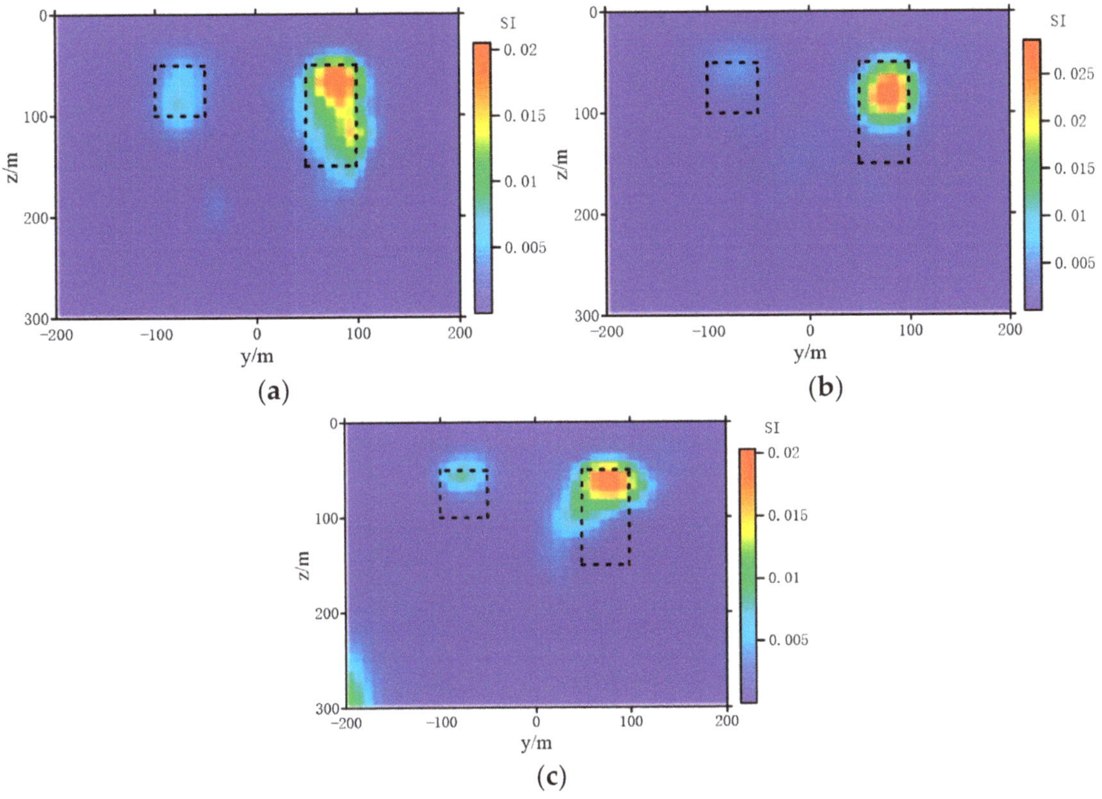

Figure 12. The inversion results with different noise level. (**a**) 1% noise, (**b**) 5% noise, (**c**) 10% noise.

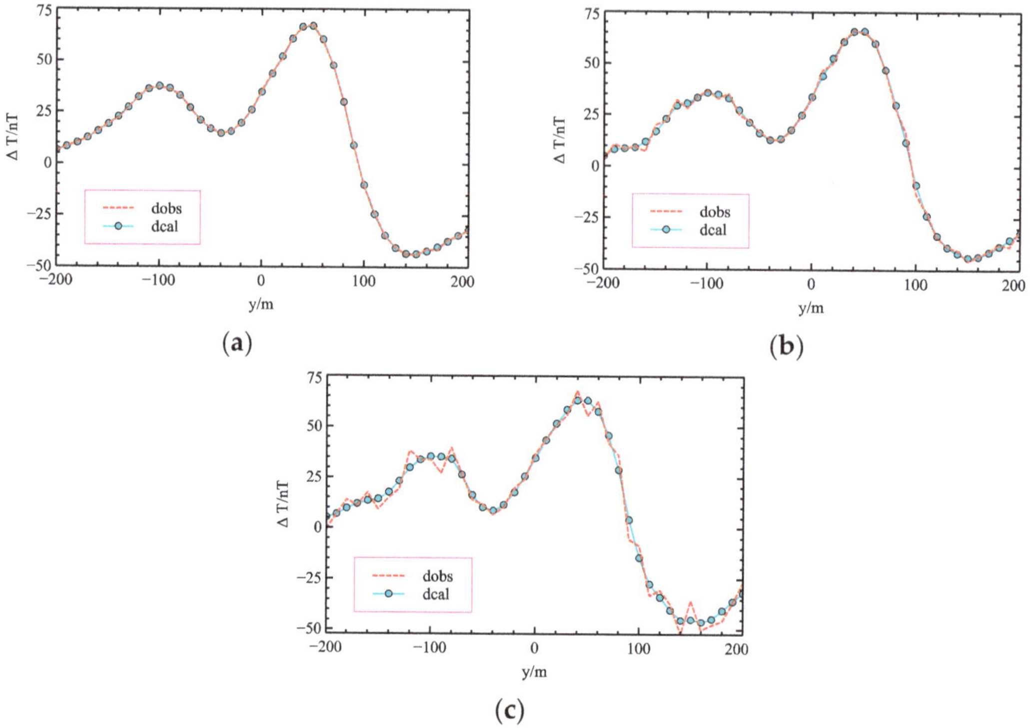

Figure 13. The comparison of observed data and predicted results. (**a**) 1% noise, (**b**) 5% noise, (**c**) 10% noise.

4. The Inversion of Real Data

The synthetic data constructed in Section 3 refers to the actual geological data and previous results, which are in line with the actual situation. From the synthetic data tests, the inversion results of synthetic models show that this method can obtain smooth solutions and describe the distribution of anomalous bodies well. The previous sections have discussed the influence of p and smoothness on the inversion results. It is concluded that $q = 2$ and $p = 1.2$ are suitable for the l_p norm smooth inversion of real data. However, the influence of the upper and lower bound constraints on the inversion results lacks discussion. Because some scholars have discussed in detail the influence of the parameters range on the results [43]. The inversion based on DE also has similar effects. The smaller the upper bound of the magnetic parameters, the larger the boundary of the recovered model, and on the contrary, the more concentrated they are. When dealing with real data, the upper and lower bound constraints need to be concerned.

The Zaohuoxi study area is located at the junction of the Qimantag suture zone and the North Kunlun magma arc zone. Most of the area is covered by Quaternary sediments with an average thickness of more than 60 m. The mineral resources in the area were mainly contact skarn-type iron polymetallic deposits. According to the work in [64], the highest magnetic susceptibility of the mineral in this region was magnetite with an average of 0.37 SI. At the same time, there was strong remanence with a maximum of 33.4 A/m in the study area. The magnetic parameters of minerals and rocks in the study area are shown in Table 1. The geomagnetic inclination I_0 and the geomagnetic field amplitude $|\boldsymbol{B}_0|$ were 53.73° and 53,739 nT, respectively.

Table 1. Magnetic parameters of minerals and rocks in the Zaohuoxi research area [64].

Rock	Lithology	Number of Specimens	Average Susceptibility χ (10^{-5} SI)	Remanent Magnetism J_r (10^{-3} A/m)
Medium strong magnetism	magnetite ore	9	3.73×10^4	3.34×10^4
	chalcopyrite magnetite ore	4	8.97×10^3	2.89×10^3
	magnetized skarn	5	2.07×10^3	8.32×10^2
Medium magnetism	granodiorite	4	5.41×10^2	1.70×10^2
	biotite granodiorite	9	4.02×10^2	3.52×10^2
	alteration granodiorite	9	2.63×10^2	1.14×10^2
Weak magnetism	marble	14	1.85×10^2	1.22×10^2
	alteration monzogranite	31	1.44×10^2	1.27×10^2
	biotite plagioclase gneiss	11	2.39×10^2	1.16×10^2
	quartz-diorite	16	1.98×10^2	1.29×10^2

The drilling results show that there was magnetite with a small distribution range in line L, which is shown in Figure 14. The magnetic anomaly of line L is shown in Figure 15. The magnetic anomaly in the central and southern part of the study area was regular in shape and distributed in a north-west-west (NWW) band. The anomaly was positive in the south and negative in the north.

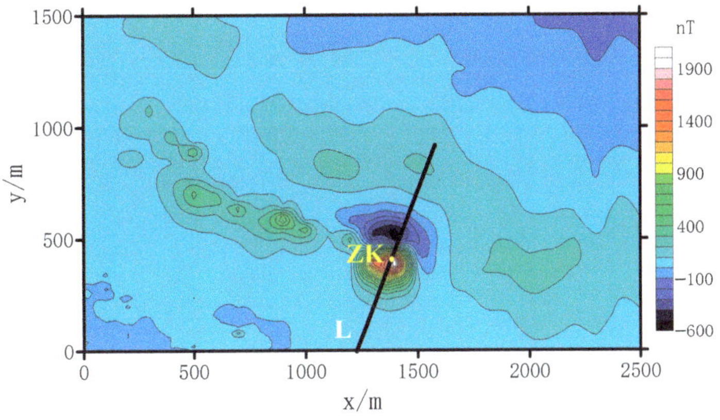

Figure 14. The magnetic contour map of ΔT in the Zaohuoxi area. (ZK is the borehole position and L is the survey line).

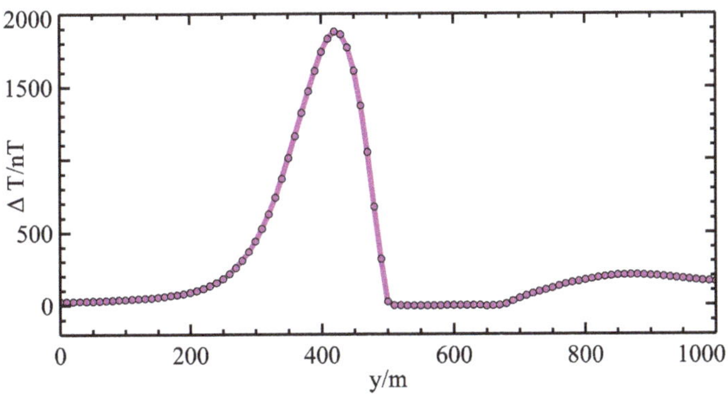

Figure 15. The magnetic anomaly of line L.

The magnetic anomaly of line L has a sharp shape (in Figure 15), high intensity (the peak value up to 1980 nT), and a steep gradient. It is the strongest magnetic anomaly zone in the region. Ou et al. [64] constructed the underground distribution of the area based on the drilling and geology information, as shown in Figure 16.

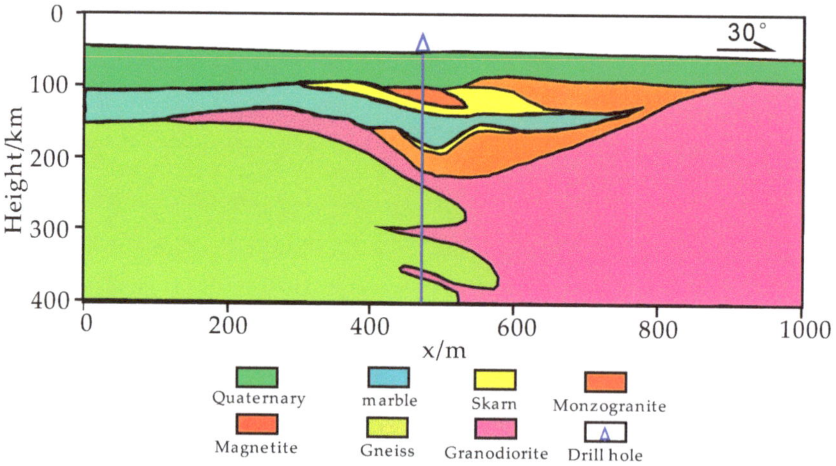

Figure 16. The underground distribution of line L [64].

In Figure 16, the magnetite with a small distribution range, buried to a depth about 100 m, x = 450 m. The high magnetic anomaly corresponding to line L in this area was mainly caused by magnetite with magnetic susceptibility $\chi = 0.37\ SI$. In the following inversion, the upper bounds of susceptibility were set as 0.37, 0.74, and 1.21 SI, respectively. The inversion results are shown in Figure 17. From Figure 17, the inverse position of the magnetite was buried at about 100 m, and the horizontal position was also consistent with the information in the drill hole. It was shown that the inversion result of this method was reliable. The inversion results with the upper bound of magnetic susceptibility can highlight the distribution of abnormal bodies better. When the upper bound of magnetic susceptibility was large, the distribution of the abnormal bodies was reduced. At the same time, the larger the upper limit of magnetic susceptibility, the dip direction of the magnetic body will be affected by oblique magnetization and deviate to the magnetization direction.

To sum up, the constrained smooth inversion of magnetic anomaly data in the study area was carried out. The results show that the constrained inversion can better reveal the distribution of the magnetic bodies. However, this kind of constraint was still rough. A more reasonable constraint was needed by using the magnetic susceptibility of different depths and different lithologies or according to the known and reliable reference model.

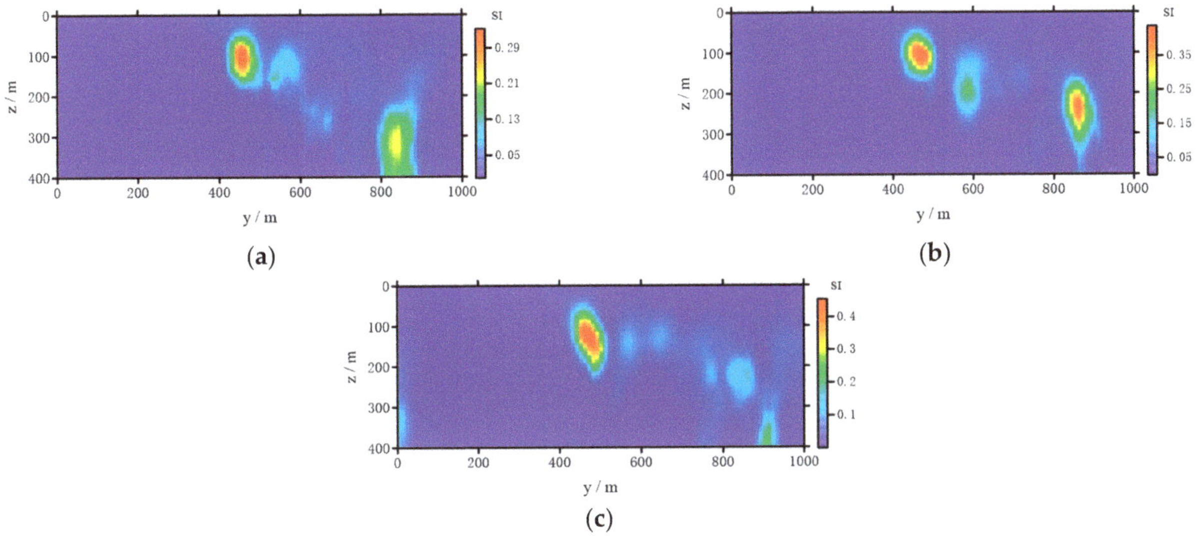

Figure 17. The susceptibility image of ΔT for line L with different susceptibility constraints in the Zaohuoxi area. (**a**) $\chi_{max} = 0.37\ SI$, (**b**) $\chi_{max} = 0.74\ SI$, (**c**) $\chi_{max} = 1.21\ SI$.

5. Conclusions

In this paper, an improved DE algorithm is proposed to solve the l_p norm inversion of 2D magnetic data. An adaptive regularization coefficient updating strategy for DE is proposed, and the mutation strategy of DE is improved by smoothing individual vectors. The proposed algorithm can obtain a smoother solution. The influence of smoothness and p in the model constraint term is analyzed. When there is more reliable constraint information, the smaller p can be used to highlight the model contour. As validated by synthetic models, this method can obtain a smoother solution and delineates abnormal bodies' distribution. Finally, the method is used in real data in the Zaohuoxi area, China, and the inversion result is consistent with the drilling results.

However, the search efficiency of the DE algorithm is not compared with other algorithms, and it is difficult to compare the DE algorithm with other techniques like gradient-based algorithms and particle swarm optimization algorithms by applying a fair criterion. This work will be completed in future research. At the same time, the 2D of magnetic inversion is based on the assumption that the strike of the anomaly body is perpendicular to the inversion profile with infinite extension. In practical problems, this assumption will restrict the application of 2D inversion. Therefore, it is necessary to extend the inversion method based on the DE algorithm to a three-dimensional case.

Author Contributions: Conceptualization, W.D. and L.C.; methodology, L.C.; software, L.C.; validation, W.D.; formal analysis, W.D.; investigation, L.C.; resources, W.D.; data curation, L.C.; writing—original draft preparation, W.D.; writing—review and editing, L.C. and Y.L.; visualization, W.D.; supervision, Y.L.; project administration, L.C.; funding acquisition, W.D. All authors have read and agreed to the published version of the manuscript.

Funding: This research was funded by the National Natural Science Foundation of China, grant number 41904129.

Institutional Review Board Statement: Not applicable.

Informed Consent Statement: Not applicable.

Data Availability Statement: The data of this study is available from the authors upon request.

Acknowledgments: This study was funded by the National Natural Science Foundation of China (grant number 41904129). The result of this study and code are available from the authors upon request.

Conflicts of Interest: The authors declare no conflict of interest.

References

1. Whitehill, D. Automated interpretation of magnetic anomalies using the vertical prism model. *Geophysics* **1973**, *38*, 1070–1087. [CrossRef]
2. Ballantyne, E.J. Magnetic curve fit for a thin dike: Calculator program (TI 59). *Geophysics* **1980**, *45*, 447–455. [CrossRef]
3. Zhdanov, M.S.; Liu, X.; Wilson, G.A.; Wan, L. Potential field migration for rapid imaging of gravity gradiometry data. *Geophys. Prospect.* **2011**, *59*, 1052–1071. [CrossRef]
4. Fedi, M.; Pilkington, M. Understanding imaging methods for potential field data. *Geophysics* **2012**, *77*, G13–G24. [CrossRef]
5. Guo, L.; Meng, X.; Zhang, G. Three-dimensional correlation imaging for total amplitude magnetic anomaly and normalized source strength in the presence of strong remanent magnetization. *J. Appl. Geophy.* **2014**, *111*, 121–128. [CrossRef]
6. Li, Y.; Oldenburg, D.W. 3-D inversion of magnetic data. *Geophysics* **1996**, *61*, 394–408. [CrossRef]
7. Pilkington, M. 3-D magnetic imaging using conjugate gradients. *Geophysics* **1997**, *62*, 1132–1142. [CrossRef]
8. Portniaguine, O.; Zhdanov, M.S. Focusing geophysical inversion images. *Geophysics* **1999**, *64*, 874–887. [CrossRef]
9. Zhdanov, M.S.; Ellis, R.; Mukherjee, S. Three-dimensional regularized focusing inversion of gravity gradient tensor component data. *Geophysics* **2004**, *69*, 925–937. [CrossRef]
10. Liu, S.; Liang, M.; Hu, X.Y. Particle swarm optimization inversion of magnetic data: Field examples from iron ore deposits in China. *Geophysics* **2018**, *83*, J43–J59. [CrossRef]
11. Storn, R.; Price, K. Differential evolution—A simple and efficient heuristic for global optimization over continuous spaces. *J. Glob. Optim.* **1997**, *11*, 341–359. [CrossRef]
12. Chen, X.; Guo, X.; Pei, J. A hybrid algorithm of differential evolution and machine learning for electromagnetic structure optimization. In Proceedings of the 32nd Youth Academic Annual Conference of Chinese Association of Automation (YAC), Hefei, China, 19–21 May 2017.
13. Qing, A.; Lee, C.K. *Differential Evolution in Electromagnetics*; Springer: Berlin, Germany, 2010.
14. Rivera-López, R.; Mezura-Montes, E.; Canul-Reich, J.; Chávez, M.A.C. A permutational-based Differential Evolution algorithm for feature subset selection. *Pattern Recognit. Lett.* **2020**, *133*, 86–93. [CrossRef]
15. Yuenyong, S.; Nishihara, A. A hybrid gradient-based and differential evolution algorithm for infinite impulse response adaptive filtering. *Int. J. Adapt. Control Signal Process.* **2014**, *28*, 1054–1064. [CrossRef]
16. Zhu, W.; Fang, J.; Tang, Y.; Zhang, W.; Du, W. Digital IIR filters design using differential evolution algorithm with a controllable probabilistic population size. *PloS ONE* **2012**, *7*, e40549. [CrossRef]
17. Karaboga, N.; Cetinkaya, B. Design of digital FIR filters using differential evolution algorithm. *Circuits Syst. Signal Process.* **2006**, *25*, 649–660. [CrossRef]
18. Shan, C.; Ma, Y.; Zhao, H.; Shi, J. Time modulated array sideband suppression for joint radar-communications system based on the differential evolution algorithm. *Digit. Signal Process.* **2020**, *97*, 102601. [CrossRef]
19. Arenas, A.; Diaz-Guilera, A.; Kurths, J.; Moreno, Y.; Zhou, C. Synchronization in complex networks. *Phys. Rep.* **2008**, *469*, 93–153. [CrossRef]
20. Shahri, A.A.; Asheghi, R.; Khorsand Zak, M. A hybridized intelligence model to improve the predictability level of strength index parameters of rocks. *Neural Comput. Appl.* **2020**, *9*, 1–14.
21. Shahri, A.A.; Moud, F.M.; Lialestani, S.P. A hybrid computing model to predict rock strength index properties using support vector regression. *Eng. Comput.* **2020**, *6*, 1–16.
22. Asheghi, R.; Hosseini, S.A.; Saneie, M.; Shahri, A.A. Updating the neural network sediment load models using different sensitivity analysis methods: A regional application. *J. Hydroinform.* **2020**, *22*, 562–577. [CrossRef]
23. Ghaderi, A.; Shahri, A.A.; Larsson, S. An artificial neural network based model to predict spatial soil type distribution using piezocone penetration test data (CPTu). *Bull. Eng. Geol. Environ.* **2019**, *78*, 4579–4588. [CrossRef]
24. Bureerat, S.; Pholdee, N. Inverse problem based differential evolution for efficient structural health monitoring of trusses. *Appl. Soft Comput.* **2018**, *66*, 462–472. [CrossRef]
25. Jena, P.K.; Thatoi, D.N.; Parhi, D.R. Differential evolution: An inverse approach for crack detection. *Adv. Acoust. Vib.* **2013**, *2013*, 321931. [CrossRef]
26. Mariani, V.C.; Neckel, V.J.; Afonso, L.D.; Coelho, L.D.S. Differential evolution with dynamic adaptation of mutation factor applied to inverse heat transfer problem. In Proceedings of the IEEE Congress on Evolutionary Computation (CEC), Barcelona, Spain, 18–23 July 2010.
27. Yu, Z.; Zhaosheng, N.; Zhige, J. An improved differential evolution algorithm for nonlinear inversion of earthquake dislocation. *Geod. Geodyn.* **2014**, *5*, 49–56. [CrossRef]
28. Mohamed, A.W.; Mohamed, A.K. Adaptive guided differential evolution algorithm with novel mutation for numerical optimization. *Int. J. Mach. Learn. Cybern.* **2019**, *10*, 253–277. [CrossRef]
29. Liu, J.; Lampinen, J. A Fuzzy Adaptive differential evolution algorithm. *Soft Comput.* **2005**, *9*, 448–462. [CrossRef]

30. Brest, J.; Greiner, S.; Boskovic, B.; Mernik, M.; Zumer, V. Self-adapting control parameters in differential evolution: A comparative study on numerical benchmark problems. *IEEE Trans. Evol. Comput.* **2006**, *10*, 646–657. [CrossRef]
31. Zhang, J.; Sanderson, A.C. JADE: Adaptive differential evolution with optional external archive. *IEEE Trans. Evol. Comput.* **2009**, *13*, 945–958. [CrossRef]
32. Tanabe, R.; Fukunaga, A. Success-history based parameter adaptation for Differential Evolution. In Proceedings of the IEEE Congress on Evolutionary Computation (CEC), Cancun, Mexico, 20–23 June 2013.
33. Tanabe, R.; Fukunaga, A.S. Improving the search performance of SHADE using linear population size reduction. In Proceedings of the IEEE Congress on Evolutionary Computation (CEC), Beijing, China, 6–11 July 2014.
34. Brest, J.; Maučec, M.S.; Bošković, B. iL-SHADE: Improved L-SHADE algorithm for single objective real-parameter optimization. In Proceedings of the IEEE Congress on Evolutionary Computation (CEC), Vancouver, British Columbia, Canada, 24–29 July 2016.
35. Awad, N.H.; Ali, M.Z.; Suganthan, P.N. Ensemble sinusoidal differential covariance matrix adaptation with Euclidean neighborhood for solving CEC2017 benchmark problems. In Proceedings of the IEEE Congress on Evolutionary Computation (CEC), Donostia, San Sebastián, Spain, 5–8 June 2017.
36. Mohamed, A.W.; Hadi, A.A.; Fattouh, A.M.; Jambi, K.M. LSHADE with semi-parameter adaptation hybrid with CMA-ES for solving CEC 2017 benchmark problems. In Proceedings of the IEEE Congress on Evolutionary Computation (CEC), Donostia, San Sebastián, Spain, 5–8 June 2017.
37. Brest, J.; Maučec, M.S.; Bošković, B. Single objective real-parameter optimization: Algorithm jSO. In Proceedings of the IEEE Congress on Evolutionary Computation (CEC), Donostia, San Sebastián, Spain, 5–8 June 2017.
38. Balkaya, Ç.; Ekinci, Y.L.; Göktürkler, G.; Turan, S. 3D non-linear inversion of magnetic anomalies caused by prismatic bodies using differential evolution algorithm. *J. Appl. Geophys.* **2017**, *136*, 372–386. [CrossRef]
39. Levent, E.Y.; Özyalın, Ş.; Petek, S.; Balkaya, Ç.; Göktürkler, G. Amplitude inversion of the 2D analytic signal of magnetic anomalies through the differential evolution algorithm. *J. Geophys. Eng.* **2017**, *14*, 1492–1508.
40. Dominique, F. A Cooperative Magnetic Inversion Method with Lp-norm Regularization. Ph.D. Thesis, University of British Columbia, Vancouver, BC, Canada, 2015.
41. Fournier, D.; Oldenburg, D.W. Inversion using spatially variable mixed ℓp norms. *Geophys. J. Int.* **2019**, *218*, 268–282. [CrossRef]
42. Meng, Z.; Xu, X.; Huang, D. Three-dimensional gravity inversion based on sparse recovery iteration using approximate zero norm. *Appl. Geophys.* **2018**, *15*, 524–535. [CrossRef]
43. Li, Z. Research on 3D Lp-norm Inversion of Gravity and Magnetic Data. Ph.D. Thesis, China University of Geosciences, Beijing, China, 2018. (In Chinese).
44. Li, Z.; Yao, C.; Zheng, Y. 3D inversion of gravity data using LP-norm sparse optimization. *Chin. J. Geophys.* **2019**, *62*, 3699–3709. (In Chinese)
45. Kazimipour, B.; Li, X.; Qin, A.K. Effects of population initialization on differential evolution for large scale optimization. In Proceedings of the IEEE Congress on Evolutionary Computation (CEC), Beijing, China, 6–11 July 2014.
46. Poikolainen, I.; Neri, F.; Caraffini, F. Cluster-Based Population Initialization for differential evolution frameworks. *Inf. Sci.* **2015**, *297*, 216–235. [CrossRef]
47. Rahnamayan, S.; Tizhoosh, H.R.; Salama, M.M.A. A novel population initialization method for accelerating evolutionary algorithms. *Comput. Math. Appl.* **2007**, *53*, 1605–1614. [CrossRef]
48. Rahnamayan, S.; Tizhoosh, H.R.; Salama, M.M.A. Quasi-oppositional differential evolution. In Proceedings of the IEEE Congress on Evolutionary Computation, Singapore, 25–28 September 2007.
49. Segredo, E.; Paechter, B.; Segura, C.; Carlos, I. González-Vila. On the comparison of initialisation strategies in differential evolution for large scale optimisation. *Optim. Lett.* **2018**, *12*, 221–234. [CrossRef]
50. Das, S.; Mullick, S.S.; Suganthan, P.N. Recent advances in differential evolution—An updated survey. *Swarm Evol. Comput.* **2016**, *27*, 1–30. [CrossRef]
51. Neri, F.; Tirronen, V. Recent advances in differential evolution: A survey and experimental analysis. *Artif. Intell. Rev.* **2010**, *33*, 61–106. [CrossRef]
52. Obeid, I.; Selesnick, I.; Picone, J. *Signal Processing in Medicine and Biology: Emerging Trends in Research and Applications*; Springer International Publishing: Cham, Switzerland, 2020; pp. 161–204.
53. Ladiray, D.; Quenneville, B. *Seasonal Adjustment with the X-11 Method*; Springer: New York, NY, USA, 2001; pp. 23–49.
54. Pitas, I.; Venetsanopoulos, A.N. *Nonlinear Digital Filters: Principles and Applications*; Springer: Boston, MA, USA, 1990; pp. 63–116.
55. Wang, N.; Zeng, W. The K nearest neighbor geometry filter based on spatial domain. In Proceedings of the 5th International Conference on Bioinformatics and Biomedical Engineering, Wuhan, China, 10–12 May 2011.
56. Šešum, V.; Tošić, D. Genetic algorithms and smoothing filters in solving the geophysical inversion problem. *Yugosl. J. Oper. Res.* **2002**, *12*, 215–225. [CrossRef]
57. Wu, J.; Ming, Y.; Zeng, R. Smooth constraint inversion technique in genetic algorithms and its application to surface wave study in the Tibetan Plateau. *Acta Seismol. Sin.* **2001**, *14*, 49–57. [CrossRef]
58. Zhou, Y.; Yi, W.; Gao, L.; Li, X. Adaptive differential evolution with sorting crossover rate for continuous optimization problems. *IEEE Trans. Cybern.* **2017**, *47*, 2742–2753. [CrossRef] [PubMed]
59. Montgomery, J.; Chen, S. An analysis of the operation of differential evolution at high and low crossover rates. In Proceedings of the IEEE Congress on Evolutionary Computation (CEC), Barcelona, Spain, 18–23 July 2010.

60. Lelièvre, P.G.; Oldenburg, D.W. Magnetic forward modelling and inversion for high susceptibility. *Geophys. J. Int.* **2006**, *166*, 76–90. [CrossRef]
61. Chen, X.; Zhao, G.; Tang, J.; Zhan, Y.; Wang, J. An adaptive regularized inversion algorithm for magnetotelluric data. *Chin. J. Geophys.* **2005**, *48*, 937–946. (In Chinese) [CrossRef]
62. Hau-San, W.; Ling, G. Application of evolutionary programming to adaptive regularization in image restoration. *IEEE Trans. Evol. Comput.* **2000**, *4*, 309–326.
63. Zhdanov, M.S. *Inverse Theory and Applications in Geophysics*, 3rd ed.; Elsevier: Amsterdam, The Netherlands, 2015; pp. 129–177.
64. Ou, Y.; Feng, J.; Zhao, Y.; Jia, D.; Gao, W. Forward modeling of magnetic data using the finite volume method with a simultaneous consideration of demagnetization and remanence. *Chin. J. Geophys.* **2018**, *61*, 4635–4646. (In Chinese)

applied sciences

MDPI

Article

3D Gravity Inversion on Unstructured Grids

Siyuan Sun [1,2], Changchun Yin [2,*] and Xiuhe Gao [1,3,*]

1 China Aero Geophysical Survey & Remote Sensing Center for Natural Resources, Comprehensive Department of Aero Geophysical Survey, Beijing 100083, China; sunsyagrs@gmail.com
2 College of Geo-Exploration Sciences and Technology, Jilin University, Changchun 130026, China
3 Key Laboratory of Airborne Geophysics and Remote Sensing Geology, Ministry of Nature and Resources, Beijing 100083, China
* Correspondence: yinchangchun@jlu.edu.cn (C.Y.); gaoxhagrs@gmail.com (X.G.); Tel.: +86-0431-8850-2426 (C.Y.); +86-010-6206-0135 (X.G.)

Abstract: Compared with structured grids, unstructured grids are more flexible to model arbitrarily shaped structures. However, based on unstructured grids, gravity inversion results would be discontinuous and hollow because of cell volume and depth variations. To solve this problem, we first analyzed the gradient of objective function in gradient-based inversion methods, and a new gradient scheme of objective function is developed, which is a derivative with respect to weighted model parameters. The new gradient scheme can more effectively solve the problem with lacking depth resolution than the traditional inversions, and the improvement is not affected by the regularization parameters. Besides, an improved fuzzy c-means clustering combined with spatial constraints is developed to measure property distribution of inverted models in both spatial domain and parameter domain simultaneously. The new inversion method can yield a more internal continuous model, as it encourages cells and their adjacent cells to tend to the same property value. At last, the smooth constraint inversion, the focusing inversion, and the improved fuzzy c-means clustering inversion on unstructured grids are tested on synthetic and measured gravity data to compare and demonstrate the algorithms proposed in this paper.

Keywords: gravity; unstructured grids; depth resolution; inversion

1. Introduction

Conventionally, gravity forward modeling is implemented by dividing the subsurface into a large number of rectangular prisms, and the gravity anomaly is calculated by integrating the gravitational effect due to each prism. However, structured grids have limited capability when modeling structures with complex shapes (such as topographic surfaces), and a staircased configuration from structured grids results in a loss of accuracy. In comparison, unstructured grids can easily be designed to conform to arbitrary straight-edged boundaries, topography, and subsurfaces. On the other hand, the local refinement or coarsening of the mesh, in regions where higher accuracy is required or lower accuracy is acceptable, can be performed without affecting the grids outside these regions. These properties can ultimately result in higher accuracy and lower computational cost.

For gravity forward modeling with unstructured grids, several methods have been developed. Talwani and Ewing [1], and Talwani [2] used numerical integration techniques by dividing models of arbitrary shape into polygonal prisms or laminas. Barnett [3] and Okabe [4] developed techniques for gravity and magnetic modeling due to homogeneous polyhedral bodies. Zhang et al. [5] and Cai and Wang [6] used the finite-element method to calculate the gravitational response of 3D rectangular grids, while May and Knepley [7] used the fast multipole method to tackle the same problem. Jahandari and Farquharson [8] used finite-volume and finite-element methods to gravity forward modeling and their results showed that the quadratic finite-element scheme is more accurate but also more computationally demanding scheme. The flexibility of unstructured tetrahedral meshes has been proved to be advantageous when complex model geometries have to be modeled. However, the numerical schemes (such as finite-volume and finite-element methods) handling such tetrahedral meshes can lead to less-accurate results in comparison with the structured, regular grids.

Three-dimensional (3D) regularized inversion has been a popular approach in right rectangular prisms [9–12]. In regularization inversion, a regularization term enables us to incorporate more information so that a smooth or sharp model can be yielded [13]. Li and Oldenburg [9,14] used an objective function that penalize both the deviation of the recovered model from a reference model and the structural complexity in three spatial directions. A focusing method is developed to reconstruct a compact and sharper image of the subsurface [11,15,16]. A fuzzy c-means clustering algorithm [17–20] is applied to induce the recovered physical properties to gather around clusters determined by some priori information. In addition, a depth weighting function is vital to recovering the model at a proper depth on structured grids. Li and Oldenburg [9] firstly introduce a depth weighting function in the Tikhonov formulation. Portniaguine and Zhdanov [16] have introduced a nonlinear depth weighting that equalizes the sensitivity of the observed data to cells located at different depths and horizontal positions. In a mathematically less rigorous way, Commer [21] directly applied a weighting function to the gradient of data misfit functional and the results revealed a great improvement in depth resolution. However, when these methods are directly used to gravity inversion on unstructured grids, a discontinuous and hollow model is more likely to be yielded. This arises because volume distinction of unstructured grids also has a considerable effect on model distribution besides depth variation of grids. This is more pronounced when the local refinement or coarsening of the unstructured mesh is used.

In this paper, by analyzing the gradient of objective functional, we find that the distribution of density is related to the conventional gradient formulation of objective functional. Subsequently, two mechanisms of depth weighting are compared and analyzed in detail: (1) a regularization-based strategy: applying a depth weighting to the regularization term [9,11,14,16]; and (2) gradient-based strategy: applying a depth weighting to the gradient of data misfit [21,22]. The comparative analysis indicates that the second method would improve gravity depth resolution more directly and effectively, and this improvement is less influenced by regularization parameters. Moreover, the regularization terms will have a restraining effect at all depths. Besides, we develop an improved fuzzy c-means clustering method combined with spatial constraints. Compared with previous works [17–20], it measures property distribution of inverted models in both the spatial domain and parameter domain, and it can provide a more continuous distribution of properties on unstructured grids.

To demonstrate the effectiveness of our algorithm, we tested both on a synthetic and a field dataset from the Vinton salt dome by Bell Geospace. The forward modeling is accomplished with the algorithm of Okabe [4] based on unstructured grids, which has been used in gravity researches [17,23,24]. We use TetGen [25] to generate tetrahedral grids. TetGen is an open-source software that generates the tetrahedral and Voronoï grids as required. The algorithms of TetGen are Delaunay-based, which ensures high quality for the whole grids by maximizing the minimum dihedral angle in the grids. The results show that the problem in gravity inversion with the anomaly concentration to the surface is largely solved. The inverted models show sharp boundaries and true positions.

2. Methods

The technical methods of this paper mainly include three issues: forward modeling algorithm on tetrahedral grids, Fuzzy c-means clustering constraints, and depth weighting. In this chapter, we will introduce a forward modeling algorithm on tetrahedral grids and Fuzzy c-means clustering constraints only, and the depth weighting will be described and analyzed in Section 3.

2.1. Gravity Forward Modeling Based on Tetrahedral Grids

In the right-handed Cartesian coordinate system, as shown in Figure 1, based on Barnett [3] and Okabe [4], by successively imposing thrice coordinate rotations on every edge of the polyhedral body, the gravity anomalies g of a homogeneous polyhedral body can be converted to the sum of the contributions from all edges, which can be expressed as

$$g = G\rho\sum_{i}^{n}\left[\cos\psi_i\sum_{j}^{m}J_j(i)\right] \tag{1}$$

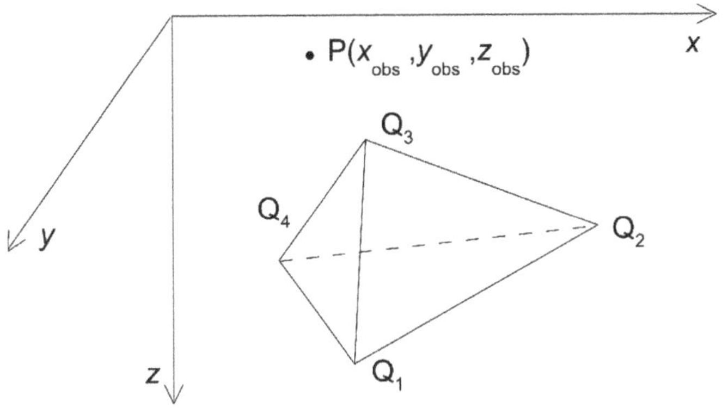

Figure 1. A tetrahedral cell in coordinate system.

In Equation (1), G is Newton's gravitational constant, and ρ is the density of the cell. n is the facet number of the polyhedral body. m is the edge number of the ith facet. For tetrahedral grids, $n = 4$ and m = 3. ψ_i is a rotation angle of the ith facet, while $J_j(i)$ is the contribution of jth edge in the ith facet, which is expressed as

$$J_j(i) = \left[(X\sin\theta_j - Y\cos\theta_j)\ln(X\cos\theta_j + Y\sin\theta_j + R) - 2Z\tan^{-1}\frac{(1+\sin\theta_j)(Y+R) + X\cos\theta_j}{Z\cos\theta_j}\right]_{(X_j,Y_j)}^{(X_{j+1},Y_{j+1})} \tag{2}$$

where

$$\begin{bmatrix} X \\ Y \\ Z \end{bmatrix} = \begin{bmatrix} \cos\psi & 0 & -\sin\psi \\ 0 & 1 & 0 \\ \sin\psi & 0 & \cos\psi \end{bmatrix} \bullet \begin{bmatrix} \cos\varphi & \sin\varphi & 0 \\ -\sin\varphi & \cos\varphi & 0 \\ 0 & 0 & 1 \end{bmatrix} \bullet \begin{bmatrix} x' - x_{obs} \\ y' - y_{obs} \\ z' - z_{obs} \end{bmatrix} \tag{3}$$

$$R = (X^2 + Y^2 + Z^2)^{1/2} \tag{4}$$

In the above equations, φ, ψ, and θ are respectively three rotation angles for the jth edge, and can be calculated as follows [4],

$$\cos\varphi = -\frac{S_{yz}}{\left(S_{yz}^2 + S_{zx}^2\right)^{1/2}} \tag{5}$$

$$sin\varphi = -\frac{S_{zx}}{\left(S_{yz}^2 + S_{zx}^2\right)^{1/2}} \tag{6}$$

$$cos\psi = -\frac{S_{xy}}{\left(S_{yz}^2 + S_{zx}^2 + S_{xy}^2\right)^{1/2}} \tag{7}$$

$$\sin\psi = \frac{S_{yz}^2 + S_{zx}^2}{\left(S_{yz}^2 + S_{zx}^2 + S_{xy}^2\right)^{1/2}} \tag{8}$$

$$\cos\theta = \frac{X_{j+1} - X_j}{\left[(X_{j+1} - X_j)^2 + (Y_{j+1} - Y_j)^2\right]^{1/2}} \tag{9}$$

$$\sin\theta = \frac{Y_{j+1} - Y_j}{\left[(X_{j+1} - X_j)^2 + (Y_{j+1} - Y_j)^2\right]^{1/2}} \tag{10}$$

In the above Equations (5)–(8), Syz, Sxz, and Sxy are the projected areas of the ith facet respectively onto the y-z-, z-x-, and x-y- planes. It should be pointed out that the vertices of the facet must be arranged clockwise in the X-Y-plane before Equation (3) is calculated. However, the generation of tetrahedral meshes is automatic and vertices are arranged and stored randomly. Therefore, we introduce a judgment and rearrangement of vertices before the gravity contribution of each facet is calculated. For more details about the rotations and judgment, please see Appendix A.

2.2. Fuzzy C-Means Clustering Inversion with Spatial Constraints

The inverse problem is solved by minimizing the Tikhonov parametric functional [13]. Fuzzy c-means clustering method (FCM) is used as the regularization term [18,19]. The objective function of gravity inversion is formulated as follows:

$$\Phi = \Phi_{\mathrm{d}} + \lambda\Phi_{\mathrm{m}} = \|\mathbf{Fm} - \mathbf{d}_{\mathrm{obs}}\|^2 + \lambda\sum_k^p \left\|\mathbf{W}_{C_k}^{f/2}(\mathbf{m} - C_k)\right\|^2 \tag{11}$$

where $\mathbf{F}$ is a forward modeling operator. $\mathbf{m}$ are density parameters of all cells. $\mathbf{d}_{\mathrm{obs}}$ are observed data. Φ_{d} is a data misfitting term, and Φ_{m} is a regularization term to constrain model structure. λ is a trade-off parameter. $\Phi_{\mathrm{m}} = \sum_k^p \left\|\mathbf{W}_{C_k}^{f/2}(\mathbf{m} - C_k)\right\|^2$ is the clustering term to classify the physical property values into distinct clusters $\mathbf{C} = \{C_k | k = 1, 2, \cdots, p\}$. In this paper, we assume that p and $\mathbf{C}$ are determined from a priori information. The parameter f, known as fuzzification parameter, is generally set to 2.0 [26]. Here, $\mathbf{W}_{Ck}$ is a diagonal matrix that measures the degree of the physical property values belonging to the kth cluster,

$$\mathbf{W}_{Ck} = diag\left[\frac{(\mathbf{m} - C_k)^{-2/(f-1)}}{\sum_k^p (\mathbf{m} - C_k)^{-2/(f-1)}}\right] \tag{12}$$

Φ_{m} can effectively classify the physical property values into distinct clusters in the parameter domain, however, it has no effect on the spatial distribution of density. As a result, the property spatial distributions may be discontinuous and discrete. So, on structured grids, Sun and Li [18,19] added the smoothness regularization term to promote continuous in the spatial domain. However, it is difficult to balance the smoothness and clustering term and reconstruct a continuous model on unstructured grids.

Thus, we combine spatial constraints into the FCM term to help recover spatial continuity of physical property. We reset component wise of $\mathbf{m}$ as follows:

$$m'_j = \left(m_j + \sum_l^{q_j} m_l\right) \Big/ (q_j + 1) \tag{13}$$

where q_j is set to be the adjacent cells number of jth cell, and it is usually equal to 4 except for the cells at boundaries. m'_j denotes the average of properties in the jth cell and its adjacent cells. The new Φ_m can measure the inverted model in both spatial domain and parameter domain. When the new Φ_m is minimized, it induces the properties of the jth cell and its adjacent cells to gather a certain cluster C_k. So, a continuous property distribution in spatial domain and a focused property distribution in parameter domain can be obtained.

In Equation (11), depth weighting function is not involved as it does in gravity inversion based on structured grids. Compared with structured grids, cell volumes are no longer the same as a constant. The dual effects of volume and depth difference of grids would result in a discontinuous and hollow model. While, the local refinement or coarsening of the mesh is an important advantage of unstructured grids, which will inevitably lead to different mesh volumes. So, it's crucial to solve this problem before gravity inversion based on unstructured grids used widely in field data. We will discuss the depth weighting in the next section.

3. Analysis of Depth Weighting on Unstructured Grids

In gravity inversion, the depth resolution is a serious issue. Different kinds of depth weighting have been proposed to counteract the decay of kernel function for improving the depth resolution. Typically, Li and Oldenburg [9,27] introduced a weighting function of cell depth,

$$\mathbf{W}_{\mathrm{L\&O}} = \frac{1}{(z+z_0)^{\beta/2}} \tag{14}$$

where z is the cell depth, β is usually 2 for gravity inversion, z_0 is a constant.

Besides, Portniaguine and Zhdanov [16] proposed a weighting function as the integrated sensitivity matrix, i.e.,

$$\mathbf{W}_{\mathrm{Zh}} = \mathrm{diag}\left(\sum_i F_{ij}^2\right)^{1/2} \tag{15}$$

where F_{ij} denotes the element of the kernel function $\mathbf{F}$.

On structured grids, the above two depth weighting functions are widely used in gravity inversions. While on unstructured grids, Figure 2 shows the variations of above 2 depth weighting functions with the depth.

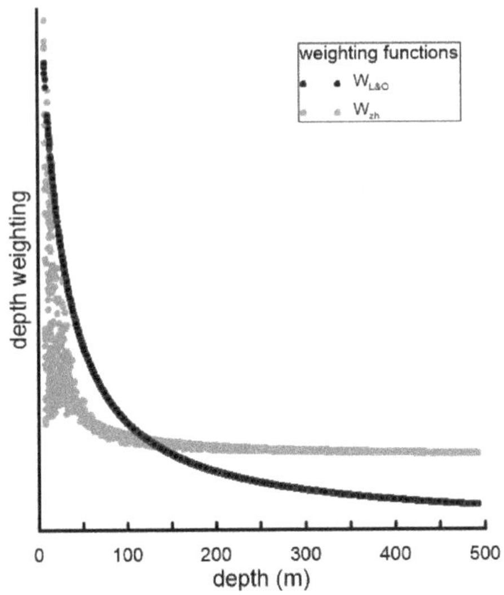

Figure 2. Depth weighting versus depth for two weighting functions on unstructured grids.

Results in Figure 2 are calculated over an area of 500 m × 500 m × 500 m composed of tetrahedral grids. z_0 is chosen to be 20 in Equation (14). As volume of grids are different for tetrahedral meshes, the depth weighting $\mathbf{W}_{\mathrm{Zh}}$ is normalized by cell volumes. In Figure 2, similar tendencies can be seen from two weightings. $\mathbf{W}_{\mathrm{L\&O}}$ is only related to depth, and all cells at the same depth have the same weightings. However, $\mathbf{W}_{\mathrm{Zh}}$ are different for cells at the same depth, especially near the surface, which is caused by difference of cell volume. On unstructured grids, if cells at the same depth have the same weightings, it will result in that the cells with larger volumes have the priority to change their densities first to fit the observed data. So based on unstructured grids, $\mathbf{W}_{\mathrm{Zh}}$ is better than $\mathbf{W}_{\mathrm{L\&O}}$ to improve resolution of gravity inversion.

Stabilized gradient-based inversion methods prove to be efficient for 3D imaging problems with highly parameterized models. Based on the principle of gradient-based inversion methods, setting the gradient to near zero can minimize the objective function efficiently [21]. The gradient of Equation (11) can be expressed as

$$\nabla\Phi = \nabla\Phi_{\mathrm{d}} + \lambda\nabla\Phi_{\mathrm{m}} = \mathbf{F}^T(\mathbf{Fm} - \mathbf{d}_{\mathrm{obs}}) + \lambda\sum_{i}^{p}\mathbf{W}_z\mathbf{W}_{Ci}^{f}(\mathbf{m} - C_i) \tag{16}$$

Firstly, let's focus on the gradient of data misfit

$$\nabla\Phi_{\mathrm{d}} = \mathbf{F}^{\mathrm{T}}(\mathbf{Fm} - \mathbf{d}_{\mathrm{obs}}) \tag{17}$$

It is well known that the gravity forward modeling operator $\mathbf{F}$ decays quickly as depth increases [9,28]. Therefore, when multiplying $\mathbf{F}^{\mathrm{T}}$ with the data misfit $\mathbf{Fm}\text{−}\mathbf{d}_{\mathrm{obs}}$, $\nabla\Phi_{\mathrm{d}}$ will be less sensitive to depth, so densities of shallow cells will first change to fit $\mathbf{d}_{\mathrm{obs}}$. As a result, the inverse model will more easily concentrate to the surface in the traditional gradient-based inversions. So, from this perspective, the key to improve the depth resolution is how to counteract the decay of $\mathbf{F}^{\mathrm{T}}$ in $\nabla\Phi_{\mathrm{d}}$.

In a conventional way, imposing depth weighting into Φ_{m} can improve depth resolution [9,14,16]. Then the gradient of the objective function can be expressed as

$$\nabla\Phi = \nabla\Phi_{\mathrm{d}} + \lambda\nabla\Phi'_{\mathrm{m}} \tag{18}$$

where Φ'_{m} denotes the regularization term after depth weighting imposed.

Then the weighted $\nabla\Phi'_{\mathrm{m}}$ decays quickly and becomes less sensitive to depth as $\nabla\Phi_{\mathrm{d}}$ does. It encourages that properties of shallow cells will first satisfy the constraint of Φ'_{m}, such that the property differences of shallow cells in smooth inversion or the relative properties of shallow cells in focus inversion tends to zero. As a result, this feature encourages the properties of shallow cells to remain unchanged and counteracts the influence of $\mathbf{F}^{\mathrm{T}}$ on shallow cells. So the depth resolution can be improved to some extent, but the improvement is affected by the balance of $\nabla\Phi_{\mathrm{d}}$ and $\nabla\Phi'_{\mathrm{m}}$, i.e., trade-off parameter λ. With a small λ, $\nabla\Phi'_{\mathrm{m}}$ is not powerful enough to offset the decay of $\mathbf{F}^{\mathrm{T}}$ in $\nabla\Phi_{\mathrm{d}}$, so the recovered model may be shallower than the true model. While with an overlarge λ, the recovered model may be deeper than the true model. What's worse, as $\nabla\Phi'_{\mathrm{m}}$ decays quickly with increasing depth, its effect on deep cells is weakened.

Another solving method is to directly apply a depth weighting function to $\nabla\Phi_{\mathrm{d}}$ instead of Φ_{m} in a mathematically less rigorous way [21]. In this way, the gradient of objective function can be expressed as

$$\nabla\Phi = \nabla\Phi'_{\mathrm{d}} + \lambda\nabla\Phi_{\mathrm{m}} = \mathbf{W}_z\mathbf{F}^{\mathrm{T}}(\mathbf{Fm} - \mathbf{d}_{\mathrm{obs}}) + \lambda\nabla\Phi_{\mathrm{m}} \tag{19}$$

where $\mathbf{W}_z$ is a depth weighting function that increases quickly as depth increases, which is just adverse to the decay of $\mathbf{F}^{\mathrm{T}}$ in $\nabla\Phi_{\mathrm{d}}$. Compared with Equation (18), $\nabla\Phi_{\mathrm{d}}$ is multiplied by $\mathbf{W}_z$ in Equation (19) instead of a depth weighted Φ'_{m}. Therefore, $\mathbf{W}_z$ can counteract the effect of $\mathbf{F}^{\mathrm{T}}$ in $\nabla\Phi_{\mathrm{d}}$ directly and effectively. Besides, different with $\nabla\Phi'_{\mathrm{m}}$, $\nabla\Phi_{\mathrm{m}}$ is not related to depth, which means that λ would have less effect on the improvement of depth resolution as it does in Equation (18).

From the perspective of objective function gradient, above analyses indicate that the gradient-based strategy in Equation (19) is more effective to improve depth resolution than the method shown in Equation (18).

4. Results

To demonstrate the effectiveness of our algorithm, we tested both on a synthetic and a field dataset from the Vinton salt dome by Bell Geospace [29] and compared the improved FCM clustering inversion with focusing inversion and conventional FCM clustering inversion in the examples. In the following examples, the open-source software TetGen [25] is used to generate tetrahedral grids. The conjugate gradient method is implemented to solve the gravity inverse problems.

4.1. Inversions of Simulation Data

The synthetic gravity data were generated for a 3D homogeneous structure with a relative density of 1 g/cm^3, as shown in Figure 3. The top bury depth is 100 m. The gravity data were calculated at the Earth's surface on a grid with 50 m spacing and a total of 320 locations. We discretized the survey area (1000 m $\times$ 800 m $\times$ 600 m) into 237,298 tetrahedral grids with maximum cell volume of 5000 m^3 when using the program TetGen [25]. Two depth weighting strategies with different inversion methods were respectively applied to recover the synclinal structure. A uniform relative density model of 0 g/cm^3 was used as the initial model. A constraint on the upper-lower limit of density [0 g/cm^3, 1.0 g/cm^3] was used in these experiments. We carried out the FCM clustering gravity inversion with $\mathbf{C}$ = {0 g/cm^3, 1 g/cm^3} and p = 2. The inversion was taken as converged when the data misfit or relative model norm update was below their respective predetermined thresholds.

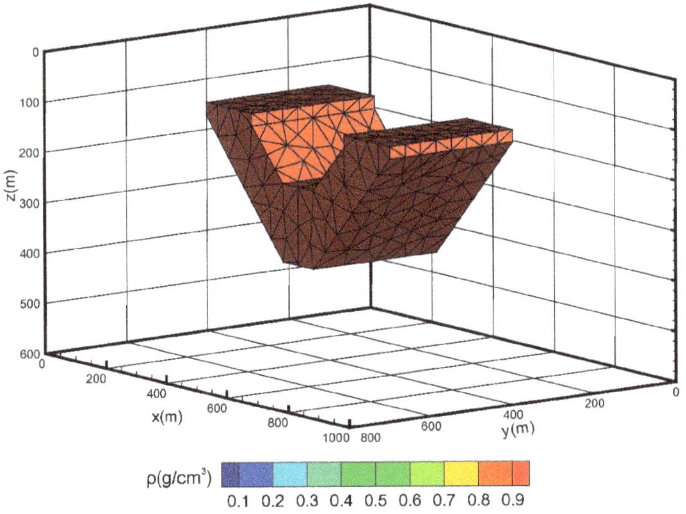

Figure 3. A V-shaped structure with a relative density of 1 g/cm^3 on unstructured grids.

The inversion results of two weighting strategies without regularization terms (λ = 0) are shown in Figure 4, where the section passing the center in y-direction is shown only. The dashed boxes in the figure outline the true boundaries of the anomalous body. With no model structure constraints, the result with a conventional strategy in Equation (16) (as shown in Figure 4a) demonstrates a concentration near the surface, while it is improved by using gradient-based strategy in Equation (17) (as shown in Figure 4b). Then using different λ, we implemented smooth inversions, and the improved FCM clustering inversions with two strategies. Respectively, the results of two inversions are respectively shown in Figures 5 and 6. It should be pointed out that, in order to better perform the influence of λ, we set λ as a constant in different inversions. In FCM clustering inversion, λ is a constant. From Figures 4–6, one can find that the depth resolution improvement of the conventional strategy is closely related to the regularization parameter. Increasing λ can improve the depth resolution to some extent, while an improper λ will result in an incorrect depth, such that a shallow model was yielded with a small λ (in Figures 5a and 6a). What's more, since the regularization terms decay as depth increases, model constraints are weakened and the inverted models become blurred and fuzzy at depth in FCM clustering inversions (as shown in Figure 5a,b and Figure 6a,b). When λ further increases, the data fit becomes deteriorated. On the contrary, in Figure 5c,d and Figure 6c,d, the model is well recovered and the top depth is well determined with gradient-based strategy in two kinds of inversion methods. It indicates that the depth resolution is less related to the regularization parameter. Besides, the constraints of regularization terms are not weakening when depth increases. These results correspond with the above analyses of depth resolution.

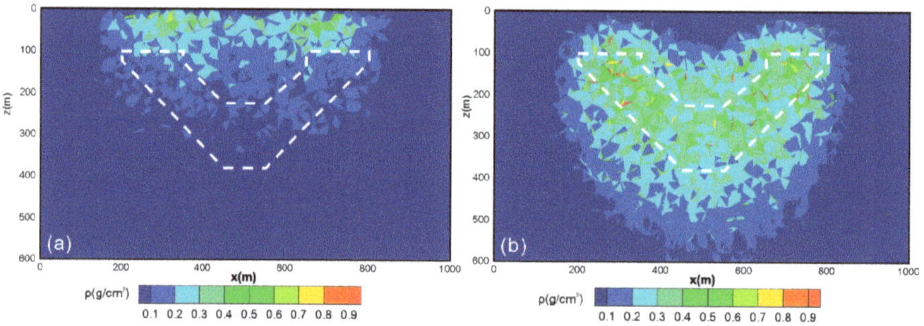

Figure 4. Inversion results with λ = 0: (**a**) conventional strategy, and (**b**) gradient-based strategy.

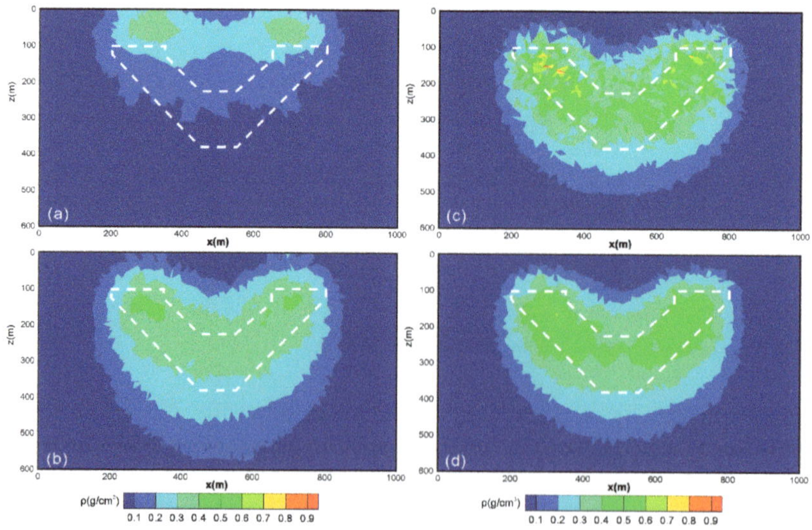

Figure 5. Smooth inversion results. (**a**,**b**) are conventional strategy with $\lambda = 1 \times 10^{-2}$ and $\lambda = 5 \times 10^{-2}$; (**c**,**d**) are gradient-based strategy with $\lambda = 1 \times 10^{1}$ and $\lambda = 5 \times 10^{1}$.

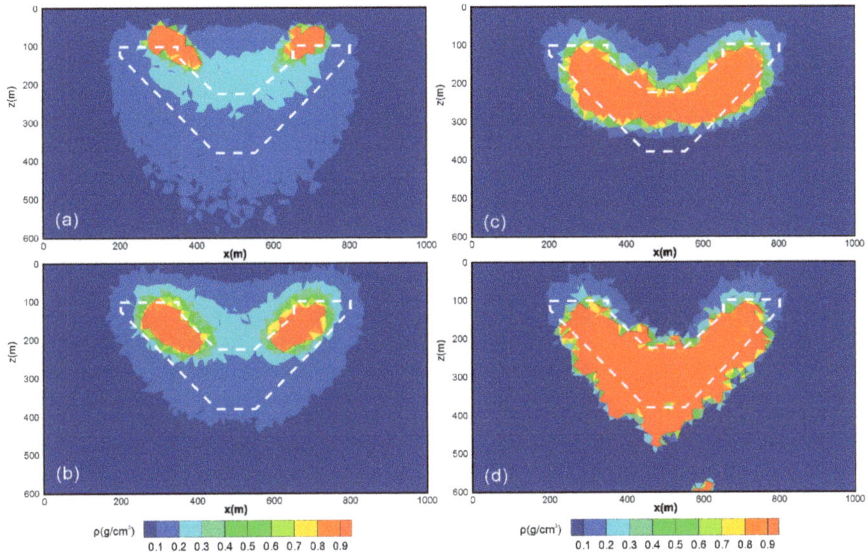

Figure 6. New FCM clustering inversion results: (**a**,**b**) are conventional strategy with $\lambda = 1 \times 10^{-3}$ and $\lambda = 3 \times 10^{-3}$; (**c**,**d**) are gradient-based strategy with $\lambda = 3 \times 10^{0}$ and $\lambda = 5 \times 10^{0}$.

In Figure 7, focusing inversion and conventional FCM clustering inversion are implemented with gradient-based strategy to compare with the improved FCM clustering inversion. We set a large initial λ, and then used the c to gradually reduce it. One can find that all these three methods can reconstruct a compact model. However, from Figure 7a,b, it is shown that focusing inversion and conventional FCM clustering inversion yield the inverted models into segments or with holes on unstructured grids. Focusing inversion is to reconstruct the model with minimum volume, and it has no constraint in spatial domain. In conventional FCM clustering inversion, one may increase smooth constraint to yield a continuous model, while the constraint of FCM clustering is weakened and the inverted model will not be focusing and compact. But in Figure 6, the models are more compact and no holes inside with the improved FCM clustering method. That is because the new method induces cells and their adjacent cells to tend to the same value except for the constraint of smooth regularization term.

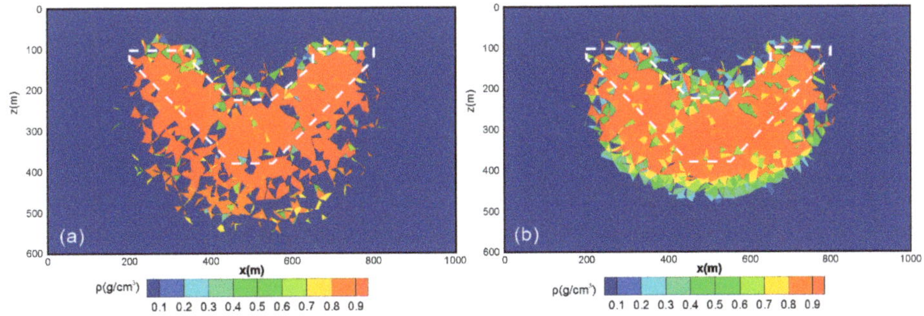

Figure 7. Inversion results with gradient-based strategy: (**a**) focusing inversion, and (**b**) conventional FCM clustering inversion.

Table 1 shows a clear perspective of inversion results with different methods, strategies, and trade-off parameters. The comparative analysis indicates that the second method would improve gravity depth resolution more directly and effectively, and this improvement is less influenced by regularization parameters. Moreover, the regularization terms will have a restraining effect at all depths.

Table 1. Inversion result list with different methods, strategies, and parameters.

Result	Inversion Method	Strategy	Trade-Off Parameter
Figure 4a	NaN	Conventional strategy	0
Figure 4b	NaN	Gradient-based strategy	0
Figure 5a	Smooth inversion	Conventional strategy	1×10^{-2}
Figure 5b	Smooth inversion	Conventional strategy	5×10^{-2}
Figure 5c	Smooth inversion	Gradient-based strategy	1×10^{1}
Figure 5d	Smooth inversion	Gradient-based strategy	5×10^{1}
Figure 6a	New FCM inversion	Conventional strategy	1×10^{-3}
Figure 6b	New FCM inversion	Conventional strategy	3×10^{-3}
Figure 6c	New FCM inversion	Gradient-based strategy	3×10^{0}
Figure 6d	New FCM inversion	Gradient-based strategy	5×10^{0}
Figure 7a	Focusing inversion	Gradient-based strategy	method of Lelièvre et al. [17]
Figure 7b	Conventional FCM inversion	Gradient-based strategy	method of Lelièvre et al. [17]

4.2. Field Data Inversion

The survey area Vinton Dome is located in southwestern Louisiana, USA. We used a subset of the survey area for this study that covers approximately 16 km^2 in the middle of the survey area, which was acquired and processed by Bell Geospace in 2008 [29]. Before we inverted the data, a second-order trend surface was removed to get rid of background, and an upward continuation of 100 m was done to eliminate the shallow local anomalies. The processed gravity data is displayed in Figure 8.

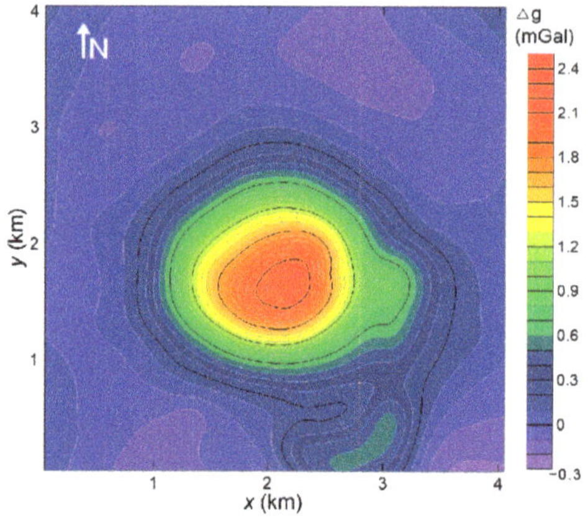

Figure 8. Field data subset from Vinton Dome area after trend surface removing and upward continuation (data courtesy of Scott Hammond from Bell Geospace).

Inversion area is 4 km × 4 km × 1 km with 169,949 cells and 1681 survey stations. The station interval was 100 m. The upper-lower limit of density was assumed to be [0 g/cm^3, 0.55 g/cm^3] and clusters $\mathbf{C}$ = {0 g/cm^3, 0.55 g/cm^3} was used in our inversions. Smooth inversion, conventional and improved FCM clustering inversion are implemented with a gradient-based strategy. We specifically point the *x*-, *y*-and *z*-axis to the East, North, and vertically downwards. Figures 9–11 show the slices of the recovered model on x-y-, x-z-, and y-z-planes of different inversion methods.

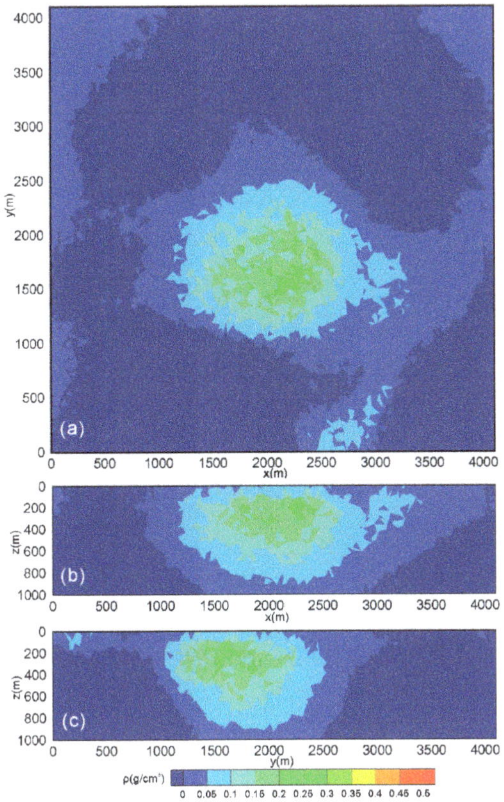

Figure 9. Smooth inversion result of Vinton Dome on (**a**) x-y-plane at z = 0.3 km; (**b**) x-z-plane at y = 1.7 km; (**c**) y-z-plane at x = 2.2 km.

Many surveys and researches have been done at the Vinton Dome. According to Coker et al. [30], the Vinton salt dome is composed of a well-defined massive cap rock extending above the salt rock. This cap rock is formed by gypsum and anhydrite that is embedded in sediments characterized by intercalated layers of sandstone and shale.

These researches indicate that the cap rock has a top depth of 160 m and a density of 2.75 g/cm^3, while the underlying media have a density of 2.20 g/cm^3 [31]. Besides, the sediment around slat dome has a density of about 2.0 g/cm^3–2.37 g/cm^3. Thus, we assumed the relative density of cap rock to be 0.55 g/cm^3, and the observed gravity data were predominantly caused by the cap rock. From previous work in this area [30–33], it is known that the extension of the anomalous body is about 1200 m in the west-east direction, 900 m in the north-south direction, 200 m in z-direction, and the top depth is about 175 m.

The result of smooth inversion is blur and fuzzy in Figure 9, and the 3D extension size of the cap rock is much larger than the real geological model. In both Figures 10 and 11, the maximum relative density is 0.55 g/cm^3, and a more compact model is yielded. The extensions of the anomalous body are well consistent with the known information. But it is more internal continuous by using the improved FCM clustering inversion in Figure 11.

In brief, the results show that the problem in gravity inversion with the anomaly concentration to the surface is largely solved. The inverted models show sharp boundaries and true positions.

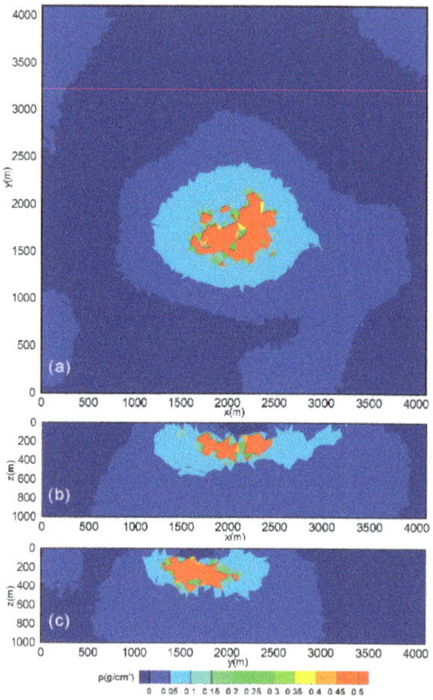

Figure 10. FCM clustering inversion result of Vinton Dome on (**a**) x-y-plane at z = 0.3 km; (**b**) x-z-plane at y = 1.7 km; (**c**) y-z-plane at x = 2.2 km.

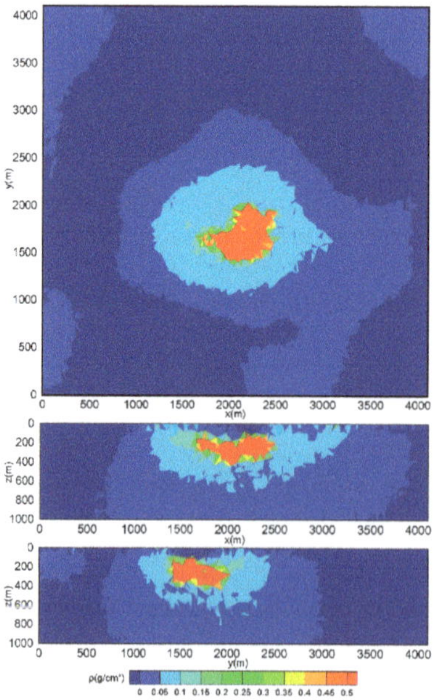

Figure 11. Improved FCM clustering inversion result of Vinton Dome on (**a**) x-y-plane at z = 0.3 km; (**b**) x-z-plane at y = 1.7 km; (**c**) y-z-plane at x = 2.2 km.

5. Conclusions

In gravity inversion based on unstructured grids, the dual effects of volume and depth difference of grids would result in a discontinuous and hollow model. However, the local refinement or coarsening of the mesh, an important advantage of unstructured grids, will inevitably lead to different mesh volumes. So, it's crucial to solve this problem before gravity inversion based on unstructured grids used widely in field data.

From the view of the objective function gradient, we analyzed the cause of poor depth resolution and two strategies of depth weighting in gradient-based inversion methods. It indicated that the key to improve the depth resolution is how to counteract the decay of kernels in the data misfit gradient. The data-based strategy can improve the depth resolution more directly and effectively than the conventional strategy.

Besides an improved FCM clustering combined with spatial constraints is developed to obtain a continuous density distribution in the spatial domain and a focused property distribution in the parameter domain. It will reconstruct a more continuous inverted model on unstructured grids. With different regularization parameters, two weighting strategies are compared by applying smooth and the new FCM clustering inversions on synthetic data sets. The results show that the depth resolution of the gradient-based strategy is independent of regularization parameters and regularization terms does not weaken with depth increases. Compared with focusing inversion and conventional FCM clustering inversion, the improved FCM clustering inversion can provide a more continuous model.

At last, we implemented the gradient-based strategy and the improved FCM clustering inversion on field data sets. The results show that a compact model can be better consistent with the known information than smooth inversion. But it is more internal continuous by using the improved FCM clustering inversion.

Author Contributions: Conceptualization, S.S. and C.Y.; methodology, validation, and visualization S.S. and X.G.; writing—original draft preparation, S.S.; writing—review and editing, C.Y. and X.G.; supervision, C.Y.; project administration, C.Y. All authors have read and agreed to the published version of the manuscript.

Funding: This research was financially supported by China Postdoctoral Science Foundation (2020M670601), the China Geological Survey Project (DD20189143) and the Basic Scientific Research Funds of the key Laboratory of Aero Geophysics and Remote Sensing Geology, Ministry of Natural Resources (2020YFL11).

Institutional Review Board Statement: Not applicable.

Informed Consent Statement: Not applicable.

Data Availability Statement: Restrictions apply to the availability of field data. Data was obtained from Bell Geospace and are available from Siyuan Sun with the permission of Scott Hammond from Bell Geospace.

Acknowledgments: We would like to thank Scott Hammond from Bell Geospace for providing their field data from Vinton salt dome.

Conflicts of Interest: The authors declare no conflict of interest.

Appendix A

As shown in Equation (1), the gravity anomalies cause by a tetrahedral is equal to the sum of the contributions from six edges. Taking the edge Q_1Q_2 in Figure 1 for example, we can describe thrice rotations as follows:

1. Rotate counterclockwise the x- and y-axes around the z-axis in Figure A1a by the angle φ until the rotated x-axis is coincident with the direction of the outward normal onto the x-y-plane.
2. Rotate counterclockwise the z- and x-axes around the y-axis by the angle ψ until the rotated z-axis is coincident with the outward normal direction of the triangle facet

($Q_1Q_2Q_3$). With previous two rotations, we obtain the target coordinate system (X, Y, Z). Thus, the polygon is in the X-Y-plane, as shown in Figure A1b.

3. Rotate counterclockwise the X-and Y-axes around the *Z*-axis by the angle θ until the rotated *Y*-axis is coincident with the direction of the outward normal on the edge Q_1Q_2 (as shown in Figure A1c).

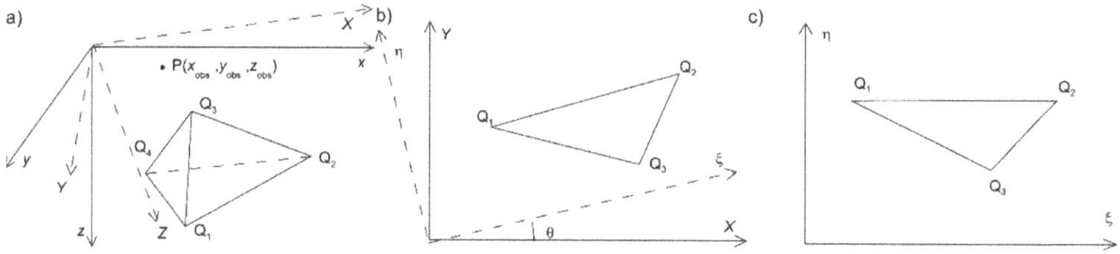

Figure A1. Process of Rotations for calculating contribution of Q1Q2: (**a**) Sketch before rotations; (**b**) Result of twice rotations; (**c**) Result of thrice rotations.

Refer to Figure A1c, $Q_1(x_1, y_1, z_1)$, $Q_2(x_2, y_2, z_2)$, and $Q_3(x_3, y_3, z_3)$ are three vertices of the triangle facet. For the three vertices, they can be either clockwise or counterclockwise in the X-Y-plane, because the grid generator stores these vertices in random order. However, the algorithm described in Section 2.1 require these vertices to be clockwise.

We use the following method to judge the rank and rearrange vertices. the normal vector **n** of the triangle facet is expressed as

$$\mathbf{n} = \begin{vmatrix} \vec{i} & \vec{j} & \vec{k} \\ x_2 - x_1 & y_2 - y_1 & z_2 - z_1 \\ x_3 - x_1 & y_3 - y_1 & z_3 - z_1 \end{vmatrix} \tag{A1}$$

$Q_4(x_4, y_4, z_4)$ is the last point of tetrahedral but outside the X-Y-plane. Define *t* as the scalar product of the normal vector n and vector $\overrightarrow{Q_1Q_4} = (x_4 - x_1, y_4 - y_1, z_4 - z_1)$, i.e.,

$$t = \mathbf{n} \cdot \overrightarrow{Q_1Q_4} \tag{A2}$$

When $t > 0$, it means that **n** and $\overrightarrow{Q_1Q_4}$ are in the same direction, the three vertices are clockwise. When $t < 0$, however, the three vertices are counterclockwise. In this case, we need to rearrange the three input vertices to make them clockwise. After rearrangement, calculate the contribution of Q_1Q_2 as described in previous Section 2.1.

Repeat the above process for every edge, and then we can obtain the gravity anomalies of a certain tetrahedral.

References

1. Talwani, M.; Ewing, M. Rapid computation of gravitational attraction of three-dimensional bodies of arbitrary shape. *Geophysics* **1960**, *25*, 203–225. [CrossRef]
2. Talwani, M. Computation with the help of a digital computer of magnetic anomalies caused by bodies of arbitrary shape. *Geophysics* **1965**, *30*, 797–817. [CrossRef]
3. Barnett, C.T. Theoretical modeling of the magnetic and gravitational fields of an arbitrarily shaped three-dimensional body. *Geophysics* **1976**, *41*, 1353–1364. [CrossRef]
4. Okabe, M. Analytical expressions for gravity anomalies due to homogeneous polyhedral bodies and translations into magnetic anomalies. *Geophysics* **1979**, *44*, 730–741. [CrossRef]
5. Zhang, J.; Wang, C.; Shi, Y.; Cai, Y.; Chi, W.; Dreger, W.; Cheng, W.; Yuan, Y. Three-dimensional crustal structure in central Taiwan from gravity inversion with a parallel genetic algorithm. *Geophysics* **2004**, *69*, 917–924. [CrossRef]
6. Cai, Y.; Wang, C. Fast finite-element calculation of gravity anomaly in complex geological regions. *Geophys. J. Int.* **2005**, *162*, 696–708. [CrossRef]

7. May, D.A.; Knepley, M.G. Optimal, scalable forward models for computing gravity anomalies. *Geophys. J. Int.* **2011**, *187*, 161–177. [CrossRef]
8. Jahandari, H.; Farquharson, C.G. Forward modeling of gravity data using finite-volume and finite-element methods on unstructured grids. *Geophysics* **2013**, *78*, G69–G80. [CrossRef]
9. Li, Y.; Oldenburg, D.W. 3-D inversion of gravity data. *Geophysics* **1998**, *63*, 109–119. [CrossRef]
10. Boulanger, O.; Chouteau, M. Constraints in 3D gravity inversion. *Geophys. Prospect.* **2001**, *49*, 265–280. [CrossRef]
11. Zhdanov, M.S.; Ellis, R.; Mukherjee, S. Three-dimensional regularized focusing inversion of gravity gradient tensor component data. *Geophysics* **2004**, *69*, 925–937. [CrossRef]
12. Silva, D.F.J.; Barbosa, V.C.; Silva, J.B. 3D gravity inversion through an adaptive-learning procedure. *Geophysics* **2009**, *7*, I9–I21. [CrossRef]
13. Tikhonov, A.N.; Arsenin, V.Y. *Solution of Ill-Posed Problems*; Wiley: New York, NY, USA, 1977; p. 258.
14. Li, Y.; Oldenburg, D.W. 3-D inversion of magnetic data. *Geophysics* **1996**, *61*, 394–408. [CrossRef]
15. Last, B.J.; Kubik, K. Compact gravity inversion. *Geophysics* **1983**, *48*, 713–721. [CrossRef]
16. Portniaguine, O.; Zhdanov, M.S. 3-D magnetic inversion with data compression and image focusing. *Geophysics* **2002**, *67*, 1532–1541. [CrossRef]
17. Lelièvre, P.G.; Farquharson, C.G.; Hurich, C.A. Joint inversion of seismic traveltimes and gravity data on unstructured grids with application to mineral exploration. *Geophysics* **2012**, *77*, K1–K15. [CrossRef]
18. Sun, J.; Li, Y. Multidomain petrophysically constrained inversion and geology differentiation using guided fuzzy c-means clustering. *Geophysics* **2015**, *80*, ID1–ID18. [CrossRef]
19. Sun, J.; Li, Y. joint inversion of multiple geophysical data using guided fuzzy c-means clustering. *Geophysics* **2016**, *81*, ID37–ID57. [CrossRef]
20. Singh, A.; Sharma, S. *Modified Zonal Cooperative Inversion of Gravity Data-a Case Study from Uranium Mineralization*; SEG Technical Program Expanded Abstracts: Houston, TX, USA, 2017; pp. 1744–1749.
21. Commer, M. Three-dimensional gravity modelling and focusing inversion using rectangular meshes. *Geophys. Prospect.* **2011**, *59*, 966–979. [CrossRef]
22. Liu, Y.P.; Wang, Z.W.; Du, X.J.; Liu, J.H.; Xu, J.S. 3D constrained inversion of gravity data based on the Extrapolation Tikhonov regularization algorithm. *Chin. J. Geophys.* **2013**, *56*, 1650–1659. (In Chinese)
23. Ren, Z.; Chen, C.; Pan, K.; Kalscheuer, T.; Maurer, H.; Tang, J. Gravity Anomalies of Arbitrary 3D Polyhedral Bodies with Horizontal and Vertical Mass Contrasts. *Surv. Geophys.* **2017**, *38*, 479–502. [CrossRef]
24. Zhao, G.; Chen, B.; Chen, L.; Liu, J.; Ren, Z. High-accuracy 2D and 3D Fourier forward modeling of gravity field based on the Gauss-FFT method. *J. Appl. Geophys.* **2018**, *150*, 294–303. [CrossRef]
25. Si, H. Tetgen: A Quality Tetrahedral Mesh Generator and a 3D Delaunay Triangulator. 2009. Available online: http://wias-berlin.de/software/index.jsp?id=TetGen&lang=1 (accessed on 25 September 2017).
26. Hathaway, R.J.; Bezdek, J.C. Fuzzy c-means clustering of incomplete data: IEEE Transactions on Systems, Man, and Cybernetics, Part B. *Cybernetics* **2001**, *31*, 735–744. [PubMed]
27. Fedi, M.; Rapolla, A. 3-D inversion of gravity and magnetic data with depth resolution. *Geophysics* **1999**, *64*, 452–460. [CrossRef]
28. Li, Y.; Oldenburg, D.W. Joint inversion of surface and three-component borehole magnetic data. *Geophysics* **2000**, *65*, 540–552. [CrossRef]
29. Dickinson, J.L.; Brewster, J.R.; Robinson, J.W.; Murphy, C.A. Imaging techniques for full tensor gravity gradiometry data. In Proceedings of the 11th SAGA Biennial Technical Meeting and Exhibition, Swaziland, South Africa, 16–18 September 2009. cp-241-00019.
30. Coker, M.O.; Bhattacharya, J.P.; Marfurt, K.J. Fracture patterns within mudstones on the flanks o f a salt dome: Syneresis or slumping? *Gulf Coast Assoc. Geol. Soc.* **2007**, *57*, 125–137.
31. Ennen, C. Mapping Gas-Charged Fault Blocks around the Vinton Salt Dome, Louisiana Using Gravity Gradiometry Data. Master's Thesis, University of Houston, Houston, TX, USA, 2012.
32. Salem, A.; Masterton, S.; Campbell, S.; Dickinson, J.D.; Murphy, C. Interpretation of tensor gravity data using an adaptive tilt angle method. *Geophys. Prospect.* **2013**, *61*, 1065–1076. [CrossRef]
33. Thompson, S.A.; Eichelberger, O.H. Vinton salt dome, Calcasieu Parish, Louisiana. *AAPG Bull.* **1928**, *12*, 385–394.

applied sciences

MDPI

Article

Prediction of Shale Gas Reservoirs Using Fluid Mobility Attribute Driven by Post-Stack Seismic Data: A Case Study from Southern China

Jing Zeng [1,*], Alexey Stovas [2], Handong Huang [1,*], Lixia Ren [3] and Tianlei Tang [4]

1 State Key Laboratory of Petroleum Resources and Prospecting, China University of Petroleum-Beijing (CUPB), Beijing 102249, China
2 Department of Geoscience and Petroleum, Norwegian University of Science and Technology (NTNU), 7031 Trondheim, Norway; alexey.stovas@ntnu.no
3 Institute of Geology, Second Oil Production Plant, North China Oilfield Branch, PetroChina Company Limited, Langfang 065700, China; cy2_dzrlx@petrochina.com.cn
4 Fourth Oil Production Plant, North China Oilfield Company, PetroChina Company Limited, Langfang 065000, China; cy4_ttl@petrochina.com.cn
* Correspondence: 2017310201@cup.edu.cn (J.Z.); hhd@cup.edu.cn (H.H.)

Featured Application: This article demonstrates the seismic fluid mobility attribute technology based on a modified matching pursuit (MP) scheme to determine shale gas reservoir.

Abstract: Paleozoic marine shale gas resources in Southern China present broad prospects for exploration and development. However, previous research has mostly focused on the shale in the Sichuan Basin. The research target of this study is expanded to the Lower Silurian Longmaxi shale outside the Sichuan Basin. A prediction scheme of shale gas reservoirs through the frequency-dependent seismic attribute technology is developed to reduce drilling risks of shale gas related to complex geological structure and low exploration level. Extracting frequency-dependent seismic attribute is inseparable from spectral decomposition technology, whereby the matching pursuit algorithm is commonly used. However, frequency interference in MP results in an erroneous time-frequency (TF) spectrum and affects the accuracy of seismic attribute. Firstly, a novel spectral decomposition technology is proposed to minimize the effect of frequency interference by integrating the MP and the ensemble empirical mode decomposition (EEMD). Synthetic and real data tests indicate that the proposed spectral decomposition technology provides a TF spectrum with higher accuracy and resolution than traditional MP. Then, a seismic fluid mobility attribute, extracted from the post-stack seismic data through the proposed spectral decomposition technology, is applied to characterize the shale reservoirs. The application result indicates that the seismic fluid mobility attribute can describe the spatial distribution of shale gas reservoirs well without well control. Based on the seismic fluid mobility attribute section, we have learned that the shale gas enrich areas are located near the bottom of the Longmaxi Formation. The inverted velocity data are also introduced to further verify the reliability of seismic fluid mobility. Finally, the thickness map of gas-bearing shale reservoirs in the Longmaxi Formation is obtained by combining the seismic fluid mobility attribute with the inverted velocity data, and two favorable exploration areas are suggested by analyzing the thickness, structure, and burial depth. The present work can not only be used to evaluate shale gas resources in the early stage of exploration, but also help to design the landing point and trajectory of directional drilling in the development stage.

Keywords: shale gas; seismic fluid mobility; matching pursuit; ensemble empirical mode decomposition

1. Introduction

The recoverable resources of shale gas in China are about 25.1×10^{12} m^3 [1] with the technological innovations of directional drilling and hydraulic fracturing. The Paleozoic marine shale in southern China provides the largest shale gas potential in China [2, 3]. At present, almost all commercial shale gas production in China comes from the Paleozoic marine shale in Sichuan Basin, southern China [4]. Previous research has also predominantly focused on the Sichuan Basin [5–8]. To meet the growing demand for shale gas, it is necessary to expand the exploration scope beyond the Sichuan Basin and find new shale gas producing areas. The western Hunan-Hubei (WHH) fold belt, adjacent to the Sichuan Basin, contains rich black marine shale. However, the strong heterogeneous reservoir, complex geological structure, and low exploration degree in the WHH fold belt greatly increase the investment risks of shale gas drilling. Seismic attribute technology is a reliable method for describing the spatial distribution of hydrocarbon reservoirs, especially in the case with few drilling data [9,10]. It can help design the direction and trajectory of directional drilling, reduce the exploration risk, and optimize the production.

Seismic attributes commonly used for hydrocarbon detection include two categories. One is the seismic attribute based on rock physics theory, such as amplitude versus offset (AVO) attribute estimated from the pre-stack data through AVO inversion [11]. Nevertheless, in areas lacking enough well controls, AVO attribute may not be adequate to assist hydrocarbon identification [12]. The other is the seismic attribute based on seismic attenuation theory, such as low-frequency shadow [13] and high-frequency attenuation [14] calculated from post-stack data. Compared with the pre-stack attribute, the post-stack one has the advantages of no need for strict well control, low computational cost, and high signal-to-noise ratio [15]. However, the post-stack attributes mentioned above can directly reflect the magnitude of seismic attenuation, but not always directly reflect the property of hydrocarbon reservoir. It is necessary to find a universal seismic attribute indicating hydrocarbon directly. In recent years, some post-stack fluid mobility attributes have been applied to hydrocarbon identification in deep-water sandstone reservoirs [7,10,15,16]. Fluid mobility is the direct reflection of the rock permeability and fluid viscosity of hydrocarbon reservoirs. Silin et al. [17] and Goloshubin and Silin [18] approximated the low-frequency seismic reflection coefficient as a function of fluid mobility in fractured reservoir. Therefore, we extract fluid mobility from low-frequency seismic reflection data to directly indicate gas in (fractured) shale reservoir.

Extracting information from low-frequency seismic data is inseparable from spectral decomposition technology [19], which is a powerful tool for obtaining time-frequency (TF) spectrum of seismic data. The bilinear Wigner–Ville distribution (WVD) can achieve a higher-resolution TF localization than the linear spectral decomposition techniques [20], but suffers from the notorious cross-term interferences. The integration of WVD and MP algorithm can produce a stable high-solution TF spectrum without cross-term interferences [10,21]. MP is a greedy algorithm for compressive sensing [22]. However, the traditional MP algorithm is difficult to determine the precise TF locations of seismic signal due to the interference effect between different seismic frequency contents [23]. More specifically, the seismic low-frequency contents of interest may be suppressed by the seismic high-frequency contents during the process of MP algorithm, which may result in an average effect in the final TF spectrum and thereby probably reduces the accuracy of hydrocarbon detection. EEMD is a nonlinear technique for adaptively decomposing a seismic signal into a finite number of narrow-band signals [24]. Performing MP over the narrow-band signals can reduce the effect of frequency interference. Therefore, a spectral decomposition technology without frequency interference is proposed by modifying MP based on EEMD.

In this paper, the spatial distribution of shale gas reservoirs in complex structural areas lacking well control is predicted by using the fluid mobility attribute based on post-stack seismic data and spectral decomposition technique without frequency interference. We first propose a new spectral decomposition technology by modifying MP based on

EEMD, and the synthetic data test shows that the new technology reduces the interference of different frequency content and obtains more accurate spectral information, especially in the low frequency domain. We then define a seismic fluid mobility attribute to detect shale gas by extracting fluid mobility measurements from the low-frequency post-stack seismic data, and real data application shows that the seismic fluid mobility attribute can describe the spatial distribution of shale gas reservoirs well. We also introduce inverted velocity to further verify the reliability of the seismic fluid mobility attribute. Finally, we predict the thickness map of shale gas reservoirs in study area by integrating seismic fluid mobility attribute and velocity data. This work not only can help design the direction and trajectory of the directional drilling, but can also be used for resource evaluation of shale gas reservoirs in the study area.

2. Target Overview

Commercial gas production has been achieved from the Lower Silurian Longmaxi shale in the eastern Sichuan Basin (ESB) [4,25]. In this paper, the research target is the Lower Silurian Longmaxi shale in the WHH fold belt, adjacent to the ESB (Figure 1a). Structurally, the WHH fold belt is composed of a series of anticlinoria and synclinoria formed by multistage tectonic deformation. Compared with eastern Sichuan Basin, the WHH fold belt belongs to a tectonically active area, which results in the strong heterogeneity of shale gas distribution [26]. The Longmaxi Formation in both the WHH and ESB areas is integrated on top of the nodular limestone of the Ordovician Baota Formation and under the yellow-green shale of the Luoreping Formation, which indicates their sealing capacities to be comparable [27]. The sedimentary environment in both WHH and ESB areas in the early Silurian was shallow water semi-closed retention basin, which made that the WHH and ESB Longmaxi shales have similar mineral composition, such as clay minerals accounting for 34.7% and 34.6% on average, feldspar accounting for 7.3% and 8.3% on average, and quartz accounting for 56.5% and 44.4% on average, respectively [25,27]. Field investigations and laboratory measurements have confirmed that the hydrocarbon generation ability and organic matter maturity of the Longmaxi shale in the WHH area are comparable to those in the ESB area [27]. The thicknesses of black Longmaxi shales with total organic carbon (TOC) greater than 2% range from 15 to 30 m in the WHH area, and these in the ESB area range from 20 to 50 m. The average kerogen reflectance of Longmaxi shales in the WHH and ESB areas are 2.68% and 2.69%, respectively, which indicates that they are both in the high maturity stage. In addition, the burial depths of the Longmaxi Formation in the WHH area range from 0 to 3300 m, which is basically in line with the favored burial depths (100–3500 m) for shale gas production [28]. Therefore, Longmaxi shale gas in the WHH fold belt is of great potential to exploration.

Specifically, the study area (blue rectangle in Figure 1a), covering an area of about 1189.7 km^2, is located between the Yidu-Hefeng anticlinorium and the Sangzhi-Shimen synclinorium. Figure 1b is the elevation plan of the bottom interface of Longmaxi Formation in study area. Only the target formation in the northwest and middle of study area is retained due to erosion. The retained formation is characterized by well-developed faults and high-steep structures (elevation range −2500–0 m). Owing to erosion and fluctuating terrain, the surface elevation of study area varies dramatically between 0 and 1300 m, which is not conducive to three-dimension (3D) seismic acquisition. Now the study area is covered by twelve 2D seismic lines (dashed lines L1–L8, C1–C4 in Figure 1b). Six wells are drilled through the Longmaxi Formation in the study area, of which S-3 is a gas-free well, S-1 is a poor gas well, and the other four are gas wells (circles with different fill colors in Figure 1b). Deviated well SX-1 and vertical wells S-2, S-3, and S-4 are located near the seismic lines. Figure 2 displays the logging data of the vertical wells S-2, S-3, and S-4. The thickness of target formation is about 100 m, and its lithology is mainly carbonaceous shale with intercalations of argillaceous siltstone in the Longmaxi Formation. Here, shale is characterized by high acoustic time difference (AC) and high natural gamma (GR). Conversely, siltstone is characterized by low AC and low GR. As marked by red rectangles, the gas zones are located at the positions with high TOC value (about >2%). Therefore, TOC data can used as an indicator of a potential gas enrichment zone. TOC data from the vertical wells implies that the vertical gas enrichment zone is located near the bottom of Longmaxi Formation. Comparing the TOC data of three wells, we can also see that the lateral distribution of shale gas enrichment zones is different. The lateral gas enrichment zone in the Longmaxi Formation is the target of interest.

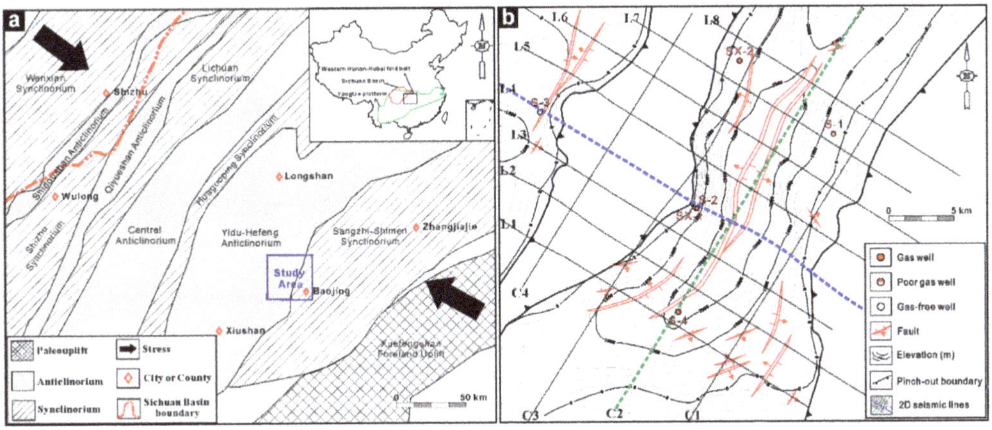

Figure 1. Location of the study area: the structural plan of Western Hunan-Hubei fold belt (**a**), and the elevation plan of Longmaxi Formation bottom interface in study area (**b**). The blue dashed line is the seismic line connecting wells S-2, S-3 and SX-1, and the green dashed line is the seismic line connecting well S-4.

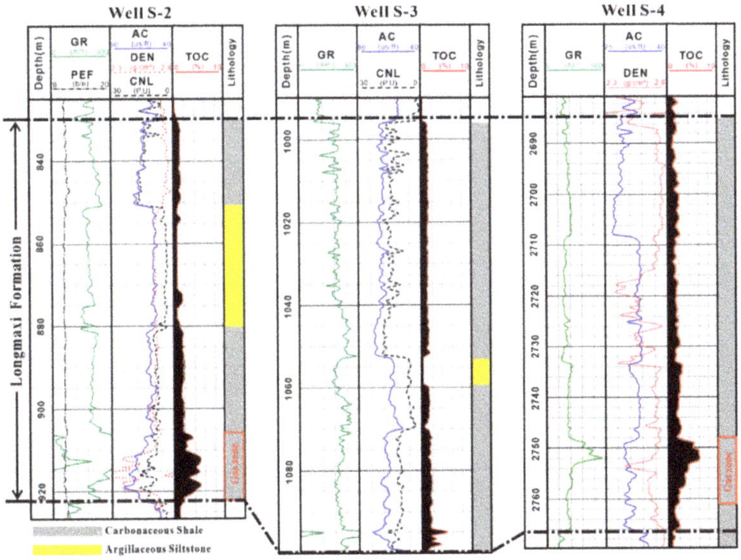

Figure 2. Logging data of wells S-2 (**left**), S-3 (**middle**) and S-4 (**right**). GR is the natural gamma curve, AC is the acoustic time difference curve, DEN is the density curve, CNL is the compensated neutron logging curve, PEF is the photoelectric absorption cross-section index curve, and TOC is the total organic carbon curve. The black dotted lines represent the interfaces of Longmaxi Formation, and the red rectangles denote the gas zones.

3. Methodology

3.1. Spectral Decomposition Technology without Frequency Interference

The key of using low-frequency seismic attribute to characterize hydrocarbon reservoir is the TF resolution of spectral decomposition algorithm. MP integrated with WVD is a high TF resolution spectral decomposition technique without the limit of Heisenberg's uncertainty principle, where MP is the crucial signal decomposition algorithm and WVD is the spectral calculation algorithm. MP decomposes a seismic signal into a linear superposition of a series of pre-defined wavelet atoms.

$$\mathbf{s} = \mathrm{Re}\left\{ \sum_{n=1}^{N} c_n \mathbf{m}_n + \mathbf{r}_N \right\}, \tag{1}$$

where $\mathbf{s}$ is the (real-valued) seismic signal, n is the iteration number, c_n is the complex coefficient at the nth iteration, $\mathbf{m}_n$ is the optimal wavelet atom at the nth iteration, $\mathbf{r}_N$ is the residual error at the Nth iteration, which is only noise and all significant features of seismic signal are recorded in the first term on the right-hand side of Equation (1), and Re{•} is the real part operator. The wavelet atom $\mathbf{m}_n$, containing the time and frequency features of seismic signal, is extracted from the atomic dictionary. Here, we choose the Morlet wavelet as the atom because it is suitable for frequency quantification and attenuation studies of seismic signal [21].

$$\mathbf{m}_n = \exp\left[-4\ln 2 f_n^2 k_n^{-2} (t-u_n)^2 + i2\pi f_n (t-u_n) \right], \tag{2}$$

where t is the time (s), u_n is the center time (s), k_n is the (dimensionless) scale, f_n is the center frequency (Hz), and i is the imaginary unit. The key parameters (u_n, k_n, f_n) of $\mathbf{m}_n$ are estimated by the following equations [10]

$$u_n = \arg\max_t [\mathbf{r}_{n-1}(t)]^{env}, \tag{3}$$

$$(k_n, f_n) = \arg \max_{k \in D_k, f \in D_f} \frac{|\mathbf{r}_{n-1}(t) \cdot \mathbf{m}_n(t, u_n, k, f)|}{\|\mathbf{m}_n(t, u_n, k, f)\|_2}, \tag{4}$$

where $\mathbf{r}_{n-1}$ is the residual error at the (n–1)th iteration, $\mathbf{r}_0 = \mathbf{s} + i\mathbf{s}^H$, superscript H is the Hilbert transform operator, superscript env is the envelope operator, • is the dot product operator, D_k and D_f is the searching interval of k_n and f_n, respectively. In general, D_k and D_f are given empirically, such as $D_k = [0.3, 6]$ and $D_f = [f_u - R_f, f_u + R_f]$, where f_u is instantaneous frequency of $\mathbf{r}_{n-1}$ at the time u_n, and R_f is the search radius of f_n, which is always set to 20 Hz [10,21]. The complex coefficient c_n, containing the amplitude and phase features of seismic signal, is expressed as follows

$$c_n = \left(\mathbf{m}_n^T \mathbf{m}_n + \varepsilon \right)^{-1} \mathbf{m}_n^T \mathbf{r}_{n-1}. \tag{5}$$

where superscript T is the transpose operator and ε is the damping constant.

According to the WVD, the TF spectrum of seismic signal is expressed as [10]

$$\mathbf{A}(t,f) = \sum_{n=1}^{N} \frac{4\left[(\mathrm{Re}\{c_n\})^2 + (\mathrm{Im}\{c_n\})^2 \right] \exp\left[-\frac{\pi^2}{2\ln 2} \frac{k_n^2}{f_n^2} (f-f_n)^2 - 8\ln 2 \frac{f_n^2}{k_n^2} (t-u_n) \right]}{1 + \exp\left(-\frac{\pi^2 k_n^2}{2\ln 2} \right) \cos\left(\arctan \frac{\mathrm{Im}\{c_n\}}{\mathrm{Re}\{c_n\}} \right)}, \tag{6}$$

where Im{•} is the imaginary part operator.

As follows from Equation (4), the accuracy of f_n affects the frequency resolution of MP algorithm. However, the interference of multi-frequency components in nonstationary seismic signals causes inaccuracy in searching f_n, and then introduce error in the TF spectrum. To reduce the effect of frequency interference, a modified MP scheme based on

EEMD (MP-EEMD) is proposed in this work. MP-EEMD mainly includes four stages. In stage 1, performing the EEMD algorithm over seismic signal yields

$$\mathbf{s} = \sum_{j=1}^{J} \boldsymbol{\varphi}_j + \boldsymbol{\eta}, \tag{7}$$

where j is the serial number of intrinsic mode function (IMF), $\boldsymbol{\varphi}_j$ is the jth IMF, and $\boldsymbol{\eta}$ is the residual. In stage 2, decomposed components with similar frequency-band characteristics are adaptively stacked according to the following principle

$$\boldsymbol{\psi}_l = \begin{cases} \boldsymbol{\varphi}_j + \boldsymbol{\varphi}_{j+1}, & \left(f_{j+1}^{max} - f_j^{min}\right) / \left(f_{j+1}^{max} - f_{j+1}^{min}\right) \geq p, \\ \boldsymbol{\varphi}_j, & otherwise, \end{cases} \tag{8}$$

where $\boldsymbol{\varphi}_{j+1} = \boldsymbol{\eta}$, $l = 1, 2, \ldots, L$ $(L \leq J + 1)$ is the serial number of the stacked components, $\boldsymbol{\psi}_l$ is the stacked component, $[f_j^{\min}, f_j^{\max}]$ is the effective frequency band of $\boldsymbol{\varphi}_j$, estimated through fast Fourier transform (FFT), and p is a constant threshold. Here, we set the threshold p to be 80%. In stage 3, the searching interval D_f^l for the stacked components $\boldsymbol{\psi}_l$ is set to be

$$D_f^l = \left[\max\left(f_l^{min}, f_l - r/2\right), \min(f_l^{max}, f_l + r/2)\right], \tag{9}$$

where f_l is the instantaneous frequency of $\boldsymbol{\psi}_l$ at the time u_n, $[f_l^{\min}, f_l^{\max}]$ is the effective frequency band of $\boldsymbol{\psi}_l$, and $r = f_l^{\max} - f_l^{\min}$ is the bandwidth of $\boldsymbol{\psi}_l$. In stage 4, we perform MP algorithm on the $\boldsymbol{\psi}_l$ to obtain the final TF spectrum. After EEMD and adaptive stack, the narrow frequency-band component $\boldsymbol{\psi}_l$ can reduce the effect of frequency interference of original seismic signal to some extent. It can be seen from Equation (9) that the searching interval D_f^l in MP-EEMD is determined by the signal characteristics of both local time (u_n) and local frequencies ($[f_l^{\min}, f_l^{\max}]$), which give more constraints to search for optimal frequency parameter f_n than MP.

To clearly illustrate the advantages of MP-EEMD, we design a synthetic signal with multi-frequency interference (blue in Figure 3a) by superimposing four wavelets W1–W4 with different frequency (black in Figure 3a). Figure 3b,c show the TF spectrum of the synthetic signal calculated by MP and MP-EEMD algorithms, respectively. As shown in Figure 3b, the spectral amplitude corresponding to the low-amplitude wavelet W1 (near dotted line) is vague and divergent due to frequency interference, and there are some significant deviations between the TF centers calculated by MP (white squares) and the true ones (black squares) of spectral amplitudes corresponding to the wavelets W2–W4. The larger the difference between interfering frequencies in synthetic signal, the larger the frequency deviation in TF spectrum. As shown in Figure 3c, the spectral amplitude corresponding to the weak wavelet W1 (near white dotted line) is clear and convergent, and the TF centers calculated by MP-EEMD (black points) are close to the true (white points) of spectral amplitudes corresponding to the wavelets W2–W4. Comparing Figure 3b,c, we can observe that the accurate and TF resolution of MP-EEMD is higher than that of MP.

Table 1. Parameters of the four wavelets (Figure 3a).

NO.	Wave Equations	Parameters
W1	$A\sin(2\pi ft)$	$A = 0.2, f = 10$ Hz
W2	$A\exp[-4\ln 2(f/k)^2(t-u)^2$ $+i2\pi f(t-u)]$	$A = 1, k = 1, f = 20$ Hz, $u = 0.1$ s
W3		$A = 1, k = 1, f = 40$ Hz, $u = 0.2$ s
W4		$A = 1, k = 1, f = 60$ Hz, $u = 0.3$ s

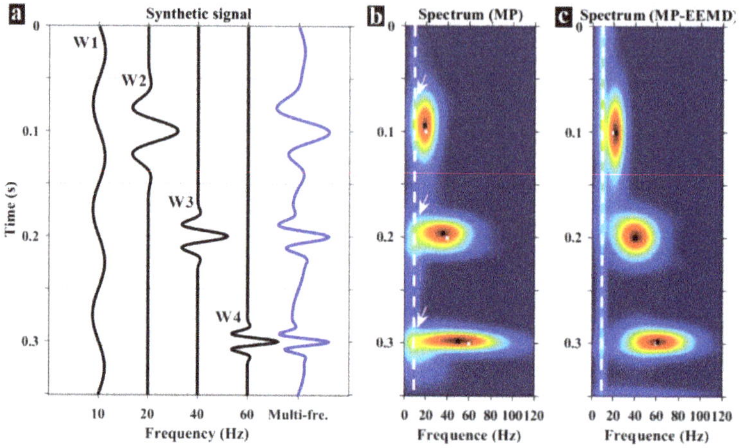

Figure 3. Synthetic signal (**a**), TF spectrum obtained by MP algorithm (**b**), and TF spectrum obtained by MP-EEMD algorithm (**c**). In panel (**a**), the synthetic signal (blue) is obtained by superimposing four wavelets W1–W4 (black), and the specific parameters of wavelets W1–W4 are shown in Table 1. In panels (**b**,**c**), the white dotted line is the true TF location of W1, the three white points are the true TF centers of W2–W4, the three black points are the TF centers of W2–W4 estimated by spectral decomposition algorithm, the cool colors represent small spectral amplitudes, and the warm colors represent large spectral amplitudes.

To further verify the feasibility of MP-EEMD, the seismic trace data near well S-2 (Figure 4a) is selected for real data test. Figure 4b,c displays the components decomposed by EEMD and their corresponding frequency spectra obtained by FFT. It can be seen that the five components have different dominant frequency and frequency band. By comparing with the frequency spectra of the real seismic signal **s** and that of the five components, we regard the component φ_1 as a high-frequency (>200 Hz) low-amplitude noise and remove it in the subsequent steps. In practice, the frequency band of seismic signal is limited and narrow (about 10–60 Hz), so we set the threshold of dominant frequency as 80 Hz in this study. The components with dominant frequency greater than this threshold are considered to be noises, and the others are the valid signals. Therefore, MP-EEMD can eliminate certain noise from the real seismic signal by analysis the spectral features of the components decomposed by EEMD. Then, after adaptive stacking (Equation (8)), the four valid components are stacked to be two sub-signals (Figure 4d), which represent the high-frequency and low-frequency information of the real seismic signal, respectively. Figure 5 displays the TF spectra of the real seismic signal using MP and MP-EEMD. As indicated in the white rectangles, MP-EEMD can avoid the obvious average effect, resulting from frequency interference of ψ_1.and ψ_2, and obtain a more focused TF spectrum than MP. In addition, the dominant frequency of MP-EEMD-based TF spectrum decreases gradually from shallow to deep, which is conforms to the propagation theory of seismic waves. Therefore, the TF spectrum obtained by MP-EEMD performs better.

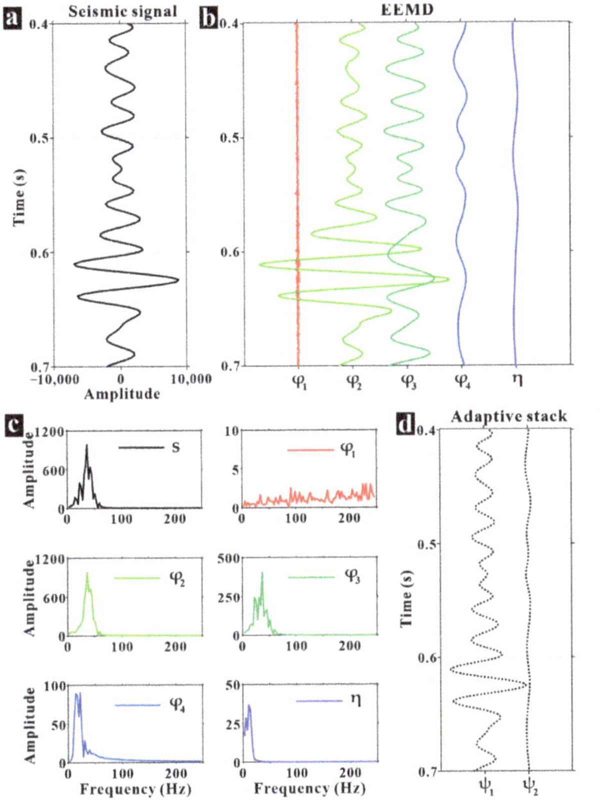

Figure 4. Real seismic signal (**a**), five components decomposed by EEMD (**b**), frequency spectra of the real seismic signal **s** and the five components (**c**), and the sub-signals obtained by adaptive stacking (**d**).

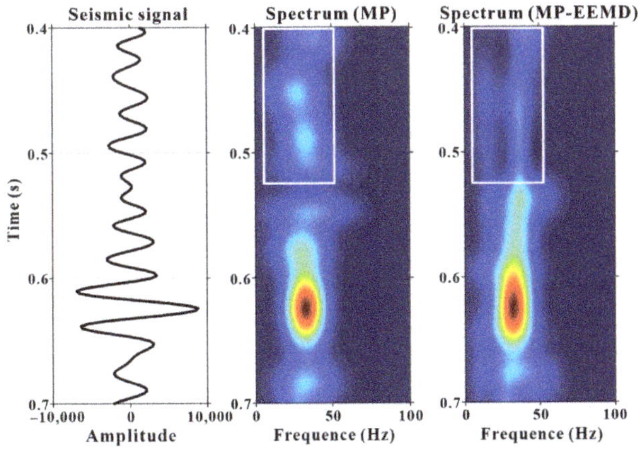

Figure 5. Real seismic signal (**left**), TF spectrum obtained by MP algorithm (**middle**), and TF spectrum obtained by MP-EEMD algorithm (**right**). The cool colors represent small spectral amplitudes, and the warm colors represent large spectral amplitudes.

3.2. Seismic Fluid Mobility Attribute

A simplified asymptotic representation of low-frequency reflection coefficient R for the normal-incidence plane P-wave in a fluid-saturated media is expressed as [17,18]

$$R(f) = R_0 + R_1\sqrt{m \cdot 2\pi f}, \tag{10}$$

where m is the fluid mobility, R_0 and R_1 are dimensionless coefficients, which depend on neither frequency nor fluid mobility. Taking the derivative of Equation (10) with respect to frequency yields

$$m = \frac{2}{\pi R_1^2}\left[\frac{\partial R(f)}{\partial f}\right]^2 f. \tag{11}$$

Equation (11) implies the fluid mobility is proportional to the product of the square of the first derivative over frequency of the reflection coefficient spectrum and the frequency. According to seismic convolution theory, the seismic amplitude spectrum $A(f)$ is proportional to the reflection coefficient spectrum $R(f)$ under the assumption of fixed seismic wavelet. Studies on the seismic amplitude spectrum change in fluid included [10,29] also show that fluid mobility reflects the first derivative of the seismic amplitude spectrum over frequency. Therefore, we define the seismic fluid mobility M as

$$M = \sum_{f \in \Omega}\left[\frac{A(f+\Delta f) - A(f)}{\Delta f}\right]^2 f. \tag{12}$$

where Δf is the frequency increment, and Ω is range of the calculational (low) frequency, which is estimated from the seismic trace data near wells and the logging interpretation data. Choosing a frequency range instead of a single frequency can improve the reliability of fluid mobility.

Figure 6 displays the seismic amplitude spectra and the corresponding frequency gradients for different lithologies with different gas-bearing conditions near S-2 versus frequency. Noted that the spectrum amplitude of gas shale is higher than that of gas-free shale and siltstone. We also can note that the frequency gradients of the gas zone 13–23 Hz (blue rectangle) are also higher than that of the gas-free zone. This means that the seismic fluid mobility attribute can effectively distinguish the gas zone and the gas-free zone.

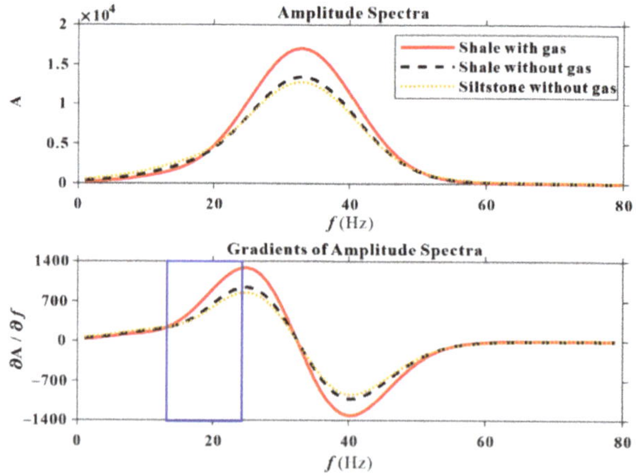

Figure 6. Seismic amplitude spectra (**top**) and the corresponding frequency gradients (**bottom**) for different lithologies with different gas-bearing conditions near well S-2 versus frequency.

Figures 7–9 display the seismic trace records and the normalized seismic fluid mobility attributes (Equation (12)) near S-2, S-3, and S-4. The seismic fluid mobility attribute extracted by using the MP-EEMD algorithm corresponds to the blue curve and extracted by using the MP algorithm corresponds to the blue dotted curve. From Figure 7, we can see that the high value of seismic fluid mobility attribute indicates the location of gas zone (red rectangle) and the low value corresponds to the gas-free zone. As shown with black arrows in Figures 8 and 9, the seismic fluid mobility based on the MP-EEMD algorithm (blue curves) can exhibit less interference for the gas-free zone at well S-3 and better indicate the gas zone at well S-4 than that based on the MP algorithm (blue dotted curves). These factors not only prove the effectiveness of seismic fluid mobility attribute, but also prove the advanced nature of the MP-EEMD algorithm.

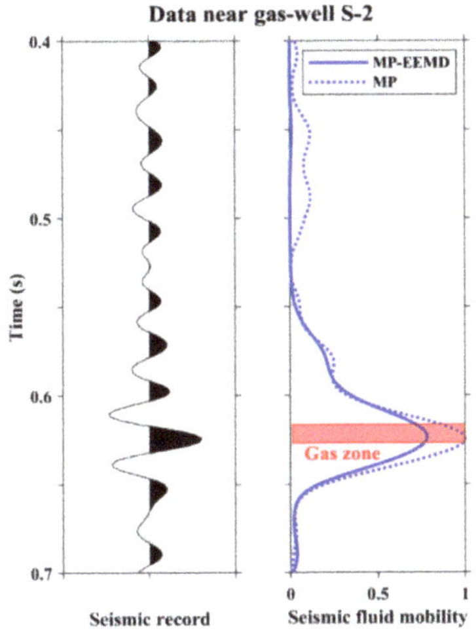

Figure 7. Seismic record (**left**) ad seismic fluid mobility attribute (**right**) near gas well S-2. According to the results of well-seismic calibration, the gas zone in well S-2 (depth 906–922.2 m) is located at time 0.617–0.625 s (red rectangle).

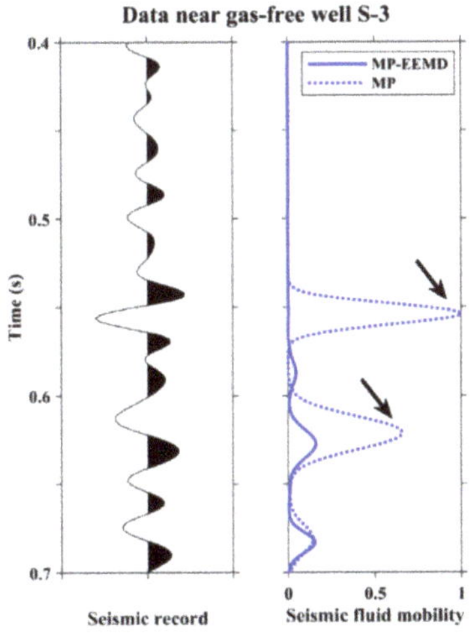

Figure 8. Seismic record (**left**) ad seismic fluid mobility attribute (**right**) near gas-free well S-3.

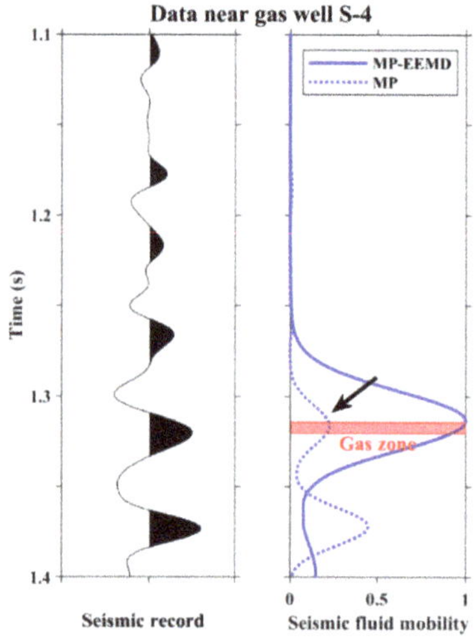

Figure 9. Seismic record (**left**) ad seismic fluid mobility attribute (**right**) near gas well S-4. According to the results of well-seismic calibration, the gas zone in well S-4 (depth 2748.3–2761.3 m) is located at time 1.313–1.319 s (red rectangle).

4. Results

The 1D seismic fluid mobility attributes have indicated that the MP-EEMD-based seismic fluid mobility technology is effective in distinguishing the gas wells from the gas-free wells. Here, the MP-EEMD-based seismic fluid mobility attribute technology is applied to predict the spatial distribution of shale gas accumulation zone in Longmaxi Formation of the whole study area (indicated with blue rectangle in Figure 1a). As shown in Figure 1b, wells S-2, S-3, and SX-1 are located near the seismic line L4 (blue dashed line), and well S-4 is located near the seismic line C2 (green dashed line). Therefore, we next focus on showing and analyzing the seismic fluid mobility attribute extracted from seismic lines L4 and C2.

Figure 10a displays the seismic section of seismic line L4 (blue dashed line in Figure 1b) near the gas wells SX-1 and S-2, Figure 10b displays the normalized seismic fluid mobility section extracted from Figure 10a, and Figure 10c displays the logging data of the deviated well SX-1. The actual well data in Figure 3 indicates a good shale gas display at the bottom section of well S-2, and the TOC curve in Figure 10c indicates a good and discontinuous shale gas display in the horizontal section (from A to B) of well SX-1. However, little information about shale gas can be seen from the post-stack seismic section (Figure 10a). It is clear from the seismic fluid mobility attribute section (Figure 10b) that the shale gas near S-2 and SX-1 is vertically concentrated at the bottom of the Longmaxi Formation, but laterally distributed in a discontinuous manner. Specifically, for the vertical well S-2, the gas-accumulation zone predicted from seismic fluid mobility attribute section is located at the bottom section with high GR and high TOC values (Figure 2). For the deviated well SX-1, the predicted gas-bearing condition from A to B has a trend of getting better, which is basically consistent with the horizontal variation of TOC curve in Figure 10c. Based on the predicted result, we have learned that the shale gas enrich areas in the study area are located near the bottom of the Longmaxi Formation. In addition, the seismic fluid mobility attribute is driven only by the post-stack seismic data, eliminating the need for excessive well control. Therefore, the spatial distribution of shale gas reservoirs predicted from seismic fluid mobility attribute can help researchers design the direction and trajectory of directional drilling, reduce the exploration risk, and optimize the production, even in the early stages of exploration and development.

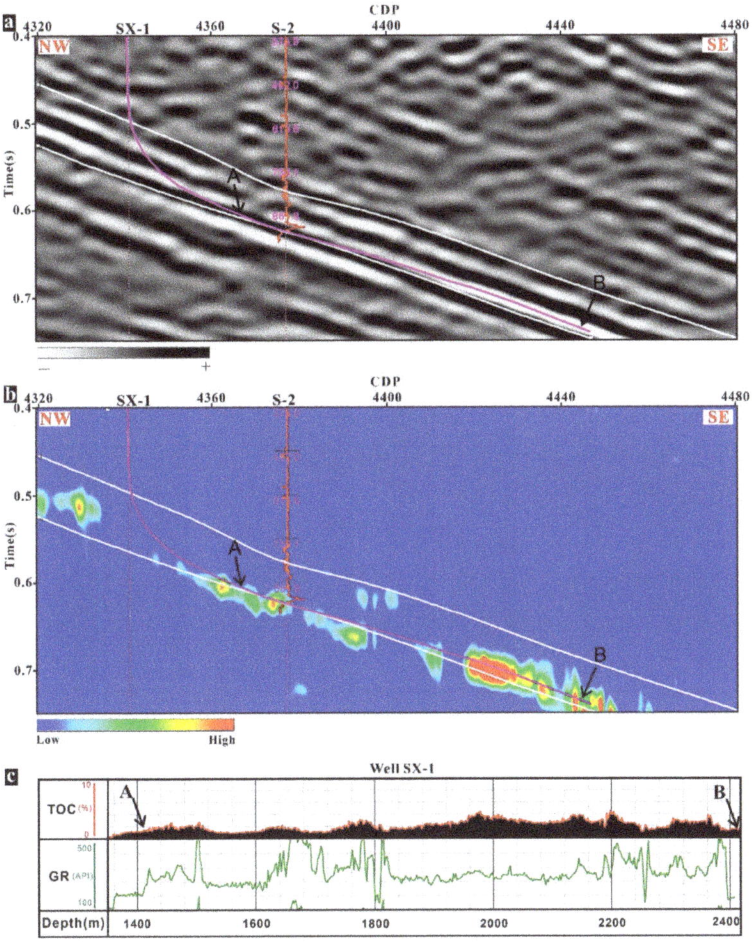

Figure 10. Seismic amplitude section (**a**), seismic fluid mobility section (**b**) connecting gas wells SX-1 and S-2, and logging data in the horizontal section (A–B) of deviated well SX-1 (**c**). In panel (**a**,**b**), NW and SE, respectively, denote the northwest and southeast direction of the study area, the two white lines are the top and bottom interfaces of Longmaxi Formation, the red curve near well S-2 is the GR logging curve, the pink scale near well S-2 is the vertical depth (m), and the pink line is the trajectory of deviated well SX-1. The cold and warm colors in panel (**b**) correspond to the low and high gas-bearing properties in Longmaxi shale.

Same as the gas wells SX-1 and S-2, the gas-free well S-3 is also located near the seismic line L4 (blue dashed line in Figure 1b). Figure 11 displays the comparison between the seismic section and the corresponding seismic fluid mobility section connecting well S-3. Compared with seismic section, seismic fluid mobility attribute section clearly describes the distribution characteristics of shale gas. Predicted result based on seismic fluid mobility attribute indicates that the area near well S-3 displays no gas, while the deeper areas far away from well S-3 (indicated with red arrows in Figure 11b) display good gas-bearing properties. To further verify the accuracy of the above shale gas prediction result, we introduce another petrophysical parameter, i.e., velocity, through post-stack seismic inversion technology (see Appendix A for detailed algorithm). The inverted velocity section connecting well S-3 is shown in Figure 12. Here the blue color represents the low-velocity pure shale, the yellow represents the middle-velocity pure siltstone, the green represents the transition between shale and siltstone in the target formation, and the red color represent the high-velocity limestone outside the target formation. The inverted velocity section reveals that the target formation mainly contains two sets of low-velocity shale reservoirs (marked by black arrows in Figure 12), which is consistent with the lithology data in Figure 2. In addition, it can be seen from the TOC data of Figure 2 that the gas content of the bottom shale reservoir is greater than that of the top shale reservoir. We can also see that the predicted two shale gas zones (marked by red arrows) in Figure 11b are located at the bottom shale reservoir in Figure 12 by comparing the inverted velocity and the seismic fluid mobility.

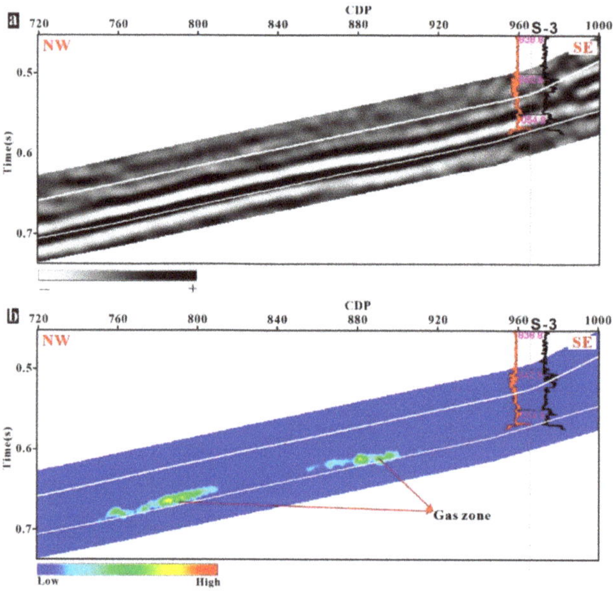

Figure 11. Seismic amplitude section (**a**) and seismic fluid mobility section (**b**) connecting gas-free well S-3. NW and SE denote the northwest and southeast directions of the study area, respectively. The two white lines are the top and bottom interfaces of Longmaxi Formation. Near well S-3, the red and black curves, respectively, are the GR and acoustic velocity logging curves, and the pink scale is the vertical depth (m).

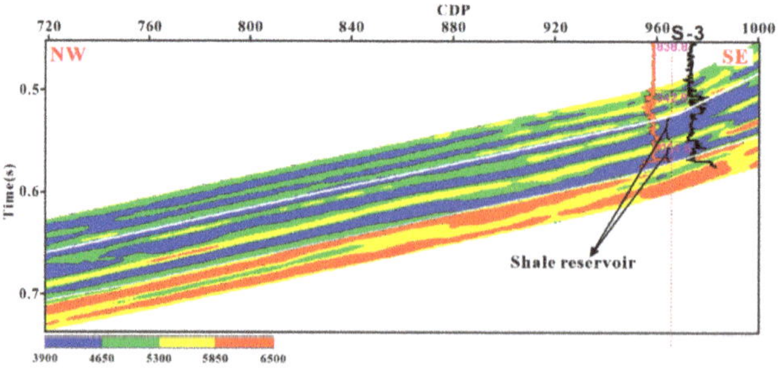

Figure 12. Inverted velocity (m/s) section connecting well S-3. NW and SE denote the northwest and southeast directions of the study area, respectively. The two white lines are the top and bottom interfaces of Longmaxi Formation. Near well S-3, the red and black curves, respectively, are the GR and acoustic velocity logging curves, and the pink scale is the vertical depth (m).

S-4 is a gas well located near the seismic line C2 (green dashed line in Figure 1b). Figure 13 displays the comparison between seismic fluid mobility section and inverted velocity section connecting well S-4. Figure 13a indicates that the gas zone in well S-4 is located near the bottom interface of the Longmaxi Formation, which is consistent with the actual well data. Figure 13b describes the distribution characteristics of shale reservoirs with low-velocity, where the blue color inside the black ellipse represents a set of low-velocity, thick, and pure shale, i.e., favorable shale reservoir. Comparing Figure 13a,b, we can see that the gas accumulation area predicted from the seismic fluid mobility attribute is located in the favorable shale reservoir interpreted from the inverted velocity data. These results confirm the validity and stability of the seismic fluid mobility attribute technique in identifying the spatial distribution of shale gas enrichment areas.

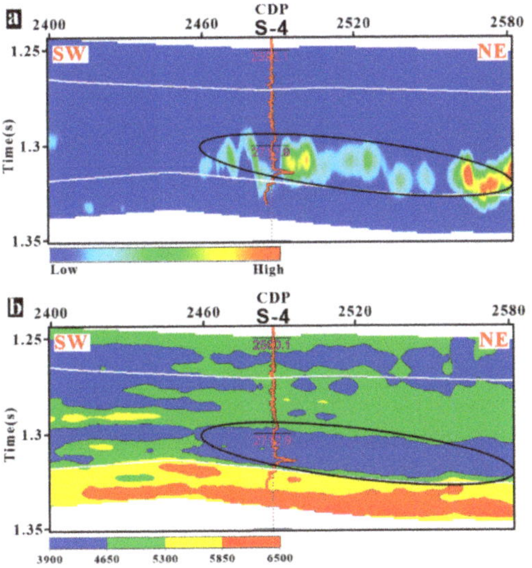

Figure 13. Seismic fluid mobility section (**a**) and inverted velocity (m/s) section (**b**) connecting gas well S-4. The two white lines are the top and bottom interfaces of Longmaxi Formation. The red curve near well S-4 is GR logging curve, and the pink scale is the vertical depth (m). SW and NE denote the southwest and northeast directions of the study area, respectively.

Figure 14 displays the thickness map of the shale gas reservoirs in Longmaxi Formation predicted by seismic fluid mobility attribute technology and seismic inversion technology. The thickness map of shale gas reservoirs is obtained in three steps: first, the time location of shale gas reservoir is estimated using a threshold value of seismic fluid mobility (i.e., $m > 0.3$), then the thickness data at the twelve seismic lines is calculated by using another threshold value of velocity (i.e., $3900 < v < 5300$ m/s), and finally the thickness in the entire study area can be obtained by interpolation technology. As shown in Figure 14, well S-3 has no gas (zero thickness), well S-1 has poor gas content (thickness less than 6 m), and the remaining four wells have good gas content (thickness greater than 10 m), which is matched with the actual drilling result. In addition, the sedimentary facies of the target formation is the transition facies from shallow water shelf in the southeast to deep water shelf in the northwest of the study area. According to the sedimentology theory, with the increase of water depth, the thickness of shale deposition and the maturity of organic matter increase as well as the thickness of shale gas reservoir. Thus, predicted thickness map is consistent with sedimentary background of marine shale. We also give two favorable exploration areas (inside the cyan dotted curves) with large thickness, few faults, and large burial depth (more than 1000 m), which can be used to guide subsequent exploration of shale gas in the Longmaxi Formation.

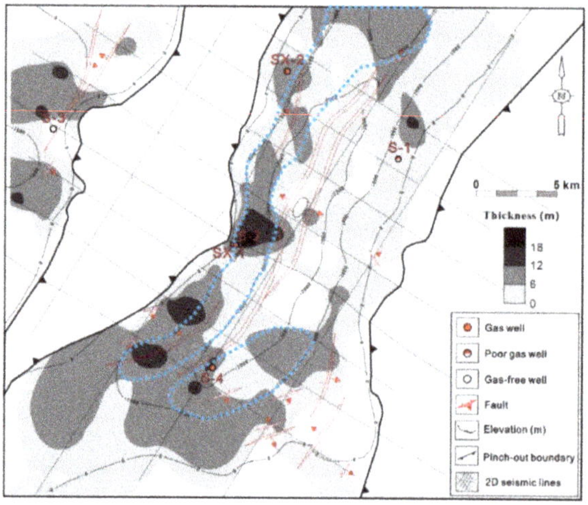

Figure 14. Predicted thickness map of the shale gas reservoirs in Longmaxi Formation.

5. Conclusions

We have predicted the spatial distribution of shale gas reservoirs in Longmaxi Formation of the study area in Southern China using a frequency-dependent seismic fluid mobility attribute, extracted from the post-stack seismic data through a novel spectral decomposition technique (i.e., MP-EEMD scheme), and we also have demonstrated its applicability for the shale reservoirs in complex structural area lacking well control. There are some significant deviations between the TF locations estimated by using the traditional MP algorithm and the true ones due to the interference of multi-frequency components in seismic signal. The synthetic data test indicates that the larger the difference between interfering frequencies, the larger the deviation between the estimated and true TF locations, and the low-amplitude component cannot be recognized on the TF spectrum. These deviations in the TF spectrum will affect the computational accuracy of frequency-dependent seismic attribute. Based on the EEMD algorithm, we propose a novel MP-EEMD scheme to minimizes the effect of frequency interference. The synthetic data test indicates that the MP-EEMD scheme provides a TF spectrum with high accuracy and resolution, which also recognize the weak component. The real data test indicates that the MP-EEMD scheme not only achieves a more focused time-frequency spectrum, but also adaptively removes noise from real seismic data. By introducing the asymptotic low-frequency reflection coefficient, the seismic fluid mobility attribute can be extracted from the low-frequency post-stack seismic data through the spectral decomposition technique. The 1D attributes near wells show that the seismic fluid mobility based on the MP-EEMD algorithm exhibits less interference at the gas-free zone and better indication at the gas zone than that based on the MP algorithm. Thus, the MP-EEMD-based seismic fluid mobility attribute technology is applied to characterize the shale reservoirs with complex subsurface structure and low exploration level in Southern China. The spatial distribution of shale gas reservoirs predicted from the seismic fluid mobility attribute section is consistent with actual well data. Based on the predicted result, we have learned that the shale gas enrich areas in the study area are located near the bottom of the Longmaxi Formation. Referring to the predicted spatial distribution pattern of shale gas reservoirs in the area can help engineers design the directions and trajectories of directional drilling during the development stage. By combining the seismic fluid mobility attribute with the inverted velocity data, the thickness map of shale gas reservoirs in target formation is obtained, which is consistent with sedimentary background of marine shale. Referring to the thickness map of shale gas reservoirs in the area can help researchers to evaluate shale gas resources in the early stage of exploration. We have also given two potential area of future research by analyzing the thickness, structure, and burial depth. This work can provide some references for the subsequent exploration and research of shale gas reservoirs.

Author Contributions: Conceptualization, J.Z.; methodology, J.Z., A.S., and H.H.; software, J.Z. and H.H.; validation, J.Z. and H.H.; formal analysis, J.Z. and H.H.; investigation, J.Z. and H.H.; resources, H.H.; data curation, J.Z.; writing—original draft preparation, J.Z.; writing—review and editing, A.S., L.R., and T.T.; visualization, J.Z. A.S., L.R., and T.T; supervision, H.H.; project administration, H.H.; funding acquisition, H.H. All authors have read and agreed to the published version of the manuscript.

Funding: This research was funded by National Natural Science Foundation of China with grant number 41974124.

Institutional Review Board Statement: Not applicable.

Informed Consent Statement: Not applicable.

Data Availability Statement: Data sharing not applicable. No new data were created or analyzed in this study. Data sharing is not applicable to this article.

Acknowledgments: The authors would like to thank Shenhua Group in China for data provided and permission to publish the results of this research.

Conflicts of Interest: The authors declare no conflict of interest.

Appendix A. Post-Stack Seismic Inversion Technology

In the petroleum industry, post-stack seismic inversion technology is a conventional and mature technology to characterize reservoirs. Ignoring multiples, transmission loss, spherical spreading, and attenuation, the objective function of post-stack seismic inversion can be expressed as

$$\min_{\mathbf{z}} \|\mathbf{d} - \mathbf{W}\mathbf{r}(\mathbf{z})\|_2^2, \tag{A1}$$

where $\mathbf{z} = [z_1, z_2, \ldots, z_g]^T$ is the acoustic impendence vector, subscript g is the sampling number, superscript T is the transpose operator, $\mathbf{d} = [d_1, d_2, \ldots, d_{g+q-2}]^T$ is the real seismic record, superscript q is the length of seismic wavelet vector $\mathbf{w} = [w_1, w_2, \ldots, w_q]^T$, and $\mathbf{W}$ is the seismic wavelet convolution matrix

$$\mathbf{W} = \begin{bmatrix} w_1 & & & \\ w_2 & w_1 & & \\ \vdots & w_2 & \ddots & \\ w_q & \vdots & \ddots & w_1 \\ & w_q & \ddots & w_2 \\ & & \ddots & \vdots \\ & & & w_q \end{bmatrix}_{(q+g-2)\times(g-1)}, \tag{A2}$$

and $\mathbf{r}(\mathbf{z})$ is the reflection coefficient vector

$$\mathbf{r}(\mathbf{z}) = [r_1, r_2, \ldots, r_{g-1}]^T = \left[\frac{z_2 - z_1}{z_2 + z_1}, \frac{z_3 - z_2}{z_3 + z_2}, \ldots, \frac{z_g - z_{g-1}}{z_g + z_{g-1}}\right]^T. \tag{A3}$$

According to Equation (A3), the reflection coefficient is controlled by the difference of the acoustic impedances of the adjacent sampling points. The acoustic impendence is the product of velocity v and density ρ. The difference of velocities of the adjacent sampling points is greater than that of the density (Figure 3). In addition, due to the thin thickness of the Longmaxi Formation, the number of time sampling points during the inversion process is about 50 (Figures 10a and 11a). Due to these properties, we approximate density of Longmaxi Formation to be a constant and then the Equation (A3) can be can be converted to be

$$\mathbf{r}(\mathbf{v}) = \left[r_1, r_2, \ldots, r_{g-1}\right]^T = \left[\frac{v_2 - v_1}{v_2 + v_1}, \frac{v_3 - v_2}{v_3 + v_2}, \ldots, \frac{v_g - v_{g-1}}{v_g + v_{g-1}}\right]^T. \quad \text{(A4)}$$

where $\mathbf{v} = \left[v_1, v_2, \ldots, v_g\right]^T$ is the velocity vector.

Due to limitation in the frequency bandwidth of seismic data, directly solving for the objective function (13) is strongly unstable. Based on the Tikhonov regularization theory and the Bayes theory, a stable and reliable impedance inversion result can be obtained [30]. The impedance update during iterative process of seismic inversion is expressed as

$$\mathbf{v}^{(y+1)} = \mathbf{v}^{(y)} + \left[\left(\mathbf{G}^{(y)}\right)^T \mathbf{G}^{(y)} + \lambda^{(y)}\right]^{-1} \left(\mathbf{G}^{(y)}\right)^T \left[\mathbf{d} - \mathbf{W}\mathbf{r}\left(\mathbf{v}^{(y)}\right)\right], \quad \text{(A5)}$$

where superscript $y = 1, 2, \ldots, Y$ is the iteration number, $\mathbf{v}^{(y)}$ is the velocity at the yth iteration, $\mathbf{v}^{(1)}$ is the initial velocity model from sonic logging data, $\mathbf{G}^{(y)}$ is the Jacobi matrix at the yth iteration

$$\mathbf{G}^{(y)} = \mathbf{W}\begin{bmatrix} \frac{-2v_2^{(y)}}{\left(v_2^{(y)}+v_1^{(y)}\right)^2} & \frac{v_1^{(y)}}{\left(v_2^{(y)}+v_1^{(y)}\right)^2} & & & \\ & \frac{-2v_3^{(y)}}{\left(v_3^{(y)}+v_2^{(y)}\right)^2} & \frac{v_2^{(y)}}{\left(v_3^{(y)}+v_2^{(y)}\right)^2} & & \\ & & \ddots & \ddots & \\ & & & \frac{-2v_g^{(y)}}{\left(v_g^{(y)}+v_{g-1}^{(y)}\right)^2} & \frac{v_{g-1}^{(y)}}{\left(v_g^{(y)}+v_{g-1}^{(y)}\right)^2} \end{bmatrix} \quad \text{(A6)}$$

$\lambda^{(y)} = \chi_{\Delta d}^{(y)} / \chi_{\Delta v}^{(y)}$ is the adaptive regularization parameter at the yth iteration, $\chi_{\Delta d}^{(y)}$ is the variance of the difference between the real and synthetic seismograms at the yth iteration, $\chi_{\Delta v}^{(y)}$ is the variance of the velocity residual before and after the yth iteration. The expression of regularization parameter implies that when the noise level in the seismic data is high, the value of regularization parameter is large, thus serving to suppress noise to obtain a relatively stable inversion result. On the contrary, when the noise approaches zero, the value of the regularization parameter is close to zero, and the inversion result is then determined mainly from the seismic data, which leads to a high-resolution solution.

References

1. Yang, R.; He, S.; Yi, J.; Hu, Q. Nano-scale pore structure and fractal dimension of organic-rich Wufeng-Longmaxi shale from Jiaoshiba area, Sichuan Basin: Investigations using FE-SEM, gas adsorption and helium pycnometry. *Mar. Pet. Geol.* **2016**, *70*, 27–45. [CrossRef]
2. Jiang, S.; Zhang, J.; Jiang, Z.; Xu, Z.; Cai, D.; Chen, L.; Wu, Y.; Zhou, D.; Jiang, Z.; Zhao, X.; et al. Geology, resource potentials, and properties of emerging and potential China shale gas and shale oil plays. *Interpretation* **2015**, *3*, SJ1–SJ13. [CrossRef]
3. Xi, Z.; Tang, S.; Wang, J. The reservoir characterization and shale gas potential of the Niutitang formation: Case study of the SY well in northwest Hunan Province, South China. *J. Pet. Sci. Eng.* **2018**, *171*, 687–703. [CrossRef]
4. Ma, X.; Xie, J. The progress and prospects of shale gas exploration and development in southern Sichuan Basin, SW China. *Pet. Explor. Dev.* **2018**, *45*, 172–182. [CrossRef]
5. Tang, X.; Jiang, Z.; Li, Z.; Gao, Z.; Bai, Y.; Zhao, S.; Feng, J. The effect of the variation in material composition on the heterogeneous pore structure of high-maturity shale of the Silurian Longmaxi formation in the southeastern Sichuan Basin, China. *J. Nat. Gas Sci. Eng.* **2015**, *23*, 464–473. [CrossRef]

6. Han, C.; Jiang, Z.; Han, M.; Wu, M.; Lin, W. The lithofacies and reservoir characteristics of the Upper Ordovician and Lower Silurian black shale in the Southern Sichuan Basin and its periphery, China. *Mar. Pet. Geol.* **2016**, *75*, 181–191. [CrossRef]
7. Wang, G.; Ju, Y.; Huang, C.; Long, S.; Peng, Y. Longmaxi-Wufeng Shale lithofacies identification and 3-D modeling in the northern Fuling Gas Field, Sichuan Basin. *J. Nat. Gas. Sci. Eng.* **2017**, *47*, 59–72. [CrossRef]
8. Fu, Y.; Jiang, Y.; Wang, Z.; Hu, Q.; Xie, J.; Ni, G.; Lei, Z.; Zhou, K.; Liu, X. Non-connected pores of the Longmaxi shale in southern Sichuan Basin of China. *Mar. Pet. Geol.* **2019**, *110*, 420–433. [CrossRef]
9. Lee, J.; Byun, J.; Kim, B.; Yoo, D. Delineation of gas hydrate reservoirs in the Ulleung Basin using unsupervised multi-attribute clustering without well log data. *J. Nat. Gas Sci. Eng.* **2017**, *46*, 326–337. [CrossRef]
10. Zeng, J.; Huang, H.; Li, H.; Miao, Y.; Wen, J.; Zhou, F. A fast complex domain-matching pursuit algorithm and its application to deep-water gas reservoir detection. *J. Geophys. Eng.* **2017**, *14*, 1335–1348. [CrossRef]
11. Veeken, P.C.H.; Rauch-Davies, M. AVO attribute analysis and seismic reservoir characterization. *First Break* **2006**, *24*, 41–52. [CrossRef]
12. Ross, C.P. Comparison of popular AVO attributes, AVO inversion, and calibrated AVO predictions. *Lead. Edge* **2002**, *21*, 244–252. [CrossRef]
13. Castagna, J.P.; Sun, S.; Siegfried, R.W. Instantaneous spectral analysis: Detection of low-frequency shadows associated with hydrocarbons. *Lead. Edge* **2003**, *22*, 120–127. [CrossRef]
14. Lichman, E.; Goloshubin, G. Unified approach to gas and fluid detection on instantaneous seismic wavelets. *SEG Tech. Program Expand. Abstr.* **2003**, 1699–1702. [CrossRef]
15. Luo, Y.; Huang, H.; Yang, Y.; Hao, Y.; Zhang, S.; Li, Q. Integrated prediction of deepwater gas reservoirs using Bayesian seismic inversion and fluid mobility attribute in the South China Sea. *J. Nat. Gas Sci. Eng.* **2018**, *59*, 56–66. [CrossRef]
16. Zhang, Y.; Wen, X.; Jiang, L.; Liu, J.; Yang, J.; Liu, S. Prediction of high-quality reservoirs using the reservoir fluid mobility attribute computed from seismic data. *J. Pet. Sci. Eng.* **2020**, *190*, 107007. [CrossRef]
17. Silin, D.B.; Korneev, V.A.; Goloshubin, G.M.; Patzek, T.W. *A Hydrologic View on Biot's Theory of Poroelasticity*; Lawrence Berkeley National Lab. (LBNL): Berkeley, CA, USA, 2004. [CrossRef]
18. Goloshubin, G.; Silin, D. *Frequency-Dependent Seismic Reflection from a Permeable Boundary in a Fractured Reservoir*; Society of Exploration Geophysicists: Tulsa, OK, USA, 2006; pp. 1742–1746. [CrossRef]
19. Naseer, T.M.; Asim, S.; Ahmad, M.N.; Hussain, F.; Qureshi, S.N. Application of Seismic Attributes for Delineation of Channel Geometries and Analysis of Various Aspects in Terms of Lithological and Structural Perspectives of Lower Goru Formation, Pakistan. *Int. J. Geosci.* **2014**, *5*, 1490–1502. [CrossRef]
20. Yuan, S.; Ji, Y.; Shi, P.; Zeng, J.; Gao, J.; Wang, S. Sparse Bayesian Learning-Based Seismic High-Resolution Time-Frequency Analysis. *IEEE Geosci. Remote Sens. Lett.* **2019**, *16*, 623–627. [CrossRef]
21. Wang, Y. Seismic time-frequency spectral decomposition by matching pursuit. *Geophysics* **2007**, *72*, V13–V20. [CrossRef]
22. Mallat, S.G.; Zhifeng, Z. Matching pursuits with time-frequency dictionaries. *IEEE Trans. Signal Process.* **1993**, *41*, 3397–3415. [CrossRef]
23. Castagna, J.P.; Sun, S. Comparison of spectral decomposition methods. *First Break* **2006**, *24*, 75–79. [CrossRef]
24. Wu, Z.; Huang, N.E. Ensemble Empirical Mode Decomposition: A Noise-Assisted Data Analysis Method. *Adv. Adapt. Data Anal.* **2011**, *1*, 1–41. [CrossRef]
25. Guo, T.; Zhang, H. Formation and enrichment mode of Jiaoshiba shale gas field, Sichuan Basin. *Pet. Explor. Dev.* **2014**, *41*, 31–40. [CrossRef]
26. Guo, T. Key geological issues and main controls on accumulation and enrichment of Chinese shale gas. *Pet. Explor. Dev.* **2016**, *43*, 349–359. [CrossRef]
27. Wan, Y.; Tang, S.; Pan, Z. Evaluation of the shale gas potential of the lower Silurian Longmaxi Formation in northwest Hunan Province, China. *Mar. Pet. Geol.* **2017**, *79*, 159–175. [CrossRef]
28. Li, Y.; Liu, H.; Zhang, L.; Lu, Z.; Li, Q.; Huang, Y. Lower limits of evaluation parameters for the lower Paleozoic Longmaxi shale gas in southern Sichuan Province. *Sci. China Earth Sci.* **2013**, *56*, 710–717. [CrossRef]
29. Goloshubin, G.M.; Korneev, V.A.; Vingalov, V.M. Seismic low-frequency effects from oil-saturated reservoir zones. *SEG Tech. Program Expand. Abstr.* **2002**, 1813–1816. [CrossRef]
30. Huang, H.; Yuan, S.; Zhang, Y.; Zeng, J.; Mu, W. Use of nonlinear chaos inversion in predicting deep thin lithologic hydrocarbon reservoirs: A case study from the Tazhong oil field of the Tarim Basin, China. *Geophysics* **2016**, *81*, B221–B234. [CrossRef]

applied sciences

MDPI

Article

Seismic AVOA Inversion for Weak Anisotropy Parameters and Fracture Density in a Monoclinic Medium

Zijian Ge [1], Shulin Pan [1,*] and Jingye Li [2]

[1] School of Earth Science and Technology, Southwest Petroleum University, Chengdu 610500, China; zingzijian@hotmail.com

[2] State Key Laboratory of Petroleum Resources and Prospecting, China University of Petroleum, Beijing 102249, China; lijingye@cup.edu.cn

* Correspondence: shulinpan@swpu.edu.cn; Tel.: +86-18502877827

Received: 24 June 2020; Accepted: 22 July 2020; Published: 26 July 2020

Abstract: In shale gas development, fracture density is an important lithologic parameter to properly characterize reservoir reconstruction, establish a fracturing scheme, and calculate porosity and permeability. The traditional methods usually assume that the fracture reservoir is one set of aligned vertical fractures, embedded in an isotropic background, and estimate some alternative parameters associated with fracture density. Thus, the low accuracy caused by this simplified model, and the intrinsic errors caused by the indirect substitution, affect the estimation of fracture density. In this paper, the fractured rock of monoclinic symmetry assumes two non-orthogonal vertical fracture sets, embedded in a transversely isotropic background. Firstly, assuming that the fracture radius, width, and orientation are known, a new form of P-wave reflection coefficient, in terms of weak anisotropy (WA) parameters and fracture density, was obtained by substituting the stiffness coefficients of vertical transverse isotropic (VTI) background, normal, and tangential fracture compliances. Then, a linear amplitude versus offset and azimuth (AVOA) inversion method, of WA parameters and fracture density, was constructed by using Bayesian theory. Tests on synthetic data showed that WA parameters, and fracture density, are stably estimated in the case of seismic data containing a moderate noise, which can provide a reliable tool in fracture prediction.

Keywords: fractures density; weak anisotropy parameters; monoclinic; Bayesian inversion

1. Introduction

In a variety of complex oil and gas reservoirs, fractured reservoirs widely exist in shale and igneous rock. Natural and induced fractures in reservoirs play an important role in determining fluid flow during production, and knowledge of the orientation and density of fractures is useful to optimize production from fractured reservoirs [1]. Areas of high fracture density may represent 'sweet spots' of high permeability, and it is important to be able to target such locations for infill drilling [2]. Recent developments in geophysics have revealed the viscous behavior of the underground strata, and demonstrated that the propagation of seismic waves has dispersion and attenuation, for fluid discrimination in porous rocks [3–5]. However, there are methods available to compensate for dispersion and attenuation. In the current implementation of the method, we also ignore the loss of seismic amplitude due to attenuation. Seismic anisotropy is defined as the dependence of seismic velocity upon angle. P-waves propagating parallel to fractures are faster than those propagating perpendicular to fractures [6]. For transversely isotropic (TI) media, empirical and analytical studies have shown that the presence of anisotropy can significantly distort conventional amplitude variation with offset (AVO) analysis [7–9], and that soil and rock anisotropy can markedly affect the dynamic and seismic response of geostructures [10–12].

Previous anisotropic models used to invert the seismic response of fractured reservoirs often assume a single set of aligned fractures, with horizontally transverse isotropic (HTI) symmetry [13,14], whereas most reservoirs contain several sets of fractures with variable orientations within a given fracture set [15–17]. Furthermore, in the HTI model, horizontal layering of the crust, leading to a polar anisotropic background (vertical transverse isotropic, VTI), is ignored. Hence, HTI is not a suitable model for the sedimentary crust. Reflection amplitudes have advantages over seismic velocities for characterizing fractured reservoirs, because they have higher vertical resolution, and are more sensitive to the properties of the reservoir. Therefore, amplitude versus offset and azimuth (AVOA) inversion is an effective method for predicting fractures [18]. The Zoeppritz equation for HTI anisotropy is obtained by linearized approximation [13]. Using the linear P-wave reflection coefficient formula, the feasibility of fracture compliance tensor inversion in orthorhombic and monoclinic media is analyzed [6,19]. In the inversion process, it is assumed that the weak anisotropy (WA) parameters of VTI background media are known. Under the assumption of HTI and orthogonal media, the elastic impedance (EI) is re-expressed in terms of the normal and tangential fracture weaknesses, and the EI inversion for fracture weaknesses is implemented, based on the linear-slip model [20,21]. Based on HTI media, many scholars have used normal move out (NMO) velocity elliptical inversion, or prestack AVOA inversion, to indirectly calculate fracture density. Scholars [22] have implemented a qualitative analysis of fracture density and orientation, based on P-wave (PP) and converted wave (PS) velocity anisotropy. They have used elliptical inversion to estimate the magnitude of anisotropy associated with fracture density, from PP data, and then qualitatively estimated fracture density, from time differences of the fast and slow PS velocities. It has been suggested that the difference in NMO times be used to estimate the axial ratio of the ellipse, which is proportional to fracture density [23]. A non-linear Bayesian inversion of the seismic amplitude versus incident and azimuthal angle (AVAZ) and dynamic production data is used to calculate the fracture density and aperture for an HTI fractured reservoir in which the fracture parameters are invariant [24]. The prestack AVAZ inversion method is used to extract the anisotropy parameters or shear-wave splitting factor, γ, which is related to fracture density [25]. The inversion algorithm has also been deeply studied by many scholars [26–28].

However, with indirect substitution there exists intrinsic errors, and the inversion methods based on simplified models such as the HTI medium can be misleading when applied to shale fractured reservoirs with a complex medium. These limit the improvement to the prestack inversion accuracy. Therefore, it cannot satisfy the accuracy requirements for obtaining fracture density.

In this paper, based upon previous research, and combined with rock physics theories, assuming that the fracture reservoir has monoclinic symmetry, and under the condition that the fracture radius, width, and orientation are known, we derived a new form of P-wave reflection coefficient, consisting of the WA parameters, and the fracture density. This was achieved by substituting the stiffness coefficients of the VTI background, and normal and tangential fracture compliances. The new form of P-wave reflection coefficient was then used to build a theoretical framework of linear AVOA inversion, based on Bayesian theory, that can avoid the assumption that the WA parameters of VTI background media are known, and avoid the intrinsic errors introduced from the indirect substitution. Finally, synthetic data, which contain Gaussian random noise with different signal-to-noise ratios (SNRs), were utilized to verify the stability of the proposed approach.

2. Methods

2.1. Elastic Compliance Tensor of Fractured Medium

In an elastic medium that contains an arbitrary number of sets of fractures, with an arbitrary orientation distribution, by using the divergence theorem and Hooke's law, it can be shown that the elastic compliance tensor of the fractured medium can be written in the following form [29,30]:

$$S_{ijkl} = S_{ijkl}{}^{0} + \Delta S_{ijkl} \tag{1}$$

where S^0 is the compliance matrix of the medium (including the effects of pores, cracks, and stress, except for those fractures explicitly included in ΔS). The subscripts i, j, k, and l are the free index. The effective excess compliance ΔS_{ijkl}, due to the presence of the fractures, with rotationally invariant shear compliance, can be written as [27]:

$$\Delta S_{ijkl} = \frac{1}{4}\left(\delta_{ik}\alpha_{jl} + \delta_{il}\alpha_{jk} + \delta_{jk}\alpha_{il} + \delta_{jl}\alpha_{ik}\right) + \beta_{ijkl}. \tag{2}$$

Here, δ_{ij} is the Krönecker delta, α_{ij} is a second-rank tensor, and β_{ijkl} is a fourth-rank tensor. If the two fracture sets are not orthogonal, but are perfectly aligned (within each set), the compliance tensors become [6]:

$$\alpha = N_{V1}\overline{A}_1\overline{B}_{T1}\begin{bmatrix} \sin^2\varphi_1 & \cos\varphi_1\sin\varphi_1 & 0 \\ \cos\varphi_1\sin\varphi_1 & \cos^2\varphi_1 & 0 \\ 0 & 0 & 0 \end{bmatrix} + N_{V2}\overline{A}_2\overline{B}_{T2}\begin{bmatrix} \sin^2\varphi_2 & \cos\varphi_2\sin\varphi_2 & 0 \\ \cos\varphi_2\sin\varphi_2 & \cos^2\varphi_2 & 0 \\ 0 & 0 & 0 \end{bmatrix} \tag{3}$$

$$\begin{aligned} \beta = {} & N_{V1}\overline{A}_1\left(\overline{B}_{N1} - \overline{B}_{T1}\right)\begin{bmatrix} \sin^4\varphi_1 & \sin^2\varphi_1\cos^2\varphi_1 & 0 & 0 & 0 & 2\sin^3\varphi_1\cos\varphi_1 \\ \sin^2\varphi_1\cos^2\varphi_1 & \cos^4\varphi_1 & 0 & 0 & 0 & 2\sin\varphi_1\cos^3\varphi_1 \\ 0 & 0 & 0 & 0 & 0 & 0 \\ 0 & 0 & 0 & 0 & 0 & 0 \\ 0 & 0 & 0 & 0 & 0 & 0 \\ 2\sin^3\varphi_1\cos\varphi_1 & 2\sin\varphi_1\cos^3\varphi_1 & 0 & 0 & 0 & 4\sin^2\varphi_1\cos^2\varphi_1 \end{bmatrix} + \\ & N_{V2}\overline{A}_2\left(\overline{B}_{N2} - \overline{B}_{T2}\right)\begin{bmatrix} \sin^4\varphi_2 & \sin^2\varphi_2\cos^2\varphi_2 & 0 & 0 & 0 & 2\sin^3\varphi_2\cos\varphi_2 \\ 0 & \cos^4\varphi_2 & 0 & 0 & 0 & 2\sin\varphi_2\cos^3\varphi_2 \\ 0 & 0 & 0 & 0 & 0 & 0 \\ 0 & 0 & 0 & 0 & 0 & 0 \\ 0 & 0 & 0 & 0 & 0 & 0 \\ 2\sin^3\varphi_2\cos\varphi_2 & 2\sin\varphi_2\cos^3\varphi_2 & 0 & 0 & 0 & 4\sin^2\varphi_2\cos^2\varphi_2 \end{bmatrix} \end{aligned} \tag{4}$$

where N_V is the number of fractures per unit volume (fracture density), $\overline{A}$ is the average area of the fractures in the set, $\overline{B}_T$ and $\overline{B}_N$ are the (area-weighted) average specific tangential and normal compliances, and φ is the azimuthal angle between the fracture strike and the survey 1-axis. The subscripts 1 and 2 denote the 1st and 2nd fracture set, respectively. If the background medium is isotropic, or transversely isotropic, and there are at least two non-orthogonal vertical fracture sets, this leads to monoclinic symmetry of the fractured rock, as shown in Figure 1.

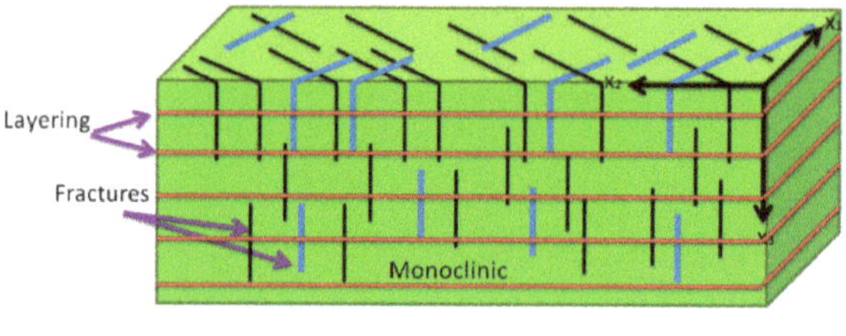

Figure 1. Two non-orthogonal vertical fracture sets with a vertical transverse isotropic (VTI) background.

The additional compliance, due to two or more sets of aligned vertical fractures not aligned with the coordinate system, has the form [6]:

$$\Delta S = \begin{bmatrix} \alpha_{11}+\beta_{1111} & \beta_{1122} & 0 & 0 & 0 & \alpha_{12}+2\beta_{1112} \\ \beta_{1122} & \alpha_{22}+\beta_{2222} & 0 & 0 & 0 & \alpha_{12}+2\beta_{1222} \\ 0 & 0 & 0 & 0 & 0 & 0 \\ 0 & 0 & 0 & \alpha_{22} & \alpha_{12} & 0 \\ 0 & 0 & 0 & \alpha_{12} & \alpha_{11} & 0 \\ \alpha_{12}+2\beta_{1112} & \alpha_{12}+\beta_{1222} & 0 & 0 & 0 & \alpha_{11}+\alpha_{22}+4\beta_{1122} \end{bmatrix} \tag{5}$$

When the fracture's additional compliance ΔS is small, one can directly calculate the stiffness matrix using the background stiffness matrix C_0 [2]:

$$C = C_0 - C_0 \Delta S C_0. \tag{6}$$

We assumed a polar anisotropic background. An isotropic background is a special case of this. Thomsen [31] proposed a simple elastic constant expression:

$$\begin{aligned} V_p &= \sqrt{C_{33}/\rho} \\ V_s &= \sqrt{C_{44}/\rho} \\ \varepsilon &= \frac{C_{11}-C_{33}}{2C_{33}} \\ \gamma &= \frac{C_{66}-C_{44}}{2C_{44}} \\ \delta &= \frac{(C_{13}+C_{44})^2-(C_{33}-C_{44})^2}{2C_{33}(C_{33}-C_{44})} \end{aligned} \tag{7}$$

2.2. PP-Wave Reflection Coefficient and Generalized Anisotropy Parameters

In this study, the elastic contrast between the overburden and the reservoir was assumed to be small. In this situation, the plane-wave/P-wave reflection coefficient for a plane separating media, with arbitrary elastic symmetry and with WA, can be written in the form [32]

$$\begin{aligned} R_{pp}(\theta,\phi) &= R_{PP}^{iso}(\theta) + \tfrac{1}{2}\Delta\varepsilon_z + \tfrac{1}{2}\begin{bmatrix} \left(\Delta\delta_x - 8\frac{\overline{V_s}^2}{\overline{V_p}^2}\Delta\gamma_x\right)\cos^2\phi + \\ \left(\Delta\delta_y - 8\frac{\overline{V_s}^2}{\overline{V_p}^2}\Delta\gamma_y\right)\sin^2\phi + \\ 2\left(\Delta X_z - 4\frac{\overline{V_s}^2}{\overline{V_p}^2}\Delta\varepsilon_{45}\right)\cos\phi\sin\phi - \Delta\varepsilon_z \end{bmatrix}\sin^2\theta + \\ &\tfrac{1}{2}\left[\Delta\varepsilon_x\cos^4\phi + \Delta\varepsilon_y\sin^4\phi + \Delta\delta_z\cos^2\phi\sin^2\phi + 2\left(\Delta\varepsilon_{16}\cos^2\phi + \Delta\varepsilon_{26}\sin^2\phi\right)\cos\phi\sin\phi\right]\sin^2\theta\tan^2\theta \end{aligned} \tag{8}$$

where $R_{PP}^{iso}(\theta)$ denotes the weak-contrast reflection coefficient at an interface separating two slightly different isotropic media, and the generalized Thomsen anisotropy parameters are given by

$$\begin{aligned} \delta_x &= \frac{A_{13}+2A_{55}-V_p^2}{V_p^2}, \delta_y = \frac{A_{23}+2A_{44}-V_p^2}{V_p^2}, \delta_z = \frac{A_{12}+2A_{66}-V_p^2}{V_p^2}, \\ X_z &= \frac{A_{36}+2A_{45}}{V_p^2}, \varepsilon_{16} = \frac{A_{16}}{V_p^2}, \varepsilon_{26} = \frac{A_{26}}{V_p^2}, \varepsilon_{45} = \frac{A_{45}}{V_s^2}, \varepsilon_x = \frac{A_{11}-V_p^2}{2V_p^2}, \\ \varepsilon_y &= \frac{A_{22}-V_p^2}{2V_p^2}, \varepsilon_z = \frac{A_{33}-V_p^2}{2V_p^2}, \gamma_x = \frac{A_{55}-V_s^2}{2V_s^2}, \gamma_y = \frac{A_{44}-V_s^2}{2V_s^2} \end{aligned} \tag{9}$$

where V_p and V_s are the P- and S-wave velocities of the background isotropic medium, respectively, and $A_{ij} = \frac{C_{ij}}{\rho}$ is the density-normalized elastic stiffness.

Incorporating Equations (5) and (7) and substituting the density-normalized elastic stiffness of Equation (9), we obtained a new expression of the generalized Thomsen anisotropy parameters, for the monoclinic medium, in terms of WA parameters and the compliance tensors. These parameters are presented in Appendix A. Under the assumption of small compliance tensors and WA parameters, we neglected the product terms of the compliance tensors and the weak anisotropic parameters.

By substituting the generalized WA parameters in Equation (8) with Equation (A1), a P-wave reflection coefficient, in terms of WA parameters and the compliance tensors, was obtained:

$$\begin{aligned} R_{pp}(\theta,\phi) &= R_{PP}^{iso}(\theta) + R_{PP}^{adi}(\theta) + F_1(\theta)\delta^{psd} + F_2(\theta)\varepsilon + F_3(\theta,\phi)\alpha_{11} + F_4(\theta,\phi)\alpha_{12} + \\ F_5(\theta,\phi)\alpha_{22} &+ F_6(\theta,\phi)\beta_{1111} + F_7(\theta,\phi)\beta_{1112} + F_8(\theta,\phi)\beta_{1122} + F_9(\theta,\phi)\beta_{1222} + F_{10}(\theta,\phi)\beta_{2222} \end{aligned} \tag{10}$$

where the pseudo weak anisotropic parameter $\delta^{psd} = \sqrt{M - \mu + 2M\delta}$. The second term on the righthand side is only a function of offset because the background is VTI, which does not depend on the azimuth. Equations for sensitivities, $\boldsymbol{F}$, are given in Appendix B.

In order to realize the direct inversion of fracture density, the linear P-wave reflection coefficient, in terms of weak anisotropic parameters and fracture density, could be obtained by combining Equations (3) and (4) and Equation (10).

$$R_{pp}(\theta,\phi) = R_{PP}^{iso}(\theta) + R_{PP}^{adi}(\theta) + F_1(\theta)\delta^{psd} + F_2(\theta)\varepsilon + a(\theta,\phi)N_{V1} + b(\theta,\phi)N_{V2} \tag{11}$$

where the sensitivities $a(\theta,\phi)$ and $b(\theta,\phi)$ for fracture density are given in Appendix C.

The interpretation of the fracture reservoir showed that the fracture apertures within hydrocarbon reservoirs were 5 mm, and the majority were 1 mm or less [1,33]. The length of the open fracture is dependent on the formation pressure. The numerical analysis results showed that an open fracture with a width of 1 mm generally does not exceed 40 cm in reservoir depth. However, it is probably reasonable to think of the 40 cm value as an average distance between weld points. In that case, the fracture length can be 1 m long or 30 m long [33]. However, fracture orientations of shale can be easily obtained by field exposure or the formation microimager (FMI) log [19]. In this paper, under the assumption that the fracture width, radius, and strike were known, the WA parameters and fracture density could be directly estimated by Equation (11).

2.3. Bayesian Inversion for Weak Anisotropy Parameters and Fracture Density

Assuming that there are two sets of fractures, we produced a matrix of data equations for the case of m incidence angles and n azimuthal angles:

$$\begin{aligned} \begin{bmatrix} d^{obs}(\theta_1,\phi_1) \\ \vdots \\ d^{obs}(\theta_m,\phi_n) \end{bmatrix} &= \begin{bmatrix} w(\theta_1,\phi_1) & & \\ & \ddots & \\ & & w(\theta_m,\phi_n) \end{bmatrix} \begin{bmatrix} R_{PP}^{iso}(\theta_1) + R_{PP}^{adi}(\theta_1) \\ \vdots \\ R_{PP}^{iso}(\theta_m) + R_{PP}^{adi}(\theta_m) \end{bmatrix} + \\ \begin{bmatrix} w(\theta_1,\phi_1) & & \\ & \ddots & \\ & & w(\theta_m,\phi_n) \end{bmatrix} & \begin{bmatrix} F_1(\theta_1) & F_2(\theta_1) & a(\theta_1,\phi_1) & b(\theta_1,\phi_1) \\ \vdots & \vdots & \vdots & \vdots \\ F_1(\theta_m) & F_2(\theta_m) & a(\theta_m,\phi_n) & b(\theta_m,\phi_n) \end{bmatrix} \begin{bmatrix} \delta^{psd} \\ \varepsilon \\ N_{V1} \\ N_{V2} \end{bmatrix} + \begin{bmatrix} n(\theta_1,\phi_1) \\ \vdots \\ n(\theta_m,\phi_n) \end{bmatrix} \end{aligned} \tag{12}$$

From the above equation, it can be seen that elastic parameters of the isotropic background should be known for the accurate inversion of weak anisotropic parameters and fracture density. The V_p, V_s, and ρ values for all layers of interest were obtained from well logs using Backus averaging [34]. The forward problem has the simple form:

$$\mathbf{d=WFm+n=Lm+n} \tag{13}$$

where the forward operator **L=WF**, **n** is random noise added into clear data, and the model parameters are:

$$\mathbf{m} = \left[\delta^{psd}, \varepsilon, N_{V1}, N_{V2}\right]^T \tag{14}$$

and the wavelet matrix is:

$$\mathbf{W} = \begin{bmatrix} w(\theta_1,\phi_1) & & \\ & \ddots & \\ & & w(\theta_m,\phi_n) \end{bmatrix}. \tag{15}$$

In Bayesian theory, the posterior probability density function (PDF) is calculated by the prior PDF and the likelihood function [35]:

$$\mathbf{p}(\mathbf{m}|\mathbf{d}) \propto \mathbf{p}(\mathbf{m})\mathbf{p}(\mathbf{d}|\mathbf{m}) \tag{16}$$

where $\mathbf{p}(\mathbf{m}|\mathbf{d})$ is the posterior PDF, $\mathbf{p}(\mathbf{m})$ is the prior PDF, and $\mathbf{p}(\mathbf{d}|\mathbf{m})$ is the likelihood function.

In the case of observed seismic data containing Gaussian random noise, the likelihood function $\mathbf{p}(\mathbf{d}|\mathbf{m})$ is given by

$$\mathbf{p}(\mathbf{d}|\mathbf{m}) = \frac{1}{\left(2\pi\sigma_e^2\right)^{\frac{N}{2}}}\exp\left\{-\sum\frac{[\mathbf{d}-\mathbf{G}(\mathbf{m})]^T[\mathbf{d}-\mathbf{G}(\mathbf{m})]}{2\sigma_e^2}\right\} \tag{17}$$

where σ_e^2 is the variance of the noise, N is the number of the input data samples, and $\mathbf{G}$ is the linear operator $\mathbf{G} = \mathbf{Lm}$.

Under the assumption of model parameters being independent of each other, the prior PDF $\mathbf{p}(\mathbf{m})$ was expressed as

$$\mathbf{p}(\mathbf{m}) = \frac{1}{(2\pi)^{\frac{N}{2}}|C_{\mathbf{m}}|^{1/2}}\exp\left(-\frac{1}{2}\mathbf{m}^T C_{\mathbf{m}}^{-1}\mathbf{m}\right) \tag{18}$$

where $C_{\mathbf{m}}$ is the variance of the model parameters.

According to Bayesian theory, we could convert the inversion problem by solving the minimum value of the objective function

$$J(\mathbf{m}) = \frac{1}{2}\exp\left\{-\frac{1}{2\sigma_{\mathbf{d}}^2}[\mathbf{d}-\mathbf{G}(\mathbf{m})]^T[\mathbf{d}-\mathbf{G}(\mathbf{m})]+\mathbf{R}(\mathbf{m})\right\} \tag{19}$$

where $\mathbf{R}(\mathbf{m}) = \frac{1}{2}\mathbf{m}^T C_{\mathbf{m}}^{-1}\mathbf{m}$, $\sigma_{\mathbf{d}}^2$ is the variance of data. By solving the derivative of the objective function with respect to the model parameters, we could obtain

$$\frac{\partial J(\mathbf{m})}{\partial \mathbf{m}} = \frac{1}{\sigma_{\mathbf{d}}^2}\mathbf{L}^T(\mathbf{Lm}-\mathbf{d}) + \frac{\partial R(\mathbf{m})}{\partial \mathbf{m}} \tag{20}$$

setting $\frac{\partial J(\mathbf{m})}{\partial \mathbf{m}}$ to zero gives the model parameters

$$\mathbf{m} = \left(\mathbf{L}^T\mathbf{L} + \frac{1}{2}\mu_h C_{\mathbf{m}}^{-1}\right)^{-1}\left(\mathbf{L}^T\mathbf{d} + \frac{1}{2}\mu_h C_{\mathbf{m}}^{-1}\right) \tag{21}$$

where super parameter μ_h allowed us to estimate the model parameters along the tuning curve, mainly through experiments. In this paper, the derived P-wave reflection coefficient, after removing the background effect, was completely linear, thus allowing us to invert the model parameters from seismic data $\mathbf{d}$.

3. Numerical Analysis

Synthetic seismic data were created by using interpretation results of well log data and vertical seismic profiling (VSP) data, which provided input for the AVOA inversion. It was assumed that the fracture medium had monoclinic symmetry, the azimuths of the two fracture sets were 10° and 70°, respectively, the average fracture width was 1 mm, the average fracture radius was 10 m, and the average area of the fracture was 0.02 m^2. Figure 2 shows two sets of fracture densities from the well log interpretation results as model parameters for inversion. Figure 3 shows the well log and VSP interpretation results, among which the isotropic elastic parameters were used as the known parameters, to calculate the reflection coefficient of isotropic background and sensitivities.

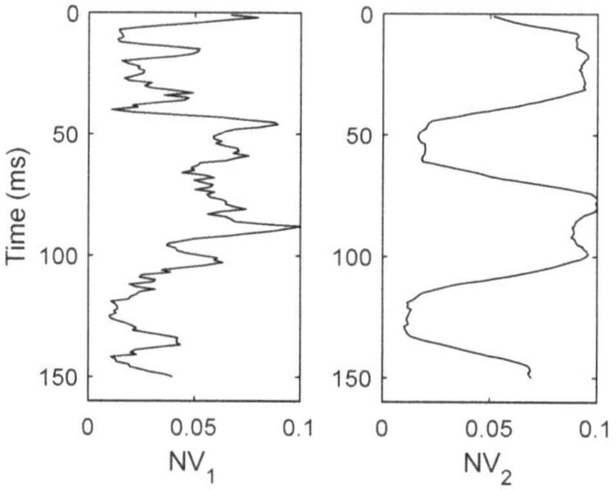

Figure 2. Interpretation results of well log data (fracture density).

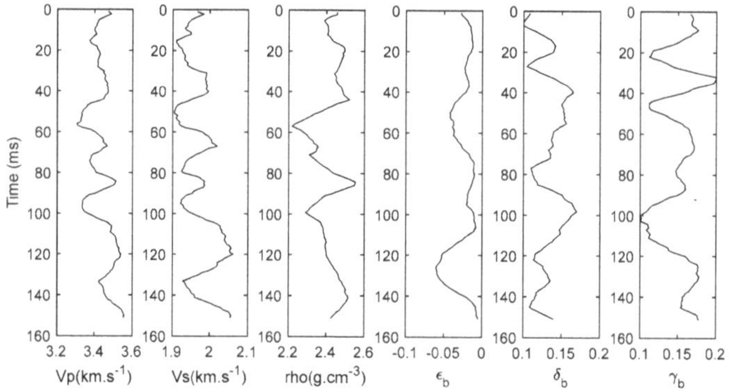

Figure 3. Interpretation results of well log and vertical seismic profiling (VSP) data (P- and S-wave velocity, density, and WA parameters).

In order to test the robustness of the proposed inversion algorithm, synthetic seismic data were created by the model parameters shown in Figures 2 and 3, as well as a 30 Hz Ricker wavelet, based on the derived PP-wave reflection coefficient, Equation (13), and the convolution model shown in Figure 4a. The incident angle had 10 angles between 0° and 45°, and the azimuths were 10°, 55°, 100°, and 145°.

Gaussian random noises (the signal-to-noise ratios (SNRs) were 5 and 2) were added into the synthetic data, as shown in Figure 4b,c. Comparisons between the inversion results (black) of the WA parameters, the fracture density, and the true values (red) are displayed in Figure 5. We can see that, in the case of the SNR being 5, the proposed inversion method could make a stable estimation of WA parameters and fracture density; however, when the SNR was 2, the inversion results of fracture density exhibited more instabilities. Random noise had some influence on the WA parameters and fracture density, but the overall agreement was good. In general, it can be concluded that the proposed inversion method is stable and valid.

(a)

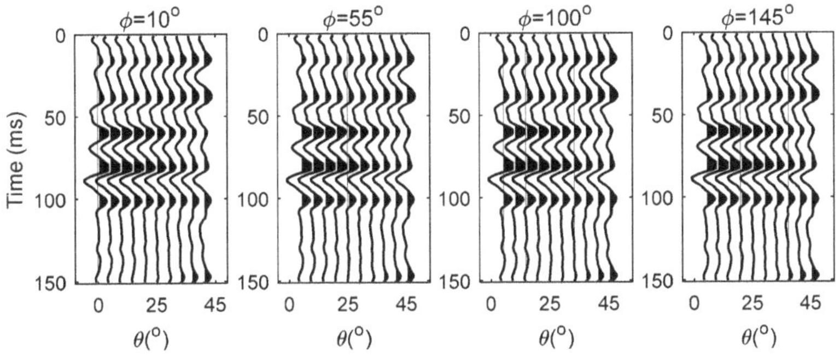

Figure 4. *Cont.*

(b)

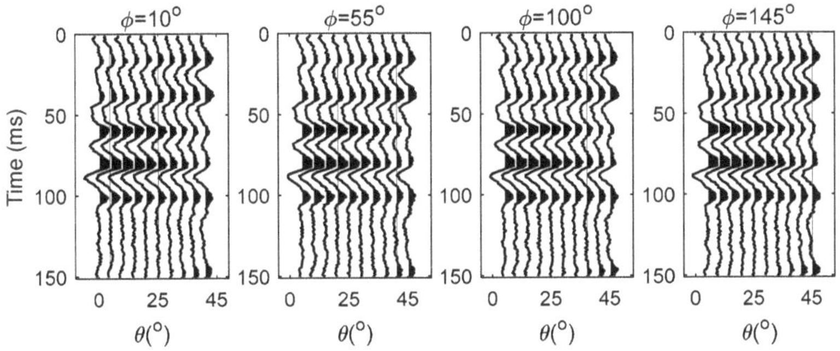

(c)

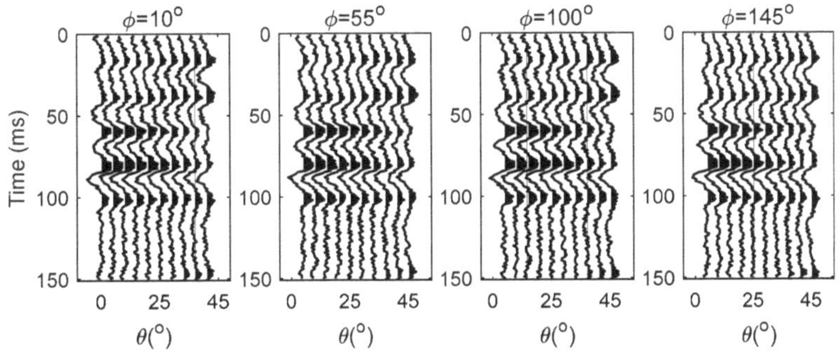

Figure 4. Synthetic seismic profiles. (**a**) noise-free, (**b**) S/N of 5, (**c**) S/N of 2.

(**a**)

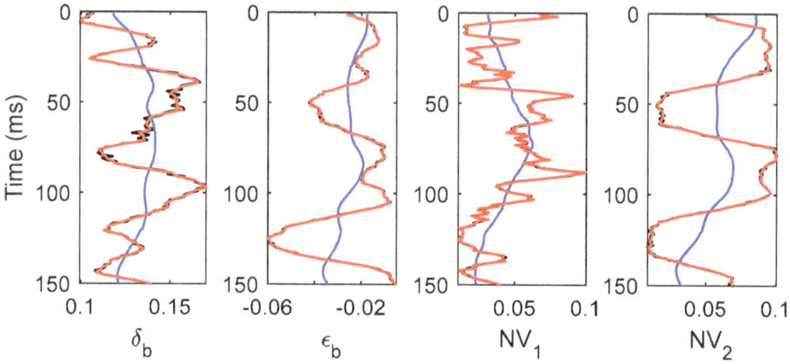

Figure 5. *Cont.*

(**b**)

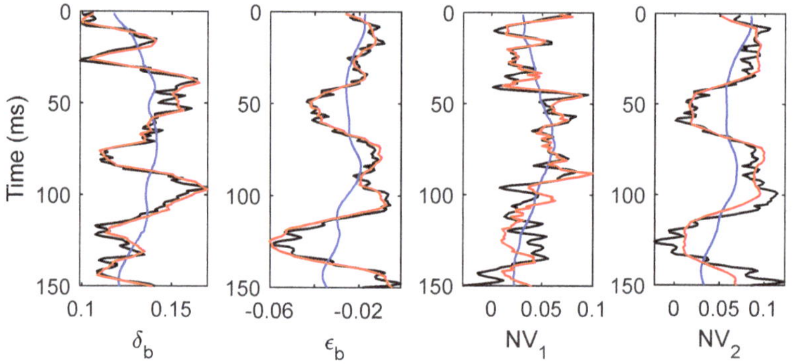

(**c**)

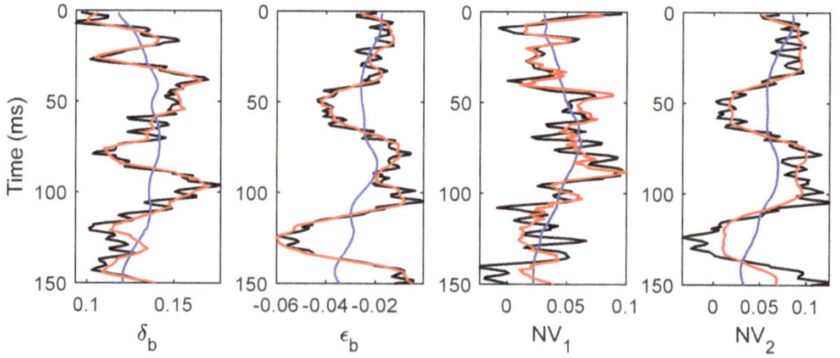

Figure 5. Comparisons between the estimated results (black curves) and the true values (red curves). The blue curves represent initial models that were constructed by smoothing the true values. (**a**) noise-free, (**b**) S/N of 5, (**c**) S/N of 2.

In the approach proposed, the fracture width, radius, and strike are required to be assumed. The dimensions of fractures are a priori functions of many factors in the medium and cannot be readily assumed in practice. In order to test the impact of this assumption (the dimensions assumed for the fractures) on the predicted results, the results of inversion, using the same synthetic data as used in Figure 4a, but with the forward operators, that fracture radius is 0.25 m and fracture width is 5 mm (wrong assumptions), are shown in Figure 6a,b, respectively. Comparing the inversion results with Figure 5a, we observed that, in the case of the fracture radius being 0.25 m, the incorrect assumption of fracture radius introduced larger errors into the prediction of WA parameters and fracture density. In general, the inverted model parameters captured the amplitudes and variability of the true models. When the fracture width was 5 mm, the incorrect assumption of fracture width almost had no impact on WA parameters; however, the accuracy and stability of the fracture density inversion was seriously influenced. More accurate information on fracture radius and width may help to improve the accuracy of WA parameters and fracture density inversion.

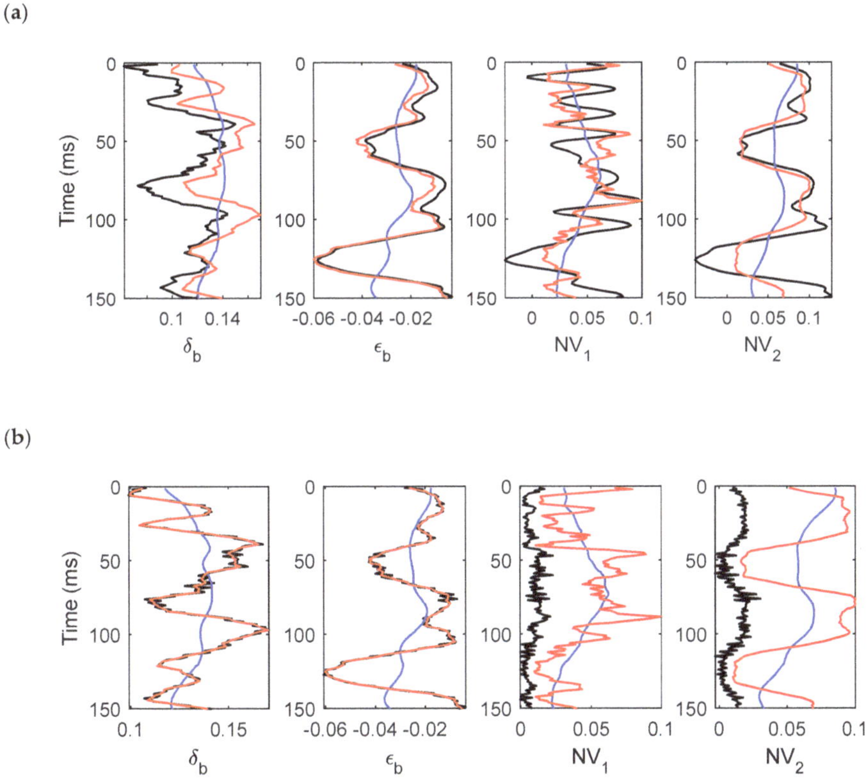

Figure 6. Inversion results using synthetic data (Figure 4a), computed with the fracture radius at 10 m and fracture width at 1 mm. (**a**) Inversion results with the forward operator that the fracture radius was 0.25 m, (**b**) Inversion results with the forward operator that the fracture width was 5 mm.

In order to demonstrate the advantage of the proposed algorithm in fracture density estimation, we made use of the same synthetic data as used in the Figure 4a and estimated anisotropy through elliptical inversion of interval velocities. The resulting NMO velocity varied ellipsoidally with the azimuth. The anisotropy intensity was defined by the axial ratio. To compare with the true value of fracture density, the anisotropy intensity, multiplied by a factor, is shown in Figure 7. We can see that the accuracy of the proposed inversion, in relative terms, is greater than that of elliptical inversion.

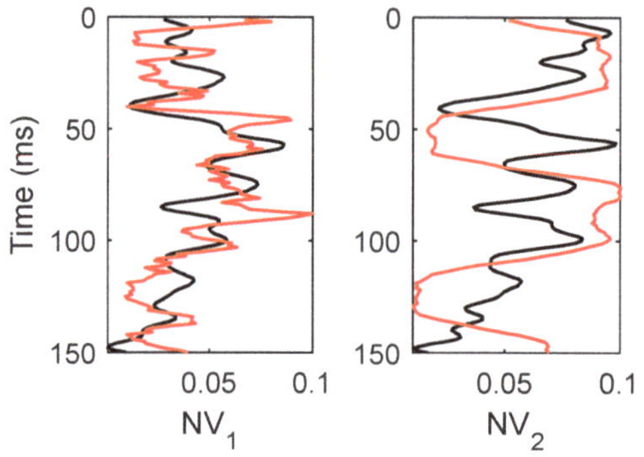

Figure 7. Fracture density (anisotropy intensity multiplied by a factor) obtained through elliptical inversion using synthetic data (Figure 4a).

4. Conclusions

Assuming that the fractured reservoir has a monoclinic symmetry of VTI background, a new form of P-wave reflection coefficient, in terms of WA parameters and fracture density, was obtained by substituting the stiffness matrix coefficient of the VTI background, and normal and tangential fracture compliances. The new expression of the reflection coefficient for a monoclinic medium can avoid calculation errors due to the assumption of a simple model (such as an HTI medium). Additionally, the inversion parameters contain WA parameters, which avoid the assumption that the parameters of the VTI background are known in the traditional form of a P-wave reflection coefficient. Finally, we achieved the direct inversion of WA parameters and fracture density, and avoided the intrinsic errors introduced by indirect substitution. The synthetic test demonstrated that this method can be used to accurately estimate WA parameters and fracture density, which significantly benefits further calculation of reservoir porosity or permeability, the prediction of the "sweet spot", and the evaluation of a reservoir's reconstruction.

In the approach proposed, the fracture width, radius, and strike are required to be assumed. Incorrect assumptions of fracture width and radius will yield erroneous results, for WA parameters and fracture density, in practice. A priori knowledge about fracture width is important in the prediction of fracture density. However, the interpretation of fracture reservoirs shows that the fracture apertures in reservoir depth are about 1 mm, and have little variation.

Author Contributions: Conceptualization, Z.G. and S.P.; methodology, Z.G. and S.P.; software, J.L.; validation, J.L.; formal analysis, S.P.; investigation, Z.G.; resources, S.P.; data curation, Z.G.; writing—original draft preparation, Z.G.; writing—review and editing, S.P. and J.L.; visualization, S.P.; supervision, J.L.; project administration, S.P.; funding acquisition, S.P. All authors have read and agreed to the published version of the manuscript.

Funding: This research received no external funding.

Acknowledgments: We appreciate the helpful comments and suggestions from editors and two anonymous reviewers that led to the substantial improvement of this article. We are grateful for the financial support of the following institutions: the Science and Technology Cooperation Project of the CNPC-SWPU Innovation Alliance and Open Projects Fund of the State Key Laboratory of Oil and Gas Reservoir Geology and Exploitation (PLN201733), and the National Natural Science Foundation of China (NSFC 41204101).

Conflicts of Interest: The authors declare no conflicts of interest.

Appendix A

In terms of the specific compliance matrices, generalized anisotropy parameters are defined as below:

$$
\begin{aligned}
\delta_x &= \frac{\frac{-M\lambda(\alpha_{11}+\beta_{1111})-\lambda^2\beta_{1122}-\lambda^2(\alpha_{22}+\beta_{2222})-M\lambda\beta_{1122}+2\left(\mu-\mu^2\alpha_{11}\right)}{M}+}{\frac{\sqrt{2M(M-\mu)\delta+(M-\mu)^2}-\mu-M}{M}} \\
\delta_y &= \frac{\frac{-\lambda^2(\alpha_{11}+\beta_{1111})-M\lambda\beta_{1122}-M\lambda(\alpha_{22}+\beta_{2222})-\lambda^2\beta_{1122}+2\left(\mu-\mu^2\alpha_{22}\right)}{M}+}{\frac{\sqrt{2M(M-\mu)\delta+(M-\mu)^2}-\mu-M}{M}} \\
\delta_z &= \frac{\frac{-M\lambda(\alpha_{11}+\beta_{1111})-\lambda^2\beta_{1122}+2\left(\mu-\mu^2(\alpha_{11}+\alpha_{22}+4\beta_{1122})\right)-M\lambda(\alpha_{22}+\beta_{2222})-MM\beta_{1122}}{M}+}{\frac{2M\varepsilon+\lambda-M}{M}} \\
X_z &= \frac{-\mu\lambda(\alpha_{12}+2\beta_{1112})-\mu\lambda(\alpha_{12}+2\beta_{1222})-\mu^2\alpha_{12}}{M} \\
\varepsilon_{16} &= \frac{-\mu M(\alpha_{12}+2\beta_{1112})-\mu\lambda(\alpha_{12}+2\beta_{1222})}{M} \\
\varepsilon_{26} &= \frac{-\mu\lambda(\alpha_{12}+2\beta_{1112})-\mu M(\alpha_{12}+2\beta_{1222})}{M}\varepsilon_{45} = -\mu\alpha_{12} \\
\varepsilon_x &= \frac{-M^2(\alpha_{11}+\beta_{1111})-\lambda^2(\alpha_{22}+\beta_{2222})-2M\lambda\beta_{1122}+2M\varepsilon}{2M} \\
\varepsilon_y &= \frac{-\lambda^2(\alpha_{11}+\beta_{1111})-\lambda\lambda\beta_{1122}-M^2(\alpha_{22}+\beta_{2222})-M\lambda\beta_{1122}+2M\varepsilon}{2M} \\
\varepsilon_z &= \frac{-\lambda^2(\alpha_{11}+\beta_{1111}+\beta_{1122})-\lambda^2(\alpha_{22}+\beta_{2222}+\beta_{1122})}{2M} \\
\gamma_x &= \frac{-\mu\alpha_{11}}{2} \\
\gamma_y &= \frac{-\mu\alpha_{22}}{2}
\end{aligned}
\tag{A1}
$$

where $M = \rho V_p^2, \mu = \rho V_s^2, \lambda = M - 2\mu$.

Appendix B

Sensitivities in Equation (10) are obtained as:

$$
\begin{aligned}
F_1(\theta) &= \frac{1}{2}\frac{\sqrt{M-\mu}}{2M}\sin^2\theta \\
F_2(\theta) &= \frac{1}{2}\sin^2\theta\tan^2\theta \\
F_3(\theta,\phi) &= \tfrac{1}{2}\tfrac{-\lambda^2}{2M} + \tfrac{1}{2}\left[\left(\tfrac{-M\lambda-2\mu^2}{M} - 8g\tfrac{-\mu}{2}\right)\cos^2\phi + \left(\tfrac{-\lambda^2}{2M}\right)\sin^2\phi + \tfrac{\lambda^2}{2M}\right]\sin^2\theta + \\
&\quad \tfrac{1}{2}\left[\tfrac{-M}{2}\cos^4\phi - \tfrac{\lambda^2}{2M}\sin^4\phi + \tfrac{-M\lambda-2\mu^2}{M}\cos^2\phi\sin^2\phi\right]\sin^2\theta\tan^2\theta \\
F_4(\theta,\phi) &= \tfrac{1}{2}\left[2\left(\tfrac{-2\mu\lambda-2\mu^2}{M} - 8g\tfrac{-\mu}{2}\right)\cos\phi\sin\phi\right]\sin^2\theta + \\
&\quad \tfrac{1}{2}\left[2\left(-\tfrac{\mu(M+\lambda)}{M}\right)\cos\phi\sin\phi\right]\sin^2\theta\tan^2\theta \\
F_5(\theta,\phi) &= \tfrac{1}{2}\tfrac{-\lambda^2}{2M} + \tfrac{1}{2}\left[\left(\tfrac{-\lambda^2}{M}\right)\cos^2\phi + \left(\tfrac{-M\lambda-2\mu^2}{2M} - 8g\tfrac{-\mu}{2}\right)\sin^2\phi + \tfrac{\lambda^2}{2M}\right]\sin^2\theta + \\
&\quad \tfrac{1}{2}\left[-\tfrac{\lambda^2}{2M}\cos^4\phi + \tfrac{-M}{2}\sin^4\phi + \tfrac{-M\lambda-2\mu^2}{M}\cos^2\phi\sin^2\phi\right]\sin^2\theta\tan^2\theta \\
F_6(\theta,\phi) &= \tfrac{1}{2}\tfrac{-\lambda^2}{2M} + \tfrac{1}{2}\left[-\lambda\cos^2\phi - \tfrac{\lambda^2}{M}\sin^2\phi + \tfrac{\lambda^2}{M}\right]\sin^2\theta + \\
&\quad \tfrac{1}{2}\left[\tfrac{-M}{2}\cos^4\phi - \tfrac{\lambda^2}{2M}\sin^4\phi - \lambda\cos^2\phi\sin^2\phi\right]\sin^2\theta\tan^2\theta \\
F_7(\theta,\phi) &= \frac{1}{2}\left[2\left(\frac{-2\mu\lambda}{M}\right)\cos\phi\sin\phi\right]\sin^2\theta + \frac{1}{2}\left[2\left(-2\mu\cos^2\phi - \frac{2\mu\lambda}{M}\sin^2\phi\right)\cos\phi\sin\phi\right]\sin^2\theta\tan^2\theta \\
F_8(\theta,\phi) &= \tfrac{1}{2}\tfrac{-\lambda^2}{M} + \tfrac{1}{2}\left[\tfrac{-M\lambda-\lambda^2}{M} + \tfrac{\lambda^2}{M}\right]\sin^2\theta + \\
&\quad \tfrac{1}{2}\left[-\lambda\cos^4\phi + \tfrac{-\lambda^2-\lambda M}{2M}\sin^4\phi + \tfrac{-\lambda^2-8\mu^2-MM}{M}\cos^2\phi\sin^2\phi\right]\sin^2\theta\tan^2\theta \\
F_9(\theta,\phi) &= \frac{1}{2}\left[2\left(\frac{-2\mu\lambda}{M}\right)\cos\phi\sin\phi\right]\sin^2\theta + \frac{1}{2}\left[2\left(-\frac{2\mu\lambda}{M}\cos^2\phi - 2\mu\sin^2\phi\right)\cos\phi\sin\phi\right]\sin^2\theta\tan^2\theta \\
F_{10}(\theta,\phi) &= \tfrac{1}{2}\tfrac{-\lambda^2}{2M} + \tfrac{1}{2}\left[\left(\tfrac{-\lambda^2}{M}\right)\cos^2\phi - \lambda\sin^2\phi + \tfrac{\lambda^2}{2M}\right]\sin^2\theta + \\
&\quad \tfrac{1}{2}\left[-\tfrac{\lambda^2}{2M}\cos^4\phi + \tfrac{-M}{2}\sin^4\phi - \lambda\cos^2\phi\sin^2\phi\right]\sin^2\theta\tan^2\theta
\end{aligned}
\tag{A2}
$$

where $g = \frac{V_S{}^2}{V_p{}^2}$ It should be noted that sensitivity **F** is only related to the elastic properties of the isotropic background.

Appendix C

Sensitivities $a(\theta,\phi)$ and $b(\theta,\phi)$ for fracture density in Equation (11) are obtained as:

$$\begin{aligned} a(\theta,\phi) &= \overline{A}_1\overline{B}_{T1}\left[F_3(\theta,\phi)\sin^2\varphi_1 + F_4(\theta,\phi)\sin\varphi_1\cos\varphi_1 + F_5(\theta,\phi)\cos^2\varphi_1\right]+ \\ \overline{A}_1\left(\overline{B}_{N1}-\overline{B}_{T1}\right)&\begin{bmatrix} F_6(\theta,\phi)\sin^4\varphi_1 + F_7(\theta,\phi)2\sin^3\varphi_1\cos\varphi_1 + F_8(\theta,\phi)\sin^2\varphi_1\cos^2\varphi_1 + \\ F_9(\theta,\phi)2\sin\varphi_1\cos^3\varphi_1 + F_{10}(\theta,\phi)\cos^4\varphi_1 \end{bmatrix} \\ b(\theta,\phi) &= \overline{A}_2\overline{B}_{T2}\left[F_3(\theta,\phi)\sin^2\varphi_2 + F_4(\theta,\phi)\sin\varphi_2\cos\varphi_2 + F_5(\theta,\phi)\cos^2\varphi_2\right]+ \\ \overline{A}_2\left(\overline{B}_{N2}-\overline{B}_{T2}\right)&\begin{bmatrix} F_6(\theta,\phi)\sin^4\varphi_2 + F_7(\theta,\phi)2\sin^3\varphi_2\cos\varphi_2 + F_8(\theta,\phi)\sin^2\varphi_2\cos^2\varphi_2 + \\ F_9(\theta,\phi)2\sin\varphi_2\cos^3\varphi_2 + F_{10}(\theta,\phi)\cos^4\varphi_2 \end{bmatrix} \end{aligned} \tag{A3}$$

For open penny-shaped cracks with radius r in a homogeneous background medium with Poisson's ratio σ and Young's modulus E:

$$B_N = \frac{16\left(1-\sigma^2\right)r}{3\pi E} B_T = \frac{32\left(1-\sigma^2\right)r}{3\pi E(2-\sigma)} \tag{A4}$$

In an isotropic medium, the relationships among P-wave moduli M, S-wave moduli μ, Poisson's ratio, and Young's modulus are shown below [36]

$$M = \mu\frac{4\mu - E}{3\mu - E} M = \mu\frac{2-2\sigma}{1-2\sigma} \tag{A5}$$

The normal and tangential compliances can be obtained by combining Equations (A4) and (A5).

References

1. Nelson, R.A. *Geologic Analysis of Naturally Fractured Reservoirs*; Gulf Publishing Company: Houston, TX, USA, 1985.
2. Sayers, C.M. Seismic characterization of reservoirs containing multiple fracture sets. *Geophys. Prospect.* **2009**, *57*, 187–192. [CrossRef]
3. Dvorkin, J.; Mavko, G.; Nur, A. Squirt flow in fully saturated rocks. *Geophysics* **1995**, *60*, 97–107. [CrossRef]
4. Rubino, J.; Holliger, K. Research note: Seismic attenuation due to wave-induced fluid flow at microscopic and mesoscopic scales. *Geophys. Prospect.* **2013**, *61*, 882–889. [CrossRef]
5. Chen, H.; Innanen, K.A.; Chen, T. Estimating P-and S-wave inverse quality factors from observed seismic data using an attenuative elastic impedance. *Geophysics* **2018**, *83*, R173–R187. [CrossRef]
6. Far, M.E.; Sayers, C.M.; Thomsen, L.; Han, D.; Castagna, J.P. Seismic characterization of naturally fractured reservoirs using amplitude versus offset and azimuth analysis. *Geophys. Prospect.* **2013**, *61*, 427–447. [CrossRef]
7. Wright, J. The effects of transverse isotropy on reflection amplitude versus offset. *Geophysics* **1987**, *52*, 564–567. [CrossRef]
8. Banik, N.C. An effective anisotropy parameter in transversely isotropic media. *Geophysics* **1987**, *52*, 1654–1664. [CrossRef]
9. Kim, K.Y.; Wrolstad, K.H.; Aminzadeh, F. Effects of transverse isotropy on P-wave AVO for gas sands. *Geophysics* **1993**, *58*, 883–888. [CrossRef]
10. Rahimian, M.; Eskandari-Ghadi, M.; Pak, R.Y.; Khojasteh, A. Elastodynamic Potential Method for Transversely Isotropic Solid. *J. Eng. Mech.* **2007**, *133*, 1134–1145. [CrossRef]
11. Shahbodagh, B.; Ashari, M.; Khalili, N. A hybrid element method for dynamics of piles and pile groups in transversely isotropic media. *Comput. Geotech.* **2017**, *85*, 249–261. [CrossRef]
12. Moghaddasi, H.; Shahbodagh, B.; Khalili, N. Lateral Vibration of Piles and Pile Groups in Nonhomogeneous Transversely Isotropic Media. *Int. J. Geomech.* **2020**, *20*, 04020124. [CrossRef]

13. Rüger, A. P-wave reflection coefficients for transversely isotropic media with vertical and horizontal axis of symmetry. *Geophysics* **1997**, *62*, 713–722. [CrossRef]
14. Sayers, C.M.; Rickett, J.E. Azimuthal variation in AVO response for fractured gas sands. *Geophys. Prospect.* **1997**, *45*, 165–182. [CrossRef]
15. Bakulin, A.V.; Grechka, V.; Tsvankin, I. Estimation of fracture parameters from reflection seismic data. Part III: Fractured models with monoclinic symmetry. *Geophysics* **2000**, *65*, 1818–1830. [CrossRef]
16. Sayers, C.M.; Dean, S. Azimuth-dependent AVO in reservoirs containing non-orthogonal fracture sets. *Geophys. Prospect.* **2001**, *49*, 100–106. [CrossRef]
17. Far, M.E. Seismic Characterization of Naturally Fractured Reservoirs. Ph.D. Dissertation, University of Houston, Houston, TX, USA, 2011.
18. Sil, S.; Davidson, M.; Zhou, C.; Olson, R.; Swan, H.; Howell, J.; Chiu, S.; Willis, M. Effect of near-surface anisotropy on a deep anisotropic target layer. In Proceedings of the 81st Annual International Meeting, SEG, Expanded Abstracts, San Antonio, TX, USA, 18–23 September 2011; pp. 305–309.
19. Far, M.E.; Hardage, B.; Wagner, D. Fracture parameter inversion for Marcellus Shale. *Geophysics* **2014**, *79*, C55–C63. [CrossRef]
20. Chen, H.; Yin, X.; Qu, S.; Zhang, G. AVAZ inversion for fracture weakness parameters based on the rock physics model. *J. Geophys. Eng.* **2014**, *11*, 65–70. [CrossRef]
21. Chen, H.; Pan, X.; Ji, Y.; Zhang, G. Bayesian Markov Chain Monte Carlo inversion for weak anisotropy parameters and fracture weaknesses using azimuthal elastic impedance. *Geophys. J. Int.* **2017**, *210*, 801–818. [CrossRef]
22. Chaveste, A.; Hall, M.; Verm, R.; Gaiser, J.E. Estimation of fracture density and orientation through azimuthal processing and interpretation of seismic data. In Proceedings of the Canadian Unconventional Resources Conference, Calgary, AB, Canada, 15–17 November 2011.
23. Cortes-Gómez, D.-M.; Agudelo-Zambrano, W.-M.; Montes-Vides, L.-A. Estimation of density and fracture orientation in HTI media through azimuthal analysis of P-waves. *CT F-Cienc. Tecnol. y Futuro* **2013**, *5*, 5–18. [CrossRef]
24. Shahraini, A.; Jakobsen, M.; Ali, A. Non linear Bayesian inversion of seismic AVAZ and production data with respect to the fracture aperture and density. In Proceedings of the 11th International Congress of the Brazilian Geophysical Society & EXPOGEF, Salvador, Bahia, Brazil, 24–28 August 2009.
25. Mahmoudian, F.; Margrave, G.F.; Wong, J.; Henley, D.C. Fracture orientation and intensity from AVAz inversion. A physical modeling study. In Proceedings of the 83rd Annual International Meeting, SEG, Expanded Abstracts, Houston, TX, USA, 22–27 September 2013.
26. Shulin, P.; Ke, Y.; Hai-Qiang, L.; José, B.; Ziyu, Q. Adaptive step-size fast iterative shrinkage-thresholding algorithm and sparse-spike deconvolution. *Comput. Geosci.* **2020**, *134*, 104343.
27. Shulin, P.; Ke, Y.; Hai-Qiang, L.; Ziyu, Q. A Bregman adaptive sparse-spike deconvolution method in the frequency domain. *Appl. Geophys.* **2019**, *16*, 463–472.
28. Shulin, P.; Ke, Y.; Hai-Qiang, L.; José, B.; Ziyu, Q. A Sparse Spike Deconvolution Algorithm Based on a Recurrent Neural Network and the Iterative Shrinkage-Thresholding Algorithm. *Energies* **2020**, *13*, 3074.
29. Hill, R. Elastic properties of reinforced solids: Some theoretical principles. *J. Mech. Phys. Solids* **1963**, *11*, 357–372. [CrossRef]
30. Sayers, C.M.; Kachanov, M. Microcrack-induced elastic wave anisotropy of brittle rocks. *J. Geophys. Res.* **1995**, *100*, 4149–4156. [CrossRef]
31. Thomsen, L. Weak elastic anisotropy. *Geophysics* **1986**, *51*, 1954–1966. [CrossRef]
32. Pšencík, I.; Martins, J.L. Properties of weak contrast PP reflection/transmission coefficients for weakly anisotropic elastic media. *Studia Geophys. Geod.* **2001**, *45*, 176–199. [CrossRef]
33. Worthington, M.H. Interpreting seismic anisotropy in fractured reservoirs. *First Break* **2008**, *26*, 57–63. [CrossRef]
34. Backus, G.E. Long-wave elastic anisotropy produced by horizontal layering. *J. Geophys. Res.* **1962**, *67*, 4427–4440. [CrossRef]

35. Buland, A.; Omre, H. Bayesian linearized AVO inversion. *Geophysics* **2003**, *68*, 185–198. [CrossRef]
36. Birch, F. The velocity of compressional waves in rocks to 10 kilobars. *J. Geophys. Res.* **1961**, *66*, 2199–2224. [CrossRef]

MDPI AG
Grosspeteranlage 5
4052 Basel
Switzerland
Tel.: +41 61 683 77 34

Applied Sciences Editorial Office
E-mail: applsci@mdpi.com
www.mdpi.com/journal/applsci

www.ingramcontent.com/pod-product-compliance
Lightning Source LLC
LaVergne TN
LVHW070540071126
851858LV00011BA/282
* 9 7 8 3 7 2 5 8 7 3 4 0 1 *